# COLOSS
# *BEEBOOK*

Volume III:
Standard Methods
for *Apis mellifera*
Hive Product Research

Edited by
Vincent Dietemann
Peter Neumann
Norman L Carreck
James D Ellis

Jointly published by: The International Bee Research Association,
a Company Limited by Guarantee, 1, Agincourt Street, Monmouth, NP25 3DZ (UK) &
Northern Bee Books, Scout Bottom Farm, Mytholmroyd, Hebden Bridge HX7 SJS (UK).

The International Bee Research Association and Northern Bee Books acknowledge permission from the Taylor & Francis Group (an Informa business) to reproduce typeset material which originally appeared in the *Journal of Apicultural Research*.

Obtainable from:
www.ibra.org.uk & www.northernbeebooks.co.uk



ISBN ISBN: 978-1-913811-05-1

# COLOSS *BEEBOOK*

## Volume III: Standard methods for *Apis mellifera* product research

The COLOSS (Prevention of Honey Bee COlony LOSSes) network was founded in 2008 as a consequence of the heavy and frequent losses of managed Western honey bee (*Apis mellifera*) colonies experienced in many regions of the world (Neumann & Carreck, 2010). The network has many accomplishments, with the COLOSS *BEEBOOK* being one of the most significant. The COLOSS *BEEBOOK* was developed to provide honey bee scientists with a single reference in which standard research methods related to various honey bee-related topics are available.

As many of the world's honey bee research teams began to address elevated colony loss rates in various areas globally, it quickly became obvious that a lack of standardized research methods was hindering scientists' ability to compare and interpret the results on colony losses obtained internationally. In its second year of activity, during a COLOSS meeting held in Bern, Switzerland, the idea of a manual of standardized honey bee research methods emerged. The manual, to be called the COLOSS *BEEBOOK*, was inspired by publications with similar purpose for fruit fly research (Roberts, 1987; Ashburner, 1989; Greenspan, 2004). Production of the *BEEBOOK* started after recruiting experts to lead the compilation of each research domain. These senior authors (first in the author list) were tasked with recruiting a suitable team of contributors to select the methods to be used as standards and then to report them in a user-friendly manner (Williams *et al*, 2012).

The first two volumes of the COLOSS *BEEBOOK* were published in 2013 and focused on general honey bee (Volume I) and honey bee pest and pathogen (Volume II) research methodologies. In keeping with this tradition, we are pleased to introduce the COLOSS *BEEBOOK* Volume III, a practical manual compiling standard methods for research on *A. mellifera* hive products.

Volume III: Standard methods for *Apis mellifera* product research contains papers on royal jelly, beeswax, propolis and brood as human food, honey, venom and pollen. These seven papers have been written by 125 authors from 23 countries. Like the previous volumes, papers in Volume III are organized according to research topics. The authors have compiled the most relevant methods for both laboratory and field research in each domain of research. We recognize that it is often necessary to use methods from several domains of research to complete a given experiment with honey bees. Whenever there is a need for such a multi-disciplinary approach, the manual describes the specific instructions necessary for a given method, and cross-references the other relevant methods from other sections of *the BEEBOOK*. The *BEEBOOK* is a tool for all who want to do research on honey bees. It was written in such a way that those new to honey bee research can use it to start research in a field with which they may not be familiar. Of course, such an endeavor is often limited by the availability of state-of-the-art and expensive equipment. However, provided access to and training on the necessary equipment are secured, the instructions provided in the *BEEBOOK* can be followed by everyone, from undergraduate student to experienced researcher.

The editors and author team hope that the *BEEBOOK* will serve as a reference tool for honey bee and other researchers globally. As with the original *Drosophila* book that evolved into a journal where updates and new methods are published, we hope that the honey bee research community will embrace this tool and work to improve it. As of August 2021, the papers from the COLOSS *BEEBOOK* have been downloaded over 180,000 times and cited 2,300+ times, and citations are beginning to appear in a large percentage of manuscripts published on honey bee related topics. We believe that this resource has evolved into a powerful tool for scientists at every level, from student to professor, from around the globe. We have no doubt that Volume III will have a similar impact. The *BEEBOOK* project is far from complete, and Volume IV: "Standard methods for *Apis cerana* research" is currently in preparation, together with other ideas and papers in the planning stage, and the revision of some of the earlier papers (Carreck *et al*, 2021). In this way, the *BEEBOOK* project will continue to develop and evolve for the foreseeable future.

We would like to thank the many international authors and numerous reviewers who wrote and refereed the *BEEBOOK* articles. We would also like to thank the International Bee Research Association and the *JAR* editorial team, Taylor & Francis Ltd, and Northern Bee Books for making its publication possible. The COLOSS Association was originally funded through the COST Action FA0803. COST (European Cooperation in Science and Technology) is a unique means for European researchers to develop their own ideas and new initiatives across all scientific disciplines through trans-European networking of nationally funded research activities. The COLOSS association is now supported by the Ricola Foundation - *Nature & Culture*, Vetopharma and the Eva Crane Trust as well as numerous local sponsors.

**Vincent Dietemann, Peter Neumann, Norman L. Carreck and James D. Ellis**

## References

Ashburner, M. (1989). Drosophila. *A laboratory handbook*. New York: Cold Spring Harbor Laboratory Press. 1331 pp.

Carreck, N.L., Dietemann, V., Ellis, J.D., Evans, J.D., Neumann, P., Chantawannakul, P. (2021) The COLOSS *BEEBOOK* evolves: hive products, 'omics research and Eastern honey bees, *Apis cerana*. *Journal of Apicultural Research*, 60(4): 1-4. https://doi.org/10.1080/00218839.2020.1739410

Greenspan, R. J. (2004). *Fly pushing: The theory and practice of* Drosophila *genetics*. New York: Cold Spring Harbor Laboratory Press. 191 pp.

Neumann, P., & Carreck, N.L. (2010). Honey bee colony losses. *Journal of Apicultural Research*, 49(1), 1-6. https://doi.org/10.3896/ IBRA.1.49.1.01

Roberts, D. B. (1987). Drosophila: *A practical approach*. Washington, DC: Oxford University Press. 295 pp.

Williams, G. R., Dietemann, V., Ellis, J. D., & Neumann, P. (2012). An update on the COLOSS network and the "*BEEBOOK*: Standard methodologies for *Apis mellifera* research". *Journal of Apicultural Research*, 51(2), 151-153. https://doi.org/10.3896/IBRA.1.51.2.01

| The COLOSS *BEEBOOK* Volume III: Standard methods for *Apis mellifera* hive product research. | |
|---|---|
| Preface | Vincent Dietemann, Peter Neumann, Norman L Carreck and James D Ellis |
| Standard methods for *Apis mellifera* royal jelly research | Fu-Liang Hu, Katarína Bíliková, Hervé´ Casabianca, Gaëlle Daniele, Foued Salmen Espindola, Mao Feng, Cui Guan, Bin Han, Tatiana Krištof Kraková, Jian-Ke Li, Li Li, Xing-An Li, Jozef Šimúth, Li-Ming Wu, Yu-Qi Wu, Xiao-Feng Xue, Yun-Bo Xue, Kikuji Yamaguchi, Zhi-Jiang Zeng, Huo-Qing Zheng, Jin-Hui Zhou |
| Standard methods for *Apis mellifera* beeswax research | Lidija Svečnjak, Lesley Ann Chesson, Albino Gallina, Miguel Maia, Marianna Martinello, Franco Mutinelli, Mustafa Necati Muz, Fernando M Nunes, Francis Saucy, Brett James Tipple, Klaus Wallner, Ewa Waś, Todd Alan Waters |
| Standard methods for Apis *mellifera* propolis research | Vassya Bankova, Davide Bertelli, Renata Borba, Bruno José Conti, Ildenize Barbosa da Silva Cunha, Carolina Danert, Marcos Nogueira Eberlin, Soraia I Falcão, María Inés Isla, María Inés Nieva Moreno, Giulia Papotti, Milena Popova, Karina Basso Santiago, Ana Salas, Alexandra Christine Helena Frankland Sawaya, Nicolas Vilczaki Schwab, José Maurício Sforcin, Michael Simone-Finstrom, Marla Spivak, Boryana Trusheva, Miguel Vilas-Boas, Michael Wilson, Catiana Zampini |
| Standard methods for *Apis mellifera* brood as human food | Annette Bruun Jensen, Josh Evans, Adi Jonas-Levi, Ofir Benjamin, Itzhak Martinez, Bjørn Dahle, Nanna Roos, Antoine Lecocq, Kirsten Foley |
| Standard methods for Apis *mellifera* honey research | Ligia Bicudo de Almeida-Muradian, Ortrud Monika Barth, Vincent Dietemann, Michael Eyer, Alex da Silva de Freitas, Anne-Claire Martel, Gian Luigi Marcazzan, Carla Marina Marchese, Carla Mucignat-Caretta, Ana Pascual-Maté, Wim Reybroeck, M Teresa Sancho, José Augusto Gasparotto Sattler |
| Standard methods for *Apis mellifera* venom research | Dirk C de Graaf, Márcia Regina Brochetto Braga, Rusleyd Maria Magalhães de Abreu, Simon Blank, Chris H Bridts, Luc S De Clerck, Bart Devreese, Didier G Ebo, Timothy J. Ferris, Margo M Hagendorens, Débora Laís Justo Jacomini, Iliya Kanchev, Zenon J Kokot, Jan Matysiak, Christel Mertens, Vito Sabato, Athina L Van Gasse, Matthias Van Vaerenbergh |
| Standard methods for pollen research | Maria G Campos, Ofélia Anjos, Manuel Chica, Pascual Campoy, Janka Nozkova, Norma Almaraz-Abarca, Lidia M R C Barreto, João Carlos Nordi, Leticia M Estevinho, Ananias Pascoal, Vanessa Branco Paula, Altino Chopina, Luis G Dias, Živoslav L j Tešić, Mirjana D Mosić, Aleksandar Ž Kostić, Mirjana B Pešić, Dušanka M Milojković-Opsenica, Wiebke Sickel, Markus J Ankenbrand, Gudrun Grimmer, Ingolf Steffan-Dewenter, Alexander Keller, Frank Förster, Chrysoula H Tananaki, |

| | |
|---|---|
| | Vasilios Liolios, Dimitrios Kanelis, Maria-Anna Rodopoulou, Andreas Thrasyvoulou, Luísa Paulo, Christina Kast, Matteo A Lucchetti, Gaëtan Glauser, Olena Lokutova, Ligia Bicudo de Almeida-Muradian, Teresa Szczęsna, Norman L Carreck |

*Journal of Apicultural Research*, 2019
Vol. 58, No. 2, 1–68, http://dx.doi.org/10.1080/00218839.2017.1286003

REVIEW ARTICLE

# Standard methods for *Apis mellifera* royal jelly research

Fu-Liang Hu[a,†*], Katarína Bíliková[b], Hervé Casabianca[c], Gaëlle Daniele[c], Foued Salmen Espindola[d], Mao Feng[e], Cui Guan[f,g], Bin Han[e], Tatiana Krištof Kraková[b], Jian-Ke Li[e], Li Li[a], Xing-An Li[h], Jozef Šimúth[b], Li-Ming Wu[i], Yu-Qi Wu[a], Xiao-Feng Xue[i], Yun-Bo Xue[h], Kikuji Yamaguchi[j], Zhi-Jiang Zeng[f], Huo-Qing Zheng[a] and Jin-Hui Zhou[i]

[a]*College of Animal Sciences, Zhejiang University, Hangzhou, China;* [b]*Department of Molecular Apidology, Institute of Forest Ecology, Slovak Academy of Sciences, Bratislava, Slovakia;* [c]*Institut des Sciences Analytiques, UMR 5280 CNRS, Université Lyon 1, ENS, Lyon, Villeurbanne, France;* [d]*Institute of Genetics and Biochemistry, Federal University of Uberlândia, Uberlândia, Brazil;* [e]*Institute of Apicultural Research/Key Laboratory of Pollinating Insect Biology, Ministry of Agriculture, Chinese Academy of Agricultural Science, Beijing, China;* [f]*Honey Bee Research Institute, Jiangxi Agricultural University, Nanchang, China;* [g]*School of Biological & Chemical Sciences, Queen Mary University of London, London, UK;* [h]*Jilin Provincial Key Laboratory of (Honey) Bee Genetics and (Queen) Breeding, Jilin Provincial Institute of Apicultural Sciences, Jilin, China;* [i]*Institute of Apicultural Research, Chinese Academy of Agricultural Science, Beijing, China;* [j]*Japan Royal Jelly Co., Ltd., Tokyo, Japan*

(*Received 20 November 2015; accepted 25 November 2016*)

Royal jelly, a honey bee secretion, plays a critical role in caste determination in honey bees because it serves as the source of nutrition for young larvae destined to become queens. It is also fed to adult queens. Royal jelly possesses numerous functional properties and thus has been used as a medication, health food, and cosmetic in many countries. In this paper, we first introduce a traditional method for producing royal jelly by artificial larvae grafting and a newly developed method that does not require grafting of larvae. We describe protocols for the storage and freeze-drying of royal jelly to preserve its biological properties. Routine methods for determination of two important quality criteria, water content and trans-10-hydroxy-2-decenoic acid content, are outlined. On a dry basis, protein, carbohydrate, and fatty acids were found to be the 3 most abundant components of royal jelly. Methods for their isolation, identification, and quantification are described. Because royal jelly is susceptible to contamination with veterinary drugs and acaricides, we also describe methods for detection and quantification of some veterinary drugs and acaricides in royal jelly.

**Métodos estándar para la investigación de la jalea real de *Apis mellifera***

La jalea real, una secreción de abejas, desempeña un papel crítico en la determinación de castas en la abeja melífera, ya que sirve como fuente de nutrición para larvas jóvenes destinadas a convertirse en reinas. También alimenta a las reinas adultas. La jalea real posee numerosas propiedades funcionales y por lo tanto se ha utilizado como un medicamento, alimento saludable y cosmético en muchos países. En este artículo, introducimos un método tradicional para producir jalea real mediante el injerto artificial de larvas y un método recientemente desarrollado que no requiere injerto de larvas. Describimos protocolos para el almacenamiento y la liofilización de la jalea real para preservar sus propiedades biológicas. Se describen métodos rutinarios para la determinación de dos importantes criterios de calidad, el contenido de agua y el de ácido trans-10-hidroxi-2-decenoico. En una base seca, proteínas, carbohidratos y ácidos grasos fueron los tres componentes más abundantes de la jalea real. Se describen métodos para su aislamiento, identificación y cuantificación. Debido a que la jalea real es susceptible a la contaminación con medicamentos veterinarios y acaricidas, también describimos métodos para la detección y cuantificación de algunos medicamentos veterinarios y acaricidas en jalea real.

西方蜜蜂蜂王浆研究标准方法

蜂王浆是蜜蜂的分泌物，作为蜂王幼虫的食物在蜜蜂级型分化过程中起到了重要作用。同时，蜂王浆也是蜂王的食物。蜂王浆有丰富的功能活性，因而在很多国家被作为药物、保健品和化妆品使用。在本章节，我们首先介绍了利用人工移虫生产蜂王浆的传统方法和一种新开发的免移虫蜂王浆生产方法。我们介绍了为保留蜂王浆生物学活性所需的蜂王浆贮存方法和冷冻干燥方法，概述了蜂王浆水分含量和10-羟基-2-癸烯酸这两个重要质量指标的检测方法。蛋白质、碳水化合物和脂肪酸是蜂王浆干物质中含量最丰富的3种成分。本章还介绍了这三种成分的分离、鉴定和定量方法。蜂王浆易受到兽药和螨药的污染，因此我们还介绍了蜂王浆中一些兽药和螨药的检测和定量方法。

**Keywords:** royal jelly; production; storage; protein; sugar; lipid; residue

*Corresponding author. Email: flhu@zju.edu.cn
†All authors except the first are listed alphabetically.
Please refer to this paper as: Hu, F-L, Bíliková, K, Casabianca, H, Gaëlle, D, Espindola, F S, Feng, M, Guan, C, Han, B, Kraková,TK, Li, J-K, Li, L, Li, X-A, Šimúth, J, Wu, L-M, Wu, Y-Q, Xue, X-F, Xue, Y-B, Yamaguchi, K, Zeng, Z-J, Zheng, H-Q, Zhou, J-H. (2019) Standard methods for *Apis mellifera* royal jelly research. In V Dietemann, P Neumann, N Carreck and J D Ellis (Eds), *The COLOSS BEEBOOK, Volume III, Part I: standard methods for Apis mellifera hive products research. Journal of Apicultural Research* 58(2): http://dx.doi.org/10.1080/00218839.2017.1286003

## 1. Introduction

Royal jelly is a substance that is secreted by the hypopharyngeal and mandibular glands of worker honey bees. It is a yellowish, creamy, and acidic substance with a slightly pungent odor and taste, composed on a wet weight basis of 60–70% water (Karaali, Meydanoğlu, & Eke, 1988; Sabatini, Marcazzan, Caboni, Bogdanov, & De Almeida-Muradian, 2009), 9–18% proteins (Karaali et al., 1988), 3–8% lipids (Karaali et al., 1988; Sabatini et al., 2009), 6–18% hydrocarbons (Daniele & Casabianca, 2012; Sesta, 2006), 0.8–3.0% minerals (Sabatini et al., 2009), and small amounts of polyphenols and vitamins. It plays a crucial role in caste determination of honey bees because larvae that are fed copious amounts of royal jelly for a longer period develop into large, fertile, and long-lived queens rather than into smaller, sterile, and short-lived workers.

Royal jelly has been demonstrated to possess numerous functional properties such as antibacterial activity, anti-inflammatory properties, vasodilatory and hypotensive activities, disinfectant action, antioxidant effects, anti-hypercholesterolaemic activity, and antitumor properties (reviewed in Ramadan & Al-Ghamdi, 2012). As a valuable bee product, royal jelly has been incorporated into traditional human medicine and is widely promoted and commercially available as a medicament, health food, and cosmetic in many countries, especially in China and Japan. Studies on royal jelly are important for our understanding of both the physiology of honey bees and royal jelly's biological activity promoting human health.

## 2. Production and harvesting of royal jelly

The development of techniques for production of royal jelly has a long history. In 1921, Sherlock Holmes simply sucked royal jelly out of a naturally produced queen cell with a syringe, to harvest this substance according to principles of vacuum physics. By the 1950s, Mexican, French, and Italian beekeepers had started small-scale production and sale of royal jelly. The method for producing royal jelly they employed involved removing the queen from the colony, so that they could take royal jelly directly from the new queen cells that the worker honey bees rapidly built after detecting the absence of the queen. Nevertheless, this method has disadvantages: the decrease in colony strength (a reduction in the number of new workers) and in honey production. To improve the situation, a queen excluder was later introduced so as to maintain the production of royal jelly in a strong colony with a queen.

Because of the improvement of royal jelly production methods, the output of royal jelly increased from 200–300 to 5000–12,000 g per colony per year. This achievement is mainly attributed to successful breeding and marketing of "royal-jelly-producing bees" (Cao, Zheng, Pirk, Hu, & Xu, 2016) and to widespread use of plastic queen cells; in other words, comprehensive methods of breeding colonies with a high yield of royal jelly have been developed (Chen, 1993; Zeng, 2013).

### 2.1. *The method for producing royal jelly by artificial larvae grafting*

During 60 years of continual development, a method for production of royal jelly has been fine-tuned. It is based on artificial larvae grafting: the transfer of larvae from worker cells to a large number of (artificial) queen cells (Chen, 1993; Zeng, 2013). In this section, the tools necessary to collect royal jelly and the management of colonies used for royal jelly production are described.

#### *2.1.1. Tools*

(1) Queen-cell-base-bar: this is made of non-toxic plastic and consists of many queen cell bases (Figure 1). A common queen-cell-base-bar has 2 rows; other models can have up to 4.

(2) Grafting pen: this is a tool used to transfer worker larvae from cells in a comb to queen cell bases. A grafting pen is made of a soft tongue made of bull horn, a plastic tube, and a spring-loaded plunger (Figure 2). Compared to the grafting needles made of metal, grafting pens are more convenient to use and increase the survival of grafted larvae.

(3) Royal jelly production frame: queen-cell-base-bars can be installed in a royal jelly production frame made of fir tree wood. The size of the periphery of a royal jelly production frame is the same as that of a comb frame. A royal jelly production frame consists of a long beam (width:

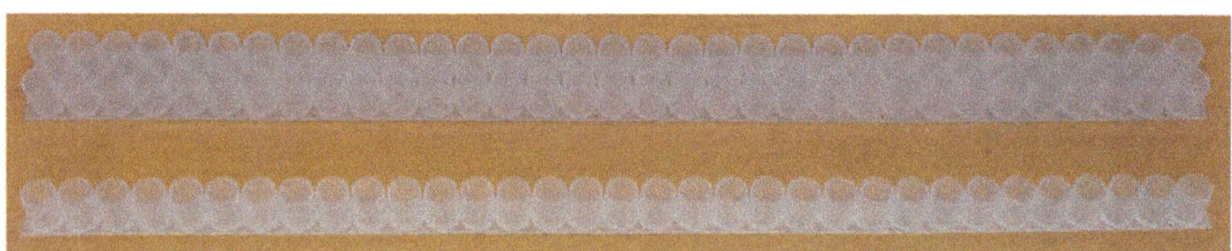

Figure 1. Plastic queen-cell-base-bar.
Photo: Linbin Zhou.

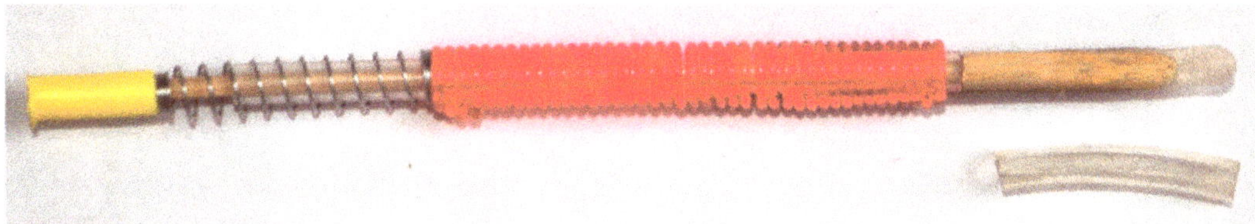

Figure 2. Grafting pen.
Photo: Qizhong Pan.

Figure 3. Royal jelly production frame.
Photo: Qizhong Pan.

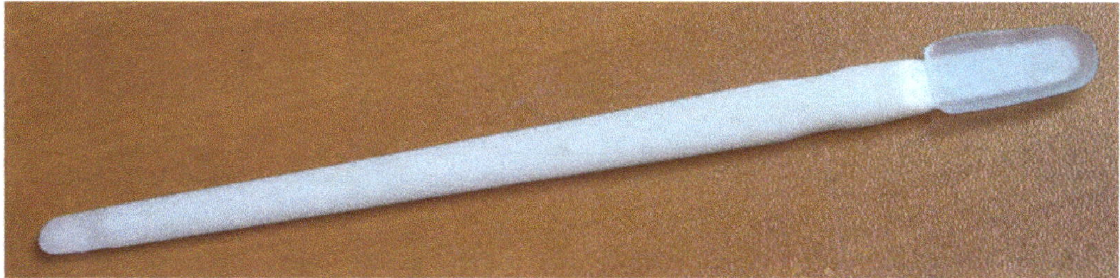

Figure 4. Royal jelly scraping bar.
Photo: Qizhong Pan.

13 mm, thickness: 20 mm), 2 sidebars (width: 13 mm, thickness: 10 mm), 5 inner plates attached to queen-cell-base-bars (width: 13 mm, thickness: 5 mm), and a bottom thicker plate (width: 13 mm, thickness: 10 mm; Figure 3).

(4) Royal jelly scraping bar: composed of a royal jelly scraping tongue and a handle (Figure 4). The scraping tongue is made of a flexible plastic or rubber sheet of high tenacity, with a flat shovel-like shape. The width of the scraping tongue piece is the same as that of the longitudinal section of a queen cell. The handle is made of rigid plastic approximately 100 mm in length.

(5) Tweezers and queen-cell-wax-cleaner: stainless steel tweezers (Figure 5 top) are used to remove the larva from a queen cell. A queen-cell-wax-cleaner has a metal piece that has a similar shape to that of a royal jelly scraping tongue and a rotatable handle sleeve. This tool is useful for scraping residual wax from the inner wall of a queen cell after royal jelly collection (Figure 5, bottom).

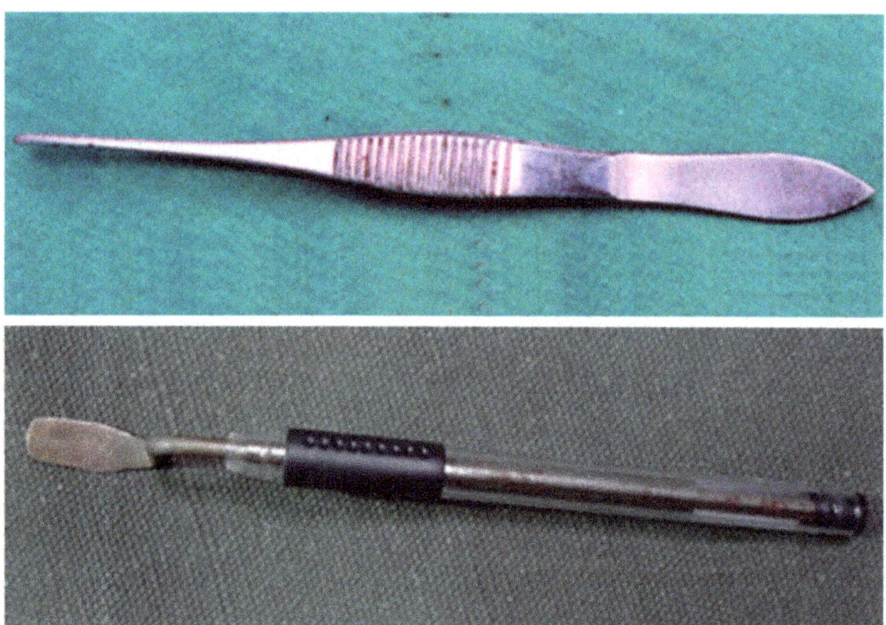

Figure 5. Tweezers (top) and queen-cell-wax-cleaner (bottom).
Photos: Zhongyin Zhang.

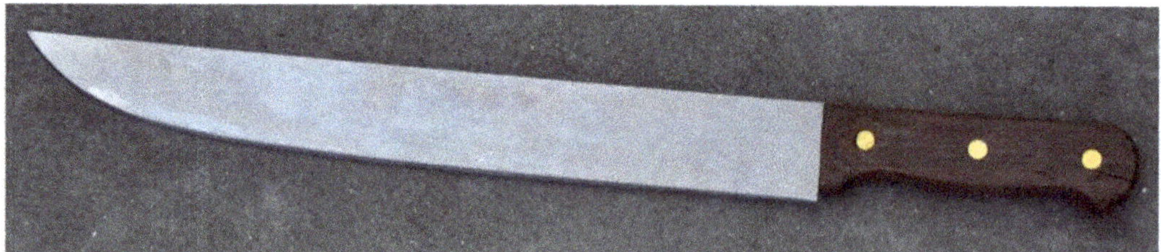

Figure 6. Stainless steel queen cell cutting blade.
Photo: Fei Zhang.

(6) Stainless steel queen cell cutting blade (Figure 6): before collection of royal jelly, a stainless steel queen cell cutting blade is used to cut off the protruding part of a queen cell built by workers above the plastic queen cup.

### 2.1.2. *Management of colonies used for producing royal jelly*

(1) Select strong colonies (standard colonies or colonies with an additional super) for the production of royal jelly.
(2) Divide the colony with a queen excluder into a propagation area (with a queen) and a production area (without a queen).

Note: The propagation area consists of 1 queen and 4–6 combs, including pollen combs, pupa combs, and empty combs for egg laying. The production area includes more than 5 combs with 1–2 pollen combs and several brood combs.
(3) Place brood combs in the middle of the frames, with pollen combs to their sides to favor brood tending by nurse bees.
(4) Insert the royal jelly production frame between 2 brood combs or between a brood comb and a pollen comb. Breed young larvae of the right age [see the *BEEBOOK* paper on queen rearing (Büchler et al., 2013)].

Note: To facilitate rapid larvae grafting, 2 queens are introduced into the same colony (with a shutter in the middle to separate them). Each separated area then includes 3–4 combs, with an empty comb close to the shutter for egg laying, and pollen combs close to the walls of a hive. Alternatively, a multiple-queen colony is organized to allow several queens to lay eggs on 1 comb [see the *BEEBOOK* paper on miscellaneous methods (Human et al., 2013) for the method to create multiple-queen colonies].

Preferably, the empty comb should be brown or light brown in order to cater to the queens' preference for used combs as well as to improve the visibility of larvae.

#### 2.1.3. *The procedure for producing royal jelly by artificially grafting larvae*

There are five steps in the process of royal jelly production, including cleaning queen cells, grafting larvae, inserting royal jelly production frames, supplementary grafting, and harvesting.

##### 2.1.3.1. *Cleaning queen cells*

(1) Clean queen cells with a detergent in warm water to remove the occasional layer of a grease-like substance on the inner wall of a new plastic queen cell.
(2) Dry the queen-cell-base-bar.
(3) Assemble the queen-cell-base-bar into a royal jelly production frame.
(4) Place the assembled frame into a colony for cleaning by the honey bees for 12–24 h.

##### 2.1.3.2. *Grafting of larvae*

(1) Put the royal jelly production frame on the comb.
(2) Rotate the queen-cell-bar so that the opening of queen cells faces upward.
(3) Insert the tongue of a grafting pen along the inner wall of a worker cell until it slides underneath the body of a larva.
(4) Lift the larva together with the royal jelly.
(5) Insert the grafting pen until the tongue rests on the bottom of a queen cell.
(6) Gently press the spring-loaded plunger to force the larva and royal jelly off the tongue.
(7) Turn the openings of queen cells downward after each queen cell on the bar is loaded with a larva.
(8) Insert the frame into a royal jelly production colony.

Note: the grafting of larvae is the most important step during the process of producing royal jelly. The grafting action has to be gentle and fast and should be completed in a single lift-and-insert operation to avoid damaging the larvae. Otherwise, the larva should be discarded and a new larva, grafted. The best larvae age for the grafting is 12–18 h after they have hatched. Larvae at this age have a curvy body, a shovel-like shape, and are bathed in royal jelly.

##### 2.1.3.3. *Inserting the royal jelly production frame*

(1) Insert the royal jelly production frame into a royal jelly production colony as soon as possible after all queen cells are loaded with larvae. Generally, a colony with 8–10 combs can support 1 or 2 royal jelly production frames, and a colony with more than 10 combs can have 2 or 3 royal jelly production frames.
(2) Open the hive 3–5 h after the royal jelly production frame was inserted into the royal jelly production colony, to check the acceptance of queen cells. If a larva in a queen cell has been accepted, it will be surrounded by nurse bees, and the queen cell will have fresh royal jelly secreted by workers. If there is damage during the transfer or rejection of a larva happens, the larvae will have been removed by workers. Supplementary grafting will be required in these situations, see Subsection 2.1.3.2.

##### 2.1.3.4. *Harvesting royal jelly*

(1) Retrieve the royal jelly production frame from the hive 68–72 h after larvae grafting.
(2) Gently shake the frame to get rid of the bees.
(3) Brush it to remove the residual bees.
(4) Use a stainless steel queen cell cutting blade to cut off the protruding part of a queen cell.
(5) Remove the larvae from the queen cells using tweezers.
(6) Collect royal jelly from queen cells using a royal jelly scraping bar.
(7) Transfer the royal jelly into a royal jelly bottle for storage.
(8) Immediately freeze the bottle (see Subsection 3.1.2) or place it temporarily in a thermos flask with ice.

### 2.2. A method for producing royal jelly without grafting larvae

A new method for royal jelly production that does not require the grafting of larvae was recently described (Human et al., 2013; Pan, Wu, Guan, & Zeng, 2013; Wu et al., 2015; Zeng, 2013). The conventional method of producing royal jelly, which is based on grafting young larvae, is time-consuming, labor-intensive, and limited by the availability of larvae and the eyesight of the technician. The new method is based on a device that obviates grafting and is therefore convenient as well as efficient. The devices consist of a plastic worker foundation with regular holes (Figure 7), plastic cell bottoms mounted on a bar that can be inserted into the comb to fill the holes (Figure 8) and into the bottomless plastic queen cups on royal jelly production bars (Figure 9).

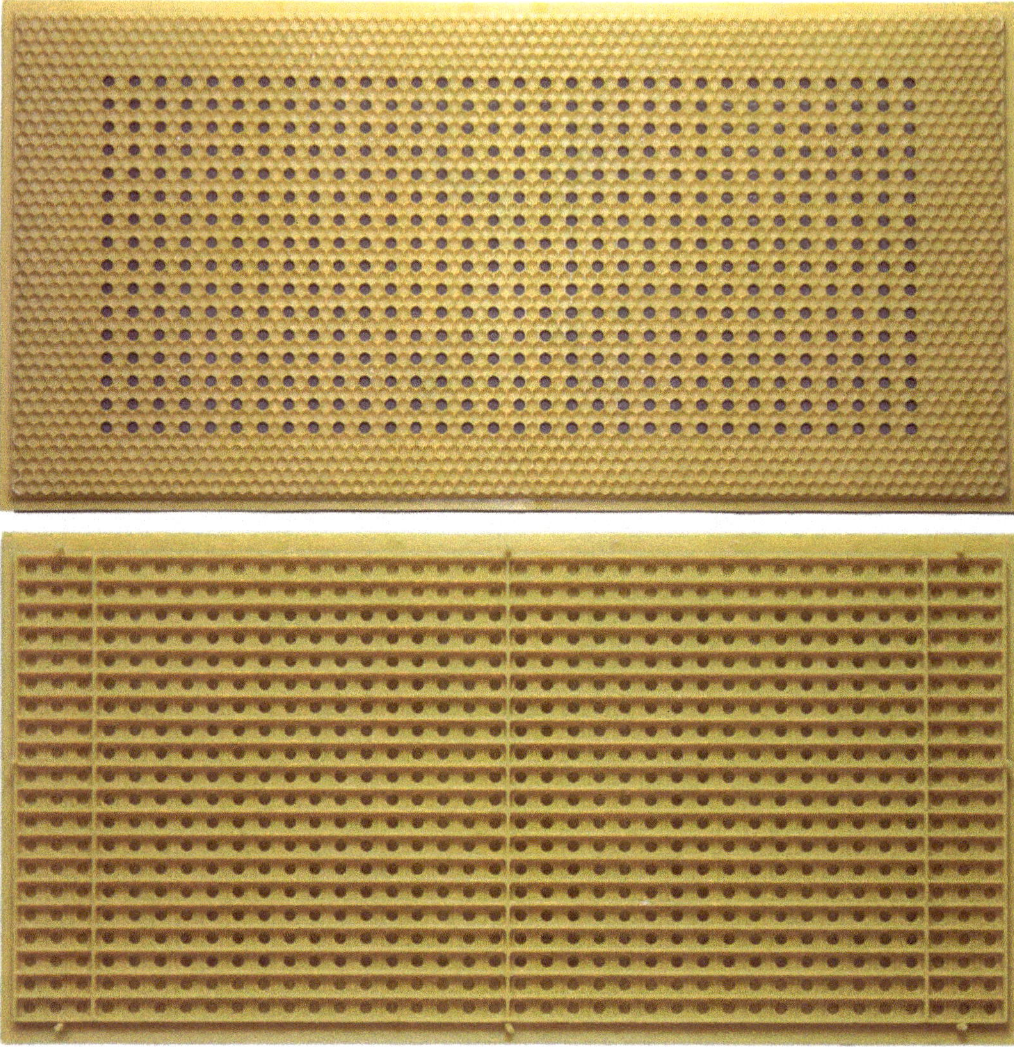

Figure 7. The plastic worker foundation with regular holes. Top: the front side. Bottom: the back side. Photo: Qizhong Pan.

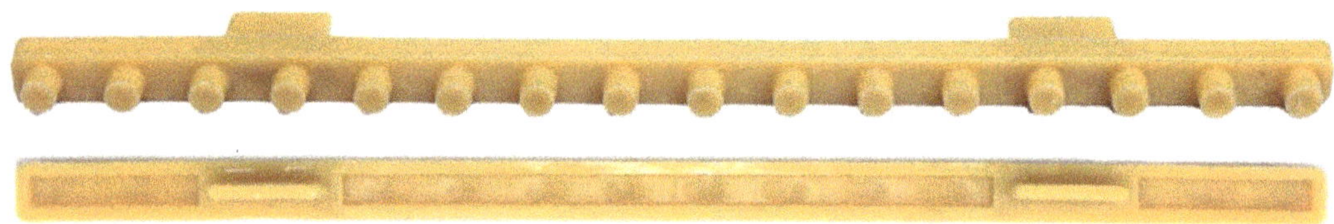

Figure 8. Cell bottom bar. Top: the front side. Bottom: the back side. Photo: Qizhong Pan.

In short, the foundation and bars with cell bottoms are assembled and placed in a colony to allow for comb building by workers. A queen is then caged on this comb to lay eggs. Four days later, when the eggs hatch into larvae, the bars with cell bottoms are taken out and inserted into the openings of queen-cup-bars. The completely assembled queen-cell-bars are placed into a colony to produce royal jelly. See Figure 10 for the combination of a royal jelly production bar and supporting larva devices and Figure 11 for the assembled worker foundation in a frame by the combination of a plastic worker foundation with regular holes, cell bottom bars, and a frame. A cover plate is added to the back side of the worker foundation (Figure 11, bottom). A cover plate (Figure 12) for an assembled queen-cell-bar is present to improve the efficiency of the method of producing royal jelly without grafting larvae. When the bottomless royal jelly production bars are assembled

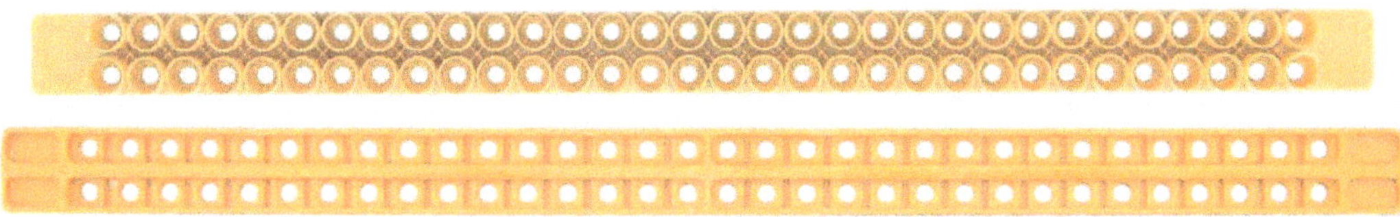

Figure 9. Royal jelly production bar. Top: the front side. Bottom: the back side.
Photo: Qizhong Pan.

Figure 10. The combination of a royal jelly production bar and cell bottom bars.
Photo: Qizhong Pan.

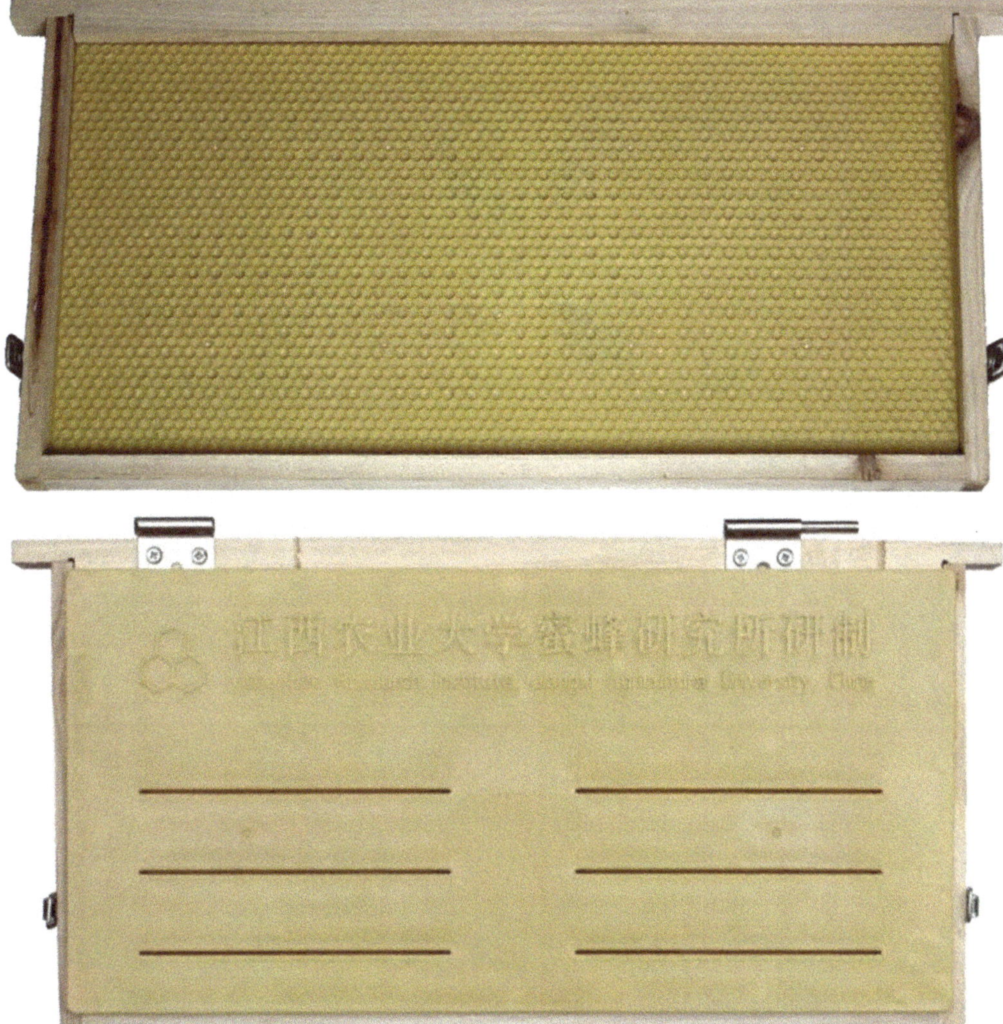


Figure 11. The assembled worker foundation in a frame by the combination of a plastic worker foundation with regular holes, cell bottom bars and a frame. Top: the front side. Bottom: the back side.
Photo: Qizhong Pan.

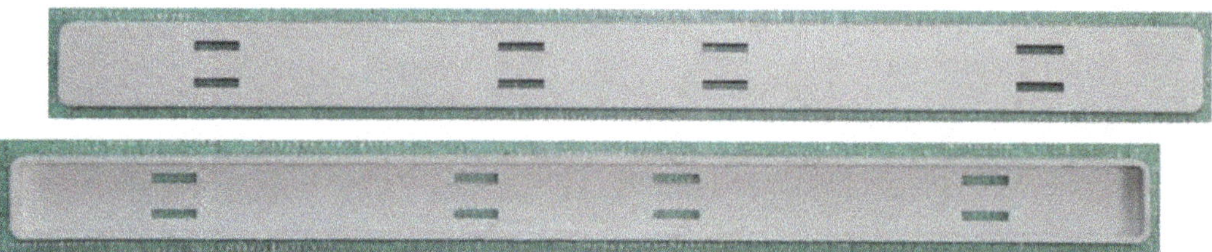

Figure 12. A cover plate for a royal jelly production bar. Top: the front side. Bottom: the back side.
Photo: Qizhong Pan.

Figure 13. Royal jelly production frame.
Photo: Qizhong Pan.

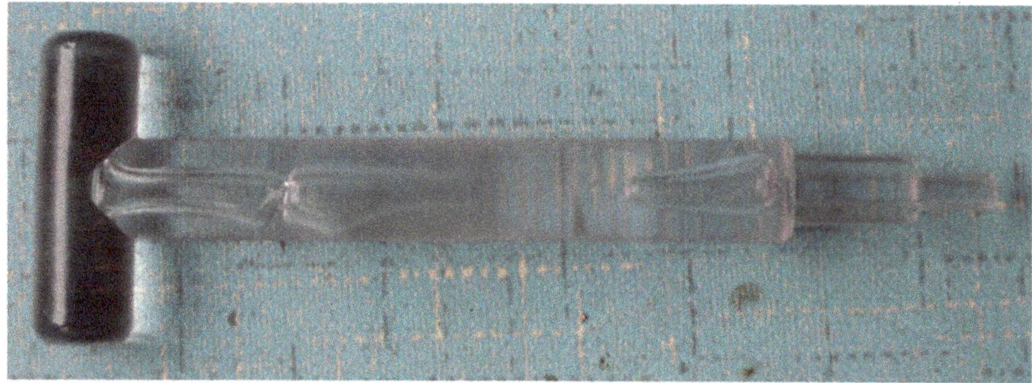

Figure 14. A queen cell cleaner for producing royal jelly without grafting larvae.
Photo: Fei Zhang.

with the cell bottom bars, there may be some hidden gaps. This cover prevents workers from secreting wax in the hidden gaps. A royal jelly production frame consists of four royal jelly production bars, cell bottom bars, and cover plates (Figure 13). See Figure 14 for a queen cell cleaner for royal jelly production without grafting larvae.

There are three main procedures: building of the comb by workers, organizing an egg-laying colony, and production of royal jelly without grafting larvae.

### 2.2.1. *Building the comb*

(1) Select a strong colony with enough honey and pollen.
(2) Reduce the number of combs to 4–5 to ensure overcrowding.
(3) Boil old combs in water.
(4) Collect the water solution.
(5) Soak the plastic worker foundation with regular holes and cell bottom bars for 24 h.

(6) Let the parts dry and then assemble the foundation and cell bottom bars together.
(7) Apply a layer of beeswax to the assembled foundation.

There are 2 ways of applying wax:

- Use a broad brush to spread a thin layer of melted beeswax on the front side of the foundation.
- Alternatively, firstly, boil beeswax with a proper amount of water to achieve appropriate wax concentration (not too high) in a large iron container, then hold one side of the hollow comb foundation by hand, and dip a half of the foundation into the melted wax and pull it out immediately and shake rapidly. Do the same again for the other half of the comb foundation. Make sure that wax is in a liquid state during the whole wax application process.

These steps are meant to facilitate acceptance of the plastic foundation and comb building.

(8) Mount the assembled foundation on a frame.
(9) Place the frame in a colony, allowing bees to build the comb.
    If the procedure is not done during a strong nectar flow, feed the colony every night to promote the comb construction.
(10) Take out the comb immediately after it is built to prevent workers from using it for storage.
(11) If workers had time to store pollen and honey, place the comb in a centrifuge to remove honey.
(12) Remove pollen from cells with a tool of appropriate size.

### 2.2.2. *Obtaining brood*

(1) Create a multiple-queen brood provider colony [see the *BEEBOOK* paper on miscellaneous methods (Human et al., 2013)]. Such colonies ensure the availability of a large number of similarly aged larvae for producing royal jelly. Alternatively Use a colony with a new queen. This approach should ensure a high egg-laying rate.
(2) Partition the colony into an egg-laying area and a hatching area, for example, with a queen excluder covered by 2/3–3/4 with a thin wood or plastic board. A partly covered queen excluder reduces the amount of worker bees going through and results in less nectar or pollen storage in the comb in the egg-laying area.
(3) Close the entrance to the egg-laying area, while the entrance to the hatching area remains open.
(4) Place a 2-sided or a single-sided comb built from an assembled plastic foundation in the egg-laying area so as to allow a new queen or several queens to lay eggs in this comb.

### 2.2.3. *Production of royal jelly*

When 4 two-sided or single-sided combs built from assembled plastic foundations are prepared, colonies can be organized for royal jelly production. There are five steps involved: cleaning up queen cells, laying of eggs, removal of larvae, insertion of frames, and harvesting of royal jelly.

#### 2.2.3.1. *Cleaning up queen cells*

(1) Boil old combs in water.
(2) Collect the water solution.
(3) Soak royal jelly production bars and cell bottom bars for royal jelly production for 24 h.
(4) Let the parts dry.
(5) Assemble the parts into a royal jelly production frame.
(6) Place the assembled frame in the royal jelly production colony to allow worker bees to clean it up for a day.

#### 2.2.3.2. *Laying eggs*

(1) Mark 4 clean built-up combs from assembled worker foundations on their beams with numbers 1, 2, 3, and 4.
(2) Before egg-laying, allow bees in a colony to clean up comb No. 1 for 5–6 h.
(3) Insert comb No. 1 into a well-organized egg-laying colony (with several queens or a new queen laying eggs).
(4) Allow queens to lay eggs for 24 h in this comb.
(5) Take the comb out, shake off the queens, and place the comb into the hatching area.
(6) Simultaneously, put comb No. 2 into a well-organized egg-laying colony for 24 h and repeat the later step (step 5) of comb No. 1.
(7) The same steps are repeated for combs No. 3 and No. 4.
(8) When comb No. 4 is placed in the hatch area for hatching, take comb No. 1 out. Eggs on the cell bottom bars of comb No. 1 now have hatched into 1-day larvae (Figure 15).
(9) Pulled out the cell bottom bars loaded with larvae on comb No. 1 for royal jelly production.
(10) Soak comb No. 1 without cell bottom bars in clean water for 5–10 min.
(11) Spin off water, larvae, and eggs from comb No. 1 in a honey centrifuge.

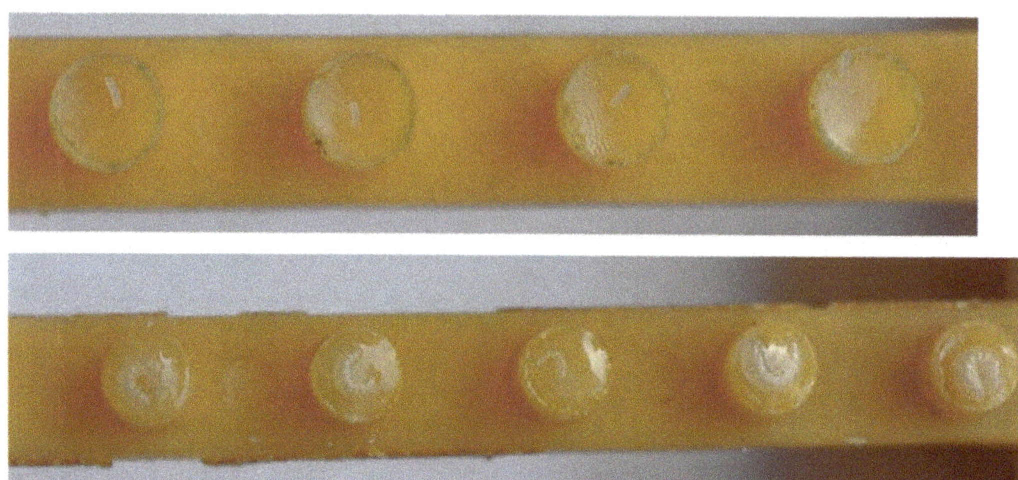

Figure 15. Eggs and larvae on cell bottom bars. Top: eggs. Bottom: larvae.
Photo: Zhijiang Zeng.

Figure 16. Royal jelly production frame covered with honey bees.
Photo: Qizhong Pan.

(12) Assemble this comb with new cell bottom bars again.
(13) Put the assembled comb in a colony for cleansing for 5–6 h.
(14) Insert it into a royal jelly production colony to allow queens to lay eggs again.

2.2.3.3. *Taking out larvae*

(1) Allow the larvae-loaded comb to hatch for 3 days.
(2) Pull out cell bottom bars loaded with larvae.
(3) Insert the cell bottom bars into a royal jelly production frame for royal jelly production.
(4) The action of pulling out cell bottom bars should be gentle, quick, and of regular intensity. Furthermore, there should be no gaps between cell bottom bars and royal jelly production bars to prevent a decrease in the acceptance rate of queen cells.
(5) Cover the assembled royal jelly production bars with cover plates.
(6) Assemble new unloaded cell bottom bars into the combs with holes in time for queens' continuously laying eggs.

2.2.3.4. *Inserting frames*

(1) Insert a royal jelly production frame between a larval comb and a pollen comb in royal jelly production colonies.

Generally, a colony with 8–11 combs can hold 1 royal jelly production frame. Nevertheless, when there are abundant nectar and pollen sources, 2 royal jelly production frames can be inserted into a colony with more than 12 combs.

#### 2.2.3.5. *Harvesting royal jelly*

(1) After 68–72 h of the frame insertion, take the royal jelly production frame out of the royal jelly production colony (Figure 16).
(2) Shake off workers remaining on the frame lightly and clean up the frame with a bee brush.
(3) Cut off the protruding part (made of wax) of queen cells (see Subsection 2.2.4.1).
(4) Pull out cell bottom bars.
(5) While harvesting royal jelly from a royal jelly production bar, clean bottomless queen cells by a queen cell cleaner (Figure 14). When bottomless queen cells in the royal jelly production bar are clean, the bar can be assembled with cell bottom bars again, and a new cycle of royal jelly production begins all over again.

### 2.2.4. *Harvesting royal jelly by a machine*

Three steps of the royal jelly production without grafting have been mechanized: cutting off the protruding part of a queen cell, removal of the larvae, and harvesting of royal jelly. The two necessary tools are a wax-cutting device and a blower for harvesting royal jelly. The process includes five steps: cutting down the protruding part of queen cells, blowing royal jelly, filtering royal jelly, collecting royal jelly, and cleaning up.

#### 2.2.4.1. *Cutting off the protruding part of queen cells*

(1) Push cutter to the left side, at the starting position (Figure 17).
(2) Draw out the royal-jelly-collecting basin along the horizontal guide line to the right (Figure 17).
A piece of warm damp cloth is used to wipe the cutter to facilitate cutting down the protruding part of a queen cell.
(3) Remove royal jelly production frames from a colony.
(4) Brush away worker bees.
(5) Pull out cell bottom bars.
The royal jelly remains in the queen cup.
(6) Wedge the royal jelly production bars from the frames in a working platform (Figure 18).
The cup openings of the royal jelly production bars and royal jelly are directed towards the bottom. The platform can accommodate 7 royal jelly production bars.
(7) Place the platform in the wax-cutting device. A wax-collecting basin is placed underneath.
(8) Hold the 2 handles of the cutter and pull away from the starting position as shown in Figure 18. The cutting blade is underneath the royal jelly production bars.

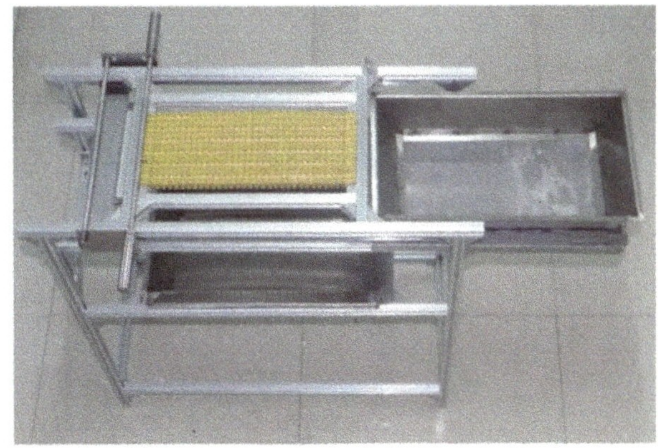

Figure 17. A wax cutting device for royal jelly production.
Photo: Qizhong Pan.

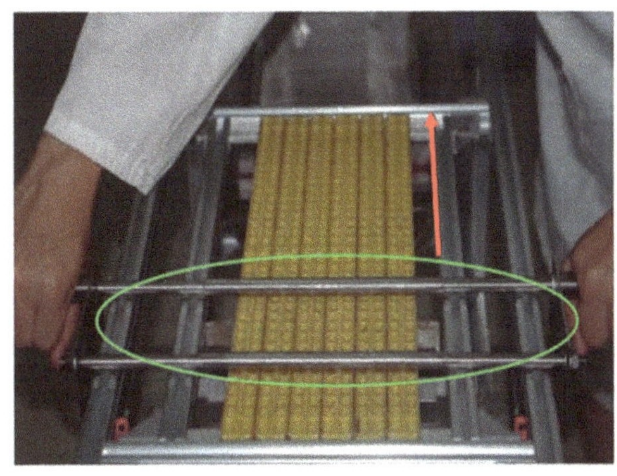

Figure 18. The operation of cutting off the protruding part of queen cells.
Photo: Qizhong Pan.

(9) Slide the cutter forward to cut the protruding part of queen cells.
(10) Slide the cutter back to the starting position.
One back-and-forth movement of these cutters can complete wax cutting of several royal jelly production bars. The removed wax will drop into the wax-collecting basin automatically.

#### 2.2.4.2. *Blowing royal jelly*

(1) Push the royal-jelly-collecting basin beneath queen cell bars.
Note: There is a royal jelly filter screen above the royal-jelly-collecting basin (Figure 19).
(2) Turn on an oil less air compressor.
(3) Aim the spray gun to bottomless openings of royal jelly production bars at 45° (Figure 19).

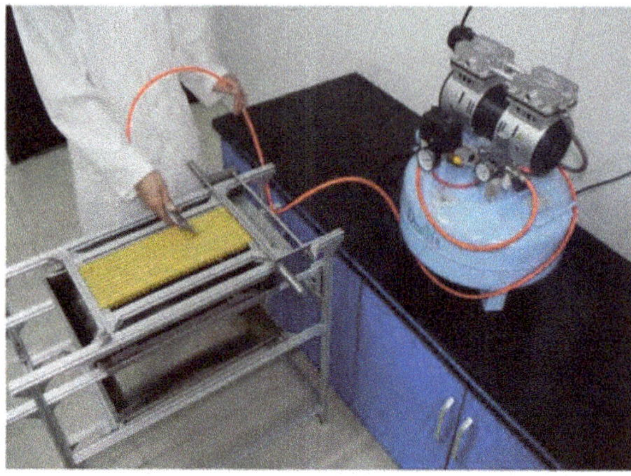

Figure 19. The operation of blowing royal jelly into the basin at the bottom.
Photo: Qizhong Pan.

Figure 20. The operation of filtering through royal jelly by a filter screen to separate royal jelly from larvae.
Photo: Linbin Zhou.

(4) Press the lever switch, and blow larvae from queen cells into the filter screen with weak airflow (3–4 kPa).
(5) Move the gun back and forth 2–3 times continuously, blowing with a stronger airflow (7–8 kPa) at the bottomless openings of royal jelly production bars, to blow royal jelly (that is attached to the queen cell wall) into the filter screen. At this stage, most of royal jelly will be filtered through the screen to the basin, and the larvae will remain above the screen.

*2.2.4.3. Filtering royal jelly*

(1) Blow the residual royal jelly found on the filter into the royal-jelly-collecting basin using weak airflow (Figure 20).
(2) Discard the larvae from the filter.
(3) Proceed to the next collection round.

*2.2.4.4. Collecting royal jelly*

(1) Collect royal jelly from the collection basin when a required amount has accumulated.
(2) Freeze or process it as required (see Subsection 3.2).

*2.2.4.5. Cleaning up.* Clean all components of the royal-jelly-collecting machine and air compressor according to instruction manuals.

See Table 1 for the pros and cons of the two methods for production of royal jelly.

## 3. Quality control of royal jelly

### 3.1. Storage of royal jelly

*3.1.1. Freshness of royal jelly*

Royal jelly is sensitive to heat, light, and air. Although there are no data on changes in its biological effectiveness for humans after inappropriate long-term storage, various changes in physical and chemical properties have been reported, such as a higher acid titre, a darker color, higher viscosity, a large insoluble-protein fraction, lower amounts of free amino acids, and less glucose oxidase (Albalasmeh, Berhe, & Ghezzehei, 2013; Kamakura, Fukuda, Fukushima, & Yonekura, 2001; Karaali et al., 1988; Takenaka, Yatsunami, & Echigo, 1986). Special precautions are thus necessary to preserve the biological properties of royal jelly during the shelf period.

*3.1.2. Freezing of royal jelly*

Refrigeration and freezing delay and reduce the chemical changes in royal jelly during storage. The following points should be considered for storage of fresh royal jelly.

(1) Transfer royal jelly into a dark and airtight container immediately after collection.
If royal jelly will be used rapidly,
(2) Refrigerate at 0–5 °C.

Table 1. Comparison of the two methods for producing royal jelly.

| Method | Advantages | Disadvantages |
|---|---|---|
| Method with grafting of larvae | Beekeepers are familiar with this technology; the acceptance rate of queen cells is high; the tools for production are simple | Artificial larvae grafting is labor-intensive; beekeepers should have good eyesight |
| Method without grafting of larvae | Beekeepers with poor eyesight can produce royal jelly; grafting of larvae is not needed | The tools for production are costlier; the egg-laying queen needs to have high fecundity |

Alternatively, if royal jelly will be stored for a longer period,

(3) Freeze at temperatures below −18 °C.
   - To protect royal jelly from light, it must be packed in dark containers.
   - To protect it from oxidation, the container must be airtight.
   - Because there are no criteria for establishing "safety" limits for product activity, storage and shelf-life should be as brief as possible.
   - After defrosting and packaging, the product should not be stored in a refrigerator for more than 12 months.
   - Repeated freeze-thaw cycles must be avoided.

### *3.1.3. Freeze-drying of royal jelly*

Freeze-drying, also known as lyophilization or cryodesiccation, is a dehydration process typically used to preserve a perishable material or make the material more convenient for transport. The process works by freezing the material and then reducing the surrounding pressure to allow the frozen water in the material to sublimate directly from the solid phase to the gas phase (Franks, 1998). Compared to fresh royal jelly, a lyophilized powder of royal jelly has the following advantages:

- Royal jelly powder is less sensitive to temperature, and can be kept at room temperature for a longer period, making longer-range transportation possible.
- Royal jelly powder can be further processed into other products, such as royal jelly tablets, or mixed with other products.

The developmental and physiological activity of fresh royal jelly and freeze-dried royal jelly were compared on the fruit fly by Kayashima, Yamanashi, Sato, Kumazawa, and Yamakawa-Kobayashi (2012). The result showed that freeze-dried royal jelly retained certain properties of fresh royal jelly, suggesting that the freeze-drying process does not alter the quality and the functions of royal jelly significantly (Kayashima et al., 2012). Generally, there are several steps to freeze-dry royal jelly, as described below.

(1) Pass freshly collected royal jelly through a 100-mesh filter, to remove potential contamination with beeswax and larvae before freeze-drying. If necessary, distilled water can be added to reduce royal jelly viscosity.
(2) Transfer the filtered royal jelly to open vessels such as glass plates and pre-freeze the sample in a freezer at −18 °C or in a freeze-dryer.
(3) Set chamber temperature of a freeze-dryer to −40 °C and place the sample into the chamber of the freeze dryer.
(4) Start to vacuumize the chamber until the chamber pressure reaches 1.33 Pa.
(5) Raise the shelf temperature to −25 °C, and keep it at this temperature for 12 h.
(6) Then raise the temperature to 30 °C and maintain it for 6 h.
(7) Package the royal jelly powder in an airtight container because freeze-dried royal jelly powder is very hygroscopic.

Because the functions and capacity of different freeze-dryers vary, for detailed operations, please refer to the corresponding operation manuals.

Freeze-dried royal jelly is certainly more stable than the fresh product, but it is still best to store it by refrigeration or freezing because we still know little about the composition and changes in biological effectiveness of royal jelly powder during processing and storage.

### *3.1.4. Freshness parameters of royal jelly*

Methods for evaluation of freshness of royal jelly are useful for the bee product industry, and identification of an indicator to measure freshness of royal jelly received a lot of attention from researchers. Trans-10-hydroxy-2-decenoic acid (10-HDA), an important quality criterion of royal jelly, has been demonstrated to be stable in royal jelly and to not be affected by the storage conditions (Antinelli et al., 2003; Kamakura et al., 2001). The enzymatic activities of superoxide dismutase (Tang & Yuan, 1999) and glucose oxidase (Wu & Zhang, 1990) were found to be affected by storage duration and conditions, but the values decreased rapidly. Chen and Chen (1995) have found that the proteins in royal jelly are affected by storage conditions (Chen & Chen, 1995; Tang & Yuan, 1999; Wu & Zhang, 1990). Kamakura et al. (2001) have stated that major royal jelly protein 1 (MRJP1, also known as royalactin) is a suitable freshness indicator because its concentration shows a good correlation with storage duration and conditions (Kamakura et al., 2001). With proteomics approaches, other MRJPs were found to be affected by storage conditions, and MRJP5 was proposed as a suitable freshness marker (Li, Feng, Zhang, Zhang, & Pan, 2008; Zhao et al., 2013). The concentration of furosine, a product of an early Maillard reaction between amino acid residues and reducing sugars, was proven to be very low in fresh royal jelly samples and increased over time with some dependence on temperature (Marconi, Caboni, Messia, & Panfili, 2002). Zheng, Wei, Wu, Hu, and Dietemann (2012) found that via an unknown mechanism, the color of the mixture of royal jelly and HCl shows a good correlation with storage duration and temperature (Zheng et al., 2012). Recently, formation of 5-hydroxymethyl-2-furaldehyde (HMF) in fresh royal jelly as a function of storage temperature and storage duration was monitored by means of an Reverse Phase-High Performance

Liquid Chromatography (RP-HPLC) method (Ciulu et al., 2015). The authors suggested the use of HMF as a possible freshness marker for royal jelly according to the finding that all the samples stored at −18 °C and 4 °C did not show any detectable levels of HMF, whereas those stored at 25 °C showed exponentially increasing amounts of HMF from the first month of storage.

Despite several criteria available as royal jelly freshness indexes, no method has been established as a standard assay of royal jelly freshness. Further research is required to clarify the variations of these criteria in different royal jelly samples, e.g., royal jelly produced in different seasons and/or areas (Wei et al., 2013) and/or by different races of bees, and to establish a threshold for each level of freshness.

## 3.2. *Quality criteria*

### 3.2.1. *Water content*

Water content is an important quality criterion of raw royal jelly. Several methods have been used for the measurement of water content of royal jelly: drying at different temperatures under reduced pressure (namely loss on drying) and refractometric analysis (Sesta & Lusco, 2008) and Karl Fischer titration (Teresa, Helena, Ewa, & Piotr, 2009). Karl Fischer titration is a classic titration method of analytical chemistry and involves colormetric or volumetric titration to determine trace amounts of water in a sample. The popularity of the Karl Fischer titration is due in large part to several practical advantages that it offers (over other methods of moisture determination), for example, accuracy, speed, and selectivity.

Presently, the titration is done with an automated Karl Fischer titrator. Elaboration of the procedures depends on the optimal sample weight and minimal homogenization time. For royal jelly, these two parameters were established at 0.02–0.05 g and 120 s, respectively, by Teresa et al. (2009).

For the cases when an automated Karl Fischer titrator is not available, the loss-on-drying method is recommended because it is a classic laboratory assay for measuring high levels of moisture in solid or semi-solid materials like royal jelly. Loss-on-drying analysis can be conducted at different temperatures (e.g., 48, 65, 90, and 105 °C) under atmospheric or reduced pressure.

A vacuum oven drying method has been described by Sesta and Lusco (2008). To enable complete water loss, the sample has to be spread in a thin layer on a wide surface. At the weighing step, however, this approach causes a visible loss of water by evaporation, producing unstable and continuously decreasing readings of weight on the scale.

To obtain an accurate measure, the method for determining the water content in royal jelly requires some care.

(1) Turn on the oven and the vacuum pump 2 h before inserting the samples.
(2) Set the temperature to 48 °C.
(3) Set the vacuum to higher than −800 mbar, resulting in a pressure lower than 200 mbar.

Due to vacuum, which accelerates water loss, a relatively low temperature (48 °C) can be employed. The 48 °C temperature was proven to be more conservative for royal jelly because it leaves the color of the sample unchanged, and proved to be sufficient for drying because it prevents formation of a superficial brown sugar crust that hampers complete evaporation (Sesta & Lusco, 2008).

(1) Weigh the Petri dish (where the sample is to be spread) on an analytical scale (precision 0.1 mg).
(2) Record the weight of the Petri dish as "Petri".
(3) Weigh the clean spatula to be used for spreading the sample.
(4) Record the weight of the spatula as "clean spatula".
(5) Transfer a homogenized royal jelly sample of approximately 3 g to the Petri dish as fast as possible.
(6) Immediately weigh it on the analytical scale.
(7) Record the weight as "Petri + R".
(8) Spread the royal jelly uniformly over the dish surface using the weighed spatula.
(9) Weigh the spatula, on which some residue of royal jelly remained.
(10) Record the weight as "dirty spatula", to take into account the weight of the residual sample left on the spreading tool.
(11) Calculate the accurate amount of royal jelly sample processed as follows: "royal jelly" = ("Petri + R" − "Petri") − ("dirty spatula" − "clean spatula").
(12) Transfer the Petri dish to a vacuum oven.
(13) Incubate for 24 h.
(14) Transfer the dried sample into a desiccator.
(15) Allow it to stand for 15 min to make sure it reaches the ambient temperature.
(16) Weigh the Petri dish on an analytical scale.
(17) Record the weight as "after drying".
(18) Calculate the water content as [("royal jelly" − "after drying")/"royal jelly"] × 100%.

In general, because the manual laboratory method is relatively slow, automated moisture analyzers based on the loss-on-drying method have been developed and can reduce the time necessary for a test from a couple hours to just a few minutes.

### 3.2.2. *10-HDA content*

10-HDA is a major fatty acid component of royal jelly. It has many pharmacological activities such as antitumor effects (Townsend, Brown, Felauer, & Hazlett, 1961), size- and lipogenesis-inhibiting activity toward the hamster ear sebaceous gland (Maeda, Kuroda, &

Motoyoshi, 1987), collagen production-promoting effects (Koya-Miyata et al., 2004), and antibiotic properties (Blum, Novak, & Taber, 1959). The presence of 10-HDA has been regarded as a marker to differentiate royal jelly from other products. Its concentration has been used as a parameter of royal jelly quality. According to the guidelines of the Ministry of Agriculture (MOA) in China and the ISO royal jelly international standard (ISO 12824:2106, 2016), the minimum concentration of 10-HDA is 1.4% for pure royal jelly.

Many different methods have been described for determination of 10-HDA in royal jelly, including high-performance liquid chromatography (HPLC) (Bloodworth, Harn, & Hock, 1995; Genç & Aslan, 1999), ultra-performance liquid chromatography (Zhou et al., 2007), gas chromatography (Anonymous, 1989), and gas chromatography–mass spectrometry (GC/MS) (see Section 6.3.1), among which, HPLC is a routine method and is thus cited in standard regulations. In particular, as a conventional method, HPLC can be carried out using standard equipment in many laboratories and is also simple, sensitive, and suitable for monitoring 10-HDA.

*Quantitative determination of 10-HDA by HPLC:*
*Requirements:*
Reagents: 10-HDA standard, methyl-4-hydroxybenzoate as the internal standard, methanol, absolute ethanol, hydrochloric acid, and phosphoric acid.
HPLC instrument: Waters 2695 with DAD detector.

*Procedure:*

Prepare a 1000 μg/ml internal standard solution (methyl-4-hydroxybenzoate) in absolute ethanol.

(1) Prepare working solutions of 0.1, 0.5, 5, 10, 20, 40, 80, 160 μg/ml 10-HDA in absolute ethanol (the concentration of the internal standard solution is 100 μg/ml).
(2) Weigh 0.5 g of royal jelly cream sample or 0.15 g of lyophilized powder (see Subsection 3.1.3) into a 50-ml volumetric flask.
(3) Add 3 ml of 1 mol/l HCl.
(4) Add 5 ml of a 1 mg/ml internal standard solution.
(5) Add 40 ml of absolute ethanol in sequence.
(6) Dissolve completely by ultrasonication with occasional shaking for 15 min.
(7) Add absolute ethanol to the marked volume (50.00 ml).
(8) Pass the solution through a 0.2-μm nylon filter before injection into the HPLC apparatus.

HPLC analytical conditions:

(1) Connect the chromatograph with a Waters Nova-pak C18 column (150 × 3.9 mm, 5 μm i.d.).
(2) Set the mobile phase to the 1/1 ratio (0.3% phosphoric acid in water/methanol).
(3) Set the flow rate to 0.8 ml / min.
(4) Set the column temperature to 30 °C.
(5) Set the injection volume to 5 μl.
(6) Set the detection wavelength to 210 nm.
(7) Set the total time to 10 min.
(8) Collect the data using Waters Empower software.

#### 3.2.3. *Protein content*

See Section 4.

#### 3.2.4. *Sugar content*

See Section 5.

## 4. Royal jelly protein research

### 4.1. *Isolation of royal jelly proteins and peptides*

#### 4.1.1. *Introduction*

The main protein components of royal jelly are major royal jelly proteins, accounting for 82% of total royal jelly protein, with molecular masses of 49–87 kDa assigned to one protein (and gene) family (Hanes & Šimuth, 1992; Malecová et al., 2003; Ohashi, Natori, & Kubo, 1997; Schmitzová et al., 1998). It has been suggested that this be denoted as MRJP, derived from major royal jelly protein, before their physicochemical properties were known (Hanes & Šimuth, 1992; Šimúth, 2001). The numerical symbol added to the acronym denotes the respective main protein. It was later found that all of these proteins are easily soluble in water, just as other albumin-like proteins: endosperm albumin, lactalbumin, ovalbumin, and serum albumin (Smith, 1997). The name albumin originates from the Latin term *albus;* it exists in nearly pure form in egg white, prompting the suggestion to rename major royal jelly proteins as *apalbumins* (prefix *ap* is derived from *Apis*). This means that apalbumin1 denotes MRJP1, apalbumin2 denotes MRJP2, and so on. These proteins are present not only in the larval food of a queen but also in larval food for the larvae of honey bee workers and drones. The term apalbumin implies the ubiquity and multifunctionality of royal jelly proteins; this name is a response to recent data that royal jelly proteins not only are present in royal jelly but also are synthesized in honey bee brain (Kucharski, Maleszka, Hayward, & Ball, 1998) and are authentic proteins of honey bee products (Bíliková & Šimúth, 2010; Di Girolamo, D'Amato, & Righetti, 2012; Šimúth, Bíliková, Kováčová, Kuzmová, & Schroeder, 2004). In this chapter, we will refer to the main protein of royal jelly as MRJP.

The gene family of MRJPs, encoding 9 closely related proteins is arranged in a 65-kb tandem array (The Honey bee Genome Sequencing Consortium, 2006). The most abundant royal jelly proteins consist of MRJP 1, 2, 3, 4, and 5 representing 90% of total proteins of royal jelly. MRJPs 6–9 are present in royal jelly in trace amounts. The molecular properties of MRJP (6–9) as authentic proteins isolated in natural form have been unknown until now. No conclusions about molecular properties of MRJPs 6–9 can be drawn from modern bioinformatics data. Other proteins such as alpha-glucosidase, glucose oxidase, and alpha-amylase have also been detected in royal jelly (Ohashi, Sawata, Takeuchi, Natori, & Kubo, 1996; Scheparts, 1965), but these enzymes are also present in honey. Proteins and peptides present in honey bee larval food play a significant role in honey bee life. They are a source of nutrition because of high concentrations of essential amino acids (Schmitzová et al., 1998). MRJPs are synthesized in cephalic glands of the honey bee (Hanes & Šimuth, 1992). Most recently, shotgun proteomics were used to identify the royal jelly proteome as well as proteomes of the hypopharyngeal gland, post-cerebral gland, and thoracic gland, from which royal jelly proteins are assumed to be derived (Fujita et al., 2013). MRJPs can be considered ubiquitous proteins, with multifunctional properties.

MRJP1 mRNA was found to be differentially expressed in heads of early emerged honey bees (Kucharski et al., 1998). The nurse brain showed increased expression of major royal jelly proteins (MRJP1, MRJP2, and MRJP7), which are related to the determination of castes during honey bee larvae differentiation (Garcia et al., 2009; Hojo, Kagami, Sasaki, Nakamura, & Sasaki, 2010). Unlike MRJPs 1–7, MRJPs 8 and 9 are expressed in the honey bee venom gland (Blank, Bantleon, McIntyre, Ollert, & Spillner, 2012).

The various health-promoting properties of royal jelly (Ramadan & Al-Ghamdi, 2012) and new proteomic data on royal jelly proteins initiated extensive research into the physiological functions of royal jelly proteins (Buttstedt, Moritz, & Erler, 2014). These proteins on a global scale are expected to have a significant impact on human immunotherapy and as potential proteinaceous antibiotics (Fontana, Mendes, Souza, & Konno, 2004; Šimúth et al., 2004; Tamura et al., 2009; Watanabe et al., 1998). The royalactin isoform of apalbumin1 is a queen determinant (Kamakura et al., 2001).

Recently, physiological properties of royal jelly and MRJPs were uncovered in another laboratory. Consequences of this fact are often problems with verification of experimental data for objective evaluation of the biological potential of royal jelly and MRJP1 from a methodological point of view. Therefore, the overview of the methods referred to in this chapter should contribute to standardization of methods used in research on royal jelly proteins.

#### *4.1.2. Isolation of major royal jelly proteins*

The systematic research on royal jelly proteins started with isolation of the individual major proteins of royal jelly using size exclusion and ion exchange column chromatography, one-dimensional sodium dodecyl sulfate-polyacrylamide gel electrophoresis (SDS-PAGE) (1DE), immunoblotting, isoelectrofocusing, cDNA cloning, and N-terminal amino acid sequencing (Bíliková et al., 2002; Schmitzová et al., 1998). These methods enabled characterization of the most abundant proteins of royal jelly, namely MRJP 1–5.

Differences in molecular and chemical properties among MRJPs are used for their isolation. The high molecular weight of the oligomeric form of MRJP1 is used for its isolation by size exclusion chromatography (see Subsection 4.1.2.2, Figure 21). MRJP1 is the most abundant protein in royal jelly and its acidic properties contribute to the acidic nature of royal jelly. MRJP2 is a slightly basic protein, and MRJP3, 4, and 5 are almost neutral proteins. These differences in polarity of royal jelly proteins are utilized in column ion exchange chromatography, where the acidic MRJP1 is eluted at 0.25 M NaCl, while the other MRJPs are present in the fraction eluted at low ionic strength (0.05 M NaCl).

These analyses formed the basis for further characterization of royal jelly proteins by proteomic methods. Before proteomic methods became available, conventional methods were applied to describe only the most abundant royal jelly proteins. Now, due to the proteomic approach, direct data on minor components at the molecular level may be derived. Latest discoveries about molecular properties and physiological function of royal jelly proteins during the last two decades help to provide new knowledge about royal jelly for proteomics and nutrigenomics and potential use for human health.

*4.1.2.1. Preparation of the water-soluble protein fraction from royal jelly.* The fraction of water-soluble proteins is prepared from fresh royal jelly according to Hanes and Šimuth (1992).

(1) Homogenize 30 g of royal jelly in a beaker (on a magnetic stirrer) containing 90 ml of phosphate buffer (50 mM $NaH_2PO_4$, 50 mM $Na_2HPO_4$, pH 7.0, 100 mM NaCl, 20 mM EDTA).

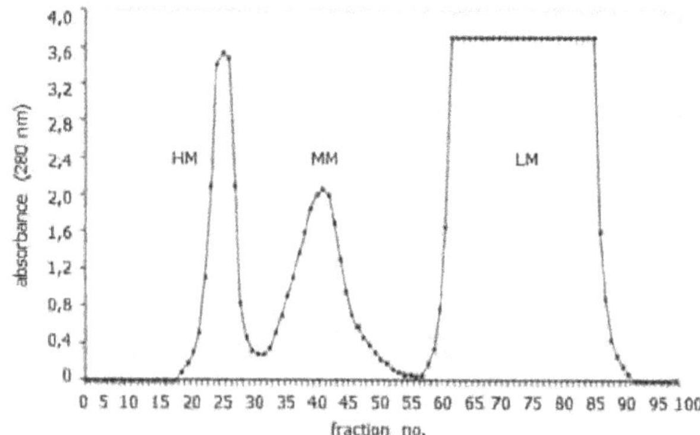


Figure 21. Fractionation of water-soluble fraction of royal jelly by size-exclusion chromatography. Sephadex G-200, elution by 50 mM NaCl in Tris/EDTA buffer, obtained high (HM), middle (MM), and low (LM) molecular weight protein fraction, respectively.

(2) Vortex the mixture for 30 min at room temperature.
(3) Centrifuge it 15,250g for 30 min, collect the supernatant.
(4) Dialyze the supernatant against 1 l of Tris-EDTA buffer (20 mM Tris–HCl pH 7.5, 1 mM EDTA) in tubing with molecular weight cut-off (MWCO) 10 kDa, at 4 °C.
(5) Change diffusates every 12 h, 4 times in total.
(6) Centrifuge at 15,250g for 30 min at 4 °C.
(7) Use the supernatant containing the water-soluble proteins for subsequent purification or concentrate the fractions by lyophilization (see Subsection 3.1.3) or by Macrosep 10 K (MWCO 10 kDa) centrifugal concentrators (Pall Gelman Sciences, Germany).
(8) Store solid protein samples at 4 °C, or protein solutions at −20 °C for further analysis.

*4.1.2.2. Fractionation of proteins of royal jelly by size exclusion chromatography.* Royal jelly proteins are purified by size exclusion chromatography from a water-soluble fraction of royal jelly (see Subsection 4.1.2.1), according to the following steps.

(1) Apply 30 ml of a sample (supernatant obtained by centrifugation of royal jelly) containing 2 g of proteins in Tris-EDTA buffer (20 mM Tris–HCl and 1 mM EDTA, pH 7) on a Sephadex G-200 (Pharmacia Biotech, Sweden) column (46 × 750 mm) at the flow rate of 1 ml/min.
(2) Wash the column with Tris-EDTA buffer.
(3) Perform elution by means of 50 mM NaCl in Tris-EDTA buffer at the flow rate of 1 ml/min.
(4) Collect 10-ml fractions.
(5) Monitor the absorbance of the fractions at 280 nm.
(6) Pool fractions corresponding to the peak with absorbance at 280 nm, in order to obtain 3 fractions: high (HM)-, middle (MM)-, and low (LM)-molecular-weight protein fraction (Figure 21).
(7) Dialyze the fractions in tubing with MWCO 3 kDa against Milli-Q water at 4 °C.
(8) Concentrate fractions by lyophilization (see Subsection 3.1.3) or using Macrosep 10 K (MWCO 10 kDa) centrifugal concentrators (Pall Gelman Sciences, Germany).
(9) Store solid protein samples at 4 °C or protein solutions at −20 °C for further analysis.

*4.1.2.3. Fractionation of royal jelly proteins by ion exchange chromatography.* Water-soluble royal jelly proteins (see Subsection 4.1.2.1) are used for purification of major royal jelly proteins by ion exchange chromatography.

(1) Load 150 ml of a dialyzed sample of water-soluble royal jelly proteins (1 mg protein per ml) at the flow rate of 1 ml/min on a DEAE (Diethylaminoethyl) cellulose column (35 × 400 mm) equilibrated with Tris-EDTA buffer (20 mM Tris–HCl pH 7.5, 1 mM EDTA).
(2) Wash the column with Tris-EDTA buffer.
(3) Use a linear gradient from 0 to 0.4 M NaCl in Tris-EDTA buffer for elution of the proteins.
(4) Collect 7-ml fractions, monitor the elution at 280 nm.
(5) Pool the fractions of the first peak eluted at approximately 0.05 M NaCl in Tris-EDTA buffer. The first peak corresponds to the weakly bound neutral and basic proteins, mainly MRJP2–MRJP5.
(6) Pool the fractions of the second peak eluted approximately at 0.2 M NaCl in Tris-EDTA buffer; it corresponds to the acidic protein MRJP1.
(7) Dialyze the protein solutions in tubing with MWCO 10 kDa against Milli-Q water.
(8) Concentrate fractions by lyophilization (see Section 3.1.3) or by means of Macrosep 10 K (MWCO 10 kDa) centrifugal concentrators with (Pall Gelman Sciences, Germany).
(9) Store the solid protein samples at 4 °C or protein solutions at −20 °C for further analysis.

*4.1.2.4. Isolation of oligomeric MRJP1 by HPLC. Preparation of a royal jelly protein sample for HPLC analysis:*

(1) Resuspend fresh royal jelly in phosphate-buffered saline (1:20, w/v) immediately before use.
(2) Dialyze 3 ml of the royal jelly suspension in cassettes with MWCO 3.5 kDa (PIERCE, Rockford, Illinois, USA) against distilled water for 1 week at 4 °C.
(3) Add 1.5 ml of distilled water to the dialyzed royal jelly.
(4) Incubate the royal jelly solution for 30 min at 37 °C.
(5) Centrifuge it at 12,500× *g* for 30 min at room temperature.
(6) Centrifuge the supernatant at 12,500× *g* for 30 min at 4 °C.
(7) Store the supernatant at 4 °C until analysis.
(8) Measure the total protein concentration in the sample using the Micro BCA protein Assay Kit (PIERCE), with human serum albumin (WAKO, Tokyo, Japan) as a standard protein.

*Size exclusion HPLC analysis:*

(1) Calibrate the column (Superose 12 column, 10 × 300 mm, GE Healthcare, Buckinghamshire, England) using Gel Filtration Calibration Kits with low-molecular-weight and high-molecular-weight proteins (GE Healthcare).
(2) Inject 100 μl of a sample into the column.

(3) Perform elution with phosphate-buffered saline (20 mM $Na_2HPO_4$, 2 mM $NaH_2PO_4 \cdot 2H_2O$, 150 Mm NaCl, pH 7.5) at the flow rate 0.5 ml/min.
(4) Collect 0.8-ml fractions and monitor HPLC elution profiles at 280 nm using the Akta Explorer System (GE Healthcare).
For anion exchange HPLC analysis, a miniQcolumn (4.6 × 50 mm, GE Healthcare) is used.
(1) Inject 1 ml of the sample in binding buffer (20 mM Tris–HCl pH 8.0) into the column.
(2) Perform elution with a buffer consisting of 20 mM Tris–HCl pH 8.0, 1 M NaCl, with NaCl gradient from 0 to 0.5 M at the flow rate 1 ml/min.
(3) Collect 1-mL fractions and monitor HPLC elution profiles at 280 nm using the Akta Explorer System (GE Healthcare).
(4) Desalt the samples by dialysis against distilled water.
(5) Concentrate the samples with Minicon (Millipore; Billerica, MA, USA).

### 4.1.3. *Isolation of royal jelly peptides*

Some honey bee antimicrobial peptides (apidaecin, abaecin, and hymenoptaecin) are induced specifically by pathogens and are released into haemolymph only after a bacterial infection, whereas others such as royalisin and apisimin are secreted into royal jelly. The molecular weight of royalisin is 5523 Da, and it consists of 51 amino acid residues, where six cysteine residues form three intramolecular disulphide linkages resulting in a compact globular structure. Royalisin belongs to the family of insect defensins: cysteine-rich cationic antimicrobial peptides.

Apisimin was found in the high-molecular-weight fraction, in the presence of MRJP1 (apalbumin1). Apalbumin1 tends to create a self-assembled structure and to form a gel. It appears that apisimin interacts with apalbumin1 and forms an oligomeric [apalbumin:apisimin] complex of yet unknown stoichiometry and structure. This complex is stable under the conditions used during size exclusion chromatography and ion exchange column chromatography. Native apisimin was obtained by separation of the high-molecular-weight fraction using ion exchange FPLC (Fast Protein Liquid Chromatography) in the presence of 100 mM glycine. In the absence of glycine, apisimin is eluted together with apalbumin at 0.2 M NaCl.

*4.1.3.1. Purification of cationic royal jelly peptides by a 2-step dialysis.* Dialysis is a simple method that can be used for separation of royal jelly proteins and peptides with appropriate buffers. Purification of royal jelly peptides according to Bíliková, Wu, and Šimúth (2001) is based on two step dialysis (Bíliková et al., 2001). The first dialysis of a royal jelly suspension in tubing with MWCO 2 kDa enables removal of sugars and the low-molecular-weight compounds. The second dialysis in tubing with MWCO 10 kDa allows for the diffusion of peptides from the retentate (royal jelly suspension) to the diffusate that are collected and concentrated. A simple scheme of 2-step dialysis of a royal jelly suspension is presented in Figure 22.

Wear gloves throughout the procedure to prevent contamination of the samples.

(1) Homogenize 30 g of royal jelly in a beaker (on a magnetic stirrer) containing 90 ml of phosphate buffer (50 mM $NaH_2PO_4$, 50 mM $Na_2HPO_4$, 100 mM NaCl, 20 mM EDTA, pH 7.0).
(2) Dialyze the royal jelly suspension in tubing with MWCO 2 kDa (Serva Electrophoresis, Heidelberg, Germany) at 15 °C against 1 l of 20 mM EDTA, pH 7.0.
(3) Change diffusates every 2 h, while monitoring the decrease in the amount of low-molecular-weight substances by measuring UV spectra in the range 210–350 nm, until the UV absorbance of diffusate at 210–350 nm cannot be observed, which means all low-molecular-weight compounds from retenate have been removed.
(4) Re-place the retenate into the dialysis tubing with a MWCO 10 kDa (Serva Electrophoresis, Heidelberg, Germany).
(5) Continue dialysis at 15 °C against 500 ml of Milli-Q water adjusted to pH 2 by addition of HCl.
(6) Change the diffusate every 12 h, 4 times.
(7) Concentrate diffusates by means of Macrosep 1 K (MWCO 1 kDa) centrifugal concentrators (Pall Gelman Sciences, Germany). Alternatively,
(8) Dry diffusates by lyophilization (see Subsection 3.1.3).
(9) Store dry peptide fractions at 4 °C or dissolve in an appropriate buffer according to the next experiments.

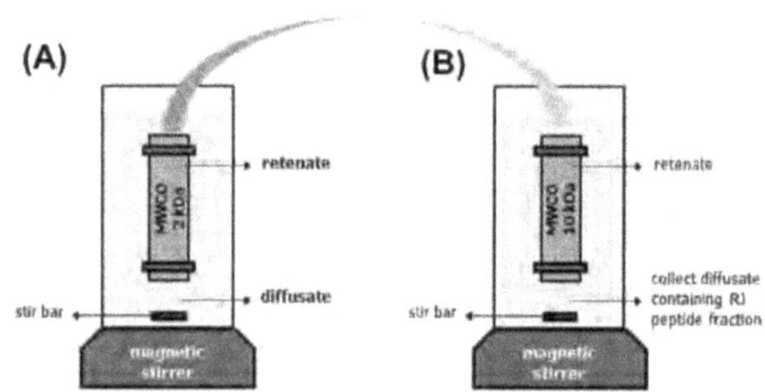


Figure 22. Scheme of two step dialysis of royal jelly suspension (Bíliková et al., 2001). (A) First dialysis performed at MWCO 2 kDa removes sugars and other low molecular weight compounds. (B) Diffusates obtained after dialysis of royal jelly suspension at MWCO 10 kDa containing peptide fraction, are collected and concentrated.

*4.1.3.2. Purification of royal jelly peptide apisimin by ion exchange FPLC.* Royal jelly peptide apisimin was purified from the HM fraction obtained by fractionation of a water-soluble fraction of royal jelly by size exclusion chromatography (see Subsection 4.1.2.2) according to Bíliková et al. (2002). All procedures for isolation of apisimin were carried out at 4 °C.

(1) Load 120 ml of the HM fraction at the concentration of 0.8 mg/ml, in Tris-glycine buffer (20 mM Tris-HCl, 100 mM glycine, pH 7.0) onto a DEAE Sepharose fast flow column (28 × 200 mm; Pharmacia Biotech, Sweden) at a flow rate of 2 ml/min.
(2) Wash the column with Tris-glycine buffer.
(3) Perform stepwise elution with 0.1 M NaCl, 0.15 M NaCl, and 0.2 M NaCl.
(4) Collect 7 ml fractions.
(5) Monitor the presence of apisimin by dot-blot analysis (see Subsection 4.2.4.1).
(6) Pool fractions containing apisimin.
(7) Concentrate and desalt the peptide fraction by means of Macrosep 1 K (MWCO 1 kDa) centrifugal concentrators (Pall Gelman Sciences, Germany).

*4.1.3.3. Electroelution of royal jelly proteins/peptides from the gel after preparative PAGE.* The royal jelly proteins or peptides can be purified in μg and/or mg quantities from the gel after preparative SDS-PAGE or native PAGE. The prepared protein or peptide can be used for immunization of experimental animals in order to raise antibodies, for characterization of the proteins, or for testing their physiological activity.

Wear gloves throughout the procedure to prevent contamination of the samples.

*Preparation of a protein sample for electroelution:*

(1) Dissolve 5 mg of the HM fraction (see Subsection 4.1.2.2) in 200 μl of Milli-Q water.
(2) Add 200 μl of 2× SDS-PAGE sample buffer and incubate for 10 min at 100 °C.
(3) Centrifuge at 18,000× g for 1 min.
(4) Load the supernatant on the gel of preparative native PAGE (160 × 140 × 1 mm).
(5) Run the electroseparation at a constant current, 12 mA.
(6) Stain the gel with 1 M KCl for 5–10 min, until white bands appear.
(7) Excise the peptide band of interest and wash the gel strip in Milli-Q water to remove the staining solution.

*Protein/peptide elution:*

(1) Cut the gel strip into smaller pieces.
(2) Place the gel pieces into the electro-eluter glass tubes with membrane caps of MWCO appropriate for the MW of the eluted protein or peptide (Figure 23).
(3) Perform electroelution using the Electro-Eluter instrument (Model 422, Bio-Rad, USA), in 1× elution buffer (Table 2) according to the instructions of the manufacturer. In general, constant current at 10 mA per glass tube for 3–5 h with vigorous stirring is sufficient. Increase elution time when using high-percentage gels or eluting high-molecular-weight proteins.
(4) Concentrate the eluate (containing the royal jelly protein or peptide) on a Microsep centrifugal column.
(5) Desalt by washing the column with Milli-Q water.

### 4.2. Characterization of major royal jelly proteins and peptides

The first necessary information for protein/peptide characterization is its concentration, which can be determined by several methods described previously in Section 1.2 of the *BEEBOOK* paper by Hartfelder et al. (2013). The principal molecular characteristics of proteins are determined by electrophoretic and immunochemical methods. Gel electrophoresis can provide information about the molecular weight, polarity, and subunit structure of proteins, and about the purity of a particular protein preparation. Isoelectric focusing (IEF) provides important information about ion properties of a protein.

Analysis by immunochemical methods is an indispensable part of characterization of bee proteins. Determination of the identity of proteins/peptides is possible by immunochemical methods with specific antibodies.

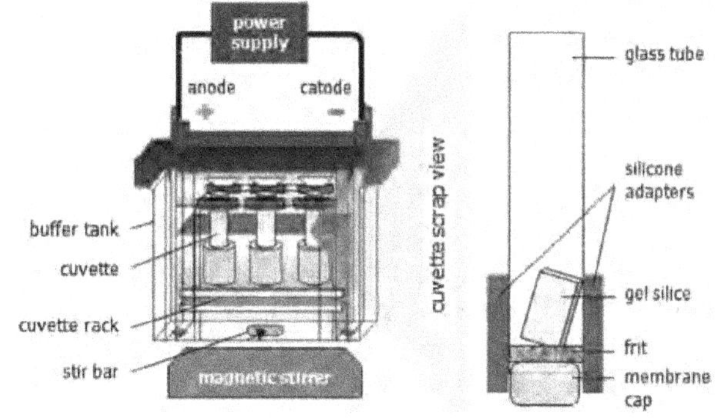


Figure 23. Electroelution performed on Elektro-Eluter (Model 422, Bio-Rad). The cuvette rack is placed into the main buffer tank. 1X elution buffer is added to the main tank until the assembly joints are covered in buffer. Likewise, 1X buffer is added to the upper buffer tank. Using a stir bar, the buffer in the main tank is gently agitated. The electrode plug is inserted into the power source. In general, each cuvette requires 10 mA of power (e.g., if three cuvettes are used, the power should be set at 30 mA).

Table 2. Composition of elution buffer for electroelution of a protein or peptide from the gel after SDS-PAGE. Store a stock solution (10×) at room temperature. Prepare a working solution (1×) by diluting the stock solution before use. In case of PAGE and electroelution under native conditions, do not add SDS to elution buffer.

| | 10× | | 1× |
|---|---|---|---|
| Elution buffer | Composition | Concentration | Concentration |
| Tris | 30 g | 0.250 M | 25 mM |
| Glycine | 144 g | 1.920 M | 192 mM |
| SDS | 10 g | 1% | 0.1% |
| Milli-Q water | To 1000 ml | | |

Note: pH of the working solution is adjusted to 8.3.

Physical spectrophotometric methods, e.g., ultraviolet (UV), circular dichroism (CD), and mass spectrometry [MS (MALDI-TOF MS and nanoLC MALDI-TOF/TOF MS)], provide important information on the secondary structure of the protein or peptide.

### 4.2.1. Determination of protein concentration

Protein quantitation is often necessary before processing of protein samples for isolation, separation, or analysis by chromatographic electrophoretic, or immunochemical techniques. Depending on the accuracy required and the amount and purity of the protein, different methods for determining protein concentration are available.

For determination of the concentration of royal jelly protein/peptides, the Bradford assay (Bradford, 1976) and bicinchoninic acid (BCA) method (Smith et al., 1985) are mostly used. These methods are described in detail in Subsection 1.2 of the article by Hartfelder et al. (2013).

### 4.2.2. One-dimensional gel electrophoresis for identification of major royal jelly proteins and their (protein) isoforms

SDS-PAGE is the most commonly practiced gel electrophoretic method for protein characterization. The typical electrophoretic profile of major proteins of royal jelly (*Apis mellifera*) is presented in Figure 24. For separation of proteins with a molecular mass lower than 30 kDa and peptides, high resolution can be achieved using Tricine SDS-PAGE (Schägger & von Jagow, 1987).

SDS-PAGE requires that proteins be denatured to their constituent polypeptide chains; therefore, this method provides limited information. In those situations where it is desirable to maintain biological activity or antigenicity, non-denaturing (native) electrophoresis systems must be employed. Nonetheless, the gel patterns from non-denaturing gels are more difficult to interpret than those from SDS-PAGE. Native or non-denaturing electrophoresis in the absence of denaturants allows for determination of the native size or subunit structure and optimal separation of proteins. Mobility depends on the size, shape, and intrinsic charge of the protein. As an example, the electrophoretic pattern of royal jelly proteins and of peptide apisimin is presented. Separation by Tricine SDS-PAGE under denaturing conditions (Figure 25(a)) shows apisimin as 1 distinct peptide band, while under non-denaturing conditions of native PAGE (Figure 25(b)), several bands of the peptide can be explained by the different degrees of oligomerization (Bíliková et al., 2002). Non-denaturing systems also provide information about the charge isomers of proteins, but this information is best obtained by IEF (see Subsection 4.2.3). An IEF run will often show heterogeneity (due to structural modifications) that is not apparent in other types of electrophoresis. Proteins thought to be a single species according to SDS-PAGE analysis are sometimes found by IEF to consist of multiple species. IEF is a method for separating molecules by differences in their isoelectric point (pI). Proteins or peptides are introduced into an immobilized pH gradient gel composed of polyacrylamide, starch, or agarose where a pH gradient has been established. A protein that is in a pH region below its pI will be positively charged and so will migrate towards the cathode (negatively charged electrode). As it migrates through a gradient of increasing pH, however, the protein's overall charge will decrease until the protein reaches the pH region that corresponds to its pI. As a result, the proteins become focused into sharp stationary bands, with each protein positioned at a point in the pH gradient corresponding to its pI. The technique is capable of high resolution,

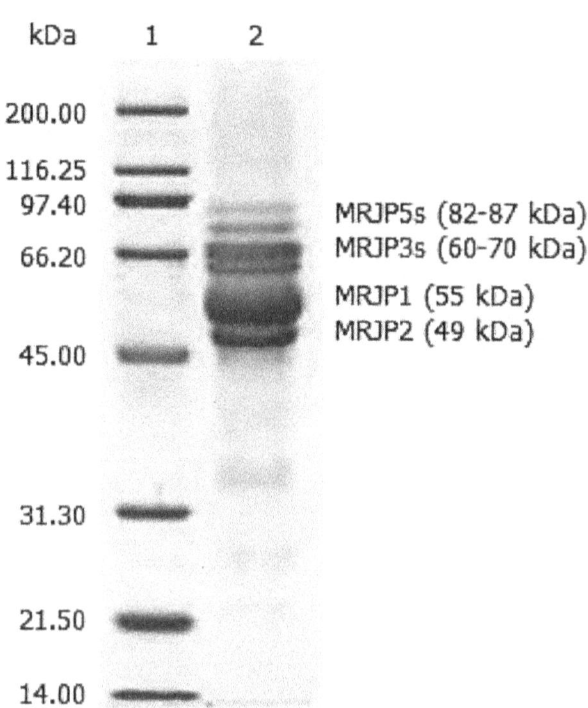


Figure 24. Electrophoretic pattern of royal jelly proteins (*Apis mellifera*) separated by SDS-PAGE (10%), lane 1: molecular mass marker; lane 2: major proteins of royal jelly. Gel stained by Coomassie Brilliant Blue.

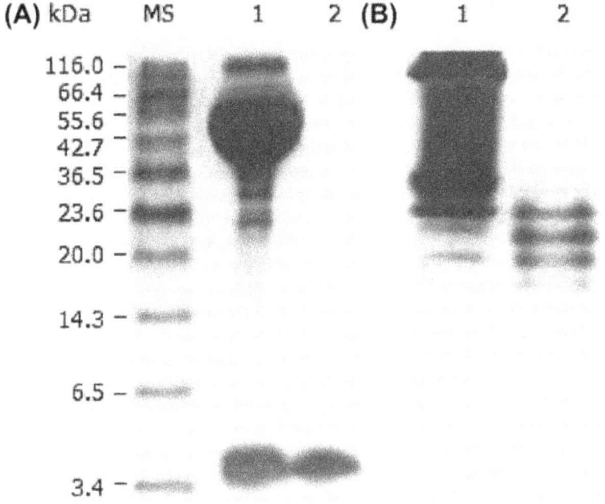


Figure 25. Protein patterns of royal jelly proteins and peptide apisimin by Tricine SDS-PAGE (A) and native Tricine PAGE (B); lane MS: protein mass standards; lane 1: royal jelly; lane 2: apisimin (From Bíliková et al., 2002. Reproduced with permission).

where proteins differing by a single charge can be fractionated into separate bands.

IEF of royal jelly proteins and peptides usually shows several isoelectrofocusing variants. In case of MRJP2, there are IEF bands in the range of pI from 7.5 to 8.5 corresponding to different posttranslational modifications of the protein (Figure 26(a)) (Bíliková, Klaudiny, & Šimúth, 1999), while in the case of royal jelly peptide apisimin, several IEF bands in the acidic range of pI (3.55–4.55) correspond to the oligomeric structures of the peptide (Figure 26(b)) (Bíliková et al., 2002).

*4.2.2.1. SDS-PAGE.* Royal jelly proteins are usually separated by SDS-PAGE in a 10 or 12% gel (Laemmli, 1970), already described in *Section* 1.3 of the *BEEBOOK* paper by Hartfelder et al. (2013).

*4.2.2.2. Tricine SDS-PAGE.* Royal jelly peptides can be separated by Tricine SDS-PAGE, which is commonly used for separation of molecules smaller than 30 kDa. Electrophoretic separation of royal jelly peptides can be performed by Tricine SDS-PAGE according to Schägger and von Jagow (1987)) (Schägger & von Jagow, 1987).

*Sample preparation*

(1) Add 10 µl of 2× Tricine sample buffer (4% SDS, 12% glycerol w/v, 50 mM Tris pH 6.8, 1.5% dithiothreitol, 0.01% Serva Blue G250) to 10 µl of a peptide sample.
(2) Incubate for 40 min at 40 °C.
(3) Centrifuge at 18,000× *g* for 1 min.

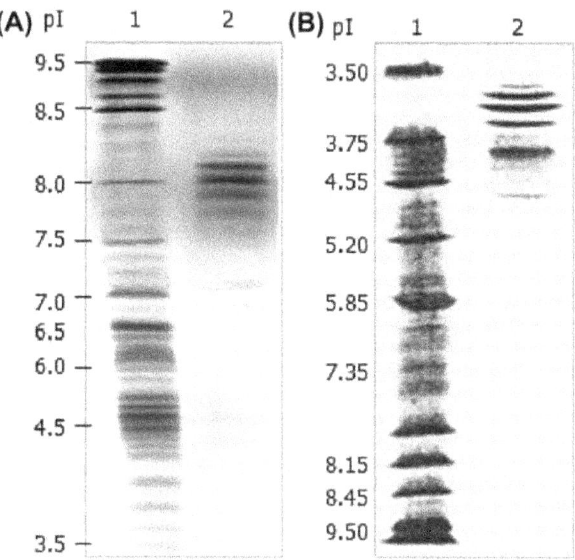


Figure 26. Isoelectrophocusing of major basic royal jelly protein MRJP2 and acidic royal jelly peptide apisimin. A: Servalyt Precotes 3–10 IEF gel, lane 1: pI standard; lane 2: MRJP2, Coomassie Blue G-250 staining. B: PhastGel Dry IEF gel, line1: pI standard; line2: apisimin, silver staining (B is from Bíliková et al., 2002. Reproduced with permission).

(4) Load 20 µl on the Tricine SDS polyacrylamide gel.

*Preparing and running vertical slab gels*

(1) Prepare stock solutions for making gels and electrolytes for Tricine SDS-PAGE according to Table 3.
(2) Prepare the stacking (4% T/3% C) and separating (16.5% T/3% C) gels for Tricine SDS-PAGE according to Table 4.
(3) Load samples onto the gel.
(4) Run electrophoresis at a constant current, 20 mA, in the presence of 2 electrolytes, the cathode and anode buffer (Table 3, Figure 27).

*Staining and destaining gels*

(1) Stain the gels with Serva Blue G250 in 10% acetic acid.
(2) Destain in 10% acetic acid.

In case of expecting a low concentration of peptides, the silver staining can be used, following Subsection 1.3.4 in the *BEEBOOK* paper by Hartfelder et al. (2013).

*4.2.2.3. Native-PAGE (n-PAGE).* Separation of the proteins under non-denaturing conditions requires the same type of equipment used for denaturing gels as described before (see Subsection 4.2.2), but without SDS in stock solutions, including the sample buffer.

Table 3. Recipes for stock solutions and electrolytes of Tricine SDS-PAGE.

| | Tris (mol $l^{-1}$) | Tricin (mol $l^{-1}$) | pH | SDS (%) |
|---|---|---|---|---|
| Anode buffer | 0.1 | – | 8.9[a] | – |
| Cathode buffer | 0.1 | 0.1 | 8.25[b] | 0.1 |
| 3× gel buffer | 3 | – | 8.45[a] | 0.3 |
| AB-mix 49.5%T/3%C | (1) Weigh 48 g of acrylamide and 1.5 g of bisacrylamide<br>(2) Dissolve in 80 ml of Milli-Q water<br>(3) Adjust volume to 100 ml<br>(4) Pass through a 0.22-μm filter<br>(5) Keep in the dark at 4 °C | | | |

[a]pH adjusted by HCl.
[b]without adjustment of pH. % T, total concentration of both monomers (acrylamide and bisacrylamide). % C, the percentage of cross-linked polymer relative to the total concentration.

Table 4. Recipes for Tricine peptide-separating and stacking gels.

| Stock solution | Stacking gel 4% T/3% C | Separating gel 16.5% T/3% C |
|---|---|---|
| AB-mix* | 5.0 ml | 0.1 ml |
| 3× gel buffer | 5.0 ml | 1.5 ml |
| Glycerol | 1.6 ml | – |
| $H_2O$ | 3.4 ml | 4.0 ml |
| 10% (APS) | 100 μl | 50 μl |
| TEMED | 10 μl | 5 μl |

*AB-mix, a stock solution of the acrylamide-bisacrylamide mixture, for composition see Table 3.

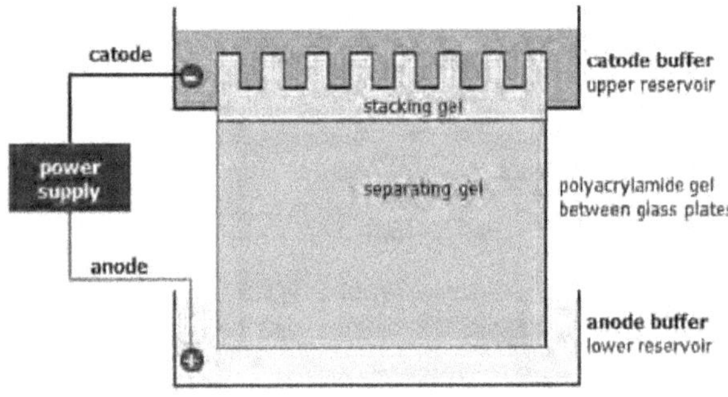


Figure 27. Setup of vertical Tricine-SDS-PAGE. The upper buffer compartment is filled by 1X cathode buffer, the lower reservoir contains 1X anode buffer.

*4.2.3. IEF*

IEF of royal jelly protein samples is performed on 5% polyacrylamide gels (0.15 mm, 125 × 125 mm, Servalyt Precotes 3–10, Serva) using SERVALYT™ ampholytes according to the instructions of producer (Serva Electrophoresis, GmbH, Germany).

(1) Run the IEF gel for 30 min at 30 V $cm^{-1}$ without samples.
(2) Load the samples and run the gel for 2 h at 120 V $cm^{-1}$.
(3) Stain the IEF gel with Coomassie brilliant blue (CBBR-250) or by silver staining. The detailed methods for staining the gels are given in Section 4.3 of the *BEEBOOK* paper by Hartfelder et al. (2013).

IEF separation of the proteins using PhastGel Dry IEF (PhastSystem-200 V, Pharmacia LKB, Uppsala, Sweden) is carried out on 5% polyacrylamide gels (0.45 mm, 40 × 50 mm, PhastGel IEF 3–9, Pharmacia Biotech, Sweden) for 30 min at 2.2 mA.

*4.2.4. Immunochemical methods*

Immunoassays are analytical methods based on highly specific binding between an antigen and an antibody. An epitope (immunodeterminant region) on the antigen surface is recognized by the antibody's binding site. The type of antibody and its affinity and avidity for the antigen determine assay sensitivity and specificity. Immunochemical methods offer simple, rapid, and sensitive methods for routine protein or peptide analyses. The most popular either quantitative or qualitative formats are enzyme linked immunosorbent assays (ELISAs), immunochromatography, dot blot, or western blot assays used to interpret data from protein analysis by gel electrophoresis.

*4.2.4.1. Dot blot analysis.* A dot blot is a simple and quick assay that can be used to determine if antibodies and a detection system are effective and can identify the appropriate starting concentration of a primary antibody for western blot analysis, as well as for detection of the protein of interest in the fractions after chromatographic separation.

(1) Assemble the dot blot equipment according to Figure 28.
(2) Spot up to 500 μl of a fraction solution (after chromatographic separation) on the wall of the dot blot equipment.
(3) Leave the sample to be soaked up by the polyvinylidene difluoride (PVDF) or nitrocellulose (NC) membrane.
 Note: An NC membrane is commonly used and cheaper but the material is brittle. PVDF is more durable and has higher chemical resistance making it ideal for reprobing and sequencing applications. A PVDF membrane has higher protein binding capacity and sensitivity; an NC membrane, on the other hand, produces less background noise.
(4) Block the membrane with 5% dry milk in TBS buffer (50 mM Tris-HCl, 150 mM NaCl, pH 7.6) for 1 h at room temperature.
(5) Decant the block buffer, but keep the membrane wet at all times for the remainder of the procedure.
(6) Incubate the membrane with a primary antibody in 5% dry milk in TBS for 1 h at room temperature.
(7) Wash the membrane 3 times (10 min each) in TBS buffer on a rocker.
(8) Incubate the membrane with a secondary antibody in 5% dry milk in TBS for 1 h at room temperature.
(9) Wash the membrane 3 times (10 min each) in TBS buffer on a rocker.
(10) Place the membrane in the chromogenic DAB/$NiCl_2$ solution (0.8 mg/ml DAB; 0.4 mg/ml $NiCl_2$; 0.1% $H_2O_2$ in 100 mM Tris–HCl pH 7.4).

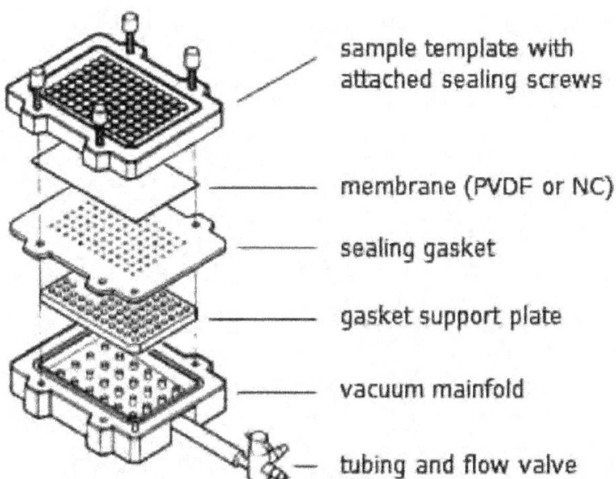


Figure 28. Set up of Dot Blot system.

(11) Incubate for 5 min for detection of the immunoactive protein/peptide spots.
(12) Wash the membrane with Milli-Q water and dry it off.
 Colored spots show us in which chromatographic fraction the protein of interest is present.

*4.2.4.2. Western blot.* As already described in Section 1.4 of the *BEEBOOK* paper by Hartfelder et al. (2013), the samples are transferred after SDS-PAGE to a PVDF or NC membrane using the tank method (MiniTrans-Blot Electrophoretic Transfer Cell, Bio-Rad Laboratories, USA). The membranes are first incubated overnight in TBS buffer containing 5% powdered non-fat milk with polyclonal rabbit antiserum against recombinant royal jelly protein or peptide at a dilution of 1:2000, then 2 h in the same buffer with a peroxidase-conjugated secondary antibody (in case of a rabbit primary *anti*-MRJP antibody, the swine or goat secondary *anti*-rabbit IgG can be used) at a dilution of 1:5000. Visualization of the immunoactive protein/peptide bands is performed by incubating the blots in the chromogenic DAB/$NiCl_2$ solution (see Subsection 4.2.4.1).

*4.2.4.3. Preparation of a polyclonal antibody.* Antibodies are serum immunoglobulins produced by a body's immune system and bind specifically to particular antigens. The methods for eliciting antibodies involve immunization of experimental animals with a purified antigen. Polyclonal antibodies are valuable for immunochemical methods such as immunoblotting (dot blot or western blot analysis), immunoprecipitation, and ELISA. Production of good antibodies (antisera) depends in large part upon the quality, purity, and amount of available antigens as well as on the specificity of the assay. For protein antigens, the material should be biochemically homogeneous and, depending on the intended use, should be in either a native or denatured conformation.

The choice of the animal for production of antibodies depends on the amount of antiserum desired. Rabbits are the usual animal to be immunized and provide as much as 25 mL of serum from each bleed. For smaller-scale experiments, the mouse may be used, where the volume of antigen suspension used for immunization is significantly less and the amount of serum obtained from a single bleed does not exceed 0.5 ml.

For preparation of antibodies against MRJPs, one can use purified royal jelly protein or peptide as an antigen. Another way is to use the water-soluble protein fraction of royal jelly for immunization of experimental animals (see Subsection 4.1.2.1) and subsequently, the individual *anti*-MRJP antibody is purified from the obtained pool of antibodies by immunoprecipitation.

A pure protein can be prepared by the chromatographic methods already described in Subsection 4.1.2. For immunization, the protein is excised from the gel strip after preparative SDS-PAGE or Tricine-PAGE directly or is purified by electroelution of the protein from the gel (see Subsection 4.1.3.3).

*4.2.4.3.1. Preparation of the MRJP antigen in an SDS-PAGE gel*

(1) Run preparative SDS-PAGE (160 × 140 × 1 mm gel; 10% separating gel; 5% stacking gel) of MRJP, at 20 mA.
(2) Stain the gel with 1 M KCl.
(3) Excise the protein band of interest.
(4) Homogenize the gel slice in 0.15 M NaCl by vortexing.
(5) Use the homogenized soft gel suspension for immunization of a rabbit or mouse directly or purify the protein from the gel by electroelution (see Subsection 4.1.3.3).

*4.2.4.3.2. Immunization with a purified protein.* Use 400 μl of the antigen (the protein of interest) and complete Freund's Adjuvant (CFA) (1:1) for the first immunization and next in incomplete Freund's Adjuvant (IFA) for 1 injection per animal.

(1) Inoculate the rabbit with 500 μg/ml antigen (the protein) in CFA.
(2) Inoculate, every 2 weeks, 3 times with 250 μg/ml protein in IFA.
(3) Collect the animal's blood a week after the last immunization.
(4) Centrifuge the blood at 4000*g* for 5 min.
(5) Collect the supernatant, where the *anti*-MRJP antibody is.
(6) Store in aliquots at −80 °C.
(7) Test antibody titre by a dot blot or western blot analysis (see Subsections 4.2.4.1 and 4.2.4.2).

*4.2.4.3.3. Immunization with WSP antigen*

(1) Administer 400 μl of the water-soluble royal jelly proteins (WSPs, see Subsection 4.1.2.1) plus Freund's complete adjuvant (1:1) as antigen samples at the concentration of 500 mg/ml to a rabbit or at 100 mg/ml to a mouse.
(2) After 15 days, administer a booster injection of 250 mg/ml (in the rabbit) or 25 mg/ml protein (in the mouse).
(3) Remove the animal's blood by cardiac puncture.
(4) Centrifuge the blood at 4,000*g* for 3 min.
(5) Store the serum at −80 °C.

*4.2.4.3.4. Antibody purification.* MRJP1 is used as an example.

(1) Separate (2 μg/μl) WSP fraction by SDS-PAGE in a 5–22% gradient gel (300 V, 2 h).
(2) Stain the gel with Coomassie brilliant blue-R250, for 1 h.
(3) Excise the bands of interest.
(4) Electrotransfer onto a nitrocellulose membrane (300 V, overnight, see Subsection 4.2.4.2).
(5) Incubate the membranes in TBS-T (50 mm Tris–HCl pH 7.4, 150 mM NaCl, 0.5% Tween 20) containing 5% of dried milk overnight at room temperature.
(6) Incubate with diluted serum (*anti*-MRJP1) in TBS (1:1) for 2 h.
(7) Wash the membranes 3 times for 5 min with TBS-T.
(8) Elute the bound antibodies (*anti*-MRJP1) with 1.4% triethylamine (1 min).
(9) Transfer the eluted antibodies to vials containing 1 M Tris–HCl (pH 8.5) at room temperature.
(10) Dialyze the antibody against TBS containing 0.1% sodium azide for 24 h.

*4.2.4.3.5. Immunoprecipitation of MRJPs*

(1) Incubate the protein extracts overnight at 4 °C with 20 mg/ml *anti*-MRJP1 and 150 mM NaCl 0.2% Triton X-100.
(2) Add the protein A Sepharose beads (2 mg/ml).
(3) Gently mix the suspension on a rocking platform for 30 min at 4 °C.
(4) Centrifuge this suspension at 18,000× *g* for 10 min to obtain the immunoprecipitation fractions.
(5) Transfer the supernatant.
(6) Wash the pellet with TBS plus 0.5 M NaCl.
(7) Repeat washing 6 times with TBS without salt addition.
(8) Dilute the pellets and the supernatant collected at step 5 with SDS sample buffer.
(9) Boil at 100 °C for 5 min.
(10) Analyze them separately for MRJPs content by immunoblotting using a rabbit *anti*-MRJPs antibody (see Section 4.2.4.2).

*4.2.5. Physical methods*

Physical methods including UV, circular dichroism (CD), and mass spectrometry (MS) provide important information on the molecular properties and secondary structure of the protein or peptide. A combination of high-resolution two-dimensional (2D) polyacrylamide gel electrophoresis (2DE), highly sensitive biological MS, and the rapidly growing protein and DNA databases have paved the way for high-throughput proteomics of MRJPs. Protein electroblotting and sequencing of N-terminal amino acids are described as a tool for *de novo* sequencing and protein identification. In the second part of this section, we highlight matrix-assisted laser desorption/ionization mass spectrometry (MALDI-MS) as one of the main contemporary analytical methods for linking gel-separated proteins to entries in sequence databases. CD is a method for rapid evaluation of the secondary structure of proteins. In the far UV region (240–180 nm), which corresponds to peptide bond absorption, the CD spectrum can be analyzed to determine regular secondary structural features (Kelly, Jess, & Price, 2005). As an example, the CD spectrum of apisimin is presented (Figure 29); this protein was isolated from royal jelly (Bíliková et al., 2002). MS is an analytical tool useful for measuring the mass-to-charge ratio (*m*/*z*) of one or more molecules present in a sample. It is used to determine the exact molecular mass of the analyzed molecule as well as for characterization of a wide variety of post-translational modifications such as phosphorylation and glycosylation; when the protein/peptide sample is pure, it is possible to determine the N-terminal amino acid sequences (Figure 30) (Bíliková et al., 2002). The *nano* LC MALDI TOF/TOF MS analysis of the peptides obtained after tryptic digest of the protein showed that apalbumin2a has 2 fully occupied N-glycosylation sites, 1 with high-mannose structure, HexNAc2Hex9, and another carrying a complex type of antennary structures, HexNAc4Hex3 and HexNAc5Hex4; this structure is different from that of MRJP2 (Figure 31). This posttranslational modification of apalbumin2a leads to changes in physiological activity of the protein. Apalbumin2a inhibits growth of *Paenibacillus larvae* sub. *larvae*, the primary honey bee pathogen of American foulbrood disease, while MRJP2 has no antimicrobial activity (Bíliková et al., 2009).

*4.2.5.1. CD analysis.* For structural studies, the CD spectra of a protein or peptide at the concentration of 0.5 µg/ml in water are recorded on a Jasco J-600 Spectrophotopolarimeter (Japan Spectroscopic Co., Ltd., Japan).

(1) Dilute a sample solution to 0.5 µg/ml in water.
(2) Transfer the sample solution to 1-cm quartz cuvettes thermostated at 25 °C.
(3) Record the spectra from 190 to 250 nm, with 3 scans at 50 nm/min, time constant 1 s, and bandwidth 1 nm.

Ellipticity and photomultiplier voltage baselines for the protein are measured using deionized water in a 1-cm cuvette. The CD spectrum for apisimin in water was obtained at peak ellipticity 50 mdeg and HT < 700 V. The obtained CD spectra are evaluated with the software provided by the instrument manufacturer, Jasco.

*4.2.5.2. N-terminal amino acid sequencing.* This sequence can be determined directly from the lyophilized sample (see Subsection 3.1.3) as well as from the individual protein/peptide bands after SDS-PAGE or Tricine SDS-PAGE.

(1) Separate the proteins or peptides by SDS-PAGE (see Section 1.3 of the *BEEBOOK* paper by Hartfelder et al. (2013)).

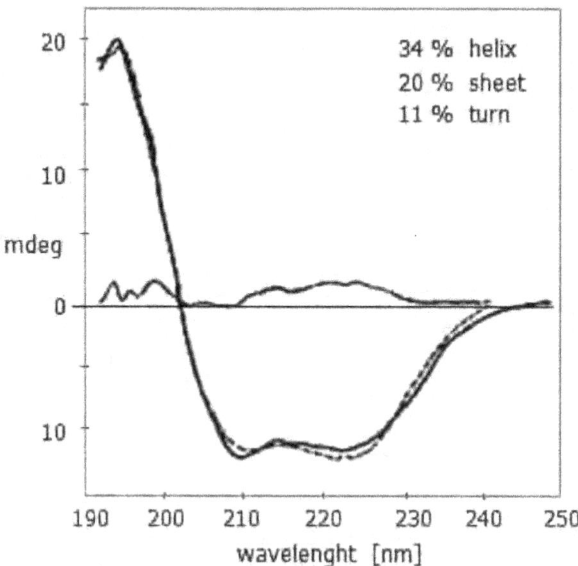


Figure 29. CD spectrum of apisimin purified from royal jelly fraction using ion-exchange FPLC. measured CD spectrum; —— calculated CD spectrum according to the software protein standards with known secondary structure; ..... differences between measured and calculated CD data. The analysis of the spectrum performed by the Jasco software showed that apisimin contains 34% K-helical, 20% *alpha*-sheet, 11% *beta* -turn, 30% random structure, that is in good agreement with the theoretically calculated parameters, typical for proteins with a predominantly helical structure. The occurrence of 65% of well-defined secondary structure in such a small peptide with only one aromatic amino acid and without disulfide bridges is unique and can be used as a new model for research into mechanisms of action of antibiotic peptides (Bíliková et al., 2002).

(2) Electroblot the proteins or peptides from the polyacrylamide gel onto a PVDF membrane (ProBlott, Applied Biosystems, USA) using the tank method of protein transfer (Mini Trans-Blot Electrophoretic Transfer Cell, Bio-Rad Laboratories, USA), in electroblotting buffer [10 mM CAPS (3-[cyclohexylamino]-1-propanesulfonic acid) in 10% methanol] according to the procedure recommended by the manufacturer [see Section 1.4 of the *BEEBOOK* paper by Hartfelder et al. (2013)].
(3) Perform visualization by Coomassie brilliant blue R250 staining (0.1% CBBR-250 in 40% methanol, 1% acetic acid).
(4) Destain the membrane with 50% methanol.
(5) Wash the membrane with Milli-Q water.
(6) Let it dry in ambient air.
(7) Excise the protein/peptide bands of interest and subject them to sequencing by automated Edman degradation on an LF3600D Protein Sequenator (Beckman, USA).
(8) Identify amino acid sequences by ExPASy BLAST program (http://br.expasy.org/tools/blast) and the Swiss-Prot database for protein searches.

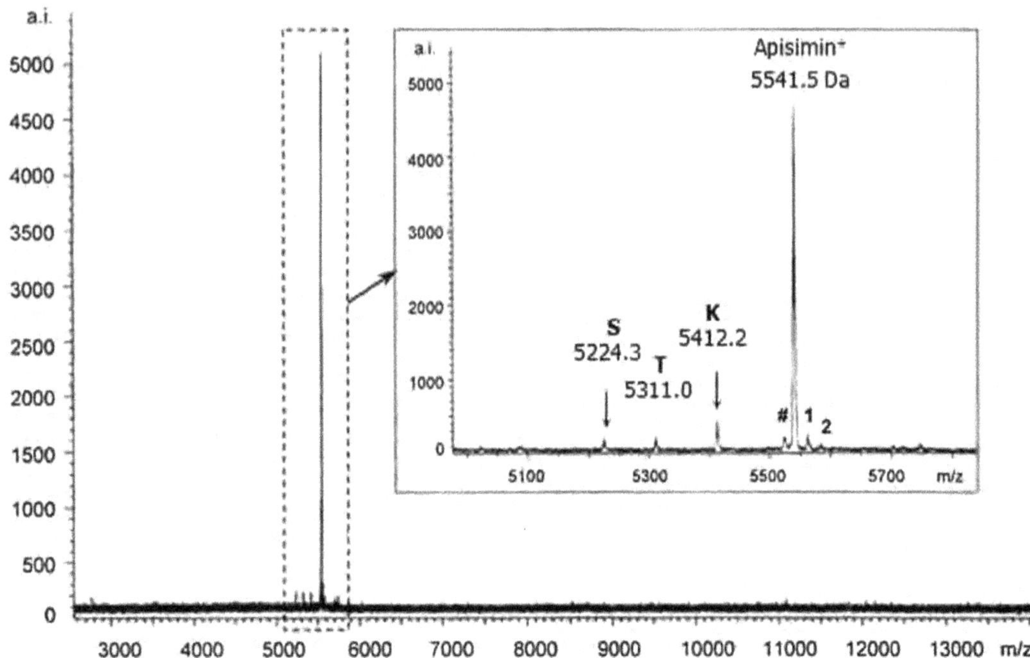


Figure 30. MALDI-TOF mass spectrum of apisimin purified from royal jelly fraction by ion-exchange FPLC. The molecular mass of apisimin determined by MALDI-TOF MS analysis (5540.5 Da) is in agreement with calculated one from amino acid sequence of the peptide (5540.4 Da), which indicates that apisimin is not post-translationally modified. The three satellite signals at 5413.2, 5312.0 and 5225.3 Da correspond to a consecutive loss of the N-terminal amino acids: Lysine, Threonine and Serine residues (From Bíliková et al., 2002. Reproduced with permission).

*4.2.5.3. Mass spectrum analysis.* See Subsection 4.3.

### *4.3. Methods for analysis of the proteome of royal jelly and its post-translational modifications*

#### *4.3.1. Introduction*

Proteins are the executors of biological functions, however, the activities of most eukaryotic proteins are modulated by posttranslational modifications (PTMs). Among hundreds of PTMs, phosphorylation and glycosylation are the most widespread and play critical roles in all parts of cellular life. Since the PTMs became a hotspot of proteomics, PTM research on royal jelly proteins has also been advancing rapidly.

Although phosphorylation is considered the reason for the high heterogeneity of MRJPs, the 2DE-based phosphoprotein fluorescent staining method only detects MRJP2, and MRJP7 carries potential phosphorylation sites without well-defined phosphorylation locations (Furusawa et al., 2008). In the research involving 2DE, shotgun analysis in combination with high-sensitivity MS and bioinformatics, phosphorylation of MRJP1, MRJP2, and apolipophorin-III-like protein were identified for the first time and a new site was localized in venom protein 2 (Zhang et al., 2012). Recently, 2 complementary phosphopeptide enrichment materials ($Ti^{4+}$-IMAC and $TiO_2$) and high-sensitivity MS have been applied to map the phosphoproteomes of royal jelly produced by *A. m. ligustica* and *A. cerana cerana*. In total, 16 phosphoproteins carrying 67 phosphorylation sites and 9 proteins phosphorylated on 71 sites were identified in royal jelly derived from Western and Eastern honey bees, respectively (Han et al., 2014). As another important protein post-translational modification, protein glycosylation mediates many important biological processes involved in cell adhesion, cell differentiation, cell growth, and immunity (Rudd, Elliott, Cresswell, Wilson, & Dwek, 2001). N-glycosylation of proteins has been reported to promote protein folding, stability, solubility, oligomerization, quality control, sorting, and transport (Lis & Sharon, 1993). Using liquid chromatography tandem mass spectrometry (LC-MS/MS) together with hydrazide chemistry and lectin enrichment technology, 13 novel proteins and 42 N-linked glycosylation sites were mapped on royal jelly proteins for the first time (Zhang et al., 2014). A 350-kDa royal jelly protein that can stimulate the cell growth (Kimura et al., 2003) is reported to bear the high-mannose type N-glycans with 4 different structures, as analyzed by NMR spectroscopy (Kimura, Washino, & Yonekura, 1995), while a 55-kDa royal jelly protein that can maintain high viability of rat liver primary cultured cells, only possesses 1 kind of N-linked sugar chain (Kimura, Kajiyama, Kanaeda, Izukawa, & Yonekura, 1996). Apalbumin2a also known as isoform MRJP2, carries 2 fully occupied N-glycosylation sites and has an antimicrobial activity (Bíliková et al., 2009).

Although the above efforts have been made to explore the unknown proteins and PTMs in royal jelly,

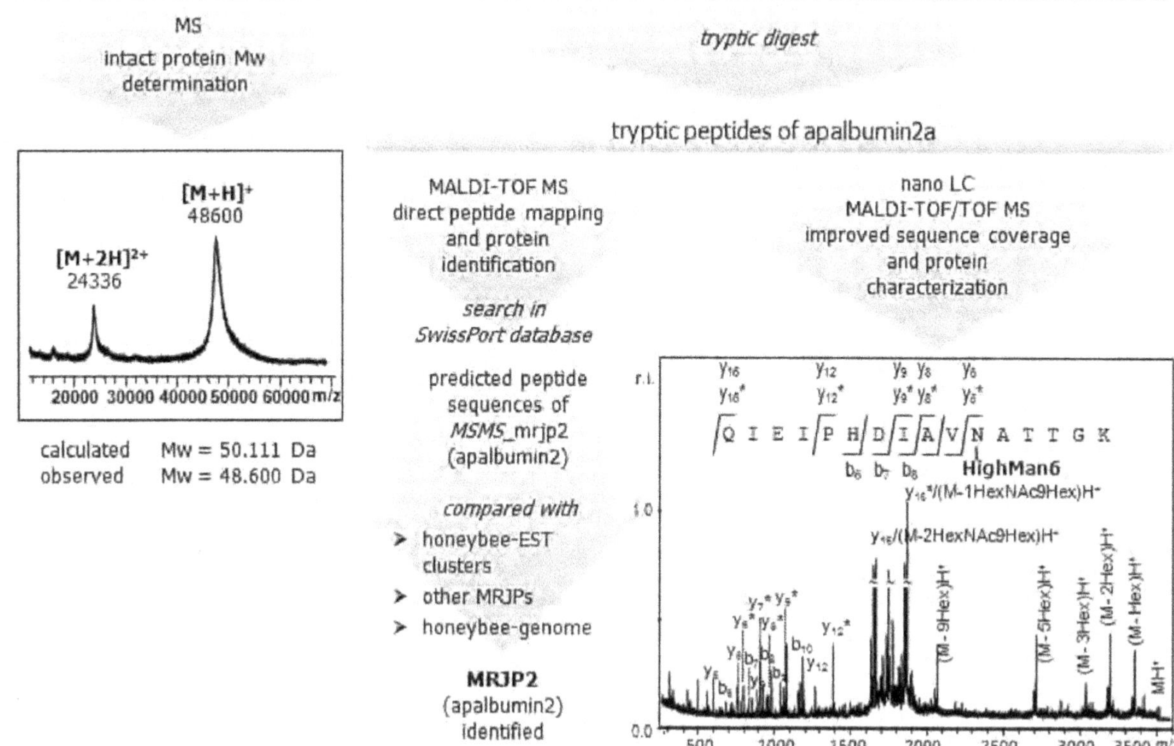


Figure 31. Simple scheme of molecular characterization of apalbumin2a, the minority homologue of major basic royal jelly protein apalbumin2 (MRJP2) using MS, MALDI-TOF MS and nanoLC MALDI-TOF MS analysis. MS analysis provide the exact molecular mass of the protein. Triptic peptides obtained after trypsin digestion of analyzed protein can be compared in protein database search to select the optimal candidate of the protein. The nanoLC MALDI TOF MS analysis of the triptic peptides showed that apalbumin2a carrying two fully occupied N-glycosylation sites, one with high-mannose structure, HexNAc2Hex9, and another carrying complex type antennary structures, HexNAc4Hex3 and HexNAc5Hex4 (Bíliková et al., 2009).

this research is still at the infancy stage. Therefore, there is a high demand for cutting-edge knowledge and state-of-the-art technology to identify the unknown proteins and the PTM status in royal jelly. This is important for gaining new insights into the new roles of such proteins both for honey bee biology and for promotion of human health.

#### 4.3.2. *Two-dimensional gel electrophoresis (2DE) for identification of major royal jelly proteins and their (protein) isoforms*

##### 4.3.2.1. *Sample preparation*

(1) Mix the royal jelly sample and lysis buffer (LB: 8 M urea, 2 M thiourea, 4% 3-[(3–151 cholamidopropyl) dimethylammonio]-1-propanesulfonate [CHAPS], 20 mM Tris base, 30 mM dithiothreitol [DTT], store at −20 °C before use) at the ratio of 1:10 (w/v).
(2) Homogenize for 30 min on ice.
(3) Sonicate 3–5 times for 10 s.
(4) Centrifuge at 12,000g, 4 °C for 10 min.
(5) Further centrifuge at 15,000g, 4 °C for 10 min.
(6) Collect the supernatant (be careful not to pipette the lipid layer).
(7) Add a 3-fold volume of acetone (pre-cooled) to the supernatant.
(8) Mix.
(9) Keep on ice for 30 min for protein precipitation.
(10) Centrifuge at 15,000g, 4 °C for 2 × 10 min.
(11) Discard the supernatant.
(12) Dry the protein pellet at room temperature.
(13) Dissolve the pellet in LB.

*4.3.2.2. Measurement of protein concentration.* Protein concentration can be determined according to previously described methods, see Section 1.2 of the *BEEBOOK* paper by Hartfelder et al. (2013).

*4.3.2.3. IEF and SDS-PAGE.* IEF is described in Section 4.2.3. SDS-PAGE is described in Section 1.3 of the *BEEBOOK* paper by Hartfelder et al. (2013). The following requirements need to be noted. IPG strips of different length require different volume of the rehydration solution. The IPG strips for the first dimensions have to be rehydrated for ~14 h in an aqueous solution consisting of 8 M urea, 2% CHAPS, 0.001% bromophenol blue,

45 mM DTT, and 0.2% Bio-lyte. After IEF, the strips should be washed with electrophoresis buffer consisting of 0.3% Tris base, 1.44% glycine, and 0.1% SDS, and then equilibrated with equilibration buffer 1 [an aqueous solution consisting of 0.375 M Tris–HCl (pH 8.8), 6 M urea, 20% glycerol, 2% SDS, 2% DTT, store at −20 °C before use] and equilibration buffer 2 [prepare an aqueous solution consisting of 0.375 M Tris–HCl (pH 8.8), 6 M urea 20% glycerol, 2% SDS, 2.5% iodoacetamide [IAA], store at −20 °C before use]. As for the second-dimension SDS-PAGE, 12% T linear gradient polyacrylamide gels are commonly used. Then, the well-equilibrated gel strips are transferred to the top of the gels and subjected to SDS-PAGE at constant voltage or current.

*4.3.2.4. Gel fixing and gel staining*

(1) Fix the gels with an aqueous solution consisting of 40% ethanol and 10% acetic acid for more than 4 h.
(2) Stain the gels using one of the following methods.

*4.3.2.4.1. Coomassie brilliant blue (CBB) G-250 staining.* This method is described in Subsection 1.3.3 of the *BEEBOOK* paper by Hartfelder et al. (2013).

*4.3.2.4.2. Fluorescent staining.* Different methods are needed for gels with different types of proteins of interest.

*Pro-Q Diamond staining for phosphorylated proteins:*

(1) Wash with double-distilled $H_2O$ ($ddH_2O$), 2 × 15 min.
(2) Stain with a Pro-Q solution for 3 h (Pro-Q: $ddH_2O$ = 1:2, V/V).
    Note: The staining solution should be kept in darkness!
(3) Destain the gel with the destaining solution consisting of 20% acetonitrile (ACN) and 5% 1 M sodium acetate (pH 4.0) for 4 × 15 min.
(4) Wash with $ddH_2O$, 2 × 5 min.

*Pro-Q Emerald staining for glycoproteins:*

(1) Add 6 ml of N,N-dimethylformamide (DMF) to the vial containing the pro-Q Emerald reagent.
(2) Mix gently and thoroughly.
(3) Store this Pro-Q Emerald stock solution at −20 °C or lower.
(4) Prepare the oxidizing solution by adding 250 ml of 3% acetic acid to the bottle containing the periodic acid.
(5) Mix until completely dissolved.
(6) Incubate the gel in an appropriate volume (according to the gel size) of the oxidizing solution with gentle agitation for 30 min.
    For a large gel, the incubation time should be 1 h.
(7) Wash the gel with 3% glacial acetic acid with gentle agitation for 10–20 min.
(8) Repeat this step 2 more times.
(9) Dilute the pro-Q Emerald stock solution 50-fold to prepare the staining buffer.
(10) Incubate the gel in an appropriate volume of staining buffer in the dark with gentle agitation for 90–150 min.
    The signal can be seen after ~20 min. Do not stain overnight.
(11) Wash the stained gel with a wash solution at room temperature for 15–20 min.
(12) Repeat this step once.
    Do not leave the gel in the wash solution for more than 2 h because the staining will start to decrease.

*SYPRO RUBY staining for total protein*

(1) Wash the gel with $ddH_2O$, 2 × 5 min.
(2) Stain with SYPRO RUBY overnight, keep in the dark.
(3) Destain the gel with a destaining solution consisting of 10% (v/v) methanol and 7% (v/v) acetic acid.
(4) Agitate for 15 min.
(5) Repeat steps 3 and 4 four times.
(6) Wash with $ddH_2O$, 2 × 5 min.

*4.3.2.4.3. Silver staining.* Silver staining is less compatible with mass spectrometric analysis. Therefore, you have to choose a MS-compatible silver stain protocol for any samples that will subsequently be subjected to MS analysis. The detailed methods for silver staining are given in Section 1.3.4 in the article by Hartfelder et al. (2013).

*4.3.2.5. Gel scanning and gel analysis.* For gels stained with CBB G-250 and a silver solution, a common gel scanner can be used.

For gels stained with Pro-Q Emerald, use an excitation maximum at ~280 nm and emission maximum at ~530 nm. Bands can be visualized using a 300-nm UV transilluminator.

For gels stained with Pro-Q Diamond, bands can be visualized using excitation at 532 nm with a 640-nm bandpass emission filter (Pharos FX Plus system, Bio-Rad), or 532 nm with a 560-nm longpass emission filter (Typhoon system, Amersham Biosciences). For gels stained with SYPRO RUBY, bands can be visualized using excitation at 582 nm with a 610 bandpass emission filter (Pharos FX Plus system, Bio-Rad). The well-scanned gels can be used for qualitative and quantitative analysis in commercially available software such as PDQuest from Bio-Rad and Imagemaster from GE. It is helpful to follow the instructions of these software programs for gel analysis.

*4.3.2.6. Protein identification*

*4.3.2.6.1. Gel cutting.* Excise the gel spots with a gel-cutting machine, or cut manually using pipette tips.

*4.3.2.6.2. Destaining. For gels stained with CBB-G250*

(1) Prepare the destaining solution: an aqueous solution consisting of 50% (V/V) ACN and 40 mM $NH_4HCO_3$.
(2) Add 100 μl of this solution.
(3) Agitate.
(4) Repeat steps 2 and 3 until the gel becomes transparent.
(5) Add 100 μl of 100% ACN.
(6) Agitate for 10 min.
(7) Remove ACN.
(8) Dry the gel spot with a Speed-Vac system.

*For silver staining*

(1) Prepare 30 mM $K_3Fe(CN)_6$ (solution A) and 100 mM $Na_2S_2O_3$ (solution B)..
(2) Wash the gel spot with $ddH_2O$ for 1 min 3 times.
(3) Mix solution A and B with equal volume to get the fresh destining solution
(4) Add 100 μl destaining solution to each gel spot.
(5) Agitate until the brown color disappears.
(6) Discard the supernatant.
(7) Rinse the gel spot with $ddH_2O$ 3 times.
(8) Discard the supernatant.
(9) Rinse the gel spot with 100 μl of 40 mM $NH_4HCO_3$ 3 times.
(10) Discard the supernatant.
(11) Add 100 μl of 100% ACN.
(12) Agitate for 10 min.
(13) Remove ACN.
(14) Dry the gel spot in a Speed-Vac system.

*4.3.2.7. Trypsin digestion*

(1) Prepare the trypsin solution with 40 mM $NH_4HCO_3$ at the final concentration 10 ng/μl.
(2) Add 10 μl of the trypsin solution to each dried gel spot with pipetting.
(3) Incubate for 60 min at 4 °C.
(4) Discard the supernatant to minimize the auto-digestion of trypsin.
(5) Incubate at 37 °C for 12–14 h keeping the Eppendorf tube upside down.
(6) Add 50% (v/v) ACN [containing 2.5% (v/v) TFA] using an appropriate volume to cover the gel spot.
(7) Incubate for 60 min at 30 °C.
(8) Pool the supernatants.
(9) Dry them by means of a Speed-Vac system.

*4.3.2.8. Mass spectrometric analysis*

The royal jelly sample is analyzed by means of an LC-MS system (QTOF G6520, Agilent Technologies). The LC-Chip (Agilent Technologies) consists of a Zorbax 300SB-C18 enrichment column (40 nl, 5 μm) and a Zorbax 300SB-C18 analytical column (75 μm × 43 mm, 5 μm). The parameter settings of liquid chromatography and the above-mentioned MS are as follows.

(1) Set the loading flow rate to 4 μl/min.
(2) Use water with 0.1% formic acid as a loading mobile phase.
(3) Perform sample elution with a binary solvent mixture composed of water with 0.1% formic acid (solvent A) and acetonitrile with 0.1% formic acid (solvent B).
(4) Use the following gradient program:
    - from 3 to 8% B in 1 min,
    - from 8 to 40% B in 5 min,
    - from 40 to 85% B in 1 min,
    - 85% B for 1 min.
(5) Set chip flow rate to 300 nl/min.
(6) MS conditions are
    - positive ion mode;
    - Vcap: 1900 V;
    - drying gas flow rate: 5 l/min;
    - drying gas temperature: 350 °C;
    - fragmentor voltage: 175 V;
    - skimmer voltage: 65 V;
    - precursors: 3.

Tandem mass spectra can be retrieved using the MassHunter software (Version B. 02. 01, Agilent Technologies).

*4.3.2.9. Database searches.* Mascot Distiller software (Matrix Science) is used to generate a peak list and to store it in a combined .mgf file.

*4.3.2.9.1. Online Mascot search.* (http://www.matrixscience.com/cgi/search_form.pl?FORMVER=2&SEARCH=MIS).

(1) Fill the Email address.
(2) Database: NCBInr.
(3) Enzyme: trypsin.
(4) Allow up to 1 missed cleavage.
(5) Taxonomy: all entries.
(6) Modifications: carbamidomethyl (C).
(7) Variable modifications: oxidation (M).
(8) Peptide tol. ± 50 ppm (change according to the accuracy of the instrument).

(9) MS/MS tol. ± 0.05 Da (change according to the accuracy of the instrument).
(10) Peptide charge: 2+, 3+, and 4+.
(11) Loading the .mgf file, start the search.

*4.3.2.9.2. In-house Mascot search.* The parameter settings are the same as for the on-line mode, except that the protein database of *Apis* can be downloaded from NCBI in FASTA format and then loaded into the Mascot software.

*4.3.2.9.3. PEAKS Studio.* The raw data of MS/MS can be searched using the PEAKS Studio software (Bioinformatics Solutions Inc.). The parameter settings are the same as those for Mascot.

According to the 2DE method described above, the proteomes of royal jelly derived from the Western honey bee (*A. mellifera ligustica*) and the Eastern honey bee (*A. cerana cerana*) were compared (Yu, Mao, & Li, 2010).

#### *4.3.3. Gel-free proteomic technology for royal jelly protein identification and quantification*

Although 2DE in combination with MS is effective at identification of royal jelly proteins, especially their isoforms, it excludes low-abundance proteins because of the restrictions of its sensitivity, as well as the extreme pI and molecular mass (Mr) of the proteins in question (Ong & Pandey, 2001). Nevertheless, gel-free proteomic technology involving high-resolution and high-mass-accuracy MS is one of the most powerful proteomic strategies for comprehensive protein identification.

##### *4.3.3.1. Sample preparation*

(1) Dissolve the protein pellet (see Subsection 4.2.1) in 100 μl of 5 M Urea.
(2) Add 400 μl of 40 mM $NH_4HCO_3$.
(3) Reduce the protein sample with 10 mM DTT for 1 h.
(4) Alkylate the protein sample with 50 mM IAA for 1 h in the dark.
(5) Digest the protein with trypsin at a 1:50 enzyme/protein ratio (w/w) at 37 °C for 14 h.
(6) Add 1 μl of formic acid to the solution to stop the reaction.
(7) Dry the solution using a Speed-Vac system.
(8) Dissolve the dried peptides with 0.1% formic acid.
(9) Centrifuge at 12,000*g* for 10 min.
(10) Transfer the supernatant into a new tube.
(11) Freeze for storage at −80 °C for further LC-MS/MS analysis.

*4.3.3.2. HPLC-MS/MS analysis.* Currently, a variety of instruments can be used to perform HPLC-MS/MS analysis. Here, we review the following system as an example: the EASY-nLC 1000 (Thermo Fisher Scientific) nano-liquid chromatography system coupled with a Q-Exactive via the nanoelectrospray source.

(1) Prepare buffer A (0.5% acetic acid) and buffer B (80% ACN in 0.5% acetic acid).
(2) Set the flow rate to 350 nl/min.
(3) The peptides are separated by the following gradient program:
    - from 3 to 8% buffer B for 5 min,
    - from 8 to 20% buffer B for 55 min,
    - from 20 to 30% buffer B for 10 min,
    - from 30 to 90% buffer B for 5 min,
    - 90% buffer B for 15 min.

The Q-Exactive is operated in data-dependent mode with survey scans acquired at the resolution of 70,000 at *m*/*z* 400. The top 10 most abundant ions with the charge ≥2 from the survey scan are selected and fragmented by higher-energy collisional dissociation with normalized collision energies of 25. Xcalibur (version 2.2, Thermo Fisher Scientific) is used to retrieve MS-MS spectra.

*4.3.3.3. Database search.* MS/MS data extracted in RAW format can be searched using such software as PEAKS (Bioinformatics Solutions Inc.), against a composite database containing protein sequences of *Apis* and a common repository of adventitious proteins (cRAP, from The Global Proteome Machine Organization). The following modifications can be applied:

- carbamidomethylation (C)/+57.02 Da is selected as a fixed modification,
- oxidation (M)/+15.99 Da is selected as a variable modification.

The other parameters are as follows:

- Parent ion mass tolerance, 15.0 ppm;
- Fragment ion mass tolerance, 0.05 Da;
- Enzyme, trypsin;
- Allow non-specific cleavage at one end of the peptide;
- Maximum missed cleavages per peptide, 2;
- Maximum allowed variable PTM per peptide, 3;
- The false discovery rate (FDR) is filtered to ≤1.0% with a target-decoy database searching strategy to distinguish positive and negative identification.

##### *4.3.3.4. Protein quantification*

*4.3.3.4.1. Quantification by labelling.* In the gel-free proteomic experiment, a number of labelling approaches can be used for quantitative comparison of protein abundance, including stable isotope labelling by amino acids in cell culture, stable isotope labelled peptides, radiolabeled amino acid incorporation, isotope-coded affinity tags (ICAT), and more recently, isobaric tags for relative and absolute quantification (iTRAQ) (Patel et al., 2009). The iTRAQ system is now commercially available and has been widely accepted as a reliable method for proteomic studies.

(1) Prepare a protein pellet as described in Subsection 4.3.2.1.
(2) Add 20 μl of dissolution buffer (0.5 M triethylammonium bicarbonate, pH 8.5) to each sample containing 100 μg of the protein pellet.
(3) Add 1 μl of the denaturant (2% SDS).
(4) Vortex.
(5) Reduce the protein sample with 10 mM DTT for 1 h.
(6) Alkylate the protein sample with 50 mM IAA for 1 h in the dark.
(7) Add 1 μl of Cysteine Blocking Reagent to each tube.
(8) Vortex to mix.
(9) Centrifuge.
(10) Incubate the tubes at RT for 10 min.
(11) Digest the protein using trypsin at a 1:50 enzyme/protein (w/w) concentration at 37 °C for 14 h.
(12) Allow each vial of iTRAQ™ Reagent to be used to reach RT.
(13) Spin each vial to bring the solution to the bottom of the tube.
(14) Add 70 μl of ethanol to each RT iTRAQ™ Reagent vial.
(15) Vortex each vial to mix the contents.
(16) Centrifuge.
(17) Transfer the contents of 1 iTRAQ™ Reagent vial to 1 sample tube.

For example, for a duplex-type experiment, transfer the contents of the iTRAQ™ Reagent 114 vial to the sample 1 protein digest tube and transfer the contents of the iTRAQ™ Reagent 117 vial to the sample 2 protein digest tube.

(18) Vortex each tube to mix.
(19) Centrifuge.
(20) Incubate the tubes at RT for 1 h.
(21) Combine the contents of each iTRAQ™ Reagent-labelled sample tube into 1 tube.
(22) Vortex to mix.
(23) Centrifuge.

After labelling, the sample is used for HPLC-MS/MS analysis. Concentration ratios of iTRAQ-labelled proteins are calculated on the basis of signal intensities of reporter ions observed in peptide fragmentation spectra, with the relative areas of the peaks corresponding to proportions of the labelled peptides.

*4.3.3.4.2. Label-free quantification.* Label-free quantitation can be subdivided into two distinct groups: (i) area under the curve (AUC) or signal intensity measurement based on precursor ion spectra, which measures the ion abundance levels at specific retention time points for the given ionized peptides without a stable isotope standard; (ii) spectral counting, which is based on counting the peptides assigned to a protein in an MS/MS experiment (Neilson et al., 2011). A large number of commercial and open-source software packages are available for label-free quantitation, such as Progenesis-LC, Scaffold, emPAI Calc, and PEAKS. The reader is directed to the respective manuals for data analyses.

### *4.3.4. Identification and quantification of phosphorylated royal jelly proteins*

Generally, the reversible phosphorylation of proteins at serine, threonine, and tyrosine residues is one of the most important and pleiotropic modifications. Protein phosphorylation plays crucial roles in enzyme activity regulation, function modulation of structural proteins, subcellular localization, protein interactions with other molecules, and in capacity for further covalent modification. Although as many as one-third of eukaryotic proteins are phosphorylated, the stoichiometry of phosphopeptides on phosphoproteins is considerably lower than that of their non-phosphorylated counterparts. Therefore, the enrichment techniques are necessary for successful phosphoproteomic experiments. A wide variety of approaches are available for phosphopeptide enrichment, among which, immobilized metal affinity chromatography (IMAC) and titanium dioxide ($TiO_2$) are the most widely used.

*4.3.4.1. Sample preparation.* Prepare a digested protein sample as described in Subsection 4.3.3.1, steps 1–7.

#### *4.3.4.2. Phosphopeptide enrichment*

*4.3.4.2.1. $TiO_2$*

(1) Prepare binding buffer consisting of 6.0% TFA, 80% ACN, and 0.2 M dihydroxy-benzoic acid (DHB).
(2) Prepare a $TiO_2$ slurry by adding 10 mg $TiO_2$ to 1 mL of binding buffer.
(3) Add 500 μl of binding buffer to the digested sample.
(4) Add 50 μl of the prepared $TiO_2$ slurry to the above mixture.
(5) Incubate at RT for 60 min with vigorous shaking.
(6) Centrifuge at 12,000g for 5 min.
(7) Remove the supernatant.
(8) Rinse the pellet in 1 ml of binding buffer for 30 min at RT with shaking.

(9) Centrifuge at 12,000*g* for 5 min.
(10) Remove the supernatant.
(11) Rinse the pellet in 1 ml of washing buffer I (0.5% TFA, 50% ACN solution) for 30 min at RT with shaking.
(12) Centrifuge at 12,000*g* for 5 min.
(13) Remove the supernatant.
(14) Rinse the pellet in 1 ml of washing buffer II (0.1% TFA, 30% ACN solution) for 30 min at RT with shaking.
(15) Elute phosphopeptides from the pellet twice with 100 μl of a 0.5 mM $K_2HPO_4$ solution.

*4.3.4.2.2. $Ti^{4+}$-IMAC*

(1) Prepare $Ti^{4+}$-IMAC
  (1.1.) Incubate 10 mg of poly-microspheres in a 100 mM $Ti(SO_4)_2$ solution at RT overnight with gentle stirring.
  (1.2.) Centrifuge at 13,000*g* for 3 min.
  (1.3.) Remove the supernatant.
  (1.4.) Wash the $Ti^{4+}$-IMAC beads 6 times with distilled water to remove the residual titanium ions.
  (1.5.) Wash the $Ti^{4+}$-IMAC beads 2 times with 200 mM NaCl.
  (1.6.) Wash the $Ti^{4+}$-IMAC beads 2 times with distilled water.
  (1.7.) Dry the beads using a Speed-Vac system.
(2) Prepare binding buffer consisting of 6.0% TFA, 80% ACN.
(3) Add 5 mg of $Ti^{4+}$-IMAC beads to 500 μl of binding buffer.
(4) Add 500 μl of the digested sample to the above mixture.
(5) Incubate at RT for 60 min with vigorous shaking.
(6) Centrifuge at 12,000*g* for 5 min.
(7) Remove the supernatant.
(8) Rinse the pellet in 1 ml of binding buffer for 30 min at RT with shaking.
(9) Centrifuge at 12,000*g* for 5 min.
(10) Remove the supernatant.
(11) Rinse the pellet in 1 ml of washing buffer I (0.5% TFA, 50% ACN solution) for 30 min at RT with shaking.
(12) Centrifuge at 12,000*g* for 5 min.
(13) Remove the supernatant.
(14) Elute phosphopeptides from the pellet twice with 100 μl of a 10% ammonia solution.

*4.3.4.3. Mass spectrometric analysis.* Refer to Subsection 4.3.3.2 "*HPLC-MS/MS analysis*".

*4.3.4.4. Data analysis*

*4.3.4.4.1. Database search and site localization.* MS/MS data extracted in RAW format are searched using in-house PEAKS software (version 6.0, Bioinformatics Solutions Inc.) against a composite database containing protein sequences of *Apis* and a common repository of adventitious proteins (cRAP, from The Global Proteome Machine Organization). The following modifications are applied:

- carbamidomethylation (C)/+57.02 Da is selected as a fixed modification,
- Oxidation (M)/+15.99 Da and Phospho (S, T, Y)/ + 79.96 Da are selected as variable modifications.

The other parameters are as follows:

- Parent ion mass tolerance, 15.0 ppm;
- Fragment ion mass tolerance, 0.05 Da;
- Enzyme, trypsin;
- Allowing non-specific cleavage at neither end of the peptide;
- Maximum missed cleavages per peptide, 2;
- Maximum allowed variable PTMs per peptide, 3.
- The false discovery rate (FDR) is filtered to ≤1.0% with a target-decoy database searching strategy to distinguish positive and negative identification.

The phosphorylation sites are assigned using Scaffold PTM (version 1.1.3; Proteome Software, Portland, OR, USA) on the basis of the Ascore algorithm. Only the site confidence more than 95% implies a mapped phosphosite.

*4.3.4.4.2. Quantitative analysis.* Refer to Subsection 4.3.3.4 "Protein quantification".

*4.3.5. Identification and quantification of glycosylated royal jelly proteins*

*4.3.5.1. Sample preparation*

(1) Prepare the protein pellet as described in Subsection 4.3.2.1.

Note: Keep the protein pellet dry without dissolving it in LB.

*4.3.5.2. N-linked glycopeptide enrichment*

*4.3.5.2.1. Hydrazide enrichment*

(1) Resuspend the protein pellet (~1 mg) in 250 μl of coupling buffer (100 mM NaAc and 150 mM NaCl, pH 5.5)
(2) Add 100 μl of 50 mM sodium periodate for glycoprotein oxidation at RT.

The solution should be protected from light.

(3) Agitate for 1 h.

(4) Hydrazide equilibration
   (4.1.) Add 250 μl of hydrazide and 750 μL of $ddH_2O$,
   (4.2.) Centrifuge at 15,000*g* for 5 min,
   (4.3.) Discard the supernatant,
   (4.4.) Repeat steps 4.2 and 4.3 thrice,
   (4.5.) Add 750 μl of coupling buffer,
   (4.6.) Centrifuge at 15,000*g* for 5 min,
   (4.7.) Discard the supernatant,
   (4.8.) Repeat steps 4.6 and 4.7 three times,
   (4.9.) Add 250 μl of coupling buffer,
   (4.10.) Agitate thoroughly.
(5) Add 500 μl of an equilibrated hydrazide solution to the dissolved protein solution.
(6) Agitate at RT for 24 h.
(7) Centrifuge at 15,000*g* for 5 min.
(8) Discard the supernatant.
(9) Desalting
   (9.1.) Add 1 ml of $ddH_2O$,
   (9.2.) Mix,
   (9.3.) Centrifuge at 15,000*g* and 4 °C for 5 min.
   (9.4.) Discard the supernatant,
   (9.5.) Repeat steps 9.1–9.5 twice,
   (9.6.) Add 1 ml of 40 mM $NH_4HCO_3$,
   (9.7.) Mix,
   (9.8.) Centrifuge at 15,000*g* and 4 °C for 5 min,
   (9.9.) Discard the supernatant,
   (9.10.) Repeat steps 9.6–9.9 twice.
(10) Resuspend the precipitate in 250 μl of 40 mM $NH_4HCO_3$.
(11) Add 50 μl of 100 mM DTT for protein reduction.
(12) Leave at RT for 30 min.
(13) Add 125 μl of 100 mM IAA for protein alkylation.
(14) Leave at RT for 30 min.
   Note: The solution should be kept in darkness.
(15) Add trypsin in the trypsin:protein ratio 1:50 (w/w).
(16) Keep at 37 °C overnight for digestion.
(17) Add 1 μl of formic acid to stop the digestion.
(18) Centrifuge at 15,000*g*, at 4 °C for 5 min.
(19) Retrieve the pellet.
(20) Wash the pellet with 1 ml of $ddH_2O$.
(21) Centrifuge at 15,000*g* and 4 °C for 5 min.
(22) Retrieve the pellet.
(23) Repeat steps 20–22 three times.
(24) Wash with 1 ml of 30% ACN.
(25) Centrifuge at 15,000*g* and 4 °C for 5 min.
(26) Retrieve the pellet.
(27) Repeat steps 24–26 three times.
(28) Wash with 1 ml of dd $H_2O$.
(29) Centrifuge at 15,000*g*, at 4 °C for 5 min.
(30) Retrieve the pellet.
(31) Repeat steps 28–30 three times.
(32) Wash with 1 ml of 40 mM $NH_4HCO_3$.
(33) Centrifuge at 15,000*g* and 4 °C for 5 min.
(34) Retrieve the pellet.
(35) Repeat steps 32–34 three times.
(36) Resuspend the pellet in 200 μl of 40 mM $NH_4HC^{18}O_3$ (prepared by dissolving $NH_4HCO_3$ in $H_2{}^{18}O$).
(37) Add 5 μl of PNGase F to release the N-linked glycopeptides.
(38) Incubate at 37 °C overnight.
(39) Add 1 μl of formic acid to stop the digestion.
(40) Centrifuge at 15,000*g* at 4 °C for 15 min.
(41) Keep the supernatant.
(42) Wash the pellet with 80% ACN (diluted with $H_2{}^{18}O$).
(43) Centrifuge at 15,000*g*, at 4 °C for 15 min.
(44) Keep the supernatant.
(45) Repeat steps 42–44 twice.
(46) Pool the supernatant and concentrate the solution by means of a Speed-Vac system for MS analysis.

#### 4.3.5.2.2. *Lectin enrichment*

(1) Resuspend the protein pellet (~1 mg) in 200 μl of UA buffer (8 M urea, 100 mM Tris–HCl, pH 8.5).
(2) Centrifuge at 14,000*g*, at 4 °C for 15 min.
(3) Pipette the supernatant onto the 0.5-ml filter unit.
(4) Add 200 μl of UA buffer.
(5) Centrifuge at 14,000*g* for 15 min.
(6) Discard the solution in the collection tube.
(7) Add 100 μl of a 100 mM IAA solution and mix for 1 min.
(8) Incubate without mixing for 30 min in the dark.
(9) Centrifuge the filter unit at 14,000*g* for 15 min.
(10) Add 200 μl of UA buffer to the filter unit.
(11) Centrifuge at 14,000*g* for 15 min.
(12) Repeat step 11 twice.
(13) Add 200 μl of 40 mM $NH_4CO_3$ to the filter unit.
(14) Centrifuge at 14,000*g* for 15 min.
(15) Repeat step 14 twice.
(16) Add trypsin (enzyme to protein ratio 1:50, w/w) to the filter unit.
(17) Place the filter unit into a new collection tube.
(18) Mix for 1 min.
(19) Incubate at 37 °C overnight.
(20) Add 1 μl of formic acid to stop the reaction.
(21) Centrifuge at 14,000*g* for 15 min.
(22) Add 50 μl of 1× binding buffer (see Table 5) to the filter unit.
(23) Centrifuge at 14,000*g* for 15 min.
(24) Repeat step 23.
(25) Transfer the solution from the collection tube to a new filter unit.
(26) Add 36 μl of lectin CWR (Table 6).
(27) Agitate for 1 h at RT.

(28) Centrifuge at 14,000*g* for 15 min.
(29) Add 200 μl of 1× binding buffer to the filter unit.
(30) Centrifuge at 14,000*g* for 15 min.
(31) Repeat step 30 four times.
(32) Add 200 μl of 40 mM $NH_4CO_3$ to the filter unit.
(33) Centrifuge at 14,000*g* for 15 min.
(34) Repeat step 33 three times.
(35) Add 50 μl of 40 mM $NH_4CO_3$ (prepared by dissolving $NH_4CO_3$ in $H_2{}^{18}O$) to the filter unit.
(36) Centrifuge at 14,000*g*, for 5 min.
(37) Add 200 μl of 40 mM $NH_4CO_3$ (prepared by dissolving $NH_4CO_3$ in $H_2{}^{18}O$) to the filter unit.
(38) Add 5 μl of PNGase F to the filter unit to release the N-linked glycopeptides.
(39) Incubate at 37 °C overnight.
(40) Add 1 μl of formic acid to stop the digestion.
(41) Transfer the filter unit to a new collection tube.
(42) Centrifuge at 14,000*g* for 15 min.
(43) Add 50 μl of 40 mM $NH_4CO_3$ (prepared by dissolving $NH_4CO_3$ in $H_2{}^{18}O$).
(44) Centrifuge at 14,000*g*, for 10 min.
(45) Collect the solution in the collection tube.
(46) Concentrate the solution by means of a Speed-Vac system for MS analysis.

*4.3.5.3. Mass spectrometric analysis.* The enriched glycopeptides are analyzed according to the method in Subsection 4.3.3.2. "HPLC-MS/MS analysis". The parameters are similar except the elution gradient program is as follows:

- from 3 to 8% buffer B for 10 min,
- from 8 to 23% buffer B for 130 min,
- from 23 to 30% buffer B for 20 min,
- from 30 to 90% buffer B for 8 min,
- 90% buffer B for 12 min.

Table 5. Composition of Binding Buffer, pH 7.6.

| Component | Concentration | |
|---|---|---|
| | 2× | 1× |
| $MnCL_2$ | 2 mM | 1 mM |
| $CaCl_2$ | 2 mM | 1 mM |
| NaCl | 1 M | 500 mM |
| Tris | 40 mM | 20 mM |

*4.3.5.4. Data analysis*

*4.3.5.4.1. Database search and glycosite assignment.* The raw data on glycoprotein MS/MS analysis are used in searches within the PEAKS Studio (Bioinformatics Solutions Inc.). The parameter settings are similar to the phosphoprotein procedure except for the selection of variable modifications:

- Oxidation (M)/+15.99 Da,
- Deamidation_$O^{18}$ (2.9883),
- Deamidation (NQ).

The glycosites are assigned using Scaffold PTM (version 1.1.3; Proteome Software, Portland, OR, USA) on the basis of the Ascore algorithm. Only those having the consensus sequence N-X-S/T ($X \neq P$) and a site confidence over 95% are considered mapped glycosites.

*4.3.5.4.2. Quantification.* Refer to Section 4.3.3.4. "Protein quantification".

*4.3.6. Conclusion*

A major challenge for honey bee proteomics is the research on low-abundance proteins in royal jelly and different organs and glands of the honey bee. The amounts of the most abundant proteins can be million-fold greater than those of the low-abundance royal jelly proteins. Many important families of honey bee and royal jelly proteins (that may be promising drug targets) such as transcription factors, protein kinases, and regulatory proteins are low-copy proteins. These proteins will not be detected in the analysis of crude soluble fractions of honey bee glands or royal jelly without some purification. Therefore, new methods must be devised for sub-proteome isolation. Despite these limitations, proteomics, when combined with other complementary technologies such as conventional separation methods, has an enormous potential in terms of new insights into physiological functions of honey bee proteins.

## 5. Royal jelly sugar research

### 5.1. Introduction

As for any food or dietary products, the knowledge about the levels and composition of carbohydrates in royal jelly offers important information on the product,

Table 6. Composition of Lectin CWR (Wisniewski et al., 2009; Zielinska et al., 2010; Zhang et al., 2014).

| Component | Concentration | Volume |
|---|---|---|
| Concanavalin A (prepare with 2× Binding buffer) | 6 mg/ml | 15 μl |
| Wheat germ agglutinin (prepare with 2× Binding buffer) | 6 mg/ml | 15 μl |
| RCA 120 agglutinin solution | | 6 μl |
| Pool together, agitate thoroughly, store at −20 °C | | |

such as its caloric content calories. Sugar content and profile can also indicate the use of exogenous sugars because of abnormal composition or too high a concentration of carbohydrates.

In contrast to honey from nectar or honeydew, few articles dealing with the sugar composition of royal jelly are available. Determination of sugar content of royal jelly is based on methods used for honey (See the *BEEBOOK* paper on honey research methods (in prep)). Nevertheless, carbohydrate physico-analytical methods applicable to honey samples are not so easily transferable to this complex matrix. Methods involved in honey analysis can show lower robustness when applied to royal jelly. The presence of higher water content (See Subsection 3.2.1), high-molecular-weight molecules (proteins, see Section 4), and lipids (triglycerides, free fatty acids, see Section 6) can cause incompatibility with specific analytical instruments or sample preparation procedures. As a result, there are fewer methods commonly employed for royal jelly carbohydrate analysis.

The relevant studies on royal jelly are mainly based on 2 approaches: either a total sugar analysis (in general by colourimetric methods) or a carbohydrate profile using chromatographic separation: HPLC, ionic chromatography (IC), capillary zone electrophoresis (CZE), or gas chromatography. These methods are described below.

### 5.2. General methods for sugar analysis applied to royal jelly

#### 5.2.1. Methods for quantification of total carbohydrates

One of the cheapest, easiest, and universally used approaches to analysis of the total carbohydrate content of a product is the colourimetric method (See the *BEEBOOK* paper on honey research methods (in prep)). It is based on the reaction between a hydrolyzed carbohydrate solution and a coloring reagent leading to a color detectable with a spectrometer in the visible range. Reagents commonly used for coloration include phenol (DuBois, Gilles, Hamilton, Rebers, & Smith, 1956), alkaline ferricyanide (Englis & Becker, 1943), alkaline cupric tartrate (Munson & Walker, 1906), and anthrone (Dreywood, 1946).

The phenol-sulfuric acid method of DuBois et al. (1956) is so far the most reliable and has been extensively used in many fields, including honey bee products.

(1) Place a clear aqueous solution of the sample in a test-tube.
(2) Add 1 ml of phenol and sulfuric acid solution (5% w/v).
    The solution turns yellow-orange as a result of the interaction between the carbohydrates and phenol.
(3) Measure the absorbance at 420 nm.
    The absorbance is proportional to the carbohydrate concentration in the sample.
(4) Prepare a calibration curve with standards of known carbohydrate concentrations for quantification.

Because sulfuric acid converts all non-reducing sugars into reducing sugars, this method determines the total sugar content.

#### 5.2.2. Methods for quantification of individual carbohydrates

*5.2.2.1. The enzymatic method.* Analytical methods based on enzymes rely on their ability to catalyze specific reactions. These methods are rapid, highly specific, and sensitive. In addition, the sample preparation is less laborious. Therefore, this approach is suitable for carbohydrate quantification in royal jelly. If samples are not liquid, dissolve them in water.

Many enzyme assay kits for the analysis of specific carbohydrates are commercially available with detailed instructions on how to carry out the analysis. Two methods are commonly used: (i) allowing the reaction to proceed to completion and measuring the concentration of the product, which is proportional to the concentration of the initial substrate; (ii) measuring the initial rate of the enzyme-catalyzed reaction because the rate is proportional to the substrate concentration. Glucose, fructose, sucrose, and malic and citric acid have been quantitated in samples of honeydew, honey, and royal jelly by enzymatic UV tests. Preliminary qualitative analysis of carbohydrates present in the analyzed samples is performed by silica gel thin-layer chromatography.

The enzymatic method has shown appreciable advantages in specificity and ease of use. Main monosaccharides (fructose and glucose) and sucrose have been quantified by this method in royal jelly by Tourn, Lombard, Belliardo, and Buffa (1980).

*5.2.2.2. The HPLC method.* Binder (1980), Nikolov, Jakovljević, and Boškov (1984), Wong-Chong and Martin (1979), Takenaka and Echigo (1980) have shown good resolution of mono- and disaccharides by liquid chromatography with refractive index detection, using an amino-linked modified silica column (Binder, 1980; Nikolov et al., 1984; Takenaka & Echigo, 1980; Wong-Chong & Martin, 1979). In royal jelly, Bogdanov et al. (2004), Sesta (2006), and Serra Bonvehi (1992) have quantified carbohydrates using HPLC (Bogdanov et al., 2004; Serra Bonvehi, 1992; Sesta, 2006).

Using HPLC with Restek Pinnacle II amino column (250 × 3.2 5 μm), Sesta (2006) quantified fructose, glucose, sucrose, and maltose in 97 royal jelly samples. The refractive index detector is used for quantification. The sample preparation consisted of protein precipitation (Currez reagent) and removal of lipids. Maximum and minimum values have been presented for each individual sugar as well as the total sugar content. In another work, Sesta (2006) studied the influence of sugar feeding by using honey, sucrose, Cerestar, and Apiinvert sources. Ninety-five royal jelly samples were analyzed by HPLC with refractive index detection. Maltose seemed to be present at higher concentrations when Cerestar syrup was used, while total sugar was found to be in agreement with previously published data. Thus, a sugar profile leads to important information regarding royal jelly quality control. The HPLC method, even with specific amino columns dedicated to sugar analysis, has some weaknesses: not enough sugars are separated during the analysis, refractive index detection is highly sensitive to temperature variations, and sensitivity is poor.

*5.2.2.3. GC method.* Carbohydrates are not usually analyzed by GC because of their low volatility and their thermal degradation. Therefore, they need to be transformed into volatile species prior to the injection step.

All methods using derivation of sugars have to take into account the moisture sensitivity of the derivation process. Preliminary drying or lyophilization of the sample is necessary (see Subsection 3.3).

Several derivatization pathways are available for the modification of the hydroxyl functions of the carbohydrates with silylating agents. Gee and Walker (1962), Bishop (1962), Fournet, Dhalliuin, Montreuil, Bosso, and Defaye (1980), Fournet, Strecker, Leroy, and Montreuil (1981) have analyzed methyl derivatives of carbohydrates (Bishop, 1962; Fournet et al., 1980, 1981; Gee & Walker, 1962). Acetylation derivatization has been applied to monosaccharide analysis by Lineback (1968) and Dutton (1973). Nevertheless, the use of trimethylsilyl derivatives is still the most popular pathway for sugar GC analysis, first introduced by Hedgley and Overend (1960) and Bayer and Witsch (1961) and broadly applied by Sweeley, Bentley, Makita, and Wells (1964) (Bayer & Witsch, 1961; Hedgley & Overend, 1960; Sweeley et al., 1964). Horváth and Molnár-Perl (1997) have achieved simultaneous GC-MS quantification of mono-, di-, and trisaccharides by means of their trimethylsilyl (TMS) ether oxime derivatives in honey samples (Horváth & Molnár-Perl, 1997).

This latter method is now widely employed for honey sugar analysis since the work of Pourtallier (1967). For royal jelly, the benchmark study was published by Lercker, Caboni, Sabatini, and Nanetti (1986). Numerous works using GC cite this study. Unlike honey, which is mainly composed of sugars (80%), royal jelly carbohydrates are present at lower concentrations (7.5–18%). The results indicate that the main carbohydrates are the monosaccharides fructose and glucose, constituting ~90% of total sugar (Bogdanov et al., 2004; Sabatini et al., 2009; Serra Bonvehi, 1992; Sesta, 2006). Between the two, fructose prevails. More carbohydrates have been identified and quantified using GC method; numerous disaccharides and trisaccharides as well as honey carbohydrate profiles have been successfully analyzed (Daniele & Casabianca, 2012; Sabatini et al., 2009; Wytrychowski, Daniele, & Casabianca, 2012; Wytrychowski et al., 2013).

Silylated derivatives of carbohydrates analyzed by GC are a powerful tool for resolution, identification, and quantification of sugars in royal jelly as described in numerous studies (Daniele & Casabianca, 2012; Lercker et al., 1986; Sabatini et al., 2009; Wytrychowski et al., 2012, 2013). In general, methyl-phenyl polysiloxane (HP 5 or DB 5) chromatographic capillary columns are used. An efficient resolution is obtained for mono-, di-, and trisaccharides. Due to calibration with a mixture of alkanes (C15–C40), retention indices were obtained (Table 7) (Cotte, 2003).

*5.2.2.3.1. Sample preparation*

(1) Lyophilise royal jelly (see Subsection 3.1.3).
(2) Weigh 40 mg of lyophilized royal jelly in a glass reactor.
(3) Add 1 mg of the internal standard (for example, sorbitol).
(4) Add 1 ml of anhydrous pyridine.
(5) Tightly close the reactor.
(6) Stir for 5 min.
(7) Add 200 μl of hexamethyldisilazane.
(8) Stir for 5 min.
(9) Add 100 μl of trimethylchlorosilane.
(10) Stir for 30 min.
(11) Leave the mixture for 20 h at room temperature with the reactor sealed.
(12) Analyze the derivatised sample by GC [see section 2.2.3 of the article by Torto et al. (2013)].

*5.2.2.3.2. GC analytical conditions*

(1) Connect the chromatograph with an HP5-MS column (30 m × 0.25 mm; 0.25 μm i.d.), a split-splitless injector, an autosampler, and a flame ionization detector (FID).
(2) Use helium (grade 5.0) as a carrier gas.
(3) Fix the injection volume at 1 μl in split mode with a ratio 1:20.
(4) Program the oven temperature as follows:
    - maintain the initial temperature (150 °C) for 5 min,
    - increase it to 325 °C at the rate of 3 °C/min,
    - maintain the final temperature for 10 min.
(5) Set the injector and detector temperatures to 280 °C.
(6) For the detector, set the hydrogen flow to 40 ml/min and the air flow to 450 ml/min.
(7) Maintain constant helium pressure at 22.04 psi.
(8) Inject 1 μl of a mixture of paraffins from C15 to C40 (at 0.2% in chloroform, w/v) prior to each batch of samples for calculation of retention indices (Cotte, 2003).
(9) Identify various sugars by means of Kovats indices instead of retention times.

Table 7. Retention indices of carbohydrates on HP5-MS.

| Sugar | Anomer-1 | Anomer-2 |
|---|---|---|
| *Monosaccharide* | | |
| Rhamnose | 1653 | |
| Arabinose | 1640 | |
| Xylose | 1737 | |
| Fructose | 1843 | 1853 |
| Mannose | 1845 | 1942 |
| Galactose | 1900 | 1944 |
| Glucose | 1931 | 2030 |
| Sorbitol | 1980 | |
| *Disaccharide* | | |
| Lactulose | 2674 | 2695 |
| Saccharose | 2707 | |
| Maltose | 2747 | 2792 |
| Cellobiose | 2756 | 2864 |
| Maltulose | 2773 | 2780 |
| Nigerose | 2784 | 2810 |
| Turanose | 2790 | |
| Trehalose | 2805 | |
| Kojibiose | 2814 | |
| Palatinose | 2824 | |
| Neo-trehalose | 2849 | |
| Laminaribiose | 2857 | 2889 |
| Leucrose | 2850 | |
| Iso-Trehalose | 2884 | 2941 |
| Melibiose | 2933 | |
| Isomaltose | 2952 | 3005 |
| Gentiobiose | 2978 | |
| *Trisaccharide* | | |
| Raffinose | 3501 | |
| Neo-Kestose | 3512 | |
| 1-Kestose | 3517 | |
| Erlose | 3549 | |
| Melezitose | 3585 | |
| Maltotriose | 3627 | |
| Panose | 3685 | |

Figure 32 shows a GC carbohydrate profile obtained for a royal jelly sample where 15 sugars are identified and quantified: fructose, glucose, sucrose, maltose, galactose, mannitol, trehalose, gentiobiose, isomaltose, turanose, erlose, maltulose, melezitose, and maltotriose (Daniele & Casabianca, 2012). The internal standard (IS) is sorbitol.

In recent years, an international working group (WG 13) elaborated a physicochemical standard for international business of royal jelly (Norm project ISO 12824 Royal-jelly: Specifications). Analysis of sugars is an important part of this project because they directly correlate with production pathways (environmental natural resources or sugar syrup bee feeding). Sesta (2006) and Daniele and Casabianca (2012) have shown important carbohydrate profile modifications (especially for di- and trisaccharides) when artificial feeding sugars were used instead of honey or natural resources (nectar). On the basis of the analysis of 800 royal jelly samples, Daniele and Casabianca (2012) and Wytrychowski et al. (2012, 2013) have proposed a discriminatory method combining the GC sugar profile with $^{13}C$ stable isotope measurements.

*5.2.2.4. Ionic chromatography.* With the development of polymeric stationary phases allowing for work with basic pH mobile phases (Dionex, 1993), important work has been done in food and agricultural chem0istry based on ionic chromatography of carbohydrates. Amperometric detection is typically used. In 1995, Goodall et al. classified honeys based on their floral origin by means of this technique (Goodall, Dennis, Parker, & Sharman, 1995). However, only a few research groups applied this method to the royal jelly matrix. In fact, unlike honey, royal jelly consists of a complex matrix containing fats and high-molecular-weight molecules (proteins) leading to difficulties with the development of robust methods. To study glucose, fructose, and sucrose in royal jelly, prior precipitation of proteins by trifluoroacetic acid is necessary, as described in the study by Lei, Peng, Ge, Mei, and Zhu (2013).

Ionic chromatography is highly sensitive to the royal jelly matrix, in contrast to routine control with honey. Right now, this method is not applicable to royal jelly due to pollution of the polymeric columns with fats and proteins and because of the amperometric detection.

*5.2.2.5. Infrared spectroscopy.* Infrared spectroscopy is a rapid, easy, and reliable method for quality analysis. In the case of honey, this method has shown good correlations among sugars, proline, free amino acids, invertase, moisture, HMF, pH, and electrical conductivity (Lichtenberg-Kraag, Hedtke, & Bienefeld, 2002; Ruoff et al., 2006a, 2006b, 2007). Using near-infrared transflectance spectroscopy, García-Alvarez, Huidobro, Hermida, and Rodríguez-Otero (2000) analyzed fructose, glucose, and moisture in 161 honey samples (García-Alvarez et al., 2000). Several studies have used this method for geographical classification and for adulteration control (Kelly, Downey, & Fouratier, 2004; Ruoff et al., 2006a, 2006b, 2007; Woodcock, Downey, Kelly, & O'Donnell, 2007). In general, an important part of the method is the statistical analysis of the numerous and multi-parametric results (principal component analysis). To our knowledge, this method has not yet been employed to study carbohydrates in royal jelly.

*5.2.2.6. Nuclear magnetic resonance.* Few studies are dealing with sugar analysis in bee products by nuclear magnetic resonance (NMR). Such a method requires considerable investment, advanced analytical skills, and time. Most of the studies involving NMR have been developed in $^{1}H$ mode especially for geographical characterization and adulteration control (Bertelli et al., 2010; Consonni & Cagliani, 2008). Specific magnetic fields are necessary to obtain higher $^{1}H$ resolution for chemical shifts because the range is smaller than that of $^{13}C$ (0–10 ppm), which represents an expensive investment and high-cost routine analysis.

Mazzoni, Bradesi, Tomi, and Casanova (1997) have conducted the first study on the qualitative and quantitative analysis of sugars in honey by $^{13}C$ NMR (40)

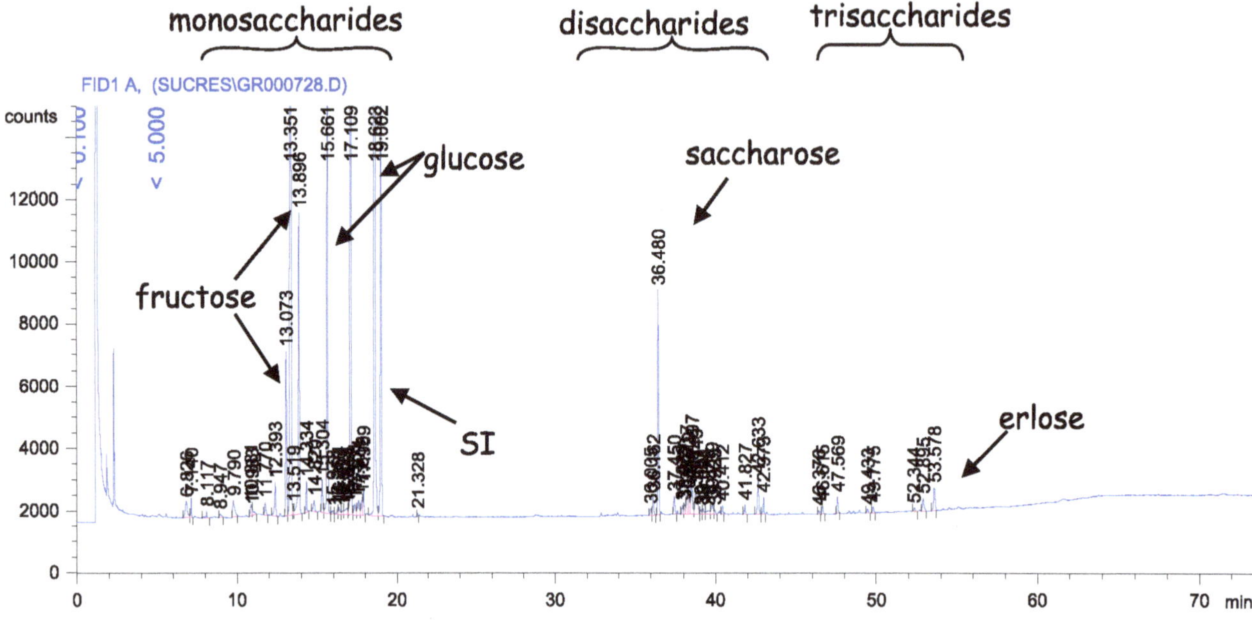


Figure 32. Royal jelly mono-, di- and trisaccharide profile obtained with GC.

(Mazzoni et al., 1997). The main advantage of $^{13}C$ NMR is that no separation is necessary for complex mixtures. $^{13}C$ chemical shifts offer an important working range allowing for identification even in complex mixtures ($^{13}C$ chemical shift from 0 to 280 ppm).

### 5.3. Discussion of methods for carbohydrate analysis as applied to royal jelly

Only little information is available in the literature on carbohydrate analysis in royal jelly. Generally, 3 kinds of methods are described in the literature: a global one to determine the total sugar content of royal jelly, a method to quantify separately the 3 main sugars, and more specific procedures to quantify the maximal number of individual sugars.

All colourimetric methods have some drawbacks: the chemical reactivity of carbohydrates with the derivatization reagent (sulfuric acid) greatly depends on whether the carbohydrates are neutral or ionic. As a result, the molar absorption coefficients can greatly vary depending on the charge of the carbohydrates analyzed (Albalasmeh et al., 2013; Mecozzi, 2005). Quisumbing and Thomas (1921) have also discussed several parameters and conditions that can affect the Fehling solution as a sucrose-reducing action, and blank or reagent auto-reduction (Quisumbing & Thomas, 1921). In many cases with a colourimetric reaction, the other constituents of the matrix can react (e.g., polyphenols, proteins, or mineral salts), resulting in overestimation. Finally, the main disadvantage of these methods is the quantification of all carbohydrates without any information on target constituents. Indeed, production involving bee feeding with syrups (cane sugar, beet sugar, inverted sugars, or hydrolyzed starch) leads to modifications of the carbohydrate profile (Daniele & Casabianca, 2012; Sesta, 2006; Wytrychowski et al., 2012, 2013). As a consequence, to ensure quality of royal jelly, chromatographic resolution of sugars is necessary.

### 5.4. Conclusion

Quality control of royal jelly requires analysis of individual carbohydrates in order to determine a production pathway corresponding to future standard requirements. This information is available only via a separation method appropriate to the specific matrix under study. As discussed above, royal jelly complexity (higher water and lower sugar contents compared to honey, and the presence of proteins and lipids) does not match traditional quantification methods for sugars. Based on carbohydrate resolution power, only GC, LC, or IC methods offer chromatographic resolution of a sufficient number of components. From a robustness point of view, the best method seems to be GC with trimethylsilyl derivation of pre-lyophilized royal jelly (see Subsection 5.2.2.3). This method is not sensitive to protein and fat content of the matrix, while LC or IC cannot offer routine methodology in this case (pollution of chromatographic support and detector). GC with capillary columns offers enough resolution for the range of carbohydrates present in the royal jelly matrix (mono-, di-, and trisaccharides) and can be coupled to several detection systems [flame ionization or mass spectrometric detectors, see Torto et al. (2013)].

## 6. Research on royal jelly fatty acids

### *6.1. Introduction*

Lipids in biological materials were described as lipid metabolites in the past but currently are known as the lipidome. The latter may be classified as a subset of metabolites, although it is typically regarded as distinct from other metabolites (Smith, Mathis, Ventura, & Prince, 2014). A new nomenclature system, Lipid Metabolites and Pathways Strategy Consortium, has been proposed for lipids in nature, which classifies them into 8 categories such as fatty acyls (i.e., free and esterified fatty acids), sterols, and glycerolipids (Fahy et al., 2005; Watson, 2006). Lipids in royal jelly, accounting for 3–8% of the fresh matter or 15–30% of the lyophilized product, are assumed to be synthesized by the mandibular glands of honey bee nurses (Li, Huang, & Xue, 2013). Although fatty acids and sterols (and even glycerolipids and glycerophospholipids) have been detected in lipid extracts from royal jelly, the lipids in royal jelly are mostly composed of (aliphatic) fatty acids (Townsend & Lucas, 1940a). Almost all of these acids are present as free fatty acids and barely any as esters (Li et al., 2013). Unlike fatty acids of most animal and plant materials, which consist mainly of triglyceride fatty acids having the main-chain length of 14–20 carbon atoms, most of fatty acids in royal jelly have a main-chain length of 6–12 carbon atoms and are therefore named as medium-chain fatty acids (MCFA) (Genç & Aslan, 1999). These fatty acids are hydroxylated at terminal and/or internal positions, terminated with mono- or dicarboxylic acid groups, and saturated or monounsaturated at the second position. Among them, the most abundant MCFA is 10-HDA, amounting to 70% of lipid extracts from royal jelly and more than 50% of free fatty acids in royal jelly (Li et al., 2013). This MCFA is considered a characteristic constituent of lipids in royal jelly and serves as an index for estimating quality of royal jelly (Chiron, 1982; Fray, Jaeger, Morgan, Robinson, & Sloan, 1961; Genç & Aslan, 1999).

10-HDA was shown to possess a variety of *in vitro* pharmacological effects such as anti-tumor, immunomodulatory, estrogen-like, collagen production-promoting, and neurogenesis-promoting effects (Li et al., 2013). These findings provide evidence that fatty acids in royal jelly make a contribution to unravelling some of the most basic processes of honey bee and offer the related knowledge on pharmacological significance of royal jelly in human health. In fact, free fatty acids in royal jelly are potential sources of bioactive compounds that function at least as nutrients required for both oogenesis in a virgin queen and early development of larvae (McFarlane, 1968; Prosser, 1978; Yanes-Roca, Rhody, Nystrom, & Main, 2009; Ziegler & Vanantwerpen, 2006), as nutritional factors in the queen-worker dimorphism (Kim, Friso, & Choi, 2009; Kucharski, Maleszka, Foret, & Maleszka, 2008; Spannhoff et al., 2011; Turner, 2000; Waterland & Rached, 2006; Zaina, 2010), hormone precursors of queen retinue pheromones (Kodai, Nakatani, & Noda, 2011), and as antimicrobial and mite-repellent agents for pathogen invasion of a queen host (Blum et al., 1959; Drijfhout, Kochansky, Lin, & Calderone, 2005; Iwanami, Okada, Iwamatsu, & Iwadare, 1979; Nazzi, Bortolomeazzi, Della Vedova, Del Piccolo, & Milani, 2009). Free acids in royal jelly are useful as preventive and supportive medicines and function, for example, as potential inhibitors of cancer growth, immune system modulators, as alternative therapies for menopause, skin-aging protectors, as neurogenesis inducers, and more (Li et al., 2013).

Just as the more established metabolomics, lipidomics is aimed at identifying and quantifying all endogenous and exogenous small metabolites with chromatography, spectrometry, and spectroscopy (Cerkowniak, Puckowski, Stepnowski, & Gołębiowski, 2013; Sandra & Sandra, 2013; Wang, Byun, & Pennathur, 2010). Methods of lipidomics are widely used for the quantification of fatty acids in royal jelly. Examples of methods are solvent extraction also known as liquid-liquid extraction (LLE), thin-layer chromatography (TLC), liquid chromatography (LC), GC, MS, optical activity (OA), attenuated total reflection-Fourier-transform infrared spectroscopy (ATR-FTIR), and nuclear magnetic resonance (NMR) as well as combined techniques. Extraction of the lipidome in royal jelly via LLE processes is usually the first step of lipidomics. Analysis of the separable fraction of lipids from royal jelly by TLC is intended to confirm the presence of the lipidome in royal jelly. Most of fatty acids in royal jelly are examined primarily by GC, in particular GC with a flame ionization detector (FID) because FID is used for analysis of organic compounds. GC-MS represents a powerful tool for the analyses of organic compounds and is a definite help in qualitative and quantitative assays of fatty acids in royal jelly. During the past decade, NMR, alone or in combination with OA and IR, has played an ever-increasing role in deciphering chemical structure of fatty acids and their derivatives and analogues in royal jelly.

### *6.2. Isolation of fatty acids from royal jelly*

To ensure reliability of lipidomic analysis, fresh royal jelly is lyophilized into a powder (see Section 3.1.3). A discontinuous LLE method was initially used to extract the lipidome from royal jelly with diethyl ether (also referred to as ether in the literature) in a dilute sodium hydroxide solution or aqueous pyridine solution (Fevold, Hisaw, & Leoard, 1931; Lercker, Capella, Conte, & Ruini, 1981; Townsend & Lucas, 1940b). Besides diethyl ether, other organic solvents are suitable for preparation of the lipidome from royal jelly by LLE methods, including trichloromethane alone or trichloromethane followed by methanol or a mixture of trichloromethane and methanol (Kodai, Umebayashi, Nakatani, Ishiyama, & Noda, 2007; Noda, Umebayashi, Nakatani, Miyahara, & Ishiyama, 2005; Weaver, Johnston, Benjamin, & Law,

1968; Weaver & Law, 1960; Weaver, Law, & Johnston, 1964), a mixture of dichloromethane and methanol (Melliou & Chinou, 2005), a mixture of trichloromethane and acetone (Kodai et al., 2011), or hexane (Terada, Narukawa, & Watanabe, 2011). This exhaustive but requisite preparative stage for lipidomic analysis is the most common choice for efficient and cost-effective extraction of the lipidome from royal jelly. As an alternative method, a solid-liquid extraction with refluxing diethyl ether was suggested by Antinelli, Davico, Rognone, Faucon, and Lizzani-Cuvelier (2002). The presence of the lipidome in crude lipid extracts of royal jelly can be monitored by TLC (Lercker et al., 1981; Melliou & Chinou, 2005). Next in importance, preparative separation of various fatty acids from the extracts by classic LC is conducted by means of poly-silicic acids (also referred to as *silicic gel* in the literature and exemplified by Cosmosil 75C18-OPN), an ion exchange resin (such as Diaion HP-20), or hydroxypropyl Sephadex gel (such as SephadexLH-20) chromatography columns (Kodai et al., 2007; Melliou & Chinou, 2005; Noda et al., 2005; Weaver et al., 1968). In many cases, however, LC is modified into an advanced technique for faster preparation of fatty acids from royal jelly, including solid-phase extraction (SPE) in a cartridge (or filter) (Terada et al., 2011), medium-pressure liquid chromatography (Melliou & Chinou, 2005), and HPLC on a reversed-phase column (Kodai et al., 2011; Noda et al., 2005). Recently, isolation of volatile fatty acids from royal jelly was carried out with headspace solid phase micro-extraction on a specific device (Isidorov, Bakier, & Grzech, 2012).

*6.2.1. Preparation of fatty acids by LLE and SPE*

*Requirements:*

- A retort, test tube, or flask: e.g., a 25-mL volumetric type or bigger ones.
- Freeze dryer: e.g., FreezeDryer FDU-1100 model (Rikakikai Co., Ltd.).
- Paper filter: e.g., Whatman™ Grade 1 qualitative cellulose filter paper.
- Extraction reagents: diethyl ether or trichloromethane and methanol.

  Note: Both of them and other reagents to be stated in the following text are high-purity organic solvents suitable for LLE, LC, LC/MS, NMR, and other demanding analytical applications.
- Centrifugal vacuum evaporator: e.g., Centrifugal Evaporator CVE-3110 model (Rikakikai Co., Ltd.).
- Vacuum elution apparatus: e.g., 12-Port Visiprep DL Vacuum Manifold model (Sigma-Aldrich Co.). This disposable Liner type is equipped with an accessory collection rack with a group of receiving tubes plugged into the device.
- SPE cartridge: e.g., Ion Exchange Cartridge Bond Elut–DEA model (Agilent Technologies, Inc.). The stationary phase of the column in this cartridge is composed of an ion exchange resin with bonded aminopropyl moieties, activated by conditioning it with an appropriate amount of hexane and is fixed in a vacuum elution apparatus.

*Procedure:*

(1) Lyophilise a royal jelly sample (see Subsection 3.1.3).
(2) Put 500–1000 mg of lyophilized royal jelly into a retort.
(3) Add 10 ml of diethyl ether or trichloromethane.
(4) Vortex for 15 min to extract lipids.
(5) Pour the extract through a paper filter.
(6) Obtain as much fatty acids as possible from the same sample by extraction repeated 2 more times.
(7) Save combined filtrates.
(8) Transfer insoluble material into another retort.
(9) Add 10 ml of methanol.
(10) Vortex for 15 min to extract lipids.
(11) Pass the extract through a paper filter.
(12) Obtain as much fatty acids as possible from the same material by extraction repeated 2 more times.
(13) Save combined filtrates.
(14) Transfer joint filtrates obtained at step 7 in a plastic tube.
(15) Concentrate them with the help of a centrifugal vacuum evaporator.
(16) Transfer combined filtrates obtained at step 13 to a plastic tube.
(17) Concentrate them by means of a centrifugal vacuum evaporator.
(18) Add 2 ml of diethyl ether or trichloromethane to dissolve the oil-like residue (left on both the bottom and walls of the tube) obtained at step 15.
(19) Add 2 ml of methanol to dissolve the oil-like residue (left on both the bottom and walls of the tube) obtained at step 17.
(20) Load the solutions obtained after steps 18 and 19 into the top space of an SPE cartridge.
(21) Elute the column with 2% acetic acid in diethyl ether or trichloromethane and save the eluate.
(22) Concentrate the eluate by means of a centrifugal vacuum evaporator. Enriched material is labelled as "test sample" and is to be used for the following analyses.

*6.2.2. Monitoring of fatty acids in the presence of lipid extracts of royal jelly by TLC*

*Requirements:*

- TLC plate: e.g., a standard TLC plate (Sigma-Aldrich Co.) or self-made TLC plate cut from cleaned and dried glass (approximately with the

dimensions 20 × 20 cm or bigger). A uniform layer (approximately 0.2-mm thickness) of Silica Gel G slurry is spread on the plate by means of a spreader, dried at room temperature, and heated in an oven at 110 °C for 30 min.
- TLC tank: rectangular TLC development tank (Sigma–Aldrich Co.).
- Developing solvent: a solution mixed with hexane and diethyl ether (1.5:1.0 v/v).
- Detection reagent: 50% sulfuric acid.
- Reference standards: some lipid compounds from royal jelly, such as 10-HDA, dissolved in 2% acetic acid in diethyl ether or trichloromethane.

*Procedure:*

(1) Add the developing solvent beforehand to the TLC tank and close it with the lid, allowing it to saturate the chamber for 10 min at RT.
(2) Take a TLC plate and draw 2 straight lines: at the top and bottom, making the first one ~2 cm from the bottom and the second one ~1 cm from the top of the plate.
(3) Subdivide the bottom line into 2-cm gaps for spotting.
(4) Re-dissolve the test sample in 3–5 ml of 2% acetic acid in diethyl ether (the most effective solution for the extraction) or 2% acetic acid in trichloromethane (a comparable alternative to the extraction).
(5) Dissolve 20 mg of each reference standard in 1 ml of the solvent mentioned above.
(6) Pipette 10 μl or 20 μl of all reference standard solutions and of the test sample solution, and spot each of them on the spotting area on the plate.
(7) Dry the plate in ambient air for 5–10 min.
(8) Pick up the TLC plate holding the top by means of forceps.
(9) Place it in the TLC tank vertically.
Note: Ensure that the solvent phase moves uniformly along the plate.
(10) TLC is carried out in a laminar flow hood.
(11) Wait until the front of the solvent phase has moved approximately to the top of the line, allowing the mobile phase to evaporate completely.
(12) Remove the plate from the chamber and place it onto an experiment table surface in ambient air.
(13) Spray the detection reagent on the plate.
(14) Place the plate in an oven at 110 °C until areas containing the test sample get charred and appear as black spots.

### 6.3. Identification of fatty acids in royal jelly

Chromatographic analysis of 10-HDA and some other dominant fatty acids, e.g., 10-hydroxydecanoic acid, in royal jelly was first performed with reference to direct determination of these compounds by GC (Weaver et al., 1964). Currently, all dominant fatty acids in royal jelly and even those of low abundance, have been analyzed by means of an automated GC/MS apparatus, a high-screen capillary column coupled with a high-resolution mass spectrometer (Isidorov, Czyżewska, Isidorova, & Bakier, 2009; Isidorov et al., 2012; Li et al., 2013). Identification of a new fatty acid in royal jelly by GC-MS is dependent on the comparison of their respective mass spectra with the related data in the 2 databases, NIST (previously known as NBT) and Wiley (Online) library, and equivalents in the literature (Melliou & Chinou, 2005).

#### 6.3.1. Quantitative determination of dominant fatty acids by GC/MS

*Requirements:*

- Derivatising reagents: pyridine, Bis(trimethylsilyl) trifluoroacetamide (BSTFA) + 1% trimethylchlorosilane (TMCS).
- Esterification reaction vial: e.g., a Reacti-Vial™ Small Reaction Vial (Thermo Fisher Scientific [China] Co., Ltd.).
- GC instrument: e.g., Agilent HP6890 GC model (Agilent Technologies, Inc.). This instrument is equipped with the split/splitless inlet.
- Chromatographic column: e.g., HP-1 ms fused with silica model (Agilent Technologies, Inc.). This non-polar capillary column has 30-m length, 0.25-mm inner diameter, and 0.25-μm film thickness.
- MS instrument: e.g., Agilent 5973 MSD model (Agilent Technologies, Inc.). This instrument is equipped with an independently heated electron-ionization source.
- Liquid sampler: e.g., Agilent HP 7673 autosampler model (Agilent Technologies, Inc.).

*Procedure:*

(1) Prepare a range of 6 application solutions, covering the range 20–2000 mg/mL of 10-HDA, by diluting each standard stock solution of dominant fatty acids with methanol.
(2) Re-dissolve the test sample in a reaction vial with 2 ml of methanol.
(3) Into a separate reaction vial, add 0.5 ml of each application solution and an equal amount of the test sample solution.
(4) After solvent evaporation, combine each of them with 80 μl of BSTFA with 1% TMCS, and 220 μl of pyridine.
(5) Seal the cap.
(6) Mix thoroughly.
(7) Heat at 60 °C for 30 min.
(8) Cool to RT.

(9) Keep a trimethylsilyl (TMS) derivative to be used for GC-MS analysis as follows.
(10) Typical GC conditions:
- Inlet temperature: 250 °C.
- Inlet mode: splitless.
- Sample value: 1 μl.
- Carrier gas: helium.
- Column flow: 1 ml/min of constant flow.
- Initial column temperature: 122 °F.
- Oven ramp: at 5 °C per min to 300 °C.

(11) Typical MS conditions:
- Ionization source: 70 eV electron ionization.
- Scan mode: full scan.
- Scan range: 41–600 atomic mass units.

(12) The TMS derivative is subjected to GC-MS analysis.
(13) A regression equation is calculated on the basis of analysis of the results.

#### 6.3.2. *Qualitative identification of fatty acids by GC-MS*

*Requirements:*

- Derivatising reagents: the same as in Subsection 6.3.1.
- Esterification reaction vial: the same as in Subsection 6.3.1.
- GC instrument: the same as in Subsection 6.3.1.
- Chromatographic columns: e.g., HP-1 ms fused silica model (as a non-polar capillary column) and HP-5 ms fused silica model (as a low-polarity capillary column) (Agilent Technologies, Inc.). They all have 30-m length, 0.25-mm inner diameter, and 0.25-μm film thickness.
- MS instrument: e.g., an Agilent 5973 MSD model (Agilent Technologies, Inc.). This instrument has an independently heated electron ionization source.
- Liquid sampler: the same as in Subsection 6.3.1.

*Procedure:*

(1) Re-dissolve a test sample in a reaction vial with 2 ml of methanol.
(2) Add 0.5 ml of the test sample solution into the reaction vial.
(3) Allow the solvent to evaporate.
(4) Combine the residue with 80 μl of BSTFA with 1% TMCS, and 220 μl of pyridine.
(5) Seal the cap.
(6) Mix thoroughly.
(7) Heat at 60 °C for 30 min.
(8) Cool to RT.
(9) Keep the TMS derivative for GC-MS analysis as follows.
(10) Typical GC conditions: the same as in Subsection 6.3.1.
(11) Typical MS conditions: the same as in Subsection 6.3.1.
(12) The TMS derivative is subjected to GC-MS analysis.
(13) The fatty acid ester is compared with data registered in NIST, Willy libraries, and previously published data.

### 6.4. *Structural characterization of fatty acids in royal jelly*

By comparing infrared absorption spectra data to those of authentic standards from the literature, it was initially confirmed that the there is a *trans* configuration of the double bond in the main chain of fatty acids in royal jelly rather than at its specific position (Brown & Freure, 1959; Melliou & Chinou, 2005). The configuration at the anomeric carbon in fatty acids in royal jelly has been assigned on the basis of specific optical rotation from empirical data (Kodai et al., 2007, 2011; Melliou & Chinou, 2005; Noda et al., 2005). NMR spectral data on the free fatty acids in royal jelly are produced to directly determine the relative amounts and partial assignment of test compounds, including $^{1}C$ spectra and $^{13}C$ spectra (Kodai et al., 2007; Noda et al., 2005). Nevertheless, the similar magnetic resonance of MCFA in royal jelly suggests that this technique has limitations for characterization of a complete pattern of fatty acids in royal jelly. In contrast, MS approaches with electrospray ionization, chemical ionization, and fast atom bombardment have proven to be the most inclusive for lipid molecular species (Kodai et al., 2007). To date, it has been well established that a combination of MS and NMR can be used in lipidomics for quantification of a wide range of fatty acids in various biofluids (Gürdeniz et al., 2013). In this section, 3 methods are presented below for quantification of fatty acids in royal jelly, including optical rotation, infrared absorption, and nuclear magnetic absorption.

#### 6.4.1. *Infrared absorption measurement of fatty acids by ATR-FTIR*

*Requirements:*

- Infrared spectrometer: Perkin-Elmer 500 spectrum model (Perkin-Elmer, Inc.).
- Sample slot: an ATR accessory.
- Solvent: methanol (to dissolve hydrophilic or polar fatty acids) or dichloromethane (to dissolve lipophilic or non-polar fatty acids) or trichloromethane (instead of trichloromethane under the same experimental conditions).

*Procedure:*

(1) Adjust spectrometer/ATR accessory settings as follows:
- Temperature during measurement: 65 °C.
- Resolution: 4 $cm^{-1}$.

- Wavelength range: 4000–400 $cm^{-1}$.

(2) Deposit uniformly a drop or several drops of a test sample solution on the crystal surface of the ATR accessory until transparent detection window in the crystal surface is fully filled.

(3) Collect infrared absorption spectra of the fatty acids versus a dry air background of an empty ATR well.

#### 6.4.2. *Optical rotation measurement of fatty acids by OA*

*Requirements:*

- Polarimeter: e.g., Perkin-Elmer 341 model (Perkin-Elmer, Inc.) or JASCO DIP-140 model (JASCO China (Shanghai) Co., Ltd.).
- Solvent: methanol.

*Procedure:*

(1) Re-dissolve the test sample in methanol until the solution is saturated. Note: It is important for the clear solution to be saturated before researchers attempt to measure its OA.

(2) Adjust polarimeter settings as follows:
- Detection mode: optical rotary dispersion.
- Wavelength range: 250–660 nm.
- Temperature during measurement: 25 °C or at RT.

(3) Use a solvent blank after each run to determine the baseline.

(4) Convert the observed value to the degree of specific rotation, $[\alpha]^{25}$, using the following formula:

$[\alpha]^{25} = \alpha/b \quad (l \times c)$, where $\alpha$ = measured value, l = sample path length in decimetres, and $c$ = concentration of a sample in grams per ml.

#### 6.4.3. *Nuclear magnetic absorption measurement of fatty acids*

*Requirements:*

- NMR instruments: e.g., JMN GX400 model (for $^{1}H$ spectra) and ECA 600SN model (for $^{13}C$ spectra) (JEOL Ltd.), or DRX 400 model (for $^{1}H$ spectra) and Bruker AC 200 model ($^{13}C$ spectra) (Bruker Biospin Co.).
- Solvent: d4-methanol (CD3OD), trichloromethane-d ($CDCl_3$).
- Internal standard: tetramethylsilane.

*Procedure:*

(1) Re-dissolve the test sample in the matrix.

(2) Transfer to NMR tubes.

(3) Typical NMR conditions:
- Probe temperature: 25–35 °C.
- Acquisition of $^{1}H$ spectra: obtained at 600 or 400 MHz.
- Acquisition of $^{13}C$ spectra: obtained at 50, 100, or 150 MHz.
- Chemical shifts: on the $\delta$ scale (ppm) with tetramethylsilane or solvent signals.
- Coupling constant: given as the J value.

(4) Record $^{1}H$ and $^{13}C$ NMR spectra on the NMR spectrometer.

## 7. Residue analysis of main veterinary drugs and acaricides in royal jelly

### 7.1. *Introduction*

Bee products can be polluted by different sources of contamination, including environmental and apicultural sources. The most important contaminants in royal jelly are veterinary drugs used against bee diseases or for prevention of outbreaks of diseases. Acaricides that are used for *Varroa* control are also important contaminants of bee products.

Although most of veterinary drugs are not authorised for the treatment of honey bees in the EU or strictly limited in other countries, veterinary-drug residues can be found in some royal jelly samples. The most important and harmful veterinary-drug residues in royal jelly are chloramphenicol, nitroimidazole, sulphonamides, fluoroquinolone, macrolides, and tetracyclines. Fluvalinate and amitraz are the main acaricides used in apiculture and are often retained in bee products.

To ensure the quality of bee products, sensitive methods for residue determination are necessary. In this section, residue analysis of the veterinary drugs and acaricides mentioned above in royal jelly is addressed, and 1 or 2 classical analytical methods for each target compound are described in detail.

### 7.2. *The LC-MS/MS method for chloramphenicol analysis in royal jelly*

Chloramphenicol (CAP) is a broad-spectrum antibiotic, showing activity against a variety of aerobic and anaerobic microorganisms. Its protein synthesis-inhibiting properties make it effective in the treatment of several infectious diseases (Forti, Campana, Simonella, Multari, & Scortichini, 2005). It is often used in beekeeping to control European and American foulbrood (Ortelli, Edder, & Corvi, 2004). Because severe side effects such as aplastic anaemia and hypersensitivity have been demonstrated in humans (Allen, 1985), the European Community banned CAP use in food-producing animals since 1994, in order to protect consumers' health. Consequently, CAP was listed in Group A of the Council Directive 96/23/EC, including those substances for which a "zero tolerance residue limit" has been established in edible tissues. Nonetheless, this drug is still illicitly used in animal farming because of its ready availability and low cost.

A method for detection of CAP in honey has been developed by Robert Sheridan and contains an acid hydrolysis step to liberate the sugar-bound sulphonamides followed by solid-phase extraction to remove possible interfering substances. Analysis was based on liquid chromatography–electrospray ionization–tandem mass spectrometry in negative mode for all 15 analytes. This MRM method generated 2 structurally significant transitions per compound, and it was designed to conform to U.S. Food and Drug Administration MS confirmation guidelines. One hundred sixteen samples from 25 countries were analyzed, and 38% were found to contain at least 1 target antimicrobial agent. Five target compounds were found in honey from 13 countries (Sheridan, Policastro, Thomas, & Rice, 2008).

Ishii, Horie, Murayama, and Maitani (2006) developed a method for detection of CAP in royal jelly samples. The quantification limit of CAP in royal jelly was 1.5 ng/g. The recovery rates of CAP from both honey and royal jelly at the quantification limits were over 92% (Ishii et al., 2006). Moreover, a liquid chromatographic/tandem mass spectrometric method was developed and validated for quantification of CAP in royal jelly. Royal jelly samples were first denatured with lead acetate solution, and the CAP was extracted with solid-phase extraction before separation by liquid chromatography. A triple-quadrupole mass spectrometer operated in the negative electrospray ionization and selected-reaction monitoring mode was used for the detection of CAP. For method validation, royal jelly samples were spiked with CAP between 0.1 and 10.0 μg/kg; at these levels, recovery values (internal standard-corrected) ranged from 93.3 to 105.0%, and the within-laboratory reproducibility (relative standard deviation) was 9.1%. The decision limit was 0.07 μg/kg, and the detection capability was 0.1 μg/kg.

#### 7.2.1. *Sample preparation*

(1) Lyophilise royal jelly (see Subsection 3.1.3).
(2) Weigh 2.00 g of lyophilized royal jelly in a centrifuge tube (50 ml).
(3) Mix the sample.
(4) Leave the mixture at room temperature for 30 min.
(5) Add 10 ml of a $Pb(AcO)_2$ solution (200 g/l).
(6) Homogenize it for 40 s using Ultra-Turrax (Suzhou, PRC).
(7) Centrifuge the mixture for 5 min at 4000g.
(8) Use a Millex filter to filter the supernatant in another centrifuge tube (50 ml).
(9) Condition the Oasis HLB 60 mg SPE cartridge with 3 ml of methanol and 5 ml of water.
(10) Add the filtered supernatant to the cartridge and elute it by gravity or vacuum (1 drop/s).
(11) Wash the cartridge twice using 3 ml of the methanol–water mixture (1:4, v/v).
(12) Elute the CAP using 5 ml of methanol.
(13) Evaporate eluate under a nitrogen stream (40 °C).
(14) Reconstitute the residue in 1.0 ml methanol–water (3:7, v/v).
(15) Filter the solution using a 0.2-μm nylon filter before injection into the LC-MS/MS apparatus.

#### 7.2.2. *LC-MS/MS analytical conditions*

(1) Connect the chromatograph with a LUNA ODS $C_{18}$ column (7.5 × 4.6 mm; Phenomenex, Torrance, CA, USA).
(2) Use isocratic mobile phase of methanol: 5 mM ammonium acetate (60:40, v/v).
(3) Set the flow rate to 0.2 ml $min^{-1}$.
(4) Fix the injection volume to 5 μl.
(5) Set the column temperature to 40 °C.
(6) Set the MS detector to negative ion mode.
(7) Heat the TurboIon Spray source to 450 °C.
(8) Set the capillary voltage to 4.5 kV with an orifice potential of 20 V.
(9) Use nitrogen as a curtain and a collision gas.

Optimize the collision energies separately for the two selected ion transitions of CAP (321→152; 321→121). Typical LC-MS/MS analysis of CAP standard, negative sample, and spiked sample is shown in Figure 33.

### 7.3. The LC-MS/MS method for nitroimidazole analysis in royal jelly

Dimetridazole (DMZ), metronidazole (MNZ), and ronidazole (RNZ) are 5-nitroimidazole-based drugs and have antibacterial and anticoccidial properties. These drugs are widely used for the treatment of infections in poultry, cattle, swine, and farmed fish (Sakamoto et al., 2011). Unfortunately, these compounds have been suspected of being human carcinogens and mutagens. The Ministry of Health, Labor and Welfare in many countries including China set "not detected" as the standard for DMZ, MNZ, and RNZ in foods, and the detection limits of DMZ, MNZ, and RNZ are 0.2, 0.1, and 0.2 μg/kg, respectively .

A method for detection of traces of MTZ, DMZ, and RNZ residues in royal jelly was developed on the basis of HPLC with tandem mass spectrometry (HPLC-MS/MS). After samples were dissolved in a sodium hydroxide solution to disassociate target analytes from the matrix, liquid-liquid extraction methods by ethyl acetate solvent were used. Matrix effects were minimized, and good quantitation results were obtained using the highly selective reaction monitoring (H-SRM)

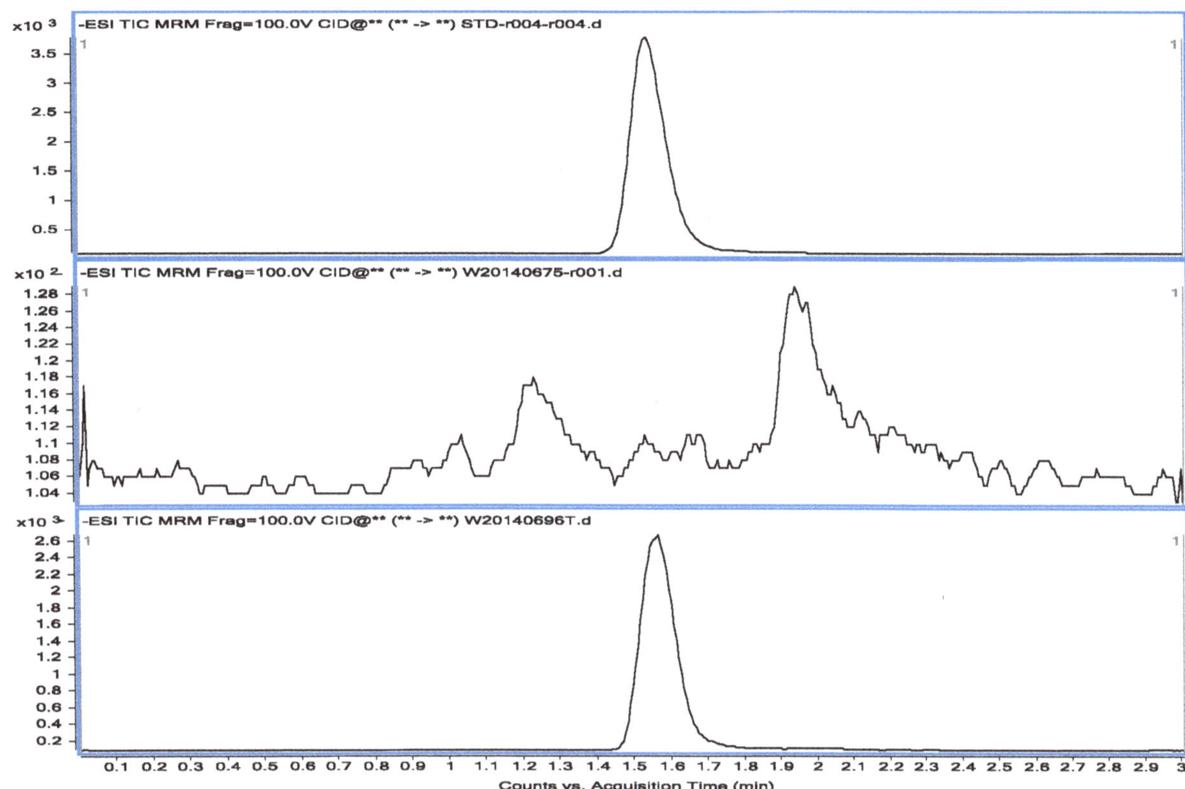


Figure 33. LC/MS/MS chromatography of chloramphenicol standard (top), spiked sample (bottom) and negative sample (middle).

technology. Limits of detection (LODs) were 1.0 mg/kg for DMZ and 0.5 mg/kg for MTZ and RNZ (Signal/Noise > 5). Limits of quantitation (LOQs) were 2.0 mg/kg for DMZ and 1.0 mg/kg for MTZ and RNZ (S/N > 10) (Ding et al., 2006).

### *7.3.1. Sample preparation*

(1) Lyophilise royal jelly (see Subsection 3.1.3).
(2) Add 5 g of royal jelly to a centrifuge tube (50 mL).
(3) Add 10 ml of 0.5 M NaOH.
(4) Mix for 15 s to dissolve.
(5) Add 10 ml of EtOAc.
(6) Mix for 30 s.
(7) Centrifuge the mixture for 3 min at 545g.
(8) Transfer the supernatant to a test tube (50 ml).
(9) Repeat extraction steps 5–8.
(10) Combine EtOAc layers obtained at step 8.
(11) Dry in a water bath at 40 °C.
(12) Dissolve the residue in 5 ml of ACN containing 10% formic acid.
(13) Pour 3 ml of methanol on top of the column for conditioning.
(14) Pour 3 ml of water on top of the column for conditioning.
(15) Load the mixture at 1–2 ml/min onto the column.
(16) Wash the column with 3 ml of water.
(17) Dry it for 5 min.
(18) Elute the column using 3 ml of methanol.
(19) Collect the eluate at 1–2 ml/min.
(20) Evaporate it.
(21) Filter the solution using a 0.2-µm nylon filter before injection into the LC-MS/MS apparatus.

### *7.3.2. LC-MS/MS analytical conditions*

(1) Set the flow rate to 0.4 ml/min, with flow ramp 2.00.
(2) Set the column temperature to 25 °C.
(3) Set the mobile phase to the 40/60 ratio (0.1% formic acid in water/0.1% formic acid in ACN).
(4) Set the injection volume to 5 µl.
(5) Set the ESI polarity to positive.
(6) Set the capillary voltage to 3 kV; set the RF lens to 0.1.
(7) Set the source temperature to 140 °C.
(8) Set the desolvation temperature to 450 °C, and desolvation gas flow to 650 l/h.
(9) Use the cone gas flow 150 l/h.
(10) Set the ion energy to 10.5.
(11) Set the entrance lens to 5.
(12) Set the collision gas flow to 18.
(13) Set the MS2 low-mass resolution to 14.5.

(14) Use the high-mass resolution of 14.5.
(15) Set the multiplier voltage to 650 V.

A typical LC-MS/MS chromatogram of a nitroimidazole standard, negative sample, and spiked sample is shown in Figure 34.

### 7.4. *The LC-MS/MS method for analysis of sulphonamides in royal jelly*

The sulphonamide family of antibiotics includes a large spectrum of synthetic bacteriostatics used against most gram-positive and many gram-negative microorganisms and protozoa. In the past decade, the irresponsible use of sulphonamide drugs in the veterinary field, for therapeutic and prophylactic purposes, as well as for treatment of human infectious diseases, has favored the development of bacterial resistance (Wegener et al., 2003), which makes it difficult to efficiently treat infections with the presently known antibiotics (Furusawa & Kishida, 2001). In bees, these antimicrobials are used to prevent and treat bacterial diseases such as American foulbrood caused by *Paenibacillus larvae* and European foulbrood caused by *Melissococcus plutonius* (Bogdanov, 2006). The method

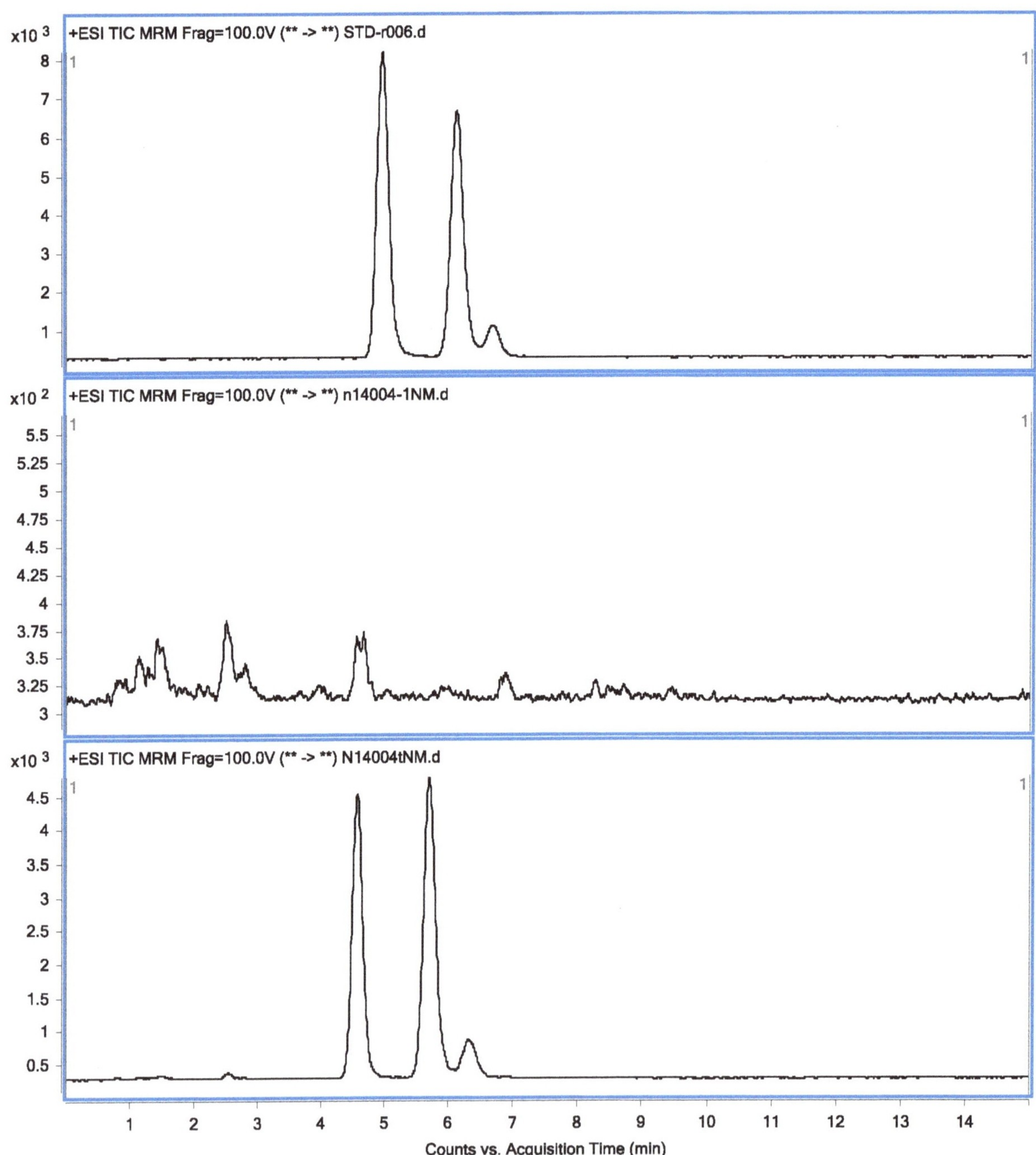


Figure 34. LC/MS/MS chromatography of nitroimidazole standard (top), negative sample (middle) and spiked sample (bottom).

commonly used for prevention is to feed the bees with a certain amount of sulphonamides in winter or early spring to improve their immunity. On the other hand, contamination of food with sulphonamide residues poses risks to human health, including an increased resistance of bacteria to antimicrobial agents, allergic reactions, and possible carcinogenicity (Enne, Livermore, Stephens, & Hall, 2001). According to European Commission (EC) Directive 2377/90, the rehabilitation of honey bees with sulphonamide antibiotics was banned in the European Union. Switzerland, Belgium, and the United Kingdom have chosen 0, 20, and 30 ng/g, respectively, as maximum residue limits (MRLs) for sulphonamide antibiotics in honey. A recent study on antibiotic residues (from sulphonamide and other antibiotic groups) in honey revealed test-positive samples exported to Europe from India, China, and Argentine that are above the MRLs (Dubreil-Chéneau, Pirotais, Verdon, & Hurtaud-Pessel, 2014).

The liquid chromatography–tandem mass spectrometry (LC-MS/MS) method for the simultaneous confirmation of 13 sulphonamides in honey was developed and fully validated by Dubreil-Chéneau in accordance with the European Commission Decision No. 2002/657/EC (Dubreil-Chéneau et al., 2014). The validation scheme was built in accordance with the target level of 50 μg kg$^{-1}$ for all analytes. The sulphonamides analyzed were the following: sulfaguanidine (SGN), sulfanilamide (SNL), sulfadiazine (SDZ), sulfathiazole (STZ), sulfamerazine (SMR), sulfamethizole (SMZ), sulfadimerazine (SDM), sulfamonomethoxine (SMNM), sulfamethoxypyridazine (SMP), sulfadoxine (SDX), sulfamethoxazole (SMX), sulfaquinoxaline (SQX), and sulfadimethoxine (SDT). Several extraction procedures were tested during the development phase. Finally, the best results were obtained with a procedure involving acidic hydrolysis and cation exchange purification. Chromatographic separation was achieved on a C18 analytical column. Matrix effects were also studied. Data acquisition implemented for the confirmatory purpose was performed by monitoring 2 MRM transitions per analyte in positive electrospray mode. Mean relative recovery ranged from 85.8 to 110.2%, and relative standard deviations were between 2.6 and 19.8% under intra-laboratory reproducibility conditions. The decision limits ranged from 1.8 to 15.5 μg/kg.

#### 7.4.1. *Sample preparation for royal jelly*

(1) Heat the fresh royal jelly in a water bath at 50 °C for 5 min.
(2) Transfer 2 g of heated royal jelly (or 0.5 g of royal jelly powder) to a test tube.
(3) Mix the sample.
(4) Leave the mixture in darkness at RT for 10 min.
(5) Condition the SCX cartridges with 4 ml of MeOH.
(6) Condition the SCX cartridges with 4 ml of ultra-pure water.
(7) Acidify the samples using 10 ml of citric acid solution (0.3 M).
(8) Mix by vortexing for 10 s.
(9) Shake the mixture immediately for 10 min.
(10) Centrifuge the mixture at 14,000× *g* and 4 °C.
(11) Transfer the upper phase to the cartridges.
(12) Elute the cartridges at a flow rate of 1 drop/s.
(13) Wash the cartridge with 4 ml of ultra-pure water.
(14) Wash the cartridge twice with 4 ml of a MeOH–ACN mixture (50:50; v/v).
(15) Elute the sulphonamides residues into a clean tube twice with 0.6 mL of a 2% ammonium hydroxide solution in methanol.
(16) Dry the extract at 40 °C under a gentle stream of $N_2$.
(17) Reconstitute the residue in 400 μl of ultra-pure water.
(18) Mix the solution.
(19) Centrifuge for 5 min at 2,500× *g* at 4 °C.
(20) Transfer the solution to auto-sampler vials for LC-MS/MS analysis.

#### 7.4.2. *LC-MS/MS analytical conditions*

(1) Use distilled water with 0.2% formic acid as mobile phase A, and mobile phase B is pure analytical-grade ACN.
(2) Use the following gradient conditions:
   - from 0 to 0.1 min, ramp linearly from 98% to 70% of mobile phase A,
   - ramp over 2.9 min to 40% of A and hold for 5 min,
   - return to initial conditions in 2 min,
   - hold for 7 min to re-equilibrate the system.
(3) Set the flow rate to 0.25 ml/min.
(4) Set the oven temperature to 25 °C.
(5) Use the injection volume of 5 μl.
(6) Operate the instrument using electrospray ionization (ESI) in positive mode.
(7) Collect the data using Xcalibur software.
(8) Set the sample tube or desolvation temperature to 350 °C.
(9) Set the spray voltage to 4,500 V.
(10) Set the sheath gas (air) to 55 arb.
(11) Set the Aux gas (air) to 20 arb.
(12) Set the Ion sweep gas pressure (air) to 10 arb.
(13) Set the Collision gas (argon) to 1.5 mTorr.
(14) Use the dwell time of 20 ms.
(15) See Table 8 for the specific MRM parameters (two transitions) for each sulphonamide.

A typical LC-MS/MS chromatogram of a sulphonamide standard, negative sample, and spiked sample is shown in Figure 35.

### 7.5. The LC-MS/MS method for fluoroquinolone analysis in royal jelly

Quinolones belong to a family of synthetic antibiotics structurally related to nalidixic acid, itself being the first quinolone used clinically in animals in the early 1960s. Because of their narrow spectrum of activity and bacterial resistance issues, the original class of quinolones was supplanted in the mid-1980s by a new generation of drugs, still structurally related to nalidixic acid, containing a fluorine covalently bound to the carbon at position 6 and a piperazine ring at carbon 7. These 6-fluoroquinolones (FQs) were shown to have a much broader spectrum of activity because they were more effective against gram-negative bacteria and moderately effective against gram-positive bacteria. In apiculture, FQs are used for the prevention and treatment of American foulbrood.

A method for detection of fluoroquinolone residues in royal jelly was developed by Zhou et al. (2009). Sample preparation includes deproteination, ultrasonication-assisted extraction with a mixed inorganic solution of monopotassium phosphate ($KH_2PO_4$) and ethylenediaminetetraacetic acid disodium salt ($Na_2EDTA$), and clean-up on a solid-phase extraction cartridge. The extraction procedure was optimized regarding the amount of an inorganic solvent and the duration of sonication for royal jelly as a complicated matrix.

#### 7.5.1. *Sample preparation for royal jelly*

(1) Weigh 5 g of fresh royal jelly in a centrifuge tube (50 ml).
(2) Add 10 ml of 0.1 M NaOH.
(3) Mix for 1 min.
(4) Centrifuge the mixture for 5 min at 5581× *g*.
(5) Transfer the supernatant to a clean test tube (50 ml).
(6) Condition the SPE ($C_{18}$ or OASIS HLB) column with 6 mL of MeOH.
(7) Condition the SPE column with 6 ml of water.
(8) Load the mixture at 1–2 ml/min.
(9) Wash the column with 6 ml of water.
(10) Dry it for 5 min.
(11) Elute the analytes using 6 ml of formic acid: methanol (1:1).
(12) Collect them at 1–2 ml/min.
(13) Evaporate at 40 °C.
(14) Dissolve the residue in 1 ml of methanol:water (1:1).
(15) Pass the solution through a 0.2-μm nylon filter before injection into the LC-MS/MS apparatus.

#### 7.5.2. *LC-MS/MS analytical conditions*

(1) Use a C18 column (3 μm, 150 × 2.1 mm) or other similar columns.
(2) Use distilled water with 0.1% formic acid as the mobile phase A and pure analytical grade MeOH as mobile phase B.
(3) Use the following gradient program:
  - from 0 to 2 min, 90% of mobile phase A,
  - from 2 to 4 min, ramp linearly from 90% to 10% of mobile phase A,
  - from 4 to 8 min, 10% of mobile phase A,
  - return to initial conditions in 1 min,
  - hold for 3 min to re-equilibrate the system.
(4) Set the flow rate to 0.2 ml/min.
(5) Set the oven temperature to 30 °C.
(6) Use the injection volume of 25 μl.
(7) Operate the instrument using electrospray ionization (ESI) in positive mode.
(8) Collect the data using the Xcalibur software.
(9) Set the sample tube or desolvation temperature to 350 °C.
(10) Use the spray voltage of 4,100 V.
(11) See Table 9 for the specific MRM parameters (2 transitions).

A typical LC-MS/MS chromatogram of a fluoroquinolone standard, negative sample, and spiked sample is shown in Figure 36.

Table 8. LC-MS/MS parameters for the sulphonamide analytes.

| Analyte | *m/z* Precursor ion | Tube lens offset | Quantification ion *m/z* (collision energy-V) | Confirmation ion *m/z* (collision energy-V) | Ion ratio (%) |
|---|---|---|---|---|---|
| SGN | 215.0 | 68 | 156.0 (18) | 108.0 (30) | 32.7 |
| SNL | 173.1 | 39 | 93.2 (22) | 76.2 (42) | 24.2 |
| SDZ | 251.0 | 78 | 156.0 (18) | 92.0 (26) | 54.5 |
| STZ | 256.0 | 76 | 156.0 (19) | 92.0 (34) | 39.4 |
| SMR | 265.1 | 89 | 172.1 (16) | 108.0 (25) | 67.1 |
| SMZ | 271.0 | 141 | 108.0 (24) | 92.0 (28) | 133.2 |
| SDM | 279.1 | 78 | 186.0 (20) | 156.0 (19) | 34.0 |
| SMNM | 281.0 | 77 | 156.0 (20) | 108.0 (35) | 41.3 |
| SMP | 281.1 | 85 | 156.0 (18) | 108.0 (28) | 42.2 |
| SDX | 311.0 | 79 | 156.0 (18) | 108.0 (30) | 31.2 |
| SMX | 254.1 | 82 | 156.0 (16) | 108.0 (26) | 49.4 |
| SQX | 301.1 | 73 | 156.0 (17) | 108.0 (36) | 26.3 |
| SDT | 311.1 | 83 | 156.0 (23) | 108.0 (35) | 25.1 |

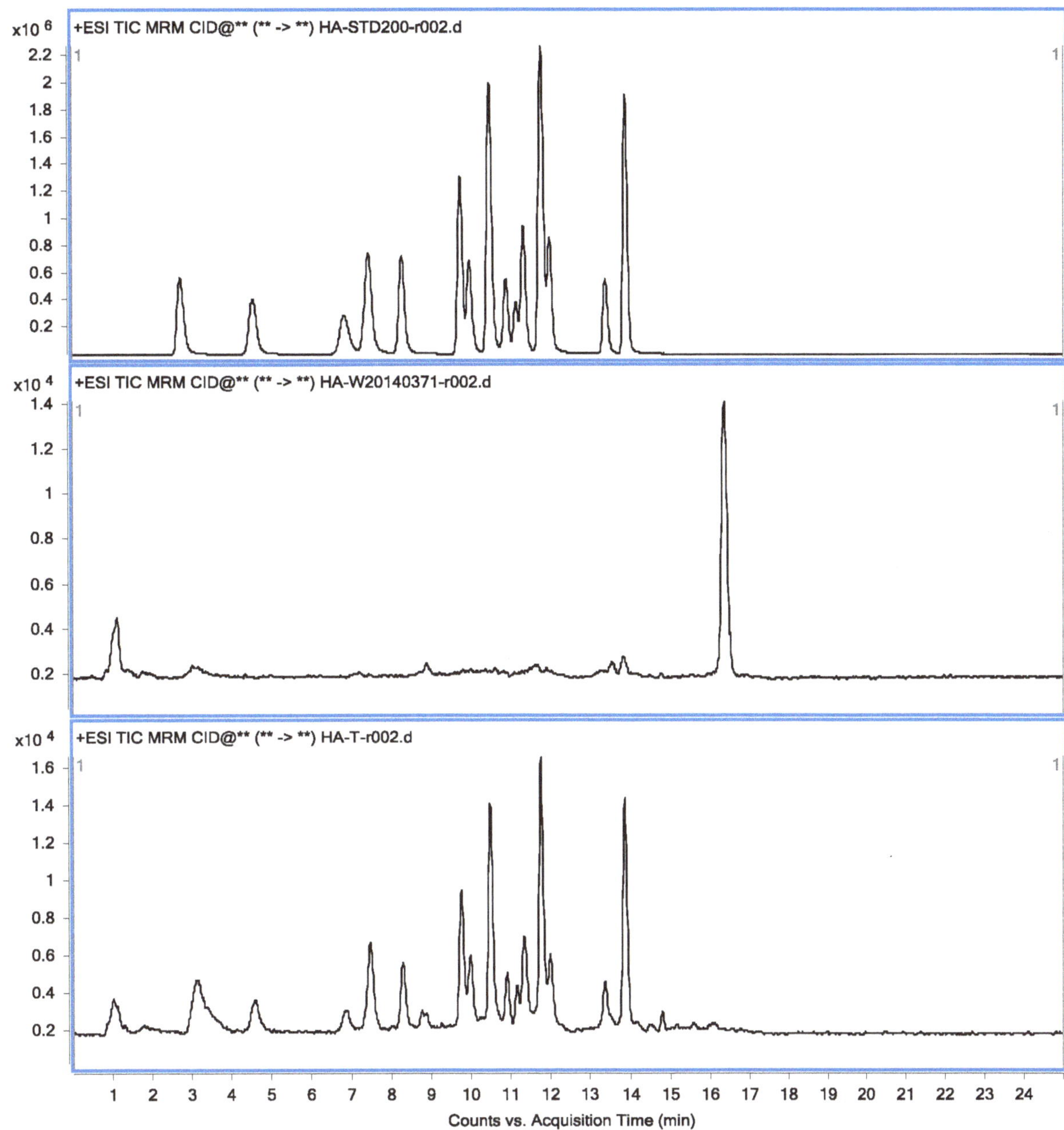


Figure 35. LC/MS/MS chromatography of sulfonamides standard (top), negative sample (middle) and spiked sample (bottom).

### *7.6. The LC-MS/MS method for tetracycline analysis in royal jelly*

Tetracyclines (TCs) are broad-spectrum antibiotics and show a strong activity against a variety of gram-positive and gram-negative microorganisms. Due to their broad spectrum of activity and cost-effectiveness, TCs are widely used in animal husbandry as veterinary drugs. The mechanism of action of TCs is inhibition of protein synthesis by binding to the small ribosomal sub-unit at the A site, which binds to the RNA. In beekeeping, TC antibiotics are used to treat bacterial brood diseases such as American Foulbrood and European Foulbrood. Because these drugs have been widely used for prevention and treatment of diseases, and often, the label directions for their use have not been followed, the resulting residues often remain in food. These antibiotic residues can lead to increased drug resistance of microbial strains in consumers and can cause allergic or toxic reactions in some hypersensitive individuals.

LC-MS/MS is the most recently adopted technique for quantification of TC residues in honey. It is a quantitative and a multi-residue method that is also characterized by increased sensitivity and accuracy. This is why LC-MS/MS is used in both screening and confirmatory tests. Almost all the TCs, including tetracycline (TET), oxytetracycline (OTC), chlortetracycline (CTC), and doxycycline (DOX), can be quantified by this method (Tarapoulouzi, Papachrysostomou, Constantinou, Kanari,

Table 9. LC-MS/MS parameters for the fluoroquinolones.

| Analyte | *m/z* Precursor ion | Quantification ion *m/z* (collision energy-V) | Confirmation ion *m/z* (collision energy-V) |
|---|---|---|---|
| ENR | 360 | 316 (17)<br>245 (25) | 316 (17) |
| CIP | 332 | 231 (16)<br>288 (34) | 231 (16) |
| NOR | 320 | 276 (22)<br>302 (16) | 276 (22) |
| OFL | 362 | 261 (35)<br>318 (26) | 261 (35) |
| DIF | 400 | 299 (31)<br>382 (26) | 299 (31) |
| SAR | 386 | 299 (31)<br>368 (23) | 299 (31) |
| SPA | 393 | 292 (24)<br>375 (22) | 292 (24) |
| DAN | 358 | 340 (23)<br>283 (27) | 340 (23) |
| FLU | 262 | 202 (29)<br>244 (18) | 202 (29) |
| FLE | 370 | 326 (21)<br>332 (16) | 326 (21) |
| MAR | 363 | 72 (21)<br>345 (14) | 72 (21) |
| ENO | 321 | 232 (35)<br>303 (18) | 232 (35) |
| ORB | 396 | 352 (19)<br>295 (25) | 352 (19) |
| PIP | 304 | 217 (30)<br>189 (22) | 217 (30) |
| PEF | 334 | 233 (25)<br>290 (20) | 233 (25) |
| LOM | 352 | 308 (30)<br>334 (215) | 308 (30) |
| NAL | 233 | 187 (28)<br>215 (24) | 187 (28) |

& Hadjigeorgiou, 2013). Xu et al. (2008) analyzed tetracycline residues in royal jelly by LC-MS/MS. The overall recovery of spiked royal jelly at the levels of 5.0, 10.0, and 40.0 µg/kg ranged from 62% to 115%, and the coefficients of variation ranged from 3.4 to 16.3% ($n$ = 6). The detection limits for TCs are under 1.0 µg/kg (Xu et al., 2008). The LC-MS/MS method is a simple, quick, and reliable assay with high sensitivity and selectivity. Furthermore, it can detect TCs in foods with good reproducibility. Thus, the LC-MS/MS method was selected for presentation here.

### *7.6.1. Sample preparation for royal jelly*

(1) Weigh 2.0 g of royal jelly into a polypropylene centrifuge tube (50 ml).
(2) Add 20 ml of 1% trichloroacetic acid extraction solution to royal jelly.
(3) Vortex for 2 min.
(4) Centrifuge the mixture at 3684× g for 5 min.
(5) Transfer the supernatant to polypropylene centrifuge tube (50 ml).
(6) Add 1.5 ml of a 1.0 mol/l $Na_2HPO_4$ solution to adjust pH to 6.0–7.0.
(7) Centrifuge the mixture at 3,684*g* for 5 min.
(8) Pass the supernatant through a filter paper.
(9) Precondition the Oasis HLB cartridges with 5 ml of methanol followed by 5 ml of Milli-Q water.
(10) Pass the supernatant through the cartridge in vacuum, at a flow rate of 1–2 ml/min.
(11) Wash the cartridge with 5 ml of Milli-Q water.
(12) Wash the cartridge with 5 ml of methanol:water mixture (2:8, v/v).
(13) Elute the TCs with 5 ml of methanol.
(14) Dry the extract under a stream of nitrogen at 45 °C.
(15) Reconstitute it in 1 ml of methanol:water solution (3:7, v/v).
(16) Pass the solution through a 0.45-µm nylon filter directly into HPLC vials for LC-MS/MS analysis.

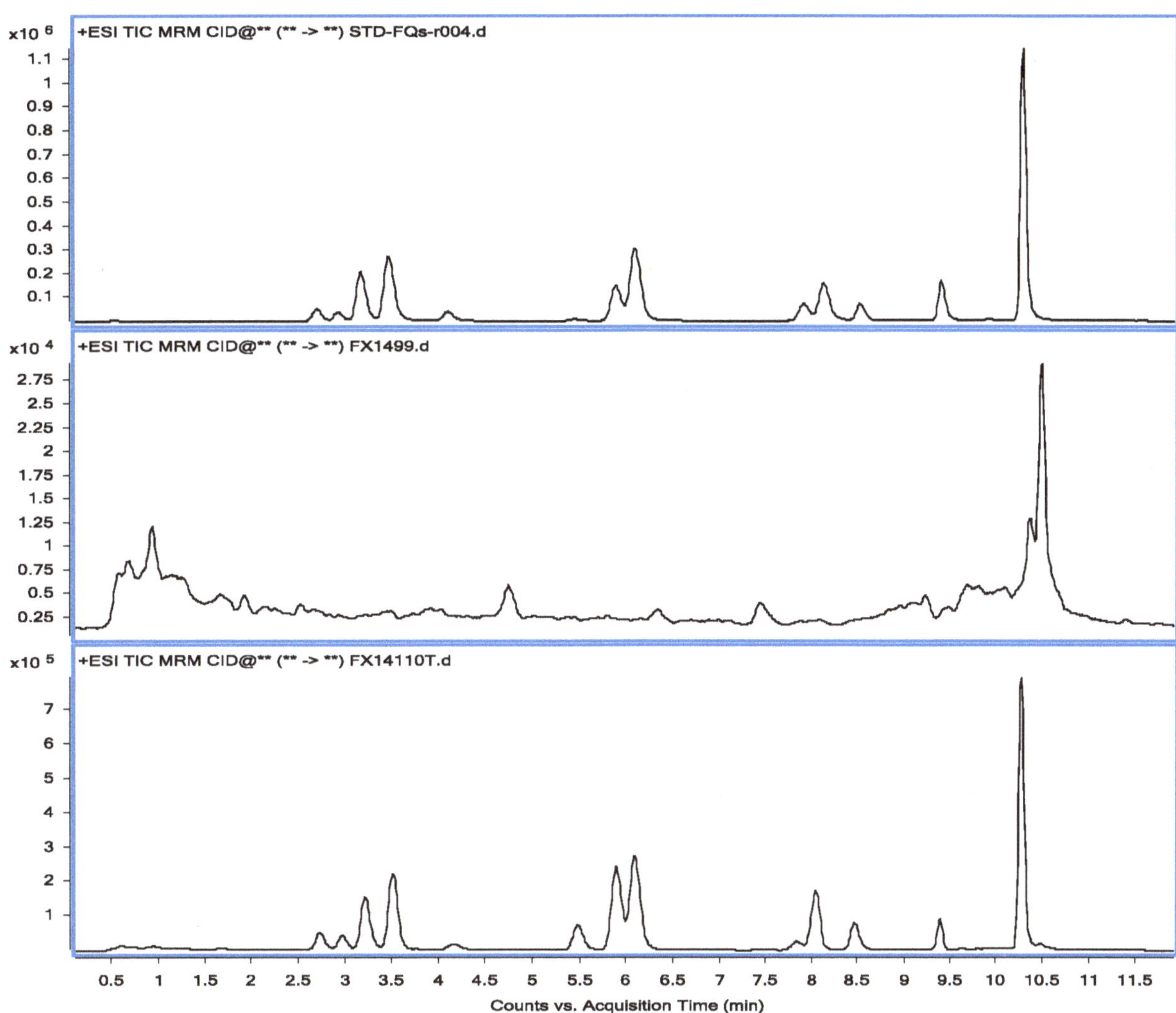


Figure 36. LC/MS/MS chromatography of fluoroquinolone standard (top), negative sample (middle) and spiked sample (bottom).

*7.6.2. LC-MS/MS analytical conditions*

(1) Connect the chromatograph with a Sunfire C18 column (150 × 2.1 mm i.d., 5.0-μm particle size) (Waters, Milford, MA, USA).
(2) Use distilled water with 0.1% formic acid as mobile phase A, and methanol as mobile phase B.
(3) Set the mobile phase gradient as follows: 0.0–5.0 min, 20–95% of mobile phase B; 5.10–7.00 min 95% of B; and 7.1–9.0 min 20% of B.
(4) Set the flow rate to 0.25 ml/min.
(5) Keep the column at room temperature.
(6) Set the injection volume to 25 μl.
(7) Detect the TC-positive electrospray ionization in multiple reaction-monitoring mode with a capillary voltage at 4.5 kV and a source temperature at 350 °C.
(8) See Table 10 for the specific MRM parameters (2 transitions).

A typical LC-MS/MS chromatogram of a TC standard, negative sample, and spiked sample is shown in Figure 37.

### *7.7. The LC-MS/MS method for macrolide analysis in royal jelly*

The macrolides are lipophilic molecules having a central lactone ring bearing 12–16 atoms, to which several amino and/or neutral sugars are bound. They are broad-spectrum antibiotics active against gram-positive bacteria and mycoplasmas, as well as some gram-negative bacteria and members of the chlamydia group. Macrolides are widely used in veterinary medicine to treat respiratory diseases and enteric infections.

It is believed that macrolide residues in food may pose a risk to consumers because of allergic reactions of individuals to the antibiotics and/or their metabolites. Incorrect use of these drugs or insufficient withdrawal time after treatment can possibly lead to the presence

Table 10. ESI MS/MS conditions of tetracycline drugs.

| Analyte | m/z Precursor ion | Quantification ion m/z (collision energy-V) | Confirmation ion m/z (collision energy-V) |
|---|---|---|---|
| TET | 445 | 410 (18)<br>427 (16) | 427 (16) |
| OTC | 461 | 426 (18)<br>443 (14)<br>337 (25) | 426 (18) |
| CTC | 479 | 444 (20)<br>462 (18) | 462 (20) |
| DOX | 445 | 428 (15)<br>410 (27) | 428 (15) |

of macrolide residues in bee products, and increases the risk to consumers in terms of allergic reactions among people sensitive to the antibiotics (Draisci, Palleschi, Ferretti, Achene, & Cecilia, 2001). Macrolides, for example, tylosin, can potentially be used for prevention and treatment of American Foulbrood disease. Nevertheless, the presence of macrolides is not allowed in honey in some countries.

Only limited data are available in the literature on the macrolide quantification in honey and royal jelly. Generally, only 2 kinds of methods are available: PremiTest method and LC-MS/MS method. Premi Test method is very simple to operate and inexpensive for analytical laboratories. However, it is a screening method without exact qualitative analysis of analytes. The LC-MS/MS method has been extensively used to quantify macrolides in honey and royal jelly, partly due to its simplicity and rapidity as well as its high sensitivity and selectivity.

#### 7.7.1. *Sample preparation for royal jelly*

(1) Weigh 2.0 g of fresh royal jelly (or 0.5 g of dried royal jelly powder) in a polypropylene centrifuge tube (50 ml) with a screw cap.
(2) Add 10 ml of a Tris solution (1 M, Ph = 6.8).
(3) Stir for 10 min.
(4) Centrifuge the mixture for 10 min at 18,000× g.
(5) Use Millex filter (0.2 μm) to filter the supernatant in another centrifuge tube (50 ml).
(6) Precondition the HLB 500 mg 60 ml SPE cartridge with 10 ml of methanol followed by 10 ml of water.
(7) Add the filtered supernatant to cartridge.
(8) Elute by gravity or vacuum (1 ml/min).
(9) Wash the cartridge with 5 ml of water.
(10) Wash with 10 ml of methanol:water (4:6, v/v).
(11) Elute the macrolides with 10 ml of methanol.
(12) Decrease the volume to 1 ml under a stream of nitrogen at 50 °C.
(13) Reconstitute the residue in 1.0 ml of ammonium acetate:water (17:3, v/v).
(14) Pass it through the Mini-UmiPrep syringeless filter vials (PVDF 0.2 μm; Whatman Inc., Clifton, NI, USA) for LC-MS/MS analysis.

#### 7.7.2. *LC-MS/MS analytical conditions*

(1) LC-MS/MS system: Alliance 2695 HPLC coupled with a MicromassOuattro Ultima Pt tandem mass spectrometer with an ESI interface and MassLynx 4.0 software (Waters Corp).
(2) Connect the chromatograph with an Atlantis C18 column (3 μm, 150 × 2.1 mm) and set the column oven temperature to 30 °C.
(3) Use distilled water with 0.1% formic acid as mobile phase A, and methanol as mobile phase B.
(4) Set the mobile phase gradient as follows:
 0.0–7.0 min, 20–95% of mobile phase B; 7.10–9.00 min 95% of B; and 9.1–15.0 min 20% of B.
(5) Program flow at 0.2 ml/min.
(6) Fix the injection volume to 20 μl.
(7) Program MS conditions as follows:
 - ionization mode, electrospray positive ion mode;
 - capillary voltage, 5.5 kV;
 - source temperature, 550 °C;
 - nebuliser nitrogen flow rate, 0.4 l/min;
 - collision gas argon pressure, 0.24 MPa;
(8) See Table 11 for the specific MRM parameters (2 transitions).

A typical LC-MS/MS chromatogram of a macrolide standard, negative sample, and spiked sample is shown in Figure 38.

### 7.8. General LC-MS/MS methods for analysis of nitrofurans in royal jelly

Furaltadone (FTD), furazolidone (FZD), nitrofurazone (NZF), and nitrofurantoin (NFT) are parent drugs of nitrofurans that contain the characteristic 5-nitrofuran ring group. For many years, this class of broad-spectrum antibacterial drugs has been widely used as additives in livestock and aquaculture feed for treatment of certain bacterial infections. In addition, nitrofurans have been used as a growth promoter in farm animal feeds and aquaculture. Nonetheless, FTD, FZD, NZF, and their metabolites have proven to have carcinogenic and muta-

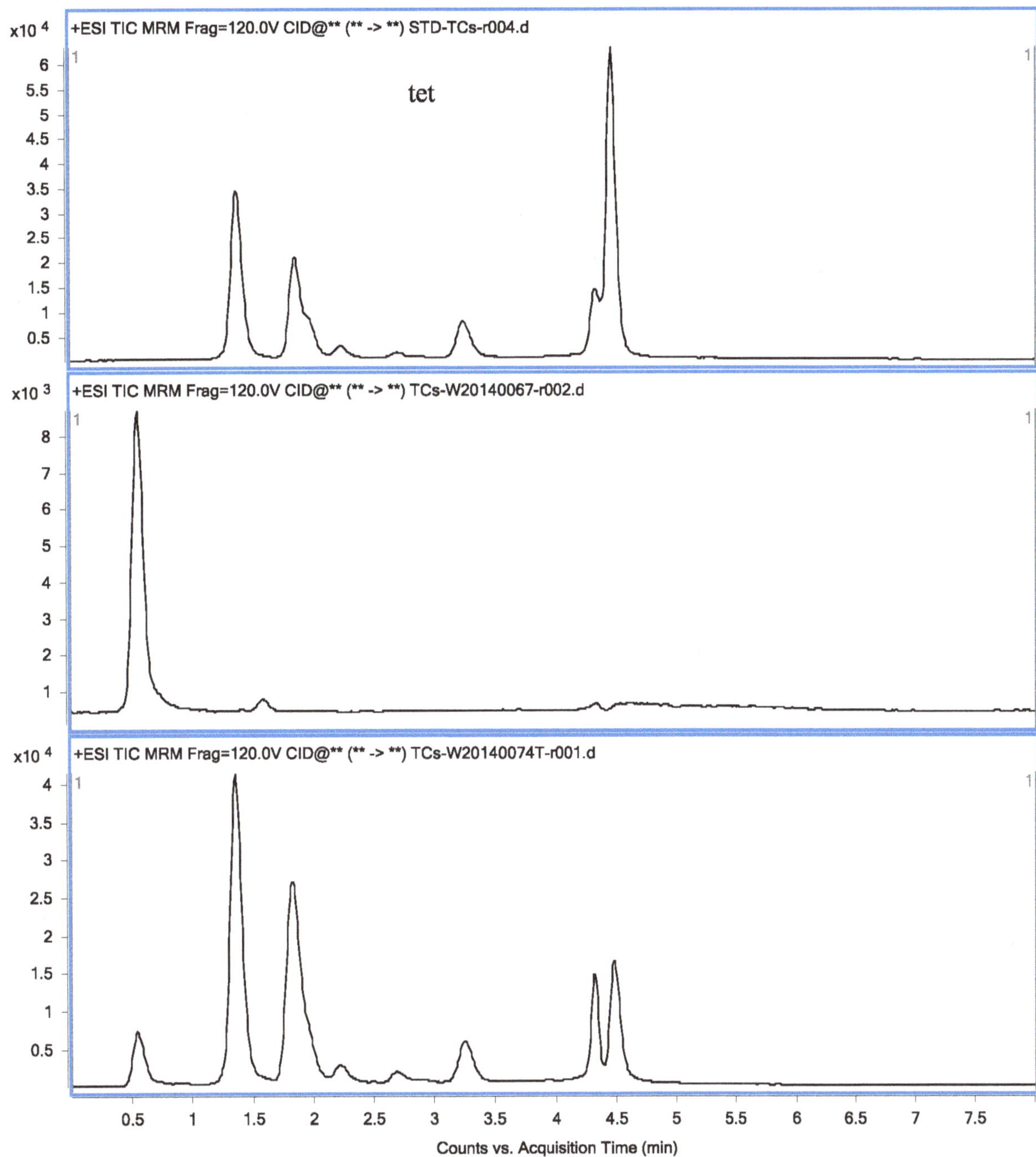


Figure 37. LC/MS/MS chromatography of tetracyclines standard (top), negative sample (middle) and spiked sample (bottom).

genic effects. As a consequence, nitrofurans have been banned from use in animal husbandry in the European Union since 1995. Because beekeepers use these nitrofurans to treat bees with bacterial infections, residues of these compounds have already been found in bee products including honey and royal jelly (Bargańska, Namieśnik, & Ślebioda, 2011; Cooper & Kennedy, 2005).

Only a few reports are available in the literature about nitrofuran detection in honey and royal jelly. Generally, two methods are available: the method of enzyme-linked immunosorbent assay (ELISA) with some modifications, and the method of LC-MS/MS. The immunoassay method has a broad detection range and high sensitivity and is a valid and cost-effective means for high-throughput monitoring of residual nitrofuran levels in many food matrices. Yet a drawback in analysis of all 4 metabolites of nitrofuran using the ELISA approach is that this approach would require 4 separate plates for testing due to the limited cross-reactivity of the antibodies. The biochip array technology is a modification of an immunochemical detection platform that offers the advantage of multiplexing several specific antibodies on a single biochip to increase the number of analytes covered. Nevertheless, the biochip method is

Table 11. ESI MS/MS conditions for macrolide drugs.

| Analyte | m/z Precursor ion | Quantification ion m/z (collision energy-V) | Confirmation ion m/z (collision energy-V) | Cone voltage V |
|---|---|---|---|---|
| Lincomycin | 407 | 126 (37)<br>359 (24) | 126 (37) | 50 |
| Erythromycin | 734 | 158 (42)<br>576 (28) | 158 (42) | 50 |
| Tilmicosin | 869 | 174 (62)<br>132 (70) | 174 (62) | 90 |
| Tylosin | 916 | 174 (54)<br>132 (70) | 174 (54) | 80 |
| Clindamycin | 425 | 126 (45)<br>377 (28) | 126 (45) | 53 |
| Spiramycin | 843 | 142 (48)<br>174 (50) | 142 (48) | 60 |
| Kitasamycin | 772 | 215 (43)<br>109 (42) | 215 (43) | 70 |
| Josamycin | 828 | 174 (45)<br>109 (45) | 174 (45) | 80 |

only a screening method. LC-MS/MS is a confirmatory and quantitative method for quantification of 4 metabolites of nitrofuran antibiotics in honey and royal jelly. Compared with other analytical methods, the sensitivity, selectivity, and specificity of the LC-MS/MS method are considerably better. In addition, LC-MS/MS provides precise results in accordance with EU requirements (2002/756/EC, 2002). Consequently, the LC-MS/MS method has been applied extensively to detect nitrofuran antibiotics in honey and royal jelly and is described below.

### 7.8.1. *Sample preparation for royal jelly*

This method is from Chinese Commodity Inspection Standards/The industry standards of entry-exit inspection and quarantine of the People's Republic of China: SN/T 2061–2008).

(1) Weigh 2.0 g of fresh royal jelly in a 50 ml centrifuge tube with a screw cap.
(2) Add 25 ml of HCl (2.0 M).
(3) Add 100 μl of 2-nitrobenzaldehyde (2-NBA).
(4) Incubate the mixture overnight at 37 °C in an oven.
(5) Add 1 ml of trichloroacetic acid.
(6) Mix and centrifuge at 1395× *g* for 5 min.
(7) Collect the supernatant.
(8) Add 1 M NaOH to adjust pH to 7.5 in the filtered supernatant.
(9) Precondition Oasis HLB solid-phase extraction (SPE) cartridge with 5 mL of methanol followed by 5 ml of deionized water.
(10) Add the adjusted supernatant mixture onto the cartridge.
(11) Add 10 ml of deionized water to clean up the cartridge.
(12) Elute the nitrofurans with 10 ml of ethyl acetate.
(13) Dry the eluate under a stream of nitrogen at 40 °C in vacuum.
(14) Reconstitute the extract with 1 ml of 0.2% acetic acid:acetonitrile (7:3, v/v).
(15) Pass the solution through a 0.45-μm filter membrane for LC-MS/MS analysis.

### 7.8.2. *LC-MS/MS analytical conditions*

(1) LC-MS/MS system: Agilent 6460 LC-MS/MS system (Santa Clara, CA, USA).
(2) Connect the chromatograph to an Agilent SB C8 column (5×150 mm, 5 μm).
(3) Set the column temperature to 30 °C.
(4) Use acetonitrile as mobile phase A; use deionized water with 5 mM ammonium acetate as mobile phase B.
(5) Program a linear gradient profile of mobile phase as follows:
   - from 0 to 7 min, ramp linearly from 30% to 90% of mobile phase A,
   - from 7 to 12 min, 90% of mobile phase A,
   - return to initial conditions in 0.1 min.
(6) Hold for 4.9 min to re-equilibrate the system (the total run time is 17 min).
(7) Set the sample injection volume to 30 μl.
(8) Use the flow rate of 0.3 ml/min.
(9) Operate the instrument using electrospray ionization (ESI) in positive mode.
(10) Use nitrogen for nebulization (42 Psi) and as a cone gas (25 Psi).
(11) Set the source temperature to 540 °C.
(12) Set the spray voltage to 4.1 kV.
(13) Use high-purity nitrogen as the collision gas.
(14) See Table 12 for the specific MRM parameters (2 transitions).

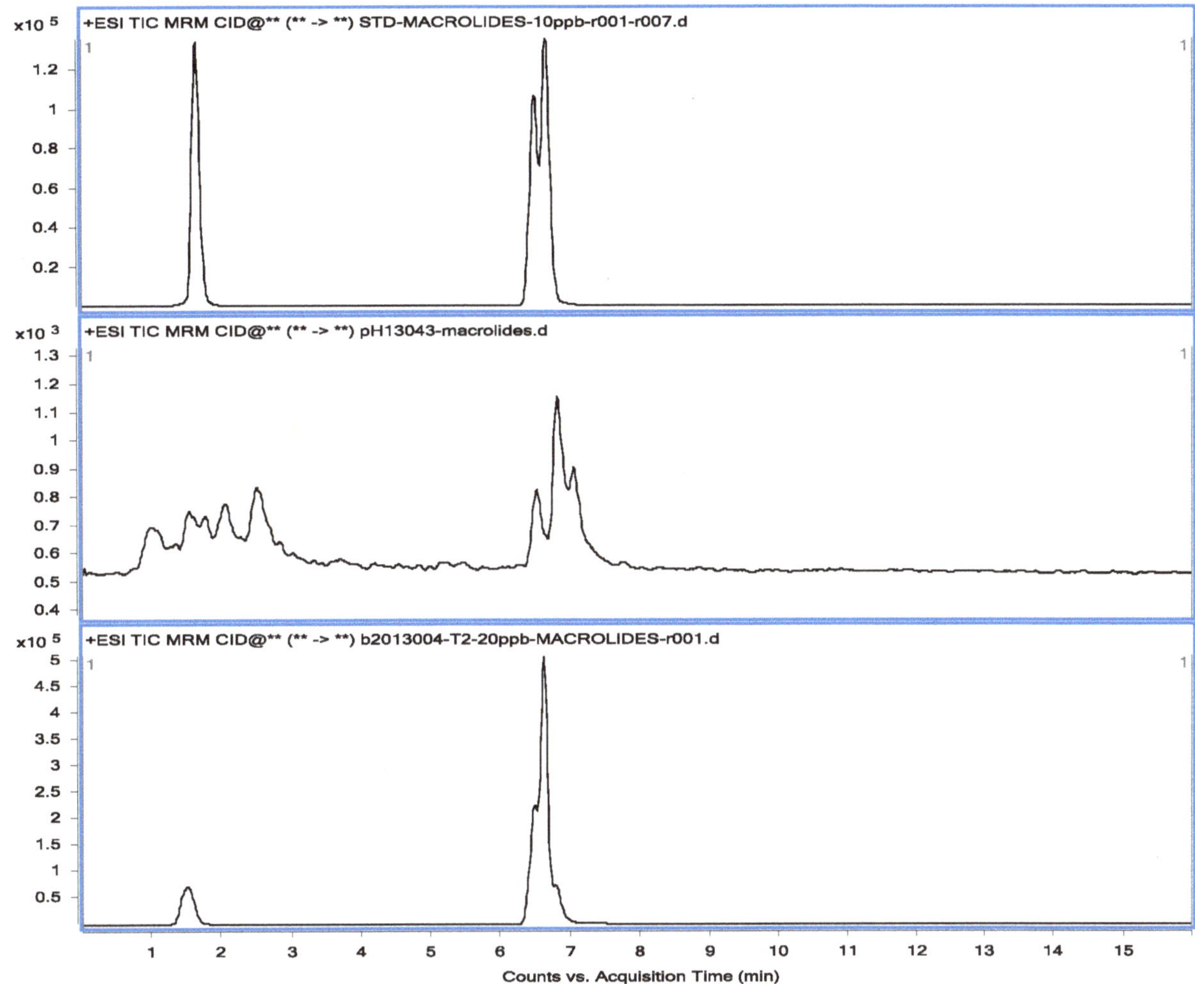


Figure 38. LC/MS/MS chromatography of macrolides standard (top), negative sample (middle) and spiked sample (bottom).

A typical LC-MS/MS chromatogram of a nitrofuran standard, negative sample, and spiked sample is shown in Figure 39.

### *7.9. GC-ECD method for fluvalinate analysis in royal jelly*

Fluvalinate has been extensively used worldwide by beekeepers since 1988 to prevent varroatosis. Lately, it became one of the most widely used acaricides. Fluvalinate residues can cause genetic mutations and cellular degradation in addition to several public health problems. These problems may occur through direct contamination as a result of beekeeping practices as well as via indirect contamination from environmental sources. The indirect contamination from the environment takes place because of the widespread use and extensive distribution of pesticides, which helped to introduce their residues into royal jelly by bees that have been consuming contaminated blossoms. This assay protocol is applicable to monitoring of fluvalinate residue in royal jelly according to national regulatory authorities and accredited labs.

#### *7.9.1. Sample preparation for royal jelly*

(1) Weigh 0.5 g of fresh royal jelly (0.15 g of royal jelly powder) in a tube.
(2) Dissolve royal jelly in 10 ml of acetonitrile:water (1:1, v/v).
(3) Homogenize the mixture in an ultrasonic bath for 15 min at 40 °C.
(4) Centrifuge the mixture for 10 min at 893× *g*.
(5) Precondition a $C_{18}$ cartridge with 5 mL of methanol followed by 5 ml of water.
(6) Pass the supernatant at a constant flow rate (1 ml/min) through the $C_{18}$ cartridge.
(7) Dry the cartridge in vacuum for 15 min.
(8) Elute the analytes with 2 mL of ethyl acetate and 2 ml of n-hexane.
(9) Evaporate the eluate in vacuum at RT.
(10) Dissolve the residue in 1 ml of isooctane.
(11) Pass it through a 0.45-μm nylon filter directly into vials for GC-ECD analysis.

Table 12. LC-MS/MS parameters for the nitrofurans.

| Analyte | m/z Precursor ion | Ion m/z (collision energy-V) | Cone energy (V) |
|---|---|---|---|
| SEM-NBA | 209.2 | 166.2 (15)<br>192.2 (17) | 60 |
| AHD-NBA | 249.2 | 134.1 (18)<br>104.1 (32) | 66 |
| AMOZ-NBA | 335.2 | 262.3 (25)<br>291.3 (17.7) | 66 |
| AOZ-NBA | 236.2 | 134.2 (18.3)<br>104.1 (32) | 66 |

#### 7.9.2. *Primary parameters of GC-ECD instruments*

(1) Connect the chromatograph to an HP-1 (30 × 0.25 mm × 0.25 μm).
(2) Set the injector temperature to 270 °C.
(3) Set the detector temperature to 300 °C.
(4) Use the nitrogen carrier gas rate of 10 ml/min.
(5) Set the make-up to 20 ml/min.
(6) Set the initial column temperature to 250 °C (3 °C/min) to 270 °C and hold it for 30 min.
(7) Use the injection volume of 1 μl.

A typical GC-ECD chromatography of fluvalinate standard is shown in Figure 40.

### 7.10. The GC-MS method for amitraz and 2,4-DMA analysis in royal jelly

Amitraz [N-methylbis(2,4-xylyliminomethyl)amine] is widely applied to beehives for control of *V. destructor*. Amitraz is a labile pesticide whose degradation products include 2,4-dimethylaniline (2,4-DMA).

Sample treatment is the key step for acaricide analysis to reduce the matrix interference and increase the sensitivity. This treatment often includes 2 main steps: extraction and clean-up. Amitraz in honey bee products is commonly extracted with an organic solvent, after diluting the samples with a buffer at different pH values because amitraz is labile in acidic media. Then, SPE cartridges with different absorbents are used to clean-up the extract. Finally, the eluent is dried under a nitrogen stream and re-dissolved in a suitable organic solvent for quantification. In the last decades, many new sample-processing methods were developed to shorten the running time and to automate the treatment. Solid-phase microextraction (SPME) (Blasco, Fernández, Picó, & Font, 2004; Rialotero, Gaspar, Moura, & Capelo, 2007), accelerated solvent extraction (ASE) (Korta, Bakkali, Berrueta, Gallo, & Vicente, 2002), and headspace solvent microextraction have also been used to detect amitraz and 2,4-dimethylaniline (2,4-DMA) in bee products (Shamsipur, Hassan, Salar-Amoli, & Yamini, 2008). These technologies are not easy to implement and require expensive facilities.

In contrast, GC with an electronic capture detector (ECD) is a low-cost, practical method for analysis of amitraz and its main metabolites. The required equipment is cheap and common in analytical laboratories (Amoli, Hasan, & Hejazy, 2009; Yu, Tao et al., 2010). It shows good performance, with an LOD of 5 μg/kg and LOQ of 10 μg/kg, but ECD is a specific detector that is used only for compounds with electronegative atoms. Thus, amitraz can be analyzed by GC-ECD, but the latter requires hydrolysis of amitraz to 2,4-DMA and derivatization with hepta-fluorobutyric acid anhydride. These steps make the method time consuming and problematic.

For some pesticides, analysis by GC is impossible due to their low thermal stability or insufficient volatility without further chemical derivatization. The most widely used HPLC detectors for acaricide analysis are diode array (DAD) and UV (Çobanoğlu & Tüze, 2008). HPLC-DAD is a simple, quick, and effective method for quantification of amitraz and its main metabolite in bee products. The sensitivity and stability of the method are suitable in most cases.

GC-MS is the most widely used method for analysis of amitraz and its metabolites, given the simple sample processing and the satisfying analytical performance (Mărghita, Bonta, Mihai, & Dezmirean, 2012; Notardonato, Avino, Cinelli, & Russo, 2014). Additionally, the HPLC tandem with MSD detection employing atmospheric pressure ionization (API), in positive and negative modes, is becoming the detection system of choice for more researchers (Gómez-Pérez, Plaza-Bolaños, Romero-González, Martínez-Vidal, & Garrido-Frenich, 2012). Nonetheless, the high price of MSD along with the expensive maintenance considerably impedes dissemination.

The GC-MS analysis for amitraz and 2,4-DMA in royal jelly is introduced in this section.

#### 7.10.1. *Sample preparation for royal jelly*

(1) Weigh 0.5 g of fresh royal jelly (0.15 g of royal jelly powder) accurately in a tube.
(2) Dissolve royal jelly in 10 ml of acetonitrile:water (1:1, v/v).
(3) Homogenize it in an ultrasonic bath for 15 min at 40 °C.

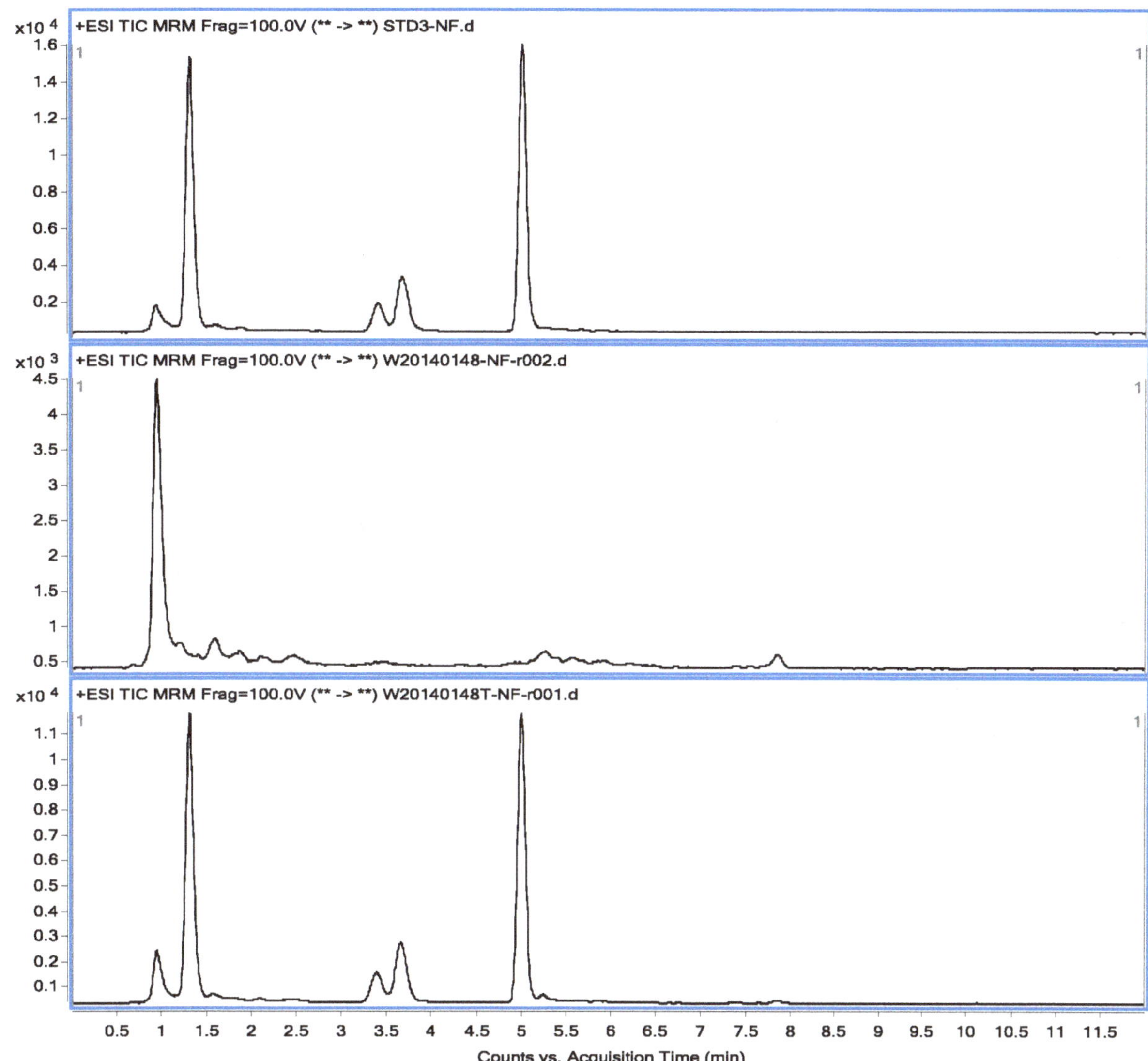


Figure 39. LC/MS/MS chromatography of nitrofuran standard (top), negative sample (middle) and spiked sample (bottom).

(4) Centrifuge the extract for 10 min at 893× *g*.
(5) Precondition a C18 cartridge with 5 ml of methanol followed by 5 ml of water.
(6) Pass the supernatant at a constant flow rate (1 ml/min) through the C18 cartridge.
(7) Dry the cartridge in vacuum for 15 min.
(8) Elute analytes with 2 ml of ethyl acetate.
(9) Elute again with 2 ml of n-hexane.
(10) Evaporate the eluate in vacuum at RT.
(11) Dissolve the residue in 1 ml of isooctane.
(12) Pass it through a 0.45-μm nylon filter directly into vials for GC-MS analysis.

#### *7.10.2. The primary GC-MS instrument parameters*

(1) Connect the chromatograph to CP-Sil 8, or SE-54, or HP-5MS, or HP-1MS (30 m × 0.25 mm × 0.25 μm).
(2) Set the injector temperature to 250 °C.
(3) Use helium as the carrier gas.
(4) Set the ion source EI to 230 °C.
(5) Set the transfer line to 270 °C.
(6) Manage the oven temperature as follows
  - start at 120 °C,
  - increase to 270 °C (20 °C/min),
  - hold at 270 °C for 7 min.
(7) Use the injection volume of 1 μl.
(8) Set the scan mode to Selected Ion Monitoring (SIM), *m/z* 293, 147, 121, 120, and 106.

### *7.11. Conclusion and outlook*

Methods such as HPLC, GC, GC/MS, and LC-MS/MS can be used for the detection of various drugs in bee products. LC-MS/MS and GC-MS presented here are so far the most sensitive, effective, and reliable assays (with good reproducibility), and have been extensively used for quantification of antibiotics and acaricides in honey (Paradis, Bérail, Bonmatin, & Belzunces, 2014; see the

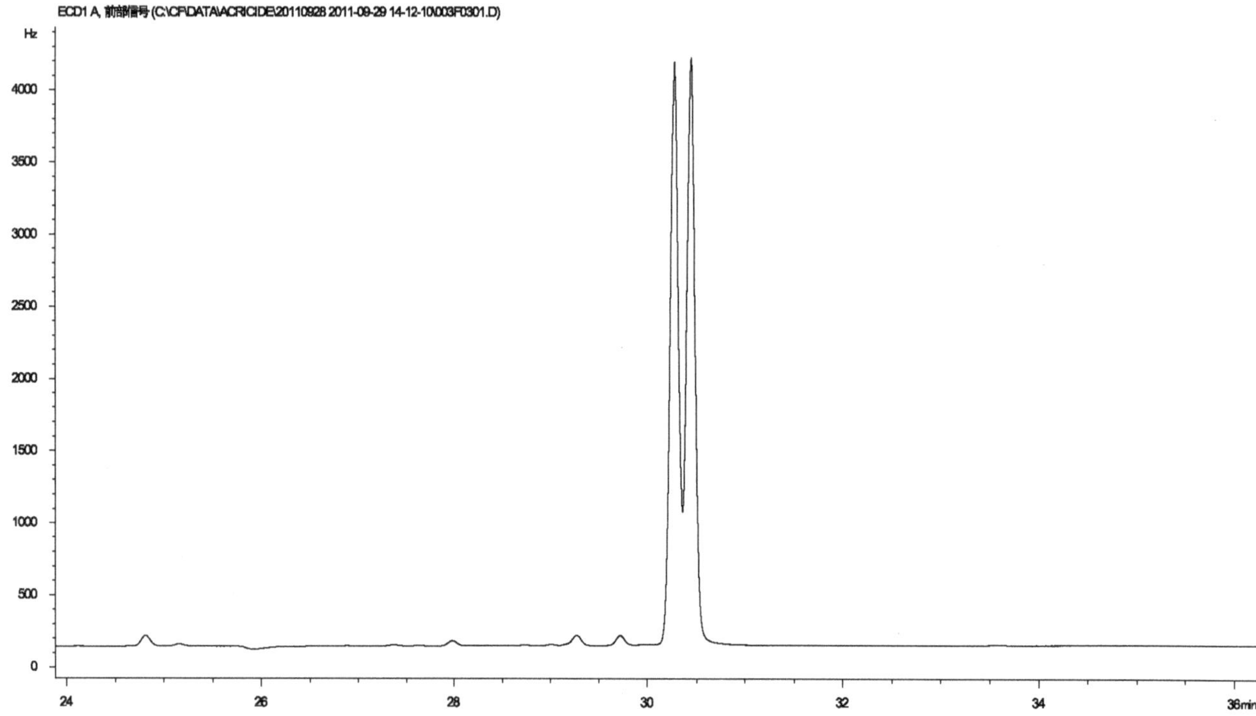

Figure 40. A typical GC-ECD chromatography of fluvalinate standard.

*BEEBOOK* paper on honey (in prep) and royal jelly. Although many new analytical techniques have been used for quality control of bee products, there is still some need for development of more advanced methods (urgent tasks are listed):

(1) To improve inadequate LC separations, as encountered for analyte classes of widely varying polarity.
(2) To develop analytical approaches for the "multi-compound class", including antibiotics and acaricides, in bee products.
(3) To find a way to adequately deal with ion-suppression effects on quantification.
(4) To enhance analyte detectability to reach LODs approximately 0.1–0.5 μg/kg for prohibited substances and drugs.
(5) To improve strategies for confirmation of analyte identity by careful perusal of IP-derived guidelines and protocols, and via evaluation of the potential of Q-TOF-MS-based detection.

## Acknowledgements

The COLOSS (Prevention of honey bee COlony LOSSes) Association aims to explain and prevent massive honey bee colony losses. It was funded through the COST Action FA0803. COST (European Cooperation in Science and Technology) is a unique means for European researchers to jointly develop their own ideas and new initiatives across all scientific disciplines through trans-European networking of nationally funded research activities. Based on a pan-European intergovernmental framework for cooperation in science and technology, COST has contributed since its creation more than 40 years ago to closing the gap between science, policy makers and society throughout Europe and beyond. COST is supported by the EU Seventh Framework Program for research, technological development and demonstration activities (*Official Journal L 412, 30 December 2006*). The European Science Foundation as implementing agent of COST provides the COST Office through an EC Grant Agreement. The Council of the European Union provides the COST Secretariat. The COLOSS network is now supported by the Ricola Foundation - Nature & Culture. Fu-Liang Hu, Jian-Ke Li, Li-Ming Wu, Yun-bo Xue and Zhi-Jiang Zeng are financially supported by the Earmarked Fund for Modern Agro-industry Technology Research System from the Ministry of Agriculture of China (CARS-45). Xing-An Li is financially supported by National Natural Science Foundation of China (31050006). Jozef Šimúth and Katarína Bíliková are financially supported by Japan Royal Jelly Co., Ltd. Tokyo, Japan. Hervé Casabianca and Gaëlle Daniele are financially supported by France Agrimer (French Agriculture ministry) and GPGR: French association of royal jelly producers.

## References

Albalasmeh, A.A., Berhe, A.A., & Ghezzehei, T.A. (2013). A new method for rapid determination of carbohydrate and total carbon concentrations using UV spectrophotometry. *Carbohydrate Polymers, 97*, 253–261.

Allen, E.H. (1985). Review of chromatographic methods for chloramphenicol residues in milk, eggs, and tissues from food-producing animals. *Journal - Association of Official Analytical Chemists, 68*, 990–999.

Amoli, J.S., Hasan, J., & Hejazy, M. (2009). Determination of amitraz residue by headspace gas chromatography in honey and beeswax samples from Iran. *American Journal of Food Technology, 4*, 56–59.

Anonymous. (1989). Turkish Standards for Royal Jelly (TS 6666).

Antinelli, J.F., Davico, R., Rognone, C., Faucon, J.P., & Lizzani-Cuvelier, L. (2002). Application of solid/liquid extraction for the gravimetric determination of lipids in royal jelly. *Journal of Agricultural and Food Chemistry, 50*, 2227–2230. doi:10.1021/jf0112466

Antinelli, J.F., Zeggane, S., Davico, R., Rognone, C., Faucon, J.P., & Lizzani, L. (2003). Evaluation of (E)-10-hydroxydec-2-enoic acid as a freshness parameter for royal jelly. *Food Chemistry, 80*, 85–89. doi:10.1016/S0308-8146(02)00243-1

Bargańska, Ż., Namieśnik, J., & Ślebioda, M. (2011). Determination of antibiotic residues in honey. *TrAC Trends in Analytical Chemistry, 30*, 1035–1041.

Bayer, E., & Witsch, H.G. (1961). *Gas chromatography*. New York, NY: Elsevier.

Bertelli, D., Lolli, M., Papotti, G., Bortolotti, L., Serra, G., & Plessi, M. (2010). Detection of honey adulteration by sugar syrups using one-dimensional and two-dimentional high-resolution nuclear magnetic resonance. *Journal of Agricultural and Food Chemistry, 58*, 8495–8501. doi:10.1021/jf101460t

Bíliková, K., & Šimúth, J. (2010). New criterion for evaluation of honey: Quantification of royal jelly protein apalbumin 1 in honey by ELISA. *Journal of Agricultural and Food Chemistry, 58*, 8776–8781. doi:10.1021/jf101583s

Bíliková, K., Klaudiny, J., & Šimúth, J. (1999). Characterization of the basic major royal jelly protein MRJP2 of honey bee (*Apis mellifera* L.) and its preparation by heterologous expression in *E.coli*. *Biologia, 54*, 733–739.

Bíliková, K., Wu, G., & Šimúth, J. (2001). Isolation of a peptide fraction from honey bee royal jelly as a potential antifoulbrood factor. *Apidologie, 32*, 275–283.

Bíliková, K., Hanes, J., Nordhoff, E., Saenger, W., Klaudiny, J., & Šimúth, J. (2002). Apisimin, a new serine-valine-rich peptide of honey bee (*Apis mellifera* L.) royal jelly: Purification and molecular characterization. *Febs Letters, 528*, 25–129.doi:10.16/S0014-5793(02)03272-6

Bíliková, K., Mirgorodskaya, E., Bukovská, G., Gobom, J., Lehrach, H., & Šimúth, J. (2009). Towards functional proteomics of minority component of honey bee royal jelly: The effect of post-translational modifications on the antimicrobial activity of apalbumin2. *Proteomics, 9*, 2131–2138. doi:10.1002/pmic.200800705

Binder, H. (1980). Separation of monosaccharides by high-performance liquid chromatography: Comparison of ultraviolet and refractive index detection. *Journal of Chromatography A, 189*, 414. doi:10.1016/S0021-9673(00)80322-2

Bishop, C.T. (1962). Separation of carbohydrate derivates by gas-liquid partition chromatography. *Methods in Biochemical Analysis, 10*, 1–42. doi:10.1002/9780470110270.ch1

Blank, S., Bantleon, F.I., McIntyre, M., Ollert, M., & Spillner, E. (2012). The major royal jelly proteins 8 and 9 (Api m 11) are glycosylated components of *Apis mellifera* venom with allergenic potential beyond carbohydrate-based reactivity. *Clinical and Experimental Allergy, 42*, 976–985. doi:10.1111/j.1365-2222.2012.03966.x

Blasco, C., Fernández, M., Picó, Y., & Font, G. (2004). Comparison of solid-phase microextraction and stir bar sorptive extraction for determining six organophosphorus insecticides in honey by liquid chromatography-mass spectrometry. *Journal of Chromatography A, 1030*, 77–85. doi:10.1016/j.chroma.2003.11.037

Bloodworth, B.C., Harn, C.S., & Hock, C.T. (1995). Liquid chromatographic determination of trans-10-hydroxy-2-decenoic acid content of commercial products containing royal jelly. *Journal of AOAC International, 78*, 1019–1023.

Blum, M.S., Novak, A.F., & Taber, S. (1959). 10-Hydroxy-delta 2-decenoic acid, an antibiotic found in royal jelly. *Science, 130*, 452–453. doi:10.1126/science.130.3373.452

Bogdanov, S. (2006). Contaminants of bee products. *Apidologie, 37*, 1–18.

Bogdanov, S., Bieri, K., Gremaud, G., Iff, D., Kanzig, A., Seiler, K., … Zurcher, K. (2004). Swiss Food Manual: Gelée royale, Chap. 23C, Bienenprodukte, BAG (Swiss Federal Office for Public Health), Berne.

Bradford, M.M. (1976). A rapid and sensitive method for the quantitation of microgram quantities of protein utilizing the principle of protein-dye binding. *Analytical Biochemistry, 72*, 248–254. doi:10.1016/0003-2697(76)90527-3

Brown, W.H., & Freure, R.J. (1959). Some carboxylic acids present in royal jelly. *Canadian Journal of Chemistry, 37*, 2042–2046. doi:10.1139/v59-296

Büchler, R., Andonov, S., Bienefeld, K., Costa, C., Hatjina, F., Kezic, N., … Wilde, J. (2013). Standard methods for rearing and selection of *Apis mellifera* queens. In V. Dietemann, J.D. Ellis, & P. Neumann (Eds.), *The COLOSS* BEEBOOK, *Volume I: Standard methods for* Apis mellifera *research. Journal of Apicultural Research, 51*(5). doi:10.3896/IBRA.1.52.1.07

Buttstedt, A., Moritz, R.F.A., & Erler, S. (2014). Origin and function of the major royal jelly proteins of the honey bee (*Apis mellifera*) as members of the yellow gene family. *Biological Reviews, 89*, 255–269. doi:10.1111/brv.12052

Cao, L.F., Zheng, H.Q., Pirk, C.W.W., Hu, F.L., & Xu, Z.W. (2016). High royal jelly-producing honey bees (*Apis mellifera ligustica*) (Hymenoptera: Apidae) in China. *Journal of Economic Entomology, 109*, 510–514.

Cerkowniak, M., Puckowski, A., Stepnowski, P., & Gołębiowski, M. (2013). The use of chromatographic techniques for the separation and the identification of insect lipids. *Journal of Chromatography B, 937*, 67–78. doi:10.1016/j.jchromb.2013.08.023

Chen, Y.C. (1993). *Apiculture in China* (61–63 pp.). Beijing: Agriculture Press.

Chen, C., & Chen, S. (1995). Changes in protein components and storage stability of Royal Jelly under various conditions. *Food Chemistry, 54*, 195–200.

Chiron, R. (1982). New synthesis of royal jelly acid. *Journal of Chemical Ecology, 8*, 709–713. doi:10.1007/BF00988312

Ciulu, M., Floris, I., Nurchi, V.M., Panzanelli, A., Pilo, M.I., Spano, N., & Sanna, G. (2015). A possible freshness marker for royal jelly: Formation of 5-hydroxymethyl-2-furaldehyde as a function of storage temperature and time. *Journal of Agricultural and Food Chemistry, 63*, 4190–4195.

Çobanoğlu, S., & Tüze, Ş. (2008). Determination of amitraz (Varroaset) residue in honey by high performance liquid chromatography (HPLC). *Tarim Bilimleri Dergisi, 14*, 169–174.

Consonni, R., & Cagliani, L.R. (2008). Geographical characterization of polyfloral and acacia honeys by nuclear magnetic resonance and chemometrics. *Journal of Agricultural and Food Chemistry, 56*, 6873–6880. doi:10.1021/jf801332r

Cooper, K.M., & Kennedy, D.G. (2005). Nitrofuran antibiotic metabolites detected at parts per million concentrations in retina of pigs—A new matrix for enhanced monitoring of nitrofuran abuse. *The Analyst, 130*, 466–468.

Cotte, J.F. (2003). *Development of method and databank applied to naturalness control of monofloral honeys* (Thesis N 33-2003). Claude Bernard University – LYON-1 France.

Daniele, G., & Casabianca, H. (2012). Sugar composition of French royal jelly for comparison with commercial and artificial sugar samples. *Food Chemistry, 134*, 1025–1029. doi:10.1016/j.foodchem.2012.03.008

Di Girolamo, F., D'Amato, A., & Righetti, P.G. (2012). Assessment of the floral origin of honey via proteomic tools. *Journal of Proteomics, 75*, 3688–3693. doi:10.1016/j.jprot.2012.04.029

Ding, T., Xu, J., Shen, C., Jiang, Y., Chen, H., Wu, B., & Liu, F. (2006). Determination of three nitroimidazole residues in royal jelly by high performance liquid chromatography-tandem mass spectrometry. *Chinese Journal of Chromatography, 24*, 331–334.

Dionex. (1993). *Technical note*. Sunnyvale, CA.

Draisci, R., Palleschi, L., Ferretti, E., Achene, L., & Cecilia, A. (2001). Confirmatory method for macrolide residues in bovine tissues by micro-liquid chromatography-tandem mass spectrometry. *Journal of Chromatography A, 926*, 97–104.

Dreywood, R. (1946). Qualitative test for carbohydrate material. *Industrial & Engineering Chemistry Analytical Edition, 18*, 499–499. doi:10.1021/i560156a015

Drijfhout, F.P., Kochansky, J., Lin, S., & Calderone, N.W. (2005). Components of honey bee royal jelly as deterrents of the parasitic varroa mite, *Varroa destructor*. *Journal of Chemical Ecology, 31*, 1747–1764. doi:10.1007/s10886-005-5925-6

DuBois, M., Gilles, K.A., Hamilton, J.K., Rebers, P.A., & Smith, F. (1956). Colorimetric method for determination of sugars and related substances. *Analytical Chemistry, 28*, 350–356. doi:10.1021/ac60111a017

Dubreil-Chéneau, E., Pirotais, Y., Verdon, E., & Hurtaud-Pessel, D. (2014). Confirmation of 13 sulfonamides in honey by liquid chromatography–tandem mass spectrometry for monitoring plans: Validation according to European Union Decision 2002/657/EC. *Journal of Chromatography A, 1339*, 128–136.

Dutton, G.G.S. (1973). Application of gas-liquid chromatography to carbohydrates. *Advances in Carbohydrate Chemistry and Biochemistry, 28*, 11–160. doi:10.1016/S0065-2318(08)60382-0

Englis, D., & Becker, H. (1943). Sugar analysis by alkaline ferricyanide method—Determination of ferrocyanide by iodometric and other procedures. *Industrial & Engineering Chemistry Analytical Edition, 15*, 262–264. doi:10.1021/i560116a011

Enne, V.I., Livermore, D.M., Stephens, P., & Hall, L.M. (2001). Persistence of sulphonamide resistance in *Escherichia coli* in the UK despite national prescribing restriction. *The Lancet, 357*, 1325–1328. doi:10.1016/s0140-6736(00)04519-0

Fahy, E., Subramanlam, S., Brown, H.A., Glass, C.K., Merrill, A.H., Murphy, R.C., ... Dennis, E.A. (2005). A comprehensive classification system for lipids. *The Journal of Lipid Research, 46*, 839–862. doi:10.1194/jlr.E400004-JLR200

Fevold, H.L., Hisaw, F.L., & Leoard, S.L. (1931). The gonad stimulating and the iuteinizing hormones of the anterior lobe of the hypophysis. *American Journal of Physiology, 97*, 291–301. Retrieved from http://ajplegacy.physiology.org/content/97/2/291.full-text.pdf+html

Fontana, R., Mendes, M.A., Souza, B.M., & Konno, K. (2004). Jelleines: A family of antimicrobial peptides from the Royal Jelly of honey bees (*Apis mellifera*). *Peptides, 25*, 919–928. doi:10.1016/j.peptides.2004.03.016

Forti, A.F., Campana, G., Simonella, A., Multari, M., & Scortichini, G. (2005). Determination of chloramphenicol in honey by liquid chromatography-tandem mass spectrometry. *Analytica Chimica Acta, 529*, 257–263.

Fournet, B., Dhalliuin, J.M., Montreuil, J., Bosso, C., & Defaye, J. (1980). Gas-liquid chromatography and mass spectrometry of oligosaccharides obtained by partial acetolysis of glycans of glycoproteins. *Analytical Biochemistry, 108*, 35–56. doi:10.1016/0003-2697(80)90690-9

Fournet, B., Strecker, G., Leroy, Y., & Montreuil, J. (1981). Gas-liquid chromatography and mass spectrometry of methylated and acetylated methyl glycosides. Application to the structural analysis of glycoprotein glycans. *Analytical Biochemistry, 116*, 489–502. doi:10.1016/0003-2697(81)90393-6

Franks, F. (1998). Freeze-drying of bioproducts: Putting principles into practice. *European Journal of Pharmaceutics and Biopharmaceutics, 45*, 221–229. doi:10.1016/S0939-6411(98)00004-6

Fray, G.I., Jaeger, R.H., Morgan, E.D., Robinson, R., & Sloan, A.D.B. (1961). Synthesis of trans-10-hydroxydec-2-enoic acid and related compounds. *Tetrahedron, 15*, 18–25. doi:10.1016/0040-4020(61)80003-3

Fujita, T., Kozuka-Hata, H., Ao-Kondo, H., Kunieda, T., Oyama, M., & Kubo, T. (2013). Proteomic analysis of the royal jelly and characterization of the functions of its derivation glands in the honey bee. *Journal of Proteome Research, 12*, 404–411. doi:10.1021/pr300700e

Furusawa, N., & Kishida, K. (2001). High-performance liquid chromatographic procedure for routine residue monitoring of seven sulfonamides in milk. *Fresenius Journal of Analytical Chemistry, 371*, 1031–1033.

Furusawa, T., Rakwal, R., Nam, H.W., Shibato, J., Agrawal, G.K., Kim, Y.S., & Yonekura, M. (2008). Comprehensive royal jelly (RJ) proteomics using one- and two-dimensional proteomics platforms reveals novel RJ proteins and potential phospho/glycoproteins. *Journal of Proteome Research, 7*, 3194–3229. doi:10.1021/pr800061j

Garcia, L., Carlos, H., Garcia, S., Calabria, K., da Cruz, G., Puentes, A., & de Sousa, M. (2009). Proteomic analysis of honey bee brain upon ontogenetic and behavioral development. *Journal of Proteome Research, 8*, 1464–1473. doi:10.1021/pr800823r

García-Alvarez, M., Huidobro, J.F., Hermida, M., & Rodríguez-Otero, J.L. (2000). Major components of honey analysis by Nearinfraed transflectance spectroscopy. *Journal of Agricultural and Food Chemistry, 48*, 5154–5158. doi:10.1021/jf000170v

Gee, M., & Walker, H.G. (1962). Gas-liquid chromatography of some methylated mono-, di-, and trisaccharides. *Analytical Chemistry, 34*, 650–653. doi:10.1021/ac60186a020

Genç, M., & Aslan, A. (1999). Determination of trans-10-hydroxy-2-decenoic acid content in pure royal jelly and royal jelly products by column liquid chromatography. *Journal of Chromatography A, 839*, 265–268. doi:10.1016/S0021-9673(99)00151-X

Gómez-Pérez, M.L., Plaza-Bolaños, P., Romero-González, R., Martínez-Vidal, J.L., & Garrido-Frenich, A. (2012). Comprehensive qualitative and quantitative determination of pesticides and veterinary drugs in honey using liquid chromatography–Orbitrap high resolution mass spectrometry. *Journal of Chromatography A, 1248*, 130–138. doi:10.1016/j.chroma.2012.05.088

Goodall, I., Dennis, M.J., Parker, I., & Sharman, M. (1995). Contribution of high-performance liquid chromatographic analysis of carbohydrates to authenticity testing of honey. *Journal of Chromatography A, 706*, 353–359. doi:10.1016/0021-9673(94)01074-O

Gürdeniz, G., Rago, D., Bendsen, N.T., Savorani, F., Astrup, A., & Dragsted, L.O. (2013). Effect of trans fatty acid intake on LC-MS and NMR plasma Profiles. *PLoS One, 8*, e69589. doi:10.1371/journal.pone.0069589

Han, B., Fang, Y., Feng, M., Lu, X., Huo, X., Meng, L., & Li, J.K. (2014). In-depth phosphoproteomic analysis of royal jelly derived from western and eastern honey bee species. *Journal of Proteome Research, 13*, 5928–5943. doi:10.1021/pr500843j

Hanes, J., & Šimuth, J. (1992). Identification and partial characterization of the major royal jelly protein of the honey bee (*Apis mellifera* L.). *Journal of Apicultural Research, 31*, 22–26. doi:10.1080/00218839.1992.11101256

Hartfelder, K., Bitondi, M.M.G., Brent, C., Guidugli-Lazzarini, K.R., SimÕes, Z.L.P., Stabentheiner, A., … Wang, Y. (2013). Standard methods for physiology and biochemistry research in *Apis mellifera*. In V. Dietemann, J.D. Ellis, & P. Neumann (Eds.), *The COLOSS* BEEBOOK, *Volume I: Standard methods for* Apis mellifera *research*. Journal of Apicultural Research, *52*(1). doi:10.3896/IBRA.1.52.1.06

Hedgley, E.J., & Overend, W.G. (1960). Trimethylsily derivates of carbohydrates. *Chemistry and Industry*, 378–380.

Hojo, M., Kagami, T., Sasaki, T., Nakamura, J., & Sasaki, M. (2010). Reduced expression of major royal jelly protein 1 gene in the mushroom bodies of worker honey bees with reduced learning ability. *Apidologie*, *41*, 194–202. doi:10.1051/apido/2009075

Horváth, K., & Molnár-Perl, I. (1997). Simultaneous quantitation of mono-, di-and trisaccharides by GC-MS of their TMS ether oxime derivatives: II. In honey. *Chromatographia*, *45*, 328–335. doi:10.1007/BF02505579

Human, H., Brodschneider, R., Dietemann, V., Dively, G., Ellts, J., Forsgren, E., … Zheng, H.Q. (2013). Miscellaneous standard methods for *Apis mellifera* research. In V. Dietemann, J.D. Ellis, & P. Neumann (Eds.), *The COLOSS* BEEBOOK, *Volume I: Standard methods for* Apis mellifera *research*. *Journal of Apicultural Research*, *52*(4). doi:10.3896/IBRA.1.52.4.10

Ishii, R., Horie, M., Murayama, M., & Maitani, T. (2006). Analysis of chloramphenicol in honey and royal jelly by LC/MS/MS. *Journal of the Food Hygienic Society of Japan (Shokuhin Eiseigaku Zasshi)*, *47*, 58–65.

Isidorov, V.A., Czyżewska, U., Isidorova, A.G., & Bakier, S. (2009). Gas chromatographic and mass spectrometric characterization of the organic acids extracted from some preparations containing lyophilized royal jelly. *Journal of Chromatography B*, *877*, 3776–3780. doi:10.1016/j.jchromb.2009.09.016

Isidorov, V.A., Bakier, S., & Grzech, I. (2012). Gas chromatographic-mass spectrometric investigation of volatile and extractable compounds of crude royal jelly. *Journal of Chromatography B*, *885–886*, 109–116. doi:10.1016/j.jchromb.2011.12.025

ISO 12824:2106. (2016). Royal jelly—Specifications.

Iwanami, Y., Okada, I., Iwamatsu, M., & Iwadare, T. (1979). Inhibitory effects of royal jelly acid, myrmicacin, and their analogous compounds on pollen germination, pollen tube elongation, and pollen tube mitosis. *Cell Structure and Function*, *4*, 135–143.

Kamakura, M., Fukuda, T., Fukushima, M., & Yonekura, M. (2001). Storage-dependent degradation of 57-kDa Protein in royal jelly: A possible marker for freshness. *Bioscience, Biotechnology, and Biochemistry*, *65*, 277–284. doi:10.1271/bbb.65.277

Karaali, A., Meydanoğlu, F., & Eke, D. (1988). Studies on composition, freeze-drying and storage of Turkish royal jell. *Journal of Apicultural Research*, *27*, 182–185. doi:10.1080/00218839.1988.11100799

Kayashima, Y., Yamanashi, K., Sato, A., Kumazawa, S., & Yamakawa-Kobayashi, K. (2012). Freeze-dried royal jelly maintains its developmental and physiological bioactivity in *Drosophila melanogaster*. *Bioscience Biotechnology and Biochemistry*, *76*, 2107–2111. doi:10.1271/bbb.120496

Kelly, J.F.D., Downey, G., & Fouratier, V. (2004). Initial study of honey adulteration by sugar solutions using midinfrared (MIR) spectroscopy and chemometrics. *Journal of Agricultural and Food Chemistry*, *52*, 33–39. doi:10.1021/jf034985q

Kelly, S.M., Jess, T.J., & Price, N.C. (2005). How to study proteins by circular dichroism. *Biochimica et Biophysica Acta (BBA) - Proteins and Proteomics*, *1751*, 119–139. doi:10.1016/j.bbapap.2005.06.005

Kim, K.C., Friso, S., & Choi, S.W. (2009). DNA methylation, an epigenetic mechanism connecting folate to healthy embryonic development and aging. *The Journal of Nutritional Biochemistry*, *20*, 917–926. doi:10.1016/j.jnutbio.2009.06.008

Kimura, Y., Washino, N., & Yonekura, M. (1995). N -linked sugar chains of 350-kDa royal jelly glycoprotein. *Bioscience, Biotechnology, and Biochemistry*, *59*, 507–509. doi:10.1271/bbb.59.507

Kimura, Y., Kajiyama, S.I., Kanaeda, J., Izukawa, T., & Yonekura, M. (1996). N -linked sugar chain of 55-kDa royal jelly glycoprotein. *Bioscience, Biotechnology, and Biochemistry*, *60*, 2099–2102. doi:10.1271/bbb.60.2099

Kimura, M., Kimura, Y., Tsumura, K., Okihara, K., Sugimoto, H., Yamada, H., & Yonekura, M. (2003). 350-kDa royal jelly glycoprotein (apisin), which stimulates proliferation of human monocytes, bears the. BETA. 1-3galactosylated N-glycan: Analysis of the N-glycosylation site. *Bioscience, Biotechnology, and Biochemistry*, *67*, 2055–2058. doi:10.1271/bbb.67.2055

Kodai, T., Umebayashi, K., Nakatani, T., Ishiyama, K., & Noda, N. (2007). Compositions of royal jelly II. Organic acid Glycosides and Sterols of the royal jelly of honey bees (*Apis mellifera*). *Chemical & Pharmaceutical Bulletin*, *55*(10), 1528–1531. doi:10.1248/cpb.55.1528

Kodai, T., Nakatani, T., & Noda, N. (2011). figThe absolute configurations of hydroxy fatty acids from the royal jelly of honey bees (*Apis mellifera*). *Lipids*, *46*(3), 263–270. doi:10.1007/s11745-010-3497-x

Korta, E., Bakkali, A., Berrueta, L.A., Gallo, B., & Vicente, F. (2002). Study of an accelerated solvent extraction procedure for the determination of acaricide residues in honey by high-performance liquid chromatography–diode array detector. *Journal of Food Protection*, *65*, 161–166.

Koya-Miyata, S., Okamoto, I., Ushio, S., Iwaki, K., Ikeda, M., & Kurimoto, M. (2004). Identification of a collagen production-promoting factor from an extract of royal jelly and its possible mechanism. *Bioscience, Biotechnology, and Biochemistry*, *68*, 767–773.

Kucharski, R., Maleszka, R., Hayward, D.C., & Ball, E.E. (1998). A royal jelly protein is expressed in a subset of kenyon cells in the mushroom bodies of the honey bee brain. *Naturwissenschaften*, *85*, 343–346. Retrieved from http://link.springer.com/article/10.1007%2Fs001140050512#page-1

Kucharski, R., Maleszka, J., Foret, S., & Maleszka, R. (2008). Nutritional control of reproductive status in honey bees via DNA methylation. *Science*, *319*, 1827–1830. doi:10.1126/science.1153069

Laemmli, U.K. (1970). Cleavage of structural proteins during the assembly of the head of bacteriophage T4. *Nature*, *227*, 680–685. doi:10.1038/227680a0

Lei, M.K., Peng, F., Ge, X.H., Mei, K., & Zhu, Z.T. (2013). Determination of Glucose, Fructose and Sucrose in royal jelly by ion chromatography coupled with dialysis as an online sample pretreatment technique. *Food Science*, *34*, 231–233. doi:10.7506/spkx1002-6630-201316046

Lercker, G., Capella, P., Conte, L.S., & Ruini, F. (1981). Components of royal jelly: I. Identification of the organic acids. *Lipids*, *16*, 912–919.

Lercker, G., Caboni, M.F., Sabatini, A.G., & Nanetti, A. (1986). Carbohydrate determination of Royal Jelly by high resolution gas chromatography (HRGC). *Food Chemistry*, *19*, 255–264. doi:10.1016/0308-8146(86)90049-X

Li, J.K., Feng, M., Zhang, L., Zhang, Z.H., & Pan, Y.H. (2008). Proteomics analysis of major royal jelly protein changes under different storage conditions. *Journal of Proteome Research*, *7*, 3339–3353. doi:10.1021/pr8002276

Li, X., Huang, C., & Xue, Y. (2013). Contribution of lipids in honey bee (*Apis mellifera*) royal jelly to health. *Journal of Medicinal Food, 16*, 96–102. doi:10.1089/jmf.2012.2425

Lichtenberg-Kraag, B., Hedtke, C., & Bienefeld, K. (2002). Infrared spectroscopy in routine quality analysis of honey. *Apidologie, 33*, 327–337. doi:10.1051/apido:2002010

Lineback, D.R. (1968). The carbohydrate-polypeptide linkages in a fungal glucoamylase. *Carbohydrate Research, 7*, 106–108. doi:10.1016/S0008-6215(00)81445-7

Lis, H., & Sharon, N. (1993). Protein glycosylation. Structural and functional aspects. *European Journal of Biochemistry, 218* (1), 1–27. doi:10.1111/j.1432-1033.1993.tb18347.x

Maeda, T., Kuroda, H., & Motoyoshi, K. (1987). Effects of royal jelly and 10-hydroxy decenoic acid on the sebaceous glands of hamster ear. *The Japanese Journal of Dermatology, 98*, 469–475.

Malecová, B., Ramser, J., O'Brien, J.K., Janitz, M., Júdová, J., Lehrach, H., & Šimúth, J. (2003). Honey bee (*Apis mellifera* L.) mrjp gene family: Computational analysis of putative promoters and genomic structure of mrjp1, the gene coding for the most abundant protein of larval food. *Gene, 303*, 165–175. doi:10.1016/S0378-1119(02)01174-5

Marconi, E., Caboni, M.F., Messia, M.C., & Panfili, G. (2002). Furosine: A suitable marker for assessing the freshness of royal jelly. *Journal of Agricultural and Food Chemistry, 50*, 2825–2829. doi:10.1021/jf0114987

Mărghita, L.A., Bonta, V., Mihai, C.M., & Dezmirean, D.S. (2012). Acaricides determination from honey through GC-MS technique using different extraction methods. *Bulletin of University of Agricultural Sciences and Veterinary Medicine Cluj-Napoca. Animal Science and Biotechnologies, 69*, 19–23.

Mazzoni, V., Bradesi, P., Tomi, F., & Casanova, J. (1997). Direct qualitative and quantitative analysis of carbohydrate mixtures using13C NMR spectroscopy: Application to honey. *Magnetic Resonance in Chemistry, 35*, S81–S90. doi:10.1002/(SICI)1097-458X(199712)35:13<S81::AID-OMR202>3.0.CO;2-E

McFarlane, J.E. (1968). Fatty acids, methyl esters and insect growth. *Comparative Biochemistry and Physiology, 24*, 377–384. doi:10.1016/0010-406X(68)90989-4

Mecozzi, M. (2005). Estimation of total carbohydrate amount in environmental samples by the phenol–sulphuric acid method assisted by multivariate calibration. *Chemometrics and Intelligent Laboratory Systems, 79*, 84–90. doi:10.1016/j.chemolab.2005.04.005

Melliou, E., & Chinou, I. (2005). Chemistry and bioactivity of royal jelly from Greece. *Journal of Agricultural and Food Chemistry, 53*, 8987–8992. doi:10.1021/jf051550p

Munson, L.S., & Walker, P.H. (1906). The unification of reducing sugar methods. *Journal of the American Chemical Society, 28*, 663–686. doi:10.1021/ja01972a001

Nazzi, F., Bortolomeazzi, R., Della Vedova, G., Del Piccolo, F., & Milani, N. (2009). Octanoic acid confers to royal jelly varroa-repellent properties. *Naturwissenschaften, 96*, 309–314. doi:10.1007/s00114-008-0470-0

Neilson, K.A., Ali, N.A., Muralidharan, S., Mirzaei, M., Mariani, M., Assadourian, G., … Haynes, P.A. (2011). Less label, more free: Approaches in label-free quantitative mass spectrometry. *PROTEOMICS, 11*, 535–553. doi:10.1002/pmic.201000553

Nikolov, Ž.L., Jakovljević, J.B., & Boškov, Ž.M. (1984). High performance liquid chromatographic separation of oligosaccharides using amine modified silica columns. *Starch - Stärke, 36*, 97–100. doi:10.1002/star.19840360307

Noda, N., Umebayashi, K., Nakatani, T., Miyahara, K., & Ishiyama, K. (2005). Isolation and characterization of some hydroxy fatty and phosphoric acid esters of 10-hydroxy-2-decenoic acid from the royal jelly of honey bees (*Apis mellifera*). *Lipids, 40*, 833–838.

Notardonato, I., Avino, P., Cinelli, G., & Russo, M.V. (2014). Trace determination of acaricides in honey samples using XAD-2 adsorbent and gas chromatography coupled with an ion trap mass spectrometer detector. *RSC Adv, 4*, 42424–42431. doi:10.1039/C4RA06822J

Ohashi, K., Sawata, M., Takeuchi, H., Natori, S., & Kubo, T. (1996). Molecular cloning of cDNA and analysis of expression of the gene for α-glucosidase from the hypopharyngeal gland of the honey bee *Apis mellifera* L. *Biochemical and Biophysical Research Communications, 221*, 380–385. doi:10.1006/bbrc.1996.0604

Ohashi, K., Natori, S., & Kubo, T. (1997). Change in the mode of gene expression of the hypopharyngeal gland cells with an age-dependent role change of the worker honey bee *Apis mellifera* L. *European Journal of Biochemistry, 249*, 797–802. doi:10.1111/j.1432-1033.1997.t01-1-00797.x

Ong, S.E., & Pandey, A. (2001). An evaluation of the use of two-dimensional gel electrophoresis in proteomics. *Biomolecular Engineering, 18*, 195–205. doi:10.1016/S1389-0344(01)00095-8

Ortelli, D., Edder, P., & Corvi, C. (2004). Analysis of chloramphenicol residues in honey by liquid chromatography-tandem mass spectrometry. *Chromatographia, 59*, 61–64.

Pan, Q.Z., Wu, X.B., Guan, C., & Zeng, Z.J. (2013). A new method of queen rearing without grafting larvae. *American Bee Journal, 153*, 1279–1280.

Paradis, D., Bérail, G., Bonmatin, J.M., & Belzunces, L.P. (2014). Sensitive analytical methods for 22 relevant insecticides of 3 chemical families in honey by GC-MS/MS and LC-MS/MS. *Analytical and Bioanalytical Chemistry, 406*, 621–633. doi:10.1007/s00216-013-7483-z

Patel, V.J., Thalassinos, K., Slade, S.E., Connolly, J.B., Crombie, A., Murrell, J.C., & Scrivens, J.H. (2009). A comparison of labeling and label-free mass spectrometry-based proteomics approaches. *Journal of Proteome Research, 8*, 3752–3759. doi:10.1021/pr900080y

Pourtallier, J. (1967). Quantitative determination of honey sugars with gas chromatography. *Bulletin Apicole, 10*, 209–212.

Prosser, C.L. (1978). *Comparative animal physiology, environmental and metabolic animal physiology*. Hoboken, NJ: John Wiley & Sons.

Quisumbing, F.A., & Thomas, A.W. (1921). Conditions affecting the quantitative determination of reducing sugars by fehling solution. Elimination of certain errors involved current in methods. *Journal of the American Chemical Society, 43*, 1503–1526. doi:10.1021/ja01440a012

Ramadan, M.F., & Al-Ghamdi, A. (2012). Bioactive compounds and health-promoting properties of royal jelly: A review. *Journal of Functional Foods, 4*, 39–52. doi:10.1016/j.jff.2011.12.007

Ramadan, M., & Al-Ghamdi, A. (2012). Bioactive compounds and health-promoting properties of royal jelly: A review. *Journal of Functional Foods, 4*, 39–52. doi:10.1016/j.jff.2011.12.007

Rialotero, R., Gaspar, E.M., Moura, I., & Capelo, J.L. (2007). Gas chromatography mass spectrometry determination of acaricides from honey after a new fast ultrasonic-based solid phase micro-extraction sample treatment. *Talanta, 71*, 1906–1914.

Rudd, P.M., Elliott, T., Cresswell, P., Wilson, I.A., & Dwek, R.A. (2001). Glycosylation and the immune system. *Science, 291*, 2370–2376. doi:10.1126/science.291.5512.2370

Ruoff, K., Luginbühl, W., Bogdanov, S., Bosset, J.O., Estermann, B., Ziolko, T., & Amado, R. (2006a). Authentication of the botanical origin of honey by near-infrared spectroscopy. *Journal of Agricultural and Food Chemistry, 54*, 6867–6872. doi:10.1021/jf060770f

Ruoff, K., Luginbühl, W., Künzli, R., Iglesias, M.T., Bogdanov, S., Bosset, J.O., ... Amado, R. (2006b). Authentication of the botanical and geographical origin of honey by mid-infrared spectroscopy. *Journal of Agricultural and Food Chemistry, 54*, 6873–6880. doi:10.1021/jf060838r

Ruoff, K., Luginbühl, W., Bogdanov, S., Bosset, J.O., Estermann, B., Ziolko, T., ... Amado, R. (2007). Quantitative determination of physical and chemical measurands in honey by near-infrared spectrometry. *European Food Research Technology, 225*, 415–423. doi:10.1007/s00217-006-0432-8

Sabatini, A.G., Marcazzan, G.L., Caboni, M.F., Bogdanov, S., & De Almeida-Muradian, L.B. (2009). Quality and standardisation of royal jelly. *Journal of ApiProduct and ApiMedical Science, 1*(1), 1–6. doi:10.3896/IBRA.4.1.01.04

Sakamoto, M., Takeba, K., Sasamoto, T., Kusano, T., Hayashi, H., Kanai, S., ... Nagayama, T. (2011). Determination of dimetridazole, metronidazole and ronidazole in salmon and honey by liquid chromatography coupled with tandem mass spectrometry. *Food Hygiene and Safety Science (Shokuhin Eiseigaku Zasshi), 52*, 51–58.

Sandra, K., & Sandra, P. (2013). Lipidomics from an analytical perspective. *Current Opinion in Chemical Biology, 17*, 847–853. doi:10.1016/j.cbpa.2013.06.010

Schägger, H., & von Jagow G. (1987). Tricine-sodium dodecyl sulfate-polyacrylamide gel electrophoresis for the separation of proteins in the range from 1 to 100 kDa. *Analytical Biochemistry, 166*, 368–379. Retrieved from http://molbiol.ru/forums/index.php?act=Attach&type=post&id=92824

Scheparts, A. (1965). The glucose oxidase of honey. II. Stereochemical substrate specificity. *Biochimica et Biophysica Acta (BBA) - Nucleic Acids and Protein Synthesis, 96*, 334–336. doi:10.1016/0005-2787(65)90597-6

Schmitzová, J., Klaudiny, J., Albert, Š., Schröder, W., Schreckengost, W., Hanes, J., & Šimúth, J. (1998). A family of major royal jelly proteins of the honey bee *Apis mellifera* L. *Cellular and Molecular Life Sciences CMLS, 54*, 1020–1030.

Serra Bonvehi, J. (1992). Sugar, acidity and pH royal jelly. *Anales de Bromatologia, 44*, 65–69.

Sesta, G. (2006). Determination of sugars in royal jelly by HPLC. *Apidologie, 37*, 84–90. doi:10.1051/apido:2005061

Sesta, G., & Lusco, L. (2008). Refractometric determination of water content in royal jelly. *Apidologie, 39*, 225–232. doi:10.1051/apido:2007053

Shamsipur, M., Hassan, J., Salar-Amoli, J., & Yamini, Y. (2008). Headspace solvent microextraction-gas chromatographic thermionic specific detector determination of amitraz in honey after hydrolysis to 2,4-dimethylaniline. *Journal of Food Composition and Analysis, 21*, 264–270. doi:10.1016/j.jfca.2007.10.004

Sheridan, R., Policastro, B., Thomas, S., & Rice, D. (2008). Analysis and occurrence of 14 sulfonamide antibacterials and chloramphenicol in honey by solid-phase extraction followed by LC/MS/MS analysis. *Journal of Agricultural and Food Chemistry, 56*, 3509–3516.

Šimúth, J. (2001). Some properties of the main protein of honey bee (*Apis mellifera* L.) royal jelly. *Apidologie, 32*, 69–80. doi:10.1051/apido:2001112

Šimúth, J., Bíliková, K., Kováčová, E., Kuzmová, Z., & Schroeder, W. (2004). Immunochemical approach to detection of adulteration in honey: Physiologically active royal jelly protein stimulating TNF-alpha release is a regular component of honey. *Journal of Agricultural and Food Chemistry, 52*, 2154–2158. doi:10.1021/jf034777y

Smith, A.D. (1997). *Oxford dictionary of bichemistry and molecular biology* (p. 23). Oxford: Oxford University.

Smith, P.K., Krohn, R.I., Hermanson, G.T., Mallia, A.K., Gartner, F.H., Provenzano, M.D., ... Klenk, D.C. (1985). Measurement of protein using bicinchoninic acid. *Analytical Biochemistry, 150*, 76–85. doi:10.1016/0003-2697(85)90442-7

Smith, R., Mathis, A.D., Ventura, D., & Prince, J.T. (2014). Proteomics, lipidomics, metabolomics: A mass spectrometry tutorial from a computer scientist's point of view. *BMC Bioinformatics, 15*, S9. doi:10.1186/1471-2105-15-S7-S9

Spannhoff, A., Kim, Y.K., Raynal, N.J., Gharibyan, V., Su, M.B., Zhou, Y.Y., ... Bedford, M.T. (2011). Histone deacetylase inhibitor activity in royal jelly might facilitate caste switching in bees. *EMBO reports, 12*, 238–243. doi:10.1038/embor.2011.9

Sweeley, C.C., Bentley, R., Makita, M., & Wells, W.W. (1964). Gas-liquid chromatography of trimethysilyl derivates of sugars and related substances. *Journal of the American Chemical Society, 85*, 2497–2507. doi:10.1021/ja00899a032

Takenaka, T., & Echigo, T. (1980). General chemical composition of the royal jelly. *Bulletin of the Faculty of Pharmacy, Tamagawa University, 20*, 71–78.

Takenaka, T., Yatsunami, K., & Echigo, T. (1986). Changes in quality of royal jelly during storage. *Journal of Japanese Society of Food Science and Technology (Japan), 33*, 1–7. doi:10.3136/nskkk1962.33.1

Tamura, S., Amano, S., Kono, T., Kondoh, J., Yamaguchi, K., Kobayashi, S., ... Moriyama, T. (2009). Molecular characteristics and physiological functions of major royal jelly protein 1 oligomer. *Proteomics, 9*, 5534–5543. doi:10.1002/pmic.200900541

Tang, C.Z., & Yuan, Y.L. (1999). Effect of temperature on activity of superoxide dismutase in royal jelly. *Apiculture of China, 50*, 10–13.

Tarapoulouzi, M., Papachrysostomou, C., Constantinou, S., Kanari, P., & Hadjigeorgiou, M. (2013). Determinative and confirmatory method for residues of tetracyclines in honey by LC-MS/MS. *Food Additives & Contaminants: Part A, 30*, 1728–1732. doi:10.1080/19440049.2013.814075

Terada, Y., Narukawa, M., & Watanabe, T. (2011). Specific hydroxyl fatty acids in royal jelly activate TRPA1. *Journal of Agricultural and Food Chemistry, 59*, 2627–2635. doi:10.1021/jf1041646

Teresa, S., Helena, R.-C., Ewa, W., & Piotr, S. (2009). Water determination in bee products using the Karl Fischer titration method. *Journal of Apicultural Science, 53*, 49–56.

The Honey bee Genome Sequencing Consortium. (2006). Insights into social insects from the genome of the honey bee *Apis mellifera. Nature, 443*, 931–947. doi:10.1038/nature05260

Torto, B., Carroll, M.J., Duehl, A., Fombong, A.T., Gozansky, K.T., Nazzt, F., ... Teal, P.E.A. (2013). Standard methods for chemical ecology research in *Apis mellifera*. In V. Dietemann, J.D. Ellis, & P. Neumann (Eds.), *The COLOSS BEEBOOK, Volume I: Standard methods for* Apis mellifera *research. Journal of Apicultural Research, 52*(4). doi:10.3896/IBRA.1.52.4.06

Tourn, M.L., Lombard, A., Belliardo, F., & Buffa, M. (1980). Quantitative analysis of carbohydrates and organic acids in honeydew, honey and royal jelly by enzymatic methods. *Journal of Apicultural Research, 19*, 144–146. doi:10.1080/00218839.1980.11100013

Townsend, G.F., & Lucas, C.C. (1940a). The chemical examination of the lipid fraction of royal jelly. *Science, 92*, 43. doi:10.1126/science.92.2376.43

Townsend, G.F., & Lucas, C.C. (1940b). The chemical nature of royal jelly. *Biochemical Journal, 34*, 1155–1162.

Townsend, G.F., Brown, W.H., Felauer, E.E., & Hazlett, B. (1961). Studies on the *in vitro* antitumor activity of fatty acids. The esters of acids closely related to 10-hydroxy-2-decenoic acid from royal jelly against transplantable mouse leukemia. *Biochemistry and Cell Biology, 39*, 1765–1770.

Turner, B.M. (2000). Histone acetylation and an epigenetic code. *BioEssays, 22*, 836–845. doi:10.1002/1521-1878(200009)22:9<836::AID-BIES9>3.0.CO;2-X

Wang, J.H., Byun, J., & Pennathur, S. (2010). Analytical approaches to metabolomics and application to systems biology. *Seminars in Nephrology, 30*, 500–511. doi:10.1016/j.semnephrol.2010.07.007

Watanabe, K., Shinmoto, H., Kobori, M., Tsushida, T., Shinohara, K., Kanaeda, J., & Yonekura, M. (1998). Stimulation of cell growth in the U-937 human myeloid cell line by honey royal jelly protein. *Cytotechnology, 26*, 23–27. doi:10.1023/A:1007928408128

Waterland, R.A., & Rached, M.T. (2006). Developmental establishment of epigenotype: A role for dietary fatty acids? *Scandinavian Journal of Food and Nutrition, 50*, 21–26. doi:10.1080/17482970601066488

Watson, A.D. (2006). Thematic review series: Systems biology approaches to metabolic and cardiovascular disorders. lipidomics: A global approach to lipid analysis in biological systems. *The Journal of Lipid Research, 47*, 2101–2111. doi:10.1194/jlr.R600022-JLR200

Weaver, N., & Law, J. (1960). Heterogeneity of fatty acids from royal jelly. *Nature, 188*, 938–939. doi:10.1038/188938b0

Weaver, N., Law, J., & Johnston, N.C. (1964). Studies on the lipids of royal jelly. *Biochimica et Biophysica Acta (BBA) - Specialized Section on Lipids and Related Subjects, 84*, 305–315. doi:10.1016/0926-6542(64)90058-7

Weaver, N., Johnston, N.C., Benjamin, R., & Law, J.H. (1968). Novel fatty acids from the royal jelly of honey bees (*Apis mellifera*, L.). *Lipids, 3*, 535–538.

Wegener, H.C., Hald, T., Wong, D.L.F., Madsen, M., Korsgaard, H., Bager, F., ... Mølbak, K. (2003). Salmonella control programs in Denmark. *Emerging Infectious Diseases, 9*, 774–780. doi:10.3201/eid0907.030024

Wei, W.T., Hu, Y.Q., Zheng, H.Q., Cao, L.F., Hu, F.L., & Hepburn, H.R. (2013). Geographical influences on content of 10-hydroxy-trans-2-decenoic acid in royal jelly in China. *Journal of Economic Entomology, 106*, 1958–1963. doi:10.1603/EC13035

Wisniewski, J. R., Zougman, A., Nagaraj, N., & Mann, M. (2009). Universal sample preparation method for proteome analysis. *Nature Methods, 6*, 359–362. doi:10.1038/Nmeth.1322

Wong-Chong, J., & Martin, F.A. (1979). Analysis of sugar cane disaccharides by liquid chromatography.I. Adsorption chromatography with flow programming. *Plant Physiology, 27*, 927–929.

Woodcock, T., Downey, G., Kelly, J.D., & O'Donnell, C. (2007). Geographical classification of honey samples by near-infrared spectroscopy: A feasibility study. *Journal of Agricultural and Food Chemistry, 55*, 9128–9134. doi:10.1021/jf072010q

Wu, C.W., & Zhang, F.X. (1990). Effect of temperature and duration on the activity of glucose oxidase in royal jelly. *Apiculture of China, 41*, 4–6.

Wu, X.B., Zhang, F., Guan, C., Pan, Q.Z., Zhou, L.B., Yan, W.Y., & Zeng, Z.J. (2015). A new method of royal jelly harvesting without grafting larvae. *Entomological News, 124*, 277–281. doi:10.3157/021.124.0405

Wytrychowski, M., Daniele, G., & Casabianca, H. (2012). Combination of sugar analysis and stable isotope ratio mass spectrometry to detect the use of artificial sugars in royal jelly production. *Analytical and Bioanalytical Chemistry, 403*, 1451–1456. doi:10.1007/s00216-012-5934-6

Wytrychowski, M., Chenavas, S., Daniele, G., Casabianca, H., Batteau, M., Guibert, S., & Brion, B. (2013). Physicochemical characterisation of French royal jelly: Comparison with commercial royal jellies and royal jellies produced through artificial bee-feeding. *Journal of Food Composition and Analysis, 29*, 126–133. doi:10.1016/j.jfca.2012.12.002

Xu, J.Z., Ding, T., Wu, B., Yang, W.Q., Zhang, X.Y., Liu, Y., ... Jiang, Y. (2008). Analysis of tetracycline residues in royal jelly by liquid chromatography-tandem mass spectrometry. *Journal of Chromatography B, 868*, 42–48. doi:10.1016/j.jchromb.2008.04.027

Yanes-Roca, C., Rhody, N., Nystrom, M., & Main, K.L. (2009). Effects of fatty acid composition and spawning season patterns on egg quality and larval survival in common snook (Centropomus undecimalis). *Aquaculture, 287*, 335–340. doi:10.1016/j.aquaculture.2008.10.043

Yu, F., Mao, F., & Li, J. (2010). Royal Jelly Proteome Comparison between *A. mellifera* ligustica and A. cerana cerana. *Journal of Proteome Research, 9*, 2207.

Yu, H., Tao, Y., Le, T., Chen, D., Ishsan, A., Liu, Y., ... Yuan, Z. (2010). Simultaneous determination of amitraz and its metabolite residue in food animal tissues by gas chromatography-electron capture detector and gas chromatography-mass spectrometry with accelerated solvent extraction. *Journal of Chromatography B, 878*, 1746–1752. doi:10.1016/j.jchromb.2010.04.034

Zaina, S. (2010). Tune into fatty acids for epigenetic programming news. *Current Opinion in Lipidology, 21*, 469–470. doi:10.1097/MOL.0b013e32833e4f00

Zeng, Z.J. (2013). *Technique for mechanized production of royal jelly* (pp. 2–46). Beijing: China Agriculture Press.

Zhang, L., Fang, Y., Li, R., Feng, M., Han, B., Zhou, T., & Li, J. (2012). Towards posttranslational modification proteome of royal jelly. *Journal of Proteomics, 75*, 5327–5341. doi:10.1016/j.jprot.2012.06.008

Zhang, L., Han, B., Li, R., Lu, X., Nie, A., Guo, L., ... Li, J. (2014). Comprehensive identification of novel proteins and N-glycosylation sites in royal jelly. *BMC Genomics, 15*, 135. doi:10.1186/1471-2164-15-135

Zhao, F.Y., Wu, Y.J., Guo, L.L., Li, X.S., Han, J.X., Chen, Y., & Ge, Y.Q. (2013). Using proteomics platform to develop a potential immunoassay method of royal jelly freshness. *European Food Research and Technology, 236*, 799–815. doi:10.1007/s00217-013-1939-4

Zheng, H.Q., Wei, W.T., Wu, L.M., Hu, F.L., & Dietemann, V. (2012). Fast determination of royal jelly freshness by a chromogenic reaction. *Journal of Food Science, 77*(6), S247–S252. doi:10.1111/j.1750-3841.2012.02726.x

Zhou, J., Zhao, J., Yuan, H., Meng, Y., Li, Y., Wu, L., & Xue, X. (2007). Comparision of UPLC and HPLC for determination of trans-10-Hydroxy-2-Decenoic acid content in royal jelly by ultrasound-assisted extraction with internal standard. *Chromatographia, 66*, 185–190.

Zhou, J., Xue, X., Chen, F., Zhang, J., Li, Y., Wu, L., ... Zhao, J. (2009). Simultaneous determination of seven fluoroquinolones in royal jelly by ultrasonic-assisted extraction and liquid chromatography with fluorescence detection. *Journal of Separation Science, 32*, 955–964. doi:10.1002/jssc.200800545

Ziegler, R., & Vanantwerpen, R. (2006). Lipid uptake by insect oocytes. *Insect Biochemistry and Molecular Biology, 36*, 264–272. doi:10.1016/j.ibmb.2006.01.014

Zielinska, D.F., Gnad, F., Wiśniewski, J.R., & Mann, M. (2010). Precision mapping of an in vivo N-glycoproteome reveals rigid topological and sequence constraints. *Cell, 141*, 897–907. doi:10.1016/j.cell.2010.04.012

*Journal of Apicultural Research*, 2019
Vol. 58, No. 2, 1–108, https://doi.org/10.1080/00218839.2019.1571556

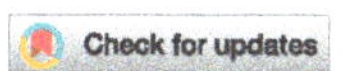

REVIEW ARTICLE

# Standard methods for *Apis mellifera* beeswax research

Lidija Svečnjak[a*], Lesley Ann Chesson[b], Albino Gallina[c], Miguel Maia[d], Marianna Martinello[c], Franco Mutinelli[c,e], Mustafa Necati Muz[f], Fernando M. Nunes[g], Francis Saucy[h], Brett James Tipple[b,i], Klaus Wallner[j], Ewa Waś[k] and Todd Alan Waters[l]

[a]*Department of Fisheries, Apiculture, Wildlife Management and Special Zoology, University of Zagreb Faculty of Agriculture, Zagreb, Croatia;* [b]*IsoForensics, Inc., Salt Lake City, UT, USA;* [c]*National Reference Laboratory for Honey Bee Health, Istituto Zooprofilattico Sperimentale delle Venezie, Legnaro, Padua, Italy;* [d]*Apismaia, Beekeeping Products & Services, Póvoa de Varzim, Portugal;* [e]*Diagnostic Services Histopathology and Parasitology Department, Istituto Zooprofilattico Sperimentale delle Venezie, Legnaro, Padua, Italy;* [f]*Department of Parasitology, Faculty of Veterinary Medicine, University of Namik Kemal, Tekirdag, Turkey;* [g]*CQ-Chemistry Research Centre, Chemistry Department, University of Tras-os-Montes e Alto Douro, Vila Real, Portugal;* [h]*Vuippens, Switzerland;* [i]*Department of Biology, University of Utah, Salt Lake City, UT, USA;* [j]*Apicultural State Institute (730), University of Hohenheim, Stuttgart, Germany;* [k]*Apiculture Department, Research Institute of Horticulture, Puławy, Poland;* [l]*Entomology Department, University of Maryland, College Park, MD, USA*

*(Received 30 April 2017; accepted 14 January 2019)*

Due to its multifunctional and complex role in the honey bee colony functioning and health (construction material allowing food storage, brood rearing, thermoregulation, mediation in chemical and mechanical communication, substrate for pathogens, toxins and waste), *Apis mellifera* beeswax has been widely studied over the last five decades. This is supported by a comprehensive set of scientific reports covering different aspects of beeswax research. In this article, we present an overview of the methods for studying chemical, biological, constructional, and quality aspects of beeswax. We provide a detailed description of the methods for investigating wax scales, comb construction and growth pattern, cell properties, chemical composition of beeswax using different analytical tools, as well as the analytical procedures for provenancing beeswax and beeswax-derived compounds based on the hydrogen isotope ratio (IRMS). Along with classical physico-chemical and sensory analysis, we describe more precise and accurate methods for detection of adulterants in beeswax (GC-MS and FTIR-ATR). Moreover, we present methods for studying the influence of beeswax (comb foundation) adulteration on comb construction. Analytical protocols for determining the pesticide residues using different chromatographic and spectroscopic techniques are also described. As beeswax is an agent of high risk for the transmission of bee diseases, we present methods for detection of pathogens in beeswax. To ensure the reproducibility of experiments and results, we present best practice approaches and detailed protocols for all methods described, as well as their advantages and disadvantages.

**Métodos estándar para la investigación de la cera de abejas *Apis mellifera***

Debido a su papel multifuncional y complejo en el funcionamiento y la salud de la colonia de abejas melíferas (material de construcción que permite el almacenamiento de alimentos, la cría, la termorregulación, la mediación en la comunicación química y mecánica, el sustrato para patógenos, toxinas y desechos), la cera de las abejas *Apis mellifera* ha sido ampliamente estudiada en las últimas cinco décadas. Esto se apoya en un amplio conjunto de informes científicos que abarcan diferentes aspectos de la investigación sobre la cera de abejas. En este artículo presentamos una visión general de los métodos que permiten el estudio químico, biológico, estructural y cualitativo de dicha cera. Proporcionamos una descripción detallada de los métodos usados para la investigación de las escamas de cera, la construcción del panal y su patrón de crecimiento, las propiedades de las celdas, y la composición química de la cera utilizando diferentes herramientas analíticas; así como los procedimientos analíticos basados en el ratio de isótopos de hidrógeno (IRMS) que permiten determinar el origen de la cera y los compuestos derivados de ella. Junto con el análisis físico-químico y sensorial clásico, describimos métodos más precisos y exactos para la detección de adulterantes en cera de abejas (GC-MS y FTIR-ATR). Además, presentamos métodos para estudiar la influencia de la adulteración de la cera de abejas (base del panal) en la construcción del panal. También se describen protocolos analíticos para determinar los residuos de pesticidas utilizando diferentes técnicas cromatográficas y espectroscópicas. Dado que la cera de abejas es un agente de alto riesgo en la transmisión de enfermedades de abejas, presentamos métodos para la detección de patógenos en dicha cera. Para garantizar la reproducibilidad de los experimentos y los resultados, presentamos modelos de buenas prácticas y protocolos detallados para todos los métodos descritos, así como sus ventajas y desventajas.

**西方蜜蜂蜂蜡研究标准方法**

蜂蜡在蜂群功能和健康中具有多种功能和复杂作用（食物储存、哺育幼虫、温度调节的筑巢材料, 化学和机械通讯的中介, 以及病原体、毒素和废物的基质）, 因此西方蜜蜂蜂蜡在过去五十年中被广泛研究。这得到了一套全面科学报告的支持, 涵盖了蜂蜡研究的不同领域。本文概述了蜂蜡的化学、生物、结构和质量的研究方法。本文详细介绍了利用不同分析工具研究蜡鳞、筑巢, 蜂蜡的生长方式、细胞特性和化学成分的方法, 以及基于氢同位素比值 (IRMS) 的蜂蜡和蜂蜡衍生化合物的来源分析方法。除了经典的物理化学和感官分析, 我们还描述了更精准的检测蜂蜡掺假的方法（GC-MS和FTIR-ATR）。此外, 我们还研究了蜂蜡（巢础）造假对蜂巢结构的影响。还介绍了使用不同色谱和光谱技术测定农药残留的分析流程。由于蜂蜡是传播蜜蜂疾病的高危因素, 我们提出了检测蜂蜡中病原体的方法。为了确保实验和结果的可重复性, 我们为所描述的所有方法提供了最佳实践方法和详细流程, 以及它们的优缺点。

*Corresponding author. Email: lsvecnjak@agr.hr

Please refer to this paper as: Svečnjak, L., Chesson, L. A., Gallina, A., Maia, M., Martinello, M., Mutinelli F., Muz, M. N., Nunes, F. M., Saucy, F., Tipple, B. J., Wallner, K., Waś, E. and Waters, T. A. (2019), Standard methods for *Apis mellifera* beeswax research. In V Dietemann, P Neumann, N Carreck and J D Ellis (Eds), *The COLOSS BEEBOOK, Volume III, Part I: standard methods for Apis mellifera hive products research. Journal of Apicultural Research* 58(2): https://doi.org/10.1080/00218839.2019.1571556

This article was originally published with errors, which have now been corrected in the online version. Please see Correction (http://dx.doi.org/10.1080/00218839.2019.1600925).

**Keywords:** *Apis mellifera*; beeswax; wax scales; comb construction; cell properties; chemical composition; beeswax provenancing; adulterants; pesticide residues; pathogens

## 1. Introduction

The wax of the western honey bee (*Apis mellifera* L.) is a complex lipid-based organic compound (natural wax) produced by worker bees using four pairs of specialized wax glands located on the inner side of the 4th to 7th abdominal sternites. Beeswax is secreted in the form of wax scales which honey bees transfer with the forelegs to their mandibles, where wax scales are chewed (salivary secretions added), and then added to the comb being constructed. The main effect of this mandibulation process is a transformation of the texturally anisotropic scale wax into isotropic comb wax (Hepburn, Pirk, & Duangphakdee, 2014). Chemically, beeswax represents a complex organic mixture of more than 300 compounds, of which the fatty acid esters (~67%), hydrocarbons (~14%), and free fatty acids (~13%) predominate (Tulloch, 1980). There are no significant differences in the basic chemical composition of wax originating from different *A. mellifera* subspecies, only small variations related to the proportion of the above mentioned predominant compounds (Beverly, Kay, & Voorhees, 1995; Fröhlich, Riederer, & Tautz, 2000; Tulloch, 1980).

The outstanding hexagonal cell structure of the honeycomb has fascinated scientists and the general public worldwide for centuries. This characteristic architectural structure reflects a complex behaviour of the honey bees in terms of their self-organization, as well as chemical and mechanical modifications of the wax during the wax scale-to-comb transformation pathway (Pirk, Hepburn, Radloff, & Tautz, 2004). Honey bees use beeswax as a construction material for building the comb (also known as honeycomb or comb wax) that serves as food (honey, pollen) storage and as brood-rearing compartment (provides infrastructure for rearing brood). Beeswax is also important for the chemical communication within a honey bee colony; its characteristic chemical composition plays an important role as a source of nest-mate recognition cues (Breed, Williams, & Fewell, 1988, et seq., D'ettorre et al., 2006; Fröhlich, Tautz, & Riederer, 2000). Furthermore, comb mediates pheromonal cues for cell capping, repairs and queen cell construction, nectar forage, colony defence and colony odour (Hepburn, 1998). It has an important role in thermoregulation and colony waste management, and serves as a humidity buffer in honey bee nests (Ellis, Nicolson, Crewe, & Dietemann, 2010; Jay, 1964). Mechanically, the comb transmits vibrational signals during the waggle dance (Kirchner, 1993; Michelsen, Kirchner, Andersen, & Lindauer, 1986; Michelsen, Kirchner, & Lindauer,

1986; Sandeman, Tautz, & Lindauer, 1996), and recruitment of new foragers (Tautz, 1996; Tautz, Casas, & Sandeman, 2001; Tautz & Lindauer, 1997).

Various practical uses of beeswax (e.g., for embalming, preserving the papyrus, protection of paintings, making of candles, figures, cult objects and ancient seals, as medicinal ingredient, and adhesive material) reach back into ancient history (Bogdanov, 2016b). Nowadays, the majority of beeswax is used for the production of comb foundations, and thus re-enters the beekeeping industry. Although it is difficult to obtain reliable figures on beeswax production and international trade statistics for products made of beeswax, it can be stated that comb foundation production is probably the major use of beeswax (Crane, 1990; Bogdanov, 2016b). Beeswax is also widely used in pharmacy, cosmetics, and in food industry as food additive E901 (Directive 2009/10/EC, 2009; EFSA, 2007; FAO, 2005; JECFA, 2005; ISO TC34/SC19, 2017).

Given the complex and important role of beeswax in the honey bee colony, it is of crucial importance that the comb foundation on which the honey bees are building their comb is genuine and uncontaminated. Nevertheless, the major beeswax quality issues nowadays include contamination of beeswax with adulterants (natural and/or synthetic substances that are deliberately added to beeswax for economic gain, such as paraffin wax as the most commonly used adulterant), and pesticide residues, as reported in numerous studies (e.g., Bernal, Jiménez, del Nozal, Toribio, & Martín, 2005; Bogdanov, 2006; Bogdanov et al., 2004; Chauzat & Faucon, 2007; Maia, Barros, & Nunes, 2013; Maia & Nunes, 2013; Ravoet, Reybroeck, & de Graaf, 2015; Serra Bonvehí & Orantes Bermejo, 2010, 2012; Svečnjak, Baranović et al., 2015; Svečnjak, Prđun, Bubalo, Matošević, & Car, 2016; Svečnjak, Prđun, Baranović, Damić, & Rogina, 2018; Wallner, 1992, 1997, 2000; Waś, Szczęsna, & Rybak-Chmielewska, 2016). In apiculture, these agents are primarily being transmitted through the comb foundation trade, as there are no regulations controlling their quality or authenticity. Additionally, commercially made comb foundation has not been incriminated as an agent for the dissemination of any bee disease although it poses a high risk for the transmission of various diseases and, to some extent, parasitic mites (Mutinelli, 2011).

In this article, we present an overview of the methods for studying beeswax in different aspects. We provide a detailed description of the methods for investigating wax scales, comb construction and its growth pattern, cell properties, chemical composition of beeswax using different analytical tools, as well as the analytical procedure for provenancing beeswax and beeswax-derived compounds based on the hydrogen isotope ratio. Along with classical physico-chemical and sensory analysis, we describe various analytical tools for determining beeswax authenticity (detection of adulterants by GC-MS and FTIR-ATR). Moreover, we present methods for studying the effects of beeswax (comb foundation) adulteration on comb construction. Methods for determination of pesticide residues using different analytical techniques are also described. As beeswax is an agent of high risk for the transmission of bee diseases, we present methods for detection of pathogens in beeswax [American Foulbrood (AFB), European Foulbrood (EFB), Small Hive Beetle (SHB), *Nosema* sp., *Ascosphera apis*]. To ensure the reproducibility of experiments and results, we present the best practice approach and detailed protocol steps for all methods described.

## 2. Beeswax sampling, processing, and storage

Sampling of the beeswax is strongly dependent on the aim of the study and the method intended to be used for its investigation. Therefore, we here provide general guidelines for collecting different types of beeswax samples from individual honey bees, from the hive or from the market (comb wax in Section 2.1.1, comb foundation in Section 2.2.2, and beeswax blocks in Section 2.2.3.) that can be followed for most of analytical methods described in this article. Specific sampling requirements related to particular methods are described in detail in corresponding sections (wax scale in Sections 3 and 4, wax caps and wax in honey in Section 5.5, and wax debris and old comb in Section 7). Beeswax samples are usually subjected to the melting (refining) prior to analysis. Thus, we also present the procedures for beeswax melting in Section 2.2, along with the requirements on beeswax storage in Section 2.3.

### *2.1. Collection of different types of beeswax specimens*

Different types of beeswax specimens can be collected for analysis. We describe the procedures for collecting the beeswax samples from the hive (comb wax, for details see Section 2.1.1), and from the market (comb foundation, see Section 2.1.2.; beeswax block, see Section 2.1.3).

#### *2.1.1. Collection of comb wax*

When aiming to analyze genuine beeswax, it is best to collect the samples of wild-built combs, i.e., newly built combs that are not constructed upon comb foundation but in empty space in the hive, as described below (steps 1–4). This allows obtaining virgin beeswax samples containing no impurities or residual contaminants that might be present when comb is constructed on the comb foundations (their origin is often questionable). To achieve this, sampling of the comb wax (wild-built combs) should be performed as follows:

1. Take out 1 or 2 frames from the hive to make space for inserting empty frames.
2. Put 1 or 2 empty frames (unwired, without comb foundation or starter) in the prepared empty space in the hive during the wax production season (spring and early summer).
3. Allow time for bees to build the required amount of comb. Note: this period can vary in duration according to local conditions.

Figure 1. Wild-built combs constructed by fulfilling an empty space in the hive after removing two frames (left) and combs collected from the hive cover board (right). Photo: Maia M. (left) and Svečnjak L. (right).

Alternatively, after taking out the frames from the hive (step 1), the empty space can be left vacant (without putting the empty frames); honey bees will build the combs in this vacant space and the comb wax samples can easily be collected (Figure 1). Also, smaller wild-build combs from the hive walls may also be collected, as presented in Figure 1.

4. Collect wild-built combs from the frames immediately after construction using beekeeping gloves (the use of laboratory gloves is required if potential contaminations need to be excluded.

Colonies used for wild-built combs collection should not have been treated with any kind of veterinary medicinal preparation previously to ensure the acquisition of pure beeswax samples, without residues of potential contaminants (unless the aim of the study requires different approach).

Make sure there are no physical impurities (propolis, pollen, nectar/honey or other hive material) on the comb wax samples before analysis. If an intact comb wax in genuine form is not required for analysis, the best way to remove impurities is melting of the comb wax that can be carried out as described below in Section 2.2. An alternative is to cut away manually the section of comb with impurities.

Old comb wax can also be collected for analysis, but it should be noted that it may contain significant amounts of other hive-derived substances (cocoons, faeces, propolis, pollen, etc.) that need to be removed by melting prior to analysis.

For the purpose of bee pathogens investigation (Section 7), (old) comb wax and wax (hive) debris can be collected. The collection of debris is an easy and non-invasive method of obtaining biological samples from apiaries during the quiet part of the year. This type of sample can be used to predict the possible occurrence of AFB in the next spring-summer.

#### 2.1.2. Comb foundation sampling

Sampling of the comb foundations can be performed by collecting the entire foundation sheet or a part of it, depending on the quantity needed for the analysis. Comb foundations are commonly purchased directly from the market (specialized beekeeping shops and/or fairs) or collected from the beekeepers. There are no particular requirements for comb foundation refining as commercial comb foundation beeswax is homogenized and free from physical impurities.

#### 2.1.3. Sampling the beeswax blocks

Beeswax blocks are obtained by melting the old combs and/or cappings. Different types of beeswax blocks (of various sizes, shapes and structures) can be found on the market, or can be collected from the beekeepers. Sampling of wax blocks depends on the sample type; the samples of wax blocks may be homogenous, or in layers/different phases, i.e., heterogeneous (this often appears due to the beeswax tendency to stratify).

- When the beeswax block is *homogeneous*, collect 3–4 aliquots from different locations, and pool all aliquots into one sample.
- When the beeswax block is stratified (*heterogeneous*), collect aliquots from the bottom, middle and upper phase, and merge all aliquots into one sample (i.e., mix the three phases collected).

When collecting aliquots, take into consideration to collect sufficient material for the analysis intended. Use a metal spatula or a spiral for sample collection. Clean the spatula with ethanol (96%) between uses to avoid contamination.

### 2.2. Beeswax melting

The simplest way to homogenize and purify the beeswax sample before analysis is melting by boiling water. This is primarily referring to the comb wax samples that usually contain hive-originating residues that have to be removed prior to analysis. Melting by boiling water will remove physical impurities from the comb wax (propolis, pollen, nectar/honey or other hive material). If the sample has already been refined (e.g., comb foundation, wax block), melting (heating) of beeswax without the water is performed (melting is required for most analyses).

The melting procedure with boiling water is performed as follows:

1. Place the beeswax sample in a glass (thermostable) or stainless steel beaker with distilled water as much to cover the sample (beakers made of other metals should not be used for melting purposes as they may darken the beeswax colour and/or contaminate it).
2. Heat at a temperature of 70–90 °C until the sample is completely molten.
3. After beeswax melted, filter the solution (or melted beeswax) through gauze or similar filtering materials, such as tightly woven cotton cloth, canvas or paper filters.
4. Leave to cool at room temperature.
5. Collect the beeswax (as a solid) from the upper layer of floating beeswax to obtain clean sample (beeswax is lighter than water and floats on its surface, and impurities can accumulate at the top of the water).
6. Dry the sample at room temperature for several hours before analysis (to remove droplets from the surface).

The melting procedure without water is performed as follows:

1. Place the beeswax sample in a glass (thermostable) or stainless steel beaker.
2. Place the beaker in an electric plate, water bath, oven or a temperature chamber.
3. Heat at a temperature of 70–90 °C until the sample is completely molten.
4. Leave to cool at room temperature.

Beeswax should not be heated at a temperature higher than 140 °C as volatile fractions begin to evaporate at this temperature (Bogdanov, 2016a). Moreover, temperatures above 150 °C may significantly affect beeswax composition (i.e., the content of hydrocarbons, free fatty acids, and long chain esters), and analytical values of beeswax (physico-chemical parameters); a prolonged heating (>24 h) at a temperature higher than 100 °C may cause the same effect (Tulloch, 1973), and it is therefore not recommended. Melting and other refining procedures commonly applied to beeswax in the beekeeping industry have been reviewed by Bogdanov (2016a).

Additional protocols for beeswax extraction and purification required for detection of pesticide residues in beeswax by LC-MS and GC-MS techniques are described in detail in Sections 6.5.2 and 6.5.3.

### 2.3. Beeswax storage

In case of performing the melting procedure, the samples must be completely dried before storage. It is not necessary to immerse the sample in any kind of preservation medium due to the natural beeswax stability. All types of beeswax specimens can be stored in the same way, as follows:

1. Store the beeswax samples in adequate packaging. Samples can be stored in one of the following packing materials:
    - plastic (polypropylene) containers or bags
    - glass containers
    - stainless steel (airtight) containers (containers made of other metals such as lead, iron, zinc, brass or copper should not be used for storage as they can make the beeswax turn dark)
    - wrapping paper or paper envelopes (suitable for storage of larger wax blocks or comb foundations)
2. Store the beeswax samples in a dark place at a temperature between 10 °C and 23 °C (note: it is advisable to store samples of brood combs or combs with bee bread below 9 °C to avoid wax moth problems).

Beeswax samples can be stored indefinitely.

## 3. Standard methods for research on wax gland cells and production of wax scales

### 3.1. Wax scales production

There were many hypotheses about the origin of beeswax throughout history, including the one of Aristotle who believed that it originates from flowers. The first correct description of beeswax origin dates back to 1744 when the German scientist Hornbostel described bees wax scales, their probable origin and use. These observations were further elaborated by Hunter in 1792, who gave additional notes on the background of beeswax secretion and comb construction (Bogdanov, 2016a; Hepburn, 1986; Hepburn et al., 2014).

Western honey bees produce wax in the form of scales used to construct their nest combs (Hunter, 1792). Secretion of wax scales is an age-related physiological process that may occur in adult honey bees at any time from day 3 to day 21 post emergence (Hepburn et al., 1991). The wax production phase primarily occurs from day 9 to 21 post emergence, peaking at day 12 (Hepburn, Hugo, Mitchell, Nijland, & Scrimgeour, 1984; Hepburn et al., 2014). In addition to age, the onset and duration of this phase is influenced by multiple biological and ecological factors including season, comb/storage space needs, honey and pollen availability, and brood-care needs (Boehm, 1965; Goetze & Bessling, 1959; Hepburn & Magnuson, 1988; Muller & Hepburn, 1992). There are eight ventrally paired wax mirrors (extremely thin sections of cuticle located underneath overlapping sternite segments on the underside of the bee's abdomen; they join adjacent sternites) on sternites 4–7, allowing bees to secrete up to eight wax scales at a time (Cassier & Lensky, 1995; Locke, 1961). Underlying fat body cells associated with the wax gland, including oenocytes and adipocytes, increase in size and undergo physiological changes prior to wax synthesis (Boehm, 1965; Claus, 1867; Hepburn, 1986; Hepburn et al., 1991). Fatty acids and hydrocarbons are taken up by oenocytes of the wax gland to produce wax precursor materials; the role of wax gland adipocytes is uncertain, but they are thought to be an important source of lipids (Hepburn et al., 1991, 2014; Piek, 1964; Sanford & Dietz, 1976). The wax precursor materials pass through the microtubule pore canals of wax mirror epithelial cells as liquid wax fractions that coalesce at the surface of the cuticle (Cassier & Lensky, 1995; Locke, 1961; Hepburn, 1986; Hepburn et al., 1991; Sanford & Dietz, 1976). At the wax mirror, subsequent layers of liquid wax exude and harden to eventually form a scale (Cassier & Lensky, 1995; Dietz & Humphreys, 1970; Hepburn et al., 1991).

### 3.2. Collection of wax scales

When observable, wax scales protrude from between sternite segments. When this is not the case, a pair of forceps can be used to pull back the preceding sternite to expose the wax scale underneath. Wax scales may be collected directly from individual bees (3–21 days post emergence) or from the floor of an experimental cage after they fall from the bee.

#### *3.2.1. Collecting workers from hives to study wax scales*

Within bee colonies, adult workers with wax scales may be found on brood frames and where new comb is being built. Please see *BEEBOOK* vol I article on standard methods for maintaining adult *A. mellifera* in cages under *in vitro* laboratory conditions, Section 4 "Obtaining adult workers for laboratory experiments" by Human et al. (2013), for more information. Caging of bees is performed to limit the influence of external factors and numerous variables that might impact the study (weather, forage resources, water availability, pests, pathogens, robbing, potential nearby pesticide use, beekeeping practice, etc.). The wax scales can be collected from bees of unknown or known age, depending on the aim of the study.

**Obtaining adult bees of unknown age**

If cohorts of known age are not needed, adult bees are removed from a hive and placed into cages as follows:

1. Removing a frame from the hive.
2. Check the frame for the queen to ensure her safety.
3. Brush bees gently off the frame into a sealable cup to bring into the lab.
4. Maintain adult bees in rearing cages in the incubator until sampling according to standard rearing methods, please see *BEEBOOK* vol I article for maintaining adult *A. mellifera* in cages under *in vitro* laboratory conditions Sections 5–7 by Williams et al. (2013).
5. Proceed to Section 3.2.2.

**Obtaining adult bees of known age**

This method is more difficult and time consuming compared to the collection of the wax scales from bees of unknown age, but provides information on the age if necessary for the research intended to be carried out. Knowing the age of bees may be important as honey bees go through significant physiological and behavioral changes in their adult lifetimes; this also includes changes in metabolism of fat stores near the cuticle and the extrusion of liquid wax. Wax scales can be collected from individual bees starting from day 3 after emergence until day 21. In addition, experience-based observations indicate that it is best to collect wax scales as early as days 5 and 6 post emergence.

If cohorts of known age are needed, please see the *BEEBOOK* vol I article on miscellaneous honey bee research methods, Section 2.5.6 "Obtaining workers of known age," by Human et al. (2013). The following is a summary of *BEEBOOK* vol I article on miscellaneous honey bee research methods, Section 2.5.6 "Obtaining workers of known age," by Human et al. (2013) and *BEEBOOK* vol I article on behavioral methods, Section 2.3 "Marking individual bees" by Scheiner et al. (2013), with modifications to aid the researcher conducting research on wax scales.

A brood frame is collected, incubated in the lab, and emerged bees are marked and returned to the hive to obtain adult bees of known age as follows:

1. Remove a frame of ample (mostly capped) brood.
2. Check the frame for the queen; if she is present, either choose another brood frame or very carefully move her to another frame.
3. Brush off adult bees back into the hive.
4. Place the selected frame into a frame cage and place in a rearing incubator (incubator settings: temp, 34.5 °C; humidity, 70%).
5. Collect and mark eclosed bees daily and return to hive according to procedures to mark bees with paint described in *BEEBOOK* vol I article on miscellaneous honey bee research methods, Section 2.5.6 "Obtaining workers of known age," by Human et al. (2013), or by following procedures for gluing tags to the bee thorax described in *BEEBOOK* vol I article on behavioral methods, Section 2.3 "Marking individual bees," Scheiner et al. (2013).
6. Come back to the hive at desired time points and collect marked bees with wide-tip featherweight forceps, gently pinching their wings, legs, or thorax, and place into a sealable container.

#### *3.2.2. Collecting wax scales from individual bees*

Wax scales can be collected directly from individual bees by using a probe or forceps to check underneath overlapping ventral sternites of the bee abdomen. Until selecting individuals for wax scales analysis, maintain bees according to standard methods found in *BEEBOOK* vol I article on maintaining adult *A. mellifera* in cages under *in vitro* laboratory conditions by Williams et al. (2013).

*3.2.2.1. Selecting and anesthetizing individual bees.* Bees are selected through a small opening in the container or cage, and then subjected to cold- anesthesia as follows:

1. Grab individual bees by the leg using wide tip forceps through a small opening in the cage and quickly remove.
2. Deposit bee into a 15 mL tube.
3. Cap the tube and place in freezer to cold-anesthetize (−2 °C for 2.5–3 min), see the *BEEBOOK* Vol I article on miscellaneous honey bee research methods, Section 2.1.2.2 "Immobilizing adults, chemical and physical immobilization" by Human et al. (2013).

If an entire cage or container of adults needs to be anesthetized to select and remove individuals for collecting wax scales,

1. Set cage in the freezer for 5 min to slow them down before selecting individuals, or
2. Place the cage in the refrigerator (4 °C) for 10–15 min.

The more bees in a cage, the longer it will take to lower their body temperatures (for a cage of 30 bees, it will take about 15 min at 4 °C). Be careful not to kill bees by leaving them in the cold for too long; check on them every 2–3 min to observe effect. Bees can also be anesthetized with carbon dioxide; however, they recover more quickly.

*3.2.2.2. Collecting wax scales from the wax mirror.* Wax scales are collected from underneath overlapping sternites of the honey bee abdomen of anaesthetized bees as follows:

1. Pinch the wings of the anesthetized bee carefully between the forefinger and thumb.
2. Pivot to display the ventral surface of the bee (Figure 2).
3. If the bee is to be killed anyway, remove the legs to the femur.
4. Prod underneath the overlapping ventral sternites with a hooked probe or fine-tip forceps to observe wax scales.
5. Rate the size of wax scales relative to the wax mirror before removal (see Section 3.3.1).
6. If using forceps, grip the sternites gently and pull outwards (away from the bee) slowly and gently to remove wax scales; this will cause enough friction between the scale and the end of the forceps to pull out the wax scale without ripping the abdomen.
7. Grab the protruding wax scale with forceps and place on a slide or into a 1.5 mL microcentrifuge tube labelled with the date, treatment group, cage number, and bee number.
8. Repeat for all eight areas where wax scales can be found: sternites 4–7, and on either side of the centerline bisecting the ventral-side of the abdomen.

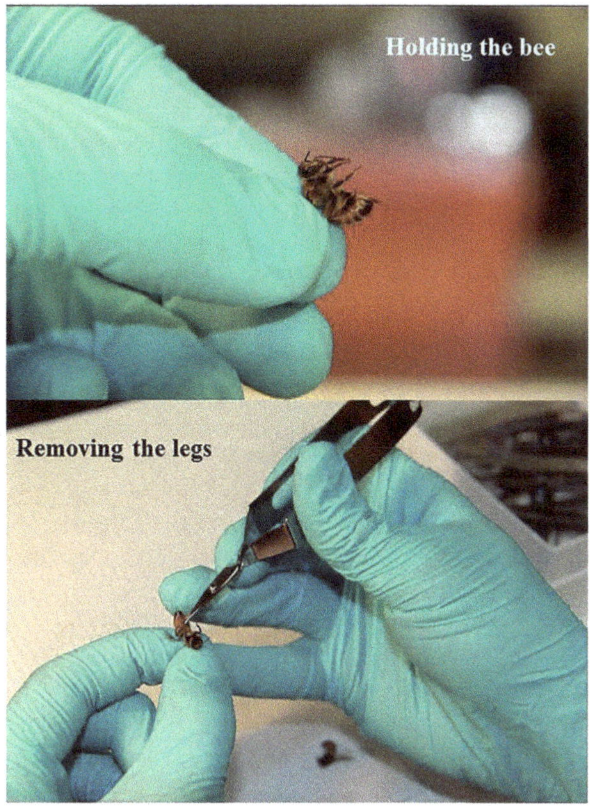


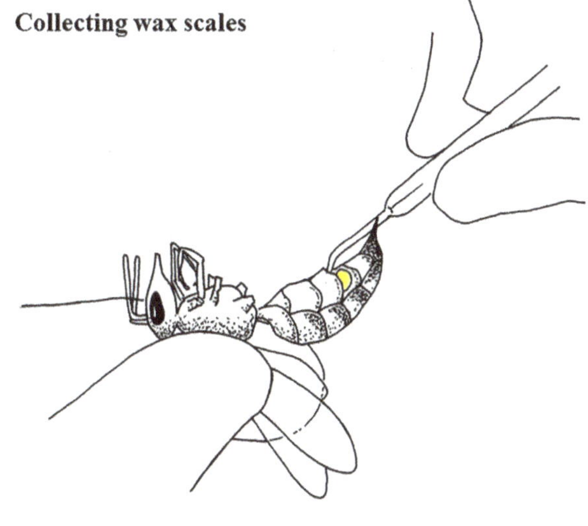


Figure 2. Demonstrates how to hold the bee, remove legs, and use a hooked probe to remove wax scales (high-lighted yellow) from the wax mirror on the underside of a honey bee abdomen. Photos and Illustration: Waters T.

9. Record the mass of collected wax scales (see Section 3.3.2).

Removing the legs is optional and this procedure may expose wax scales to the risk of contamination from hemolymph. Collecting wax scales from individual bees is more time consuming than retrieving fallen scales in cages; however, there is a decreased chance of contamination with cage debris.

*3.2.3. Collecting fallen wax scales from cages*

The easiest way to collect wax scales is to wait until after they have fallen off the bees, using wooden cages or plastic cup cages with raised wire mesh floors. The mesh needs to be sturdy enough to serve as a stable enclosure for the bees but large enough to allow wax scales to fall through (#6, 3 mm opening, steel wire mesh).

*3.2.3.1. Modifying plastic cup cages.* Procedure to modify plastic cup cages to insert a wire mesh false floor:

1. Cut a circle out of a Solo Cup® lid leaving the rim intact (Figure 3(a,b)).
2. Use cuticle scissors to remove the bottom of the cup (Figure 3(c)).
3. Cut steel wire mesh into a circle approximately 7 cm in diameter for a 473 mL clear Solo cup (this just so happens to be the same diameter as a wide-mouth Ball® mason jar lid that can be used as a model/gauge, Figure 3(d,e)).
4. Place wire mesh circle into the cup – the mesh circle should be small enough to fit snugly in the cup leaving about 2 cm of space between the mesh and the bottom of the cup as shown (Figure 3(f,g)).
5. Cut approximately 15 cm × 15 cm piece of fine-mesh cloth.
6. Place fine-mesh cloth square over cup and snap lid on top to secure (Figure 3(h)).
7. Poke a hole in the fine mesh with an razor blade or snip with dissection scissors, to allow top feeder to be placed through hole without falling through (Figure 3(h)).
8. Set the cup onto a petri dish to collect fallen scales without disturbing the bees (Figure 4).

*3.2.3.2. Collecting fallen wax scales from cages designed with wire mesh floors.* For *in vitro* adult rearing cages designed with wire mesh floors, remove fallen wax scales as follows:

1. Place a flat removable surface underneath the cage to collect fallen wax scales (such as a strip of plastic from a notebook divider, with edges bent up to form a rim) (Figure 5).
2. Collect wax scales at set time-course.
3. Record the number of wax scales collected from each cage.
4. Measure the mass of wax scales (see Section 3.3.2).

Feeder drips, bee feces, mold, and other contaminants are a problem in most caged experiments. Compared to rearing cages with closed or solid floors, the wire-mesh floors appear to improve cage ventilation and prevent food and feces build-up on the floor of the cage, reducing mold problems and improving cage health. Additionally, the wire mesh floors facilitate wax collection while minimizing disturbance of bees. Despite the raised floor, wax scales may still be contaminated by cage debris.

### 3.3. Wax scale measurements

*3.3.1. Recording scale size*

Wax scales are flat oval-shaped flakes of raw wax (Figure 6) that can be up to 3 mm wide. When collected directly from bees, their size can be compared to the wax mirror as described in Jordan (1962) (as cited in Ferguson & Winston, 1988; Ledoux et al., 2001; Otis, Winston, & Taylor, 1981). In this way, scale size can be rated as follows:

0, no wax
1, very small scales, not easily removable
2, medium scales, not extending beyond the overlapping sternite
4, very large scale or irregularly shaped clumps of wax (covering the entire wax mirror and extending well beyond the overlapping sternite).

Described rating method represents a simplified procedure for measuring the wax scales collected directly from the bees. However, if wax scales need to be sized or when fallen wax scales are collected from the cages, their size (mm) can be determined using an eye-piece ocular micrometer fixed within a stereomicroscope (alternative: digital photographs in combination with various image analysis programs).

*3.3.2. Measuring scale mass*

The relative quantity of wax secreted by different age groups can be estimated by measuring the weight of wax scales recovered per bee or per group. For measuring the mass of wax secreted by cohorts of bees (e.g., Hepburn et al., 1991), a microscale readable to 1 mg should be sufficient when large numbers (dozens) of wax scales are pooled together. To weigh wax scales of individual bees (e.g., Hepburn et al., 1991) a microscale readable to 0.01 mg will be needed:

1. Tare microscale with clean, small (0.4–1.5 mL), microcentrifuge tube.

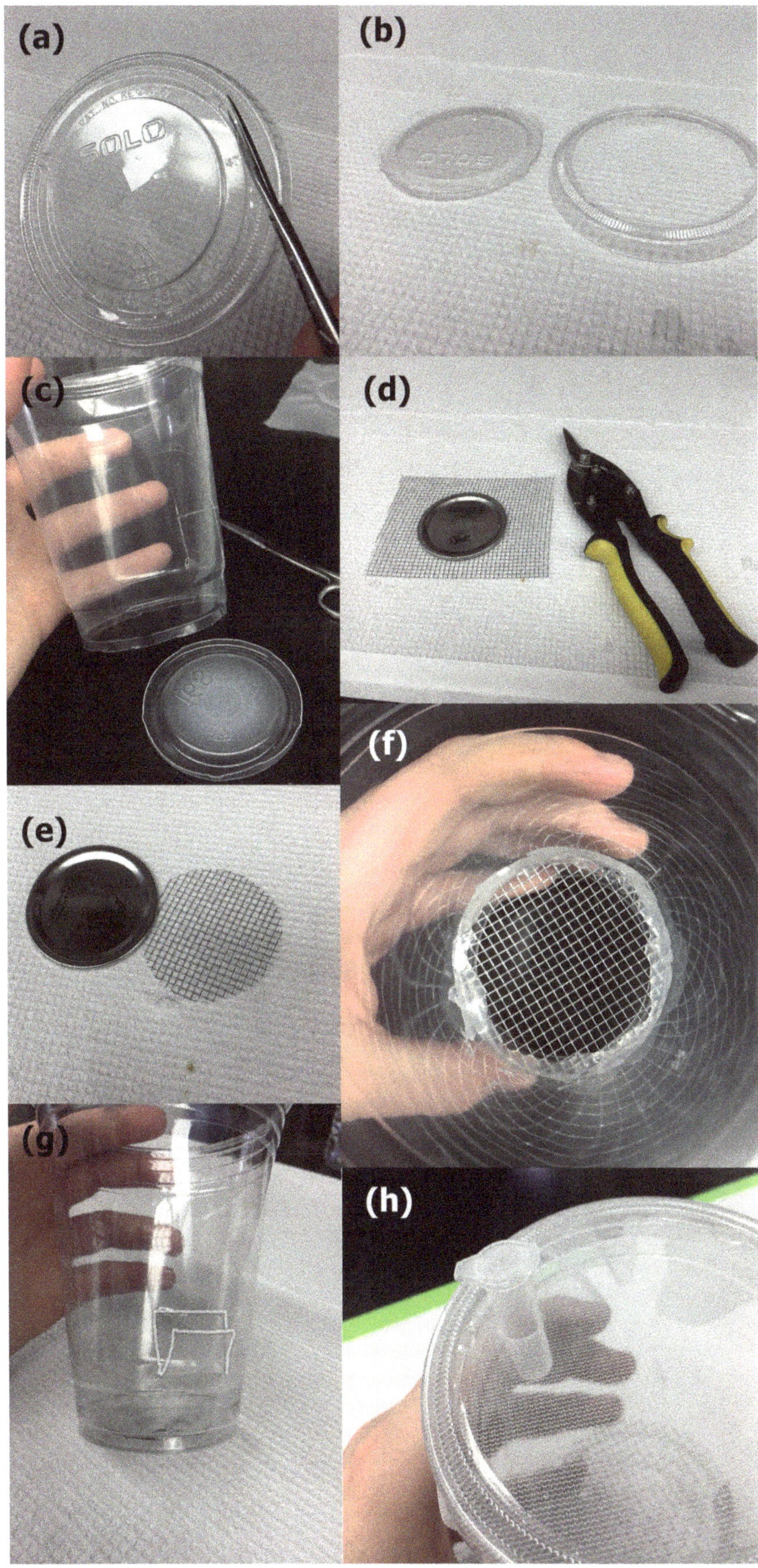


Figure 3. Steps to modify plastic cup rearing cage to collect wax scales (a–h). Photo: O'Grady M.

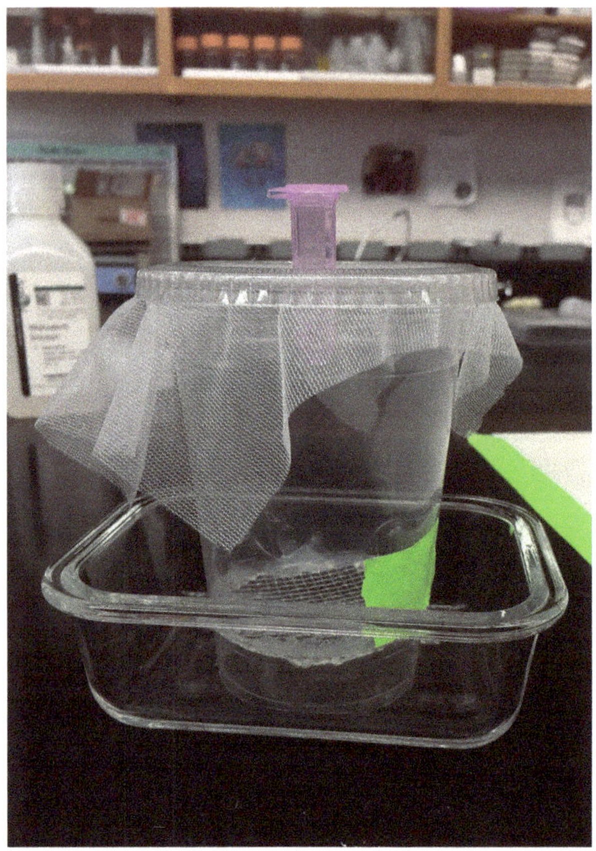

Figure 4. Modifying a Solo Cup® rearing cage to collect wax scales by inserting a wire mesh false floor, removing the bottom of the cup, and placing over dish. Photos: O'Grady M.

2. Place wax scales in small microcentrifuge tube.
3. Measure and record the weight of the scales.
4. Freeze at −20 °C to preserve until analysis.

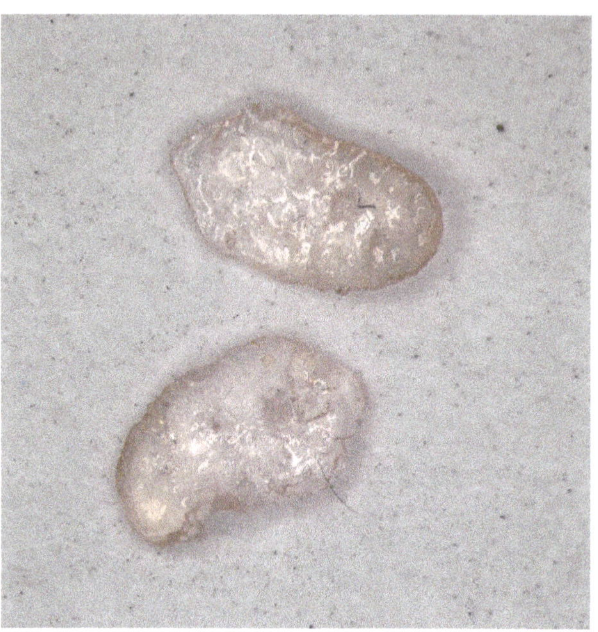

Figure 6. Wax scales viewed under a dissection scope. Photo: Waters T.

## 4. Methods for investigating honeycomb cell properties and comb construction

### *4.1. The two-dimensional structure of the hexagonal cell*

Despite the fact that the honeybee cells might originate from initially circular tubes, cells exhibiting a hexagonal structure as a final result remains universally accepted. Stating this assertion is essential for the rest of the chapter, since adequate measurements depend crucially on the model of the cell that we adopt. From a

Figure 5. Plastic placed under rearing cages with wire mesh floors to collect wax scales. Photo: Waters T.

geometrical point of view, the honey bee cell can be described as follows (Figure 7, Saucy, 2014).

1. The dimensions of a regular hexagon can be approached by two circles, i.e., an inscribed circle (i) and a circumscribing circle (c), of radius $r_i$ (the apothem) and $r_c$, respectively.
2. The hexagon can be decomposed in 6 isosceles triangles, i.e., triangles with edges of equal lengths.
3. The small diameter of the inscribed circle $d_i$ is $d_i = 2*r_i$.
4. The large diameter $d_c$ of the circumscribed circle is: $d_c = 2*r_c$.
5. Since the triangles are isosceles, the length of each edge of the hexagon is equal to the radius of the circumscribed circle $r_c$.
6. If follows that the height of each triangle is equal to $r_i$.
7. The area of each isosceles triangle is given by: $A = r_c*r_i/2$.
8. The area of a cell is therefore: $A_{cell} = 6*r_c*r_i/2 = 3*r_c*r_i$.
9. Each of the six triangles of the hexagon being also equilateral triangles, the length of each side of the hexagon is equal to the radius of the circumscribed circle $r_c$, while the height of each triangle is equal to $r_i$. Applying the Pythagorean theorem we deduce $r_i^2 = r_c^2 - (r_c/2)^2 = 3r_c^2/4$, from which the following relationships between $r_i$ and $r_c$:

$$r_i = \frac{\sqrt{3}}{2} r_c \text{ and } r_c = \frac{2}{\sqrt{3}} r_i,$$

10. Therefore, the area of a cell can be expressed as:

$$A_{cell} = 3*\frac{2}{\sqrt{3}} r_i * r_i = 2\sqrt{3} r_i^2$$

11. Or, expressed in $d_i$ units: $A_{cell} = \frac{\sqrt{3}}{2} d_i^2$
12. Finally, the cell density (expressed as the number of cells per area unit) is given by: $D_{(1side)} = \frac{1}{\frac{\sqrt{3}}{2} d_i^2}$
13. Taking the two sides of the comb into account yields: $D_{(2sides)} = \frac{4}{\sqrt{3} d_i^2}$

*4.1.1. Linear measurements along the small diameters ($d_i$)*

The linear measurements were formalized by Baudoux (1933). His results have been summarized as follows in de Meyer (1938): "wax foundation should be measured in three directions and each time along the double apothem of the cells. To minimize error measurements, they should be conducted in several series of 10 cells" (Figure 8). To estimate cell densities, Baudeaux developed a simplified formula for field work (de Meyer, 1938):

1. take several measurements of 10 cells (in mm)
2. calculate the average of these measurements
3. divide the number 2,309,467 by the square of the averaged measurements

The number 2,309,467 corresponds to the square of a measurement of 10 cells in a comb of cell density = 1000 cells/dm$^2$, which in turn corresponds to a cell width of 4.805691417 mm. Beaudoux's simplified formula is correct, still valid and in full accordance with the geometrical and mathematical properties of a two-sided comb made of hexagonal cells, as described in the formula $D_{(2sides)} = \frac{4}{\sqrt{3} d_i^2}$.

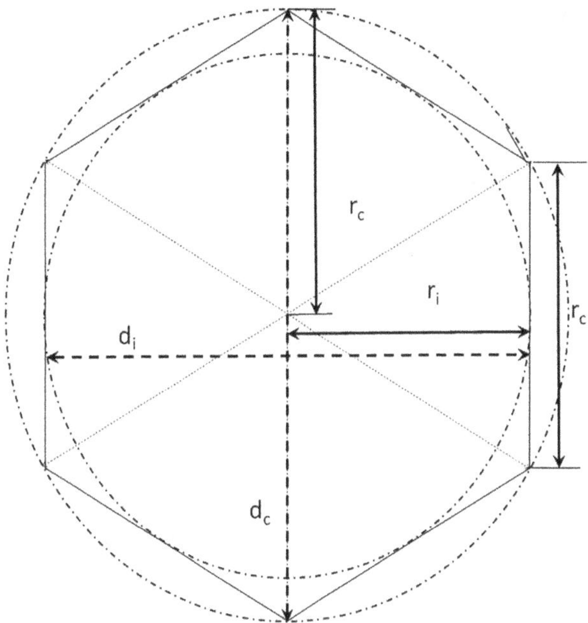

Figure 7. Two-dimensional model of the cell as a regular hexagon with respect to its inscribed (i) and circumscribing (c) circles. $r_i$, radius of the inscribed circle (apothem); $d_i$, diameter of the inscribed circle; $r_c$, radius of the circumscribing circle; $d_c$, diameter of the circumscribing circle.

Figures 8–10 illustrate various ways of taking measurements of 10 cells in different configurations, for instance within a hexagon encompassing exactly 100 full cells (Figure 9). Figure 10 shows how the same hexagon of small diameter 10d$_i$ can be transformed into a rhomb of same area $A_{hexagon} = A_{rhomb} = 100\frac{\sqrt{3}}{2} d_i^2$.

Linear measurements should be taken along rows of a minimum of 10 aligned cells through their small diameters, i.e., through the diameters of their inscribed circles ($d_i$), either horizontally, either laterally at 45° angles (Figure 8), through the middles of the external walls of the most extreme cells of the row:

1. Identify an area of cells homogenous in size and shape (worker cells, drone cells, honey cells).
2. Identify a row of n aligned cells (at least 10).
3. Mark a cell, then count n cells and mark the next cell.
4. Measure the distance from the 1st to the nth cell, from the middle of the external walls of the 1st and nth cell using a calliper (Figure 11).
5. Calculate the average cell width ($d_i$) by dividing the measurement by *n*, the number of cells of the measurement.

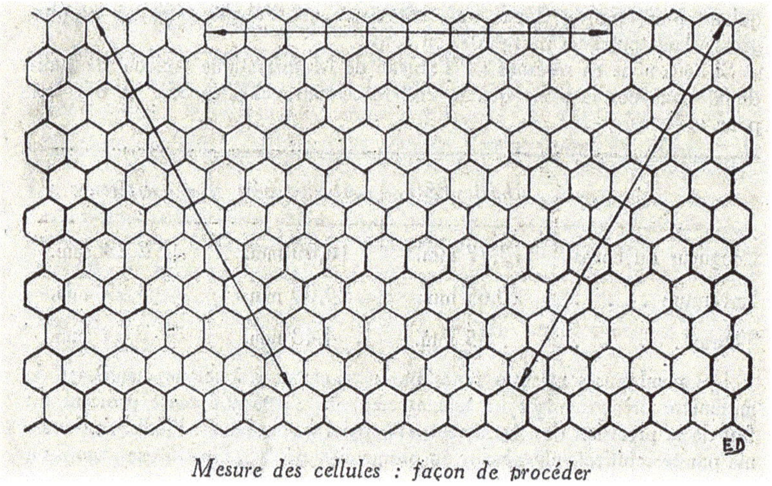

*Mesure des cellules : façon de procéder*

Figure 8. Measurement of the cells: way to proceed according to Baudoux (De Meyer, 1938).

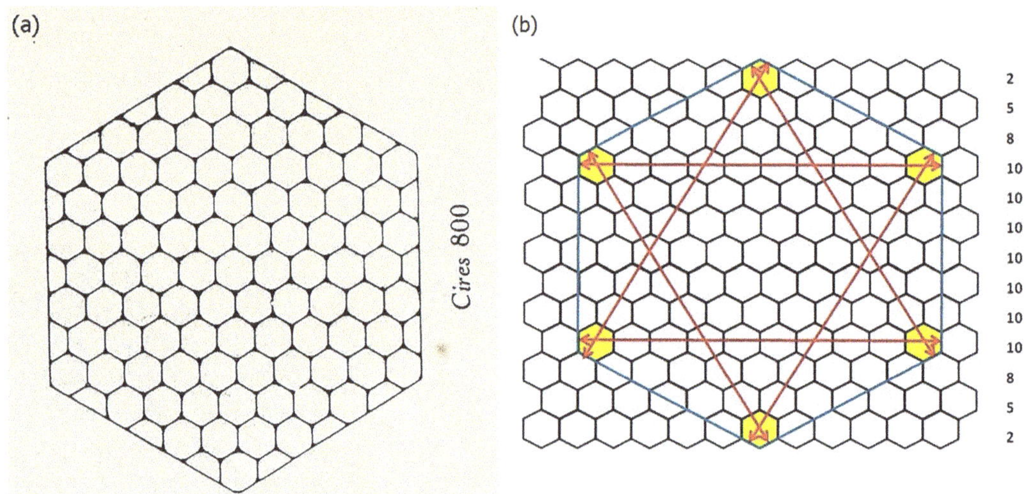


Figure 9. (a) Example of a hexagonal subset of wax foundation with a density of 800 cells/dm$^2$ encompassing the area of hundred full cells (according to Baudoux, in De Meyer, 1938); (b) diagram of the same hexagonal subset of a comb exhibiting that 7 rows encompass an area of 10 full cells in 3 directions. The red lines indicate 6 possible measurements of rows of 10 cells within the hexagon. The figures on the right give the area (in full hexagonal cells) of each horizontal row.

6. Calculate cell density using formula (13): $D_{(2sides)} = \frac{4}{\sqrt{3}d_i^2}$, or read it from Table 1.
7. Replicate the same measurements to get statistical estimates of measure errors.
8. Repeat measurements at regular intervals on the comb to get statistical estimates of variability of cell sizes.

*4.1.2. Sampling the comb*

*4.1.2.1. Sampling the comb using linear measurements along the small diameters ($d_i$).* Since a comb may contain different kinds of cells of different sizes (worker, drone or honey cells), sampling the comb is usually necessary to seize its cell size diversity.

The method of linear measurements can be extended in different ways:

1. Three linear measurements of 10 cells in a triangular pattern around a topic cell (Figure 13(a)).
2. Three linear measurements of 10 cells crossing a topic cell in a star pattern (Figure 13(b)).
   In both cases (a and b), the average of the three measurements gives an estimate of cell sizes over approximately 1/4 of dm$^2$. Five such measurements on each side of the comb should be sufficient to assess the comb's diversity.
3. Linear measurements at regular intervals (e.g., every each 5th row) from the top to the bottom of the comb or from the right to the left of the comb.

*4.1.2.2. Sampling the comb using rectangular measurements along the small and large diameters.* Combs built on artificial foundation are expected to exhibit little variability, but sampling the comb using rectangular measurements may be a useful approach to assess the regularity of wax foundation sheets. Measurements should be conducted along both dimensions of the cells, i.e., along the

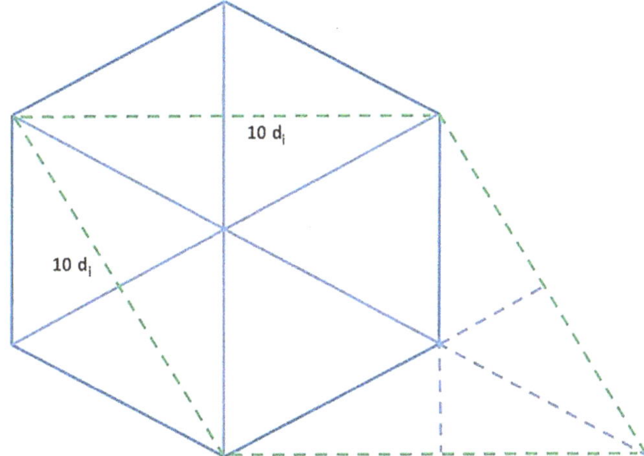


Figure 10. Picture showing how a hexagon of diameter $10d_i$ can be transformed into a rhomb of length $10d_i$ and of same area as the hexagon, i.e., $A_{hexagon} = A_{rhomb} = 100\,\frac{\sqrt{3}}{2}d_i^2$. Note that each of the 6 isocele triangles of the hexagon can be divided in 2 right triangles, yielding 12 such triangles of same area for both the hexagon and the rhomb. A square of side $10d_i$, of area $100d_i^2$, would encompass 16 such rright triangles. The ratio $H/S = \frac{\text{area of the hexagon}}{\text{area of the square}}$ is therefore $\frac{12}{16} = \frac{3}{4} = \frac{\sqrt{3}}{2}$.

Figure 11. Measuring 10 cells in a row using a caliper. (Source: http://parkerbees.com/beespace.html)

small and large diameters, $d_i$ and $d_c$ (Figure 12). The measurements and calculations are slightly more complicated, since linear measurements through rows of cells pass alternatively through large diameters and large radii of the cells. The average height ($h$) of a row is therefore:

$$h_{1row} = \frac{3}{4}d_c = \frac{\sqrt{3}}{2}d_i.$$

In practice, it is recommended to measure an even number of rows (e.g., 10), so that the average height is estimated based on an equal number of large diameters and large radii.

1. Mark a row, count $n$ (10) rows vertically along the comb, mark the $n+1$th row (11th row).
2. Take a measurement of the $n$ (10) rows.
3. Divide your measurement by $n$ (10) to get $h_{1row}$.
4. Calculate $d_c = \frac{4}{3}h_{1row}$ and $d_i = \frac{\sqrt{3}}{2}d_c$.

On this basis, it is therefore easy to estimate the regularity of the cells. For instance, a linear horizontal measurement of 53 mm along the small diameters ($d_i$) of 10 cells corresponds to an average 5.3 mm cell width. If cells are regular, we expect for a vertical measurement of 10 rows a figure of $10*\frac{\sqrt{3}}{2}d_i$, i.e., 45.9 mm. A larger figure indicates a vertical distortion of the hexagons, with $d_c$ larger than expected from $d_i$, while a smaller figure is indicative of a horizontal distortion of the cell.

Table 2 gives examples of such linear cell measurements conducted on wax and plastic foundation. In this case, the longest possible rows of full cells were measured in both horizontal and vertical directions and compared to samples of 10-cell measurements using a ruler

Table 1. Cell density estimates based on measures of the diameter of the inscribed circle $d_i$, according to the formula: $D_{(2sides)} = \frac{4}{\sqrt{3}d_i^2}$.

| Cell width (diameter $d_i$ in mm) | Density (number of cells/$dm^2$) |
|---|---|
| 4.5 | 1140.4 |
| 4.6 | 1091.4 |
| 4.7 | 1045.5 |
| 4.8 | 1002.3 |
| 4.9 | 961.8 |
| 5.0 | 923.8 |
| 5.1 | 887.9 |
| 5.2 | 854.1 |
| 5.3 | 822.1 |
| 5.4 | 792.0 |
| 5.5 | 763.4 |
| 5.6 | 736.4 |
| 5.7 | 710.8 |
| 5.8 | 686.5 |
| 5.9 | 663.4 |
| 6.0 | 641.5 |
| 6.1 | 620.6 |
| 6.2 | 600.8 |
| 6.3 | 581.9 |
| 6.4 | 563.8 |
| 6.5 | 546.6 |
| 6.6 | 530.2 |
| 6.7 | 514.5 |
| 6.8 | 499.4 |
| 6.9 | 485.1 |
| 7.0 | 471.3 |

or a calliper. The results indicate that using callipers (Column J; Table 2) yields closer estimates of cell widths than using a ruler (I) for rows of 10 cells with regard to cell width estimated using measurements of whole cell rows (L). Columns P and Q give estimates of horizontal and vertical distortion ranging from 0.2% for plastic foundation (third row) to 1.1% for wax foundation (first row).

*4.1.2.3. Measurements and analyses conducted on photographs.* Taking measurements on combs is not only a time-

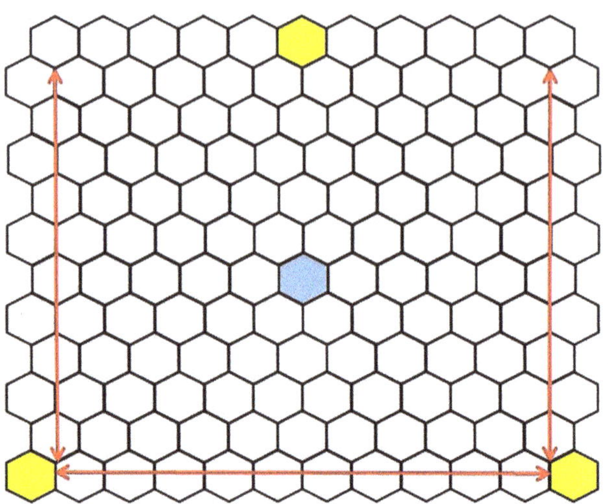

Figure 12. Rectangular measurements of 10 rows of cells to assess the two-dimensional regularity of the cells.

consuming task which may be postponed for further analysis in the laboratory, but may also be complex because of the bees activity during field work, as well as fragility of the wax. Taking pictures of the combs is an alternative and efficient way of speeding up field work. In all cases, one should label the subject and use reference objects (e.g., a ruler) to be able to calibrate all further measurements. Such reference tags should be located as close as possible of the areas of interest and should be disposed at different locations of the comb to be able to assess possible distortions due resulting from different angles of view of the camera with respect to the picture's subject. Taking measurements on photographs has the advantage that measurements can be postponed, can be replicated and that pictures can be shared with colleagues everywhere in the world. In addition, pictures may be enlarged, allowing increased precision measurements. They are particularly suited if the subject has to be enlarged, e.g., in the case of cell width measurements. Various software are available for the analyses. Examples given in this section were operated using ImageJ (Fereira & Rasband, 2012), an open source software widely used among scientists:

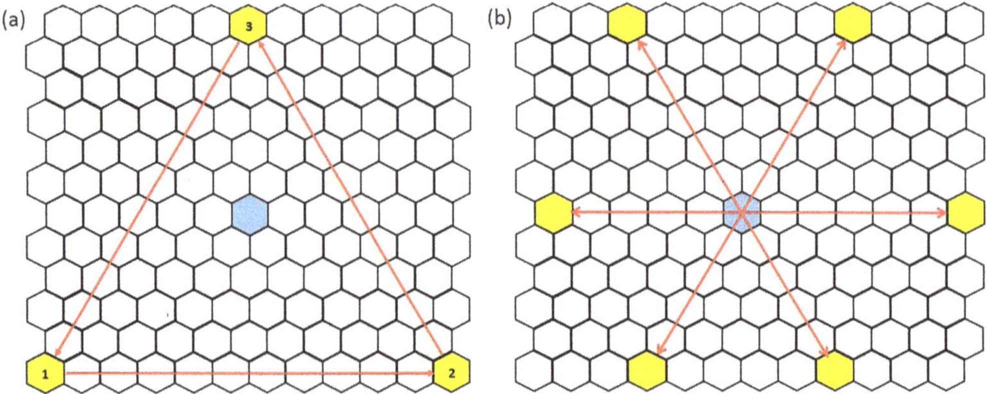


Figure 13. Linear measurements of three rows of: (a) 10 cells in a triangle pattern and (b) of three rows of 10 cells in a star pattern (blue: topic cell; yellow: cells adjacent to each row of 10 cells). Note that the triangle pattern is easier to implement since it only requires marking 3 cells, instead of 7 (including the topic cell) in the star pattern. Moreover, the latter displays an asymmetrical pattern with either 4 or 5 cells on each side of the topic cell for each measurement.

Table 2. Examples of linear measurements conducted on artificial foundation from three different suppliers.

| A | B | C | D | E | F | G | H | I | J | K | L | M | N | O | P | Q |
|---|---|---|---|---|---|---|---|---|---|---|---|---|---|---|---|---|
| Type of foundation | Number of cells per (horizontal) row | Number of (horizontal) rows | Length of (horizontal) rows (mm) | Height of (horizontal) rows (mm) | Area (calculated from col. D & E) ($dm^2$) | Number of cells on measured area (calculated from col. B & C) | Cell density (number of cells/$dm^2$ calculated from col. F & G) (2 sides) | Measurement of 10 cells using a ruler along the small diameter $d_i$ (cm) | Measurement of 10 cells using a caliper along the small diameter $d_i$ (cm) | Average cell width $d_i$ (calculated from col. D & B) (mm) | Average row height $h_{row}$ (calculated from col. E & C) (mm) | verage cell large diameter $d_c$ (calculated as dc= $h_{row}$*4/3 (mm) | $d_i$ estimated from $d_i = d_c*\sqrt{3}/2$ | $d_c$ estimated from $d_c = 2d_i/\sqrt{3}$ | Horizontal distortion (ratio $d_i$ measured/$d_i$ estimated after $d_c$ (col. K & N) | Vertical distortion (ratio $d_{ic}$ measured/$d_c$ estimated after $d_i$ (col. M & O) |
| Dadant (body) wax | 76 | 56 | 407 | 257 | 10.5 | 4 256 | 814 | 5.5 | 5.4 | 5.355 | 4.589 | 6.119 | 5.299 | 6.184 | 1.011 | 0.990 |
| Dadant (super) wax | 77 | 28 | 408 | 128 | 5.2 | 2 156 | 826 | 5.3 | 5.3 | 5.299 | 4.571 | 6.095 | 5.279 | 6.118 | 1.004 | 0.996 |
| Dadant (super) plasic | 72 | 27 | 410 | 135 | 5.5 | 1 944 | 702 | 5.8 | 5.7 | 5.694 | 5.000 | 6.585 | 5.703 | 6.575 | 0.998 | 1.002 |

1. Tag the comb using a reference for measurements (e.g., a ruler).
2. Take pictures at different time intervals.
3. Document pictures (number, label, date and time, remarks).
4. Measure reference tags.
5. Set the scale (e.g., transform pixels in mm).
6. Take measurements of comb features (cell widths, wall widths, series of cells, etc).
7. Save measurements.

*4.1.2.4. Addressing the regularity of the cell.* The honey bee comb, composed of hexagonal cells, is commonly viewed as a model of regularity and perfection in nature. This comes both from an idealization of the honey bee construction abilities and from the fascination for the comb's two-dimensional structure. However, this regularity partly results from visual impressions and is enforced by the use of wax foundation. However, it is often challenged and is rarely observed on natural combs. On the contrary, the combs exhibit many irregularities with cells of different shapes, often far from regular hexagons, including squared, pentagonal and heptagonal cells (Figure 15). Irregularities may result from different processes:

a. Transition from worker to drone cells during comb construction.
b. Junction from comb sections started at different distances during the construction process.
c. Repair of damaged combs.

Examining how bees handle transitions from worker to drone cells and junctions of combs started at different locations or how they repair damaged combs is perhaps even more fascinating than understanding how they build apparently regular cells. During these processes, bees may exhibit solutions which preserve the function and the overall integrity of the comb, generating numerous irregularities. On some combs, it is hardly possible to find a typical hexagonal structure.

Heaf (2012) has published several pictures and cell size measurements from natural combs (http://www.dheaf.plus.com/warrebeekeeping/cell_size_measurements.htm). As Figure 16 shows, trying to find regular rows of cells as they commonly appear on wax foundation is a hopeless goal. Analyses of Heaf's measurements display a clear trend of decreasing cell size from the top to the bottom of the comb (Figure 17(a)), as well as little correlation between the two sides of the comb with respect to cell size (Figure 17(b)).

To take measurements addressing the regularity of the comb:

1. Apply a 10 cm × 10 cm grid on the top row centred on the middle of the comb.
2. Measure the length and the height of the comb along the horizontal and vertical lines of the grid.

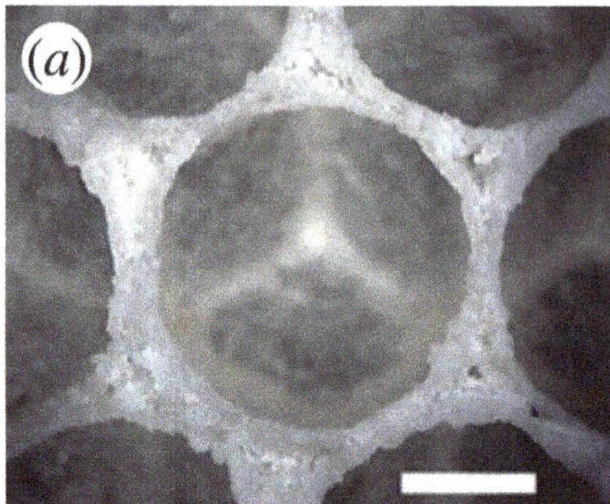


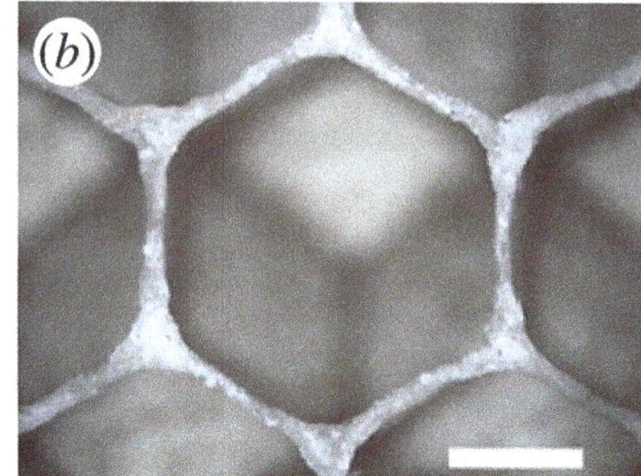


Figure 14. Photographs showing the process of construction of the honey bee cells exhibiting different wall widths during the "maturation" process of the cells: (a) cell at "birth", (b) 2-days old cell. The scale is given by the reference tag (115 pixels = 2 mm), which for the 2-days old cell yields a width of 5.32 mm between wall centers (306 pixels) (from Karihaloo et al., 2013).

Figure 15. Picture of a natural comb exhibiting irregularities; note a heptagonal cell in the centre and "filling" cells. A regular hexagonal cell can hardly be found.

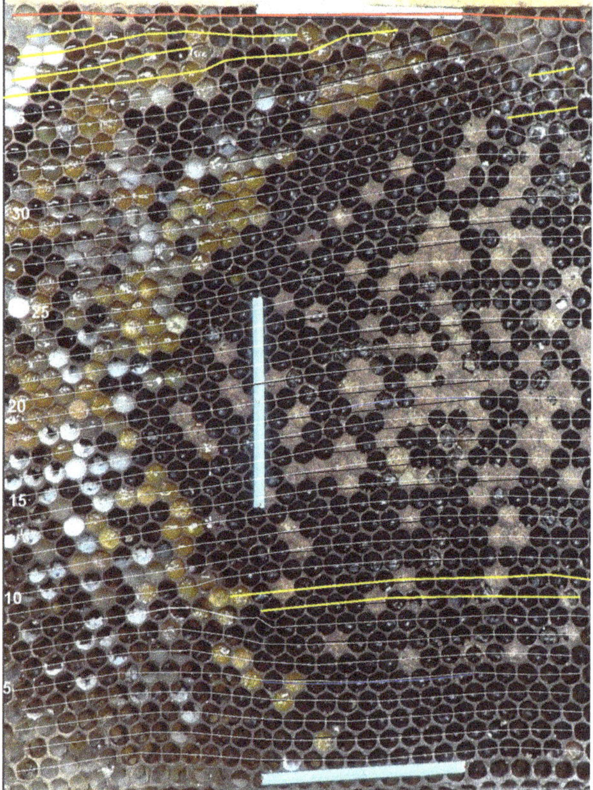


Figure 16. Identification of "horizontal" rows of cells on a natural comb exhibiting a large variability of the two-dimensional cell structure.

3. Count the total number of rows from the top to the bottom along the vertical lines of the grid.
4. Count the total number of cells from the right to the left along the horizontal lines of the grid.
5. Take measurements of rows of 10 cells at regular intervals (e.g., every single, 2nd or 5th row) along the vertical lines of the grid (or at each node of the grid) from the top to the bottom on both sides of the comb.

Recent methods of photograph analyses allow much more sophisticated analyses of the two-dimensional structure of the comb than simple lengths measurements. For instance, Kaatz, Bultheel, and Egami (2008) conducted a study in which they compared natural combs and combs built on wax foundation against an ideal hexagonal structure (Figure 18).

*4.1.2.5. Advantages and disadvantages of different measurement techniques.* Table 3 gives a summary of the principal characteristics, strengths and weaknesses of the different methods and techniques of measurements described in this section. For instance, measurements of rows of 10 cells are clearly best suited for field work, because the method is simple and measurements quick to conduct. Linear measurements of all full cells in a row are particularly well suited for artificial foundation: it is much quicker than repeated measures of series of 10 cells in the same row. While there is only one approach suited for estimating cell distortion in horizontal or vertical direction, among the various approaches to sample the comb in different directions, the triangular pattern is clearly the quickest. If measurements in a

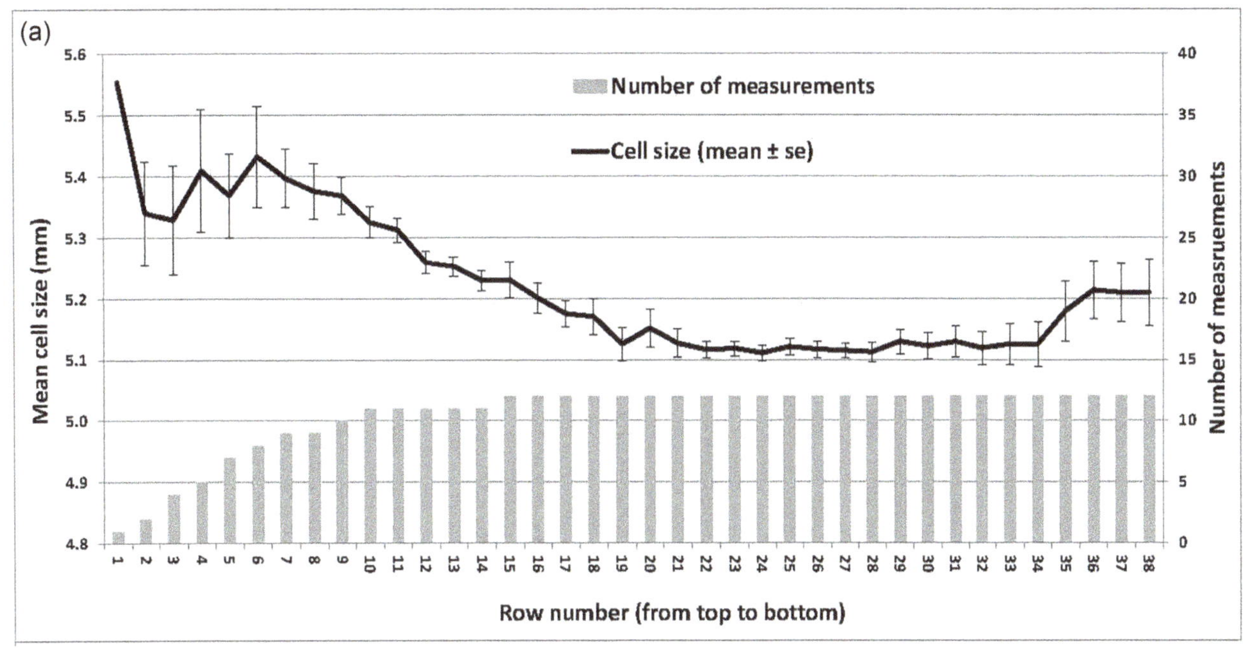


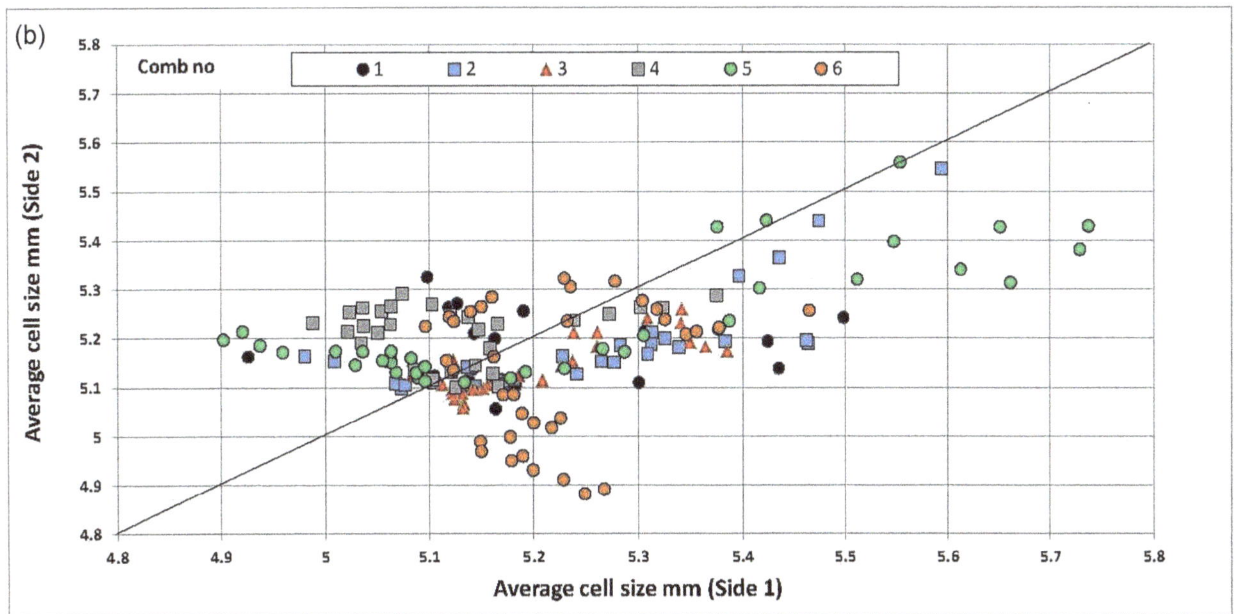


Figure 17. Example of natural cell size variations among combs. Measurements of 6 combs (12 sides) from one Warré box (topbar hive without wax foundation; data from David Heaf: http://www.dheaf.plus.com/warrebeekeeping/cell_size_measurements.htm; Warré 1, measured 04.12.2011). (a) Average cell size (±SE) from top to bottom in the middle of the comb (average of twelve measurement of 10 cells per row; missing data correspond to accidentally damaged top of combs); (b) scatterplot of cell size measurements (mm) for corresponding rows on both sides of the combs (same data as above) exhibiting variation between and within combs. The oblique line corresponds to equal cell size measurements on both sides of the comb.

star pattern are basically equivalent, the approach is more time consuming (7 cells to identify). In addition, from an esthetical point of view, it is geometrically unsatisfactory and would be best suited for measurements of rows of odd numbers of cells (e.g., 9 or 11). In terms of precision, callipers should be preferred to rulers. As long as optical distortions are controlled, taking measurements on pictures is a satisfying approach. It is a quick method in field work and measurements can be reproduced. In conclusion, the best approach depends on the goals, means and resources of the study. What really matters is to take into account the peculiar properties of the hexagonal cell properly, to avoid crude errors (Saucy, 2014).

*4.1.2.6. Investigating interspecific and intraspecific cell size variability.* Cell size varies between *Apis* species and is in direct relationship with the size of the adult workers. It also varies within species, between subspecies, but also within subspecies. It has long been known that smaller

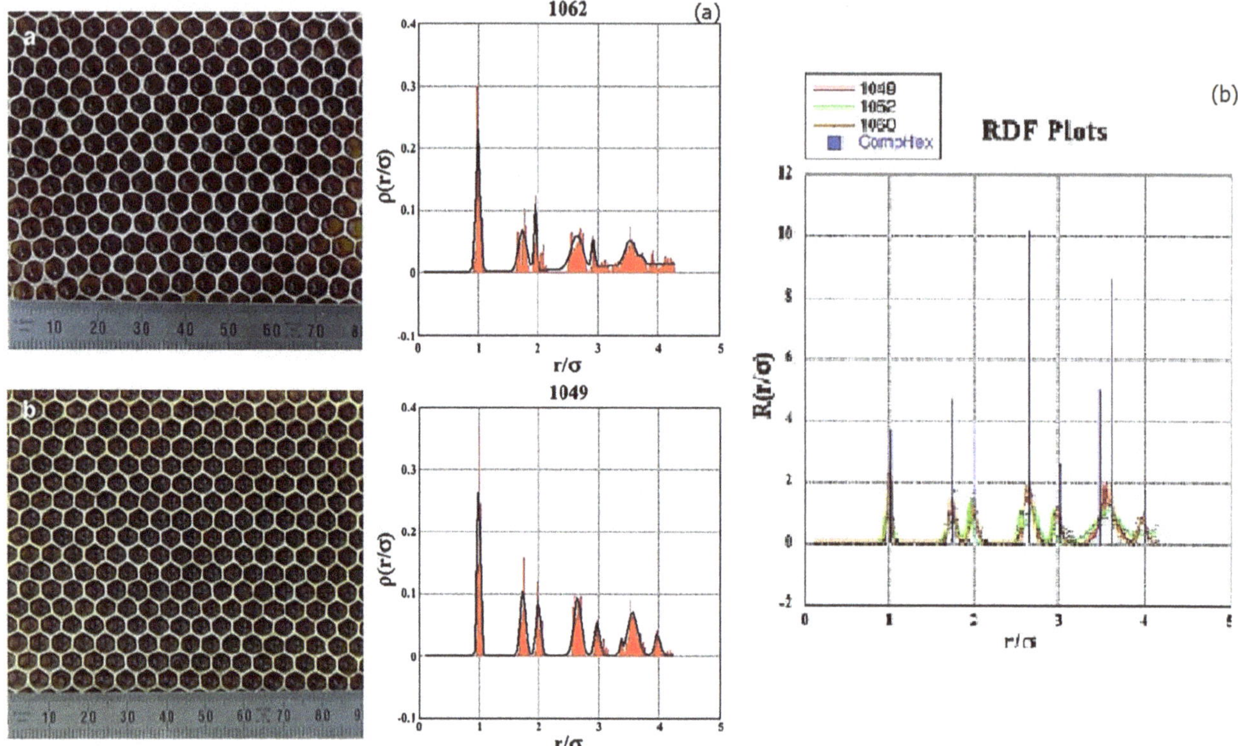


Figure 18. (a) Analysis of the comb structure (from Kaatz et al., 2008): left: comb built on artificial foundation (bottom) and natural comb (top), right: plots of radial diffusion functions; (b) plots of radial diffusion functions (RDF) of 3 combs compared to a perfect hexagonal array (blue lines).

workers emerge from old brood comb because silk residues accumulate in the cells and that smaller or larger workers can be produced depending on the size of wax foundation (Baudoux, 1933; Honegger, 1937; Vogt, 1911). In the early 1990s, uncertainties arose regarding the natural size of cells (Erickson, Lusby, Hoffmann, & Lusby, 1990a, 1990b) that led to a debate with repercussions on beekeeping management, honeybee biology, physiology and health issues (Heaf, 2011, 2012; Saucy, 2014; Stever, 2003; Zeissloff, 2007), an issue which highlighted the need for proper methods.

*4.1.2.7. Interpreting historical data on cell size.* To interpret historical data on cell width properly, the following steps should be conducted:

1. Find a copy of the original publication.
2. Read the publication in the original language or get a certified translation.
3. Correctly identify the unit(s) used for the measurements.
4. Establish the correspondence between the historical measure units and metric units.
5. Convert the measurements of the publication in metric units.

Point 3 is of particular importance. Before the introduction of the metric system, the same words were often used to refer to locally different units of measure. For instance, foot and inch were different in Great Britain, France, some provinces of Germany, Switzerland, the Netherlands and Eastern European countries. Definitions might also change with time. Failing to understanding these peculiarities properly may lead to errors of interpretation (Saucy, 2014 for a review).

Several examples of erroneous interpretations have been reported (Saucy, 2014). For instance, Huber (1814), probably based on the measurements of de Castillon presented in 1781 at the Academy of Sciences of Berlin and which were the only replicated measurements avaible at Huber's times, gives the same cell size as de Castillon, i.e., an average cell size of 2 2/5 lines. Properly interpreted from the Parisian foot used by the French speaking authors before the introduction of the metric system (de Castillon, 1781; Huber, 1814; Maraldi, 1712), this corresponds to a cell size of 5.41 mm, and not to 5.08 as incorrectly transposed from the modern Anglo-American foot defined in the 19th century.

#### *4.1.3. The three-dimensional structure of the comb*

With its hexagonal cells, the bottom of each facing three cells on the other side (Figures 19 and 20), understanding the three-dimensional structure of the honeybee comb has been a practical and theroretical issue since the antiquity (review in Saucy, 2014). The three-rhombic configuration of the bottom of the cell,

Table 3. Comparison of the different approaches to measuring various comb parameters.

| Characteristics | Speed | Use in field work | Precision | Natural comb | Artificial foundation | Reproducibility of measures | Remarks, advantages and recommendations |
|---|---|---|---|---|---|---|---|
| Linear measurements of (horizontal) rows of 10 cells | G | G | S | S | S | S | Best approach for field work |
| Linear measurements of all cells in a row | Nr | Nr | B | Na | B | S | Best approach for artificial foundation |
| Linear measurements of the whole comb | R | R | G | G | G | G | Best for estimates of whole comb area |
| Three measurements of 10 cells in a triangle pattern | R | R | R | R | R | G | Best for comb sampling in field work |
| Three measurements of 10 cells in a star pattern | Nr | S | S | S | G | G | Good for comb sampling in field work |
| Measurements in horizontal and vertical directions | Nr | S | S | Nr | G | G | For estimates of cell distortion |
| Ruler | S | S | S | S | S | S | Lowest precision |
| Caliper | S | G | G | G | G | G | Best precision in field |
| Picture | B | B | B | B | B | B | Best reproducibility |

B: best; G: good; Na: not applicable; Nr: not recommended; R: recommended; S: satisfactory.

widely accepted since Maraldi (1712), has been questioned by Pirk et al. (2004), but later confirmed by Hepburn et al. on the basis of polyester resin mouldings. It can therefore be accepted as an established fact.

*4.1.3.1. Measuring the depth and estimating the internal volume of the cell.* Because of its particular geometry, measuring the depth of the honey bee cell is not an easy task. Three different measurements can be taken, depending on the purposes of the study:

a. along JR (the longest possible measurement: $h_1$) (Figure 19),
b. along CO, AN or EP (the smallest side lengths: $h_2$) and
c. along BH, DK or FM (the longest side lengths: $h_3$).

From the geometry, $h_3 = (h_1 + h_2)/2$. The volume of the cell is therefore given by:

$$V_{cell} = A_{cell} * h_3 = h_3 * \frac{\sqrt{3}}{2} d_i^2$$

Although cell depth is commonly measured as $h_1$, one should measure $h_3$ to compute estimates of the cell volume. In addition, one should also take into account the effect of the 10–14 cell's angle with the horizontal line perpendicular to the comb's vertical direction, which makes measurements quite complicated.

In practice, cell depth is measured as $h_1$, from the top to the bottom of the cell using the calliper's depth probe.

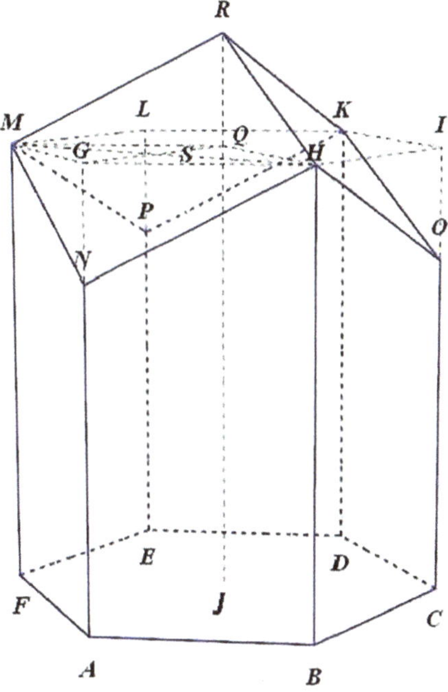


Figure 19. A three-dimensional view of the honeycomb cell (from Huber, 1814).

Alternatively, one could fill a row of 10 cells and measure their capacity using a pipette or weighing their water content. A proper method with estimations of sources of error is still to be developed.

*4.1.3.2. Investigating the bottom of the cells.* As already mentioned (Section 4.1.3; Figure 20), the bottoms of the cells are made of three rhombs connected to each

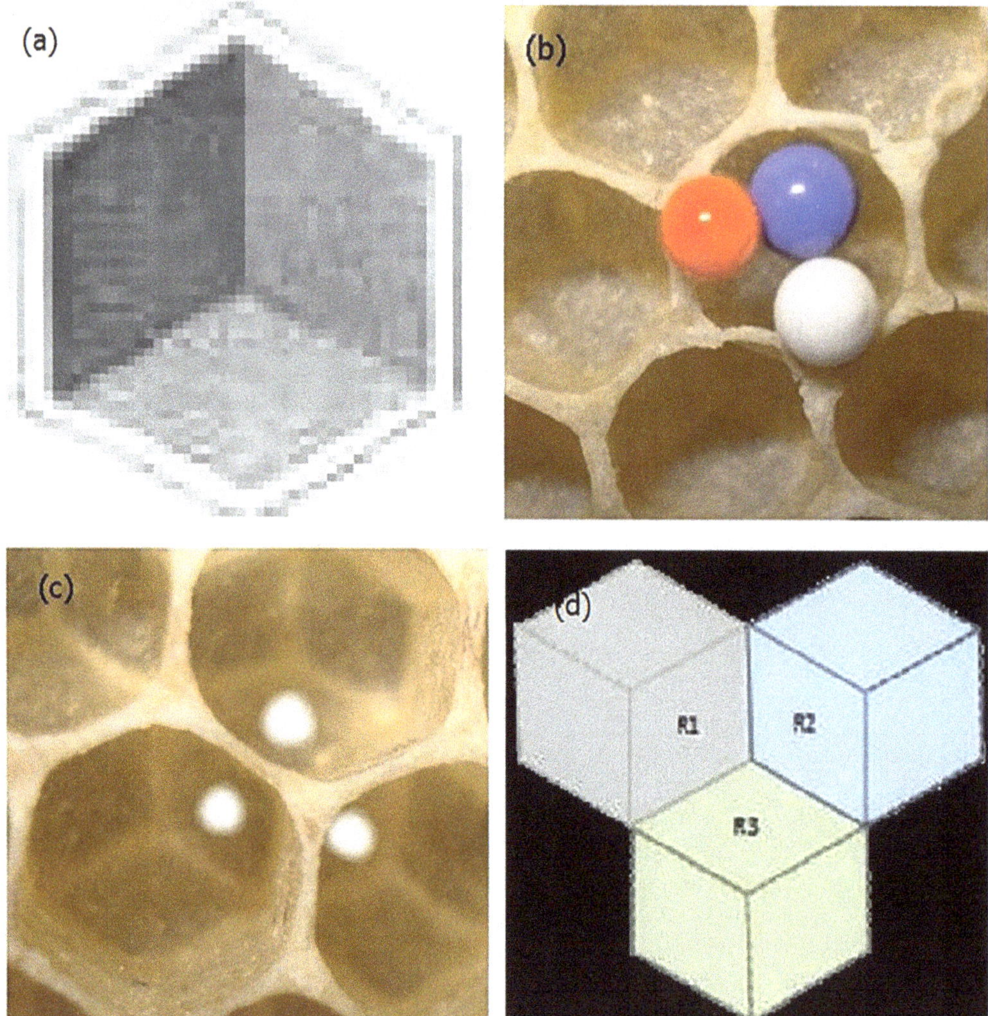


Figure 20. (a) Bottom of the cell showing the three rhombs and the reversed Y (λ) pattern resulting from their connections; (b) three pins in each rhomb of one cell; (c) produce 1 hole in 3 different cells on the other side of the comb; (d) three-dimensional model showing the reversed Y (λ) at the junction of R1, R2, and R3 and the upside Y in each of the three cells on the other side of the cell (from Saucy, 2014).

other on one edge and facing each other at angles of approximately 120°. This structure can be studied using mouldings with plaster or artificial resins (Hepburn, Muerrle, & Radloff, 2007; Vogt, 1911). Measurements can then be conducted on the dry moldings according to Vogt (1911).

*4.1.3.3. Estimating the width and the external volume of the comb.* In practice, since estimations refer to both sides of the comb, the difficulty of measuring the cell's depth is overcome using a measure of the average width of the comb based on several measurements (e.g., 1/4, 1/2, 3/4 of the length of the comb from the top to the bottom) using a calliper. The volume of the comb is then obtained as the area of the comb multiplied by its average width.

*4.1.3.4. Estimating the capacity of the comb.* The effective capacity of the comb, i.e., its internal volume excluding the volume of the wax, is of course smaller than its external volume. Moreover, it is reduced in old brood combs in which cocoons accumulate. Content can be estimated by filling them with water:

1. Clean the comb in a water bath (4 h; pollen might be difficult to remove).
2. Remove water from the comb.
3. Dry the comb overnight in a ventilated oven at 25–30 °C.
4. Weigh the comb.
5. Fill both sides with water using a shower.
6. Weigh the filled comb.
7. Subtract the weight of the dry comb from the weight of the filled comb to obtain the weight of the water.
8. Transform weight of the water into volume (1 kg = 1 L).

*4.1.3.5. Documenting transitions between different cell types or junctions of combs.* Transitions from one type of cell to another (e.g., from worker to drone cells) appears not only from the two dimensional perspective, but also affect the three-dimensional aspects of the comb (Hepburn et al., 2014). Such transitions have already long been described (Huber, 1814) and documented as natural processes maintaining the function and integrity of the comb. They are difficult to quantify and should be described qualitatively. Some examples are depicted in Figure 21. Figure 21(a) shows the transition from worker to drone cells. Transitions as changes in cell size, bottom cell structure, as well as numbers of rows affected should be documented. For instance, Figure 21(b) shows a complete transition from the *Y* to the $\lambda$ pattern across a range of 11 adjacent cells, with intermediate cells displaying four bottom plates facing 4 cells (instead of three) on the opposite side of the comb, while cell size remains unaffected.

### 4.2. Perspectives for methods in comb construction research

The architecture of the comb has fascinated the best human minds for centuries (reviews in Hales, 2001; Hepburn et al., 2014; Huber, 1814; Réaumur, 1742; Saucy, 2014; Thompson, 1945). According to some authors, bees are seen as architects able to master the geometry of the hexagonal cell and able to measure distances and angles (e.g., Bauer & Bienefeld, 2013; Huber, 1814; Nazzi, 2016; Réaumur, 1742), while others suggested that the hexagonal structure results from self-organizing processes depending on physical forces applied on originally cylindrical cells built in a malleable medium (e.g., Buffon, 1753; Darwin, 1859; Hepburn et al., 2007; Karihaloo, Zhang, & Wang, 2013; Pirk et al., 2004; Figure 14).

Understanding the structure of the comb and its building rules has practical consequences, e.g., for the production of comb foundation, as well as in health issues of the honey bee. For instance, the significance of the bees forming living chains while building comb (known as festoons) remains poorly understood. Despite interesting and promising studies, no standard methods have emerged yet.

#### *4.2.1. Investigating comb construction and its growth pattern*

To be able to observe every single bee active in the building process, Huber (1814) induced the bees to construct their combs from the bottom to the top in an upward direction. Although such combs seem undistinguishable from those built downwards, Huber's approach cannot be recommended as a standard method for other purposes. Usually researchers install and observe bees building downwards from the top. They smoke them out at times to observe and periodically record the progress of their work. Investigations the comb's growth pattern have been reviewed in great detail by Hepburn et al. (2014). The many experimental approaches reported suggest that the methodology should be developed according to the questions under scrutiny. For instance, honey bee combs may exhibit various cell shapes (e.g., pentagons, squares, heptagons) aside from the typical hexagonal cell, various sizes as well as amazing plasticity in their organization (horizontal, oblique, vertical rows, rosettes, etc.). Investigating these patterns involves experimentally manipulate cell sizes or patterns on embossed wax foundation, submit bees to restrictive conditions to assess their plasticity limits (Hepburn et al., 2014).

#### *4.2.2. Investigating timing and type of comb construction*

Under natural conditions, comb construction depends strongly on the time of the year and on the availability of nectar, with wax production reaching its climax when vegetation blooms and nectar becomes superabundant (Bogdanov, 2016a, 2016b). Many other factors may affect or stimulate wax production, such as outside temperature, the presence/absence of an egg-laying queen, pollen availability, ratio of brood combs, ratio of unoccupied combs, population of workers or drones, as well as available space in the nest cavity (Hepburn et al., 2014). These factors are often intermingled and correlated (Pratt, 2004). For instance, in temperate climates of both southern and northern hemispheres, the vegetation bloom is correlated with increasing sunlight and average temperature, as well as abundance of pollen and nectar. To address the importance and the relative contribution of these factors, we recommend apply classical methodological approaches involving a strong experimental design linked to appropriate statistical analyses according to Pirk et al. (2013). Construction may be measured counting the number of built cells of each type (worker, drones, honey), measuring comb area and weighing comb produced at the end of the experiment. Timing can be investigated experimentally manipulating external factors (such as day duration, food abundance) under laboratory conditions or replicating the experiments at various times of the year, or at different locations. The following approach should be applied:

1. Clearly define the goals of the study and identify the factors to test.
2. Identify major possible confounding factors to control or to monitor.
3. Build a solid experimental design taking account of all these factors.
4. Define the statistical methods to use before starting the experiments.
5. Conduct the experiments, collect the data and perform the statistical analyses.

## 5. Standard methods for beeswax chemical characterization

Numerous studies employing different gas chromatography (GC) and/or gas chromatography–mass

(a)

(b)

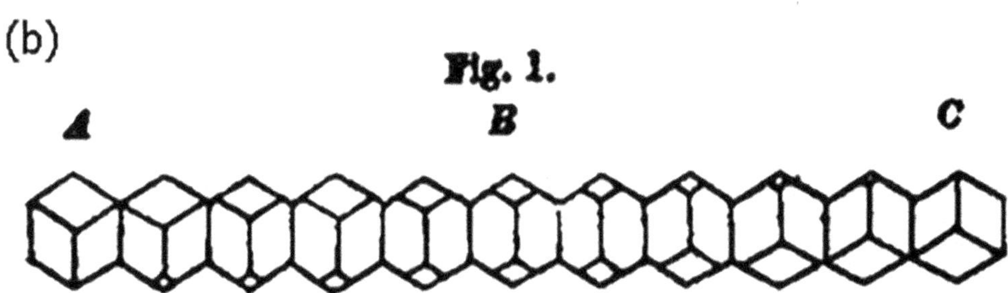

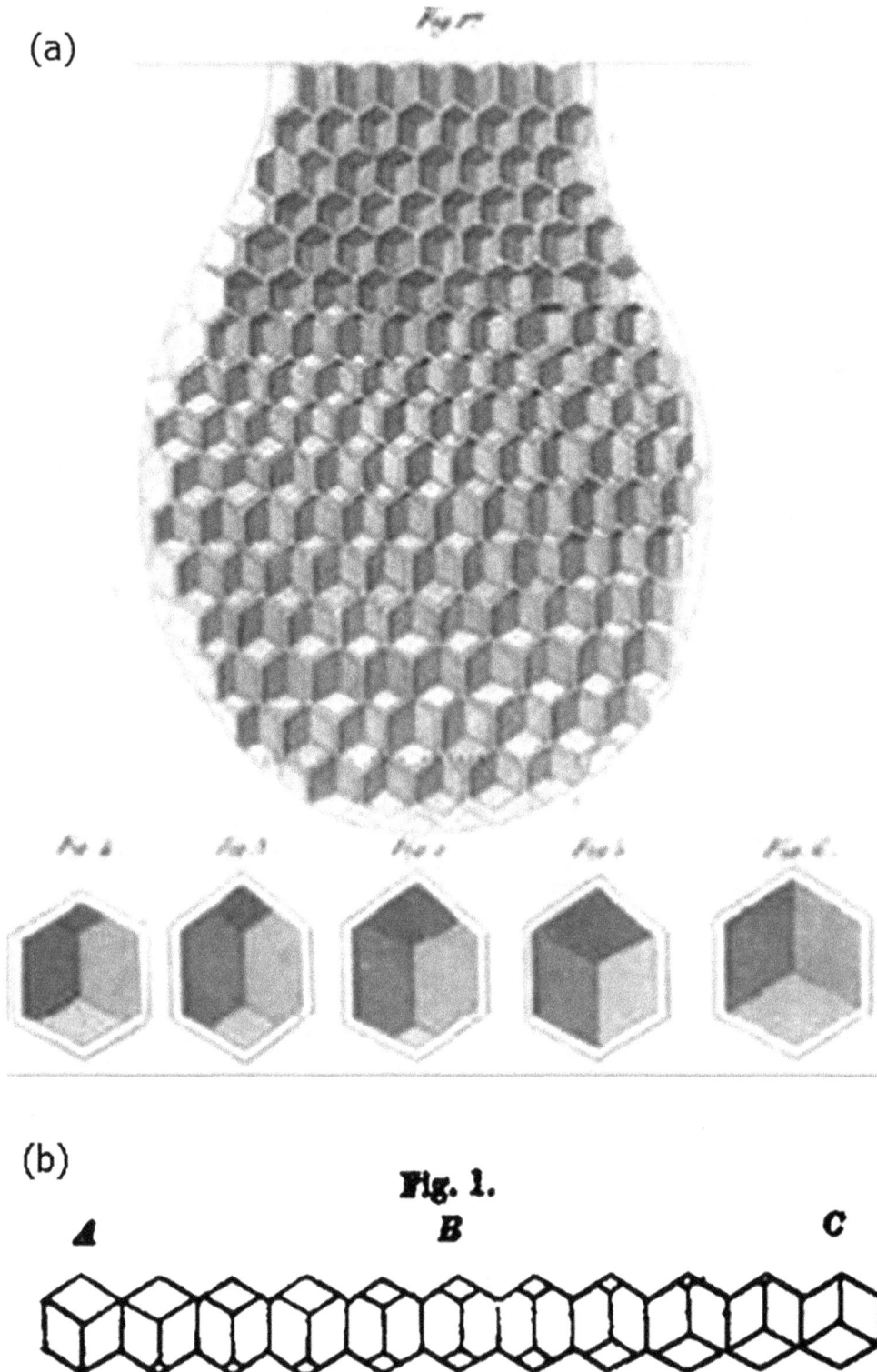


Figure 21. (a) Transition from worker to drone cells (Huber, 1814); (b) transition from an upside *Y* to a reversed *Y* ($\lambda$) cell pattern over 11 cell rows (Wyman, 1866).

spectrometry (GC-MS)-coupled analytical methods allowed detailed insight into beeswax chemical composition (Aichholz & Lorbeer, 1999, 2000; Jiménez, Bernal, Aumente, Toribio, & Bernal, 2003; Jiménez, Bernal, Aumente et al., 2004; Jiménez, Bernal, del Nozal, Martín, & Bernal, 2006; Maia & Nunes, 2013; Serra Bonvehí & Ornantes Bermejo, 2012; Tulloch, 1980; Waś, Szczęsna, & Rybak-Chmielewska, 2014a, 2014b).

Besides GC-coupled techniques, other analytical tools have also been employed for investigating beeswax, but focusing on more specific aspects of beeswax research. Hence, liquid chromatography (LC) techniques were mostly used for determination of pesticide residues in beeswax (Bernal, del Nozal, Toribio, Jimènez, & Atienza, 1997; Jiménez, Bernal, del Nozal, & Alonso, 2004; Pirard et al., 2007; Yáñez, Martín, Bernal, Nozal, & Bernal, 2014), while Fourier transform infrared spectroscopy (FTIR) was employed for the chemical fingerprinting of beeswax, as well as successful detection of adulterants in beeswax (Maia et al., 2013; Svečnjak, Baranović et al., 2015). Furthermore, isotope ratio mass spectrometry (IRMS) was found to be very useful in establishing the geographical origin of beeswax (Chesson, Tipple, Erkkila, Cerling, & Ehleringer, 2011; Tipple, Chesson, Erkkila, Cerling, & Ehleringer, 2012). All of the above methods are described in the following sub-sections of Sections 5 and 6. Due to their complexity and very recent application for beeswax analysis, FTIR-ATR and IRMS methods are described in more detail in Sections 5.3 and 5.5, respectively.

### 5.1. Chemistry of beeswax

Understanding the chemical composition of beeswax is of crucial importance when applying chromatographic and spectroscopic analytical methods, which are commonly used for the chemical investigation of beeswax. Here, we present an overview of the major constituents of beeswax to enable easier interpretation of results obtained by the methods described in Sections 5 and 6.

#### 5.1.1. Beeswax constituents

Beeswax derived from *A. mellifera* comb wax is a lipid-based complex mixture of more than 300 constituent fractions; it consists mainly of esters (67%), hydrocarbons (14%), free acids (12%), free alcohols (1%), and other (unidentified) constituents (6%), as presented in Table 4 (Tulloch, 1980). Specific chemical composition of beeswax holds a significant importance for a honey bee colony given that comb wax mediates the acquisition of nestmate recognition cues in honey bees (Breed, Williams et al., 1988, et seq.), and due to its capturing of toxins. The major compound families in beeswax are defined as those exceeding 5% (w/w) of the total beeswax composition (alkanes, alkenes, free fatty acids, monoesters, diesters and hydroxymonoesters); those of lesser abundance are regarded as minor constituents representing <5% of the total composition (fatty alcohols) (Aichholz & Lorbeer, 1999, 2000; Hepburn et al., 2014). However, according to Tulloch (1980), only several individual constituent fractions constitute more than 5% of the total beeswax composition, namely C40 monoester (6%), C46 monoester (8%), C48 monoester (6%) and C24 (lignoceric/tetracosanoic) acid (6%). Aichholz and Lorbeer (1999, 2000) also observed that each compound family in beeswax represents a series of homologues differing in chain length by two carbon atoms. The list of major and minor compound families, and their chain lengths (number of carbon atoms), are summarized in Table 5.

*Hydrocarbons*. The total content of hydrocarbons determined in beeswax ranges between 12.3 and 17.8%, as summarized by Waś et al. (2014b). Beeswax is mostly comprised of straight chain (unbranched) saturated hydrocarbons (12.8%), with a predominant chain length of C27–C31. More specifically, the most numerous group of beeswax hydrocarbons are linear saturated hydrocarbons (n-alkanes), accounting for ca. 67% of all hydrocarbons occurring in beeswax, while branched alkanes occur in much smaller amounts (0.2%) (Streibl et al., 1966). Moreover, the percentage share of the odd-numbered n-alkanes is significantly higher in comparison to even-numbered alkanes and amounts to about 95% (Waś et al., 2014a, 2014b).

Numerous studies revealed that the number of carbon atoms in unbranched alkanes in beeswax is generally ranging from C17 to C35 (Aichholz & Lorbeer, 1999, 2000; Downing, Kranz, Lamberton, Murray, & Redcliffe, 1961; Jiménez, Bernal, Aumente et al., 2004; Maia & Nunes, 2013; Serra Bonvehí and Ornantes Bermejo (2012; Tulloch, 1980; Waś et al., 2014a, 2014b; White, Reader, & Riethof, 1960).

The alkane C27 (heptacosane) is the most abundant hydrocarbon in beeswax, and comprises 6.2% of the total composition of *A. mellifera* wax (Aichholz & Lorbeer, 1999). The presence of even higher amounts of heptacosane was reported in recent study by Maia and Nunes (2013); the average percentage of C27 alkane determined in virgin beeswax was 13%, ranging from 9.19 to 16.8%, followed by nonacosane (C29) and hentriacontane (C31). The most recent study investigating hydrocarbon composition of *A. mellifera* beeswax collected from light and dark coloured combs (Waś et al., 2014b), confirms previously reported results; the highest contents were determined for C27, C29, C31, and C25, with an average of 34.0, 23.5, 20.3, and 8.6%, respectively, of the total content of unbranched alkanes in the light virgin beeswax (dark comb wax showed slightly different values, 34.5, 22.5, 18.4, and 10.1%, respectively). The proportion of less abundant (2.9%)

Table 4. Chemical composition of European *A. mellifera* comb wax (Tulloch, 1980).

| Constituent fraction | % (w/w) |
|---|---|
| Hydrocarbons | 14 |
| Monoesters | 35 |
| Diesters | 14 |
| Triesters | 3 |
| Hydroxy monoesters | 4 |
| Hydroxy polyesters | 8 |
| Acid esters | 1 |
| Acid polyesters | 2 |
| Free acids | 12 |
| Free alcohols | 1 |
| Unidentified | 6 |
| Total | 100 |

Table 5. Compound families identified in *A. mellifera* comb wax (Aichholz & Lorbeer, 1999, 2000).

| Compound family | % (w/w) | Chain length (number of C atoms) |
|---|---|---|
| Hydrocarbons | 15.7 | C23–C35 |
| Saturated (alkanes) | (12.8) | C23–C35 |
| Unsaturated (alkenes) | (2.9) | C29–C33 |
| Fatty acids | 18 | C20–C36 |
| Lignoceric (tetracosanoic) acid | (5.8) | C24:0 (saturated) |
| | | C14–C36[a] |
| Fatty alcohols | 0.6 | C33, C35 |
| | | C24–C34[a] |
| Monoesters (palmitates, oleates) | 40.8 | C38–C54 |
| | (6.6) | C40 |
| | (11.9) | C46 |
| | (9) | C48 |
| Hydroxymonoesters | 9.2 | C40–C50 |
| Diesters | 7.4 | C54–C64 |
| Esters total | 57.4 | C38–C64 |
| Total | 91.7 | C14–C64 |

[a]Reported by other authors: Serra Bonvehí and Orantes Bermejo (2012) and Jiménez et al. (2003, Jiménez, Bernal, Aumente et al., 2004, Jiménez et al. 2006)

unsaturated hydrocarbons (alkenes) increases with the chain length (>C29), while dienes (C31–C35) were reported at only low, almost insignificant levels (Giumanini, Verardo, Strazzolini, & Hepburn, 1995; Jiménez, Bernal, Aumente et al., 2004, Jiménez et al., 2006; Jiménez, Bernal, del Nozal, Toribio, & Bernal, 2007; Maia & Nunes, 2013; Waś et al., 2014a, 2014b).

None of the major structural hydrocarbons of honey bees (i.e., n-alkanes) produces a positive result in a recognition bioassay, nor do these compounds differ significantly in relative concentration among families of bees. However, hydrocarbons that are present in smaller quantities, alkanes hexadecane and octadecane, and alkenes heneicosene and tricosenee, yielded positive results as nestmate recognition signals (Breed, 1998; Breed, Garry et al., 1995).

*Fatty acids. A. mellifera* beeswax contains remarkably high levels of free fatty acids (18%) in comparison to beeswaxes of other species studied, *A. cerana* (3.6%), *A. florea* (0.8%), *A. andreniformis* (2.6%), *A. dorsata* (4.9%), *A. laboriosa* (4.3%), respectively (Aichholz & Lorbeer, 1999). It mostly contains saturated, unbranched long-chain fatty acids (ca. 85%) with an even number of carbon atoms (C14–C36). The most abundant free fatty acid of *A. mellifera* beeswax is tetracosanoic (lignoceric) acid (Aichholz & Lorbeer, 1999, 2000; Jiménez et al., 2003; Serra Bonvehí and Ornantes Bermejo (2012; Tulloch, 1980), comprising more than 5% of the total beeswax composition (Aichholz & Lorbeer, 1999; Tulloch, 1980). Serra Bonvehí and Ornantes Bermejo (2012) gave a detailed overview of beeswax fatty acids profile and reported C16 (palmitic) as the main acid among saturated group, with a concentration of about 14.8%, followed by tetracosanoic (3.24%) and octacosanoic (C28) acid (1.80%). In relation to the unsaturated acids, C18:1 (oleic) represents major compound (3.90%), followed by C18:2 (linoleic). Lower quantities of other unsaturated fatty acids (primarily C16:1–C36:1), such as hexadecenoic C16:1 acid (palmitoleic), docosenoic C22:1 and eicosenoic C20:1, were also reported (Jiménez et al., 2003).

Fatty acids play an important role in honey bee colony as a source of nestmate recognition cues, thus allowing a discrimination of nestmate vs. non-nestmate individuals (along with other, queen derived and environmentally acquired "colony odours") (Breed, Williams et al., 1988, et seq.; Hepburn et al., 2014). The pattern of nestmate recognition within a honey bee colony has been widely studied by Breed, Williams et al. (1988), Breed, Stiller, and Moor (1988), Breed and Stiller (1992), Breed, Garry et al. (1995), Breed, Page, Hibbard, and Bjostad (1995), Breed, Leger, Pearce, and Wang (1998), and Breed, Diaz, and Lucero (2004) in a series of bioassays. The results of these studies revealed that fatty acids, rather than hydrocarbons, represent a key element in honey bee nestmate recognition which was also confirmed by other authors (Brockmann, Groh, & Fröhlich, 2003; D'ettorre et al., 2006; Fröhlich, Riederer et al., 2000). Accordingly, fatty acid recognition cues include four unsaturated acids (palmitoleic, oleic, linoleic, and linolenic) and two saturated fatty acids (palmitic and lignoceric). These acids are present in beeswax in substantial quantities and provide a chemical signature on the surface of bees that identifies to which colony they belong. Although very abundant in the free fatty acid fraction of beeswax, stearic acid proved to be inactive as a recognition cue (Breed, 1998). The role of fatty acids in the mechanical properties of beeswax (particularly resilience and stiffness) confirmed by Buchwald, Greenberg, and Breed (2005), Buchwald, Breed, Greenberg, and Otis (2006), and Buchwald, Breed, Bjostad, Hibbard, and Greenberg (2009), should also be mentioned here.

*Fatty alcohols.* Free fatty alcohols (~1% of the total beeswax composition) with a chain length of C28–C35 are minor compounds. Saturated fatty alcohols C30 and

C32 have been reported in higher concentrations by Serra Bonvehí and Ornantes Bermejo (2012), while Aichholz and Lorbeer (1999, 2000) identified C33 and C35 fatty alcohols in *A. mellifera* alcohol fraction.

*Esters*. Monoesters are the most abundant fraction of beeswax. The total content of the monoesters comprises more than 35% of *A. mellifera* beeswax. Beeswax monoesters are linear wax esters almost entirely derived from palmitic and oleic acid (palmitates, oleates), and comprise two main types of structures: saturated esters, which are predominantly alkylpalmitates (C38–C52), and unsaturated esters, mostly alkyloleates (C46–C54) (Aichholz & Lorbeer, 1999, 2000; Tulloch, 1980). These findings are consistent with results reported in other studies (Jiménez et al., 2003; Jiménez, Bernal, Aumente et al., 2004; Maia & Nunes, 2013; Serra Bonvehí and Ornantes Bermejo (2012). The content of hydroxymonoesters and diesters is significantly smaller in comparison to monoesters, 9.2 and 7.4%, respectively (Aichholz & Lorbeer, 1999). Two types of hydroxyesters structure dominate in beeswax: long chain alcohols, esterified by a hydroxy acid (mainly 15-hydroxypalmitic acid), and diols esterified with palmitic acid (Aichholz & Lorbeer, 1999, 2000; Maia & Nunes, 2013; Tulloch, 1971). The presence of smaller quantities of diesters, triesters and higher esters was determined by the same authors.

### 5.2. Investigation of beeswax composition by gas chromatography-mass spectrometry (GC-MS) and other GC-coupled techniques

***Definitions, acronyms:***

- GC: gas chromatography
- MS: mass spectrometry
- GC-MS: gas chromatography–mass spectrometry (gas chromatography with mass detector technique)
- IS: internal standard
- SPE: solid phase extraction
- FID: flame ionization detector
- GC-FID: gas chromatography coupled with flame ionization detector
- EI: electron impact ionization

Gas chromatography with different detectors (FID, MS) is commonly used in studies of beeswax composition (Aichholz & Lorbeer, 1999, 2000; Giumanini et al., 1995; Jiménez et al., 2003, 2004, 2006, 2007; Maia & Nunes, 2013; Namdar, Neuman, Sladezki, Haddad, & Weiner, 2007; Serra Bonvehí and Ornantes Bermejo (2012; Waś et al. 2014a, 2014b). The majority of compounds belonging to certain homologous series have been characterized by different researchers (Table 6). The GC-MS is one of the latest techniques used in testing the chemical composition of beeswax. In many cases GC-MS and GC-FID complement each other. The GC-MS allows identification of unknown compounds based on the mass spectra without using standards, as described by Torto et al. (2013) in the *BEEBOOK* chapter on chemical ecology. It is mainly used for qualitative analysis of beeswax composition (Jiménez et al., 2003, Jiménez, Bernal, Aumente et al., 2004; Maia & Nunes, 2013; Serra Bonvehí and Ornantes Bermejo (2012), while GC-FID is preferable for quantification (Jiménez, Bernal, Aumente et al., 2004, Jiménez et al., 2007; Maia & Nunes, 2013; Serra Bonvehí and Ornantes Bermejo (2012).

The GC-MS method elaborated by Waś et al. (2014a) allows in a single run for identification of beeswax hydrocarbons (alkanes, alkenes and dienes) as well as for quantification of n-alkanes, the biggest group of hydrocarbons occurring in beeswax.

This method is described in Section 5.2.1 and is recommended as standard method for analysis of beeswax hydrocarbons. However, for determination of other beeswax constituents such as esters, acids and alcohols, other methods elaborated by different authors should be taken into consideration (Jiménez et al., 2003; Jiménez, Bernal, Aumente et al., 2004; Jiménez et al., 2006; Maia & Nunes, 2013; Serra Bonvehí and Ornantes Bermejo (2012).

In the following subsections we also present the methods for simultaneous determination of hydrocarbons and monoesters (see Section 5.2.2), as well as fatty acids and fatty alcohols (see Section 5.2.3) by GC-MS and GC-FID according to Maia and Nunes (2013).

#### *5.2.1. Determination of beeswax hydrocarbons by GC-MS*

The most important advantage of the proposed GC-MS method according to Waś et al. (2014a) is that it allows the identification of beeswax hydrocarbons (alkanes, alkenes and dienes) and quantification of n-alkanes in a single run. Quantitative analysis of n-alkanes is conducted by the method of internal standard with squalane used as the internal standard. For extraction of hydrocarbons from beeswax the SPE technique with columns filled with neutral aluminum oxide is used (Waś et al., 2014a). There are many advantages of using SPE, which include mainly decreasing the time of chromatographic analysis and substantial simplification of the qualitative and quantitative analysis by improving detection and increasing selectivity of the determined compounds. Applying the SPE have the most significant effect for determining n-alkanes with even numbers of carbon atoms in a molecule, which are difficult to detect because they occur in beeswax in a very small amounts. In the case of determination of hydrocarbons without the preliminary purification of a sample, the n-alkanes peaks, especially even-numbered n-alkanes, overlap with the peaks of esters. This overlap can lead to inaccurate results of the quantitative analysis. Application of GC-MS method in analysis of beeswax hydrocarbons according to Waś et al. (2014a) allowed

Table 6. Investigation of beeswax composition with gas chromatography methods.

| Identified groups of compounds | Method | Quantitative analysis | References |
|---|---|---|---|
| Hydrocarbons (dienes) | GC-MS (EI) | Relative abundances of parent ion | Giumanini et al. (1995) |
| Hydrocarbons (alkanes, alkenes, dienes), free fatty acids, monoesters, hydroxy monoesters, diesters, hydoxy diesters, fatty alcohols | GC-FID, GC-MS (CI) | Estimation of share (%) of individual compounds based on the ratio of the certain peak area to the sum of the areas of peaks for all of analyzed compounds ( method of internal normalization) | Aichholz and Lorbeer (1999, 2000) |
| Hydrocarbons (alkanes, alkenes, dienes), monoesters, alcohols (monoalcohols, diols, propanetriols), fatty acids (saturated, unsaturated), hydroxyacids | GC-MS (EI) | Relative amounts (%) calculated from the peak areas and in relation to the peak of the most abundant compound | Jiménez et al. (2003, Jiménez, Bernal, Aumente et al., 2004) |
| Hydrocarbons (alkanes, alkenes, dienes), esters, acids, hydroxyacids, alcohols, monoesterified propanetriols | GC-FID | Expression in weight percentage referred to one standard | Jiménez et al. (2006, 2007) |
| Hydrocarbons (n-alkanes) | GC-FID, GC-MS (EI) | Relative peak areas normalized to the most abundant alkane | Namdar et al. (2007) |
| Hydrocarbons (alkanes, alkenes), monoesters, acids, alcohols | HTGC-FID/MS (EI) | Comparisons peaks areas with those of a reference external standard (eicosane) with the assumption the response of all compounds is equal | Serra Bonvehí and Orantes Bermejo (2012) |
| Hydrocarbons (alkanes, alkenes, dienes), monoesters | GC-FID, GC-MS (EI) | Percentage area method | Maia and Nunes (2013) |
| Hydrocarbons (alkanes, alkenes, dienes) | GC-MS (EI) | Internal standard method with using standard mixture of n-alkanes and squalane used a IS | Waś et al. (2014a, 2014b) |

EI, electron ionization; CI, chemical ionization.

identifying these compounds and determining the ranges of their amounts in natural beeswax (Waś et al., 2014b). These values can be used as concentration guide-value to distinguish between pure and adulterated beeswax. The GC-MS method presented here is characterized by good linearity as well as satisfactory repeatability and within-laboratory reproducibility (Waś et al., 2014a). The disadvantage of proposed method may be that it only allows the analysis of hydrocarbons, but from the point of view of detection of beeswax adulteration with hydrocarbons of foreign origin, e.g., paraffin or ceresin, substances the most frequently used for this purpose, the method is very important and helpful (Waś, Szczęsna, & Rybak-Chmielewska, 2015; Waś et al., 2016).

The procedure for determining hydrocarbons in beeswax according to Waś et al. (2014a) is described in the following subections. Typical GC-MS chromatograms of beeswax hydrocarbons are presented in Figure 22(a–c).

*Equipment required:*

- GC-MS instrument such as Shimadzu GCMS-QP2010 Plus system, optionally equipped with autosampler
- Non-polar column ZB-5HT INFERNO (20 m × 0.18 mm × 0.18 μm, Phenomenex) or other with similar characteristics
- Incubtaor Shaker such as Innowa 40 (New Brunsfic Scientific)
- SPE Column Processor (J.T. Baker)
- Vortex
- Analytical balance (with accuracy ±0.001 g)
- Erlenmayers flasks of 100 mL with PTFE stopper
- Volumetric flasks of 10 and 100 mL
- Vials of 3 mL
- Autosampler vials of 2 mL
- Automatic pipettes (20–200 μL, 100–1000 μL, 1000–5000 μL).

***Materials and reagents required:***

All analytical standards and reagents used must be for gas chromatography (≥98% purity). The all reagents listed below comparable purity can be used from other companies.

- Standard mixtures of n-alkanes $C_8H_{18}$–$C_{20}H_{42}$ in hexane and $C_{21}H_{44}$–$C_{40}H_{82}$ in toluene at 40 mg/l

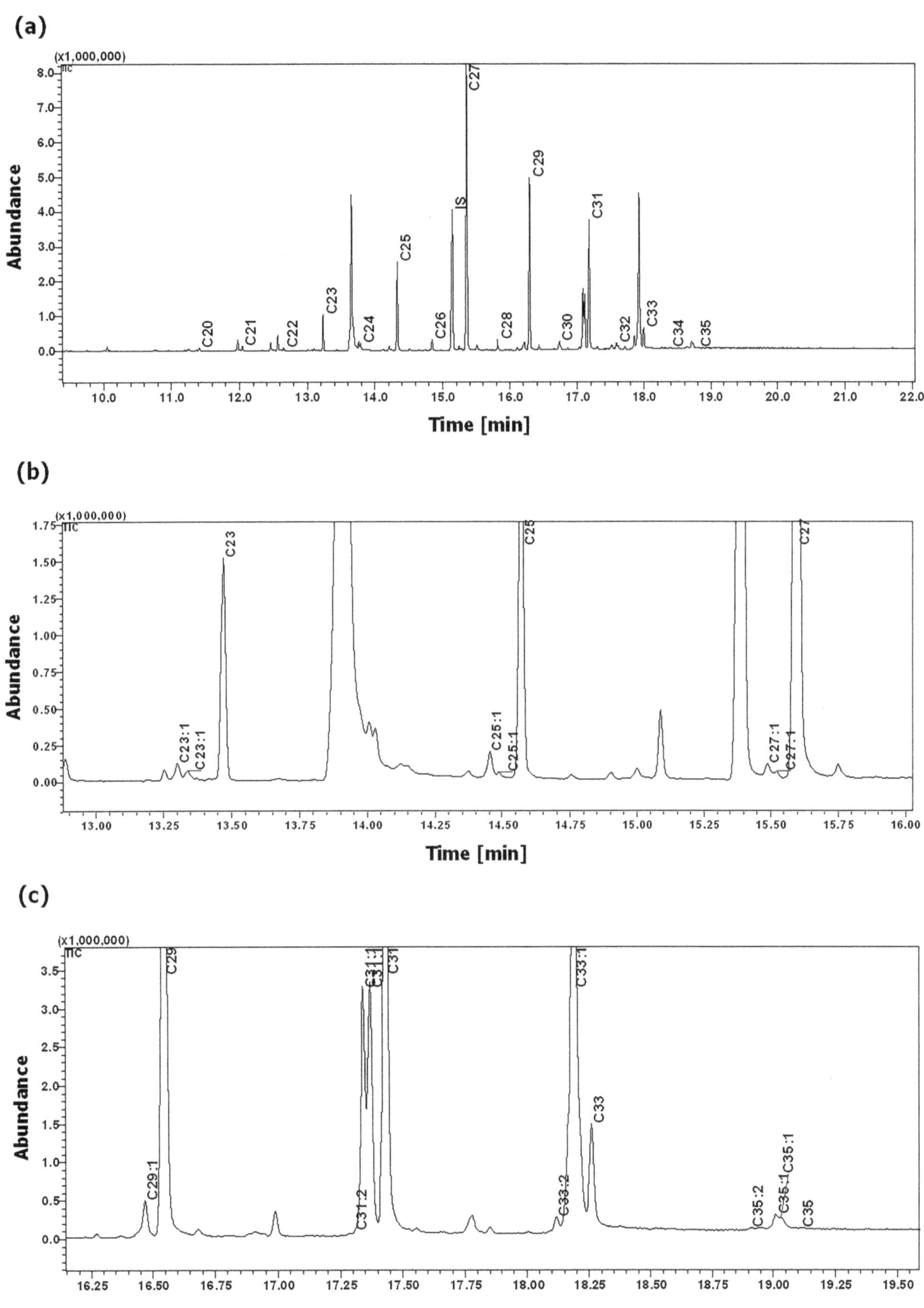


Figure 22. GC-MS chromatogram of hydrocarbons in beeswax: (a) saturated hydrocarbons, C20–C35 – n-alkanes with the formula $C_{20}H_{42}$–$C_{40}H_{82}$; IS – internal standard ($C_{30}H_{62}$); (b, c) unsaturated hydrocarbons (alkenes and dienes): 1,2 – number of double bonds in the molecule.

concentration of each (Fluka, Buchs, Switzerland; Saint Louis, MO, USA)

- Squalane, 99.9% purity (Supelco, Bellefonte, PA, USA)
- Hexane SupraSolv®, ≥98% purity (Merck, Darmstadt, Germany)
- Heptane anhydrous, ≥98.5% purity (Sigma-Aldrich, Steinheim, Germany)
- SPE cartridges (Alumina–N, 1000 mg, 6 mL (Agela Technologies, Wilmington, DE, USA).
- Helium, 99.9999% purity (Air Products, Warsaw, Poland).

*5.2.1.1. Preparation of standard solution.*

1. Prepare the stock solution of squalane ($c = 400$ mg/L).
   a. Weight the 40 mg of squalane ($C_{30}H_{62}$) with an accuracy of 0.2 mg.
   b. Dissolve it in a small amount (about 10 mL) of heptane.
   c. Transfer to a volumetric flask of 100 mL.
   d. Fill the flask to a 100 mL volume with heptane.
2. Prepare the solution of squalane ($c = 20$ mg/L).
   a. Pipette of 1 mL of squalane stock solution ($c = 400$ mg/L) into a volumetric flask of 10 mL.
   b. Fill to a 10 mL volume with heptane.
   c. Pipette of 1 mL solution into a vial of 3 mL.
   d. Add 1 mL heptane.
   e. Mix the solution on vortex.
3. Prepare the standard solution of 10 mg/L of n-alkanes ($C_8H_{18}$–$C_{40}H_{82}$) with IS ($C_{30}H_{62}$).
   a. Use a commercially available reference mixtures of n-alkanes ($C_8H_{18}$–$C_{20}H_{42}$) and ($C_{21}H_{44}$–$C_{40}H_{82}$) at 40 mg/L concentration of each.
   b. Pipette of 200 µL of each above standard mixtures of n-alkanes to the autosampler vial.
   c. Add of 400 µL of IS solution at 20 mg/L concentration.
   d. Mix the standard solution on vortex.

The standard solution of n-alkanes ($C_8H_{18}$–$C_{40}H_{82}$) with IS ($C_{30}H_{62}$) at 10 mg/L concentration is used for further GC-MS analysis.

*5.2.1.2. Beeswax sample preparation.* The preparation procedure for the samples of beeswax combs should begin with melting down the pieces of beeswax combs at a temperature of 70–75 °C and purifying on a filter made of gauze. Homogenized and free from mechanical impurities, beeswax should be kept in a dry and dark place (as described in Section 2.3) until further analysis. For the other beeswax samples (wax blocks or comb foundations), the sample preparation can be started from the preparation of beeswax solution for solid phase extraction (SPE) which is described below.

1. Prepare the beeswax solution ($c = 5$ mg/mL) for SPE.
   a. Weigh the 0.05 ± 0.001 g of beeswax.
   b. Add 7.5 mL of heptane.
   c. Shake the solution at 50 °C for about 12 min until the beeswax dissolved.
   d. Cool the solution to room temperature.
   e. Add 2.5 mL of squalane (IS) at 400 mg/L concentration.
   f. Mix a beeswax solution with IS on vortex.
2. Perform the SPE extraction of hydrocarbons from beeswax with neutral aluminum oxide (Alumina-N, 1000 mg, 6 mL).
   a. Wash the sorbent with 2 mL of hexane.
   b. Inject of 1 mL of beeswax solution in heptane ($c = 5$ mg/mL).
   c. Eluate of hydrocarbons from the intergrain spaces with 3 mL of hexane (in 3 portions of 1 mL).
   d. Collect the hydrocarbon fraction into a 10 mL flask.
   e. Fill the flask to a 10 mL volume with heptane.
   f. Transfer the solution to the autosampler vial.
   g. Use the solution to carry out chromatography analysis (see Section 5.2.1.3).

The SPE procedure used for isolation of hydrocarbons from beeswax is also schematically illustrated by Waś et al. (2014a).

*5.2.1.3. Conditions of GC-MS analysis.* The hydrocarbon composition of beeswax is analyzed by GC-MS after extraction using SPE described in Section 5.2.1.2. The GC-MS analysis should be conducted with appropriate instrument such as Shimadzu GCMS-QP2010 Plus system equipped with turbo molecular pomp, the GCMSsolution software, a non-polar column ZB-5HT INFERNO (20 m × 0.18 mm × 0.18 µm, Phenomenex) or other with similar characteristics, and optionally equipped with autosampler. The following steps describe in detail the parameters of GC-MS that could be applied for the hydrocarbon analysis of beeswax. The GC-MS condition settings can depend on the instrument with consideration to the optimization of the chromatographic separation.

1. Set the initial temperature of column at 80 °C (1 min), and then apply a temperature gradient of 15 °C/min until 340 °C, at which further separation of hydrocarbons is carried out.
2. Use the helium as the carrier gas, and set its flow through the column at 1.0 mL/min.
3. Set the temperature of the injector at 320 °C.
4. Inject the sample of 1 µL in the splitless mode and under 300 kPa (High Pressure Injection Mode).
5. Use the electron source (EI) with a standard ionization energy (70 eV), and the temperature of the ion source adjusted at 250 °C.
6. Set the interface temperature at 348 °C.
7. Use the value of voltage on the detector equal to the voltage obtained during autotune.
8. Set the range of mass scanning at 50–700 U.

*5.2.1.4. Qualitative analysis of beeswax hydrocarbons.* Identification of individual hydrocarbons of beeswax (n-

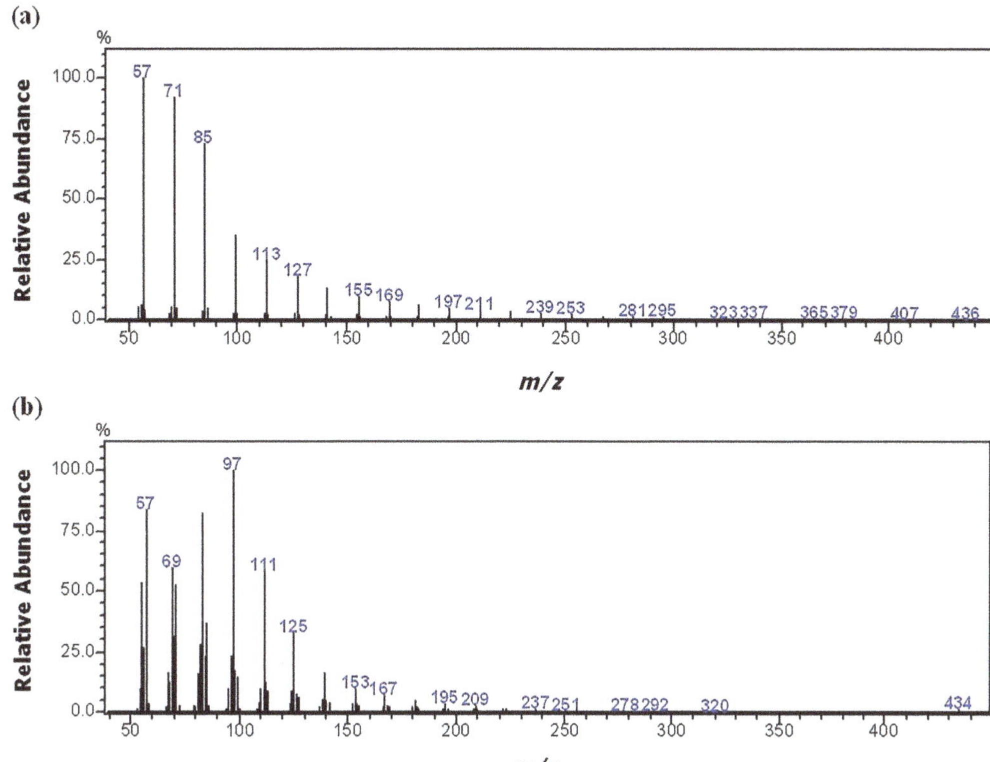


Figure 23. Electron impact spectrum of hydrocarbons: (a) hentriacontane ($C_{31}H_{64}$) with molecular ion of $m/z = 436$; (b) heneitriacontene ($C_{31}H_{62}$) with molecular ion of $m/z = 434$.

alkanes, alkenes and dienes) can be performed by comparison of the obtained mass spectra of n-alkanes with the mass spectra collected in commercial libraries (e.g., NIST 05 Mass Spectra Library). If no reference spectra are available, identification can be performed based on the fragmentation pathway and the presence of ions characteristic for the particular group of hydrocarbons in the following way. In mass spectra of n-alkanes and alkenes, certain groups of peaks should be identified when the *m/z* values differed by 14 Da. The peak of the highest intensity occurred in every group. In the case of n-alkanes, the peak corresponded to the ions $[C_nH_{2n+1}]^+$ and occurred when $m/z = 57, 71, 85, 99$, etc.; in the case of alkenes, the highest intensity had the peaks of the $C_nH_{2n}$ and $C_nH_{2n-1}$ ions. In addition, mass spectra of these compounds differ in molecular ions. Examples of the mass spectra of n-alkanes and alkenes are presented in Figure 23(a,b).

The qualitative analysis of n-alkenes can also be performed by comparison of the retention times of individual n-alkanes in the standard solution and the analyzed beeswax solution. If the retention time of the compound in beeswax sample corresponds to the retention time of the compound in the standard solution, it should be identified as the same compound.

The rules of identification of unknown compounds using GC-MS were also described in details by Torto et al. (2013).

*5.2.1.5. Quantitative analysis of n-alkanes in beeswax.* The quantitative analysis of n-alkanes is conducted using an internal standard method. General advantages resulting from using of internal standard in quantitative analysis are listed by Torto et al. (2013). The standard mixture of n-alkanes ($C_8H_{18}$–$C_{40}H_{82}$) is used with the internal standard - squalane ($C_{30}H_{62}$) at 10 mg/L concentration. Preparation of the standard solution is described in details in Section 5.2.1.1.

*5.2.1.6. Advantages and disadvantages of GC-MS anylsis.*

**Advantages:**

- precise, accurate and selective method
- good linearity as well as satisfactory repeatability and reproducibility
- suitable for the characteristics of beeswax hydrocarbons
- allows identification of the beeswax hydrocarbons (alkanes, alkenes and dienes) and quantification of n-alkanes (the most numerous group of hydrocarbons occurring in beeswax) in a single run
- preliminary purification of a sample with SPE, eliminates the overlapping of peaks, especially even-numbered n-alkanes with the peaks of esters

**Disadvantages:**

- allows only analysis of hydrocarbons
- expensive due to a very specialized equipment

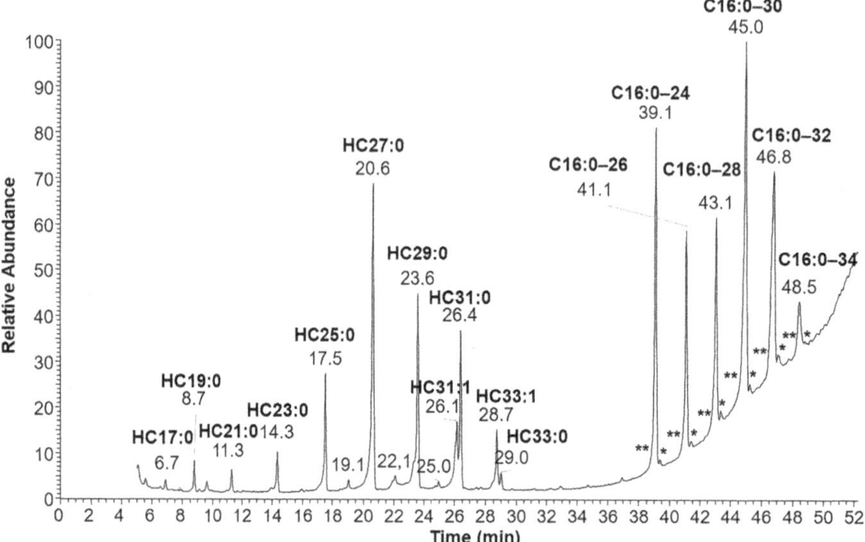


Figure 24. Typical GC-MS chromatogram of beeswax odd number hydrocarbons HC, even number hydrocarbons (unlabeled peaks between odd number hydrocarbons), palmitate monoesters C, oleate monoesters ** and 15-hydroxypalmitate monoesters * (Nunes & Maia, 2017, unpublished data).

- time-consuming at initial stage (preparation of sample using SPE)

*5.2.2. Simultaneous analysis of monoesters and hydrocarbons by GC-MS and GC-FID*

GC coupled with Flame Ionization Detector (GC-FID) is commonly used method for detection of volatile organic compounds. General analytical properties of CG-FID technique is provided in the *BEEBOOK* chapter on chemical ecology research in *A. mellifera*; for details see Section 2.2.4. "Detection and analysis of volatiles" (Torto et al., 2013). Using the methods described by Jiménez et al. (2004), and Maia and Nunes (2013) for the analysis of beeswax by GC-MS and GC-FID, odd number hydrocarbons (with 17–35 carbon atoms), even number hydrocarbons (with 22–34 carbons), and monounsaturated hydrocarbons with an odd number of carbon atoms (from 21 to 35) containing isomers with a different position of the double bond (although from the mass spectra the position of the double bond cannot be identified) can be determined qualitatively and quantitatively in a single run. Also, the highly abundant monoesters with carbon numbers from 34 to 50 corresponding to the esterification of palmitic acid (C16) with fatty alcohols containing 18 (1-octadecanol) to 34 (1-tetratriacontanol) carbon atoms) can be identified (Figure 24). Not all of the monoester components can be completely resolved under these chromatographic conditions, but their presence can be detected using the extracted ion chromatograms with specific ions for each monoester series: *m/z* 256 for the palmitate monoesters, *m/z* 264 for the oleate monoesters, *m/z* 236 for the hydroxypalmitate monoesters. The characteristic EI mass spectrum of palmitate monoesters is shown in Figure 25(a) for triacontyl palmitate, the most abundant palmitate ester found in beeswax, and in Figure 25(b) for the hydroxypalmitate monoester.

***Equipment required:***

- Gas-chromatograph (GC) equiped with a mass spectrometer (MS) with electron impact ionization (EI)
- Gas-chromatograph (GC) equiped with a flame ionization detector (FID)
- Balance with a 0.01 mg precision
- Horinzontal shaker

***Materials and reagents required:***

- mL glass tube with screw cap with a teflon faced rubber
- Chloroform p.a.
- Stearylstearate p.a.
- High temperature resistant apolar column 30 m column with 0.25 mm ID and 0.25 μm film thickness (for example, ZB-5 Inferno column, Phenomenex, or an equivalent high temperature resistant apolar column)
- μL syringe
- Helium 99.999%
- Air 99.999%
- Hydrogen 99.999%

*5.2.2.1. Preparation of beeswax solution.*

1. Weigh 3 mg of beeswax.
2. Add 4 mL of chloroform (if quantitative analysis is needed, replace 2 mL of chloroform with 2 mL of the internal standard solution of stearylstearate at 2 mg/mL).
3. Shake the solution mechanically until all the sample is dissolved.

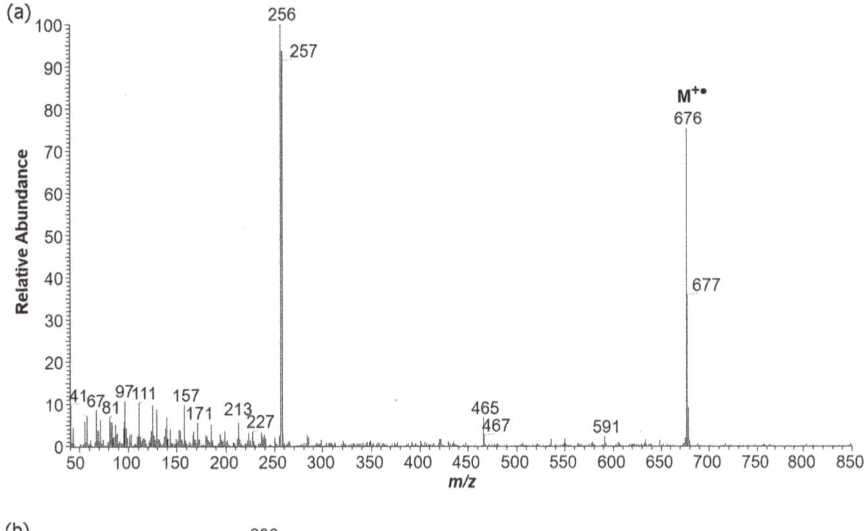


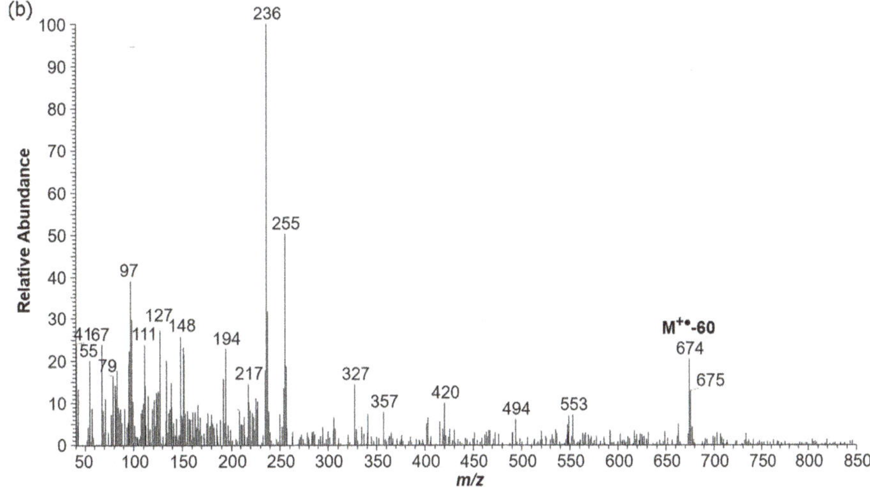


Figure 25. EI-MS of (a) triacontyl palmitate; (b) triacontyl hydroxypalmitate (Nunes & Maia, 2017, unpublished data).

4. Analyze the sample by GC-MS for qualitative analysis and GC-FID for quantitative analysis according to Sections 5.2.2.2 and 5.2.2.3 described below (for hydrocarbons analysis only see Section 5.2.1).

*5.2.2.2. Qualitative analysis by GC-MS.*

1. Use a 30 m ZB-5 Inferno column (Phenomenex) with 0.25 mm ID and 0.25 lm film thickness or an equivalent high temperature resistant apolar column.
2. Set the injector temperature at 325 °C.
3. Inject 1 µL of sample in a splitless mode with 2 min splitless time.
4. Use the following oven temperature program: initial temperature 50 °C, held for 3 min, a 50 °C/min ramp to 180 °C and held for 1 min, then a 3 °C/min ramp to 390 °C held for 5 min.
5. Use the carrier gas (He) at constant flow-rate of 1 mL/min.
6. Set the transfer line temperature at 350 °C.
7. Set the MS ion source temperature to 220 °C.
8. Set the scan range to *m/z* 40–850 with a total scan time of 0.72 s with a maximum ion time of 25 ms.
9. Use a solvent delay of 5 min.

*5.2.2.3. Quantitative analysis by GC-FID.*

1. Use a 30 m ZB-5 Inferno column (Phenomenex) with 0.25 mm ID and 0.25 µm film thickness or equivalent high temperature resistant apolar column for beeswax analysis.
2. Set the injector temperature at 325 °C.
3. Inject 1 µL sample in a splitless mode with 2 min splitless time.
4. Use the following oven temperature program: initial temperature 50 °C, held for 3 min, a 50 °C/min ramp to 180 °C and held for 1 min, then a 3 °C/min ramp to 390 °C held for 5 min.
5. Use the carrier gas (He) at a constant flow-rate of 1 mL/min.
6. Set the detector temperature at 400 °C.
7. Use hydrogen (20 mL/min) and synthetic air (200 mL/min) as auxiliary gases for the flame ionization detector.

After integration of all hydrocarbons peaks and internal standard in the obtained chromatogram, perform semi-quantification of each peak using the response factor calculated for the internal standard (IS) as:

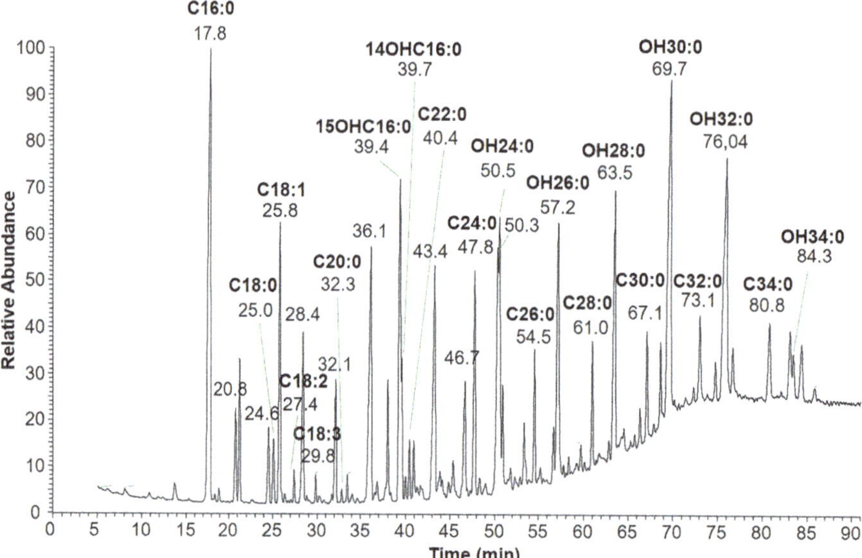


Figure 26. Typical GC-MS chromatogram of beeswax fatty acids and fatty alcohols analyzed by GC-MS after derivatization: fatty acids C; hydroxyl-fatty acids OHC; fatty alcohols OH. Unlabeled peeks correspond to hydrocarbons (Nunes & Maia, 2017, unpublished data).

$$\text{Response factor of IS} = \frac{\text{IS area}}{\text{weight of IS in sample injected}}$$

*5.2.3. Simultaneous analysis of fatty acids and fatty alcohols by GC-MS and GC-FID*

The fatty acids and fatty alcohols are two of the main beeswax components. Fatty acids are present in beeswax in the free form or esterified with fatty alcohols in the monoesters, mainly palmitates. Fatty alcohols are mainly present in monoesters with low amount present in the free form (Figure 26). For their characterization in beeswax, the determination of the total amount of fatty acids and fatty alcohols can be performed by GC-MS for qualitative analysis, and by GC-FID for quantitative analysis (Jiménez et al., 2003). According to the procedure described below, the method employed involves an esterification with methanol of the carboxyl groups of fatty acids followed by acetylation of the free hydroxyl groups of fatty alcohols and also of the hdroxyl fatty acids.

***Equipment required:***

- Gas-chromatograph (GC) equiped with a mass spectrometer (MS) with electron impact ionization (EI)
- Gas- chromatograph (GC) equiped with a flame ionization detector (FID)
- Balance with a 0.01 mg precision
- Dry-block heater
- Horinzontal shaker
- Centrifuge

***Materials and reagents required:***

- mL glass tube with screw cap with a teflon faced rubber
- Chloroform p.a.
- Methanol p.a.
- 14% $BF_3$ (boron trifluoride) solution in methanol p.a.
- Heptadecanoic acid p.a.
- Acetic anhydride p.a.
- Pyridine p.a.
- 30 m polar column with 0.25 mm ID and 0.25 μm film thickness (for example Supelcowax 10M column, Supelco).
- μL syringe
- Helium 99.999%
- Air 99.999%
- Hydrogen 99.999%

*5.2.3.1. Preparation of beeswax solution and derivatization.*

1. Weight 30 mg of beeswax for a 10 mL glass tube with screw cap with a teflon faced rubber.
2. Add 4 mL of chloroform containing 2 mg/mL of internal standard (Heptadecanoic acid C17:0).
3. Add 2 mL of methanol and 2 mL of a 14% $BF_3$ solution in methanol as catalyst (commercially available).
4. Close tightly the screw cap and heat during 1 h at 90 °C in a dry-block heater.
5. After cooling to room temperature, transfer 2 mL of the solution to a new glass tube.
6. Add 2 mL of acetic anhydride and 0.2 mL of pyridine.
7. Close tightly the screw cap and heat during 2 h at 90 °C in a dry-block heater.
8. Cool the tubes in an ice bath during 15 min.
9. Add 2 mL of water and mix the contents thoroughly.
10. Centrifuge at 3000 rpm during 5 min to separate the phase and discard the upper aqueous phase.
11. Repeat steps 9 and 10 twice.
12. Analyze the sample by GC-MS for qualitative analysis and GC-FID for quantitative analysis according to Sections 5.2.3.2 and 5.2.3.3 described below.

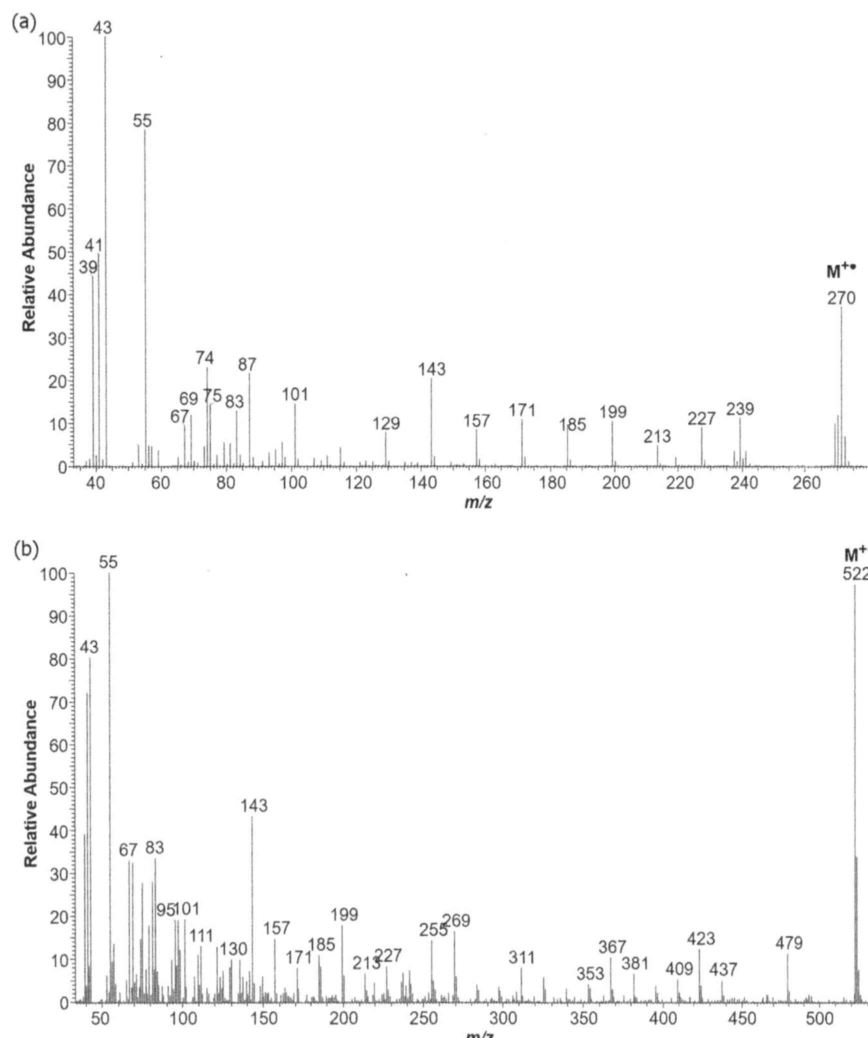


Figure 27. EI-MS of (a) palmitic acid methyl ester (C16:0), and (b) tetratriacontanoic acid methyl ester (C34:0) (Nunes & Maia, 2017, unpublished data).

The fatty acids methyl esters and acetyl fatty acids can be readily identified using an available standard or alternatively by comparison of the mass spectra available in several data bases. Fatty acids methyl esters can be easily identified by the presence of the molecular ion in their spectra (Figure 27). For the fatty alcohols the spectra are less characteristic and contain less distinguishing elements, contrary to fatty acids (Figure 28). The acetyl hydroxyl fatty acids methyl esters also present the molecular ion on their mass spectra although with a lower intensity, being also present the ion at *m/z* corresponding to the loss of an acetyl group (−60 Da) (Figure 29). For the acetyl fatty alcohols, the molecular ion is of very low intensity or absent for all practical purposes. Also present with low intensity is the ion resulting from the loss of an acetic acid (−60 Da). One of the most distinctive ions present in the mass spectra of acetyl fatty acids is the ion at *m/z* 61 presumed to be a protonated acetate moiety (Figure 28).

*5.2.3.2. Qualitative analysis by GC-MS.*

1. Use a 30 m Supelcowax 10M column (Supelco) with 0.25 mm ID and 0.25 μm film thickness or equivalent polar column for beeswax analysis.
2. Set the injector temperature at 280 °C.
3. Inject 1 μL of sample in splitless mode with a 2 min splitless time.
4. Use the following oven temperature program: initial temperature 140 °C, held for 1 min, a 2 °C/min ramp to 280 °C and held for 20 min.
5. Use a constant carrier gas (He) flow-rate at 1 mL/min.
6. Use a transfer line temperature of 280 °C.
7. Set the MS ion source temperature to 220 °C.
8. Set the scan range from *m/z* 33–650 with a total scan time of 0.72 s with a maximum ion time of 25 ms.
9. Use a solvent delay of 5 min.

*5.2.3.3. Quantitative analysis by GC-FID.*

1. Use a 30 m Supelcowax 10M column (Supelco) with 0.25 mm ID and 0.25 μm film thickness or equivalent polar column

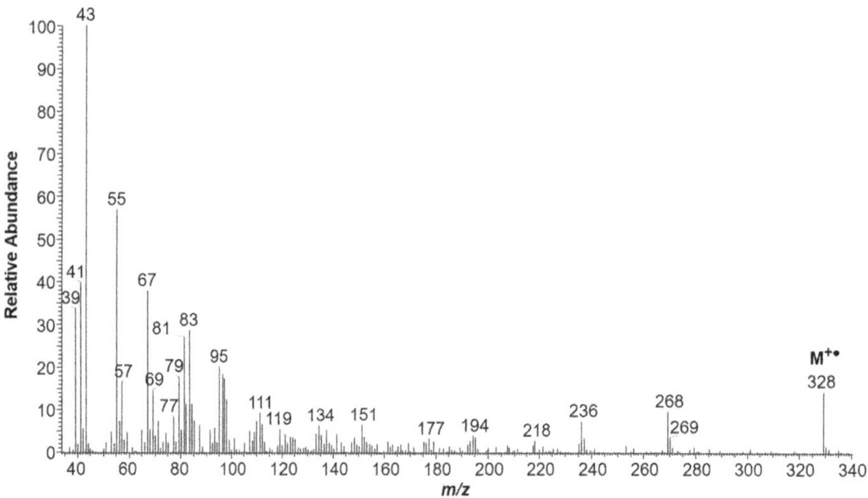


Figure 28. EI-MS of acetyl 1-triacontanol (OH30:0) (Nunes & Maia, 2017, unpublished data).

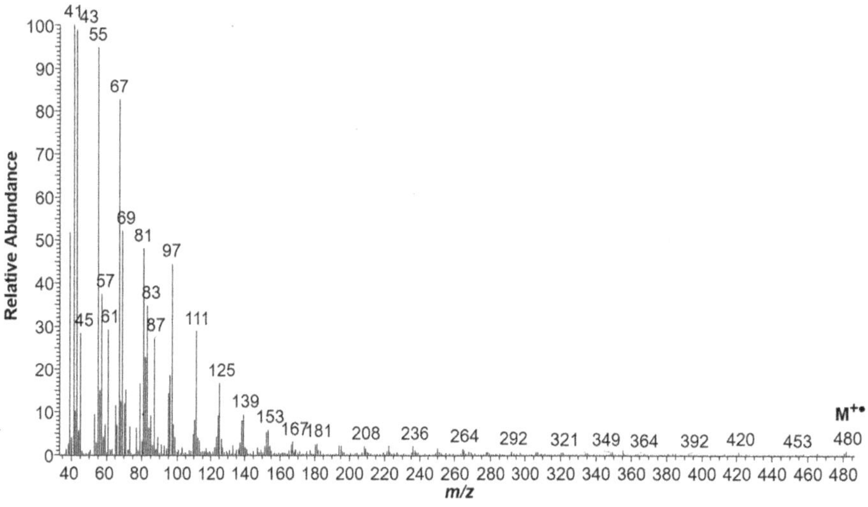


Figure 29. EI-MS of 15-hydroxypalmitic acid methyl ester (15OHC16:0) (Nunes & Maia, 2017, unpublished data).

2. Set the injector temperature at 280 °C.
3. Inject 1 μL of sample in a splitless mode with 2 min splitless time.
4. Use the following oven temperature program: initial temperature 140 °C, held for 1 min, a 2 °C/min ramp to 280 °C and held for 20 min.
5. The carrier gas (He) flow-rate is constant at 1 mL/min.
6. Set the detector temperature at 350 °C.
7. Use hydrogen (20 mL/min) and synthetic air (200 mL/min) as auxiliary gases for the flame ionization detector.

After integration of all fatty acids and fatty alcohols peaks and internal standard in the obtained chromatogram, perform semi-quantification of each peak using the response factor calculated for the internal standard (IS) as:

$$\text{Response factor of IS} = \frac{\text{IS area}}{\text{weight of IS in sample injected}}$$

#### 5.2.4. *Advantages and disadvantages of simultaneous analyses by GC-MS and GC-FID*

**Advantages:**

- provide broader chemical characterization of the chemical composition of beeswax
- simultaneous analysis of hydrocarbons and esters or fatty acids and fatty alcohols in a single run
- minimum sample preparation requirments
- sensitive and selective method

**Disadvantages:**

- expensive due to a very specialized equipment
- require advanced chemical analysis training

### *5.3. Investigation of beeswax by infrared (IR) spectroscopy*

***Definitions, acronyms:***

- IR: infrared
- $\lambda$: wavelenght
- $\nu$: frequency
- $cm^{-1}$: reciprocal centimeter
- $\tilde{\nu}$: wavenumber
- mid-IR (mid-infrared region): spectral region between 4000 and 400 $cm^{-1}$
- fingerprint region: spectral region between 1500 and 500 $cm^{-1}$ (approximatelly)
- FTIR: Fourier transform infrared (spectroscopy/spectrometer)
- ATR: Attenuated Total Reflectance (recording technique/accessory)
- T: transmittance (the amount of IR radiation passed through the sample)
- A: absorbance (the amount of IR radiation absorbed by the sample)
- a.u.: absorbance units
- DTGS: deuterated triglycine sulfate (detector)
- CsI: sesium iodide
- ZnSe: zinc selenide

### *5.3.1. Interaction of IR radiation with the sample*

Infrared (IR) spectroscopy is a method that utilizes infrared radiation as a physical medium for investigating the molecular structure of the sample. The basic principle of IR spectroscopy involves an interaction of IR radiation with the sample where molecules are being excited by the radiation. When IR radiation is absorbed by the sample, the molecules begin to vibrate with greater amplitudes (molecules gain energy) and experience a wide variety of vibrational motions characteristic for their component atoms and functional groups.

In the spectrum of electromagnetic radiation, an infrared portion (invisible, harmless thermal radiation) ranges from 0.8 to 1000 μm (800 nm–1 mm) in the wavelength ($\lambda$), i.e., from the nominal red edge of the visible spectrum to the microwave radiation. Frequency ($\nu$) is another parameter often encountered in IR spectroscopy, and represents the number of cycles/waves per second; in this context, the infrared portion extends from 375 THz to 300 GHz. However, a third unit, wavenumber ($\tilde{\nu}$), is the most widely used in IR spectroscopy. Wavenumber is the reciprocal of the wavelength ($1/\lambda$) expressed in reciprocal centimetres ($cm^{-1}$), which represents the spatial frequency of a wave, more specifically, the number of waves in a length of one centimetre. The most important argument for the application of the wavenumber in IR spectroscopy is its proportionality with the frequency of the electromagnetic field, and thus, its linearity with energy (the energy associated with a photon of IR light is directly proportional to its frequency). IR radiation induces vibrational excitation of covalently bonded atoms and groups within a molecule which helps in identifying the sample compounds due to strong relationship between vibrational energy and molecular structure.

According to the wavenumber, the infrared portion of the electromagnetic radiation covers a range from 12,500 to 10 $cm^{-1}$. The IR spectrum is further divided into three main regions: the far-infrared (<400 $cm^{-1}$), the mid-infrared (4000–400 $cm^{-1}$), and the near-infrared (12,500–4000 $cm^{-1}$), named after their relation to the visible spectrum. Mid-IR is most frequently used for biological applications as the majority of organic compounds exhibits characteristic absorption bands in this region. This is particularly referring to the spectral region between ~1500 and 500 $cm^{-1}$ (considered as the *fingerprint region*) which typically contains a series of absorption bands appearing in the IR spectrum mostly due to bending and skeletal vibrations of atoms within a molecule.

IR spectroscopy represents a valuable analytical tool for identification and analysis of numerous compounds in various organic specimens. As a type of vibrational (molecular) spectroscopy, where molecular vibrations are being analyzed, IR spectroscopy provides information on the total chemical composition of a sample, and permits acquisition of the absorption IR spectra of compounds that are a unique reflection of their molecular structure. IR spectrum, therefore, represents a unique chemical "fingerprint" of a sample, which is readily distinguished from the absorption patterns of all other compounds; only optical isomers absorb IR radiation in exactly the same way (Skoog & Leary, 1992). Interaction of vibrations of all atoms in the sample gives extremely complex but very characteristic absorption of IR radiation which prominent in the IR spectrum. Intensive IR signals, as well as weak bands, indicate a specific functional group positions and intensities associated with specific molecules in the specimen being analyzed. Details on the mechanisms of IR spectroscopy can be found in numerous contemporary scientific literature, such as Stuart (2004), Günzler and Gremlich (2002), and Barth and Haris (2009).

### *5.3.2. Analysis of beeswax by IR spectroscopy*

The utilization of IR spectroscopy for the beeswax analysis was first employed by Birshtein and Tul'chinskii (1977), who investigated basic beeswax composition and associated impurities. Further investigations of molecular structure and dynamics in beeswax, as well as general interpretation of beeswax IR spectrum, were reported in only several spectroscopic studies (Edwards, Farwel, & Daffner, 1996; Muscat, Tobin, Guo, & Adhikari, 2014; Zimnicka & Hacura, 2006). A notable progress in spectroscopic investigation of beeswax was brought by Maia et al. (2013), who introduced a single-reflection attenuated total reflectance (ATR) recording technique (coupled with FTIR spectrometer) in beeswax analysis, aiming to detect adulterants in beeswax. The application of FTIR-ATR method was further improved and modified by Svečnjak, Baranović et al. (2015) who demonstrated an analytical procedure for reliable routine detection of beeswax adulteration.

*5.3.2.1. FTIR-ATR recording technique.* Due to the numerous benefits elaborated further in this section, FTIR spectroscopy coupled with a diamond ATR accessory (FTIR-ATR recording technique) has become an important analytical tool for beeswax research as it was successfully applied for the chemical characterization of beeswax, as well as the qualitative and quantitative detection of adulterants in beeswax (Maia et al., 2013; Svečnjak, Baranović et al., 2015; Svečnjak, Prđun et al., 2015; Svečnjak et al., 2016, 2018).

Fourier transform infrared spectrometers (FTIR) are mainly used to measure absorption of the mid-infrared radiation (4000–400 $cm^{-1}$). FTIR spectroscopy has many advantages over conventional (dispersive) spectroscopy. Most importantly, FTIR spectrometers employ an interferometer based on the mathematical process of Fourier-transformation, and thus, reduce the measuring time and provide an acquisition of high quality spectra within seconds.

The IR spectrum of a sample can be measured either as transmittance (*T*, expressed as percent transmittance – %*T*) or absorbance (*A*, expressed in absorbance units – a.u.) which are commonly labelled on the *y*-axis of the IR spectrum, while the wavenumbers ($cm^{-1}$) that measure the position of an infrared absorption, are labelled on the *x*-axis. Absorbance is defined as the amount of IR radiation absorbed by a sample, and transmittance as the amount of IR radiation passed through that sample. Researchers use absorbance mode rather than transmittance, especially when it comes to quantitative analysis in biological applications. For the same reason, the absorbance mode is preferable for investigating beeswax samples.

An addition of ATR instrumental accessory further simplifies the measurement procedure, i.e., enables samples to be examined directly as obtained (in the solid or liquid state), without previous sample preparation. Unlike conventional transmittance measurements, ATR measurements are not dependent on the Beer–Lambert law which expresses the relationship of the IR radiation absorbed by the sample to the concentration of the targeted component and to the path length of the sample. ATR technique involves low and effective depth of penetration of the IR beam into the sample, typically between 0.5 and 5 μm, regardless of the sample thickness. This also ensures that only a small amount of the sample is sufficient for analysis. ATR spectroscopy utilizes the phenomenon of total internal reflection; a beam of radiation entering a crystal undergoes total internal reflection when the angle of incidence at the interface between the sample and crystal is greater than the critical angle, where the latter is a function of the refractive indices of the two surfaces (Stuart, 2004). This internal reflectance creates an evanescent wave that extends beyond the surface of the crystal into the sample being held in contact with the crystal. The crystals used for ATR accessories are made from optically dense materials with a high refractive index [such as diamond, germanium (Ge) or zinc selenide (ZnSe)]. Diamond is considered the best ATR crystal material due to its robustness, durability and chemical inertness.

*5.3.2.2. FTIR-ATR spectra acquisition.* Here, we describe a method for investigating beeswax samples by FTIR-ATR spectroscopy introduced by Maia et al. (2013) and modified by Svečnjak, Baranović et al. (2015). Different types of beeswax specimens can be analyzed by FTIR-ATR spectroscopy (wax scales, comb wax, comb foundations, beeswax blocks) depending on the aim of the study; sampling and refining should be carried out as described in Sections 2.1 and 2.2., respectively. Beeswax samples should be analyzed directly as obtained, without sample pre-preparation (purification or the usage of any kind of chemical reagents). The IR spectra of beeswax samples should be recorded using FTIR spectrometer coupled with a single-reflection high temperature heated diamond ATR system (alternatively, a heated germanium crystal plate can be used).

We present the method using the example of Cary 660 Fourier transform mid-infrared spectrometer (Agilent Technologies) with a DTGS detector and CsI optics, coupled with Golden Gate high temperature (200 °C) heated single-reflection ATR accessory (Specac) with a diamond as internal reflection element (the depth of the beam penetration into the sample is ~2 μm).

***Equipment required:***

- FTIR spectrometer with accompanying software (spectrometers with a DTGS detector and CsI optics are recommended, such as Agilent Technologies's Cary 660 FTIR spectrometer)
- High temperature (200 °C) heated single-reflection ATR accessory (ATR accessory with a diamond as internal reflection element is recommended, such as Specac's Golden Gate ATR)
- ATR thermocontroller (comes as a supplement to the ATR accessory listed above)
- Analytical balance (accuracy ±0.001 g)
- Laboratory (nitrile) gloves
- Metal spatula
- 96% ethanol (alternatively, acetone or isopropanol)
- Cotton wool
- Soft paper tissue

***Procedure:***

IR spectra of the beeswax samples are acquired as follows:

1. Heat the ATR plate to 75 °C (by adjusting ATR thermocontroller).

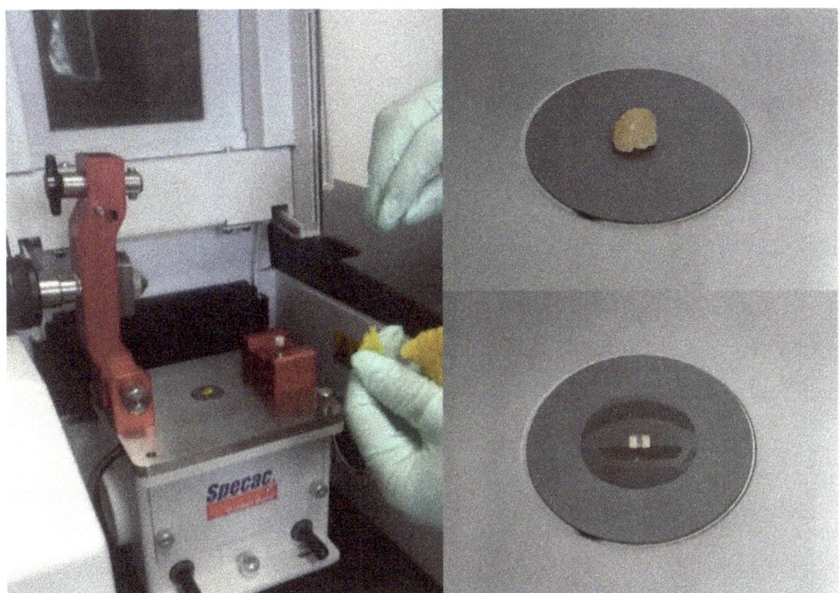


Figure 30. Beeswax sample on the ATR crystal (diamond) prepared for the acquisition of its IR spectrum (after melting at 75 °C). Photo: Svečnjak L.

2. Collect a background spectrum (spectrum of blank ATR crystal).
3. Place a small amount of a sample (~0.03 g; 0.02–0.05 g) directly onto the ATR sampling crystal using laboratory gloves or metal spatula, depending on the type of the specimen investigated (this amount of the sample should be sufficient to cover the internal reflection element/ATR crystal.
4. Leave the sample on the ATR crystal for 1 min to allow melting, homogenisation, and stabilisation of the sample (samples are analyzed in the liquid state) (Figure 30).
5. Collect sample IR spectrum (acquire two replicate spectra of each sample using the same aliquot).
   The following instrumentation parameters should be used to acquire the spectra:
   - *spectral mode*: absorbance
   - *spectral range*: mid-infrared spectral region (4000–400 cm$^{-1}$)
   - *resolution*: 4cm$^{-1}$ (gives a distance between two points in the resulting spectrum $\Delta\tilde{\nu} = 2cm^{-1}$)
   - *number of scans per spectrum*: 64
6. Store raw spectral data as SPC or ASCII file (this enables easier spectral data manipulation given that classical spectral formats are acceptable for numerous commercial software products commonly used for spectral data analysis (such as Origin, Matlab, The Unscrambler, Spekwin 32, etc), unlike instrument software-specific file formats, such as BSP in case of ResolutionsPro.
7. Clean the ATR sampling crystal thoroughly using 96% ethanol (it is best to use cotton wool soaked in alcohol for this purpose).
   Alternatively, other organic solvents can be used for cleaning the plate: acetone or isopropanol.
8. Dry the crystal with soft paper tissue.
9. Ensure the ATR crystal is completely dry and clean before the acquisition of another spectrum (scan an empty ATR crystal to confirm the absence of sample residues).

The analytical procedure for acquiring spectra described above is optimized by means of both time efficiency and reproducibility; analysis of beeswax samples in liquid state directly on the heated ATR plate using above listed instrumentation parameters enables easy and quick analysis and obtaining of high quality spectra.

After cleaning, the ATR plate is ready for the collection of another spectrum, starting with the background scan (new background spectrum should be acquired before each sample). It is recommended to use the live spectrum monitoring option (if possible) during the whole period of spectra processing, including cleaning of the ATR element. This enables an observation of the potential presence of impurities or other deviations, and ensures a clean ATR plate prior to the collection of the following spectrum.

Modern FTIR spectrometers ensure reproducible and accurate measurements. However, to validate the accuracy of measurement data obtained by specific methodology, a statistical analysis that provides information on the repeatability and reproducibility of the method can be performed by following ISO 5725-2,6:1994 standards (Accuracy (trueness and precision) of measurement methods and results – Part 2, Part 6, 1994). The repeatability of the IR measurements should be determined by ten-fold measurement of different aliquots of one beeswax sample. For this, the mean absorbance, standard deviation, coefficient of variation ($c_v$) and/or repeatability limit (95% confidence limits) using the absorbance at 2921 cm$^{-1}$ (band with the highest absorption intensity) can be calculated. Examples for calculating $c_v$/repeatability are provided in sections on detecting the adulterants in beeswax, i.e., Sections 6.2.5.3.3.1 and 6.2.5.3.3.2.

*5.3.2.3. Spectral data processing.* Raw spectral data can be stored and pre-analyzed using different software packages. Any software suitable for analysis of spectral data can be used. We present a procedure where spectra are recorded and stored using Resolutions Pro version 5.3.0 software package (Agilent Technologies):

1. Import spectral data in favourable software.
2. Exclude the noisy parts of the spectrum by cutting the spectrum to 3200–650 cm$^{-1}$ (use simple *cut* or *truncate* options).
3. Select the region of interest for further qualitative, quantitative, or chemometric and statistical modeling.

Given that the most characteristic spectral features in beeswax sample appear in the fingerprint region (1800–900 cm$^{-1}$), this region is usually being studied. Selection of the spectral region of interest and appropriate chemomeric models for investigating the beeswax spectra primarily depend on the aim and research design (qualitative or quantitative IR analysis, investigation of spectral variations in beeswax, detection of adulterants, monitoring of temperature-depended structural changes in beeswax, etc.).

*5.3.3. Characteristic FTIR-ATR spectrum of beeswax*

When analysing beeswax by FTIR-ATR spectroscopy, the first step represents qualitative interpretation of spectral features (positions and intensities of absorption bands) appearing in the beeswax IR spectrum. Visual inspection can be performed using different software packages specialized for spectral data analysis (e.g., Origin version 8.1 or other/Origin Lab Corporation, Spekwin 32), and complemented with the use of electronic spectral libraries, spectral atlases or books (e.g., Socrates, 2001), and/or available scientific literature. Contrary to the simplicity from an analytical point of view, the interpretation of IR spectrum (assignation of underlying molecular vibrations) is demanding and time consuming process.

The characteristic FTIR-ATR spectrum of *A. mellifera* beeswax is dominated by the spectral features due to esters (mainly monoesters), free fatty acids and hydrocarbons (Figure 31). The other less abundant components can be considered here as traces. In Section 5.3.3.1, we provide a detailed assignation of molecular vibrations occurring in FTIR-ATR spectrum of genuine *A. mellifera* beeswax sample (derived from melted wild-built comb wax, as described in Sections 2.1.1 and 2.2). As demonstrated below in Section 5.3.3.1, IR spectrum of *A. mellifera* beeswax should be characterized by unique spectral features; an appearance of absorption bands other than listed below indicate a questionable genuinity of the sample, i.e., spectral effects related to the presence of adulterants (see Section 6.2.5).

Minor hemical variability occurring in different specimens of *A. mellifera* beeswax should be considered when analysing beeswax by FTIR-ATR spectroscopy (this may be important for other methods as well, depending on the aim of the study). There are several biological factors affecting chemical differences that naturally occur among different types of beeswax samples. These chemical variations are summarized by Hepburn et al. (2014) presenting comparative compositional changes (in hydrocarbon and fatty acid content) observed in oenocytes, epidermis, scale wax and comb wax samples of *A. mellifera capensis*. Notable species-related differences were observed in beeswaxes originating from different species (Aichholz & Lorbeer, 1999). However, there are no significant differences in the chemical composition of wax originating from different *A. mellifera* subspecies; it has the same basic composition with only small variations related to the proportion of particular constituents, i.e., hydrocarbons, esters and fatty acids (Beverly et al., 1995; Fröhlich, Riederer et al., 2000; Tulloch, 1980). As reported by Beverly et al. (1995), African, Africanized, and European waxes showed differences in the relative amounts of several compounds; but no unique biomolecules were found that were distinctive for a particular type of wax. Chemical composition of comb waxes varies between colonies; small interfamily variations are mostly related to the content of hydrocarbons, esters and alcohols (Breed, Page et al., 1995, Svečnjak, Baranović et al., 2015; Svečnjak, 2017, unpublished data). An investigation of the differences between the waxes of different age (wax scales, new, middle aged, and old waxes) within the same honey bee colony, also revealed small variations; analyzed samples could only be distinguished after chemometric treatment (based on the content of hydrocarbons and esters) (Fröhlich, Riederer et al., 2000).

*5.3.3.1. Assignation of underlying molecular vibrations in beeswax IR spectrum.* When exposed to IR radiation (mid-IR spectral range), beeswax exhibits characteristic molecular (functional group) vibrations that can be identified and assigned to particular compounds. We present an assignation of major and minor molecular vibrations (synonyms: analyte signal, absorption band, absorption, vibration, peak) observed in the characteristic IR spectrum of beeswax. Assignation was conducted based on Socrates's tables and charts of IR spectra/characteristic group frequencies (2001), and scientific literature (Maia et al., 2013; Pielichowska et al., 2008; Svečnjak, Baranović et al., 2015; Zimnicka & Hacura,

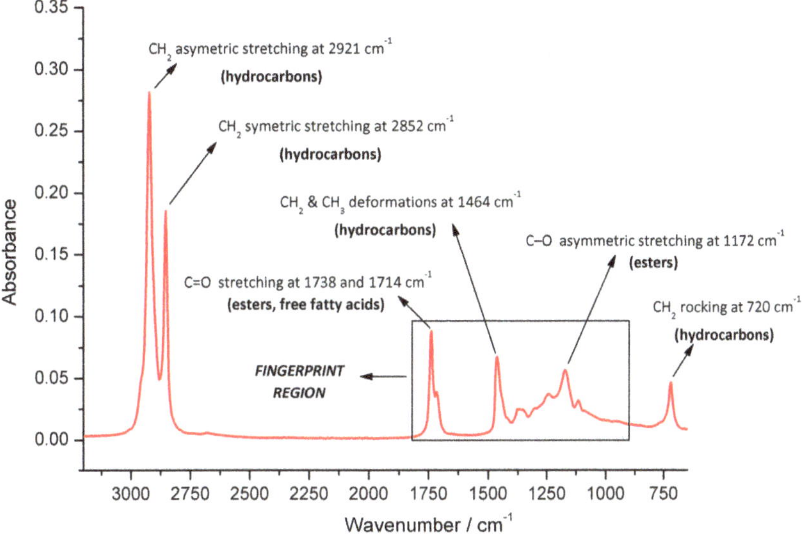


Figure 31. Characteristic FTIR-ATR spectrum of beeswax with assigned underlying absorption bands (whole spectrum – spectral region 3200–650 cm$^{-1}$) (modified from Svečnjak, Baranović et al., 2015).

2006). Respective molecular vibrations are given in Table 7 and various figures (noted below).

As shown in Figure 31, hydrocarbons are represented by very intensive absorptions observed at 2921 and 2852 cm$^{-1}$, medium band at 1464 cm$^{-1}$, and the weaker one arising at 720 cm$^{-1}$. These four bands are characteristic for all types of wax samples, being similar or almost of equal intensity and position.

Spectral region between 1800 and 800 cm$^{-1}$ (fingerprint region) is populated by a number of absorption bands. The most prominent spectral feature in the fingerprint region is the absorption band occurring at 1738 cm$^{-1}$ due to the carbonyl group (C=O) stretching vibrations of the of the ester bond. Given that saturated aliphatic esters typically absorb at 1750–1725 cm$^{-1}$, this absorption is attributed to the beeswax monoesters, which are the major ester component of beeswax (~40%). A weaker band peaking at 1714 cm$^{-1}$ is assigned to the C=O stretching vibrations of the carboxyl groups of free fatty acids present in beeswax (Figure 32).

The most complex absorption envelope of the IR spectrum of beeswax arises in the region between 1400 and 970 cm$^{-1}$. The most prominent absorption in this region is a broad band in the 1150–1050 cm$^{-1}$ with absorption maximum observed at 1172 cm$^{-1}$. This absorption corresponds to the C–O asymmetric stretching vibration of esters related to long-chain aliphatic acids (oleates, palmitates, stearates, etc.).

The spectral region between 1100 and 950 cm$^{-1}$ is characterized by broad signal (1090–1020 cm$^{-1}$) due to a number of overlapping bands which can be attributed to several vibrations; this spectral envelope is also the most variable part of beeswax spectra. For details on molecular vibrations observed in the fingerprint region, see Table 7.

For assays where researchers are focused on determining the chemical alterations and corresponding spectral variations naturally occurring in various beeswax samples, we present an example of IR spectra of 20 different beeswax samples (wild-built combs) collected from different *A. mellifera* colonies situated in different geographical regions in Croatia. It is observable that integral spectral features appear to be very similar among different samples; only small changes in absorbance intensities are noticeable (Figure 33(a)). An exception is spectral region between 1090 and 1015 cm$^{-1}$ characterised by different absorbance intensities, which indicate variability related to hydrocarbons (skeletal vibrations), esters and alcohols (Figure 33(b)). This region can be further subjected to spectral data mining and chemometric modelling, as discussed in Section 5.3.3.2.

*5.3.3.2. Spectral data preprocessing and chemometric modelling.* Along with basic investigation of beeswax molecular structure, the proposed FTIR-ATR methodology may also serve as an analytical tool for identifying chemical alterations and associated spectral variations occurring in beeswax due to different factors (such as different geographical origin, *A. mellifera* subspecies, comb age, type of beeswax specimen, etc.). However, such investigations are strongly dependent on the aim of the study, as well as the structure and complexity of the spectral data set attempted to be investigated. They also require more demanding analytical approach and employment of comprehensive chemometric assessment involving spectral data preprocessing and transformation (such as normalization, alignment, Savitzky–Golay filtering/smooting – SG, deconvolution, standard normal variate – SNV, multiplicative signal correction – MSC, and others), and multivariate statistical analyses (such as PCA, LDA, PLS-DA, FDA, ANN, etc.) to enhance the spectral variations between beeswax samples by

Table 7. Beeswax IR absorption bands with assignation of underlying molecular vibrations.

| Band position (wavenumber, $cm^{-1}$) | Functional group assignment and type of vibration | Absorption intensity | Associated beeswax chemical compound |
|---|---|---|---|
| 2957 | $CH_3$ stretching, asymmetric | Weak, shoulder | Hydrocarbons |
| 2921 | $CH_2$ stretching, asymetric | Strong, sharp | Hydrocarbons |
| 2852 | $CH_2$ stretching, symetric | Medium-strong, sharp | Hydrocarbons |
| 1738 | C = O stretching | Medium | Monoesters – saturated aliphatic esters, saturated free fatty acids |
| 1714 | C = O stretching | Weak, shoulder | Free fatty acids |
| 1464 | $CH_2$ scissoring, $CH_3$ asymmetric deformation (overlapped) | Medium | Hydrocarbons |
| 1420 | $CH_2$ deformation | Weak, shoulder | Esters |
| 1376 | $CH_3$ symmetric deformation | Weak | Hydrocarbons |
| 1368 | C–H deformation | Weak | Hydrocarbons |
| 1353 | C–H deformation | Weak | Hydrocarbons |
| 1302 | $CH_2$ wagging | Weak | hydrOcarbons |
| 1242 | C–O–C asymmetric stretching | Medium-weak, broad | Aliphatic esters |
| 1172 | C–O asymmetric stretching | Medium, broad | Esters of long-chain aliphatic acids (stearates, oleates, etc.) |
| 1115 | C–O–C symmetric stretching | Weak | Aliphatic esters |
| 1080–1020 | $CH_3$ rocking | Weak-to-medium, broad | Esters |
| 1065 | (unsat.)–$CH_3$ | | Hydrocarbons |
| 1050 | CCO stretching | | Alcohols |
| 720 | $CH_2$ rocking | Medium-weak, | Hydrocarbons |

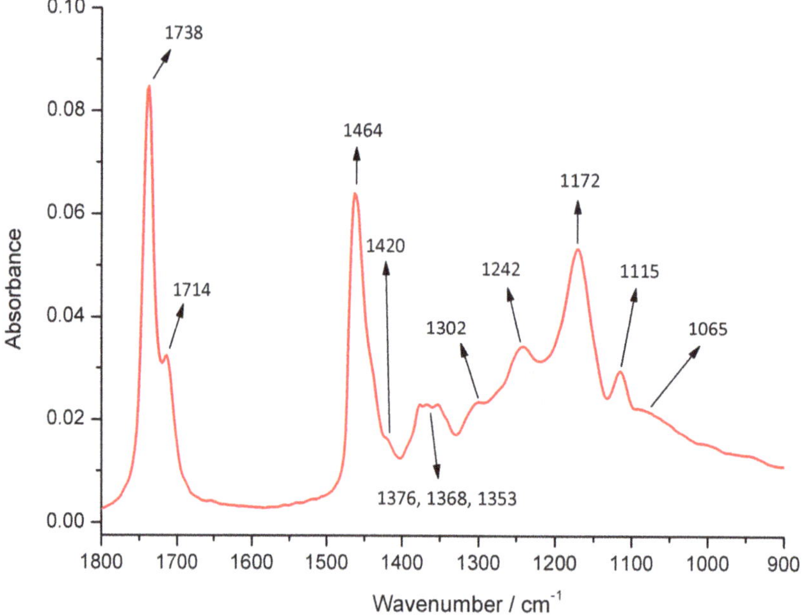


Figure 32. Fingerprint region (1800–900 $cm^{-1}$) of beeswax FTIR-ATR spectrum and underlying absorption bands.

enhancing the analyte signals. Therefore, a methodology for the chemometric modelling cannot be standardized. To determine the best chemometric model for particular data set, refer to the recent research notes on the optimization of SG parameters for improving spectral resolution and quantification reported by Zimmermann and Kohler (2013). As emphasised by the authors, calculating derivatives of spectral data by the SG numerical algorithm can be used as a preliminary pre-processing step to resolve overlapping signals, enhance signal properties, and suppress unwanted spectral features that arise due to nonideal instrument and sample properties. However, make sure to use the proper data treatment (preprocessing) and appropriate statistical models given that model with large number of latent variables (most often involved in spectral data mining) may lead to overfitting and perplexity of the statistical model (Toher, Downey, & Murphy, 2007). For general multivariate statistical approaches, refer to the respective Sections 5.4, 5.5, 6, 7, and 8 of the *BEEBOOK* paper on statistics (Pirk et al., 2013).

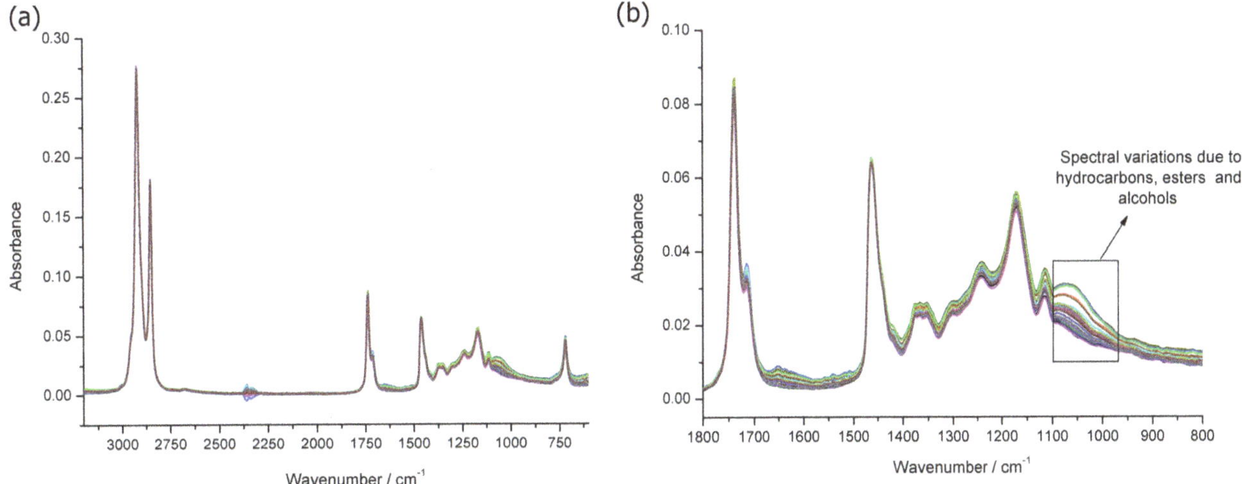


Figure 33. IR spectra and spectral variations between different beeswax samples ($n = 20$) originating from different *A. mellifera* colonies: whole spectra (3200–650 cm$^{-1}$) (a); fingerprint region (1800–900 cm$^{-1}$) (b) (modified from Svečnjak, Baranović et al., 2015).

#### *5.3.4. Advantages and disadvantages of FTIR-ATR measurements*

**Advantages:**

- Provide information on the total chemical composition of a sample.
- Speed advantage (after optimization of the analytical procedure as described above, it is possible to obtain the results in only few minutes).
- Cheap (consumables required only for the cleaning of ATR crystal) and easy-to-use (undemanding from the analytical point of view).
- Non-destructive and reagent-free method (requires no sample preparation).
- Only a small amount of a sample is sufficient to acquire the IR spectra (<0.5 g).
- Good accuracy, repeatability, and reproducibility of measurement results.
- Suitable for analysis of different types of beeswax specimens (wax scales, comb wax, comb foundations).
- Accompanying software and optional accessories for FTIR spectrometers (e.g., different ATR plates, liquid and gas transmission cells, FTIR microscope) enable implementation of various experimental designs using different measurement and chemometric techniques that can be used to complement and/or expand the beeswax research.

**Disadvantages:**

- Minor beeswax compounds cannot be detected easily by FTIR-ATR spectroscopy due to an overlap with the absorption bands of predominant beeswax compounds (esters, hydrocarbons and fatty acids).
- Only suitable for qualitative analysis since quantitative methods (analytical procedures for quantification of individual beeswax compounds) still have to be developed.

### *5.4. Determination of ash content and mineral composition*

The ash content and mineral composition of beeswax have generally been poorly studied. Ash content is a measure of the total amount of minerals present in the wax, whereas mineral content is a measure of the amount of specific inorganic components present in it. Determination of the ash and mineral content of wax may be useful to evaluate the composition and quality of the product (Bernal et al., 2005; KEBS, 2013), and its age (Taha, Manosur, & Shawer, 2010).

#### *5.4.1. Determination of ash content*

Ash is the inorganic residue remaining after the water and organic matter have been removed by heating in presence of oxidizing agents, which provides a measure of the total amount of minerals within a food. The most widely used laboratory methods rely on the fact that minerals are not destroyed by heating, and that they have a low volatility compared to other components.

***Equipment required:***

- Muffle furnace
- Porcelain capsules
- Desiccator
- Analytical balance (accuracy ±0.001 g)
- Bunsen burner

***Procedure:***

1. Heat the porcelain capsules for 30 min in the muffle furnace set at 600 °C.
2. Allow the capsules to cool in a desiccator to room temperature.
3. Determine their weight on analytical balance to the nearest 0.001 g.

4. Transfer 2 of wax in a capsule.
5. Weigh on analytical balance to the nearest 0.001 g.
6. Heat gently on a Bunsen burner till the material is completely charred.
7. Transfer and heat in muffle at 600 °C for 1.5 h.
8. Transfer and cool the capsule with the ashes in a desiccator until constant weight.
9. Repeat incineration, cooling and weighing until the difference between two successive weighing is less than one milligram.

The ash percentage was determined as follows:

$$\text{Ash }\% = 100 \times \frac{W_1 - W_2}{w}$$

where $W_1$ is the mass, in g, of the porcelain capsule with the ash content; $W_2$ is the mass, in g, of the empty capsule; and $w$ is the mass, in g, of the beeswax sample.

*5.4.2. Determination of mineral composition*

Various spectroscopic techniques may be used to determine the mineral content in the beeswax.

The older techniques, but still widely used, are atomic absorption spectroscopy (AAS), to determine metallic elements, and atomic emission spectroscopy (AES), to determine alkaline and alkaline-earth elements. The most recent and performing technique is inductively coupled plasma combined with mass spectrometry (ICP-MS), which allows the simultaneous determination of many elements, replacing both AES and AAS.

Whatever the instrumental technique, beeswax has to be wet digested to eliminate the organic part with the following procedure:

1. Weight 0.5 g of beeswax into a burning cup.
2. Add 15 mL of pure $HNO_3$.
3. Digest the sample in a microwave oven at 200 °C.
4. Dilute the solution to the desired volume with water.
5. Determine the concentrations with a spectroscopic technique.

To determine the amount of an element present in the wax, a calibration function will be calculated for each analyte using the relative values of the signals generated by the analysis of standard solutions with known and increasing concentration of these elements.

### 5.5. Standard methods for sampling and analysing beeswax for hydrogen isotope ratios

**Definitions, acronyms:**

- $^1H$: hydrogen-1
- $^2H$: hydrogen-2 or deuterium
- $\delta^2H$: stable hydrogen isotope value
- IRMS: Isotope ratio mass spectrometer
- *R*: ratio of a less common isotope form to the more common isotope form ($^2H/^1H$)
- $\delta$-notation: parts per thousand (‰) difference from an accepted standard reference point; $\delta = (R_{sample}/R_{standard} - 1)$
- V-SMOW: Vienna Standard Mean Ocean Water
- QA: quality assurance (QA standard)
- QC: quality control (QC standard)
- RM: reference material

The $\delta^2H$ values of beeswax are related to hydrogen incorporated by a honey bee through consumption of foods and water (Chesson et al., 2011; Tipple et al., 2012). The most significant source of hydrogen to animals is meteoric waters (Bowen, Wassenaar, & Hobson, 2005) and the $\delta^2H$ values of meteoric waters vary in systematic and highly predictable patterns across land surfaces (Bowen & Revenaugh, 2003; Craig, 1961). In general, more positive $\delta^2H$ values of meteoric waters are observed at lower elevations and locations nearer to the coasts, while more negative $\delta^2H$ values are observed at higher elevations and inland areas. Geographic patterns in water hydrogen isotope signatures create *isotope landscapes* or *isoscapes* (Bowen et al., 2009). These spatial patterns in the $\delta^2H$ values of waters are subsequently transferred to organisms, as organisms incorporate isotopic signals from their local waters. Thus, the $\delta^2H$ values of beeswax can be useful in identifying the region-of-origin (Figure 34). Stable isotope analysis and the isoscape method have been applied to bee products (Chesson et al., 2011, 2014a; Cho et al., 2012; Schellenberg et al., 2010; Tipple et al., 2012).

*5.5.1. Methods for sampling and handling beeswax for hydrogen isotope analysis*

Here, we present the methods applied by Chesson et al. (2011, 2014b). These methods are relatively standard procedures and can be modified for particular laboratory settings. Different types of beeswax can be analyzed for stable hydrogen isotope ratios. Sampling beeswax for hydrogen isotope analysis may include:

1. Wax scales (as described in Section 3.2)
2. Comb wax (as described in Section 2.1.1)
3. Wax caps (as described below in Section 5.5.1.1)
4. Dissolved wax in liquid honey (as described below in Section 5.5.1.2).

*5.5.1.1. Sampling honeycomb wax caps.* In both commercial beekeeping, as well as in the natural environment, the interior wax that makes up the cell walls of the honeycomb is often recycled year after year. For this reason, the wax cap covering the honey within the honeycomb is preferable for studies investigating the provenance of honey and honeycomb using isotope

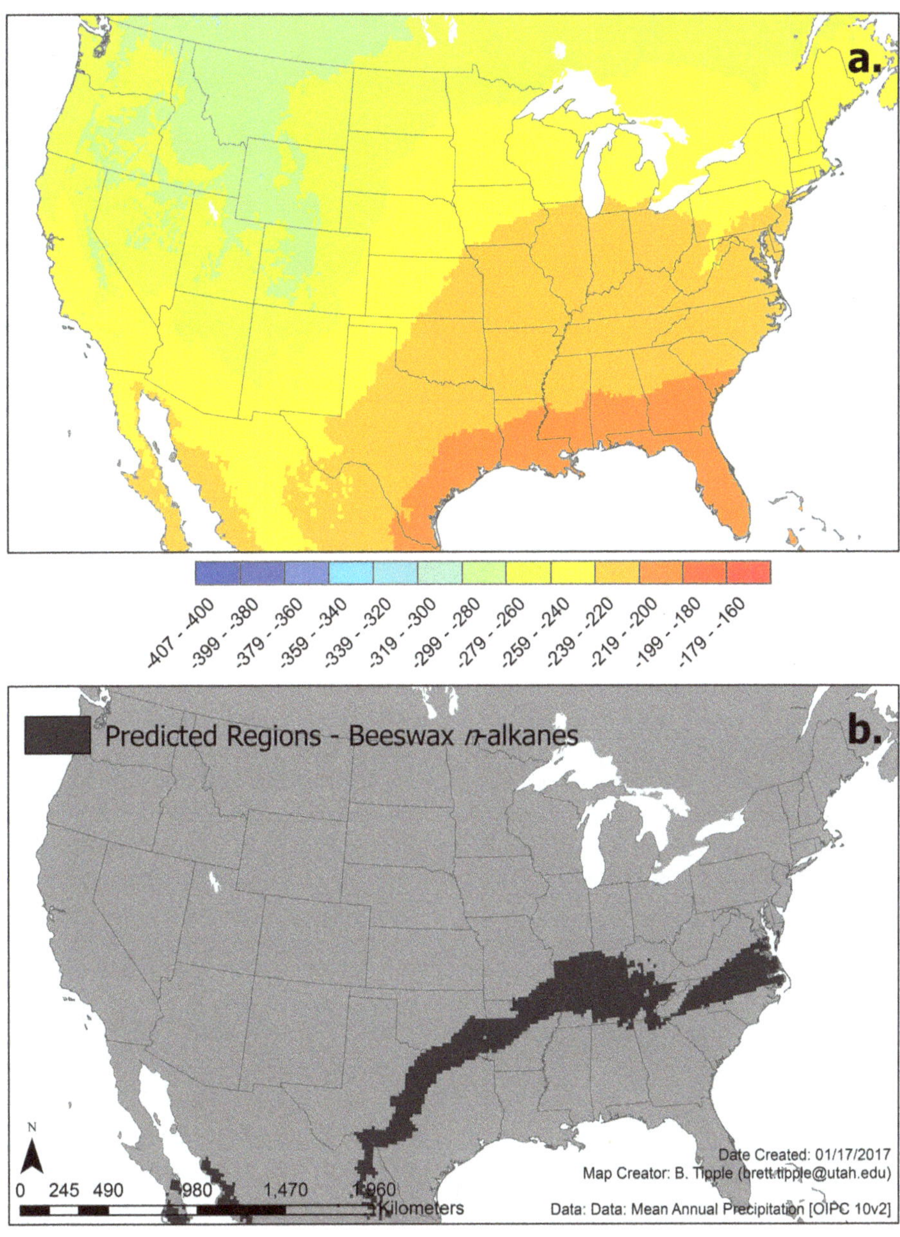


Figure 34. (a) Isoscape projection of beeswax wax cap *n*-alkanes as described by relationship from Tipple et al. (2012). (b) Regions consistent with a beeswax wax cap *n*-alkane $\delta^2$H value of $-227 \pm 2.5‰$.

analysis. Here, the exclusive use of wax caps ensures that the associated honey and waxes reflect the same production season and geographic location. Thus, prior to any isotopic analysis, the wax caps must be isolated from the comb and free of any residual honey. The method below outlines the use of forceps and razor blades to isolate wax caps from the surrounding cells of the honeycomb (Figure 35). Following isolation from the honeycomb, techniques are described to clean any residual honey from the wax cap sample.

***Equipment required:***

- Ashed (i.e., baked at 500 °C for 4 h) 4-mL glass vial with Teflon-lined caps
- Balance with precision of ±1 mg
- Nitrile gloves
- Metal spatula
- Metal forceps
- Razor blade
- "Solvent residue" waste container
- Sonicator
- 50-mL glass beaker
- Ashed aluminium foil
- Drying apparatus

***Reagents required:***

- Laboratory soap
- Deionised or distilled water in squeeze bottle
- HPLC-grade (or better) methanol in Teflon® squeeze bottle

Figure 35. Representative honey and beeswax samples in ashed canning jars fitted with ashed aluminium foil cap liners. From left to right, orange, tupelo, sage, wildflower, and sourwood honeys.

***Procedure:***

1. Prepare a 4-mL vial (also known as a 1-dram vial) by weighing empty vial and cap while working in a fume hood and wearing nitrile gloves.
2. Wash spatula, forceps, and razor blade with laboratory soap and water: Rinse three times with tap water, followed by three rinses with deionised or distilled water from squeeze bottle.
   This step removes surface contaminants and soap residue from the tools.
3. Squirt clean spatula, forceps, and razor blade with methanol three times within a fume hood. Collect residual methanol in solvent waste container. This step removes any organic contamination from the tools.
4. Remove the honeycomb from storage container using forceps to a clean surface, such as a sheet of aluminium foil.
5. Remove 2–5 honeycomb cell caps using forceps and razor blade
6. Place in the pre-weighed 4-mL vial.
7. Label vial.
8. Fill 4-mL vial containing wax caps with ~3.5 mL deionised or distilled water.
   This is the *wash* water.
9. Cap vial.
10. Sonicate vial containing wax caps for 10 min at room temperature.
11. Decant and discard *wash* water from vial.
12. Repeat steps 8 and 10 two additional times, for a total of three washes.
13. Decant final *wash* water.
14. Dry wax caps within vial under a gentle stream of nitrogen gas or purified air at room temperature.
15. Cap vial once washed wax caps and vial are completely dry.
16. Determine weight of wax caps by weighing vial with cap and subtracting the mass of the empty vial with cap.
17. Place in dry storage area (see Section 2.3 on beeswax storage for details). Wax cap samples can be stored indefinitely.

While these steps describe sampling and cleaning wax caps from honeycomb, this method could be easily modified for sampling and cleaning of interior cells of a honeycomb.

*5.5.1.2. Extraction of beeswax from liquid honey.* Here, we present methods applied by Tipple et al. (2012). Solvent-extractable organic compounds can be isolated from liquid honey using a two-phase liquid-liquid extraction. Liquid-liquid extraction methods are particularly useful when visible wax is not present in a honey sample, such as filtered honeys (Tipple et al., 2012). The procedures presented here are those used for isotope analysis of *normal*-alkanes (*n*-alkanes) and different liquid-liquid extraction methods may be used for other analytical techniques.

***Equipment required:***

- Nitrile gloves
- Ashed aluminium foil
- Balance with precision of ±1 mg
- Ashed 20-mL glass scintillation vial with Teflon-lined cap
- 5.75" ashed, borosilicate glass Pasteur pipettes
- Vortex mixer (optional, but recommended)
- Ashed round-bottom culture tube
- Drying apparatus
- Ashed 4-mL glass vial with Teflon-lined cap

***Reagents required:***

- Solvent-extracted deionised or distilled water in squeeze bottle
- HPLC-grade (or better) hexanes in Teflon® squeeze bottle

***Procedure:***

1. Select a liquid honey sample.
   Either filtered honey or comb-honey can be used.
2. Weigh out 10 g of honey in a 20-mL glass scintillation vial while wearing nitrile gloves. Record honey mass.
3. Label vial.

4. Add 5 mL of distilled and deionised water to the scintillation vial with honey sample with a pipette.
5. Gently agitate to dissolve honey.
6. Add 10 mL of hexane to scintillation vial.
7. Seal vial with solvent-rinsed Teflon-lined cap. Close cap tightly.
8. Agitate vial for 2 min either by shaking in hand or using a vortex mixer.
9. Allow solvent and honey/water layers to separate (~10–20 min).
10. Remove top layer (i.e., hexanes) using a pipette to a culture tube.
11. Label tube.
12. Repeat steps 6–11 two additional times, combining extracts in culture tube, for a total of three extractions.
13. Dry extract under a gentle stream of nitrogen gas or purified air at room temperature using a drying apparatus. Concentrate to ~2–3 mL
14. Transfer concentrate to 4-mL glass vial.
15. Label vial.
16. Dry concentrate completely under a gentle stream of nitrogen gas or purified air at room temperature using a drying apparatus.
17. Cap with Teflon-lined cap.
18. Dried extracts can be stored at room temperature indefinitely (see Section 2.3 for details).

*5.5.2. Methods for isolating beeswax compounds for compound-specific hydrogen isotope ratio analysis*

Here, we apply the methods of Tipple et al. (2012). Prior to hydrogen isotope analysis of specific compounds from beeswax, organic compounds must be isolated from the complex mixtures of bulk beeswax or beeswax extracted from liquid honey.

*5.5.2.1. Dissolution of bulk beeswax.* A mixture of 2:1 dichloromethane:methanol is used to dissolve sample wax in a sonicator. When the sample has dried, the remaining residue is the total lipid extract.

***Equipment required:***

- Nitrile gloves
- Metal spatula
- Metal forceps
- Razor blade
- Ashed aluminium foil
- Balance with precision of ±1 mg
- Ashed 4-mL glass vial with Teflon-lined cap
- 50-mL glass beaker
- Sonicator
- Drying apparatus

***Reagents required:***

- HPLC-grade (or better) dichloromethane
- HPLC-grade (or better) methanol in Teflon® squeeze bottle
- Deionised or distilled water

***Procedure:***

1. Select a clean and dry wax cap.
2. Remove from storage vial with spatula or forceps
3. Place on aluminium foil.
4. Cut a single wax cap into quarters with a clean and solvent-rinsed razor blade.
5. Weigh an empty 4-mL borosilicate glass vial.
6. Label vial.
7. Place a 1/4 of wax cap into the weighed, labelled vial.
8. Add additional pieces of wax cup until mass is approximately 5–10 mg.
9. Record final mass.
10. Mix a 2:1 solution of dichloromethane:methanol.
11. Add 0.5 mL 2:1 dichloromethane:methanol to the 4-mL vial containing wax cap sample of known weight.
12. Seal vial with solvent-rinsed, Teflon-lined cap.

    Close cap tightly.

13. Fill a 50-mL beaker with ~25 mL deionised water.
14. Place wax-containing vial in beaker.

    Ensure water in the beaker does not reach the threads of the capped vial.

15. Sonicate beaker and vial for 10–15 min or until all wax is dissolved.
16. Remove vial from beaker and uncap.

    At this point the dissolved wax cap can be quantitatively divided into separate aliquots using a graduated syringe. Aliquots can either be used at this point or dried for storage.

17. Dry dissolved wax within vial under a gentle stream of nitrogen gas or purified air at room temperature.
18. Cap vial once sample is completely dry.
19. Place in cool, dry storage area. Wax samples can be stored indefinitely (see Section 2.3 on beeswax storage for details).

*5.5.2.2. Silica gel column chromatography.* The preferred method for preparation of beeswax for compound specific isotope analysis follows Torto et al. (2013). This method uses Pasteur pipette-packed silica gel columns (i.e., the stationary phase) with organic solvents (i.e., the mobile phase) to separate organic compounds from beeswax into multiple fractions (see Section 4.2.2. in the *BEEBOOK* article on standard methods for chemical ecology research in *A. mellifera*, Torto et al., 2013). The separation of the fractions is based on compound and solvent polarity. This method can be used prior to any gas chromatographic analysis to simplify chromatogram baseline. This is particularly important when using gas chromatography-isotope ratio mass spectrometry of beeswax compounds, as a clean baseline is mandatory (see Section 5.5.3). Here, dissolved wax from Section 5.5.2.1 is added to the top of the Pasteur pipette column as described in Torto et al. (2013).

*5.5.3. Isotope ratio mass spectrometry and measurement systems*

Here, we provide methods applied in Chesson et al. (2011, 2014a), Cho et al. (2012), Schellenberg et al. (2010), and Tipple et al. (2012). Isotope ratio mass spectrometry (IRMS) is used to analyze the stable hydrogen isotope abundance at natural levels (Hoef, 2004). With an IRMS instrument, materials must be converted to $H_2$ gas prior to analysis. The most common conversion methods are either offline pyrolysis or the use of an online pyrolysis system. Traditionally, offline methods were applied where organic material, such as beeswax, was reacted to produce $H_2$ gas initially, and then analyzed with a duel-inlet IRMS system (Bigeleisen, Perlman, & Prosser, 1952; Friedman, 1953; Schimmelmann & DeNiro, 1993; Sofer & Schiefelbein, 1986; Stuermer, Peters, & Kaplan, 1978; Wong & Klein, 1986). The main advantage of offline methods to measure the stable hydrogen isotope ratio of beeswax is high precision. The disadvantages include low sample throughput, long instrument dwell times, and numerous and costly consumables. Another significant disadvantage for offline methods is the requirement for highly-trained IRMS technicians. In addition, offline methods are typically non-standard and must be tailored to specific laboratories.

Most laboratories now perform stable hydrogen isotope measurements using continuous-flow IRMS systems with online preparative methods to produce $H_2$ gas (Begley & Scrimgeour, 1996; Burgoyne & Hayes, 1998; Hilkert, Douthitt, Schluter, & Brand, 1999; Prosser & Scrimgeour, 1995; Tobias & Brenna, 1997). This online $H_2$ gas preparation occurs by thermal decomposition (i.e., pyrolysis) and the $H_2$ gas is swept into the IRMS instrument with helium (He) as a carrier gas. The main advantages of continuous-flow systems are increased throughput, fewer consumable at lower costs, and relatively standard methodologies across laboratories. The main disadvantage is slightly less precise measurements.

*5.5.3.1. Hydrogen isotope analysis of bulk beeswax and specific compounds.* The relative abundances of stable isotopes within a material are described as the ratio ($R$) of a less common isotope form to the more common isotope form (i.e., $R = {}^2H/{}^1H$). Since these ratios are very small numbers, values are expressed in $\delta$-notation, which is the parts per thousand (‰) difference from an accepted standard reference point, where:

$$\delta = \left(R_{sample}/R_{standard} - 1\right) \qquad (5.5.1)$$

As defined in Equation (5.5.1), the stable hydrogen isotope value ($\delta^2H$) is:

$$\delta^2H = \left[\left({}^2H/{}^1H_{sample}\right)/\left({}^2H/{}^1H_{standard}\right) - 1\right] \qquad (5.5.2)$$

The standards used to express isotope abundances in $\delta$-notation are internationally recognized reference materials. Primary reference materials have absolute isotope compositions. In the case of hydrogen, $\delta^2H$ values are reported relative to Vienna Standard Mean Ocean Water (V-SMOW), which has a ${}^2H/{}^1H$ of 0.00015576.

Laboratories also have access to a variety of secondary reference materials that have been calibrated to the primary reference material (Table 8). A recent effort to develop reference materials has specified 16 new or updated secondary reference materials for hydrogen isotope ratio measurements (Schimmelmann et al., 2016). Often, secondary reference materials are used to calibrate in-house reference materials or standards (Dunn & Carter, 2018). In-house standards can be produced in significant quantities to allow for day-to-day analysis and long-term usage. The "principle of identical treatment" requires in-house standards to be chemically matched to the samples being analyzed to produce robust hydrogen isotope ratios. Two-point normalisation of measured data to the international $\delta$ scale is also a requisite to report accurate $\delta^2H$ values (Coplen, 1988). In-house standards of contrasting isotope ratios are used in daily operation both for normalising results to the international $\delta$ scale, but also for accumulating long-term quality assurance data to assess system performance. A variety of in-house standards can be used for normalising measured sample data and compiling quality assurance information (e.g., Table 9).

The following descriptions are general considerations for hydrogen isotope ratio measurement of beeswax using either online or offline IRMS systems. We strongly advocate working with an established laboratory with the knowledge of and facilities in the area of stable hydrogen isotope ratio analysis (see list in Schimmelmann et al., 2016), as the specific methods described below may slightly vary between laboratories. Detailed procedures for making hydrogen isotope ratio measurements are beyond the scope of this contribution and chapter; however, specific IRMS guidelines and methodologies are thoroughly discussed elsewhere (Dunn & Carter, 2018). Detailed procedures for compound-specific IRMS operations are also available (e.g., Burgoyne & Hayes, 1998; Hilkert et al., 1999). Here, we briefly discuss quality assurance and quality control methods and data normalisation procedures.

*5.5.4. Quality assurance and quality control reference materials*

Both quality assurance (QA) and quality control (QC) standards or reference materials (RMs) must be

Table 8. Some commercially available secondary reference materials for hydrogen isotope measurement (compiled October 2016).

| Reference material | Description | Material | $\delta^2H_{VSMOW}$ |
|---|---|---|---|
| NBS-30 | Lakeview Tonalite | Biotite | −65.7 |
| USGS-42 | Human Hair from India | Keratin | −78.5 |
| USGS-43 | Human Hair from Tibet | Keratin | −50.3 |
| USGS-67 | $n$-$C_{16}$, Hexadecane | $n$-alkane | −166.2 |
| USGS-68 | $n$-$C_{16}$, Hexadecane | $n$-alkane | −10.2 |
| USGS-69 | $n$-$C_{16}$, Hexadecane | $n$-alkane | +381.4 |
| NBS-22 | Mineral oil | Oil | −117.2 |
| IAEA-CH-7 | Polyethylene Foil (formerly PEF-1) | Polyethylene | −99.2 |
| GISP | Greenland Ice Sheet Precipitation | Water | −189.7 |
| V-SLAP | Vienna Standard Light Antarctic Precipitation | Water | −428 |

Table 9. In-house standards and hydrogen isotope values used for hydrogen isotope measurement (as of October 2016).

| Reference material | Description | Material | $\delta^2H_{VSMOW}$ |
|---|---|---|---|
| SQ | Squalane | Isoprenoid | −177 |
| C28 | $n$-$C_{28}$, Octacosane | $n$-alkane | −251 |
| C36 | $n$-$C_{36}$, Hexatriacontane | $n$-alkane | −240 |
| C24 | $n$-$C_{24}$, Tetracosane | $n$-alkane | −36 |
| SM | Arndt Schimmelmann's Mix A ($n$-$C_{16}$, $n$-$C_{24}$… $n$-$C_{29}$, $n$-$C_{30}$) | $n$-alkanes | Varies by compound and production run |
| PARA | Parafilm M® | Paraffin wax film | −101 |
| ANDRO | 5a-Androstane | Steroid | −253 |

analyzed alongside beeswax samples analyzed for $\delta^2H$ values to ensure reproducible and accurate measurements. Primary and secondary RMs are used to calibrate in-house RMs (see discussion above in Section 5.5.3.1), which are in turn used for isotope data normalisation (i.e., QA reference materials, see Section 5.5.4.1) and quality control (i.e., QC reference materials, see Section 5.5.4.2) step. All reputable stable isotope analysis laboratories maintain primary and secondary reference materials from NIST and IAEA as a means of initially calibrating in-house RMs to the international $\delta$ scales, as well as to periodically reconfirm those isotope values.

QA RMs are the basis upon which measured isotope data for beeswax samples are normalised to international $\delta$ scale (i.e., V-SMOW for $\delta^2H$ values). QA RMs must be analyzed alongside samples or in samples, in the case of compound-specific measurements, within a given analytical sequence. This allows for systematic in-run variations to be captured, monitored, and corrected as needed. Further, QA RMs should have a wide isotopic range that brackets the expected isotopic composition of the beeswax samples (Table 9).

QC RMs must always be included in an analytical sequence to assess the accuracy of the isotopic normalisation (see Sections 5.5.4.1 and 5.5.4.2) generated by QA RMs. QC RMs should ideally match the chemical matrix of the type of sample being analyzed (Table 9).

*5.5.4.1. Normalisation of hydrogen isotope values to international scale.* Normalisation is the process of converting measured hydrogen isotope values of a sample to the international $\delta$ scale. For hydrogen isotope measurements, all $\delta^2H$ values are reported relative to V-SMOW. Normalisation of hydrogen isotope values is compulsory as it is the only method to ensure $\delta^2H$ values from different laboratories and instruments are comparable. Normalisation should be done in each analytical sequence. The normalised $\delta^2H$ value of a sample is calculated using:

$$\delta^2H_{Normalised} = m * \delta^2H_{Measured} + b \qquad (5.5.3)$$

The slope ($m$) and intercept ($b$) are determined by comparing the measured and known $\delta^2H$ values of two QA RMs. Table 10 shows an example normalisation of measured $\delta^2H$ values using in-house RMs.

*5.5.4.2. Additional quality control metrics.* Normalised $\delta^2H$ values of QC materials (see Section 5.5.4.1) are compared to the known $\delta^2H$ values of those materials. Before any normalised $\delta^2H$ data of beeswax samples can be accepted, the normalised $\delta^2H$ values of the QC RMs must meet accepted limits. Typically, for bulk hydrogen isotope analysis this limit is ±2‰ (Chesson et al., 2011). If the QC RMs results do not meet this limit, $\delta^2H$ values from the beeswax samples are deemed unacceptable and must be re-analyzed.

Using best practices, each sample should be analyzed in replicate to provide robust metrics for reproducibility (Dunn & Carter, 2018). However, this is not always possible, particularly in commercial laboratories, due to the cost of preparing and analysing these additional samples. Alternatively, several samples within an analytical sequence may be analyzed multiple times and the reproducibility of these selected samples may be applied to other samples in the sequence. Nonetheless, either of these approaches to determine reproducibility is acceptable.

Table 10. Example two-point $\delta^2H$ normalization using in-house RMs.

| ID | $\delta^2H_{Measured}$ | $\delta\ ^2H_{Known}$ | $\delta\ ^2H_{Normalized}$ |
|---|---|---|---|
| C24 | −30 | −36 | – |
| C28 | −244 | −251 | – |
| PARA | −94.7 | −101 | −101 |
| Sample 1 | −151 | – | −158 |
| Sample 2 | −178 | – | −185 |
| Sample 3 | −203 | – | −210 |

## 6. Standard methods for beeswax authenticity and quality control

### 6.1. Legislation and quality control of beeswax

The quality control of beeswax worldwide includes official statutory documents referring to beeswax used in pharmaceutical and food industry, as pharmaceutical grade beeswax and food additive E901, respectively (national Pharmacopoeias, FAO legislation, EU regulations, U.S. FDA regulations, etc.).

There are different international and national Pharmacopeias (International Pharmacopeia, European Pharmacopeia, United States Pharmacopeia, British, French, German, Swiss, Slovak, etc.; with only small differences regarding beeswax criteria values) that give a definition of beeswax, and provides reference measurement guidelines for determination of its general physical and chemical characteristics.

Alongside Pharmacopeias where beeswax is described in terms of a pharmaceutical substance specification (yellow beeswax – *cera flava*, white beeswax – *cera alba*), beeswax is also an authorised food additive. Identified as additive E901, beeswax is permitted as a glazing agent, and as carrier for colours and flavours (Directive 95/2/EC, 2005; Directive 2009/10/EC, 2009; EFSA, 2007; FAO, 2005; ISO TC34/SC19, 2017; JECFA, 2005); two types of beeswax are specified: yellow beeswax (C.A.S No. 8006-40-4), and white beeswax (C.A.S. No. 8012-89-3).

The methods defining pure beeswax quality criteria based on 10 physico-chemical parameters determined in accordance with European Pharmacopoeia (EP, Council of Europe, 2007), and the Methods of German Society for Fat Science (Methods of Deutsche Gesellschaft für Fettwissenschaft: DGF-M-V2,3,6, 1957), have been proposed as general quality criteria for beeswax by the International Honey Commission (IHC) (Bogdanov, 2016a), as presented in Table 11 (comparative overview of beeswax quality parameters according to FAO, European legislation and IHC are presented in Table 12). However, the IHC material remained in the form of a proposal to date. As beeswax material used in the apiculture sector (primarily referring to the wax blocks and comb foundations) is considered as apiculture by-product not intended for human consumption, it is still not covered by any effective legislation nor subjected to any kind of authenticity or quality control prior to its placement on the market. This has resulted in a growing beeswax adulteration trend (Maia et al., 2013; Serra Bonvehí & Orantes Bermejo, 2012; Svečnjak, Baranović et al., 2015; Svečnjak et al., 2016, 2018; Waś et al., 2016). The quality of beeswax is further jeopardized by residues that mostly originate from the pesticides commonly used in the beekeeping technology (primarily acaricides), as well as environmental commercial pesticides widely utilized in agriculture (Bogdanov, 2006; Boi et al., 2016; Bonzini, Tremolada, Bernardinelli, Colombo, & Vighi, 2011; Chauzat & Faucon, 2007; Jiménez, Bernal, del Nozal, & Martín, 2005; Ravoet et al., 2015; Serra Bonvehí & Orantes Bermejo, 2010; Tremolada, Bernardinelli, Colombo, Spreafico, & Vighi, 2004; Wallner, 1992, 1993, 1997, 1999, 2000). This implies an urgent need for the implementation of mandatory regulations defining the beeswax quality criteria and standardized analytical methods for its testing, as well as routine quality/authenticity control of beeswax prior to its placement on the market. The methods for detection of adulterants and pesticide residues in beeswax are therefore described in the following sections.

Table 11. Quality criteria parameters for routine authenticity testing of beeswax and reference range values defining pure beeswax according to the methods prescribed by European Pharmacopoeia (2007/2008) and the Methods of German Society for Fat Science (DGF, 1957) – proposal of the International Honey Commission (IHV) (Bogdanov, 2016a).

| Quality criteria | Value | Method |
|---|---|---|
| Water content | <1% | DGF-M-V-2 |
| Refractive index (75 °C) | 0.4398–1.4451 | EP |
| Melting point | 61–65 °C | EP |
| Acid number | 17–22 | EP |
| Ester number | 70–90 | EP |
| Ester/acid ratio | 3.3–4.3 | EP |
| Saponification number | 87–102 | EP |
| Mechanical impurities, additives | Absent | DGF-M-V-3 |
| Glycerols, polyols, fatty acids fats | Absent | EP |
| Hydrocarbons max. | 14.5%[a] | DGF-M-V-6 |

DGV, V2,3,6 – Methods of Deutsche Gesellschaft für Fettwissenschaft.
EP – European Pharmacopoeia 7th Edition, 2008, 56.
[a]Wax from African and Africanized bees: max. 13.8%

### 6.2. Standard methods for beeswax adulteration detection

#### 6.2.1. Beeswax adulteration

Adulteration has been one of the main quality issues of beeswax production and represents a long-term and increasing problem worldwide. Nevertheless, there are still no internationally standardised analytical methods for routine authenticity control of beeswax used in the apiculture sector (honeycombs, wax blocks and comb foundations). Moreover, beeswax and its products are classified as animal by-products (ABP) not intended for human consumption, and are usually marketed

Table 12. Comparative overview of beeswax quality parameters according to FAO (2005), European legislation (2009/10/EC; EP, 2007) and International Honey Commission (Bogdanov, 2016a).

| Parameter | FAO (2005) | 2009/10/EC | EP 5.0 (2007) Yellow beeswax | EP 5.0 (2007) White beeswax | IHC (2016) |
|---|---|---|---|---|---|
| Water Content | | | | | <1% |
| Melting range (°C) | 62–65 | 62–65 | 61–66 | 61–66 | 61–65 |
| Specific gravity ($D_{2020}$) | | 0,96 | 0.960 | 0.960 | |
| Refractive index (75 °C) | | | | | 1.4398–1.4451 |
| Solubility | Insoluble in water; sparingly soluble in alcohol; very soluble in ether | Insoluble in water; sparingly soluble in alcohol; very soluble in chloroform and ether | Practically insoluble in water, partially soluble in hot alcohol (90% *V/V*) and completely soluble in fatty and essential oils. | Practically insoluble in water, partially soluble in hot alcohol (90% *V/V*) and completely soluble in fatty and essential oils. | |
| Acid value (mg KOH/g): | 17–24 | 17–24 | 17–22 | 17–24 | 17–22 |
| Saponification value (mg KOH/g) | 87–104 | 87–104 | 87–102 | 87–104 | 87–102 |
| Ester value (mg KOH/g) | | | 70–80 | 70–80 | 70–90 |
| Ester/acid ratio | | | | | 3.3–4.3 |
| Peroxide value (mM $H_2O_2$/kg): | <5 | <5 | | | |
| Glycerol and other polyols: | <0,5% (as glycerol) | <0,5% (as glycerol) | Absent | Absent | Absent |
| Carnuba wax | Passes test | | | | |
| Ceresin, paraffins and other waxes: | Passes test | Absent | Absent | Absent | Absent |
| Fats, Japan wax, resin and soaps | Passes test | Absent | Absent | Absent | Absent |
| Arsenic | | <3 mg/kg | | | |
| Lead | <2 mg/kg | <5 mg/kg | | | |
| Mercury | | <1 mg/kg | | | |

(imported) as category 3 material which includes ABPs that do not present a potential risk for the food chain as they must not contain residues of other (foreign) substances and environmental contaminants (Reg. (EC) No. 1069/2009, Reg. (EU) No 142/2011, USDA/APHIS regulations – most of other regulations are in compliance with mentioned EC legislation for export/import purposes). However, beeswax is frequently marketed as "safe" category 3 even when it contains substances of questionable origin and chemical background (such as most commonly used adulterants, paraffin and stearic acid/stearin, as well as pesticide residues), due to the lack of obligatory legal regulations controlling its authenticity (FFN, 2017; Svečnjak et al., 2018). This problem is not referring only to the aspects of uncontrolled contamination via comb foundation trade (deliberate addition of adulterants into comb foundations that are not subjected to authenticity testing prior to their placement on the market) but also to the fact that adulterants are being accumulated in the apiculture sector through the comb foundation production process as comb foundations re-enter beekeeping technology (Svečnjak et al., 2016). Consequently, adulterated beeswax may enter other industries, such as cosmetic, pharmacy and food industry. Furthermore, given that beeswax (honeycomb built upon comb foundation) comes into contact with honey during the honey production process, food safety is another concerning aspect of beeswax adulteration. The negative aspects and potential risks of beeswax adulteration for animal and public health were recently brought to the attention of the European Commission (EC) by the EU Food Fraud Network (FFN, 2017).

Nowadays, there are more than 15 different natural (petroleum-derived, mineral, animal, and plant waxes) and industrial waxes used as beeswax adulterants worldwide (Bogdanov, 2004b, 2016a). Among them, paraffin waxes represent the greatest problem due to their wide availability and low price. Additionally, physico-chemical properties of paraffin (it is a chemically inert, almost odourless, and white or colourless substance) makes it almost "ideal" for adulteration. Other adulterants, such as stearic acid, stearin, tallow, microcrystalline wax, and others, are observed sporadically. The presence of high amounts of paraffin (> 50%; up to 93%), as well as the sporadic appearance of stearic acid and/or stearin in comb foundations collected from the market has been reported in several recent studies (Maia et al., 2013; Serra Bonvehí & Orantes Bermejo, 2012; Svečnjak, Baranović et al., 2015, 2016, 2018; Waś et al., 2016). The negative effects of beeswax adulterated with stearin on the brood development have also been recently reported (Reybroeck & Van Nevel, 2018).

The disconcerting situation on the comb foundation market indicated by the results obtained in these studies implies an urgent need for routine beeswax authenticity control. This kind of outcome is a consequence of a larger-scale problem: general deficit of beeswax and "chronic" accumulation and circulation of paraffin (sporadically, other adulterants as well) in the comb foundation production process. The lack of regulations and standardized analytical methods for beeswax authenticity testing definitely contribute to this issue (Svečnjak, Baranović et al., 2015; Waś et al., 2015, 2016). We therefore provide a detailed description of the methods for beeswax adulteration detection using classical physico-chemical parameters, sensory analysis and spectroscopic techniques (GC-MS and FTIR-ATR).

#### 6.2.2. *Beeswax authentication by classical analytical methods (physico-chemical parameters)*

Determination of physico-chemical parameters describing both physical and chemical properties of different organic substances have been widely used to establish the reference values defining quality (authenticity) criteria for particular substance, including beeswax. Over the course of the last six decades, considerable research has been invested in the attempt to build a reliable reference physico-chemical model defining pure beeswax (Bogdanov, 2016a; Tulloch, 1973, 1980; White, Riethof, & Kushnir, 1960).

Among the list of analytical methods and physico-chemical parameters defining beeswax quality criteria finally proposed by IHC based on pharmaceutical regulations (Bogdanov, 2016a; Table 11), the melting point, acid value, saponification value, ester value and ester/acid ratio, are the most commonly used to evaluate the beeswax quality and to detect possible adulteration. Nevertheless, the use of these parameters for detection of beeswax adulteration has some drawbacks in detecting the common adulterants, (paraffin, tallow, stearic acid and carnauba wax); as presented in Table 13, the minimum amount of mentioned adulterants that can be detected by corresponding methods is relatively high (high detection thresholds: 5–50%, depending on the type of adulterant).

Although the majority of proposed analytical range values characterizing pure beeswax are generally accepted and widely utilized, corresponding methods have not yet been officially employed for general beeswax authentication and quality control due to unresolved legislation issues mentioned in previous section. Another possible reason why these methods are not implemented in sectors other than pharmacy and food industry might be related to analytical deviations (values of acid, saponification, and ester value, as well as ester/acid ratio outside the proposed ranges) observed and reported in numerous studies in the past and current literature (Bennett, 1944; Bernal et al., 2005; Maia & Nunes, 2013; Serra Bonvehí, 1990; Svečnjak, Baranović et al., 2015; Tulloch, 1973). An overview of the ranges of physico-chemical parameters reported in the literature for authentic beeswax are summarized in Table 14 (Bennett, 1944; Bernal et al., 2005; Maia & Nunes, 2013; Nunes and Maia, 2017, unpublished data; Poncini, Poncini, & Prakash, 1993; Puleo & Rit, 1992; Serra

Table 13. Minimum adulteration percentages detected in beeswax by the measurement of reference physico-chemical parameters (Bernal et al., 2005).

| Parameter | Paraffin (54–74 °C) | Stearic acid | Tallow | Carnauba wax |
|---|---|---|---|---|
| Melting point | 30–50% | 30% | 40% | 5% |
| Acid value | 10% | 2% | 10% | 20% |
| Saponification value | 10% | 3% | 15% | [a] |
| Ester value | 5% | 5% | 10% | [a] |
| Ester/acid ratio | 10% | 15% | 10% | 40% |
| Iodine value | 15% | 15% | 15% | [a] |

[a]Not useful for beeswax adulteration adulteration.

Bonvehí, 1990; Serra Bonvehí & Orantes Bernejo, 2012; Svečnjak, Baranović et al., 2015; Tulloch, 1973; Tulloch & Hoffman, 1972). The major factor affecting the analytical range values (primarily saponification value, and consequently, ester value and ester/acid ratio) is an exposure of beeswax to high heat treatment (Tulloch, 1973) which represents an integral part of the comb foundation production process given that beeswax used for comb foundation production is commonly subjected to different high heat treatments (121–140 °C) necessary to kill the spores of the heat-resistant *Paenibacillus larvae*. Thus, anomalous range values can be partially explained by the heat treatment applied (>100 °C) during beeswax recycling and processing (Bogdanov, 2016a; Svečnjak, Baranović et al., 2015; Tulloch, 1973): However, this does not explain the anomalous range values reported for virgin beeswax samples collected directly from the beehives (Bernal et al., 2005; Maia & Nunes, 2013; Svečnjak, Baranović et al., 2015). Certain deviations may also arise from a different geographical origin of the beeswax (Beverly et al., 1995), but the overall analytical deviation effects are not yet fully explained. Mentioned deviations of the range values proposed as beeswax quality standards do not affect standard methods for determining the respective parameters, but should be considered when performing physico-chemical tests (especially on comb foundations) for authentication purposes.

Along with proposed analytical measurands mentioned above, other physico-chemical parameters have been utilized sporadically for the purpose of beeswax authentication, namely, density, peroxide value (Bernal et al., 2005; Bogdanov, 2004a, 2004b), ash content and iodine number (Bernal et al., 2005; Puleo & Rit, 1992; Serra Bonvehí, 1990).

The methods for determination of all physico-chemical parameters listed above, as well as the challenges related to their use in beeswax adulteration detection, are described in the following sections (with an exception of ash content determination which has already been presented in Section 5.4.1).

*6.2.2.1. Drop point.* The drop point of beeswax is the temperature at which a drop of beeswax grease is extruded from the bottom of a special cup under the conditions of this test. The range of values defined by the European Pharmacopoeia and EU legislation is 61–66 °C.

It should be noted that determination of drop point and melting point (described in the following section) are not reliable analytical methods for beeswax adulteration detection if adulterants of the same or similar melting point as beeswax (60–65 °C) were used (often in the case of beeswax adulteration with paraffin wax).

An apparatus for the determination of drop point presented in Figure 36 (dimensions in mm) consists of 2 metal sheaths (A) and (B) screwed together. Sheath (A) is fixed to a mercury thermometer. A metal cup (F) is loosely fixed to the lower part of sheath (B) by means of 2 tightening bands (E). Fixed supports (D) 2 mm long determine the exact position of the cup in addition to which they are used to centre the thermometer. A hole (C) pierced in the wall of sheath (B) is used to balance the pressure. The draining surface of the cup must be flat and the edges of the outflow orifice must be at right angles to it. The lower part of the mercury thermometer has the form and size shown in the figure; it covers a range from 0 to 110 °C and on its scale a distance of 1 mm represents a difference of 1 °C. The mercury reservoir of the thermometer has a diameter of 3.5 ± 0.2 mm and a height of 6.0 ± 0.3 mm. The

Table 14. Ranges of the physico-chemical parameters reported in the literature for authentic beeswax.

| Parameter | Melting point | Acid value | Saponification value | Ester value | Ester/acid ratio | Peroxide value | Iodine number |
|---|---|---|---|---|---|---|---|
| Bennett (1944) | 61–65 | 16–23 | 85–101 | 72–79 | 3.6–4.3 | | 4.0–12.0 |
| Tulloch and Hoffman (1972) | 63.4–65.1 | 17.4–21.8 | | 70.3–75.4 | 3.38–4.12 | | |
| Tulloch (1973) | | 19.1 | | 73.5 | 3.84 | | |
| Serra Bonvehí (1990) | 61.9–64.1 | 17.4–19.8 | 90.1–90.8 | 70.3–79.0 | 3.54–4.34 | | 9.6–17.3 |
| Puleo and Rit (1992) | 61–65 | 17–24 | 87–104 | 70–80 | | | 7–12 |
| Bernal et al. (2005) | 64–66 | 17.1–21.9 | 82.8–147.1 | 62.7–74.8 | 3.09–7.08 | <0.01 | 7.6–13.1 |
| Serra Bonvehí and Orantes Bermejo (2012) | 61.9–64.1 | | 90.1–98.3 | | | | |
| Maia and Nunes (2013) | 63.0–67.3 | 14.4–23.0 | 65.5–124.2 | | | | |
| Svečnjak, Baranović et al. (2015) | 60–65 | 20.7–30.2 | 57.5–134.0 | 31.1–112.2 | 1.18–5.14 | | |
| Nunes and Maia (2017, unpublished data) | | | | 46.4–103.3 | 2.3–5.3 | 0–19.9 | 5.9–14.3 |

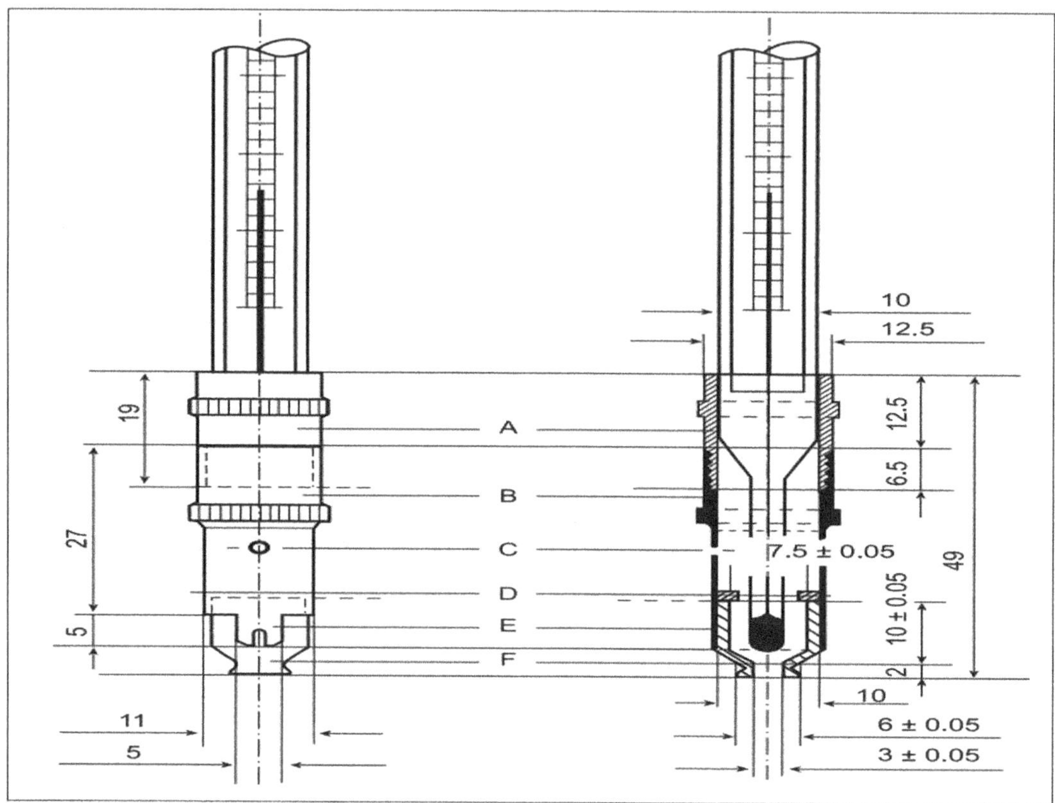


Figure 36. Apparatus for the determination of beeswax drop point (dimensions in mm); metal sheath fixed to a mercury thermometer (A), lower part of sheath (B), hole pierced in the wall of sheath (C), fixed supports (D), 2 tightening bands (E), metal cup fixed to the lower part of the sheath (F).

apparatus is placed in the axis of a tube about 200 mm long and with an external diameter of about 40 mm. It is fixed to the test-tube by means of a stopper through which the thermometer passes, and is provided with a side groove. The opening of the cup is placed about 15 mm from the bottom of the test-tube

***Equipment required:***

- Water bath with controlled temperature
- Drop point apparatus
- Thermometer
- 1L beaker
- Heat and stirring plate
- Stirrer

***Procedure:***

1. Melt the beeswax (10 g) by heating on a water-bath
2. Pour onto a glass plate.
3. Allow cooling to a semi-solid mass.
4. Fill the metal cup (Figure 36) by inserting the wider end into the beeswax.
5. Repeat the procedure until beeswax extrudes from the narrow opening.
6. Remove the excess with a spatula.
7. Insert the thermometer immediately.
8. Remove the beeswax displaced.
9. Allow to stand at room temperature for at least 12 h before determining the drop point.
10. Place the apparatus in the water-bath (the whole device is immersed in a beaker with a capacity of about 1 L) filled with water.
    The bottom of the test-tube is placed about 25 mm from the bottom of the beaker. The water level reaches the upper part of sheath (A). A stirrer is used to ensure that the temperature of the water remains uniform.
11. Heat the water-bath.
12. When the temperature is at about 10 °C below the presumed drop point, adjust the heating rate to about 1 °C/min.
13. Note the temperature at the fall of the first drop.
14. Carry out at least 3 determinations, each time with a fresh beeswax sample. The difference between the readings must not exceed 3 °C. The mean of three readings is the drop point of the beeswax sample.

*6.2.2.2. Melting point.* This index is determined following the "capillary tube method" as described by Bernal et al. (2005).

***Equipment required:***

- Water bath with controlled temperature
- 10 cm long 62 mm internal diameter thin-wall hollow capillary tube

***Procedure:***

1. Melt beeswax (see Section 2.2)
2. Insert melted beeswax in a 10 cm long 62 mm internal diameter thin-wall hollow capillary tube, up to about 1 cm.
3. Allow sample to solidify.
4. Keep at RT for 24 h once solidified.
5. Put the capillary tube containing beeswax into a water bath.
6. Slowly warm the water bath: 1–2 °C/min
   The temperature is checked with a thermometer whose bulb had to be as close as possible to the capillary tube with beeswax.
7. The melting temperature is that at which the beeswax is completely molten: the beeswax liquid is entirely transparent without turbidity.
8. Repeat steps 1–7 twice so that the analysis is performed in triplicate.

Alternatively, the capillary tube-based instruments for automatic determination of melting point (or slip melting point – SMP) employing simple and fast protocols based on microprocessor-controlled temperature ramping up system and in some models, a built-in digital camera that allows the observation of beeswax melting, can be used (for example, MP 55 Melting Point System, Mettler Toledo).

6.2.2.3. *Acid value.* The acid value is defined as the amount of KOH in milligrams needed to neutralize 1 g of beeswax. This method is intended to give a measure of free fatty acids present in beeswax, although the presence of other interfering acid substances can affect this relation.

The acid value determination is one of the most efficient and simple methods for detecting beeswax adulteration with paraffin and stearic acid. Acid value decreases with the addition of paraffin (Bernal et al., 2005; Maia & Nunes, 2013; Svečnjak, Baranović et al., 2015), and increases with the addition of stearic acid (Bernal et al., 2005) (Figures 37 and 38, respectively).

***Equipment required:***

- Balance with a 0.1 mg precision
- 250 mL conical flask
- Reflux condenser
- Glass beads
- Electric plate
- Burette

***Materials and reagents required:***

- Xylene p.a.
- Ethanol p.a.
- Phenolphthalein p.a.
- KOH p.a. or available standardized 0.5 M KOH solution

***Procedure:***

1. Weigh 2 g of beeswax in a 250 mL conical flask fitted with a reflux condenser (***m***, g).
2. Add 40 mL of xylene.
3. Add a few glass beads.
4. Heat on an electric plate at 60–70 °C until the material is completely dissolved.
5. Add 20 mL of ethanol.
6. Add 0.5 mL of phenolphthalein solution (1% phenolphthalein in 95% ethanol).
7. Titrate the hot solution with a standardized c.a. 0.5 M potassium hydroxide (***C***, mol/L) in ethanol until the change in colour of the indicator (appearance of a persistent red colour for at least 10 s), (***v1***, mL).
8. Repeat steps 2–7 to create the reference (blank) solution without the addition of sample, (***v2***, mL).
9. Calculate the acid value using the formula:

$$\text{Acid value} = \frac{(v1 - v2) \times C \times 56.1}{m}$$

6.2.2.4. *Saponification value.* The saponification value is defined as the amount of KOH in milligrams needed to

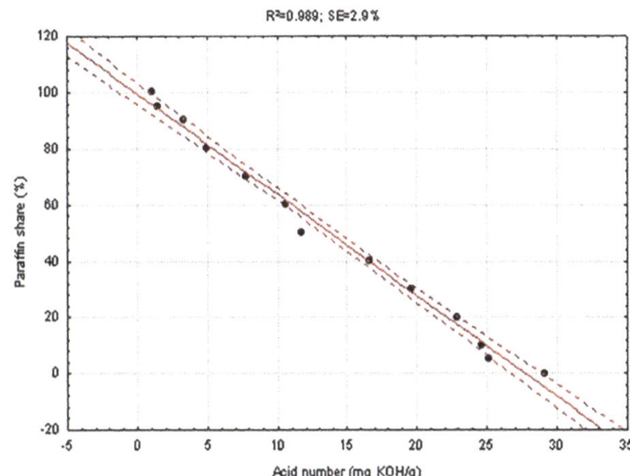


Figure 37. Effect of different levels of beeswax contamination with paraffin on the acid value of beeswax (Svečnjak, Baranović et al., 2015).

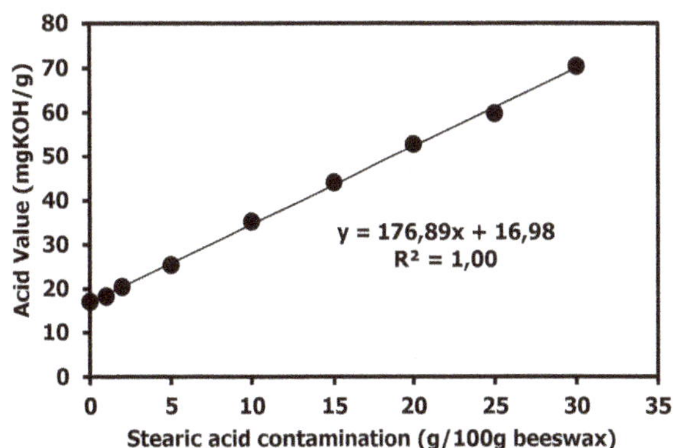


Figure 38. Effect of different levels of beeswax contamination with stearic acid on the acid value of beeswax (Nunes & Maia, 2017, unpublished data).

neutralize 1 g of beeswax after saponification, i.e., alkaline hydrolysis. This method is intended to give a measure of the total fatty acids present in beeswax, either free or esterified, although the presence of other interfering acid substances can affect this relation (any acid or ester present but not associated directly with the sample being analyzed, including adulterant-originated compounds).

Determination of saponification value (and consequently, ester value and ester/acid ratio) is not a reliable for detecting adulterants in comb foundations due to analytical anomalies that may be caused by high heat treatment applied during the comb foundation production process. Also, it is a time-consuming analysis, not suitable for routine testing.

***Equipment required:***

- Balance with a 0.1 mg precision
- 250 mL conical flask
- Reflux condenser
- Glass beads
- Electric plate
- Burette

***Materials and reagents required:***

- Xylene p.a.
- Ethanol p.a.
- Phenolphthalein p.a.
- KOH p.a or available standardised 0.5 M KOH solution
- HCl p.a. or available standardised 0.5 M HCl solution

***Procedure:***

1. Weigh 2 g of beeswax in a 250 mL conical flask fitted with a reflux condenser (***m***, g).
2. Add 30 mL of a mixture of ethanol:xylene (1:1 v/v)
3. Add a few glass beads.
4. Heat until the material is dissolved.
5. Add 25.00 mL of standardized c.a. 0.5 M potassium hydroxide (***C***, mol/L)
6. Heat under reflux during 3 h.
7. Titrate the hot solution immediately with a standardized solution of c.a. 0.5 M of HCl, using 1 mL of phenolphthalein solution (1% phenolphthalein in 95% ethanol) as indicator (***v3***, mL).
8. Reheat the solution to boiling several times (usually three to four times) during the titration to dissolve the material.
9. Repeat steps 2–8 to create the reference (blank) solution without the addition of sample, (***v4***, mL).
10. Calculate the saponification value using the formula:

$$\text{Saponification value} = \frac{(v4 - v3) \times C \times 56.1}{m}$$

*6.2.2.5. Ester value.*

$$\text{Ester value} = \text{saponification value} - \text{acid value}$$

The ester value is defined as the amount of KOH in milligrams needed to neutralize 1 g of ester linked acids of beeswax being calculated by subtracting the saponification value (Section 6.2.2.4) by the acid value (Section 6.2.2.3). This method is intended to give a measure of esterified fatty acids in beeswax.

*6.2.2.6. Iodine number.* The iodine number is defined as the amount in grams of iodine absorbed by 100 g of lipids. The iodine value gives an indication of the average degree of unsaturation, i.e., the number of double bonds of a lipid. There are various methods available in the literature for determining the iodine number of a lipid; the two more common methods are those of Wijs and Hanus. The method mostly used for determination of the iodine number of beeswax is the Hanus method.

The iodine number can also be advantageously used to detect adulteration with paraffin, tallow, and stearic acid (Bernal et al., 2005). An example for iodine number increase in case of adulteration with tallow, and its decrease in case of paraffin adulteration is presented in Figures 39 and 40, respectively.

***Equipment required:***

- Balance with a 0.1 mg precision
- 250 mL conical flask with stopper
- Burette

***Materials and reagents required:***

- Chloroform p.a.
- Iodine p.a.
- Glacial acetic acid p.a.
- Bromine water p.a.
- Potassium iodide p.a.
- Sodium thiosulfate p.a or available standardised 0.1 N sodium thiosulfate solution
- Starch

***Procedure:***

1. Prepare Hanus solution:
    a. Dissolve 18.2 g of iodine in 1 L of glacial acetic acid
    b. Add 3 mL of bromine water for doubling the halogen content.
    Alternatively, there are commercially available Hanus solutions from various distributors.
2. Weigh the appropriate amount of beeswax (***m***, g) in a 250 mL conical flask.

    The amount of material used in iodine value determination is dependent on its expected iodine value

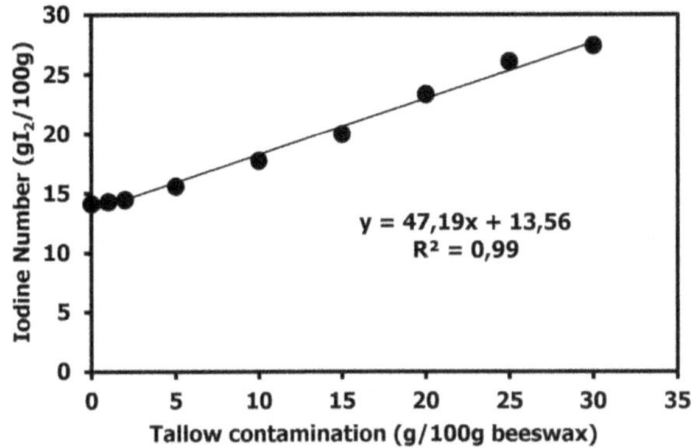


Figure 39. Effect of different levels of beeswax contamination with tallow on the beeswax iodine number (Nunes & Maia, 2017, unpublished data).

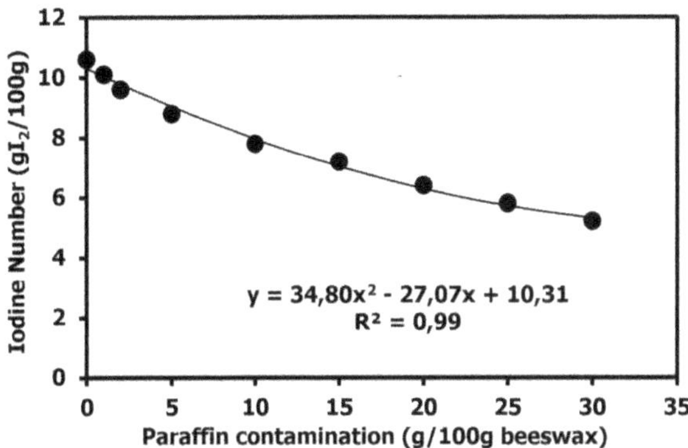


Figure 40. Effect of different levels of beeswax contamination with paraffin on the beeswax iodine number (Nunes & Maia, 2017, unpublished data).

and should be determined by a previous titration of the sample before carrying out the analytical determination, as presented in Table 15.

3. Add 10 mL of chloroform.
4. Add 25mL of Hanus solution by draining in a definite time (the time of draining should be kept constant between analysis).
5. Close the flask with a glass stopper.
6. Mix thoroughly.
7. Allow standing in the dark for 30 min with occasional shaking.
8. Add 10 mL of 15% potassium iodide solution.
9. Shake thoroughly.
10. Add 100 mL of freshly boiled and cooled distilled water, washing down any free iodine on the stopper.
11. Titrate against standardized 0.1 N sodium thiosulfate (**N**, equivalent $L^{-1}$) solution until the yellow solution turns almost colourless.
12. Add 2–3 drops of starch solution (1%); a deep blue colour must be formed.
13. Shake the stoppered flask vigorously (so that any iodine remaining in the chloroform is taken up by the potassium iodide solution).
14. Continue with titration until the blue colour disappears (**v5**, mL).

Table 15. Amount of sample to be weighed in the iodine number determination as a function of the expected iodine number.

| Expected iodine number | Mass of sample to be used (g) |
|---|---|
| 0–30 | 0.8000 |
| 30–50 | 0.5000 |
| 50–100 | 0.2500 |
| 100–150 | 0.1700 |
| 150–200 | 0.1200 |

15. Repeat steps 3 to 14 to create the reference (blank) solution without the addition of sample, (**v6**, mL).
16. Calculate the iodine number using the formula:

$$\text{Iodine number} = \frac{(v6 - v5) \times N \times 12.69}{m}$$

6.2.2.7. *Peroxide value.* The peroxide value of a sample gives a measure of the amount of hydroperoxides present, being a measure of the oxidation of a sample, expressed commonly as milliequivalents of peroxide per kilogram of sample (AOAC 965.33, 2000).

The method has two main sources of errors, the oxidation of the reagents by air, accelerated by light, and the re-absorption of the liberated iodine by the material, which can reduce the peroxide value obtained and these sources of errors should be taken into account when interpreting the results.

Determination of peroxide value is commonly used for beeswax purity testing in the frame of chemical and technical assessment of beeswax as food additive (FAO, 2005; JECFA, 2005).

***Equipment required:***

- Balance with a 0.1 mg precision
- 250 mL conical flask with stopper
- Burette

***Materials and reagents required:***

- Chloroform p.a.
- Glacial acetic acid p.a.
- Potassium iodide p.a.
- Sodium thiosulfate p.a or available standardised 0.1 N sodium thiosulfate solution
- Starch

***Procedure:***

1. Transfer approximately 5 g of the sample (**m**, g) into a 250 mL Erlenmeyer flask with glass stopper.
2. Add 30 mL of the glacial acetic acid: chloroform 3:2 (v/v).
3. Swirl to dissolve.
4. Add 0.5 mL of saturated potassium iodide solution, freshly prepared (dissolve excess KI in freshly boiled water, excess solid must remain, store in the dark).

5. Shake thoroughly for 60 ± 1 s while reaction is occurring.
6. Add 30 mL of water.
7. Shake vigorously.
8. Slowly titrate with 0.1 N sodium thiosulfate solution (**N**, equivalent $L^{-1}$) with vigorous shaking until the solution presents a pale yellow colour.
9. Add 0.5 mL of 1% aqueous starch solution and continue titration with vigorous shaking to release all iodine from the organic chloroform layer, until the colour disappears. (**v7**, mL).
10. If less than 0.5 mL of the 0.1 N sodium thiosulfate solution is used, repeat the determination using the standardized 0.01 N sodium thiosulfate solution (step 8).
11. Repeat steps 2–10 to create the reference (blank) solution without the addition of sample (the volume used should be less than 0.1 mL of the 0.1 N thiosulfate solution) (**v8**, mL).
12. Calculate the peroxide value using the formula:

$$\text{Peroxide value} = \frac{(v7 - v8) \times N}{m} \times 1000$$

*6.2.2.8. Detection of carnauba wax.*

***Equipment required:***

- Balance with a 0.1 mg precision
- Test tubes
- Water bath with controlled temperature
- Beaker
- Light microscope

***Materials and reagents required:***

- *n*-butanol p.a.

***Procedure:***

1. Weigh 100 mg of the sample into a test tube.
2. Add 20 mL of *n*-butanol.
3. Heat the test tube in a boiling water bath.
4. Shake the mixture gently until the sample dissolves completely.
5. Transfer the test tube to a beaker of water at 60 °C.
6. Allow the water to cool to room temperature.
7. A loose mass of fine, needlelike crystals separates from clear mother liquor. Under the microscope, the crystals appear as loose needles or stellate clusters, and no amorphous masses are observed, indicating the absence of carnauba wax.

*6.2.2.9. Detection of ceresin, paraffin and other waxes.*

***Equipment required:***

- Balance with a 0.1 mg precision
- Round-bottomed flask
- Reflux condenser
- Thermometer
- Water bath with controlled temperature

***Materials and reagents required:***

- KOH p.a.
- Ethanol p.a.

***Procedure:***

1. Weigh 3 g of material in a round-bottomed flask.
2. Add 30 mL of a 40 g/L solution of KOH in aldehyde-free ethanol.
3. Boil gently during 2 h under reflux by attaching a condenser.
4. Remove the condenser and immediately insert a thermometer.
5. Place the round-bottomed flask in a water-bath at 80 °C.
6. Allow cooling by swirling the solution continuously.
7. For pure beeswax no precipitate is formed up to 65 °C, although the solution may be slightly opalescent. If above 65 °C there is a formation of precipitates, this indicates the presence of ceresin, paraffin and other waxes. At 59 °C, the solution of pure beeswax is cloudy.

*6.2.2.10. Detection of glycerol and other polyols.*

***Equipment required:***

- Balance with a 0.1 mg precision
- Round-bottomed flask
- Reflux condenser
- G4 glass filter
- Test tubes
- Thermometer
- Water bath with controlled temperature

***Materials and reagents required:***

- KOH p.a.
- Ethanol p.a.
- Sulfuric acid p.a.
- Glycerol p.a.
- Sodium periodate p.a.
- Fuchsin p.a.
- Sodium sulphite p.a.
- HCl p.a.
- Activated charcoal

***Procedure:***

1. Add 10 mL of ethanolic potassium hydroxide solution (0.5 N) to 0.20 g of beeswax.
2. Heat on a water-bath under a reflux condenser for 30 min.
3. Add 50 mL of dilute sulphuric acid (98 g/L of $H_2SO_4$)

4. Cool.
5. Filter through a G4 glass filter.
6. Rinse the flask and the filter with dilute sulphuric acid.
7. Combine the filtrate and washes.
8. Dilute to 100.0 mL with dilute sulphuric acid.
9. Prepare a 10 mg/L glycerol standard in diluted sulphuric acid (98 g/L of $H_2SO_4$).
10. Place 1.0 mL of the test solution in a test-tube.
11. Place 1 mL of the glycerol standard in another test tube.
12. Add 0.5 mL of a 10.7 g/L solution of sodium periodate to each tube
13. Mix thoroughly.
14. Allow to stand for 5 min.
15. Prepare decolorized fuchsin solution by dissolving 0.1 g of basic fuchsin in 60 mL of water.
    a. Allow the fuchsin solution to stand protected from light for at least 12 h.
    b. Add a solution of 1 g of anhydrous sodium sulfite (reagent grade) in 10 mL of water.
    c. Add 2 mL of hydrochloric acid slowly and with continuous shaking of the solution.
    d. Dilute to 100 mL with water.
    e. Decolorize with activated charcoal.
    f. Filter through a G4 glass filter.
    g. If the solution becomes cloudy, filter again before use.
    h. If by standing the solution becomes violet, decolorize again by adding activated charcoal.
    i. Store protected from light.
16. Add 1.0 mL of decolorized fuchsin solution to each tube.
17. Mix thoroughly (any precipitate disappears).
18. Place the tube in a beaker containing water at 40 °C.
19. During cooling observe for 10–15 min.
20. The violet blue colour developed by the pure beeswax containing solution should not be more intense than that of the glycerol standard prepared and analyzed simultaneously [using 1.0 mL of a 0.001% (w/v) solution of glycerol in dilute sulphuric acid].

*6.2.2.11. Advantages and disadvantages of physico-chemical methods.*

**Advantages:**

- Inexpensive equipment, material and reagents.
- Easy to implement.
- No need for advanced chemical analysis training.
- Existence of reference values for pure beeswax for the majority of physico-chemical parameters (though analytical anomalies that may occur should be taken into consideration).

**Disadvantages:**

- Questionable reliability of certain physico-chemical methods for beeswax adulteration detection (determination of melting point, saponification and ester value, and ester/acid ratio) in relation to the type of beeswax specimen or adulterant type.
- High detection thresholds: 5–50% (depending on the type of adulterant and parameter used for its detection).
- Methods require high amounts of the sample for analysis (compared to GC-MS and FTIR-ATR method).
- Analyses are destructive.
- Provide no chemical information about the beeswax constituents.
- Performance of more than one analysis is required to determine the presence of adulterants.
- For quantification purposes (and sometimes reliability), the beeswax adulteration should be confirmed by more reliable analytical tools - spectroscopic techniques (GC-MS or FTIR-ATR).

*6.2.2.12. Conclusion.* Considering all the challenges associated with the use of physico-chemical methods for beeswax adulteration detection, we recommend the following approach to be applied when assessing the authenticity of beeswax using physico-chemical parameters:

- Always combine a set of several physico-chemical measurements (at least 3; an exclusive measurement of only one physico-chemical parameter should not be used for detecting adulterates in beeswax; the most reliable combination of physico - chemical measurements to discriminate between pure and adulterated beeswaxes is determination of acid value, iodine number and peroxide value (Bernal et al., 2005).
- For reliable confirmation of adulterants presence (detection of in average <5–15% of adulterants) and quantification of adulterants in beeswax, physico-chemical measurements should be complemented with spectroscopic analysis (GC-MS and/or FTIR-ATR technique described below in Sections 6.2.4 and 6.2.5, respectively).

### *6.2.3. Sensory analysis of beeswax*

Determination of the sensory characteristics of beeswax is not sufficient evidence of adulteration, but in some cases can give hints on possible adulteration. In this way, this kind of analysis can be important to beekeepers because they are fast and cheap and can be a useful as a preliminary analysis of the beeswax quality in industry. The negative control for sensory assessment is authentic beeswax (collected in accordance with guidelines described in Section 2.1.1).

To obtain the beeswax sample from comb wax, comb foundation or wax blocks for sensory assessment, follow the procedure described in Section 2.1. Beeswax melting and storage should be performed as described in Sections 2.2 and 2.3, respectively.

Figure 41. Diversity of colours in unadulterated comb foundations due to mixing of different wax blocks and absence of wax bleaching. Photo: Maia M.

*6.2.3.1. Organoleptic characteristics.*

*6.2.3.1.1. Colour of beeswax.* The colour of the freshly produced beeswax is white, later it turns to yellow. This yellow colour originates from propolis and pollen colorants (Bogdanov, 2004a). The colour of the beeswax is an important quality parameter (point of view of refiners and industry), but should not be taken as a criterion for establishing wax purity (Barros, Nunes, & Maia, 2009). Figure 41 shows comb foundation of naturally different colours (without adulterants).

In the industry, the colour of beeswax can be altered by bleaching or by the addition of colorants (Bianchi, 1990). Bleaching is often done with chemicals. The use of these chemicals can be problematic from an environmental and toxicological point of view. Beeswax can be brightened simply by exposing it to the sun, or by chemicals using the following methods.

1. Use 2–3 g of oxalic acid per kg wax and 1 L of water. Add acid to water and not *vice versa* (Bogdanov, 2016a).
2. Use 1 mL concentrated of sulphuric acid to 1 L of water per kg wax.
3. Add concentrated hydrogen peroxide solution (about 35% in basic *milieu*) to hot wax (100 °C). There is no point in using more than the amount of hydrogen peroxide suggested, and any excess may cause too much frothing and a boiling over of wax (Brown, 2009). An excess of hydrogen peroxide can cause problems in the manufacture of creams and ointments.

*6.2.3.1.2. Aroma and taste of beeswax.* The odour is very important to characterize the beeswax sensory profile and/or to indicate an addition of certain adulterants, such as oils, tallow, paraffin or resins. Beeswax has a characteristic odour, originating from the bees, propolis and honey (Bogdanov, 2004b). Odour is best assessed by placing the wax sample in a test tube (Bianchi, 1990) and should be pleasant and honey-like (Bogdanov, 2016a). Some factors can contribute to the change of aroma of wax, like the use of combs containing fermented honey and the use of some solvents (Bogdanov, 2004b). The taste of beeswax is normally pleasant and is not specific – any unpleasant taste is a sign of quality deterioration due to foreign matter (Bogdanov, 2016a). See Table 16 to compare the odour and taste of pure beeswax and adulterated beeswax. It should be noted that it is not recommendable to assess the taste of beeswax in case of suspicious beeswax samples given the unknown chemical background of various foreign substances that might be used for beeswax adulteration.

*6.2.3.1.3. Consistency of beeswax.* Consistency level is an important test to indicate the presence of some adulterants in beeswax (Bianchi, 1990). Consistency test should be performed as follows:

1. Weigh 1 g of beeswax and melt in a test tube.
2. Put into a watch glass and leave 24 h at room temperature.
3. After this period, take the sample between index finger and thumb and make a ball.
4. See Table 16 for differences between pure beeswax and mixtures of beeswax with different adulterants.

Table 16. Differences between pure beeswax and beeswax mixture with some adulterants.

| Substance | Odour/taste | Consistency |
|---|---|---|
| Beeswax | Should have a honey-like smell | The mass of beeswax is slightly sticky to fingers and has a homogeneous, transparent and dull appearance |
| Paraffin | It also has a peculiar odour of petroleum (the more intense the smell, the greater the percentage of paraffin) | Adulteration with paraffin is recognised by the appearance of an intense brightness in the moulded beeswax sample between the fingers, and become slippery between the fingers |
| Tallow | Fat-like odour | Elastic, easily mouldable |
| Stearin/stearic acid | Hard to characterize specific odour/not distinctive | Like porcelain bright, with visible crystalline appearance, undoes easily between fingers |

*6.2.3.2. Qualitative methods for rapid detection of beeswax adulteration.* The qualitative tests are quick and interesting because they can indicate, with some assurance, the level of adulteration, especially with regard to paraffin (Barros et al., 2009). The description of these tests is modified from Bianchi (1990).

*6.2.3.2.1. Detection of paraffin.*

***Procedure:***

1. The alcohol solution of 12% KOH should be prepared on the same day.
2. Weigh 5 g of beeswax in 250 mL beaker.
3. Add 25 mL of 12% KOH alcohol solution.
4. Heat to a pasty consistency (at ~ 70–80 °C).
5. Quickly dissolve with 20 mL of hot glycerine.
6. Add hot distilled water.
7. In the presence of paraffin or ceresin, a milky liquid is produced.
8. Can be converted into a precipitate if there are high amounts of paraffin.

It is possible to designate wax samples as adulterated with paraffin in excess of 15%. We also found that in case of adulteration with 10% of paraffin, there was a slight increase in the opacity of the solution. However, we think it may not be convincing enough to confirm that a wax sample is adulterated, because this type of analysis is highly dependent on the operator's observation. For samples with high percentages of paraffin (> 20%), this methodology is quite persuasive, independently of the operator, because the milky appearance of the liquid is very typical (Figure 42). Results of this test are related with acid value, and these two methods are complementary to each other for the detection of paraffin. The qualitative method for rapid detection of paraffin generates less than 5% false positives; the absence of paraffin was confirmed by FTIR-ATR (Maia & Nunes, 2017, unpublished data).

*6.2.3.2.2. Detection of tallow.*

***Procedure:***

1. In a beaker (50 mL), bring to the boil 1 g of beeswax with 10 mL of sodium carbonate solution at anhydride 30%, until its dilution.
2. Leave it to cool at room temperature.
3. Allow standing for another 3 min.
4. Carefully pour the liquid into the test tube.
5. Add 5 mL of water.
6. In the presence of fat, a milky liquid rapidly form some lumps that rise to the surface after ±10 min (Figure 43).

*6.2.3.2.3. Detection of stearin.*

***Procedure:***

1. Heat 1 g of beeswax in a beaker (50 mL) until it is melted.
2. Add 7 mL of ethyl alcohol and over 3 mL of water until its dissolution.
3. Cool to 20 °C.
4. Filter into a test tube (filter used: Whatman 114).
5. Add 4 mL of distilled water to the filtrate.
6. In the presence of stearin, a milky white precipitate is observed (Figure 44).

*6.2.3.2.4. Detection of resin.*

***Procedure:***

1. Heat 1 g of beeswax and 4 mL of concentrated nitric acid (65%) in a beaker (50 mL) until it is melted.
2. Boil for 1 min.
3. Leave it to cool at room temperature.
4. Dilute with 4 mL of water.
5. Rapidly neutralize with with few drops of ammonia.
6. In the presence of resins, reddish vapours are formed during the heating of the blend and the final solution is a reddish-brown color.

*6.2.3.2.5. Detection of starch.*

***Procedure:***

1. Place 1g beeswax in a 50 mL beaker.
2. Heat (at 62–65 °C, until melting starts)
3. Add 12 mL of essence of turpentine until its dissolution.
4. Add an aqueous solution of iodine if a white deposit was formed.
5. If the deposit becomes blue, the sample contained starch or flour.

Figure 42. Detection of paraffin by qualitative methods: (a) sample without paraffin; (b) sample containing 15% of paraffin; (c) sample containing 20% of paraffin; (d) sample containing 30% of paraffin.

Figure 43. Formation of lumps characteristic for tallow adulteration.

*6.2.3.2.6. Detection of minerals.*

***Procedure:***

1. Heat 1 g of beeswax in a beaker (50 mL) until it is melted.
2. Add 12 mL of chloroform until dilution of the beeswax dilution.
3. In the presence of minerals, an insoluble residue is formed.

*6.2.3.3. Advantages and disadvantages of sensory analysis.*

**Advantages:**

- Cheap and fast methods.
- Can be related to other analytical parameters (physico-chemical, spectroscopic) and complement the confirmation of adulteration.
- Low percentage of false positives.

**Disadvantages:**

- Sensory analysis comprises both objective and subjective evaluation
- Subjective measures
- High detection limit for some adulterants (paraffin: 10–15%).
- For some parameters, the detection of adulteration may depend on the operator experience.

*6.2.4. Adulteration detection by GC-MS technique*

The GC-MS method elaborated by Waś et al. (2014a) for the determination of beeswax hydrocarbons is also suitable for the detection of beeswax adulteration with hydrocarbons of foreign origin, e.g., paraffin or ceresin

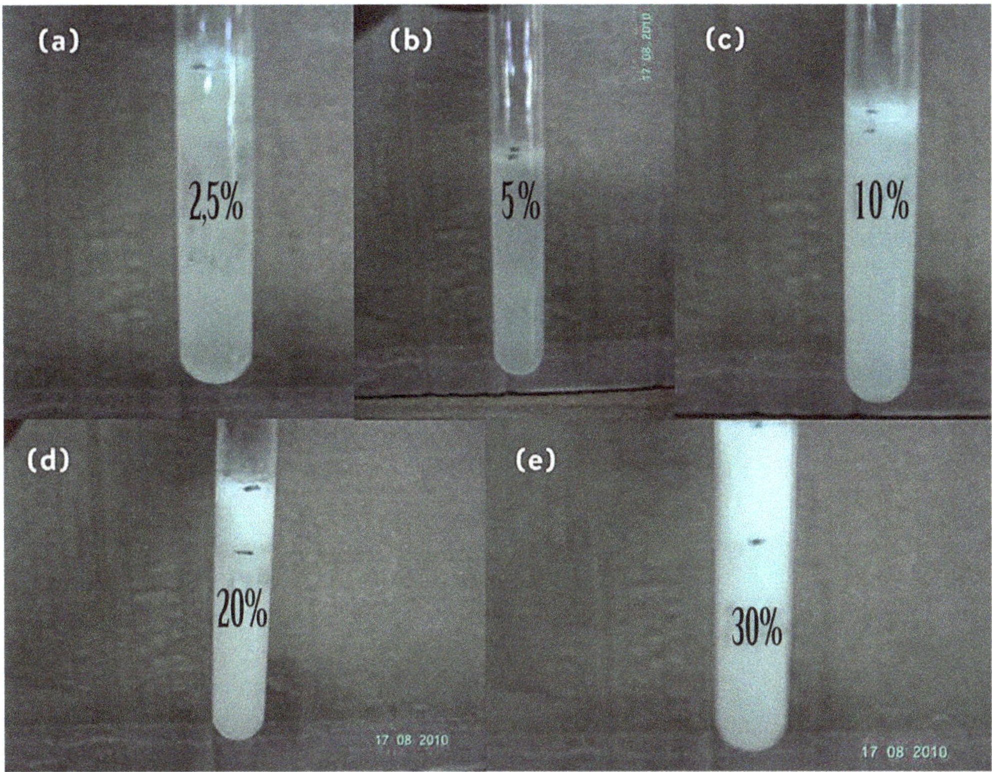


Figure 44. Adulteration with stearin: (a) sample containing 2.5% of stearin; (b) sample containing 5% of stearin; (c) sample containing 10% of stearin; (d) sample containing 20% of stearin; (e) sample containing 30% of stearin.

(Waś et al., 2015, 2016). The method is schematically illustrated in Figure 45.

The GC-MS method which is described in details in Section 5.2.1 focuses on identification of hydrocarbons (alkanes, alkenes, and dienes) and quantitative analysis of n-alkanes. These compounds naturally occur in beeswax, but they can also come from adulteration with substances such as paraffin or ceresin. The routine analysis of detection of beeswax adulteration with hydrocarbons of foreign origin can be carried out rapidly by comparing the chromatograms (fingerprints) of hydrocarbons in pure and adulterated beeswax as is presented in Figure 46. The proposed method also enables the quantitative analysis of n-alkanes, which could be used for estimation a degree of beeswax adulteration with paraffin. The minimum estimated percent of paraffin detectable using the GC-MS technique is 3%. According to the criteria established by Waś et al. (2016) the adulteration of beeswax with paraffin is indicated by the presence of hydrocarbons containing over 35 atoms of carbon in the molecule, and by the higher contents of individual n-alkanes ($C_{20}H_{42}$–$C_{35}H_{72}$) as well as a higher content for the total of these compounds, in comparison to the maximum contents determined in beeswax (Waś et al., 2014b). Similar criteria, but with slightly different critical values for the contents of beeswax hydrocarbons, were recommended by other authors (Jiménez et al., 2007).

*6.2.4.1. Detection of beeswax adulteration with foreign hydrocarbons.* The analytical procedure for detection of beeswax adulteration with hydrocarbons of foreign origin (substances such as paraffin or ceresin) is the same as procedure for determination of hydrocarbons in pure beeswax. This procedure is described step by step in Section 5.2.1 (along with the list of equipment and materials/reagents required):

1. Preparation of standard solution (5.2.1.1).
2. Beeswax sample preparation (5.2.1.2).
3. Conditions of GC-MS analysis (5.2.1.3).
4. Qualitative analysis of beeswax hydrocarbons (5.2.1.4).
5. Quantitative analysis of n-alkanes in beeswax (5.2.1.5).

*6.2.4.2. Quality assessment of beeswax using GC-MS.* In assessing the quality of commercial beeswax, the following criteria indicating its adulteration with hydrocarbons of foreign origin (e.g., paraffin) should be applied:

1. Presence of hydrocarbons containing over 35 carbon atoms in the molecule.
2. Higher contents of individual n-alkanes ($C_{20}H_{42}$–$C_{35}H_{72}$).
3. Higher content for the total of n-alkanes in comparison to the maximum values determined in pure beeswax.

In the routine preliminary quality control of beeswax is recommend the comparative analysis of the hydrocarbon chromatograms, an example of which is given in

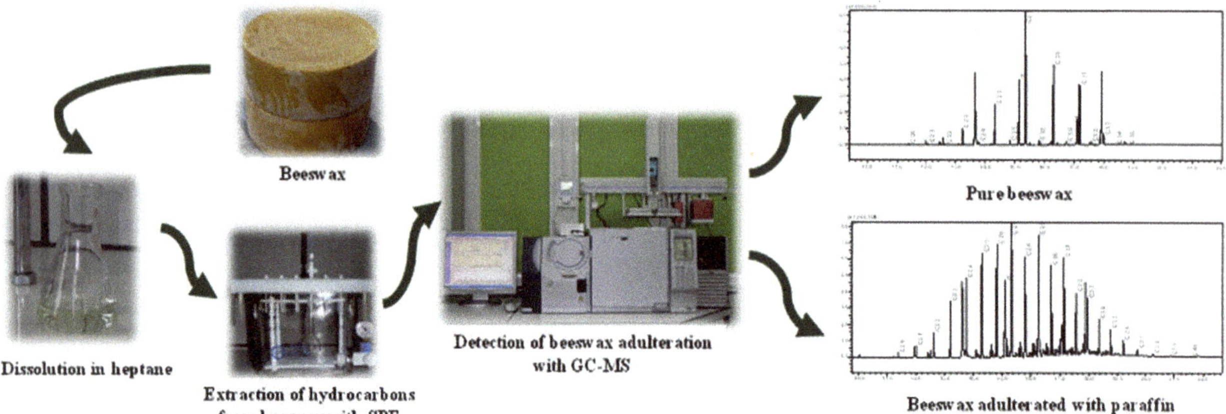


Figure 45. General procedure of the detection of beeswax adulteration by GC-MS.

Figure 46. For the compaerative analysis of chromatograms the same amounts of beeswax samples should be taken. The GC-MS chromatogram of hydrocarbons presented in Figure 46(a) is proposed as standard fingerprint of beeswax originate from *A. mellifera*. By comparing of the chromatograms of hydrocarbons in pure beeswax (Figure 46(a)) and adulterated with paraffin (Figure 46(b,c)), it can be easily noted that the intensity of the peaks of individual alkanes in beeswax increases when the paraffin is added. The increase of the peak intensities is particularly visible for the alkanes with even numbers of carbon atoms in the molecule ($C_{24}H_{50}$, $C_{26}H_{54}$, $C_{28}H_{58}$, $C_{30}H_{62}$, $C_{32}H_{66}$, $C_{34}H_{70}$), even with only a 3% addition of paraffin (Figure 46(b)). Moreover, the addition of substances such as paraffin or ceresin to beeswax resulted in the occurrence of alkanes containing over 35 atoms of carbon in the molecule (Figure 46(b,c)), which are not detected in pure beeswax (Figure 46(a)).

In the next step, it is advisable to confirm the qualitative results based on chromatograms by quantitative analysis performed using the internal standard method (described in Section 5.2.1.5). Then, n-alkane contents determined in unknown sample must be compared with the maximum contents determined for pure beeswax, proposed as the concentration – guide values to distinguish between pure and adulterated beeswax (Waś et al., 2014b) shown in Table 17. According to these requirements, the maximum concentrations accepted for pure beeswax amounted to 11.7 g/100 g (for total n-alkanes $C_{20}H_{42}$–$C_{35}H_{72}$) and 1.0 g/100 g (for total even-numbered n-alkanes from $C_{20}H_{42}$ to $C_{34}H_{70}$).

The most frequently used adulterant is paraffin. However, various types of paraffin available on the market differ not only in physico-chemical properties (such as melting point), but also in the hydrocarbon composition. The results of studies investigating different types of paraffin and their mixtures with beeswax were presented by Waś et al. (2015, 2016). In all investigated types of paraffin, the homologous series of n-alkanes much longer than those in beeswax were found. Also the amounts of n-alkanes were different depending on the type of paraffin. These results might be helpful and useful in detection of beeswax adulteration with paraffin. Depending on the type and amount of paraffin used for beeswax adulteration, chromatograms and determined contents of hydrocarbons in beeswax adulterated with paraffin may be different. Nonetheless, the same criteria and rules defined in this paragraph for assaying the quality of unknown beeswax sample should be applied.

*6.2.4.3. Advantages and disadvantages of GC-MS method for beeswax adulteration detection.*

**Advantages:**

- more precise, accurate and selective in comparisons to classical physico-chemical methods
- efficient in detection of beeswax adulteration with hydrocarbons of alien origin e.g., paraffin, ceresin (allows detection of about 3% of paraffin added to beeswax)
- preliminary purification of a sample with SPE, eliminates the overlapping of peaks and prevents inaccurate results of the quantitative analysis
- Suitable for a routine control of beeswax carried out rapidly by comparing the chromatograms (fingerprints); possible without SPE procedure.

**Disadvantages:**

- allows only the detection of adulteration with hydrocarbons of alien origin (paraffin, ceresin)
- expensive due to a very specialized equipment

*6.2.5. Detection of beeswax adulteration by FTIR-ATR spectroscopy*

***Definitions, acronyms*** (in addition to those listed in Section 5.3):

- RS: reference (calibration) standard/s
- ABM: adulterant–beeswax mixtures
- *R*: coefficient of correlation

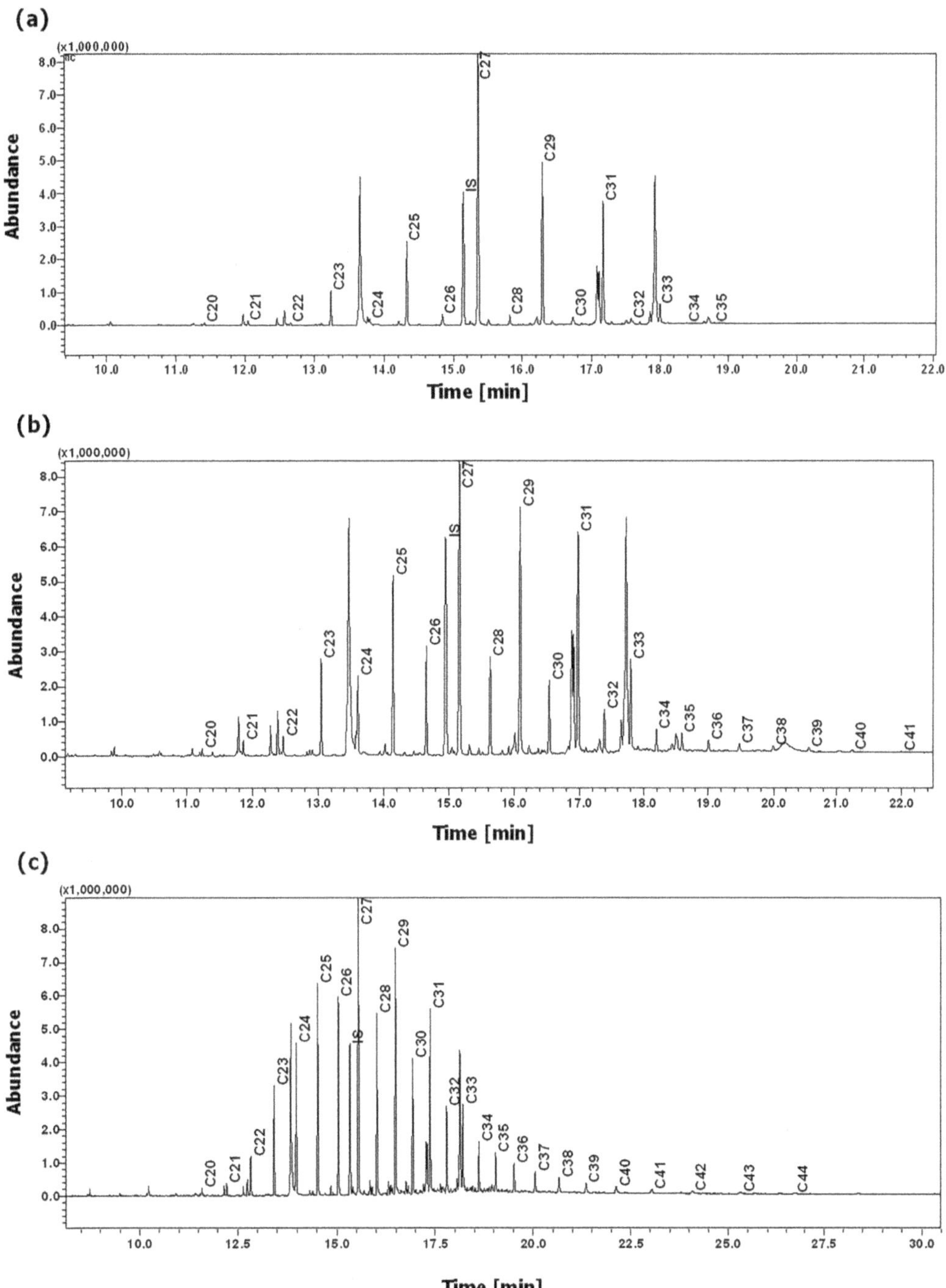


Figure 46. GC-MS chromatogram of hydrocarbons in beeswax: (a) pure beeswax originated from *A. mellifera*, (b) beeswax with 3% addition of paraffin, (c) beeswax with 30% addition of paraffin; C20–C44 – n-alkanes with the formula $C_{20}H_{42}$–$C_{44}H_{90}$; IS – internal standard ($C_{30}H_{62}$).

- $R^2$: coefficient of determination
- SE: standard error
- $c_v$: coefficient of variation
- RSD: relative standard deviation

FTIR spectroscopy is generally well-recognized as very reliable or promising method in analysis of various food systems (such as honey, milk, olive oil, juice, etc.) for authentication purposes (Casale & Simonetti, 2014; Cozzolino, Corbella, & Smyth, 2011; Karoui, Downey, & Blecker, 2010; Rios-Corripio, Rojas-López, & Delgado-Macuil, 2012), and even more demanding systems such as those involved in forensic sciences (Yu & Butler, 2015). However, it was not used for the authentication of beeswax until the second decade of this century. In 2013, Maia et al. demonstrated a feasibility study of FTIR-ATR for detection of beeswax adulterants with good detection limits (≤5%) of the four most commonly used beeswax adulterants: paraffin, microcrystalline wax, tallow, and stearic acid. Shortly after, Svečnjak, Baranović et al. (2015) established an analytical procedure for routine detection of beeswax adulteration by

Table 17. Contents of hydrocarbons (min-max) proposed by different authors as concentration guide-values to distinguish between pure and adulterated beeswax originated from *A. mellifera*.

| Compound | Jiménez et al. (2007)[a] | Waś et al. (2016)[b] |
|---|---|---|
| Heptadecane | 0.03–0.22 | Not detected |
| Nonadecane | 0.08–0.64 | Not detected |
| Eicosane | 0.03–0.13 | 0.01–0.06 |
| Heneicosane | 0.19–0.61 | 0.03–0.10 |
| Docosane | 0.02–0.13 | 0.02–0.09 |
| Tricosene | 0.03–0.18 | Not determined quantitatively |
| Tricosane | 0.69–1.33 | 0.12–0.68 |
| Tetracosane | 0.04–0.18 | 0.03–0.13 |
| Pentacosene | 0.05–0.18 | Not determined quantitatively |
| Pentacosane | 1.26–1.78 | 0.42–1.47 |
| Hexacosane | 0.18–0.35 | 0.06–0.22 |
| Heptacosene | 0.05–0.12 | Not determined quantitatively |
| Heptacosene isomer | 0.09–0.21 | Not determined quantitatively |
| Heptacosane | 2.55–3.20 | 2.44–4.40 |
| Octacosane | 0.14–0.37 | 0.06–0.19 |
| Nonacosene | 0.06–0.52 | Not determined quantitatively |
| Nonacosane | 1.87–2.68 | 1.68–2.73 |
| Triacontene | 0.02–0.07 | Not detected |
| Triacontene isomer | 0.02–0.05 | Not detected |
| Triacontane | 0.11–0.31 | 0.05–0.19 |
| Hentriacontadiene | 0.01–0.08 | Not determined quantitatively |
| Hentriacontene | 0.61–0.89 | Not determined quantitatively |
| Hentriacontene isomer | 1.05–1.58 | Not determined quantitatively |
| Hentriacontane | 1.62–2.45 | 1.53–2.64 |
| Dotriacontene | 0.06–0.12 | Not detected |
| Dotriacontane | 0.04–0.14 | 0.01–0.12 |
| Tritriacontadiene | 0.01–0.08 | Not determined quantitatively |
| Tritriacontene | 1.19–1.82 | Not determined quantitatively |
| Tritriacontene isomer | 0.06–0.38 | Not detected |
| Tritriacontane | 0.34–0.72 | 0.31–0.76 |
| Tetratriacontane | Not detected | <0.025[c]–0.03 |
| Pentatriacontene | 0.03–0.16 | Not determined quantitatively |
| Pentatriacontene isomer | 0.02–0.08 | Not determined quantitatively |
| Pentatriacontane | 0.01–0.09 | <0.025[c]–0.03 |
| Total | | 8.27–11.66 |

[a]GC-FID method, concentration in wt%, related to the internal standard (octadecyl octadecanoate).
[b]GC-MS method, concentration in g/100 g, internal standard method with using standard mixture of n-alkanes ($C_{20}H_{42}$-$C_{40}H_{82}$) and squalane ($C_{30}H_{62}$) used as IS.
[c]Limit of determination.

generating the calibration standards and curves for detection of various adulterants.

Considering the advantages of this spectroscopic technique and observations reported in studies mentioned above, we here present a simple and reliable FTIR-ATR method that enables distinguishing between authentic and adulterated beeswax samples, including the procedure for the quantification of adulterants in contaminated beeswax samples [method by Svečnjak, Baranović et al. (2015) is complemented and slightly modified for this purpose]. The general list of equipment and materials required is given in Section 5.3.

*6.2.5.1. Generating IR spectral database of reference samples.* Prior to infrared analysis aiming to detect the presence of adulterants in beeswax, it is necessary to conduct detailed IR characterization of beeswax and adulterant samples (as described in Section 5.3). To detect and quantify foreign substances in beeswax, a reference IR spectral database of authentic (genuine) beeswax, adulterants (as much different types as possible), and their mixtures should be generated. This requires preparation of in-house quality control materials, as described in the following subsections. Generating IR spectral database of the reference samples involves several steps described in detail bellow:

1. Sampling of the reference specimens (genuine beeswax, adulterants) – see Section 6.2.5.1.1.
2. Preparation of adulterant–beeswax mixtures/reference (calibration) standards – see Section 6.2.5.1.2.
3. Acquisition of FTIR-ATR spectra of prepared samples – see Section 6.2.5.1.3.

*6.2.5.1.1. Sampling of reference specimens.*

*6.2.5.1.1.1. Genuine beeswax – sampling and storage conditions.* To obtain the reference IR spectra of genuine beeswax, virgin beeswax samples (wild-built combs) should be collected directly from the hives maintained in controlled conditions, as described in Section 2.1.1. Sampling, melting and storage of comb wax and other types of beeswax specimens is performed as described in Section 2.

It is recommendable to collect as much wild-built combs as possible (>10) from different hives to determine minor chemical variations naturally occurring among different beeswaxes in particular geographical region (this is important to reduce the level of uncertainty when evaluating test samples, as explained in Section 6.2.5.2.3.3.).

*6.2.5.1.1.2. Adulterants – purchasing and storage conditions.* Adulterants of interest can be purchased on the national or international market depending on their availability in different countries. Only certified products purchased from the commercial suppliers (companies and/or specialized shops) that guarantee product's authenticity should be used for generating the IR spectral database of the reference adulterant specimens.

For example, crude paraffin wax with melting point 56 °C (*paraffinum solidum*, Ph. Eur. 7.8) and stearic acid (*acidum stearicum*, Ph. Eur. 8.1) can be purchased from the local pharmaceutical supplier. Different types of high melting point paraffin waxes, commonly used in the beekeeping technology for hive waxing, can also be purchased for this purpose if they have a product specification. Animal fats are available at different market places, but it is preferable to obtain beef and/or mutton tallow from the local butcher shop.

The majority of wax-based adulterants can be stored in original packaging with lids and stored according to the manufacturer's instructions or as described in Section 2. However, due to the lower stability (biological degradation tendency) of particular animal-derived specimens, such as beef or mutton tallow, these must be melted and filtered before storing. Melting and filtering are necessary to remove needless tissues (venation) that are naturally present in animal fat. Commercial paper filters or canvas filters can be used for this purpose. Afterwards, samples should be stored in an airtight glass containers at +4 °C to prevent oxidation and decomposition.

*6.2.5.1.2. Preparation of adulterant–beeswax mixtures (reference standards).* To obtain reference samples exhibiting IR spectral features specific for particular level of adulteration, and ensure precise prediction and quantification of the same levels in test samples, mixtures of genuine beeswax containing different proportions of particular adulterant (adulterant–beeswax mixtures – ABM) should be prepared as follows (as an example, we present a preparation description for 5 g mixtures):

1. Weigh the material (beeswax and adulterant) for the preparation of ABM containing different proportions of adulterant (w/w) in accordance with the recipe given in Table 18 (recipe for 5 g ABM; higher amounts can also be prepared if required).
   Note: prepared mixtures are to be used for the establishment of calibration (standard) curves exhibiting a relationship between the FTIR instrument response and the known concentrations of the analyte (adulterant). It is therefore of crucial importance to conduct an accurate weighing a precise analytical balance (precision: ±0.001 g).
2. Place the weighed specimens in small glass thermostable storage containers.
3. Place the containers in a temperature chamber for 3 h at 90 °C for melting and homogenisation (stir the mixtures gently several times during this period for better homogenization).
   Note: complete set of material prepared must be subjected to the same temperature treatment (including pure beeswax and pure adulterant, i.e., mixtures 1 and 13 from Table 18). 90 °C temperature treatment must be applied to ensure melting of adulterants with higher melting point (such as carnauba wax: 83 °C, some paraffin waxes: >75 °C).
4. Re-solidify the mixtures by allowing to cool to room temperature (without any additional cooling treatment aiming to decrease the temperature rapidly) (Figure 47).
   Note: some adulterants have a tendency to precipitate after melting (tallow for instance) – stir those mixtures continuously to the solid state to avoid a formation of separated layers.
5. Store the mixtures (with a lid on) in a dark place at room temperature, as described in Section 2.3 on beeswax storage.

The source of the material used to prepare the RS requires careful consideration regarding its authenticity. For example, more than 20 wild-built combs collected from different hives were used as beeswax material for the preparation of mixtures presented here, while adulterants were purchased as specified in Section 6.2.5.1.1.2. Furthermore, make sure a sufficient quantity of material for the preparation of the whole set of mixtures (13) is provided.

Along with the preparation of mixtures by following usual 10% increasing sequence of adulterant addition (10, 20, 30, …, 90%), we propose a preparation of two additional mixtures containing 5% and 95% of adulterant to obtain more precise results based on additional marginal range values. This can be useful in case of "borderline" test samples (5%) or almost pure adulterants (95%). Also, mixtures can be prepared to cover a wider range of adulteration level (5, 10, 15, 20, 25%, … sequence) and decrease the calibration error. Nonetheless, we here describe the methodology using adulterant proportions (Table 18) that provide simple application and reliable detection of adulterants with good detection limits (≤3%).

Table 18. Recipe for the preparation of 5 g adulterant–beeswax mixtures (w/w).

| Mixture | Adulterant:beeswax ratio (%) | Adulterant (g) | Beeswax (g) |
|---|---|---|---|
| 1 | 0:100 (pure beeswax) | – | 5 |
| 2 | 5:95 | 0.25 | 4.75 |
| 3 | 10:90 | 0.5 | 4.5 |
| 4 | 20:80 | 1 | 4 |
| 5 | 30:70 | 1.5 | 3.5 |
| 6 | 40:60 | 2 | 3 |
| 7 | 50:50 | 2.5 | 2.5 |
| 8 | 60:40 | 3 | 2 |
| 9 | 70:30 | 3.5 | 1.5 |
| 10 | 80:20 | 4 | 1 |
| 11 | 90:10 | 4.5 | 0.5 |
| 12 | 95:5 | 4.75 | 0.25 |
| 13 | 100:0 (pure adulterant) | 5 | – |
| Total | | 32.5 | 32.5 |

*6.2.5.2. Acquisition of IR spectra of reference standards.* The IR spectra of all sample types (beeswax, adulterants, ABM) should be recorded in accordance with the procedure described in Section 5.3.2.2. Given the solid state of the RS after melting and re-solidification, as well as their storage in containers, placing the samples on the ATR crystal is best done by using a laboratory scoop or scraper (this helps to tranfer small aliquots that will melt on the crystal prior to spectral analysis). In case of recording the spectra of samples with naturally higher melting point, the temperature of the ATR crystal should be increased to 80–90 °C during the spectra acquisition.

*6.2.5.3. Detection of adulterants using IR spectral data.* Detection of adulterants based on IR spectral data involves several basic steps:

I. Visual exploration of spectral similarities (and dissimilarities) between beeswax and adulterants and identification of adulterant-specific spectral region(s) of interest in ABM (see Section 6.2.5.3.1 for details).

   Visual exploration represents the first step towards selecting a suitable spectral region of interest indicative for adulteration detection (further used for calibration purposes). Each adulterant type exhibits different spectral features as both individual sample and when mixed with beeswax. We describe examples of spectral alteration based on four adulterant types and respective adulterant–beeswax mixtures (RS):

   1. paraffin wax–beeswax mixtures (petroleum-derived adulterant)
   2. tallow–beeswax mixtures (adulterant of animal origin)
   3. stearic acid–beeswax mixtures (synthetic adulterant)
   4. carnauba wax–beeswax mixtures (adulterant of plant origin)

II. Construction of a calibration curve using the spectral data of RS (see Section 6.2.5.3.2 for details).

   Construction of a calibration curve using the spectral data of pre-analyzed reference (calibration) standards is conducted by following several steps:

   1. Estimate the coefficient of correlation (R) to determine the strength and direction of a relationship between the instrument response and known adulterant proportions in the indicative spectral regions identified in RS (see Section 6.2.5.3.2.1 for details).
   2. Select the indicative spectral region (targeted peak areas) showing the best correlation effects.
   3. Generate (plot) a calibration curve based on the best correlation results

      The spectral regions with reference peaks exhibiting the best correlation parameters ($R \geq 0.998$; $SE < 0.05\%$), are used for the construction of a calibration curve and further quantification of adulterants in test samples (see Section 6.2.5.3.2.2 for details).
   4. Carry out statistical (linear regression) analysis to estimate the prediction strength and

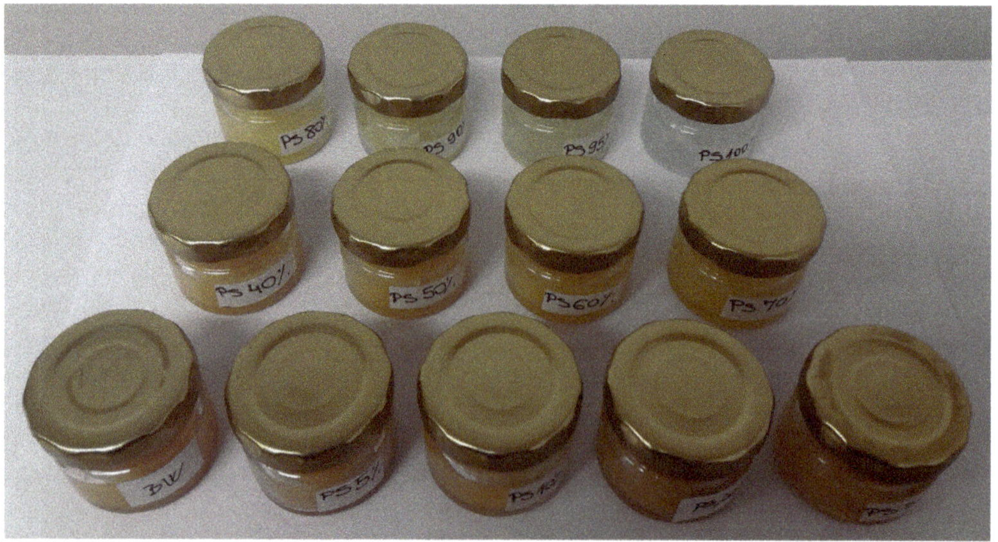


Figure 47. Containers with prepared adulterant (paraffin/*paraffinum solidum* – PS)–beeswax mixtures after melting. Photo: Svečnjak L.

prediction error of selected calibration model for adulteration level detection (determination of the coefficient of determination – R2, and standard error – SE). This is essential for the final interpretation of results which includes both the predicted value (% of adulterant in beeswax) and measurement uncertainty in terms of prediction accuracy (see Section 6.2.5.3.2.3 for details).

III. Detection of adulterants in test samples (see Section 6.2.5.3.3 for details).
The quantification of adulterants in the test samples includes the following steps:
   1. Acquire the spectrum of test sample under the same measurement conditions as RS.
   2. Estimate the measurement uncertainty associated with the instrument and the sample being analyzed (see Sections 6.2.5.3.3.1 and 6.2.5.3.3.2 for details).
   3. Determine (quantify) the exact amount of adulterant in test sample (see Section 6.2.5.3.3.3 for details).

*6.2.5.3.1. Identification of adulterant-specific spectral regions of interest in adulterant–beeswax mixtures.* As mentioned, each adulterant type exhibits different spectral features (Figure 48) as both individual sample and when mixed with beeswax. Thus, IR spectra of prepared ABM should reveal a unique trend of spectral alterations (increasing and/or decreasing absorption intensities) which follow the increasing amounts of adulterant added to beeswax. These alterations represent spectral regions of interest that are further used for calibration purposes.

Comparative spectral features of beeswax versus selected adulterants (paraffin, beef tallow, stearic acid, and carnauba wax) are described and presented below in the following sections along with IR spectra of the reference standards (ABM) and identification of spectral regions indicative for adulteration.

*6.2.5.3.1.1. Paraffin–beeswax mixtures.* In comparison to the beeswax IR spectrum, a typical FTIR-ATR spectrum of *paraffin* is characterised by a simple molecular structure related to hydrocarbon absorption bands (at 2921, 2852, 1464, and 720 $cm^{-1}$), as shown in Figure 48. The main spectral differences between pure beeswax and paraffin (as well as other adulterants) are observed in the fingerprint region. The analyte signals related to the ester and free fatty acids vibrations (at 1738, 1714, and 1172 $cm^{-1}$) are not present in the IR spectrum of paraffin, and therefore, represent valuable and indicative spectroscopic data for detection of beeswax adulteration.

When prepared and recorded in a satisfactory manner as described above, IR spectra of *paraffin–beeswax mixtures* should reveal a clear linear decrease in the absorption intensities of the lipid components following the increasing percentages of the paraffin (Figure 49). These effects are observed in the spectral region between 1760 and 1700 $cm^{-1}$ (esters and free fatty acids with absorption maximums at 1738 $cm^{-1}$ and 1714 $cm^{-1}$, respectively), and between 1260 and 1100 $cm^{-1}$ (esters with absorption maximum at 1172 $cm^{-1}$).

Paraffin waxes with different melting point exhibit the same IR spectra (Figure 50). Thus, all types of paraffin waxes can be used for calibration; there are no effects on the reliability of results.

*6.2.5.3.1.2. Tallow–beeswax mixtures.* Contrary to the lack of particular lipid-based signals in the fingerprint region of a paraffin spectrum, a characteristic IR spectrum of *beef tallow* is dominated by the strong intensive absorption bends that correspond to the esters vibrations, occurring at 1744 and 1160 $cm^{-1}$ (Figure 48). These signals are considerably stronger and characterised by the absorption maxima shifts in comparison to beeswax signals.

Tallow-beeswax mixtures also exhibits linear behaviour trend, but in the opposite direction in comparison to paraffin; IR absorption intensities (most prominent in the spectral ranges between 1770–1710 $cm^{-1}$ and 1260–1070 $cm^{-1}$) increased linearly for the increasing percentages of the tallow added to beeswax (Figure 51). This increasing trend is withal accompanied by the characteristic absorption maximum shifts in comparison to beeswax signals (from 1738 to 1744 $cm^{-1}$, and from 1172 to 1160 $cm^{-1}$). Different types of tallow might give spectra with slightly different signal properties, especially in the region 1770–1710 $cm^{-1}$ due to a different composition of free fatty acids. However, these spectral data are equally valid and reliable for the purposes of further calibration.

*6.2.5.3.1.3. Stearic acid–beeswax mixtures.* Stearic acid exhibits the most distinctive IR spectrum, with numerous absorption bands that are not observable in the IR spectrum of beeswax, more specifically, absorptions at 1710, 1412, 1281, and 929 $cm^{-1}$ (Figure 48).

Stearic acid–beeswax mixtures exhibit unique and complex spectral features with numerous regions that are indicative for adulteration (Figure 52). The appearance of a strong band at 1710 $cm^{-1}$ in the IR spectrum of stearic acid can be attributed to the C=O stretching vibrations of saturated aliphatic carboxylic acids as dimer given that these bands characteristically appear in the region between 1730 and 1700 $cm^{-1}$. Such spectral effect can be explained by the dimer association of stearic acid in different solutions and mixtures; the crystallization process of stearic acid in various media may lead to different polymorphic forms depending on the different crystallization parameters (Garti, Sato, Schlichter, & Wellner, 1986). Pielichowska et al. (2008) reported the similar appearance of a band at 1710 $cm^{-1}$ in stearic acid, belonging to C=O stretching in dimers in liquid state. The fingerprint region contains a series of bands that are very useful for detection of stearic

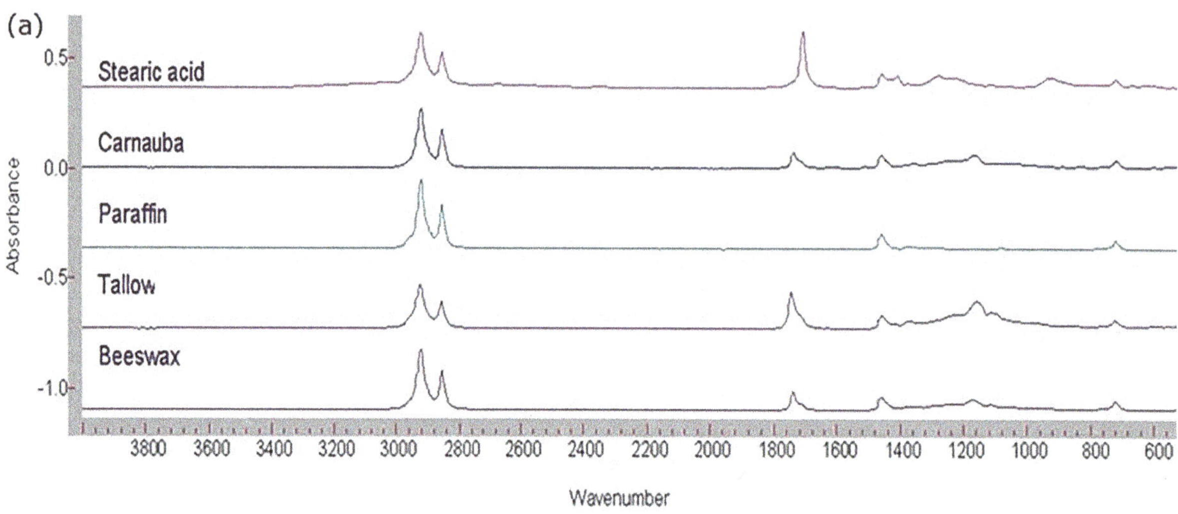


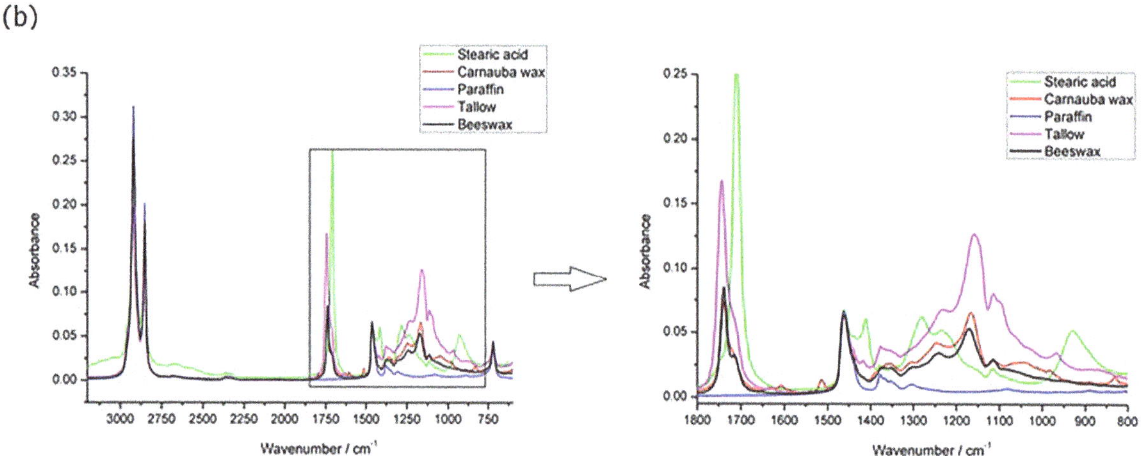


Figure 48. Comparative spectral features of beeswax and four commonly used adulterants (paraffin, tallow, stearic acid, and carnauba wax): (a) individual spectra and, (b) combined spectra representation (modified from Svečnjak, Baranović et al., 2015).

acid in beeswax; the most indicative are vibrations associated with carboxylic acids observed at 1281 cm$^{-1}$ (C–O stretching vibrations) and 929 cm$^{-1}$ (O–H bending vibration) (Pielichowska et al., 2008; Socrates, 2001).

*6.2.5.3.1.4. Carnauba wax–beeswax mixtures.* Unlike other adulterants that show a unique spectral fingerprint which can easily be distinguished from beeswax, the *carnauba wax* spectrum is very similar to the beeswax spectrum with only minor spectral differences (Figure 48), which makes a detection of carnauba wax the most challenging. Differences can be observed in the region where variations among different genuine beeswax samples also appear (1090–1015 cm$^{-1}$ in particular). Therefore, these spectral regions should not be used for further calibration modelling. However, specific small-intensity absorption bands that are observable at 1633, 1607, 1514, and 830 cm$^{-1}$, specify IR spectrum of carnauba wax and carnauba wax–beeswax mixtures (Figure 53). An integrated visual identity of these absorptions should be considered when detecting the presence of carnauba wax in beeswax. However, these small bands do not necessarily show an equal linear increasing trend following increasing amonts of carnauba wax in carnauba-beeswax mixtures. In case of the example presented here, linear trend is well represented for absorbance at 1514 cm$^{-1}$, while other two peaks show a desultory in linearity (Figure 53). The same goes for the region from 840 to 820 cm$^{-1}$.

The indicative regions (and corresponding distinctive absorption bands/peak areas) determined for a particular adulterant of interest are further used in the calibration process, starting with an evaluation of a correlation coefficient between IR measurements (absorbance intensities) and known (real) proportions of adulterant (described below in the following subsections).

*6.2.5.3.2. Construction of a calibration curve using the spectral data of reference standards.*

*6.2.5.3.2.1. Estimation of the correlation coefficient (R).* The correlation coefficient (R) is one of the statistical parameters commonly used in analytical measurements prior

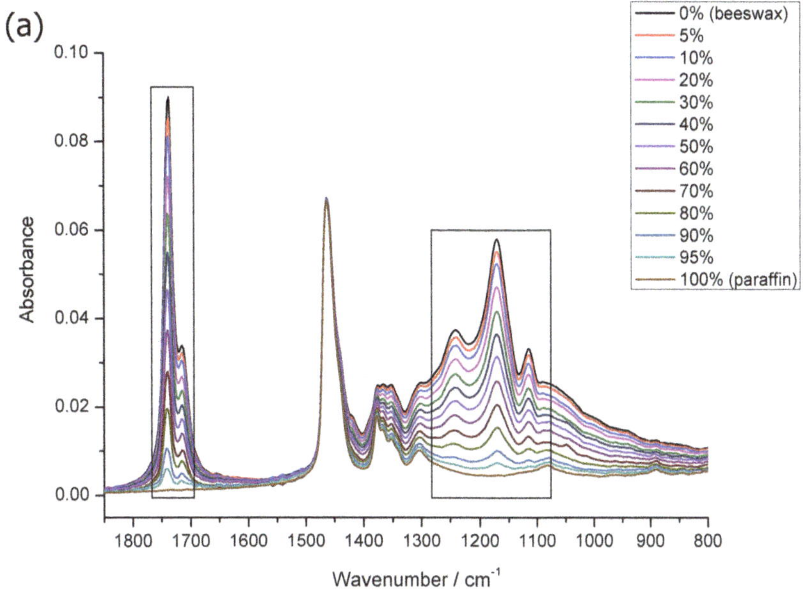


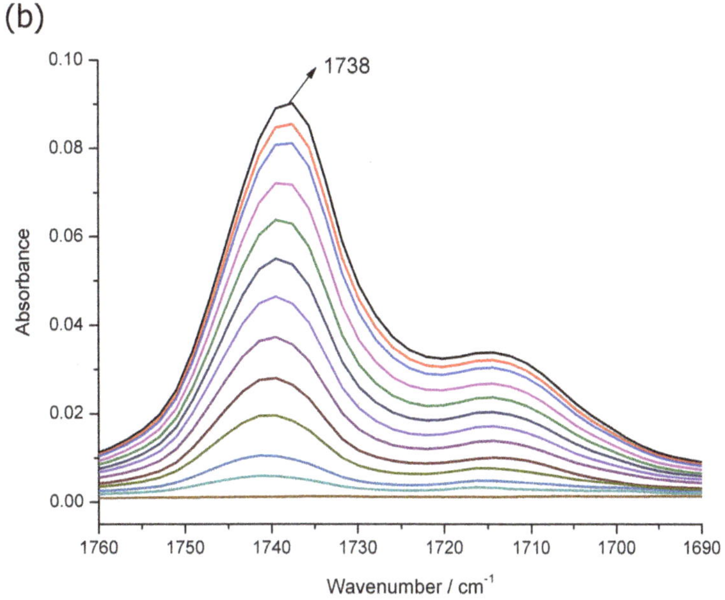


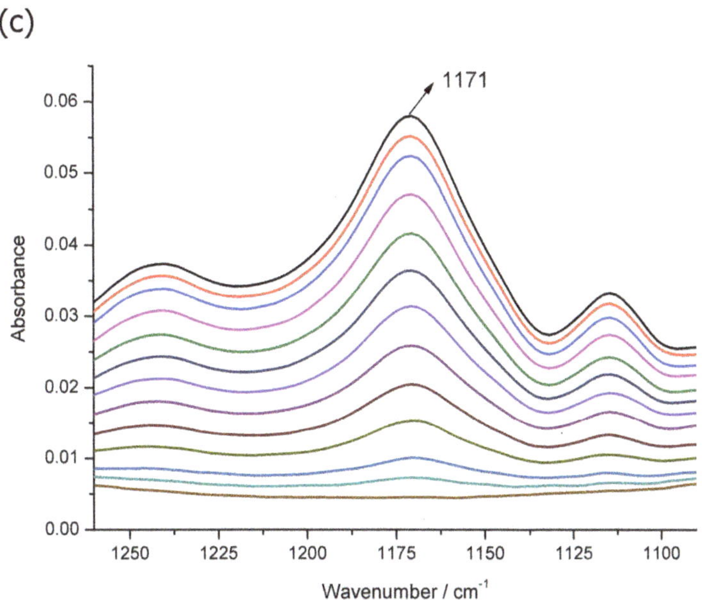


Figure 49. IR spectra of beeswax, paraffin, and paraffin–beeswax mixtures (calibration standards) containing different proportions of paraffin; an emphasis on the spectral regions indicative for paraffin detection: (a) fingerprint region – 1800–800 cm$^{-1}$; (b) spectral region from 1760 to 1700 cm$^{-1}$; (c) spectral region from 1250 to 1100 cm$^{-1}$ (modified from Svečnjak, Baranović et al., 2015).

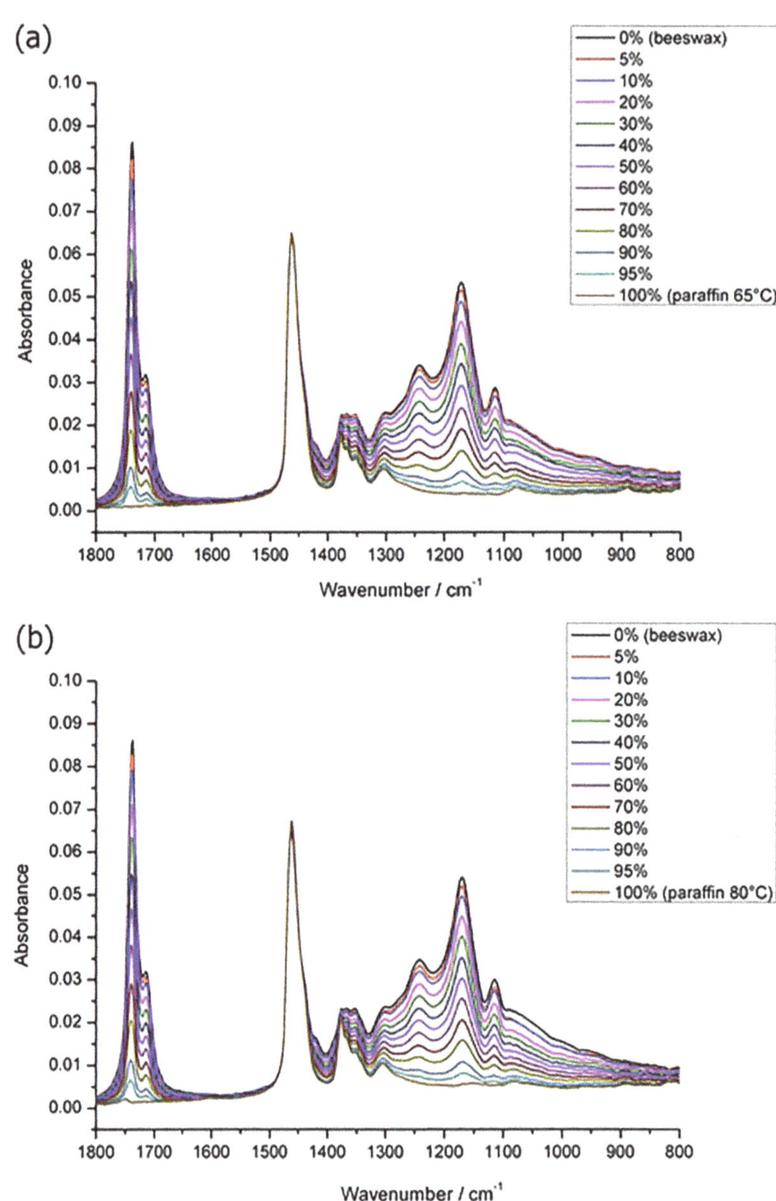


Figure 50. IR spectra of paraffins with different melting point: (a) 65 °C; (b) 80 °C.

to setting up a calibration curve. The correlation coefficient represents the strength and direction of a relationship (degree of correlation) between the FTIR instrument response and the known (real) concentrations of the analyte (amount of adulterant).

The overall correlation results related to the target peaks of indicative spectral regions for four different adulterant (paraffin, beef tallow, stearic acid, and carnauba wax)–beeswax mixtures are given in Table 19. The results representing the best correlation between the instrument response and known adulterant proportions in the adulterant–beeswax RS are shown in Figures 54–57, for paraffin, tallow, stearic acid, and carnauba wax, respectively. In these figures we show the validation plots representing the instrument response for each adulterant–beeswax RS (mixtures containing 0–100% of adulterant; marked as 1–13 point in the plots), i.e., its indicative spectral regions. We recommend these spectral regions (absorbances at selected reference peak areas) to be used for further calibration process for corresponding adulterants (recommended peak areas are also highlighted with asterisks in Table 19). Also, we recommend estimating and selecting the best *R* value for other adulterants (not presented here) in like manner. It should be noted that stearic and palmitic acid, as well as commercially available stearin (often present on the market as a mixture of stearic and palmitic acid) exhibit almost equal IR spectral features and therefore, the same indicative spectral regions as those recommended for stearic acid, can be used to detect palmitic acid and stearin in beeswax. The same goes for commercial stearic acid which is also often available on the market as a mixture of stearic and palmitic acids.

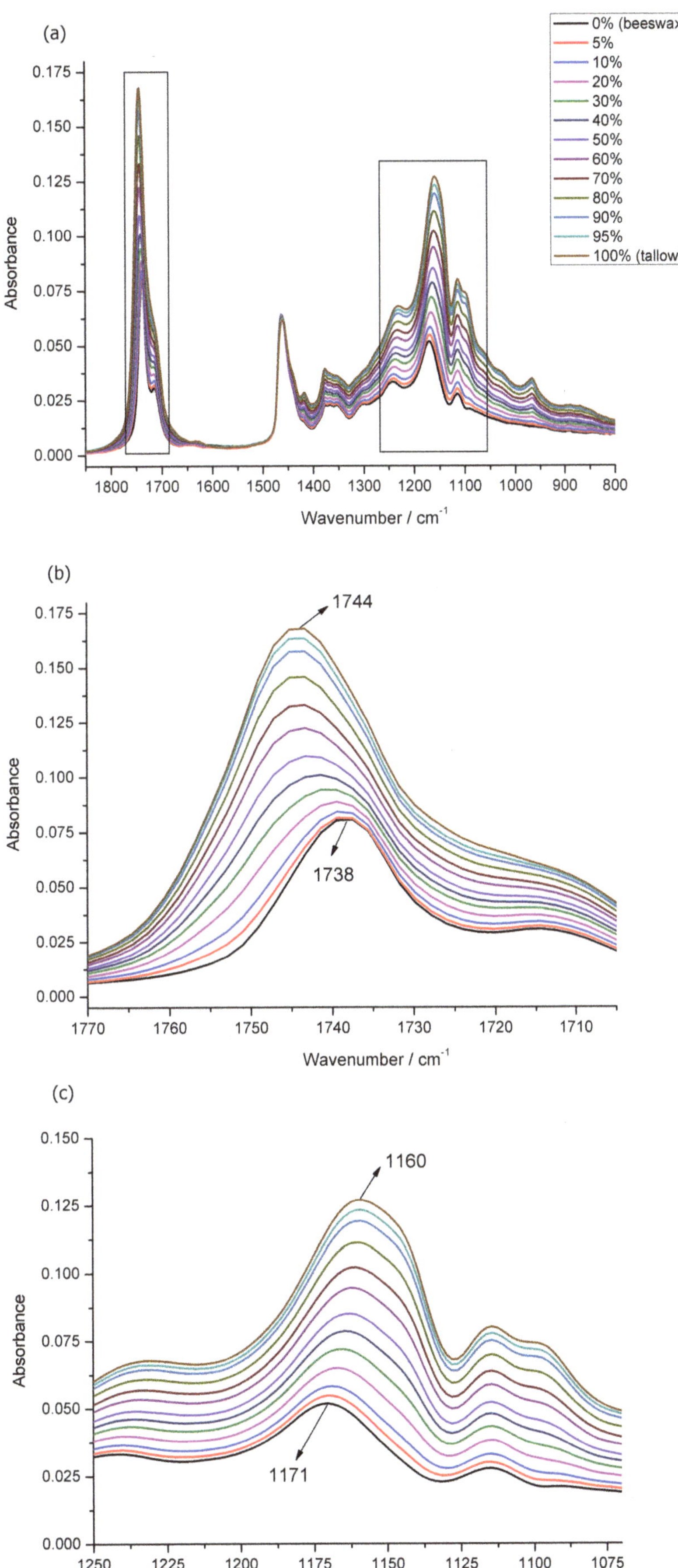


Figure 51. IR spectra of beeswax, tallow, and tallow–beeswax mixtures (calibration standards) containing different proportions of tallow; an emphasis on the spectral regions indicative for tallow detection: (a) fingerprint region – 1800–800 cm$^{-1}$; (b) spectral region from 1770 to 1710 cm$^{-1}$; (c) spectral region from 1250 to 1075 cm$^{-1}$ (modified from Svečnjak, Baranović et al., 2015).

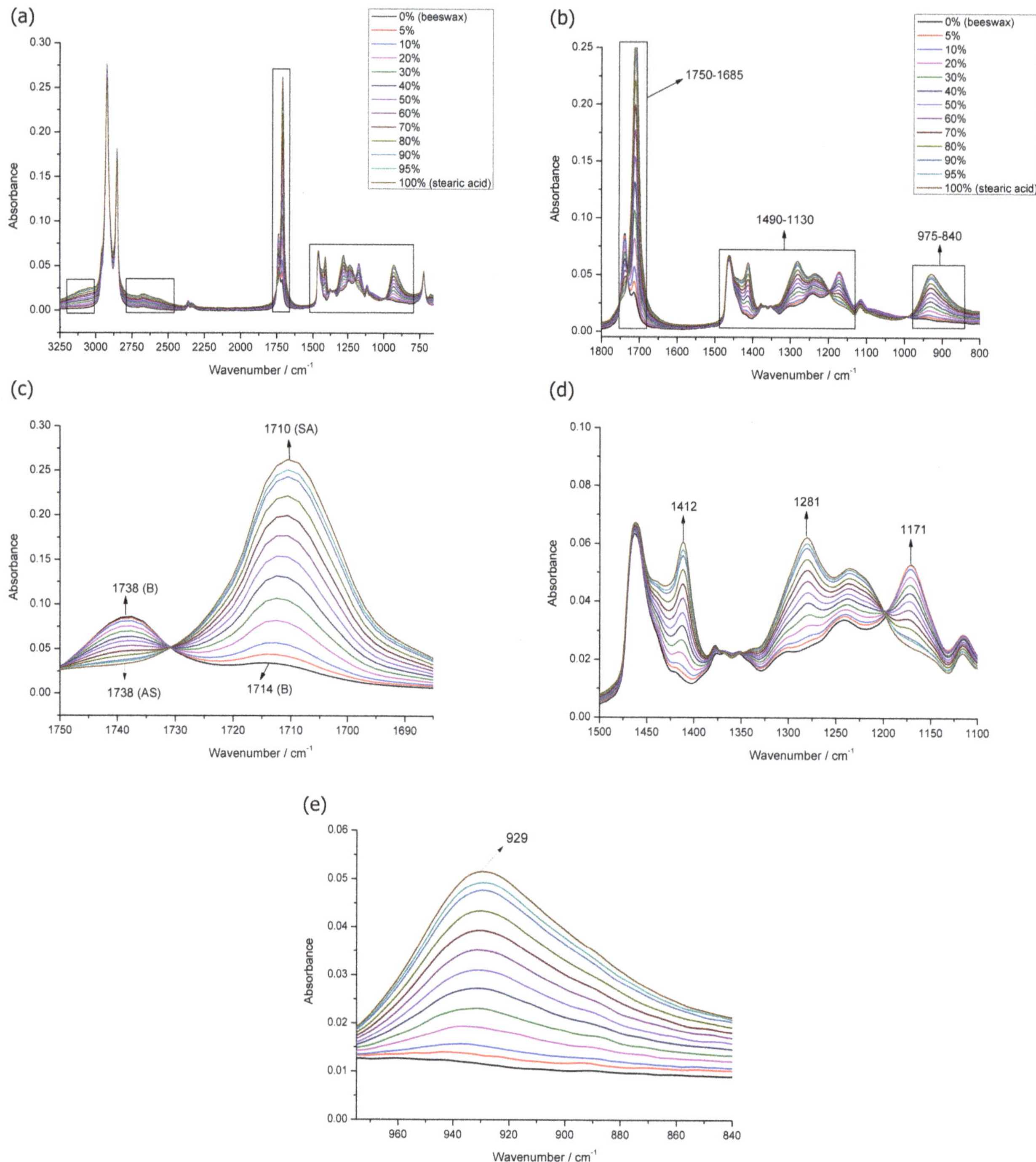


Figure 52. IR spectra of beeswax, stearic acid, and stearic acid–beeswax mixtures (calibration standards) containing different proportions of stearic acid; an emphasis on the spectral regions indicative for stearic acid detection: (a) whole spectrum – 3250–700 $cm^{-1}$; (b) fingerprint region – 1800–800 $cm^{-1}$; (c) spectral region from 1750 to 1690 $cm^{-1}$; (d) spectral region from 1500 to 1100 $cm^{-1}$; (e) spectral region from 970 to 840 $cm^{-1}$ (Svečnjak, 2018, unpublished data).

The correlation assessment presented here was performed using the Resolutions Pro version 5.3.0 software package (Agilent Technologies). Commonly used statistical packages (such as SPSS, SAS, R, Statistica-StatSoft line of softwares, etc.), as well as those specialized for the spectral data manipulation (such as Origin) can also be used for this purpose as they provide simple instructions for calculating the coefficient of correlation (*R*). *R* value can also be determined using Microsoft Excel spreadsheet. For more details, see Section 7 in the *BEEBOOK* article on statistics (Pirk et al., 2013).

*6.2.5.3.2.2. Generating a calibration curve.* The spectral regions with reference peak areas exhibiting the best correlation parameters ($R \geq 0.998$) are used for the construction of a calibration curve that will be further utilized for the quantification of adulterants in test samples. Most of modern FTIR spectrometers have

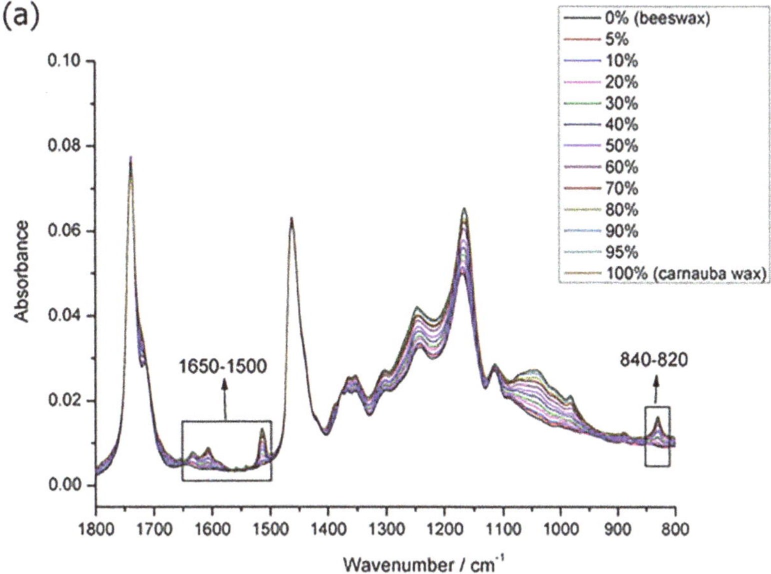


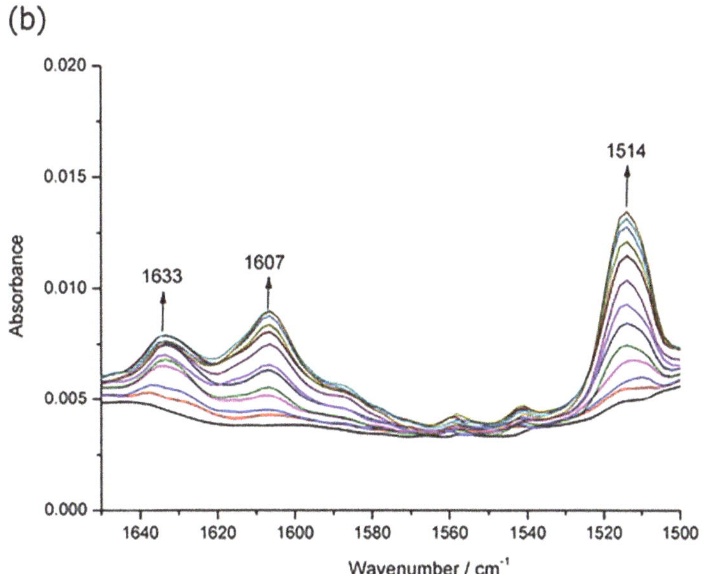


Figure 53. IR spectra of beeswax, carnauba wax, and carnauba wax–beeswax mixtures (calibration standards) containing different proportions of carnauba wax; an emphasis on the spectral regions indicative for carnauba wax detection: (a) fingerprint region – 1800–800 cm$^{-1}$; (b) spectral region from 1650 to 1500 cm$^{-1}$ (Svečnjak, 2017, unpublished data).

accompanying software that enables default quantification of the analyte of interest after recording the spectra of the reference (calibration) standards. To detect and quantify the analyte of interest (adulterant) in the samples of unknown chemical background (test samples), the calibration data (targeted peak areas of reference/calibration standards showing the best *R* value) must be previously stored as special (calibration) file formats (e.g., as BSQ file format in case of Agilent/Cary 660 FTIR spectrometer and accompanying software Resolutions Pro). Afterwards, the quantitative prediction of adulterant amount in test samples (based on calibration curve – known adulterant amounts in RS) is performed by a simple click on respective quantification toolbar button.

If there is no specialzed software for default quantification available, other statistical packages (such as those listed in Section 6.2.5.3.2.1) can be used to perform a factor-based Partial Least Squares Regression (PLSR) method for determining the adulteration level based on ABM (calibration set) and validation (test) set of samples (see Svečnjak, Baranović et al., 2015 for details).

*6.2.5.3.2.3. Estimation of the prediction performance parameters.* Interpretation of the results and complete analytical report on the amount of adulterant in beeswax test samples should include both the best estimate/predicted value (% of adulterant in beeswax based on generated calibration curves) and level of measurement uncertainty in terms of prediction accuracy (±%). Therefore, after estimating the *R* values, the spectral regions with reference peaks exhibiting the best correlation parameters (used for calibration purposes) are further subjected to linear regression modelling.

Table 19. Determination of a correlation coefficient conducted on the target peaks of indicative spectral regions of four different adulterant (paraffin, beef tallow, stearic acid and carnauba wax)–beeswax mixtures (Svečnjak, Baranović et al., 2015; Svečnjak, 2017, unpublished data).

| Reference standard (adulterant–beeswax) | Reference peak ($cm^{-1}$) | Peak area ($cm^{-1}$) | Predominant beeswax compound associated to reference peak | Correlation coefficient (*R*) |
|---|---|---|---|---|
| Paraffin | 1738 | 1750–1727 | Monoesters | 0.9999[a] |
| | 1171 | 1195–1147 | Esters of aliphatic acids | 0.9998[b] |
| Tallow | 1738 | 1753–1724 | Monoesters | 0.9969 |
| | 1171 | 1195–1148 | Esters of aliphatic acids | 0.9995[a] |
| Stearic acid | 1738 | 1747–1730 | Monoesters | 0.9971 |
| | 1710 | 1721–1707 | Free fatty acids | 0.9982 |
| | 1412 | 1423–1400 | Esters (shoulder) | 0.9989 |
| | 1281 | 1308–1253 | Free fatty acids | 0.9996[a] |
| | 929 | 978–880 | None | 0.9983 |
| Carnauba wax | 1633 | 1638–1628 | None | 0.2609 |
| | 1607 | 1610–1603 | None | 0.9859 |
| | 1514 | 1523–1506 | None | 0.9995[a] |
| | 830 | 740–820 | None | 0.9889 |

[a]The highest R values determined for each adulterant (used for further calibration process).
[b]Reference peak 1171 $cm^{-1}$ can also be used for further calibration due to almost equally good correlation results as for 1738 $cm^{-1}$.

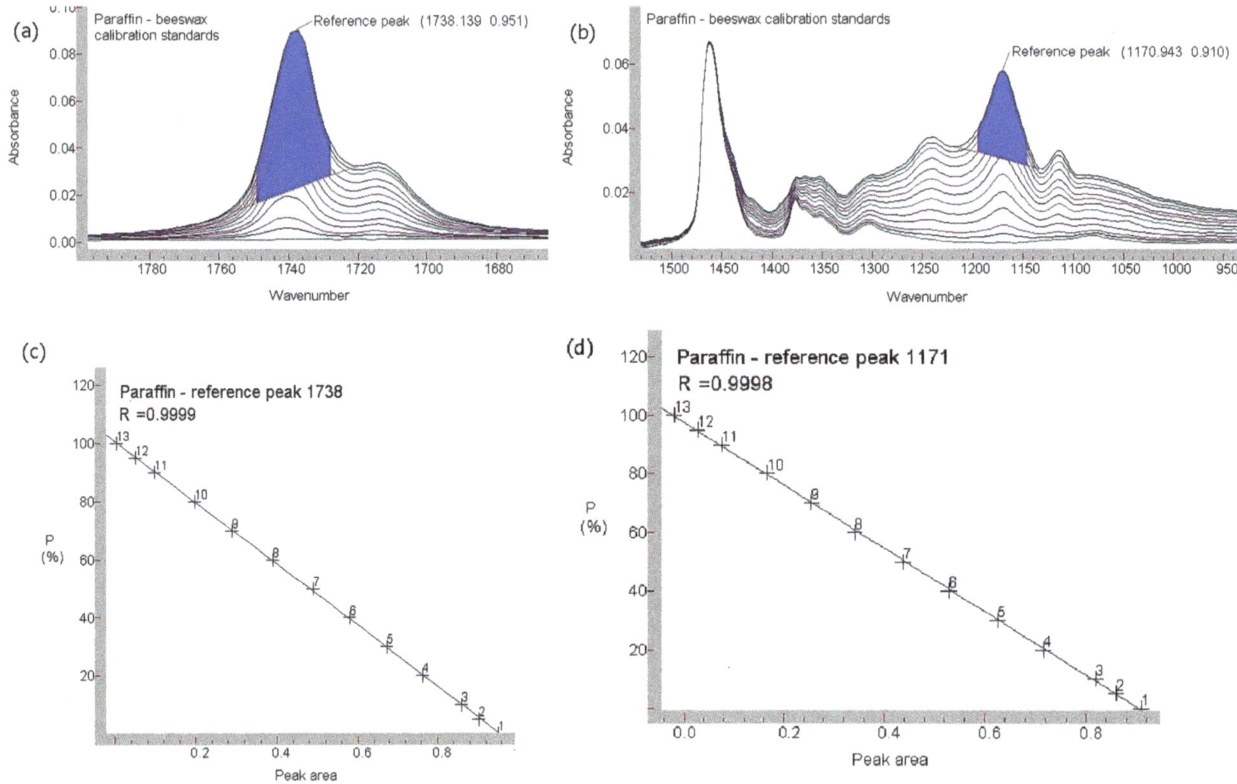


Figure 54. Reference peak areas: (a) absorbance at 1738 $cm^{-1}$, and (b) absorbance at 1171 $cm^{-1}$ exhibiting (c, d) the best results of correlation between the instrument response and known paraffin proportions in the paraffin–beeswax reference standards (Svečnjak, 2017, unpublished data).

The aim of linear regression is to estimate the prediction strength and prediction error of selected calibration curve for adulteration level detection, and this is achieved by calculating the coefficient of determination ($R^2$) and standard error (SE), respectively. $R^2$ can be considered here as adjusted *R*, while SE estimates the measurement uncertainty in terms of accuracy (closeness of agreement between measured adulterant share values and true adulterant share values). An example of prediction performance parameters of the calibration curve constructed for determination of stearic acid share in beeswax based on stearic acid-beeswax mixtures is shown in Figure 58). For reliable quantification of adulterants in beeswax, it is best to hold to the following performance parameters: $R^2 \geq 0.997$, SE $< 0.05\%$.

The statistical analysis presented here (Figure 58) was performed using the statistical package Origin

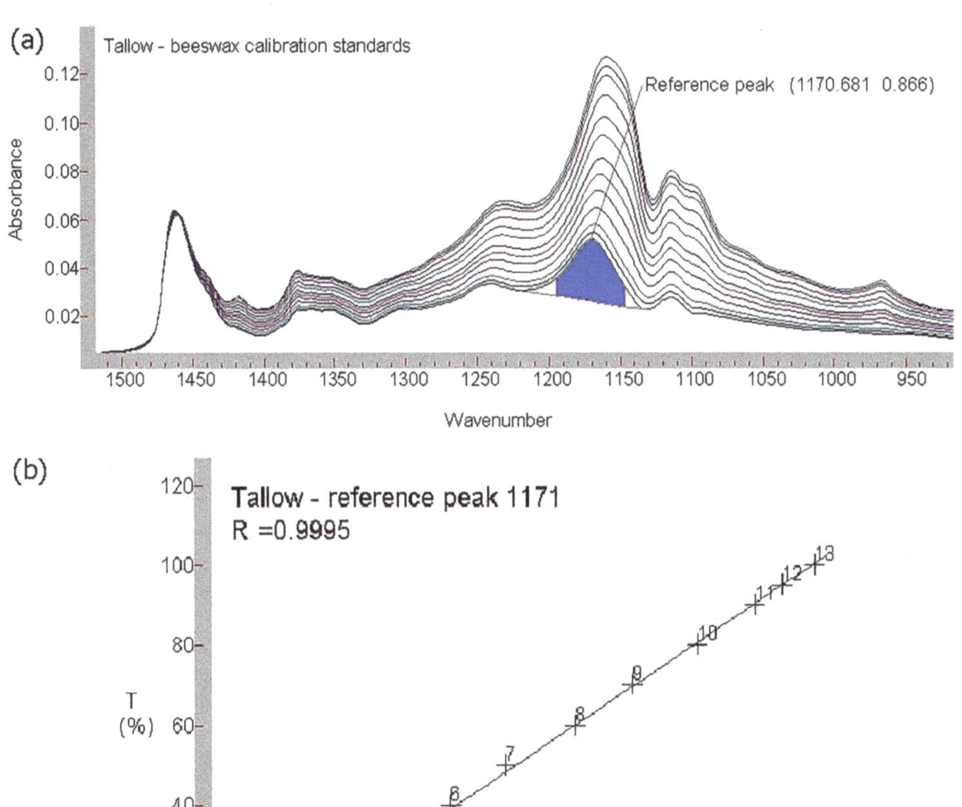

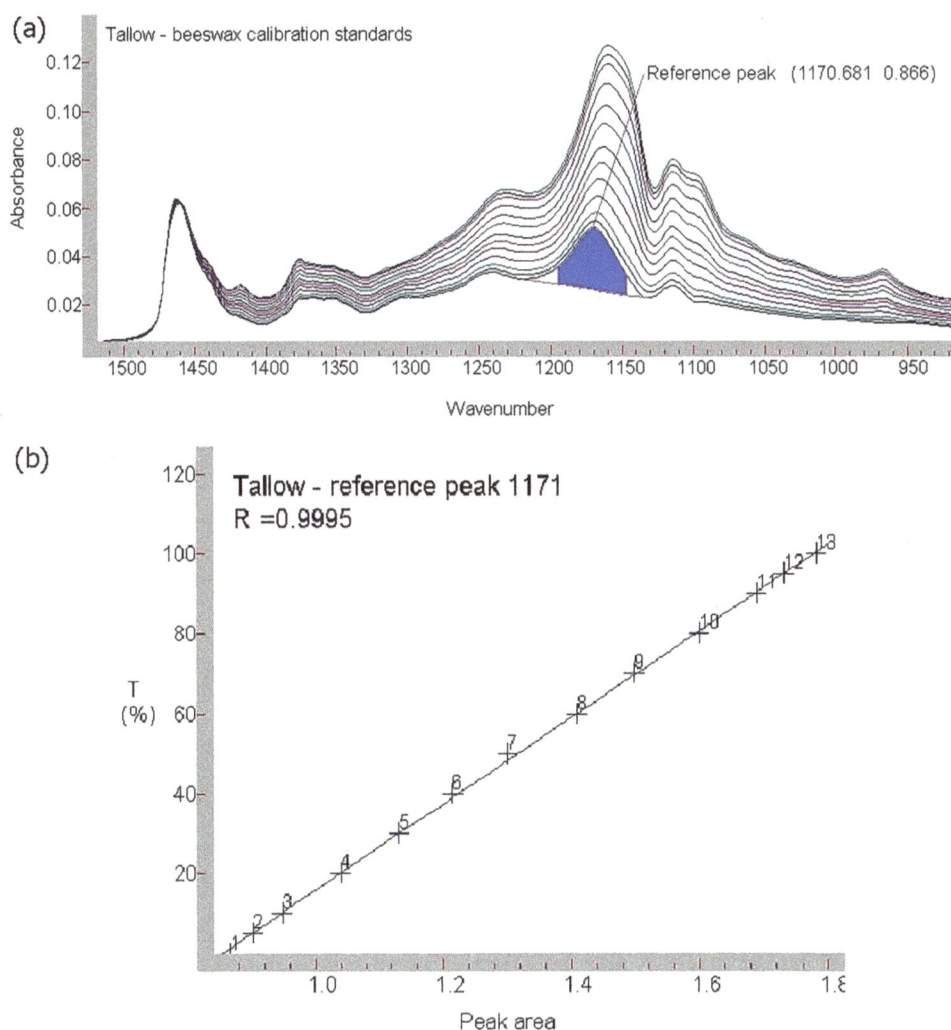

Figure 55. (a) Reference peak area (absorbance at 1171 cm$^{-1}$), (b) exhibiting the best results of correlation between the instrument response and known tallow proportions in the tallow–beeswax reference standards (Svečnjak, 2017, unpublished data).

version 8.1 (OriginLab Corporation), but other statistical packages may also be used (examples of respective packages are provided in Section 6.2.5.3.2.1). For general guidelines on linear regression models, see Sections 5 and 5.2 in the *BEEBOOK* paper on statistics (Pirk et al., 2013).

*6.2.5.3.3. Detection of adulterants in test samples.* In practice, the test samples are usually comb foundations of unknown chemical background (origin) collected from the market, and they often contain adulterants.

Quantification of the analytes of interest (adulterants) in test samples is performed by comparing the unknown samples to a set of reference/calibration standards of known adulteration level (based on the generated calibration curve). After calibration curves are constructed and evaluated to be satisfactory ($R^2 \geq 0.997$, SE $< 0.05\%$), they can be used to estimate the proportion (%) of the targeted analyte (adulterant) in test samples. This requires each test sample to be analyzed (two replicate spectra) under the same measurement conditions as RS (see Section 6.2.5.2), and the result (predicted value/best estimate) is obtained based on the generated calibration curve, as described in Section 6.2.5.3.2.2. However, the complete result on the proportion of adulterants in test samples should include an estimate of the level of confidence associated with the estimated value, which is commonly represented as:

$$\text{adulterant share} = \text{predicted value} \pm \text{measurement uncertainty } \%$$

Therefore, prior to determining the adulteration level in test samples, it is necessary to estimate several aspects of measurement uncertainty. Along with the measurement uncertainty (in terms of prediction accuracy) determined by means of SE (see previous section), the measurement uncertainty associated with the measuring instrument and the sample being analyzed should also be estimated.

*6.2.5.3.3.1. Measurement uncertainty arising from the sample.* When comparing minor spectral variations identified in 20 different genuine *A. mellifera* beeswax samples (presented in Section 5.3.3.1; Figure 33) with absorption changes observed in paraffin–beeswax mixtures, it is

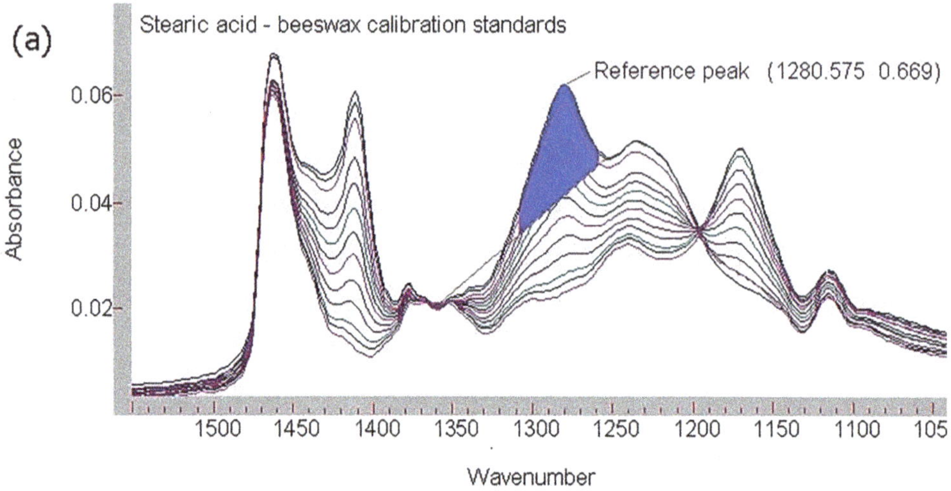


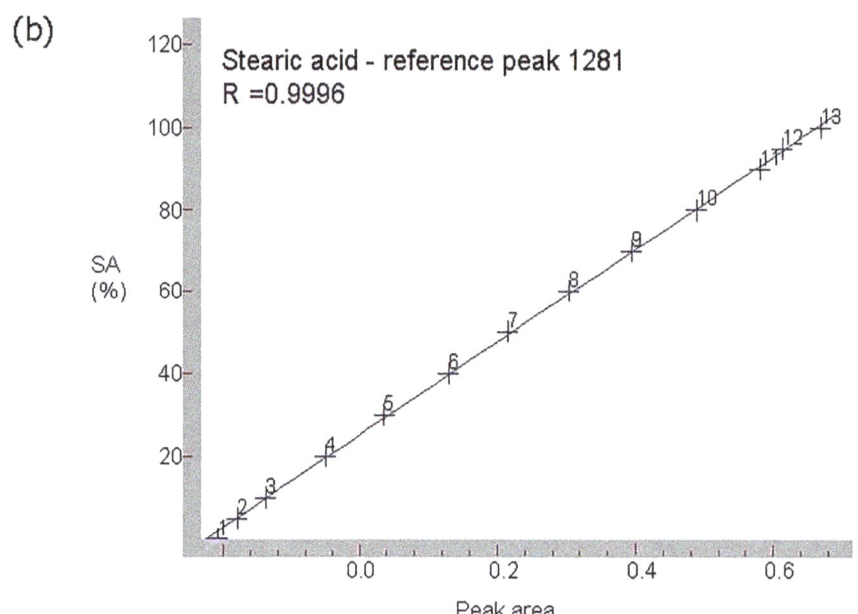


Figure 56. (a) Reference peak area (absorbance at 1281 cm$^{-1}$), (b) exhibiting the best results of correlation between the instrument response and known stearic acid proportions in the stearic acid–beeswax reference standards (Svečnjak, 2017, unpublished data).

obvious that the most intensive spectral variations determined in different beeswaxes (observed between 1090 and 1015 cm$^{-1}$) do not intercept those crucial for detection of beeswax adulteration with paraffin (absorbance at 1738 and 1171 cm$^{-1}$; Figure 50). However, these minor differences may slightly affect the results for test samples. When detecting paraffin in beeswax, we detect the components whose share decreases by addition of paraffin. This reflects in the lower absorbance intensities of bands associated with esters and fatty acids. Thus, it can be concluded that paraffin in some respects "dilutes" the beeswax. Minor spectral variations that occur among different comb wax samples may also reflect slightly lower absorbance intensities of mentioned absorption bands throughout spectrum (Figures 33 and 59). To avoid uncertainty and potential declaration of false negative results for test samples (especially in cases of "borderline" test samples containing ≤5% of paraffin), and determine the exact amount of paraffin in beeswax, it is necessary to quantify spectral variations occurring in different genuine beeswax samples, at a spectral region that overlaps with the paraffin calibration pattern (with maximum absorbance at 1738 cm$^{-1}$, or 1171 cm$^{-1}$ if used for calibration). This is achieved by calculating the coefficient of variation ($c_v$), also known as relative standard deviation (RSD). $c_v$ should be determined using the absorbance at 1738 cm$^{-1}$ (or 1171 cm$^{-1}$). Using an example of spectra obtained from 20 different beeswax samples presented in Section 5.3.3.1 (Figure 33), we present a calculation of the following: mean absorbance at 1738 cm$^{-1}$ ($\mu_A$= 0.5665), with standard deviation $\sigma$= 0.0197. The coefficient of variation $c_v$, defined as:

$$c_v = \frac{\sigma}{\mu_A}$$

is 0.0348 (3.48%), which means that 3.48% of variation

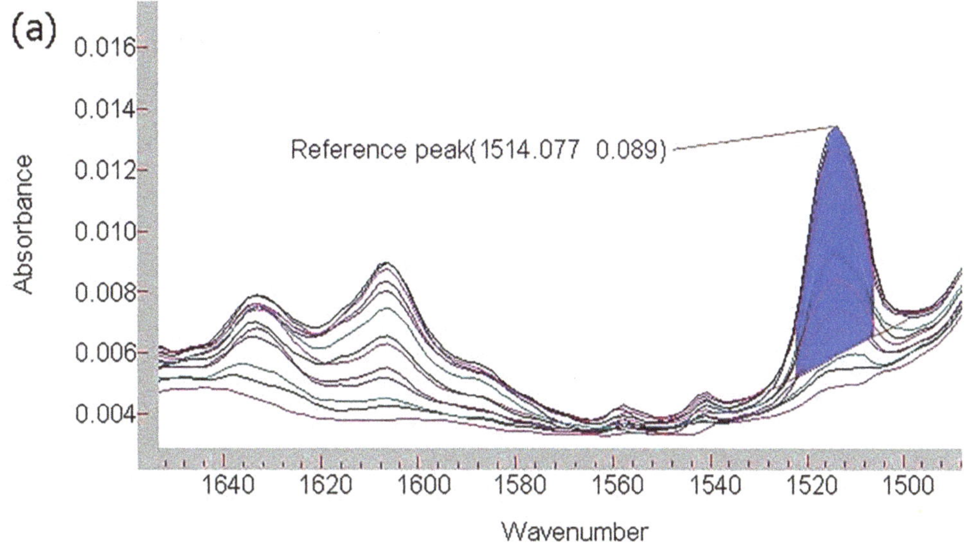


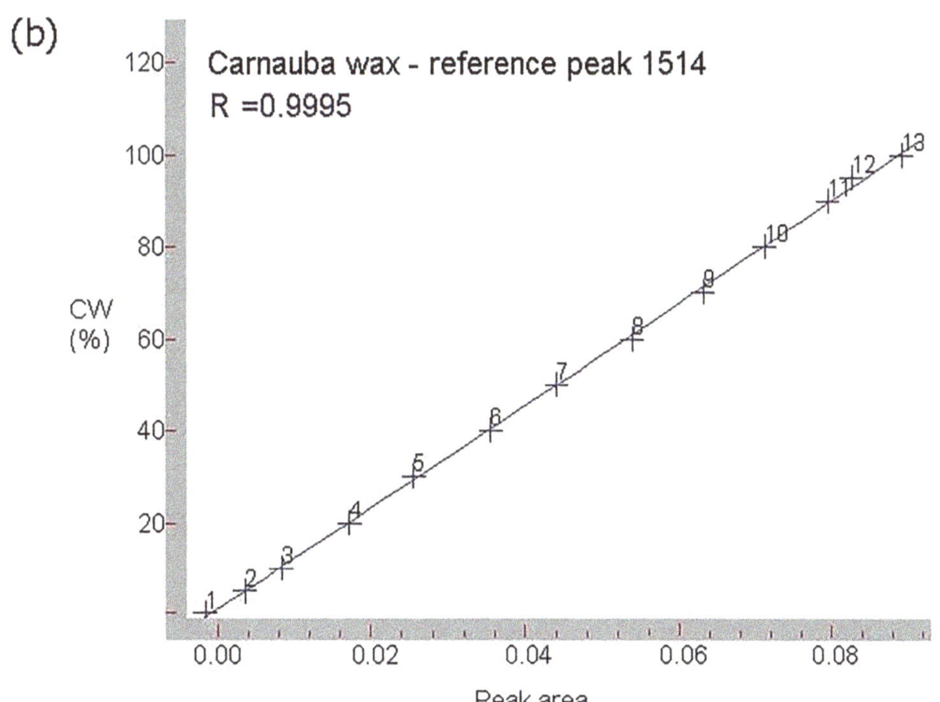


Figure 57. (a) Reference peak area (absorbance at 1514 cm$^{-1}$), (b) exhibiting the best results of correlation between the instrument response and known carnauba wax proportions in the carnauba wax–beeswax reference standards (Svečnjak, 2017, unpublished data).

can be assigned to spectral alterations naturally occurring in authentic beeswax samples (Figure 59).

Obtained $c_v$ value should be considered in the final result.

The uncertainty test procedure related to the natural minor variations occurring in different beeswax samples ($c_v$ – sample) should be applied when detecting adulterants whose calibration curves involve reference peaks associated with the compounds present in the beeswax (e.g., it should be applied in the case of paraffin, tallow, and stearic acid detection, while can be omitted in the case of carnauba wax; see Table 19. for details).

*6.2.5.3.3.2. Instrument related measurement uncertainty.* Measurement uncertainties can also come from the measuring instrument (FTIR spectrometer) so the precision (repeatability) of a measurement system has to be assessed. This is achieved by determining the $c_v$ value based on ten-fold measurement of different aliquots of the same beeswax sample, following the calculation steps described in previous section. For comparison, $c_v$ value (repeatability) determined by ten-fold measurement of different aliquots of genuine beeswax by using the maximum absorbance (at 2921 cm$^{-1}$) is 0.3% (generally, it should not exceed 0.5%).

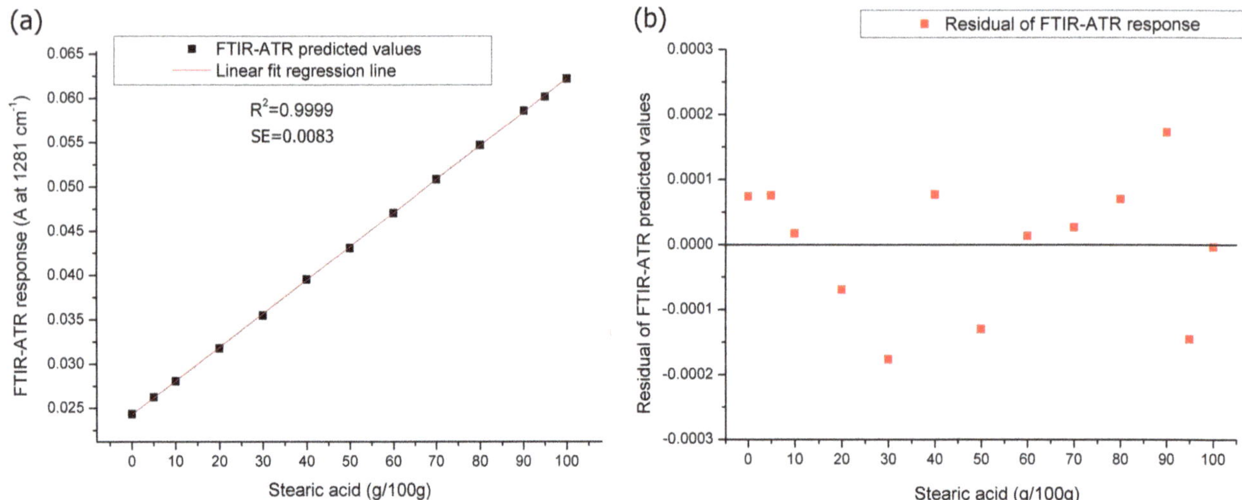


Figure 58. (a) Prediction performance parameters of the calibration curve constructed for determination of the stearic acid share in beeswax: a scatter plot of instrument response data (FTIR-ATR predicted values) versus real stearic acid share values; (b) residuals of FTIR-ATR prediction (Svečnjak, 2018, unpublished data).

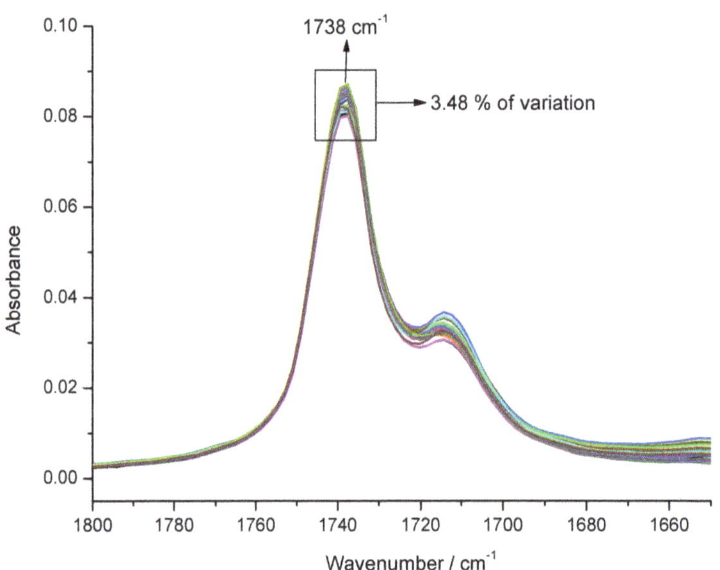


Figure 59. IR spectra and spectral variations ($c_v$ = 3.48%) between different beeswax samples ($n$ = 20) originating from different *A. mellifera* colonies observed at 1798 cm$^{-1}$ (Svečnjak, 2017, unpublished data).

*6.2.5.3.3.3. Interpretation of results.* For a proper interpretation of the results, the complete measurement result on the adulterant share in the test sample should account for all measurement uncertainties determined in above sections, i.e., in Section 6.2.5.3.2.3 (SE/prediction accuracy), Section 6.2.5.3.3.1 ($c_v$ – sample/natural variations in genuine beeswax), and Section 6.2.5.3.3.2 ($c_v$ – repeatability/measurement precision). Thus, if for example 19.3% of paraffin has been detected in the test sample, the final result should be presented as follows:

$$\text{paraffin share} = 19.3 \pm 3.8\%$$

where the final uncertainty value of 3.8% is consisted of 0.023% (SE determined for paraffin), 3.48% ($c_v$ – sample), and 0.3% ($c_v$ – repeatability).

*6.2.5.4. Advantages and disadvantages of FTIR-ATR for beeswax adulteration detection.* Compared to other analytical tools used for detection of adulterants in beeswax described above, FTIR-ATR analytical approach is characterized by the following:

**Advantages:**

- Numerous instrumental and measurement advantages as presented in 5.3.4. 'Advantages and disadvantages of FTIR-ATR measurements' (summarized: fast, cheap, reliable - good accuracy and precision, reagent-free, and easy-to-use method).
- Suitable for qualitative and quantitativedetection of all adulterant types.

- High temperature treatment commonly used in comb foundation production process (~125 °C) has no influence on the spectral values.

**Disadvantages:**

- The method relies entirely on the in-house preparation of reference standards, i.e., adulterant–beeswaxmixtures (time-consuming at initial stage).
- Relatively high price of FTIR spectrometer coupled with ATR accessory (but cheaper compared to GC-MS and IRMS instrumentation).
- Hard to distinguish microcrystalline wax from paraffin wax, and stearic acid from palmitic acid and stearin (although irrelevant from the viewpoint of authenticity given that mentioned substances represent foreign matters that should not be present in beeswax).

*6.2.6. Investigating brood survival on adulterated comb foundations*

Field studies on the acceptance or rejection of adulterated comb foundation by the honey bee colony are scarce (Castro, Medici, Sarlo, & Eguaras, 2010; Darchen & Lavie, 1958; Medici, Castro, Sarlo, Marioli, & Eguaras, 2012; Ribeiro da Silva, Ribeiro, Toledo, & Toledo, 2002; Semkiw & Skubida, 2013; Toledo, 1991). As reported in several studies (Maia et al., 2013; Serra Bonvehí & Orantes Bermejo, 2012; Svečnjak, Baranović et al., 2015), comb foundation without any adulterants are rare and most contain 5–20% of paraffin. Foundations with significantly higher percentages, i.e., > 50% of adulterants, have occasionally been found on the European market. These results indicate the variability of beeswax quality, probably due to various factors, such as the recycling of old combs and accumulation of residual amounts of paraffin (Maia & Nunes, 2013; Svečnjak, Baranović et al., 2015; Svečnjak, Prđun et al., 2015), the use of 10% of paraffin during the comb foundation production process to improve the mold of beeswax (Castro et al., 2010), mixing of beeswax from different countries, and also, the different types of paraffin wax and other substances used for adulteration (Serra Bonvehí & Orantes Bermejo, 2012).

This market situation implies a need for implementing bioassays to understand the consequences of particular biological factors on honey bee colonies (Section 6.2.6.2), as well as economic losses for the beekeepers (see Section 6.2.6.3). However, it is necessary to homogenize productive, environmental, management and technical factors that may influence the behaviour of honey bee colonies against pure and adulterated comb foundations. In Section 6.2.6.1, we present several points that justify the necessity of the bioassays having the same conditions of the factors indicated above.

*6.2.6.1. General factors important for bioassay standardization.*

1. For general guidelines to equalize colony size see Section 3.1 or 3.2 (setting up experimental colonies of uniform strength) in the *BEEBOOK* paper on *A. mellifera* strength parameters (Delaplane, van der Steen, & Guzman, 2013)
2. The races of honey bees: different races can produce different quantity of beeswax in equal conditions (Hepburn, 1986; Toledo, 1991).
3. Age of a queen: colonies with new queen can have more capacity to produce more beeswax (Hepburn et al., 2014).
4. Age of honey bee and physiological state of the colony: pre-swarming colony have more potential to construct comb wax (Hepburn, 1986), and queen loss and swarming tend to be the most disruptive factors to colony populations (Delaplane et al., 2013).
5. Temperature values (mean, maximum and minimum) and humidity (%): these factors can also influence the construction of the comb (Szabo, 1977).
6. Nectar/pollen/beebread availability: the production of beeswax varies with the quantity and quality of food (Pratt, 2004) and inadequate pollen nutrition can interfere with normal wax gland development (Hepburn, 1986).
7. Condition of the combs: honey bee colonies with more honey reserves produce more wax than colonies with scarce reserves, even during the nectar flow period (Pratt, 2004). If possible, use manual or electronic hive scales for weighing full hives and to compare the development between colonies. See Section 4.1. "Research methods at the colony level" in the *BEEBOOK* paper on miscellaneous methods for *A. mellifera* research (Human et al., 2013).
8. Type of artificial diet (sugar and protein): normally, beekeepers stimulate honey bee colonies with sugar syrup at a ratio of 1:1 (by weight). In case of proteins, it is better to use a commercial product because botanical origin of pollen can differ in nutritional value between different geographical regions.
9. Sanitary management (type of acaricide). Varroa control treatment conducted before bioassay is important to avoid the death of brood. Some organic acaricides can interfere with brood development (Rademacher & Harz, 2006).
10. Type of beehive or nucleus: different volume of the hive and colony size can have some interactions.
11. Quality of adulterants (especially paraffin with different melting point): the beeswax industry uses paraffin with different melting points taking into account two points of view: paraffin with low melting point (<58 °C) because it is cheaper, and paraffin with higher melting points (60–65 °C) because of the similarity to the melting point of the beeswax. The adulteration impact on the bee colonies may depend on

the type of the paraffin used, especially in relation to its melting point (Semkiw & Skubida, 2013). For instance, the use of low melting point paraffin wax can result in comb deformations and/or collapses.

12. Thickness of the comb foundation: the thickness of the comb foundation can influence the comb construction (Toledo, 1991): (a) if the foundation is thicker, the honey bees spend more time working on building the combs; (b) larger areas of honeycomb are constructed on the comb foundation with the finest thickness.
13. Beeswax origin: it is important that beeswax has the same origin since different concentration of residues and adulterants can significantly influence test results, especially if the wax comes from different suppliers (Büchler et al., 2013).

*6.2.6.2. Bioassay: brood survival versus beeswax adulteration.* For biological factors, it is important to consider the dynamics of the honey bee colonies because some studies showed that the addition of adulterants may decrease the area built on the comb foundation, reduce the queen acceptance for oviposition, cause the appearance of irregular brood patterns, and may cause other negative effects (Castro et al., 2010; Darchen & Lavie, 1958; Medici et al., 2012; Semkiw & Skubida, 2013; Toledo, 1991). The brood survival on adulterated comb foundation experiment should be conducted as follows:

1. Use pure comb foundation (made of genuine beeswax) and comb foundation adulterated with paraffin (or other adulterant of interest)
2. Adulteration levels should be similar to the situation on the beeswax market (the adulterant amounts can range from 5 to >90%).
3. Adulterated beeswax can be obtained from a stock proven (or suspected) to be adulterated to a certain percentage or adulteration can be performed by the investigator. Adulterated wax foundations can be obtained as follows:
   a. Choose the adulterant of interest (note that the most commonly used/available solid paraffin wax has melting point ranging from 56 to 58 °C; if different melting point want to be tested, they can be found in specialized shops).
   b. Melt the beeswax and paraffin wax (or other adulterant) in the same container. Take into account the proportions of each to obtain the levels of adulteration desired, and the total quantity required (see Section 2.2. for details on melting for general guidelines).
   c. When the wax mixture melts, place a smooth wet table board in cold water.
   d. Dip the board in the wax mixture 2–3 times to obtain a sheet of wax.
   e. Allow to cool.
   f. Remove the smooth wax sheet.
   g. For each sheet obtained always wet the board to facilitate separation
   h. Place the wax sheet in press machine.(alternatively, put the melted wax in the portable beeswax machine).
4. Insert experimental frames in the hives: one frame with comb foundation made of pure beeswax and another frame with a sheet containing certain percentage of paraffin (or other adulterant) in each beehive. Repeat this operation on 3 hives (minimum) (Figure 60).
5. To avoid any effect of the frame position, two experimental frames are placed in contact with the brood nest in each hive; one on the left and the other on the right side of the brood.
6. To equalize colony size see above Section 6.2.6.1, step 1.
7. The measurements of the brood survival can be carried out in two ways:
   a. Compare the survival of brood in each hive and each experimental frame, the one with particular percentage of adulterant, and with pure beeswax (see Section 4.1. "Measuring colony strength at end of experiment" in the BEEBOOK paper on *A. mellifera* strength parameters' (Delaplane et al., 2013). Survival of brood can be quantified by first selecting a relatively contiguous patch of brood in the late larval/capped stage, and overlaying on the patch a 10-cm horizontal transect and a 10 cm vertical transect intersecting at the centre (Delaplane et al., 2013).
   b. Methodology by Medici et al. (2012) can also be applied: count the number of worker honey bee cells with eggs (day 0) over a rectangular area of 7 x 6 $cm^2$ marked on each wax section on both sides. This is repeated at intervals of ten days (day 10), counting the number of pupae. The brood survival rate in each treatment is estimated by the following formula:

$$\text{Brood survival} = \left(\frac{\text{N of pupae}}{\text{N of eggs}}\right) \times 100$$

8. The experiment should last at least 7 weeks to cover two generations.

*6.2.6.3. Bioassay: colony development versus beeswax adulteration.* This methodology has the objective of verifying the economic impact of adulteration on the beekeeping by assessing the colonies that develop on pure compared to adulterated combs. The rejection or bad acceptance of comb foundation by the honey bees is frequently observed by the beekeepers, and it is cosely related to the poor quality of commercial comb foundations. Comb foundation adulteration may cause an economic loss in the beekeeping because it delays the development of the honey bee colonies. This usually result in the delay in placing the honey supers (negative

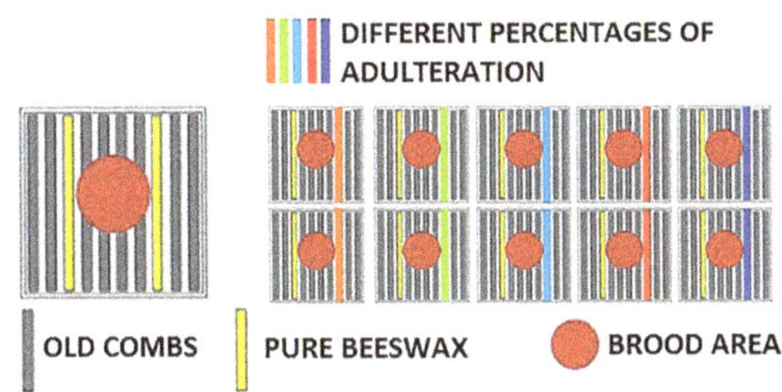


Figure 60. Bioassay for assessing biological factors.

impact on honey production) and compromises the production of swarms. Also, given that good apicultural practices recommend replacing at least 2–3 old combs per colony with comb foundations each year (Bogdanov et al., 2009), it is important that the honeybees are given wax that they can use as fast as possible to build their combs (Johansson & Johansson, 1978):

Bioassay tests for economic factors should be performed as follows:

1. Use pure comb foundation and comb foundation adulterated with paraffin or other adulterant of interest (see step 1. and 2. in Section 6.2.6.2).
2. Place comb foundation with the same percentage of paraffin (or other adulterant) in the beehive. This option consists of performing a method as close as possible to the operations in normal bee management (usually beekeepers acquire beeswax at the same supplier).
3. Use three test frames to replace frames with old combs.
4. Put the frames between the last brood comb and the first comb with honey and pollen (Figure 61).
5. To transfer a nucleus (five frames) or a package of bees to a beehive (10 frames), put five frames with bees, open and closed brood and food reserves (pollen and honey) into a beehive (Figure 62).
6. Put marks (numbers or letters) or spray the top of the experimental frames (pure and adulterated comb foundation) with different colours.
7. Before use in experiment, weigh experimental frames (Figure 63).
8. Use 1 L of sugar syrup at a ratio of 1:1 (by weight) to stimulate colonies in the first day. Remember: this bioassay is performed when there is a nectar flow and it is not economical to use more syrup. Feed with more syrup during the experiment period only in case of unfavourable climatic conditions (prolonged rain, the wind that dries the production of nectar during the flowering season, etc.).
9. Check the development of the colonies weekly. Use manual or electronic hive scales to weigh full hives (see Section 4.1 "Research methods at the colony level" in the *BEEBOOK* article Miscellaneous standard methods for *Apis mellifera* research' (Human et al., 2013).

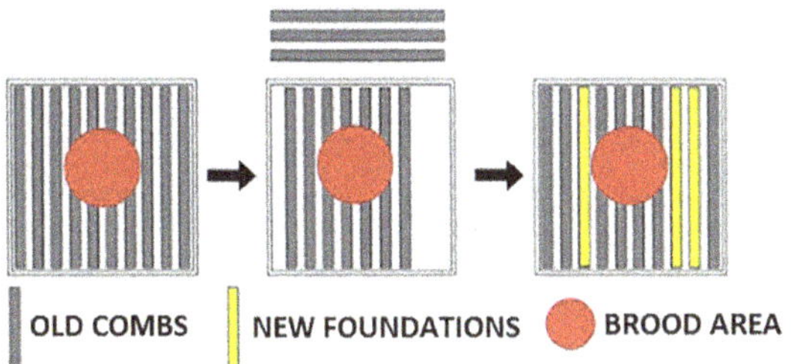


Figure 61. Replacement of the old combs (generally three frames).

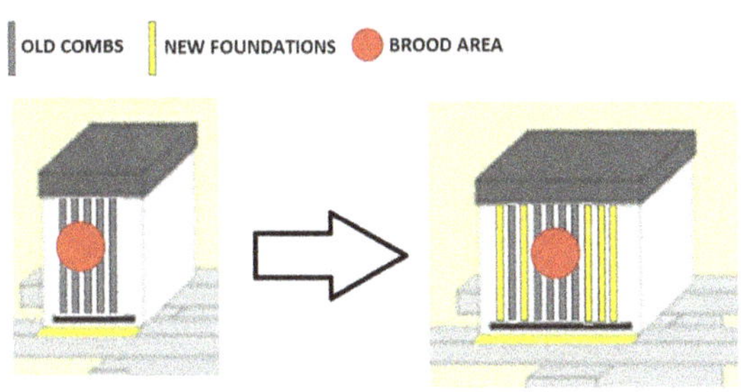


Figure 62. Transferring of a nucleus (five frames) or a package of bees to a beehive (ten frames).

10. Each time a beehive is checked, weigh the experimental frames (pure and adulterated comb foundation) with the same scale.
11. Compare the weigh between colonies and each experimental frame (built upon pure and adulterated comb foundation) (Figure 63). In this case, the weight includes the production of wax, the storage of honey, the honey bees and the brood.
12. Exchange the fully drawn frames from the middle (near the frames with brood) with the initially or partially built frames from outside positions (near nectar and pollen frames) during the weekly visits (step 9), immediately after weighing the combs. This operation is only performed if this is required (if the comb foundations are pulled out with wax).
    *Caution*: do not insert the comb foundation into the hive without pulling in the middle of brood to avoid development of some brood diseases. The placement of experimental frames between the brood area can divide the colony; consequently, a part of the brood may not receive enough heat and some brood diseases, such as fouldbrood and chalkbrood, may develop.
13. A comb is considered built when the cells are more than 0.5 cm deep (Szabo, 1977) and when honey bees have constructed cells on more than 80% of the comb foundation area.
14. The duration of the study is 3 weeks (sufficient time for the colony to pull the pure wax out). However, this time period need to be adjusted according to climatic conditions and various nectar flow. It is important to compare the development of the colonies with different ratios of adulterated wax in the same environmental conditions, and at the same time.

Figure 63. Weighing of the experimental frames in the field.

*6.2.6.4. Advantages and disadvantages.*

**Advantages:**

- Better understanding of the consequences of beeswax adulterants on biological and economic factors in beekeeping technology.
- Potential for conducting complementary studies related to the adulterants (especially paraffin wax) that act as diluents of residual contaminants present in comb foundations; for instance, for verifying diluent action of the paraffin with the acaricides and consequent contamination of honey and brood (Medici et al., 2012; Wilmart et al., 2016).

**Disadvantages:**

- Possible need for a considerable number of colonies to carry out this kind of study.
- For some parameters, there may be difficulties in the measurement procedure, such as the evolution of the colony weight and relating economic factors (it might be necessary to purchase scales and instruments that may not be economically viable).

### *6.3. Standard methods for detection of pesticide residues in beeswax*

***Definitions, acronyms:***

- MRL: maximum residue limits
- GC: gas chromatography
- ECD: electron capture detector
- NPD: nitrogen-phosphorus detector
- FPD: flame photometric detector
- MS: mass spectrometry
- LC: liquid chromatography
- PES: polyethersulfone
- DE: diatomaceous earth
- OCLLE: on-column liquid-liquid extraction using diatomaceous earth
- SPE: solid-phase extraction
- QuEChERS: quick easy cheap effective rugged safe
- PSA: primary secondary amine
- CDL: curved desolvation line
- ESI: electrospray ionisation
- LOD: limits of detection
- LOQ: limits of quantification

*6.3.1. Residues in beeswax*

Data on beeswax contamination by pesticides are less available and mostly come from nationwide surveys, rather than from regular inspections, simply because beeswax is not directly consumed by humans. Therefore, an official maximum residue limits (MRLs) are not set. On the other hand, beeswax is not only recycled almost continuously in the form of comb foundation, but it is also processed for pharmaceutical purposes or for cosmetics industries, and in the food industry as a glazing agent, for the surface treatment of certain fruits and as a carrier for flavour and colours (EFSA, 2007).

Beeswax is often contaminated by persistent lipophilic acaricides (Bogdanov, 2006; Jiménez et al., 2005). Moreover, many substances can easily migrate from wax to honey (Tremolada et al., 2004; Wallner, 1999), hence pesticide residues even at trace levels are problematic. The contamination level of residues in beeswax has to be controlled and the accumulation process driven by wax recycling should be limited.

Gas chromatography (GC), initially in combination with selective detectors (electron capture detector (ECD), nitrogen-phosphorus detector (NPD), flame photometric detector (FPD) and subsequently in combination with mass spectrometry (MS) analyser, has been the main instrumental technique used in the analysis of pesticide residues since the 70s. GC and liquid chromatography (LC) coupled with MS detection are currently the most widely used techniques for multiresidue analysis of pesticides in bee products (Wiest et al., 2011). The choice of the chromatographic technique depends mainly on the physicochemical characteristics of the pesticides of interest. Volatile, semi-volatile and thermally stable compounds can be determined by GC technique, while non-volatile and/or thermally unstable compounds are usually determined by LC technique (Kujawski et al., 2014). The number of active ingredients that are determined by LC has increased significantly in the last years, since the pesticides used today are more polar, thermally unstable or not easily vaporizable, which makes them more suitable for this technique (Albero, Sanchez-Brunete, & Tadeo, 2004; Wille et al., 2011).

Over last recent years many studies have been focused in the multiple determination of pesticides in bee matrices covering insecticides, acaricides, fungicides and herbicides which can be divided into distinct chemical classes and mechanism of action such as organophosphates, organochlorurates, carbamates, phenylpyrazoles, pyrethroids, neonicotinoids, spynosins, benzoylureas and more. The next sections will describe the most used methods for wax extraction and purification for the subsequent analysis with LC-MS (see Section 6.3.2.1) and GC-MS techniques (see Section 6.3.3.1), and related chromatographic and MS conditions (see Sections 6.3.2.2 and 6.3.3.2).

*6.3.2. Trace-level determination of pesticides in beeswax by LC-MS*

*6.3.2.1. Extraction and purification.* The extraction of pesticide residues and the partial or total removal of the interfering substances are two critical issues in the analysis of complex matrices like beeswax. These operations affect performance parameters of sensitivity, selectivity and trueness of the method.

There is no official method for sample preparation for pesticide detection in beeswax. There are many methods that can be used and they are usually chosen based on the materials and equipment present in the laboratory. The extraction and clean up protocols more used can be distinguished into three types: on-column liquid-liquid extraction using diatomaceous earth, solid phase extraction (SPE) and modified QuEChERS (Quick Easy Cheap Effective Rugged Safe) protocol. The QuEChERS technique is the most commonly used method as it is able to determine a broad spectrum of analytes and it is certainly faster than using diatomaceous earth (OCLLE) or other SPE cartridges, despite a lower sample purification.

To get the best sample homogeneity of wax and improve the extraction of residues is always recommended to:

1. Reduce the wax in small pieces.
2. Place them in the milling chamber.
3. Freeze the wax pouring liquid nitrogen in the milling chamber.
4. Grind it finely with the analytical mill.

*6.3.2.1.1. OCLLE protocol. (Nguyen et al., 2009):*

1. Weigh 2 g of homogenized wax.
2. Add 15 mL of n-hexane mixture/isopropanol (8:2 v/v).
3. Heat to 50 °C for 3 min until wax is dissolved.
4. Add 10 mL of water.
5. Centrifuge for 5 min at 50 °C and 700 rpm.
6. Transfer the aqueous phase in diatomaceous earths cartridge.
7. Elute the analytes after 15 min with 20 mL of acetone, collecting the extract into a glass flask.
8. Dry on the rotary evaporator.
9. Reconstitute with 1 mL of acetonitrile/water (50:50 v/v).
10. Filter (polyethersulfone (PES) syringe filter, diameter 13 mm, pore size 0.45 μm).
11. Transfer into vial.

*6.3.2.1.2. SPE protocols. (Yáñez, Bernal, Nozal, Martín, & Bernal, 2013):*

1. Weigh 0.5 g of homogenized wax.
2. Add 1.5 mL of water.
3. Add 2.5 mL of ethanol.
4. Stir for 1 min.
5. Centrifuge for 20 min.
6. Condition the ChemElut SPE cartridge with 1 mL NaCl 20%.
7. Transfer the supernatant solution in the cartridge.
8. After 15 min, twice elute the analytes with 10 mL of n-hexane-dichloromethane (50:50 v/v).
9. Dry at 30 °C under gentle stream of nitrogen.
10. Reconstitute with 200 μL of acetonitrile-water (50:50 v/v).
11. Filter and transfer into vial.

*SPE protocol (Wallner, 1993):*

1. Weigh 0.2 g of homogenized wax.
2. Add 6 mL of hexane.

3. Heat to 70 °C.
4. Centrifuge at −10 °C for 5 min.
5. Transfer the supernatant solution in a glass tube.
6. Repeat the steps 2, 3, 4 and 5 with 3 mL hexane two times.
7. Condition a 500 mg Florisil column with 6 mL of hexane.
8. Run the collected solution through the column.
9. Elute with 1.5 mL of acetone/hexane (1:1).
10. Transfer 0.5 mL into vial.

*6.3.2.1.3. Modified QuEChERS protocol. (Herrera Lòpez, Lozano, Sosa, Hernando, & Fernàndez-Alba, 2016) according to the European EN 15662 (CEN 2008) method:*

1. Weigh 10 g of homogenized sample.
2. Add 10 mL of acetonitrile.
3. Stir for 1 min.
4. Add the premixed extraction salts, consisting of 4 g of Magnesium sulphate anhydrous, Sodium chloride 1 g, 1 g of Trisodium citrate dihydrate and 0.5 g of Disodium hydrogencitrate sesquihydrate.
5. Stir for another min.
6. Centrifuge for 5 min at 4000 rpm.
7. Transfer 6 mL of the liquid phase in a polypropylene tube.
8. Add 150 mg of Primary Secondary Amine (PSA) and 900 mg of magnesium sulphate anhydrous.
9. Stir for 1 min.
10. Centrifuge for 2 min at 6000 rpm.
11. Transfer 1 mL of extract in a test tube with 220 µL of acetonitrile.
12. Vortex and transfer into vial.

*6.3.2.2. LC-MS detection.* This subsection presents a short review of parameters relative to LC-MS detection of pesticides in beeswax performed by different studies (summarized in Table 20). The used parameters depend on the equipment and instrumentation available to the laboratory, and the investigated analytes. The expexted result consists on a chromatogram with acceptable peaks shape and resolution.

Chromatographic separation of pesticides was more often achieved by reversed-phase chromatography that includes any chromatographic method that uses a hydrophobic stationary phase, and the most used column is the octadecyl carbon chain (C18)-bonded silica, such as:

- Polaris C18-A HPLC column (3 µm, 2.0 × 150 mm) (Pirard et al., 2007).
- Agilent Zorbax SB-C18 column (3.5 µm, 2.1 x 150 mm) (Mullin et al., 2010).
- Fused-core type column Kinetex C18 (2.6 µm, 4.6 x 150 mm) (Yáñez et al., 2013, 2014).
- Halo C18, Eksigent column (2.7 µm, 0.5 x 50 mm) (Herrera Lòpez et al., 2016).

Kinetex Phenyl-Hexyl column (2.6 µm, 2.1 x 100 mm) (Jabot et al., 2015) as it displays a higher affinity for aromatic and moderately polar analytes, and RP-Amide Ascentis Express column (2.7 µm, 2.1 x 100 mm) (Porrini et al., 2016) that provides enhanced selectivity for samples containing acidic and basic compounds that can act as a hydrogen-bond donor, have also been used.

Chromatographic conditions:

- Column oven temperatures settings were quite variable depending on the research study, ranging from 25 to 60 °C. In detail, 25 °C (Yáñez et al., 2014), 30 °C (Herrera Lòpez et al., 2016), 35 °C (Yáñez et al., 2013), 40 °C (Pirard et al., 2007; Porrini et al., 2016), and 60 °C (Jabot et al., 2015) were the different temperatures used.
- Chromatographic mobile phases consisted of an acidified aqueous solution (mobile phase A) and an organic phase (mobile phase B) as acetonitrile or methanol, acidified or not. In detail, different combinations have been used:

  - (A) water and (B) acetonitrile, both acidified with 0.1% of acetic acid (Pirard et al., 2007) or with 0.1% formic acid (Herrera Lòpez et al., 2016).
  - (A) 0.1% formic acid in water and (B) acetonitrile (Yáñez et al., 2013, 2014).
  - (A) 0.04 mM ammonium acetate in water with 0.01% formic acid and (B) methanol (Jabot et al., 2015).
  - (A) 5 mM ammonium formiate in water with 0.1% formic acid and (B) 5 mM ammonium formiate in methanol with 0.1% formic acid (Porrini et al., 2016).
  - Only in one case an isocratic elution mode was described (Yáñez et al., 2014), but usually different gradient modes were performed. The flow rate varied from 0.4 to 0.5 mL/min.

Electrospray ionisation mass spectrometry (ESI-MS) is the elective technique. ESI uses electrical energy to assist the transfer of ions from solution into the gaseous phase before they are subjected to mass spectrometric analysis. In positive ion mode, as performed for pesticide detection in beeswax, the analyte is sprayed at low pH to encourage positive ion formation.

The instrumental parameters vary in the different papers:

- Jabot et al. (2015): capillary voltage 3200 V, desolvation temperature 450 °C, source temperature 150 °C and nitrogen desolvation and nebulizer gas flows 900 and 150 L/h, respectively.
- Pirard et al. (2007): capillary and cone voltages 3 kV and 35 V, respectively, temperature source 125 °C, desolvatation temperature 250 °C. Nitrogen used as cone and desolvating gas at a flow rate of 100 and 680L/h, respectively.

Table 20. Summary of LC coupled to MS methods for detection of pesticides in beeswax.

| Number of chemicals detected | Pesticides extraction | Instrument | Chromatographic column | Mobile phases | LOQ | Validation | Advantages/ disadvantages | BeeBook section number | References |
|---|---|---|---|---|---|---|---|---|---|
| 120 | QuEChERS method | Microflow-LC-ESI-MS/MS | Halo C18 (50 × 0.5 mm × 2.7 μm) | (A) Water (B) ACN both with 0.1% formic acid | 5 μg/kg (for most compounds) | Yes | Well described method parameters | 6.3.2.1.3., 6.3.2.2. | Herrera Lòpez et al. (2016) |
| 128 | QuEChERS method | LC-ESI-MS | RP-Amide Ascentis Express (100 × 2.1 mm × 2.7μm) | (A) Water (B) ACN both with 0.1% formic acid and 5 mmol ammonium formiate | n.r. (5 μg/kg LOD) | n.r. | LOQ not available | 6.3.2.2. | Porrini et al. (2016) |
| 13 | On-column liquid–liquid extraction (OCLLE) using diatomaceous earth | LC-ESI-MS/MS | Kinetex phenyl-hexyl (100 × 2.1 mm × 2.6 μm) | (A) 0.1% formic acid and 0.04 mmol ammonium acetate in water (B) MeOH | 1–40 μg/kg | Yes | Few chemicals detected | 6.3.2.2. | Jabot et al. (2015) |
| 300 | QuEChERS method | LC-ESI-MS/MS | n.r. | n.r. | 10 μg/kg (for most compounds) | n.r. | Lacking critical information | | Ravoet et al. (2015) |
| 99 | Solid phase extraction | LC-ESI-MS/MS | n.r. | n.r. | 100 μg/kg (for most compounds) | n.r. | Lacking critical information | 6.3.2.2. | Simon-Delso et al. (2015) |
| 2 | Solid-liquid extraction | LC-ESI-MS | Kinetex C18 (150 × 4.6 mm × 2.6 μm) | (A) Water with 0.1% formic acid (B) ACN | 0.4–0.7 μg/kg | Yes | Few chemicals detected | 6.3.2.2. | Yáñez et al. (2014) |
| 7 | OCLLE using diatomaceous earth | LC-ESI-MS | Kinetex C18 (150 × 4.6 mm × 2.6μm) | (A) Water (B) ACN both with 0.1% formic acid | 1.5–7 μg/kg | Yes | Few chemicals detected | 6.3.2.1.2., 6.3.2.2. | Yáñez et al. (2013) |
| 200 | QuEChERS method | LC-ESI-MS/MS | Agilent Zorbax SB-C18 column (150 × 2.1 mm × 3.5 μm) | n.r. | n.r. (0.1–50 ng/kg LOD) | n.r. | Chemicals detected by LC and GC techniques, not mentioned which and how many compounds for each technique. | 6.3.2.2. | Mullin et al. (2010) |
| 1 | OCLLE using diatomaceous earth | LC-ESI-MS/MS | n.r. | n.r. | 0.5 μg/kg | n.r. | Few chemicals detected | 6.3.2.1.1., 6.3.2.2. | Nguyen et al. (2009) |
| 19 | OCLLE using diatomaceous earth | LC-ESI-MS/MS | Polaris C18 (150 × 2.0 mm × 3 μm) | (A) Water (B) ACN both with 0.1% acetic acid | 0.0002–1.232 μg/kg (CC$\beta$) | Yes | Few chemicals detected | 6.3.2.2. | Pirard et al. (2007) |

ACN, acetonitrile; MeOH, methanol; n.r., not reported.

- Porrini et al. (2016): ESI interface voltage 1.5 kV, curved desolvation line (CDL) temperature 250 °C, block heater temperature 200 °C. Nitrogen was the nebulising gas with a flow rate of 1.5 L/min.
- Yáñez et al. (2013, 2014): fragmentor voltage 140 V, drying gas ($N_2$) flow 9 L/min, drying gas ($N_2$) temperature 340° C, nebulizer gas ($N_2$) pressure 40 or 60 psi, capillary voltage 2500 or 3000 V.

Limits of detection (LOD) of the methods ranged from 0.05 to 50 ng/kg and limits of quantification (LOQ) from 0.5 to 70 μg/kg, strictly depending on the analyte and on the studies considered, as summarised in Table 20.

*6.3.3. Chromatographic multiresidue analysis by GC-MS*

In the scientific literature concerning determination of pesticides in wax by GC-MS, prevails the analyses of residues of active ingredients linked to pharmacological treatments in the control of *Varroa* mite (miticides). Few publications describe analysis methods for pesticides determination in wax (Bogdanov, Kilchenmann, & Imdorf, 1998; Boi et al., 2016; Lodesani, Costa, Serra, Colombo, & Sabatini, 2008; Mullin et al., 2010; Wallner, 1993, 1999).

The next sections will describe the most used methods for wax extraction and purification for the subsequent analysis with GC-MS (see Section 6.3.3.1) and related chromatographic and MS conditions (see Section 6.3.3.2).

*6.3.3.1. Extraction and purification.* Similarly to what has already been written for the phases of extraction and purification in LC-MS paragraph (see Section 6.3.2.1), also in GC analysis these phases assume a relevant importance, due to the chemical complexity of the wax.

To obtain the best homogeneity of the sample is recommendable to freeze the wax with liquid nitrogen and then to grind it finely (as described in Section 6.3.2.1.).

The extraction and purification protocols can be distinguished into three types according to the use of diatomaceous earth (DE), SPE or QuEChERS method.

*6.3.3.1.1. DE protocol. (Boi et al., 2016):*

1. Weigh 3 g of wax in a beaker.
2. Melt at 80 °C in oven.
3. Add 5 mL of hexane.
4. Add 15 g of diatomaceous earth.
5. Vortex the powdered mixture.
6. Fill an empty 80 mL cartridge with the resulting powdered mixture.
7. Elute the sample with approximately 120 mL of acetonitrile.
8. Freeze the extract at -18 °C, when the elution is complete.
9. Dry the extract in rotary evaporator at low pressure with water bath at 55 °C.
10. Dissolve the residue with 3 mL of 2,2,4-trimethylpentane.
11. Filter (PES syringe filter, 13 mm, 0.45 μm) the sample in vials for GC-MS analysis.

**Advantages:**

- performance of good extractions and purifications

**Disadvantages:**

- requires large volumes of solvent and large freezers

*6.3.3.1.2. SPE protocol. (Bogdanov et al., 1998):*

1. Weigh 1 g of wax into a centrifugation tube.
2. Add 10 mL hexane.
3. Extract in ultrasonic bath for 45 min.
4. Cool the n-hexane solution in a freezer for 1.5 h.
5. Centrifuge for 18 min at 10,000 rpm at −5 °C.
6. Concentrate the supernatant to a volume of 1 mL using a Rotor evaporator.
7. Transfer to a new centrifuge tube.
8. Repeat twice freezing and centrifugation.
9. Load the final supernatant (1 mL) into a Florisil (60/100 mesh) column (1.5 g Florisil + 0.5 g anhydrous $Na_2SO_4$ in a 6 mm i.d. glass column).
10. Add 20 mL of n-hexane.
11. Discard this fraction.
12. Add 10 mL of n-hexane/acetone (1:1).
13. Concentrate this fraction by Rotor evaporator.
14. Dry under a gentle nitrogen stream.
15. Re-suspend purified extract in 0.5 mL of n-hexane for GC-MS quantification.

**Advantages:**

- performance of good extractions and purifications

**Disadvantages:**

- time-consuming analysis

*6.3.3.1.3. QuEChERS protocol. (Mullin et al., 2010):*

1. Weigh 3 g of wax into a 50 mL centrifuge tube.
2. Add 100 μL of process control spiking solution.
3. Prepare an adequate volume of extraction solution by mixing water:acetonitrile:glacial acetic acid, 44:55:1 (v/v).
4. Add 27 mL of extraction solution.
5. Melt the sample at 80 °C in water bath.
6. Cool the sample to room temperature.

7. Add the premixed extraction salts, consisting of 6 g of $MgSO_4$ and 1.5 g of sodium acetate.
8. Seal tube.
9. Shake vigorously for 1 min.
10. Centrifuge for 1 min.
11. Add to a dual layer SPE cartridge (containing 250 mg graphitized carbon black and 500 mg PSA) about 80 mg of anhydrous magnesium sulfate.
12. Condition cartridge by adding 4.0 mL of acetone/toluene (7:3, v:v).
13. Elute solvent to waste, using a positive pressure or vacuum manifold.
14. Add 2 mL of supernatant obtained after centrifugation (step 10), to the top of the cartridge.
15. Elute cartridge using 3–4 mL of acetone/toluene (7:3, v:v) into a 15 mL graduate glass centrifuge tube.
16. Dry eluate to a final volume of 0.4 mL in evaporator at 50 °C.
17. Sample is ready for GC-MS analysis.

**Advantages:**

- quick and simple technique

**Disadvantages:**

- technique does not provide an excellent purification of the analytes

*6.3.3.2. GC-MS detection.* Most of the scientific literature (Boi et al., 2016; Bonzini et al., 2011; Serra Bonvehí & Orantes Bermejo, 2010) agrees that a DB5 MS column 30 m × 0.25 mm × 0.25 μm is the best choice for the chromatographic separation of pesticides by GC.

Instrumental parameters settings:

- Injection volume: 1 μL.
- Injector: splitless mode (60 s).
- Injector temperature: 280 °C.
- Carrier gas: helium.
- Flow rate: 1 mL $min^{-1}$.
- Temperature program:
  - 60 °C for 2min.
  - 20 °C $min^{-1}$ to 150 °C.
  - 4 °C $min^{-1}$ to 260 °C.
  - 15 °C $min^{-1}$ to 320 °C.
  - 320 °C for 7 min.
- Transfer line temperature: 325 °C.
- Electron ionization: 70 eV.
- Source temperature: 230 °C.
- Quadrupole temperature 150 °C.
- Scan range: 50–600 mass unit.

The limits of detection (LOD) and quantification (LOQ) vary greatly in relation to the monitored active ingredient (see Table 21 for some examples, the conditions they were acquired in and the advantages and disadvantages of each).

*6.3.4. Other methods for pesticides detection in beeswax*

Other techniques different from LC or GC-MS are currently not used commonly for the detection of pesticides in beeswax, except for the pesticides group of acaricides, mostly used to treat honey bee colonies to control *Varroa* mite infestation.

The methods for acaricide detection in beeswax are mainly based on the determination by GC with different detectors (apart from MS analyser, described in Section 6.3.3) such as flame photometric detector (FPD), ion trap detector (ITD), flame ionization detector (FID) and electron capture detector (ECD), but only allow for the detection of a restricted number of analytes.

Only a few researchers used LC, and then only for the determination of a limited number of active principles, mainly acaricides. A widely used LC detection is the fluorescence detector but its use is restricted to analytes that are fluorescent or are converted to fluorescent derivatives.

Given the limited number of analytes and the relatively high limits of detection, these methods are to be considered of secondary importance compared to multiresidual studies previously described in Sections 6.3.2., 6.3.3., and 6.3.4., and these have been summarized in Table 22. They are outdated and of very limited interest and can therefore be considered as historical literature, not recommended for future studies.

## 7. Standard methods for detection of pathogens in beeswax

Beeswax is an important bee product which returns partly into the hive in the form of comb foundation (Máchová, 1993). Consequently, investigation methods applicable to beeswax herein described aim at detecting and quantifying honey bee pathogens and consequently guaranteeing sterility or at least the absence of specific pathogens from comb foundation. When beeswax is processed for producing foundations, cleaning and disinfection could necessitate chemicals, e.g., 0.5% $H_2SO_4$, and a temperature of 80–90 °C for 30 min. This procedure is sufficient to kill vegetative stages of *Paenibacillus larvae* (Rose, 1969), but does not destroy *P. larvae* spores; their heat resistance can even be increased at these temperatures (Gerhardt & Marquis, 1989). A temperature of 121 °C at a pressure of 101.325 kilopascal (kPa) for 20–30 min to remove *P. larvae* spores from beeswax is recommended (Hornitzky & Wills, 1983; Plessis, Du-Rensbrug, & Van-Johannsmeier, 1985; Shimanuki, Herbert, & Knox, 1984). Temperature not combined with pressure could also be effective but longer application is required (Savov & Arsenov, 1963). Methods for isolation of *P. larvae* spores in beeswax had been already elaborated in the 70s by Kostecki and

Table 21. Summary of GC coupled to MS methods for detection of pesticides in beeswax.

| Number of chemicals detected | Pesticides extraction | Chromatographic column | LOD | Validation | Advantages/disadvantages of the method | Beebook section number | References |
|---|---|---|---|---|---|---|---|
| 5 | DE | J&W DB 5MS 30 m, 0.25 mm, 0.25 μm | 1 μg/kg | Yes | Few active ingredients | 6.3.3.1.1. | Boi et al. (2016) |
| 200[a] | QuEChERS method | J&W DB 5MS 30 m, 0.25 mm, 2 μm | 0.1–50 μg/kg | n.r. | Number of chemicals detected by GC not indicated | 6.3.3.1.3. | Mullin et al. (2010) |
| 5 | SPE | HP 5 30 m, 0.25 mm, 0.25 μm | n.r. | n.r. | Few active ingredients | 6.3.3.2. | Bonzini et al. (2011) |
| 14 | SPE | J&W DB-17MS 30 m, 0.25 mm, 0.25 μm | 7–38 μg/kg | Yes | Few active ingredients | 6.3.3.2. | Serra Bonvehí and Orantes Bermejo (2010) |

n.r., not reported.
[a]Used in combination with LC-MSMS technique.

Orlowski (1975), Kostecki and Jelinski (1977) and then by Hansen and Rasmussen (1991). More recently wax (hive) debris have been investigated as indicator of the possible presence of spores of *P. larvae* and to predict the onset of AFB in the colony in the next season (Bzdil, 2007; Forsgren & Lauden, 2014; Ryba, Titera, Haklova, & Stopka, 2009; Titěra & Haklová, 2003). Beeswax is the most relevant component of hive debris and it is where pathogens can accumulate. Other than *P. larvae* here we report on the protocols to determine other pathogens from beeswax and wax (hive) debris, i.e., *Melissococcus plutonius, Ascosphaera apis, Nosema* spp. and small hive beetle (SHB). Furthermore, the reader is referred to *COLOSS BEEBOOK Volume II* for more general information on the different pathogens of honey bees.

### 7.1. Early detection of American foulbrood by beeswax analysis

AFB is a devastating brood disease of the honey bee caused by the spore-forming, Gram-positive rod-shaped bacterium *P. larvae* (de Graaf et al., 2013; Hansen & Brodsgaard, 1999). AFB is one of the bee diseases listed in the OIE (Office International des Epizooties – the World Organization for Animal Health) Terrestrial Animal Health Code (OIE, 2018a) and internationally recognised diagnosis protocols are available in the OIE Manual for diagnostic tests and vaccines for terrestrial animals (OIE, 2018b). The possible role of beeswax in the spread of pathogens through trade in honey bees and their products has been considered by Mutinelli (2011). An AFB infected dead honey bee larva transforms in a dark scale adherent to the cell wall that contains up to 2.5 billion spores of *P. larvae* (Figure 64) and can contaminate beeswax. Methods of detecting *P. larvae* in beeswax have implications in both sterility of foundations and prognostic value for the development of the disease.

We provide these methods in a simple and easy-to-be-applied form to facilitate their implementation in the laboratory. However, reference to original papers is provided. Furthermore, readers are addressed to COLOSS *BEEBOOK Volume II Standard methods for American foulbrood research* (de Graaf et al., 2013) that covers a broad set of technical information on *P. larvae*.

#### *7.1.1. Determination of* P. larvae *spores in beeswax*

*7.1.1.1. Choice of solvent.* Wax samples should be dissolved in Tween 80 that is today considered the most appropriate solvent (Bzidil, 2007) (Section 7.1.1.2.2). An historical overview of different solvents used to extract *P. larvae* spores from beeswax and wax (hive) debris is presented in Table 23.

All the solvents should be stored in a flammable cabinet and used under chemical safety hood using personal protective equipment, i.e., laboratory suit, gloves and googles.

Table 22. List of publications concerning the research of pesticides in beeswax using techniques other than liquid and gas chromatography coupled with mass spectrometry.

| Analytes | Technique | LOD (μg/kg) | Reference |
|---|---|---|---|
| Malathion, coumaphos | GC-FPD | – | Thrasyvoulou and Pappas (1988) |
| Amitraz | GC-ITD | 1 | Leníček et al. (2006) |
| Para-dichlorobenzene | GC-FID | 700 | Bogdanov et al. (2004) |
| Amitraz, fluvalinate, bromopropylate | GC-ECD | 1, 1, 50 | Lodesani et al. (1992) |
| Bromopropylate, coumaphos, flumethrin, fluvalinate | GC-ECD | 100–300 | Bogdanov et al. (1998) and Bogdanov, Kilchenmann, and Bütikofer (2003) |
| Fluvalinate | GC-ECD | 100 | Tsigouri, Menkissoglu-Spiroudi, Thrasyvoulou, and Diamantidis (2000) |
| Amitraz and metabolites (DPMF, DMA) | GC-ECD | 80 | Jiménez, Bernal, del Nozal, and Alonso (2004b) |
| Lindane, chlorpyriphos, z-chlorfenvinphos, endosulfan A and B, 4,4′-DDE, 4,4′-TDE, acrinathrine, bromopropylate, tetradifon, coumaphos, fluvalinate | GC-ECD | 5–40 | Jiménez, Bernal, del Nozal, and Alonso (2004a) |
| Benomyl, carbendazim | LC-FL | 10–50 | Bernal et al. (1997) |
| Coumaphos | LC-FL | 2 | Van Buren, Marien, Velthuis, and Oudejans (1992) |
| Amitraz | LC-UV | – | Korta et al. (2001) |
| Fluvalinate, coumaphos, bromopropylate and 4,4′-dibromobenzophenone | LC-DAD | 2–100 | Adamczyk, Làzaro, Pérez-Arquillué, and Herrera (2007) |

FPD, Flame Photometric Detector; ITD, Ion Trap Detector; FID, Flame Ionization Detector; ECD, Electron Capture Detector; FL, Fluorimetric Detector; UV, Ultra-Violet detector; DAD, Diode Array Detector.

The following characteristics are examined visually providing direct information about the solvent efficacy:

- solubility of wax samples.
- formation of mist or sediment with solvent.

In the latter, mist should be removed by replicated dissolving and centrifugation until beeswax is completely dissolved.

According to the results obtained by Bzdil (2007), the Tween method is statistically significantly more effective in detecting *P. larvae* spores than toluene. This might be explained by high aggressiveness of toluene that might kill a part of spore-forming microorganisms in one of the sporulation stages. It is also possible that toluene residues on culture media plates negatively influence the germination and growth of microorganisms.

The viable *P. larvae* spores trapped in beeswax and extracted by solvent is determined by sample plating on MYPGP medium.

*7.1.1.2. Method of isolation of* P. larvae *spores from beeswax and wax (hive) debris.*

***Definitions, acronyms:***

- Spiked sample: sample to be examined spiked with spores of *P. larvae*.
- Positive control: reference strain of *P. larvae*, ATCC 9545.
- Negative control: matrix without spores.
- $H_2O_2$: hydrogen peroxide

***Equipment required:***

- Bunsen or Laminar flow cabinet
- Sterile 50 mL tubes
- Sterile Pasteur pipette
- 50–200 μL micropipette
- 200μL sterile tips
- Sterile disposable loops
- Bench centrifuge
- Disposable L-shaped spatula
- Microscopic slides
- Incubator +37 °C
- Dry heat oven 90 °C
- Freezer -18 °C
- Refrigerator +2 °C/+8 °C
- Jar for microaerobic conditions.
- Water bath +50 °C/+80 °C
- Microscope with objective 100x

***Reagents required:***

- Sterile distilled water stored at RT
- MYPGP agar stored at 5 ± 2 °C:
    - 10 g Mueller-Hinton broth (Oxoid CM0405)
    - 15 g yeast extract
    - 3 g $K_2HPO_4$
    - 1 g Na-pyruvate
    - 20 g agar
    - Autoclave at 121 °C/15 min.
    - Add 20 mL 10% glucose (autoclaved separately).
    - Cool the media to 50 °C and add the antibiotics to a final concentration of 20 μg/mL for nalidixic acid and 10 μg/mL for pipemidic acid:

Figure 64. Comb with typical signs of AFB: dark scales adhering to the wall of the cells; perforated caps with brown dense material draining from the cells. Once dried this material becomes scale.

- Nalidixic acid stock solution (1 mg/mL) is prepared by dissolving 0.1 g in 2 mL of 1 M NaOH and diluting to 100 mL with 0.01 M phosphate buffer (pH 7.2).
- Pipemidic acid stock solution (2 mg/mL) is prepared by dissolving 0.2 g in 2 mL of 1 M NaOH and then diluting to 100 mL with 0.01 M phosphate buffer (pH 7.2).
- Both antibiotic solutions are filter sterilized.

The medium is poured (20 mL) into sterile Petri dishes and plates are dried before use (15 min).

- Campygen (Oxoid) gas pack for generating microaerobic conditions stored at RT
- 3% oxigen peroxide solution stored at 5 ± 2 °C
- Tween 80 solution stored at RT
- Reagents for GRAM staining (Hucker's technique modified): safranine, cristal violet and Lugol's solution stored at RT
- Saline solution stored at RT
- Positive control stored at -20 °C
- Negative control stored at RT

*7.1.1.2.1. Spiked beeswax preparation to be used as* Paenibacillus larvae *positive control.*

1. Weigh 1 g of beeswax.
2. Cut in small pieces in a 50 mL tube.
3. Add an AFB scale (Figure 64), previously tested positive for *P. larvae* (see 7.1.1.2.5.).

*7.1.1.2.2. Culture.*

1. Weigh 1 g of beeswax or wax debris.
2. Cut in small pieces (<3 mm) in a 50 mL tube.
3. Add 8.5 mL sterile distilled water.
4. Add 0.5 mL Tween 80 previously heated for 30 min in water bath at 70 ± 2 °C to reduce its viscosity.
5. Put the sample for 30 min in water bath at 70 ± 2 °C.
6. Stir in longitudinal direction every 10 min (if beeswax is not properly melt or the beeswax size is >3 mm, put in water bath for additionally 30 min). The result is a grey-brown pulp.
7. Cool at RT for 2–4 h to clearly separate liquid/solid phase.
8. Transfer 2–5 mL of liquid in a sterile tube (15 mL).
9. Add the same volume of sterile distilled water.
10. Stir for 5 min.
11. Put in incubator at 90 ± 2 °C for 10 min.
12. Cool at RT.
13. Smear 200 µL of the suspension on MYPGP agar using a spatula.
14. Incubate at +37 ± 1 °C in a jar with CampyGen OXOID gas pack for 5–8 days.
15. After 5–8 days, evaluate the presence of colonies compatible with *P. larvae*.
16. Small, regular, rough, flat or convex, whitish to beige colonies should be considered suspect.
17. Colony morphology is not conclusive but might serve to select the bacterial colonies for further identification through catalase test or Gram staining. Molecular methods can also be applied for identification and characterisation (see COLOSS *BEEBOOK VOLUME II*).

*7.1.1.2.3. Catalase test.*

1. Transfer a colony to a clean microscope slide.
2. Add a drop of 3% hydrogen peroxide on the colony.
3. Observe the drop with the naked eye.
4. *P. larvae* colonies do not produce bubbly foam or only delayed and weak (Haynes, 1972).

*7.1.1.2.4. Gram staining* (OIE, *2018b*).

1. Flood (cover completely) slide of a heat fixed smear with crystal violet.
2. Leave for about 60 s.

Table 23. Solvents used to extract *P. larvae* spores from beeswax and hive debris.

| References | Rose (1969) and Wilson (1972) | Máchová (1993) | Ritter and Metzinger (1998) and Ritter (2003) | Titěra and Haklová (2003) | Bzdil (2007) | Ryba et al. (2009) | Forsgren and Laugen (2014) |
|---|---|---|---|---|---|---|---|
| Matrix/solvents | Beeswax | Beeswax | Beeswax | Hive debris | Beeswax/ hive debris | Hive debris | Hive debris |
| Benzene | Was found to be the best | Was found to be the best | – | – | – | – | – |
| Ether also named diethyl ether | – | Mixed with wax, combined to form a sediment which did not disappear even after repeated centrifugation. After plating, a wax crust was formed in which the spores did not germinate. There was a reduction in the number of spores in replicated dissolution, sedimentation and decantation in the lower part of the jelly supernatant. | Only this solvent was used and found appropriate. | – | – | – | – |
| Chloroform | – | As above | – | – | – | – | – |
| Toluene | – | An excessive growth of additional microflora in the Petri dishes was observed in an experiment with toluene. | – | Only this solvent was used and found appropriate. | – | Toluene (common in cultivation tests) gave consistent results. | – |
| Tween 80 | – | – | – | – | Tween method is statistically significantly more effective in detecting *P. larvae* spores than the Toluene method (Titěra & Haklová (2003)). | According to Bzdil (2007). | According to Bzdil (2007). |
| Isopropanol and ethanol | – | – | – | – | – | They showed the average value of efficiency. | – |
| Water + Tween 80 | – | – | – | – | – | Extraction into water with Tween 80 did not show any differences compared with water alone. | – |
| Water | – | – | – | – | – | Showed the highest efficiency. | – |

3. Wash the slide for 5 s with water. The smear should appear blue-violet when observed with the naked eye.
4. Flood the slide with the iodine solution.
5. Leave for about 60 s.
6. Rinse the slide with water for 5 s and immediately proceed. At this point, the smear should still be blue-violet.
   Note: The next step involves addition of the decolouriser, ethanol. This step is somewhat subjective because using too much decolouriser could result in a false Gram (–) result. Likewise, not using enough decolouriser may yield a false Gram (+) result.
7. To be safe, add the ethanol drop-wise until the blue-violet colour is no longer emitted from the specimen.
8. Rinse with water for 5 s.
9. Flood the slide with the counter-stain safranin.
10. Let this stand for about 60 s to allow the bacteria to incorporate the dye.
11. Rinse with water for 5 s to remove any excess of dye.
12. Blot the slide gently with bibulous paper or allow it to air dry before viewing it under the microscope (Figure 65).

*7.1.1.2.5. Bacterial DNA extraction from wax (hive) debris.*

1. Weight 1 g of debris.
2. Mix 1 g of debris with 5 mL of sterile, distilled water.
3. Shake at RT for 1 h.
4. Immediately subject 100 µL of the suspension to DNA extraction using the QIAamp® genomic DNA isolation mini kit for Gram-positive bacteria (Qiagen) (Ryba et al., 2009).
5. Elute the DNA with 100 µL elution buffer and store at -20 °C until further use.

*7.1.1.2.6. Quantitative real-time PCR for* P. larvae *16s RNA sequence.* The quantitative real-time PCR assay is modified from Martinez, Simon, Gonzalez, and Conget (2010):

1. The reaction contained 10 µL of SsoFast™ EvaGreen® Supermix (Bio-Rad), 0.4 µM each of primers PL2-Fw and PL2-Rev, 2 µL of template.
2. The final reaction volume should be adjusted to 20 µL with nuclease free water.
3. A negative control containing water instead of DNA template is included in each run.
4. All real time qPCR reactions (standards, unknown samples and controls) are performed in duplicate in neighbouring wells on the sample plate.
5. The amplification and data acquisition is carried out using a CFX Connect® (Bio-Rad) real-time PCR machine under the following cycling condition.
6. Enzyme activation step, 98 °C for 2 min, PCR cycling (40 cycles of 98 °C for 5 s (denaturation), 58 °C for 5 s (annealing/extension) and data collection.
7. The identity of the amplified product is confirmed using a melting curve analysis, by raising the temperature from 55 to 95 °C in 0.5 °C increments with a hold of 1 s at each increment.
8. The assay specificity and the presence/absence of non-specific amplification products is determined through the melting temperature(s) ($T_m$) of the amplified product(s).
9. Each set of PCR assays included serial dilutions of DNA extracted from known concentrations of *P. larvae* spores (total microscopic count) as external standards, for relating the qPCR data to spore counts (Forsgren, Stevanovic, & Fries, 2008).
10. Standard curves are prepared by using serial dilutions of target DNA ranging from $10^2$ to $10^8$ as quantification standards in every run.
11. The quantification data are normalized to tissue or debris weight and converted to bacterial cells per bee or bacterial cells per gram debris.

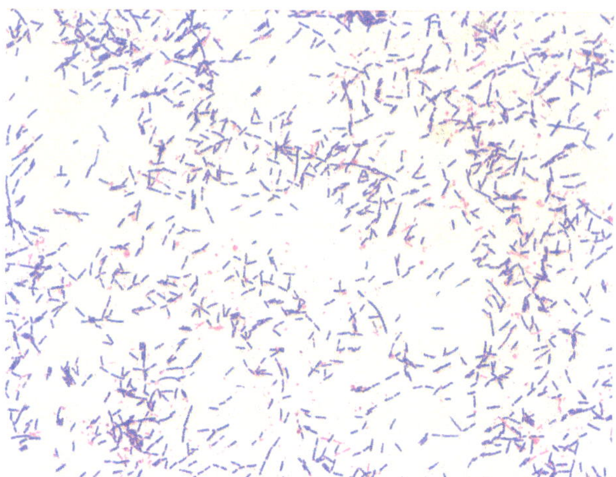

Figure 65. Gram + staining of *P. larvae*. Gram staining, 1000×, immersion oil.

This PCR method on debris samples revealed the highest bacterial levels if compared to honey bee samples (Forsgren & Laugen, 2014). Eventually, sub-clinical levels of *P. larvae* can be easily detected by this PCR in hive debris irrespective of disease symptoms.

### *7.2. Detection of other pathogens/pests in beeswax and wax (hive) debris*

#### *7.2.1. European foulbrood*

There is no reference about investigation of *Melissococcus plutonius*, the causative agent of European foulbrood, in beeswax except for McKee, Djordjevic, Goodman, and Hornitzky (2003) who developed a hemi-nested PCR to improve the detection of *Melissococcus plutonius* in honey bees and their products starting from a previous attempt by Djordjevic, Noone, Smith, and Hornitzky (1998). Briefly, to obtain brood comb cell washes from colonies to be screened for the presence of *M. plutonius*:

1. Soak a sterile swab in sterile distilled water.
2. Rotate it in a single comb cell containing a freshly laid egg.

3. Agitate the swab in 1 mL of sterile distilled water for 30 s.

The application of this hemi-nested PCR to brood comb cell washes could be useful to provide information about the possible presence of *M. plutonius* on beeswax (McKee et al., 2003).

Further information on *M. plutonius* is available in COLOSS *BEEBOOK* volume II Standard methods for European foulbrood research (Forsgren, Budge, Charrière, & Hornitzky, 2013).

The lack of references about the detection of *M. plutonius* in beeswax is likely related to its non-spore forming nature. Hence, it is not considered relevant as beeswax contaminant.

*7.2.2. Ascosphaera apis*

Beeswax, provisions, brood and excreta are rich in nutrients and serve for fungal growth when sufficient moisture is available. Fungi associated with honey bee hives (including yeasts) may be parasites, commensals or mutualistics. *Ascosphaera apis* (Maasen ex Claussen) L.S. Olive and Spiltoir is a pathogen that affects honey bee brood and causes chalkbrood (Johnson et al., 2005). This species occurs widely in temperate regions, being the most widespread infectious fungal disease (Gilliam, Taber, Lorenz, & Prest, 1988). Due to the intimate contact between larvae and comb cells (beeswax) it is reasonable to hypothesize that fungal spores are embedded in beeswax. Foundations artificially contaminated with ground sporulated mummies (mummy is the name and typical clinical presentation of infected larvae covered by *A. apis* iphae and/or mycelium) were able to infect honey bee larvae (Flores, Spivak, & Gutierrez, 2005).

Here we present the protocol to analyse the fungal community from brood combs (beeswax) and wax (hive) debris according to Rivas and Bettucci (2007).

*7.2.2.1. Protocol for determination of A. apis in comb wax and wax (hive) debris.*

1. Place brood comb and bottom board debris on Petri dishes containing MEA 20% sucrose.
   Malt extract agar (for 1 L of MEA + 20% sucrose):
   - Glucose 20 g,
   - Malt extract 20 g,
   - Sucrose 200 g,
   - Peptone 1 g, Agar 20 g,
   - Distilled water1 L.

   Preparation: dissolve ingredients and sterilize at 121 °C for 15 min.
   Final pH: 5.0–5.5.
   Store at 4 °C.
2. Incubate at 25 °C until the fungal colonies develop.
3. Identify grown colonies based on COLOSS *BEEBOOK Volume II* (Jensen Bruun et al., 2013).

Detailed information on further methods to study fungal disease of honey bee is available in COLOSS *BEEBOOK Volume II Standard methods for fungal brood disease research* (Jensen Bruun et al., 2013).

*7.2.3.* Nosema *spp.*

An extensive literature review exists on *Nosema* spp. in COLOSS *BEEBOOK Volume II Standard methods for Nosema research* (Fries et al., 2013). However, detection of *Nosema* spores in beeswax is a rarely studied subject. The nature of beeswax could make retrieving of spores difficult and during wax separation spore viability and genomic DNA could be affected. Different methods may be suitable for spore separation from wax, spore purification and DNA extraction. Here, we present a rapid method developed for a successful separation of *Nosema* spores from beeswax (Muz, Özdemir, & Mutinelli, 2017).

*7.2.3.1. Beeswax melting and spore extraction by water.*

1. Heat 100 mL of distilled water to 81 °C on a digital hot-plate with magnetic stirrer device combined with an encapsulated magnetic stir bar to mix the solution in a borosilicate quality glass chamber.
2. Gently add 10 g of the sampled beeswax to heated water.
3. Close the glass cover well for 30 min.
4. Cool the mixture beeswax-water very slowly at a rate of 1 °C/2 min.
   This facilitates the separation of *Nosema* spores from beeswax.
5. Stop magnetic stirring.
6. Remove the magnetic stir when temperature reaches 66 °C.
7. Let the temperature of the hot-plate decrease at a rate of 0.5 °C/2 min until 60 °C.
8. Gently aspirate the water in a centrifuge tube.
9. Centrifuge at 14,000 rpm for 3 min.
10. Gently discard the supernatant and get *Nosema* spore pellet ready for DNA extraction (Fries et al., 2013).

If spore purification is needed, this can be done by Percoll® centrifugation (Pertoft, Laurent, Laas, & Kagedal, 1978).

*7.2.3.2. Beeswax melting and spore extraction by organic solvents.* If the expected spore number is very low, different chemicals like dichloromethane, acetonitrile, acetone, ethyl acetate, hexane, toluene and xylene should replace water:

1. Use a microtube heating block.
2. Add 0.25 g of beeswax to 1 mL of any of the solvents above to the microtube.
3. Close microtube.
4. Melt at 81 °C in a fume hood.

5. Gently vortex the tube for 5 min.
6. Incubate for 30 min.
7. Cool heating block until 60 °C.
8. Put a drop of solvent containing dissolved beeswax and spores on a microscopy slide.
9. Cover the slide with a cover slip.
10. Check under light microscope (400x) for *Nosema* spore.
11. If the drop contains spores, quickly transfer the drop to a glass tube.
12. Keep the tube open and wait until the solvent evaporates completely in fume hood.
13. Add 1 mL purified water into the glass tube
14. Vortex gently.

The beeswax separated and purified *Nosema* spore suspension is ready to use in any commercial gDNA extraction kit or any traditional method like cetyl trimethylammonium bromide (CTAB), phenol-chloroform or DNAzol methods. See also Fries et al. (2013) for further study methods.

Laboratory trials demonstrated that acetonitrile, acetone, ethyl acetate and xylene caused a reduction in the total number of spores of 34, 25, 19 and 10% respectively and their use is then discouraged. Toluene and hexane had instead no noteworthy effect on the total number of *Nosema* spores recovered (Muz et al., 2017).

*7.2.4. Small hive beetle.* The small hive beetle (SHB), *Aethina tumida*, is a parasite and scavenger of honey bee colonies (Neumann, Pettis, & Schäfer, 2016). It has recently become an invasive exotic species creating the need for an efficient and reliable detection method. Ward et al. (2007) developed a method to screen wax (hive) debris for the presence of SHB using real-time. This method (modified) was included in COLOSS *BEEBOOK volume II Standard methods for small hive beetle research*, and recommended to screen imported hives for SHB (Neumann et al., 2013).

*7.2.4.1. Protocol for determination of A. tumida in wax (hive) debris in the context of monitoring programs.* Hive debris were collected from 398 colonies in a monitoring study on SHB carried out during 2010 and 2011 in Spain (Cepero, Higes, Martínez-Salvador, Meana, & Martín-Hernández, 2014). Following the protocol of mentioned study, hive debris should be collected and processed as follows:

1. Collect the debris of the bottom of six hives randomly selected (8.5 g per the six colonies as average) in a plastic jar (125 mL).
2. To prepare the pooled samples, introduce 5 g from each sample of debris into a double bag strainer (Seward, BA6040).
3. Add 18 mL of milliQ water (MQW).
4. Crush in a Stomacher 80 (Biomaster) for 120 s.
5. Add 9 mL of MQW.
6. Homogenise the suspension again for 60 s.
7. Collect the macerate obtained in a 50 mL tube.
8. Centrifuge at 1751xg for 10 m.
9. Discard the supernatant.
10. Add 1 mL of MQW to the pellet.
11. Re-suspend an aliquot of 50 mg (on average) from the pellet in 3 mL of MQW and homogenise.
12. Add 400 µl to a 96-well plate and process for DNA extraction as described by Botías et al. (2012); include some wells with water alone as negative controls.
13. Freeze the plates at −20 °C until use.

Real-time PCR should be performed in a LightCycler 480 (Roche) using the LightCycler 480 Probes Master mix (Roche, 04887301001), and PCR is carried out as described by Ward et al. (2007).

### 7.3. Conclusions

We have listed the available protocols to determine pest and pathogens analysing beeswax and wax (hive) debris (of which beeswax is the most relevant component). Available literature is most concentrated on protocols to extract *P. larvae* spores from beeswax and more recently from wax (hive) debris based on the relevance and the impact this bacterium has on the apiculture industry. According to the results presented, the Tween method (Bzdil, 2007) (Section 7.1.1.2.2.) appeared statistically significantly more effective in detecting *P. larvae* spores than the toluene method (Titěra & Haklová, 2003) as well as the others listed in Table 23. In addition, Ryba et al. (2009) and Forsgren and Laugen (2014) successfully used the Bzdil's method on hive debris. Hive debris gained importance during the last decade since they have a predictive value on the possible development of a disease condition of the colony, particularly for AFB. Protocols to identify this bacterium are well established (de Graaf et al., 2013; OIE, 2018b), but as for other matrices and pathogens, it is important to make available fast and sensitive methods of detection from beeswax and wax (hive) debris based on PCR that do not require time-consuming culture methods. Protocols to determine other pathogens from beeswax and wax (hive) debris, i.e., *A. apis*, *Nosema* sp. and SHB have also been provided based on PCR at least for the latter two.

## 8. Overview of other methods and perspectives

There are numerous methods covering different aspects of beeswax research that are not presented here in detail as it was not possible to encompass such an extensive beeswax research area, i.e., to standardize all methods available in the literature. However, these methods should also be mentioned here in terms of their perspectives for standardization in the future. We highlight the following ones:

- Methods for beeswax synthesis and secretion research covering investigation of wax gland complex (Hepburn et al., 1991).
- Investigation of biodegradability of beeswax by adapted $CO_2$ evolution test (Hanstveit, 1992).
- Methods for investigating cell cappings (Goetz & Koeniger, 1992; Hepburn, 1998).
- Method for research on mechanical properties of beeswax, such as electromechanical test system methodology (Buchwald et al., 2006, 2009), investigation of beeswax viscoelasticity and resistance to deformation (Shellhammer, Rumsey, & Krochta, 1997), determination of fracture toughness (Kurstjens, Hepburn, Schoening, & Davidson, 1985), investigation of tensile properties (Zhang, Duan, Karihaloo, & Wang, 2010), honeycomb vibratory signals tested by laser - Doppler vibrometers (Sandeman et al., 1996; Tautz & Lindauer, 1997; Tautz et al., 2001), research on crystal texture and alterations (Davidson & Hepburn, 1986; Hepburn & Kurstjens, 1988; Kurstjens et al., 1985; Kurstjens, Mcclain, & Hepburn, 1990), and optical and X ray diffraction studies (Mellema, 2009; Nikolova, Panchev, Kovacheva, & Pashova, 2009).
- Methods for investigating wax role in nestmate recognition (Breed, Garry et al., 1995, Breed, Page et al., 1995, Breed et al., 1998; Breed, 1998; D'ettorre et al., 2006).
- Molecular structure analysis of beeswax by nuclear magnetic resonance (solid-state $^{13}C$ NMR) (Kameda, 2004).
- Methods for investigating cell structure: polyester resin cast method described by Pirk et al. (2004), and silicone-based moulding rubber method for investigation of cell bases described by Hepburn et al. (2007).
- Determination of honeycomb order parameters from image analysis (Kaatz et al., 2008).
- Investigation of ultrasonic acoustic resonances exploited by the bees during comb construction, and accuracy of the hexagonal symmetry (Kadmon, Ishay, & Bergman, 2009).
- Investigation of comb integrity by environmental scanning electron microscope (ESEM) (Zhang, Duan et al., 2010).
- Thermographic analysis of comb and hexagonal cell building (Bauer & Bienefeld, 2013).
- Methods for honey bee silk research employing various molecular, spectroscopic and chromatographic analytical tools (Campbell et al., 2014; Sutherland et al., 2007, 2011, 2014; Walker, Warden, Trueman, Weisman, & Sutherland, 2013;; Weisman et al., 2010; Wittmer et al., 2011), and ESEM (Zhang, Si, Duan, & Wang, 2010).
- Method for research on thermal properties of beeswax based on classical melting point determination, determination of the heat of fusion, differential scanning calorimetry (DSC), and thermal conductivity determination (Buchwald et al., 2005; Buchwald, Breed, & Greenberg, 2008; Timbers & Gochnauer, 1982; Timbers, Robertson, & Gochnauer, 1977).

**Acknowledgements**

We would like to acknowledge the Ricola Foundation - Nature & Culture, and COST Action FA0803 for funding the COLOSS (Prevention of honey bee COlony LOSSes) network and making the *BEEBOOK* possible. We would also like to express our gratitude to Editor Vincent Dietemann for his kind support all the way, and for providing helpful and constructive suggestions.

## Disclosure statement

No potential conflict of interest was reported by the authors.

## ORCID

*Lidija Svečnjak* http://orcid.org/0000-0002-7833-8891
*Lesley Ann Chesson* http://orcid.org/0000-0002-8457-9341
*Marianna Martinello* http://orcid.org/0000-0001-9811-6308
*Franco Mutinelli* http://orcid.org/0000-0003-2903-9390
*Brett James Tipple* http://orcid.org/0000-0001-8994-8452
*Ewa Waś* http://orcid.org/0000-0001-9464-8286

## References

Adamczyk, S., Làzaro, R., Pérez-Arquillué, C., & Herrera, A. (2007). Determination of synthetic acaricides residues in beeswax by high-performance liquid chromatography with photodiode array detector. *Analytica Chimica Acta*, *581*(1), 95–101. doi:10.1016/j.aca.2006.07.085

Aichholz, R., & Lorbeer, E. (1999). Investigation of combwax of honeybees with high-temperature gas chromatography and high-temperature gas chromatography-chemical ionization mass spectrometry I. High-temperature gas chromatography. *Journal of Chromatography A*, *855*(2), 601–615. doi:10.1016/S0021-9673(99)00725-6

Aichholz, R., & Lorbeer, E. (2000). Investigation of combwax of honeybees with high-temperature gas chromatography and high-temperature gas chromatography-chemical ionization mass spectrometry. II: High-temperature gas chromatography-chemical ionization mass spectrometry. *Journal of Chromatography A*, *883*(1–2), 75–88. doi:10.1016/S0021-9673(00)00386-1

Albero, B., Sanchez-Brunete, C., & Tadeo, J. L. (2004). Analysis of pesticides in honey by solid-phase extraction and gas chromatography-mass spectrometry. *Journal of Agricultural and Food Chemistry*, *52*(19):5828–5835. doi:10.1021/jf049470t

AOAC 965.33. (2000). Peroxide value of oils and fats. In *AOAC official methods of analysis* (17th ed.). Washington, DC: Association of Official Analytical Chemists.

Barros, A. R. N. A., Nunes, F. M., & Maia, M. (2009). *Manual de boas práticas na produção de cera de abelha: Princípios gerais*. Bragança, Portugal: Artegráfica Brigantina, 63 pp.

Barth, A., & Haris, P. I. (2009). *Biological and biomedical infrared spectroscopy: Vol. 2. Advances in biomedical spectroscopy*. Amsterdam: IOS Press, pp. 1–46.

Baudoux, U. (1933). The influence of cell size. *Bee World*, *14*(4), 37–41. doi:10.1080/0005772X.1933.11093211

Bauer, D., & Bienefeld, K. (2013). Hexagonal comb cells of honeybees are not produced via a liquid equilibrium process. *Naturwissenschaften*, *100*, 45–49. doi:10.1007/s00114-012-0992-3

Begley, I. S., & Scrimgeour, C. M. (1996). On-line reduction of $H_2O$ for $\delta^2H$ and $\delta^{18}O$ measurement by continuous-flow isotope ratio mass spectrometry. *Rapid Communications in Mass Spectrometry*, *10*(8), 969–973. doi:10.1002/(SICI)1097-0231(19960610)10:8 < 969::AID-RCM532 > 3.0.CO;2-O

Bennett, H. (Ed.). (1944). *Commercial waxes, natural and syntetic, a symposium and compilation*. New York, NY: Chemical Publishing Co., 583 pp.

Bernal, J. L., del Nozal, M. J., Toribio, L., Jimènez, J. J., & Atienza, J. (1997). High-performance liquid chromatographic determination of benomyl and carbendazim residues in apiarian samples. *Journal of Chromatography A*, *787*(1–2), 129–136. doi:10.1016/S0021-9673(97)00633-X

Bernal, J. L., Jiménez, J. J., del Nozal, M. J., Toribio, L., & Martín, M. T. (2005). Physico-chemical parameters for the characterization of pure beeswax and detection of adulterations. *European Journal of Lipid Science and Technology*, *107*(3), 158–166. doi:10.1002/ejlt.200401105

Beverly, M. B., Kay, P. T., & Voorhees, K. J. (1995). Principal component analysis of the pyrolysis mass spectra from African, Africanized hybrid and European beeswax. *Journal of Analytical and Applied Pyrolysis*, *34*(2), 251–263. doi: 10.1016/0165-2370(95)00891-H

Bianchi, E. N. (1990). *Control de calidad de la miel y de la cera*. Rome: FAO Agricultural Services Bulletin 68/3.

Bigeleisen, J., Perlman, M. L., & Prosser, H. C. (1952). Conversion of hydrogenic materials to hydrogen for isotopic analysis. *Analytical Chemistry*, *24*(8), 1356–1357. doi: 10.1021/ac60068a025

Birshtein, V. Y., & Tul'chinskii, V. M. (1977). Determination of beeswax and some impurities by IR spectroscopy. *Chemistry of Natural Compounds*, *13*(2), 232–235. doi: 10.1007%2FBF00563956

Boehm, B. (1965). Beziehungen zwischen Fettkörper, Oenocyten und Wachsdrüsenentwicklung bei *Apis mellifica* L. *Zeitschrift für Zellforschung und Mikroskopische Anatomie*, *65*(1), 74–115. doi:10.1007/BF00319220

Bogdanov, S. (2004a). Quality and standards of pollen and beeswax. *Apiacta*, *38*, 334–341.

Bogdanov, S. (2004b). Beeswax: Quality issues today. *Bee World*, *85*(3), 46–50. doi:10.1080/0005772X.2004.11099623

Bogdanov, S. (2006). Contaminants of bee products. *Apidologie*, *37*: 1–18. doi:10.1051/apido:2005043

Bogdanov, S. (2009). Beeswax: Production, properties, composition and control. Beeswax book. Bee Product Science, pp. 1–17.

Bogdanov, S. (2016a). Beeswax: Production, properties, composition and control. Beeswax book (Chapter 1). Bee Product Science, pp. 1–18. Retrieved from http://www.bee-hexagon.net/wax/

Bogdanov, S. (2016b). Beeswax: History, uses and trade. Beeswax book (Chapter 2). Bee Product Science, pp. 1–17. Retrieved from http://www.bee-hexagon.net/wax/

Bogdanov, S., Kilchenmann, V., & Bütikofer, U. (2003). Determination of acaricide residues in beeswax: Collaborative study. *Apiacta*, *38*, 235–245.

Bogdanov, S., Kilchenmann, V., & Imdorf, A. (1998). Acaricide residues in some bee products. *Journal of Apicultural Research*, *37*(2), 57–67. doi:10.1080/00218839.1998.11100956

Bogdanov, S., Kilchenmann, V., Seiler, K., Pfefferli, H., Frey, T. H., Roux, B., … & Noser J. (2004). Residues of paradichlorobenzene in honey and beeswax. *Journal of Apicultural Research*, *43*(1), 14–16. doi:10.1080/00218839.2004.11101102

Boi, M., Serra, G., Colombo, R., Lodesani, M., Massi, S., & Costa, C. (2016). A 10 year survey of acaricide residues in beeswax analysed in Italy. *Pest Management Science*, *72*, 1366–1372. doi:10.1002/ps.4161

Bonzini, S., Tremolada, P., Bernardinelli, I., Colombo, M., & Vighi, M. (2011). Predicting pesticide fate in the hive (part 1): Experimentally determined τ-fluvalinate residues in bees, honey and wax. *Apidologie*, *42*(3), 378–390. doi:10.1007/s13592-011-0011-2

Botías, C., Martín-Hernández, R., Garrido-Bailón, E., González Porto, A., Martínez-Salvador, A., De La Rúa, P., … & Higes, M. (2012). The growing prevalence of *Nosema ceranae* in honey bees in Spain, an emerging problem for the last decade. *Research in Veterinary Science*, *93*(1), 150–155. doi:10.1016/j.rvsc.2011.08.002

Bowen, G. J., & Revenaugh, J. (2003). Interpolating the isotopic composition of modern meteoric precipitation. *Water Resources Research*, *39*(10), 1299. doi:10.1029/2003WR002086

Bowen, G. J., Wassenaar, L. I., & Hobson, K. A. (2005). Global application of stable hydrogen and oxygen isotopes to wildlife forensics. *Oecologia*, *143*(3), 337–348. doi:10.1007/s00442-004-1813-y

Bowen, G. J., West, J. B., Vaughn, B. H., Dawson, T. E., Ehleringer, J. R., Fogel, M. L., … & Still, C. J. (2009). Isoscapes to address large-scale earth science challenges. *EOS: Transactions of the American Geophysical Union*, *90*(13), 109–111. doi:10.1029/2009EO130001

Breed, M. D. (1998). Recognition pheromones of the honey bee. *Bioscience*, *48*, 463–447. doi:10.2307/13132440

Breed, M. D., Diaz, P. H., & Lucero, D. (2004). Olfactory information processing in honeybee, *Apis mellifera*, nestmate recognition. *Animal Behaviour*, *68*, 921–928. doi:10.1016/j.anbehav.2003.10.033

Breed, M. D., Garry, M. F., Pearce, A. N., Hibbard, B. E., Bjostad, L. B., & Page, R. E. (1995). The role of wax comb in honey-bee nestmate recognition. *Animal Behaviour*, *50*(2), 489–496. doi:10.1006/anbe.1995.0263

Breed, M. D., Leger, E. A., Pearce, A. N., & Wang, Y. J. (1998). Comb wax effects on the ontogeny of honey bee nestmate recognition. *Animal Behaviour*, *55*, 13–20. doi: 10.1006/anbe.1997.0581

Breed, M. D., Page, R. E., Hibbard, B. E., & Bjostad, L. B. (1995). Interfamily variation in comb wax hydrocarbons produced by honey bees. *Journal of Chemical Ecology*, *21*, 1329–1338. doi:10.1007/BF02027565

Breed, M. D., & Stiller, T. M. (1992). Honeybee, *Apis mellifera*, nestmate discrimination – Hydrocarbon effects and the evolutionary implications of comb choice. *Animal Behaviour*, *43*, 875–883. doi:10.1016/S0003-3472(06)80001-1

Breed, M. D., Stiller, T. M., & Moor, M. J. (1988). The ontogeny of kin discrimination cues in the honey bee, *Apis mellifera*. *Behavior Genetics*, *18*, 439–448. doi:10.1007/BF01065513

Breed, M. D., Williams, K. R., & Fewell, J. H. (1988). Comb wax mediates the acquisition of nest-mate recognition cues in honey bees. *Proceedings of the National Academy of Sciences of the United States of America*, *85*, 8766–8769.

Brockmann, A., Groh, C., & Fröhlich, B. (2003). Wax perception in honeybees: Contact is not necessary. *Naturwissenschaften*, *90*, 424–427. doi:10.1007/s00114-003-0442-3

Brown, R. H. (2009). *Beeswax*. Burrowbridge, UK: Bee Books New & Old, 81 pp.

Büchler, R., Andonov, S., Bienefeld, K., Costa, C., Hatjina, F., Kezic, N., … & Wilde, J. (2013). Standard methods for rearing and selection of Apis mellifera queens. In V Dietemann, J. D. Ellis, & P. Neumann (Eds.), *The coloss beebook, Vol. I: Standard methods for Apis mellifera research.*

*Journal of Apicultural Research, 51*(5), 1–30. doi:10.3896/IBRA.1.52.1.07

Buchwald, A. R., Breed, M. D., Bjostad, L., Hibbard, B. E., & Greenberg, A. R. (2009). The role of fatty acids in the mechanical properties of beeswax. *Apidologie, 40*(5), 585–594. doi:10.1051/apido/2009035

Buchwald, R., Breed, M. D., & Greenberg, A. R. (2008). The thermal properties of beeswaxes: Unexpected findings. *The Journal of Experimental Biology, 211*, 121–127. doi:10.1242/jeb.007583

Buchwald, R., Breed, M. D., Greenberg, A. R., & Otis, G. (2006). Interspecific variation in beeswax as a biological construction material. *The Journal of Experimental Biology, 209*, 3984–3989. doi:10.1242/jeb.02472

Buchwald, R., Greenberg, A. R., & Breed, M. D. (2005). Biomechanical perspective on beeswax. *American Entomologist, 51*(1), 39–41. doi:10.1242/jeb.007583

Buffon, G. L. L. (1753). *Discours sur la nature des animaux, in Histoire naturelle* (Vol. IV), p. 3. Retrived from http://www.buffon.cnrs.fr/ice/ice_book_detail-fr-text-buffon-buffon_hn-4-1.html

Burgoyne, T. W., & Hayes, J. M. (1998). Quantitative production of $H_2$ by pyrolysis of gas chromatographic effluents. *Analytical Chemistry, 70*, 5136–5141. doi:10.1021/ac980248v

Bzdil, J. (2007). Detection of *Paenibacillus larvae* spores in the debris and wax of honey bee by the Tween 80 method. *Acta Veterinaria Brno, 76*, 643–648. doi:10.2754/avb200776040643

Campbell, P. M., Trueman, H. E., Zhang, Q., Kojima, K., Kameda, T., & Sutherland, T. D. (2014). Cross-linking in the silks of bees, ants and hornets. *Insect Biochemistry and Molecular Biology, 48*, 40–50. doi:10.1016/j.ibmb.2014.02.009

Casale, M., & Simonetti, R. (2014). Review: Near infrared spectroscopy for analysing olive oils. *Journal of Near Infrared Spectroscopy, 22*, 59–80. doi:10.1255/jnirs.1106

Cassier, P., & Lensky, Y. (1995). Ultrastructure of the wax gland complex and secretion of beeswax in the worker honey bee *Apis mellifera* L. *Apidologie, 26*(1), 17–26. doi:10.1051/apido:19950103

Castro, A. V., Medici, S. K., Sarlo, E. G., & Eguaras, M. J. (2010). Agregado de parafina en ceras estampadas y su efecto sobre el labrado de panales y viabilidad de las crías de *Apis Mellifera*. *Zootecnia Tropical, 28*(3), 353–361.

Cepero, A., Higes, M., Martínez-Salvador, M., Meana, A., & Martín-Hernández, R. (2014). A two year national surveillance for *Aethina tumida* reflects its absence in Spain. *BMC Research Notes, 7*, 878. doi:10.1186/1756-0500-7-878. Retrived from http://www.biomedcentral.com/1756-0500/7/878

Chauzat, M. P., & Faucon, J. P. (2007). Pesticide residues in beeswax samples collected from honey bee colonies (*Apis mellifera* L.) in France. *Pest Management Science, 63*, 1100–1106. doi:10.1002/ps.1451

Chesson, L. A., Tipple, B. J., Erkkila, B. R., Cerling, T. E., & Ehleringer, J. R. (2011). B-HIVE: Beeswax hydrogen isotopes as validation of environment. Part I: Bulk honey and honeycomb stable isotope analysis. *Food Chemistry, 125*, 576–581. doi:10.1016/j.foodchem.2010.09.050

Chesson, L. A., Tipple, B. J., Howa, J. D., Bowen, G. J., Barnette, J. E., Cerling, T. E., & Ehleringer, J. R. (2014a). Stable isotopes in forensics applications. In H. D. Holland & K. K. Turekian (Eds.), *Treatise on geochemistry* (2nd ed., pp. 285–317). Oxford: Elsevier.

Chesson, L. A., Tipple, B. J., Howa, J. D., Bowen, G. J., Barnette, J. E., Cerling, T. E., & Ehleringer, J. R. (2014b). Stable isotopes in forensics applications. In T. E. Cerling (Ed.), *Treatise on geochemistry* (2nd ed., pp. 285–317). Oxford: Elsevier.

Cho, Y. J., Kim, J. Y., Chang, M. I., Kang, K. M., Park, Y. C., Kang, I., … & Oh, J. H. (2012). A study on stable isotope ratio of circulated honey in Korea. *Korean Journal of Food Science and Technology, 44*(4), 401–410. doi:10.9721/KJFST.2012.44.4.401

Claus, C. (1867). Über die Wachbereitenden Hautdrüsen der Insekten. SB Ges Beford Naturwiss Marburg.

Coplen, T. B. (1988). Normalization of oxygen and hydrogen isotope data. *Chemical Geology, 72*(4), 293–297. doi:10.1016/0168-9622(88)90042-5

Council of Europe. (2007). Beeswax, white (01/2008-0069); Beeswax yellow (01/2008-0070). In European Pharmacopoeia (6th ed., Vol. 2, pp. 1085–3308). Strasbourg, France: EDOM – European Directorate for the Quality of Medicine & Healthcare.

Cozzolino, D., Corbella, E., & Smyth, H. E. (2011). Quality control of honey using infrared spectroscopy: A review. *Applied Spectroscopy Reviews, 46*, 523–538. doi:10.1080/05704928.2011.587857

Craig, H. (1961). Isotopic variations in meteoric waters. *Science, 133*, 1702–1703. doi:10.1126/science.133.3465.1702

Crane, E. (1990). *Bees and beekeeping: Science, practice and world resources*. Ithaca, NY: Cornell University Press.

D'ettorre, P., Wenseleers, T., Dawson, J., Hutchinson, S., Boswell, T., & Ratnieks, F. L. W. (2006). Wax combs mediate nestmate recognition by guard honeybees. *Animal Behaviour, 71*(4), 773–779. doi:10.1016/j.anbehav.2005.05.014

Darchen, R., & Lavie, P. (1958). Étude préliminaire de quelques qualités des cires gaufrées présentées aux abeilles. *Les Annales de l'Abeille, 1*(1), 11–18. doi:10.1051/apido:19580102

Darwin, C. (1859). *On the origin of species by means of natural selection or the preservation of favoured races in the struggle for life*. London: John Murray, 556 pp. Retrieved from http://darwin-online.org.uk/converted/pdf/1859_Origin_F373.pdf

Davidson, B. C., & Hepburn, H. R. (1986). Transformations of the acylglycerols in comb construction by honeybees. *Naturwissenschaften, 73*(3), 159–160. doi:10.1007/BF00367408

de Castillon. (1781). Lhuillier, S. Mémoire sur le minimum de cire des alvéoles des abeilles et en particulier sur un minimum minimorum relatif à cette matière. Nouveaux Mémoires de l'Académie Royale des Sciences et belles-lettres, Berlin, Germany, pp. 277–300.

de Graaf, D. C., Alippi, A. M., Antúnez, K., Aronstein, K. A., Budge, G., De Koker, D., … & Genersch, E. (2013). Standard methods for American foulbrood research. In V. Dietemann, J. D. Ellis, & P. Neumann (Eds.), The coloss beebook, Vol. II: Standard methods for *Apis mellifera* pest and pathogen research. *Journal of Apicultural Research, 52*(1), 1–28. doi:10.3896/IBRA.1.52.1.11

de Meyer, E. (1938). *L'oeuvre de Baudoux: La cellule plus grande*. Chambre syndicale belge d'apiculture, comité national (ed), 55 pp.

Delaplane, K. S., van der Steen, J., & Guzman, E. (2013). Standard methods for estimating strength parameters of *Apis mellifera* colonies. In V. Dietemann, J. D. Ellis, P. Neumann (Eds.), The coloss beebook, Vol. I: Standard methods for *Apis mellifera* research. *Journal of Apicultural Research, 52*(1), 1–12. doi:10.3896/IBRA.1.52.1.03

DGF-M-V-2,3,6. (1957). DGF - Einheitsmethoden - Abteilung M - Wachse. German Standard - Beeswax. In Deutsche Einheitsmethoden zur Untersuchung von Fetten, Fettprodukten, Tensiden und verwandten Stoffen (2016). Retrived from http://www.dgfett.de/methods/inhaltsverzeichnis.pdf

Dietz, A., & Humphreys, W. J. (1970). Scanning electron microscopy study of the structure of honey bee wax scales. *Journal of Georgia Entomological Society, 5*(2), 1–6.

Djordjevic, S. P., Noone, K., Smith, L., & Hornitzky, M. A. Z. (1998). Development of a hemi-nested PCR assay for the specific detection of *Melissococcus pluton*. *Journal of Apicultural Research*, *37*(3), 165–174. doi:10.1080/00218839.1998.11100968

Downing, T. D., Kranz, Z. H., Lamberton, J. A., Murray, K. E., & Redcliffe, A. H. (1961). Studies in waxes. XVIII. Beeswax: A spectroscopic and gas chromatographic examination. *Australian Journal of Chemistry*, *14*, 253–263. doi:10.1071/CH9610253

Dunn, P., & Carter, J. F. (2018). *Good practice guide for isotope ratio mass spectrometry* (2nd ed.). Bristol, UK: The FIRMS Network, 84 pp.

Edwards, H. G. M., Farwel, D. W., & Daffner, L. (1996). Fourier-transform Raman spectroscopic study of natural waxes and resins. *Spectrochimica Acta Part A: Molecular and Biomolecular Spectroscopy*, *52*(12), 1639–1648. doi:10.1016/0584-8539(96)01730-8

Ellis, M. B., Nicolson, S. W., Crewe, R. M., & Dietemann, V. (2010). Brood comb as a humidity buffer in honeybee nests. *Naturwissenschaften*, *97*(4), 429–433. doi:10.1007/s00114-010-0655-1

Erickson, E. H., Lusby, D. A., Hoffmann, G. D., & Lusby, E. W. (1990a). On the size of the cells. Speculations on foundation as a colony management tool. *Gleaning in Bee Culture*, *118*(2), 98–101.

Erickson, E. H., Lusby, D. A., Hoffmann, G. D., & Lusby, E. W. (1990b). On the size of cells. Speculations on foundation as a colony management tool. *Gleanings in Bee Culture*, *118*(3), 173–174.

EU. (1995). Council Directive 95/2/EC. Food additives other than colours and sweeteners. *Official Journal of the European Community*, *L61*, 41, 46. Retrieved from https://www.fsai.ie/uploadedFiles/European_Parliament_and_Council_Directive_No_95_2_EC.pdf

EU. (2009). Council Directive 2009/10/EC. Specific purity criteria on food additives other than colours and sweeteners. *Official Journal of the European Union*, *L44*, 63, 76–77. Retrieved from http://eur-lex.europa.eu/legal-content/EN/TXT/PDF/?uri=CELEX:32009L0010&rid=1

EU. (2009). Regulation (EC) No 1069/2009 of the European Parliament and of the Council laying down health rules as regards animal by-products and derived products not intended for human consumption and repealing Regulation (EC) No 1774/2002 (Animal by-products Regulation). *Official Journal of the European Union*, *L300*, 1–33. Retrieved from https://eur-lex.europa.eu/LexUriServ/LexUriServ.do?uri=OJ:L:2009:300:0001:0033:EN:PDF

EU. (2011). Commission Regulation (EU) No 142/2011 implementing Regulation (EC) No 1069/2009 of the European Parliament and of the Council laying down health rules as regards animal by-products and derived products not intended for human consumption and implementing Council Directive 97/78/EC as regards certain samples and items exempt from veterinary checks at the border under that Directive. *Official Journal of the European Union*, *L54/13*, 18, 101. Retrieved from https://eur-lex.europa.eu/legal-content/EN/TXT/?uri=uriserv:OJ.L_.2011.054.01.0001.01.ENG

EU. (2012). Commission Regulation (EU) No 231/2012 laying down specifications for food additives listed in Annexes II and III to Regulation (EC) No 1333/2008 of the European Parliament and of the Council. *Official Journal of the European Union*, *L83*, 250–251. Retrieved from https://www.fsai.ie/uploadedFiles/Consol_Reg231_2012.pdf

EU. (2017). Food Fraud Network (FFN), European Commission, Beeswax intended for honey production adulterated with paraffin and stearin (draft material under evaluation of the European Commission). Retrieved from https://ec.europa.eu/food/sites/food/files/safety/docs/food-fraud_succ-coop_beeswax.pdf

European Committee for Standardisation (CEN) BS EN 15662:2008. (2008). Foods of plant origin – Determination of pesticide residues using GC-MS and/or LC-MS/MS following acetonitrile extraction/partitioning and clean-up by dispersive SPE – QuEChERS-method. Retrived from http://www.chromnet.net/Taiwan/QuEChERS_Dispersive_SPE/QuEChERS_%E6%AD%90%E7%9B%9F%E6%96%B9%E6%B3%95_EN156622008_E.pdf

European Food Safety Authority (EFSA). (2007). Beeswax (E901) as a glazing agent and as carrier for flavours. *The EFSA Journal*, *615*, 1–28. Retrieved from http://www.efsa.europa.eu/sites/default/files/scientific_output/files/main_documents/615.pdf

Fereira, T., & Rasband, W. (2012). The ImageJ user guide. Retrieved from https://imagej.nih.gov/ij/docs/guide/user-guide.pdf

Ferguson, L. A., & Winston, M. L. (1988). The influence of wax deprivation on temporal polyethism in honey bee (*Apis mellifera* L.) colonies. *Canadian Journal of Zoology*, *66*(9), 1997–2001. doi:10.1139/z88-292

Flores, J. M., Spivak, M., & Gutierrez, I. (2005). Spores of *Ascosphaera apis* contained in wax foundation can infect honeybee brood. *Veterinary Microbiology*, *108*, 141–144. doi:10.1016/j.vetmic.2005.03.005

Food and Agricultural Organisation of the United Nations (FAO). (2005). Compendium of food additive specifications Addendum 13. Revised specifications prepared at the Joint FAO/WHO Expert Committee on Food Additives (JECFA) 65th meeting (2005) and published in FNP (FAO Food and Nutrition Paper) 52 Add. 13. pp. 11–13. Retrieved from http://www.fao.org/3/a-a0044e.pdf

Forsgren, E., Budge, G. E., Charrière, J.-D., & Hornitzky, M .A. Z. (2013). Standard methods for European foulbrood research. In V. Dietemann, J. D. Ellis, & P. Neumann (Eds.), The coloss beebook, Vol. II: Standard methods for *Apis mellifera* pest and pathogen research. *Journal of Apicultural Research*, *52*(1), 1–14. doi:10.3896/IBRA.1.52.1.12

Forsgren, E., & Laugen, A. T. (2014). Prognostic value of using bee and hive debris samples for the detection of American foulbrood disease in honey bee colonies. *Apidologie*, *45*(1), 10–20. doi:10.1007/s13592-013-0225-6

Forsgren, E., Stevanovic, J., & Fries, I. (2008). Variability in germination and in temperature and storage resistance among *Paenibacillus larvae* genotypes. *Veterinary Microbiology*, *129*(3–4), 342–349. doi:10.1016/j.vetmic.2007.12.001

Friedman, I. (1953). Deuterium content of natural waters and other substances. *Geochimica et Cosmochimica Acta*, *4*(1–2), 89–103. doi:10.1016/0016-7037(53)90066-0

Fries, I., Chauzat, M.-P., Chen, Y.-P., Doublet, V., Genersch, E., Gisder, S., … & Williams, G. R. (2013). Standard methods for Nosema research. In V. Dietemann, J. D. Ellis, & P. Neumann (Eds.), The coloss beebook, Vol. II: Standard methods for *Apis mellifera* pest and pathogen research. *Journal of Apicultural Research*, *52*(1), 1–28 doi:10.3896/IBRA.1.52.1.14

Fröhlich, B., Riederer, M., & Tautz, J. (2000). Comb-wax discrimination by honeybees tested with the proboscis extension reflex. *Journal of Experimental Biology*, *203*(10), 1581–1587.

Fröhlich, B., Tautz, J., & Riederer, M. (2000). Chemometric classification of comb and cuticular waxes of the honeybee *Apis mellifera carnica*. *Journal of Chemical Ecology*, *26*(1), 123–137. doi:10.1023/A:1005493512305

Garti, M., Sato, K., Schlichter, J., & Wellner, E. (1986). The dimer association of stearic acid in solution. *Crystal Research and Technology*, *21*(5), 653–656. doi:10.1002/crat.2170210518

Gerhardt, P., & Marquis, R. E. (1989). Spore thermoresistance mechanisms. In I. Smith, R. A. Slepecky, & P. Setlow (Eds.), *Regulation of procaryotic development* (pp. 43–64). Washington, DC: American Society of Microbiology.

Gilliam, M., Taber, S., Lorenz, B. J., & Prest, D. B. (1988). Factors affecting development of chalk brood disease in colonies of honey bees, *Apis mellifera*, fed pollen contaminated with *Ascosphaera apis*. *Journal of Invertebrate Pathology*, *52*(2), 314–325. doi:10.1016/0022-2011(88)90141-3

Giumanini, A. G., Verardo, G., Strazzolini, P., & Hepburn, H. R. (1995). Rapid detection of high-molecular-mass dienes in beeswax. *Journal of Chromatography A*, *704*, 224–227. doi: 10.1016/0021-9673(95)00145-D

Goetz, B., & Koeniger, N. (1992). Structural features trigger capping of brood cells in honey bees. *Apidologie*, *23*, 211–216. doi:10.1051/apido:19920303

Goetze, G., & Bessling, B. K. (1959). Die Wirkung vershiedener Fütterung der Honigbiene Wachserzeugung und Bautätigkeit [The effect of different feeding on honeybee wax production and building activity]. *Zeitschrift Bienenforsch*, *4*, 202–209.

Günzler, H., & Gremlich, H-U. (2002). *IR spectroscopy: An introduction*. Weinheim, Germany: Wiley-VCH, 374 pp.

Hales, T. C. (2001). The honeycomb conjecture. *Discrete and Computational Geometry*, *25*(1), 1–22. doi:10.1007/s004540010071

Hansen, H., & Brodsgaard, C. J. (1999). American foulbrood: A review of its biology, diagnosis and control. *Bee World*, *80*(1), 5–23. doi:10.1080/0005772X.1999.11099415

Hansen, H., & Rasmussen, B. (1991). Empfindlichkeit des Faulbrutbakteriums *Bacillus larvae* gegenüber Hitzebehandlung. *Die Biene*, *3*, 129–131.

Hanstveit, A. O. (1992). Biodegradability of petroleum waxes and beeswax in an adapted $CO_2$ evolution test. *Chemosphere*, *25*(4), 605–620. doi:10.1016/0045-6535(92)90291-X

Haynes, W. C. (1972). The catalase test. An aid in the identification of Bacillus larvae. *American Bee Journal*, *112*, 130–131.

Heaf, D. (2011). Do small cells help bees cope with Varroa. A review. *The Beekeepers Quarterly*, *104*, 39–45.

Heaf, D. (2012). Natural cell size. Retrieved from http://www.dheaf.plus.com/warrebeekeeping/natural_cell_size_heaf.pdf

Hepburn, H. R. (1986). *Honeybees and wax: An experimental natural history*. Berlin: Springer-Verlag, 171 pp. doi:10.1007/978-3-642-71458-0

Hepburn, H. R. (1998). Reciprocal interactions between honeybees and combs in the integration of some colony functions in *Apis mellifera* L. *Apidologie*, *29*, 47–66 doi:10.1051/apido:19980103

Hepburn, H. R., Bernard, R T. F., Davidson, B. C., Muller, W. J., Lloyd, P., Kurstjens, S. P., & Vincent, S. L. (1991). Synthesis and secretion of beeswax in honeybees. *Apidologie*, *22*(1), 21–36. doi:10.1051/apido:19910104

Hepburn, H. R., Hugo, J. J., Mitchell, D., Nijland, M. J. M., & Scrimgeour, A. G. (1984). On the energetic costs of wax production by the African honeybee (*Apis mellifera adansonii* Latreille). *South African Journal of Science*, *80*(8), 363–368.

Hepburn, H. R., & Kurstjens, S. P. (1988). The combs of honeybees as composite materials. *Apidologie*, *19*, 25–36. doi: 10.1051/apido:19880102

Hepburn, H. R., & Magnuson, P. C. (1988). Nectar storage in relation to wax secretion by honeybees. *Journal of Apicultural Research*, *27*(2), 90–94. doi:10.1080/00218839.1988.11100786

Hepburn, H. R., Muerrle, T., & Radloff, S. E. (2007). The cell bases of honeybee combs. *Apidologie*, *38*(3), 268–271. doi: 10.1051/apido:2007005

Hepburn, H. R., Pirk, C. W. W., & Duangphakdee, O. (2014). *Honeybees nests: Composition, structure, function*. New York, NY: Springer.doi:10.1007/978-3-642-54328-9

Herrera Lòpez, S., Lozano, A., Sosa, A., Hernando, M. D., & Fernàndez-Alba, A. R. (2016). Screening of pesticide residues in honeybee wax comb by LC-ESI-MS/MS. A pilot study. *Chemosphere*, *163*, 44–53. doi:10.1016/j.chemosphere.2016.07.008

Hilkert, A. W., Douthitt, C. B., Schluter, H. J., & Brand, W. A. (1999). Isotope ratio monitoring gas chromatography/mass spectrometry of D/H by high temperature conversion isotope ratio mass spectrometry. *Rapid Communications in Mass Spectrometry*, *13*, 1226–1230. doi:10.1002/(SICI)1097-0231(19990715)13:13 < 1226::AID-RCM575 > 3.0.CO;2-9

Hoef, J. (2004). *Stable isotope geochemistry*. Berlin: Springer, 285 pp.

Honegger, A. (1937). Großzellen ja oder nein? *Schweizerische Bienenzeitung*, *60*(3), 149–152.

Hornbostel, H. C. (1744). Neue Entdeckung, wie das Wachs von den Bienen entsteht. *Vermis Bibliothek Hamburg*, *2*, 45–62.

Hornitzky, M. A. Z., & Wills, P. A. (1983). Gamma radiation inactivation of *Bacillus larvae* to control American foulbrood. *Journal of Apicultural Research*, *22*(3), 196–199. doi:10.1080/00218839.1983.11100587

Huber, F. (1814). *Nouvelles observations sur les abeilles (2nd ed.). Vol. II: Revue, corrigée et considérablement augmentée.* Paris & Genève: J J Paschoud, 479 pp. Retrived from https://dl.dropboxusercontent.com/u/6639518/Nouvelles_observations_sur_les_abeilles_1814_tomeII.pdf

Human, H., Brodschneider, R., Dietemann, V., Dively, G., Ellis, J., Forsgren, E., … & Zheng, H.-Q. (2013). Miscellaneous standard methods for *Apis mellifera* research. In V. Dietemann, J. D. Ellis, & P. Neumann (Eds.), The coloss beebook, Vol. I: standard methods for *Apis mellifera* research. *Journal of Apicultural Research*, *52*(4), 1–53. doi: 10.3896/IBRA.1.52.4.10

Hunter, J. (1792). Observations on bees. *Philosophical Transactions of the Royal Society B: Biological Sciences*, *82*, 128–196. doi:10.1098/rstl.1792.0011

ISO TC34/SC19. (2017). *Bee products/Bee Products Secretariat*. International Organization for Standardization. Retrieved from https://www.iso.org/committee/6716626.html https://www.apiservices.biz/documents/articles-en/iso_standards_bee_products.pdf

ISO 5725-2:1994. (1994). *Accuracy (trueness and precision) of measurement methods and results – Part 2: Basic method for the determination of repeatability and reproducibility of a standard measurement method*. Geneva, Switzerland: International Organisation for Standardisation.

ISO 5725-6:1994 (1994). *Accuracy (trueness and precision) of measurement methods and results – Part 6: Use in practice of accuracy values*. Geneva, Switzerland: International Organisation for Standardisation.

Jabot, C., Fieu, M., Giroud, B., Buleté, A., Casabianca, H., & Vulliet, E. (2015). Trace-level determination of pyrethroid, neonicotinoid and carboxamide pesticides in beeswax using dispersive solid-phase extraction followed by ultra-high-performance liquid chromatography-tandem mass spectrometry. *International Journal of Environmental Analytical Chemistry*, *95*(3), 240–257. doi:10.1080/03067319.2015.1016011

Jay, S. C. (1964). The cocoon of the honeybee, *Apis mellifera* L. *The Canadian Entomologist*, *96*(5), 784–792. doi:10.4039/Ent96784-5

Jensen Bruun, A., Aronstein, K., Flores, J. M., Vojvodic, S., Palacio, M. A., & Spivak, M. (2013). Standard methods for fungal brood disease research. In V. Dietemann, J. D. Ellis, & P. Neumann (Eds.), The coloss beebook, Vol. II: Standard methods for *Apis mellifera* pest and pathogen research.

*Journal of Apicultural Research, 52*(1), 1–20. doi:10.3896/IBRA.1.52.1.13

Jiménez, J. J., Bernal, J. L., Aumente, S., del Nozal, M. J., Martín, M. T., & Bernal, J. (2004). Quality assurance of commercial beeswax. I. Gas chromatography-electron impact ionization mass spectrometry of hydrocarbons and monoesters. *Journal of Chromatography A, 1024*(1–2), 147–154. doi:10.1016/j.chroma.2003.10.063

Jiménez, J. J., Bernal, J. L., Aumente, S., Toribio, L., & Bernal, J. (2003). Quality assurance of commercial beeswax II. Gas chromatography-electron impact ionization mass spectrometry of alcohols and acids. *Journal of Chromatography A, 1007*(1–2), 101–116. doi:10.1016/S0021-9673(03)00962-2

Jiménez, J. J., Bernal, J. L., del Nozal, M. J., & Alonso, C. (2004a). Liquid-liquid extraction followed by solid-phase extraction for the determination of lipophilic pesticides in beeswax by gas chromatography-electron-capture detection and matrix-matched calibration. *Journal of Chromatography A, 1048*(1), 89–97. doi:10.1016/j.chroma.2004.07.034

Jiménez, J. J., Bernal, J. L., del Nozal, M. J., & Alonso, C. (2004b). Extraction and clean-up methods for the determination of amitraz total residues in beeswax by gas chromatography with electron capture detection. *Analytica Chimica Acta, 524*(1–2), 271–278. doi:10.1016/j.aca.2004.03.039

Jiménez, J. J., Bernal, J. L., del Nozal, M. J., & Martín, M. T. (2005). Residues of organic contaminants in beeswax. *European Journal of Lipid Science and Technology, 107*, 896–902. doi:10.1002/ejlt.200500284

Jiménez, J. J., Bernal, J. L., del Nozal, M. J., Martín, M. T., & Bernal, J. (2006). Sample preparation methods for beeswax characterization by gas chromatography with flame ionization detection. *Journal of Chromatography A, 1129*(2), 262–272. doi:10.1016/j.chroma.2006.06.098

Jiménez, J. J., Bernal, J. L., del Nozal, M. J., Toribio, L., & Bernal, J. (2007). Detection of beeswax adulterations using concentration guide-values. *European Journal of Lipid Science and Technology, 109*(7), 682–690. doi:10.1002/ejlt.200600308

Johansson, T. S. K., & Johansson, M. P. (1978). *Some important operations in bee management.* Cardiff, UK: Northern Bee Books, 145 pp.

Johnson, R. N., Tauheed Zaman, M., Decelle, M. M., Siegel, A. J., Tarpy, D. R., Siegel, E. C., & Starks, P. (2005). Multiple micro-organisms in chalkbrood mummies: Evidence and implications. *Journal of Apicultural Research, 44*(1), 29–32. doi:10.1080/00218839.2005.11101143

Joint FAO/WHO Expert Committee on Food Additives (JECFA). (2005). Beeswax Chemical and Technical Assessment (CTA), 65th JECFA meeting. Retrieved from: http://www.fao.org/fileadmin/templates/agns/pdf/jecfa/cta/65/beeswax.pdf

Jordan, R. (1962). Anomale Wachsplattchen. *Bienenvater, 63*, 299–301.

Kaatz, F. H., Bultheel, A., & Egami, T. (2008). Order parameters from image analysis: A honeycomb example. *Naturwissenschaften, 95*(11), 1033–1040. doi:10.1007/s00114-008-0418-4

Kadmon, J., Ishay, J. S., & Bergman, D. J. (2009). Properties of ultrasonic acoustic resonances for exploitation in comb construction by social hornets and honeybees. *Physical Review E, 79*(6), 061909. doi:10.1103/PhysRevE.79.061909

Kameda, T. (2004). Molecular structure of crude beeswax studied by solid-state $^{13}$C NMR. *Journal of Insect Science, 4*(29), 1–5. doi:10.1673/031.004.2901

Karihaloo, B. L., Zhang, K., & Wang, J. (2013). Honeybee combs: How the circular cells transform into rounded hexagons. *Journal of the Royal Society Interface, 10*(86), 20130299. doi:10.1098/rsif.2013.0299

Karoui, R., Downey, G., & Blecker, C. (2010). Mid-infrared spectroscopy coupled with chemometrics: A tool for the analysis of intact food systems and the exploration of their molecular structure–quality relationships – A review. *Chemical Reviews, 110*, 6144–6168. doi:10.1021/cr100090k

Kenya Bureau of Standards (KEBS). (2013). Specification for natural beeswax. Kenya Standard CD/ 05-1279: 2013 ICS 67.120. Retrived from http://www.puntofocal.gov.ar/notific_otros_miembros/ken379_t.pdf

Kirchner, W. H. (1993). Vibrational signals in the tremble dance of the honeybees, *Apis mellifera. Behavioral Ecology and Sociobiology, 33*, 169–172. doi:10.1007/BF00216597

Korta, E., Bakkali, A., Berrueta, L. A., Gallo, B., Vicente, F., Kilchenmann, V., & Bogdanov, S. (2001). Study of acaricide stability in honey. Characterization of amitraz degradation products in honey and beeswax. *Journal of Agricultural Food Chemistry, 49*(12), 5835–5842. doi:10.1021/jf010787s

Kostecki, R., & Jelinski, M. (1977). Investigations on the sterilization of beewax for foundation production. *Bulletin of the Veterinary Institute Pulawy, 21*(1–2), 6–9.

Kostecki, R., & Orlowski, J. (1975). Ausspülen von Bacillus larvae-Sporen aus dem für Mittelwände bestimmten Wachs. In Proceedings of 25th International Apicultural Congress, Grenoble, France, 8–14 September 1975, pp. 402–403.

Kujawski, M. W., Barganska, A., Marciniak, K., Miedzianowska, E., Kujawski, J. K., Slebioda, M., & Namiesnik, J. (2014). Determining pesticide contamination in honey by LC-ESI-MS/MS. Comparison of pesticide recoveries of two liquid-liquid extraction based approaches. *LWT-Food Science and Technology, 56*(2), 517–523. doi:10.1016/j.lwt.2013.11.024

Kurstjens, S. P., Hepburn, H. R., Schoening, F. R. L., & Davidson, B. C. (1985). The conversion of wax scales into comb wax by African honeybees. *Journal of Comparative Physiology B, 156*(1), 95–102. doi:10.1007/BF00692930

Kurstjens, S. P., Mcclain, E., & Hepburn, H. R. (1990). The proteins of beeswax. *Naturwissenschaften, 77*(1), 34–35. doi:10.1007/BF01131795

Ledoux, M. N., Winston, M. L., Higo, H., Keeling, C. I., Slessor, K. N., & Leconte, Y. (2001). Queen and pheromonal factors influencing comb construction by simulated honey bee (*Apis mellifera* L.) swarms. *Insectes Sociaux, 48*(1), 14–20. doi:10.1007/PL00001738

Leníček, J., Sekyra, M., Novotná, A. R., Vášová, E., Titěra, D., & Veselý, V. (2006). Solid phase microextraction and gas chromatography with ion trap detector (GC-ITD) analysis of amitraz residues in beeswax after hydrolysis to 2,4-dimethylaniline. *Analytica Chimica Acta, 571*(1), 40–44. doi:10.1016/j.aca.2006.04.039

Locke, M. (1961). Pore canals and related structures in insect cuticle. *Journal of Biophysical and Biochemical Cytology, 10*(4), 589–618. doi:10.1083/jcb.10.4.589

Lodesani, M., Costa, C., Serra, G., Colombo, R., & Sabatini, A. G. (2008). Acaricide residues in beeswax after conversion to organic beekeeping methods. *Apidologie, 39*(3), 324–333. doi:10.1051/apido:2008012

Lodesani, M., Pellacani, A., Bergomi, S., Carpana, E., Rabitti, T., & Lasagni, P. (1992). Residue determination for some products used against Varroa infestation in bees. *Apidologie, 23*(3), 257–272. doi:10.1051/apido:19920309

Máchová, M. (1993). Resistance of *Bacillus larvae* in beeswax. *Apidologie, 24*(1), 25–31. doi:10.1051/apido:19930103

Maia, M., Barros, A. R. N. A., & Nunes, F. M. (2013). A novel, direct, reagent-free method for the detection of bees-wax adulteration by single-reflection attenuated total reflectance mid-infrared spectroscopy. *Talanta, 107*, 74–80. doi:10.1016/j.talanta.2012.09.052

Maia, M., & Nunes, F. M. (2013). Authentication of beeswax (*Apis mellifera*) by high-temperature gas chromatography and chemometric analysis. *Food Chemistry, 136*(2), 961–968. doi:10.1016/j.foodchem.2012.09.003

Maraldi, G. P. (1712). *Observations sur les abeilles. Histoire de l'Académie royale des sciences Mémoires de l'Année 1712.* Paris, France, pp. 297–331. Retrived from http://www.biodiversitylibrary.org/item/86705#page/445/mode/1up

Martinez, J., Simon, V., Gonzalez, B., & Conget, P. (2010). A real-time PCR-based strategy for the detection of *Paenibacillus larvae* vegetative cells and spores to improve the diagnosis and the screening of American foulbrood. *Letters in Applied Microbiology, 50*(6), 603–610. doi:10.1111/j.1472-765X.2010.02840.x

Mckee, B. A., Djordjevic, S. P., Goodman, R. D., & Hornitzky, M. A. (2003). The detection of Melissococcus pluton in honey bees (*Apis mellifera*) and their products using a hemi-nested PCR. *Apidologie, 34*(1), 19–27. doi:10.1051/apido:2002047

Medici, S. K., Castro, A., Sarlo, E. G., Marioli, J. M., & Eguaras, M. J. (2012). The concentration effect of selected acaricides present in beeswax foundation on the survival of *Apis mellifera* colonies. *Journal of Apicultural Research, 51*(2), 164–168. doi:10.3896/IBRA.1.51.2.03

Mellema, M. (2009). Co-crystals of beeswax and various vegetable waxes with sterols studied by X-ray diffraction and differential scanning calorimetry. *Journal of the American Oil Chemists' Society, 86*, 499–505. doi:10.1007/s11746-009-1385-4

Michelsen, A., Kirchner, W. H., Andersen, B. B., & Lindauer, M. (1986). The tooting and quacking vibration signals of honeybee queens: A quantitative analysis. *Journal of Comparative Physiology A, 158*, 605–611. doi:10.1007/BF00603817

Michelsen, A., Kirchner, W. H., & Lindauer, M. (1986). Sound and vibrational signals in the dance language of the honeybee, *Apis mellifera. Behavioral Ecology and Sociobiology, 18*, 207–212. doi:10.1007/BF00290824

Muller, W. J., & Hepburn, H. R. (1992). Temporal and spatial patterns of wax secretion and related behavior in the division of labour of the honeybee (*Apis mellifera* capensis). *Journal of Comparative Physiology, 171*(1), 111–115. doi:10.1007/BF00195966

Mullin, C. A., Frazier, M., Frazier, J. L., Ashcraft, S., Simonds, R., van Engelsdorp, D., & Pettis, J. S. (2010). High levels of miticides and agrochemicals in North American apiaries: implications for honey bee health. *PLoS One, 5*(3), e9754. doi:10.1371/journal.pone.0009754

Muscat, D., Tobin, M. J., Guo, Q., & Adhikari, B. (2014). Understanding the distribution of natural wax in starch-wax films using synchrotron-based FTIR (S-FTIR). *Carbohydrate Polymers, 102*, 125–135. doi:10.1016/j.carbpol.2013.11.004

Mutinelli, F. (2011). The spread of pathogens through trade in honey bees and their products (including queen bees and semen): Overview and recent developments. *Revue Scientifique et Technique de l'Office international des Epizooties, 30*(1), 257–271. Retrived from http://www.oie.int/doc/en_document.php?numrec=4021703

Muz, M. N., Özdemir, N., & Mutinelli, F. (2017). A method for detection of *Nosema* sp. spores in beeswax. In Proceedings of 45th Apimondia International Apicultural Congress, Istanbul, Turkey, 29 September–3 October 2017, p. 64.

Namdar, D., Neuman, R., Sladezki, Y., Haddad, N., & Weiner, S. (2007). Alkane composition variations between darker and lighter colored comb beeswax. *Apidologie, 38*: 453–461. doi:10.1051/apido:2007033

Nazzi, F. (2016). The hexagonal shape of the honeycomb cells depends on the construction behavior of bees. *Scientific Reports, 6*, 28341. doi:10.1038/srep28341

Neumann, P., Evans, J. D., Pettis, J. S., Pirk, C. W., Schäfer, M. O., Tanner, G., & Ellis, D. D. (2013). Standard methods for small hive beetle research. In V. Dietemann, J. D. Ellis, & P. Neumann (Eds.), The coloss beebook, Vol. II: Standard methods for *Apis mellifera* pest and pathogen research. *Journal of Apicultural Research, 52*(4), 1–32. doi:10.3896/IBRA.1.52.4.19

Neumann, P., Pettis, J. S., & Schäfer, M. O. (2016). Quo vadis *Aethina tumida*? Biology and control of small hive beetle. *Apidologie, 47*(3), 427–466. doi:10.1007/s13592-016-0426-x

Nguyen, B. K., Saegerman, C., Pirard, C., Mignon, J., Widart, J., Thirionet, B., … & Haubruge, E. (2009). Does imidacloprid seed-treated maize have an impact on honey bee mortality? *Journal of Economic Entomology, 102*(2), 616–623. doi:10.1603/029.102.0220[pubmedMismatch]

Nikolova, K. R., Panchev, I., Kovacheva, D., & Pashova, S. (2009). Thermophysical and optical characteristics of bee and plant waxes. *Journal of Optoelectronics and Advanced* Materials, *11*(9), 1210–1213.

OIE (World Organisation for Animal Health). (2018a). Chapter 9.2. Infection of honey bees with *Paenibacillus larvae* (American foulbrood). In *Terrestrial animal health code* (25th ed., Vol. 2), Paris, France. Retrived from http://www.oie.int/index.php?id=169&L=0&htmfile=titre_1.9.htm

OIE (World Organisation for Animal Health). (2018b). Chapter 2.2.2. American foulbrood of honey bees (infection of honey bees with *Paenibacillus larvae*) (NB: Version adopted in May 2016). In *Manual of diagnostic tests and vaccines for terrestrial animals* (Vol. 1). Retrived from http://www.oie.int/fileadmin/Home/eng/Health_standards/tahm/2.02.02_AMERICAN_FOULBROOD.pdf

Otis, G. W., Winston, M. L., & Taylor, O. R. (1981). Engorgement and dispersal of Africanized honey bee swarms. *Journal of Apiculture Research, 20*(1), 4–12. doi:10.1080/00218839.1981.11100464

Pertoft, H., Laurent, T. C., Laas, T., & Kagedal, L. (1978). Density gradients prepared from colloidal silica particles coated by polyvinylpyrrolidone (Percoll). *Analytical Biochemistry, 88*(1), 271–282. doi:10.1016/0003-2697(78)90419-0

Piek, T. (1964). Synthesis of wax in the honeybee (*Apis mellifera* L.) *Journal of Insect Physiology, 10*(4), 563–572. doi:10.1016/0022-1910(64)90027-7

Pielichowska, K., Głowinkowski, S., Lekki, J., Binias, D., Pielichowski, K., & Jenczyk, J. (2008). PEO/fatty acid blends for thermal energy storage materials. Structural/morphological features and hydrogen interactions. *European Polymer Journal, 44*: 3344–3360. doi:10.1016/j.eurpolymj.2008.07.047

Pirard, C., Widart, J., Nguyen, B. K., Deleuze, C., Heudt, L., Haubruge, E., … & Focant, J. F. (2007). Development and validation of a multi-residue method for pesticide determination in honey using on-column liquid-liquid extraction and liquid chromatography-tandem mass spectrometry. *Journal of Chromatography A, 1152*(1–2), 116–123. doi:10.1016/j.chroma.2007.03.035

Pirk, C. W. W., de Miranda, J. R., Fries, I., Kramer, M., Paxton, R., Murray, T., … & van Dooremalen, C. (2013). Statistical guidelines for *Apis mellifera* research. In V. Dietemann, J. D. Ellis, & P. Neumann (Eds.), The coloss beebook, Vol. I: Standard methods for *Apis mellifera* research. *Journal of Apicultural Research, 52*(4), 1–24. doi:10.3896/IBRA.1.52.4.13

Pirk, C. W. W., Hepburn, H. R., Radloff, S. E., & Tautz, J. (2004). Honeybee combs: Construction through a liquid equilibrium process? *Naturwissenschaften, 91*(7), 350–353. doi:10.1007/s00114-004-0539-3

Plessis, T. A., Du-Rensbrug, H. J., & Van-Johannsmeier, M. F. (1985). The gamma sterilization of beewax. *South African Bee Journal, 57*(3), 54–57.

Poncini, L., Poncini, A., & Prakash, D. (1993). The effects of washing on the fluorescent impurities and chemical properties of Fijian beeswax from *Apis mellifera* L. *Apiacta, 28*(2), 42–51.

Porrini, C., Mutinelli, F., Bortolotti, L., Granato, A., Laurenson, L., Roberts, K., … & Lodesani, M. (2016). The status of honey bee health in Italy: Results from the nationwide bee monitoring network. *PLoS One 11*(5), e0155411. doi: 10.1371/journal.pone.0155411

Pratt, S. C. (2004). Collective control of the timing and type of comb construction by honey bees (*Apis mellifera*). *Apidologie, 35*(2), 193–205. doi:10.1051/apido:2004005

Prosser, S. J., & Scrimgeour, C. M. (1995). High-precision determination of $^2H/^1H$ in $H_2$ and $H_2O$ by continuous-flow isotope ratio mass spectrometry. *Analytical Chemistry, 67*(13), 1992–1997. doi:10.1021/ac00109a014

Puleo, S., & Rit, T. P. (1992). Natural waxes: Past, present and future. Lipid Technology, *4*, 82–90.

Rademacher, E., & Harz, M. (2006). Oxalic acid for the control of varroosis in honey bee colonies – A review. *Apidologie, 37*(1), 98–120. doi:10.1051/apido:2005063

Ravoet, J., Reybroeck, W., & De Graaf, D. C. (2015). Pesticides for apicultural and/or agricultural application found in Belgian honey bee wax combs. *Bulletin of Environmental Contamination and Toxicology, 94*(5), 543–548. doi:10.1007/s00128-015-1511-y

Réaumur, R. A. F. (1742). *Mémoires pour servir à lhistoire des Insectes* (Vol. 5). Paris, France, 728 pp. Retrived from http://gallica.bnf.fr/ark:/12148/bpt6k99227k/f811.image

Reybroeck, W., & Van Nevel, J. (2018). Effect of beeswax adulterated with stearin on the development of worker bee brood: Results of a field trial. In Program & Abstracts Book EurBee 8, 8th Congress of Apidology, Ghent, Belgium, 18–20 September 2018, p. 115.

Ribeiro da Silva, C., Ribeiro, L. R., Toledo, V. A. A., & Toledo, J. O. A. (2002). Uso da parafina incorporada à cera alveolada em colônias de abelhas *Apis mellifera* L. africanizadas para produção de mel. *Acta Scientiarum, 24*(4), 875–879. doi:10.4025/actascianimsci.v24i0.2336

Rios-Corripio, M. A., Rojas-López, M., & Delgado-Macuil, R. (2012). Analysis of adulteration in honey with standard sugar solutions and syrups using attenuated total reflectance-Fourier transform infrared spectroscopy and multivariate methods. *CyTA – Journal of Food, 10*, 119–122. doi: 10.1080/19476337.2011.596576

Ritter, W. (2003). Early detection of American foulbrood by honey and wax analysis. *Apiacta, 38*, 125–130. Retrived from http://www.apimondia.com/apiacta/articles/2003/ritter_1.pdf

Ritter, W., & Metzinger, S. (1998). Evidence of *Paenibacillus larvae larvae*, the trigger of American foulbrood, in wax. *Apidologie, 29*(5), 420–421. doi:10.1051/apido:19980502

Rivas, F., & Bettucci, L. (2007). Frequency and abundance of *Ascosphaera apis* and other fungal species in *Apis mellifera* hives from Uruguay. *Journal of Apicultural Research, 46*(1), 1–2. doi:10.1080/00218839.2007.11101358

Rose, R. I. (1969). *Bacillus larvae*, isolation, culturing and vegetative thermal death point. *Journal of Invertebrate Pathology, 14*(3), 411–414. doi:10.1016/0022-2011(69)90171-2

Ryba, S., Titera, D., Haklova, M., & Stopka, P. (2009). A PCR method of detecting American foulbrood (*Paenibacillus larvae*) in winter beehive wax debris. *Veterinary Microbiology, 139*(1–2), 193–196. doi:10.1016/j.vetmic.2009.05.009

Sandeman, D. C., Tautz, J., & Lindauer, M. (1996). Transmission of vibration across honeycombs and its detection by bee leg receptors. *Journal of Experimental Biology, 199*, 2585–2594.

Sanford, M. T., & Dietz, A. (1976). The fine structure of the wax glands of the honeybee (*Apis mellifera* L). *Apidologie*, 7(3), 197–207. doi:10.1051/apido:19760301

Saucy, F. (2014). On the natural cell size of European honey bees: A "fatal error" or distortion of historical data? *Journal of Apicultural Research, 53*(3), 327–336. doi:10.3896/IBRA.1.53.3.01

Savov, D., & Arsenov, L. (1963). Dejstvijeto na njakom chimicni i fizicni sredstva srescu amerikanskaja gnilec. *Izv Veterin Int Zarazni Parazit Bolezsti, 8*, 195–200.

Scheiner, R., Abramson, C. I., Brodschneider, R., Crailsheim, K., Farina, W., Fuchs, S., … Thenius, R. (2013). Standard methods for behavioural studies of Apis mellifera. In V. Dietemann, J. D. Ellis and P. Neumann (Eds) The COLOSS BEEBOOK, Volume I: standard methods for Apis mellifera research. *Journal of Apicultural Research, 52*(4). doi:10.3896/IBRA.1.52.4.04

Schellenberg, A., Chmielus, S., Schlicht, C., Camin, F., Perini, M., Bontempo, L., … & Horacek, M. (2010). Multielement stable isotope ratios (H, C, N, S) of honey from different European regions. *Food Chemistry, 121*(3), 770–777. doi: 10.1016/j.foodchem.2009.12.082

Schimmelmann, A., & DeNiro, M. J. (1993). Preparation of organic and water hydrogen for stable isotope analysis: Effects due to reaction vessels and zinc reagent. *Analytical Chemistry, 65*(6), 789–792. doi:10.1021/ac00054a024

Schimmelmann, A., Qi, H., Coplen, T. B., Brand, W. A., Fong, J., Meier-Augenstein, W., … & Werner, R. A. (2016). Organic reference materials for hydrogen, carbon, and nitrogen stable isotope-ratio measurements: Caffeines, n-alkanes, fatty acid methyl esters, glycines, L-valines, polyethylenes, and oils. *Analytical Chemistry, 88*(8), 4294–4302. doi:10.1021/acs.analchem.5b04392

Semkiw, P., & Skubida, P. (2013). Comb construction and brood development on beeswax foundation adulterated with paraffin. *Journal of Apicultural Science, 57*(1), 75–83. doi: 10.2478/jas-2013-0009

Serra Bonvehí, J. (1990). Estudio de la adulteración de la cera de abejas (*Apis mellifera* L.). *Grasas y Aceites, 41*(1), 69–72.

Serra Bonvehí, J., & Orantes Bermejo, J. (2010). Acaricides and their residues in Spanish commercial beeswax. *Pest Management Science, 66*(11), 1230–1235. doi:10.1002/ps.1999

Serra Bonvehí, J. S., & Orantes Bermejo, F. J. (2012). Detection of adulterated commercial Spanish beeswax. *Food Chemistry, 132*(1), 642–648. doi:10.1016/j.foodchem.2011.10.104

Shellhammer, T. H., Rumsey, T. R., & Krochta, J. M. (1997). Viscoelastic properties of edible lipids. *Journal of Food Engineering, 33*, 305–320. doi:10.1016/S0260-8774(97)00030-7

Shimanuki, H., Herbert, E. W., Jr., & Knox, D. A. (1984). High velocity electron beams for bee disease control. *American Bee Journal, 124*, 865–867.

Simon-Delso, N., San Martin, G., Bruneau, E., Minsart, L.-A., Mouret, C., & Hautier, L. (2014). Honeybee colony disorder in crop areas: The role of pesticides and viruses. *PLoS One, 9*(7), e103073. doi:10.1371/journal.pone.0103073

Skoog, D. A., & Leary, J. J. (1992). *Principles of instrumental analysis* (4th ed.). New York, NY: Saunders College Publishing, 700 pp.

Socrates, G. (2001). *Infrared and Raman characteristic group frequencies tables and charts.* New York, NY: Wiley, 366 pp.

Sofer, Z., & Schiefelbein, C. F. (1986). Hydrogen isotope ratio determinations in hydrocarbons using the pyrolysis preparation technique. *Analytical Chemistry, 58*(9), 2033–2036. doi: 10.1021/ac00122a023

Stever, T. (2003). Verkleinerte Bienen - Irrweg der Züchtung oder Wunderwaffe gegen Varroamilben? *Bienenpflege, 3*, 93–95.

Streibl, M., Stransky, K., & Sorm, F. (1966). Über einige neue Kohlenwasserstoffe in Wachs der Honigbiene (Apis mellifera L.). *Fette Seifen Anstrichmittel, 68*, 799–805.

Stuart, B. (2004). *Infrared spectroscopy: Fundamentals and applications*. Chichester, UK: Wiley, 244 pp. Retrieved from http://www.kinetics.nsc.ru/chichinin/books/spectroscopy/Stuart04.pdf

Stuermer, D. H., Peters, K. E., & Kaplan, I. R. (1978). Source indicators of humic substances and protokerogen: Stable isotope ratios, elemental compositions and electron spin resonance spectra. *Geochimica et Cosmochimica Acta, 42*(7), 989–997. doi:10.1016/0016-7037(78)90288-0

Sutherland, T. D., Church, J. S., Hu, X., Huson, M. G., Kaplan, D. L., & Weisman, S. (2011). Single honeybee silk protein mimics properties of multi-protein silk. *PLoS One, 6*(2), e16489. doi:10.1371/journal.pone.0016489

Sutherland, T. D., Sriskantha, A., Church, J. S., Strive, T., Trueman, H. E., & Kameda, T. (2014). Stabilization of viruses by encapsulation in silk proteins. *ACS Applied Materials & Interfaces, 6*(20), 18189–18196. doi:10.3390/ijms17071170

Sutherland, T. D., Weisman, S., Trueman, H. E., Sriskantha, A., Trueman, J. W., & Haritos, V. S. (2007). Conservation of essential design features in coiled coil silks. *Molecular Biology and Evolution, 24*(11), 2424–2432. doi:10.1093/molbev/msm171

Svečnjak, L., Baranović, G., Vinceković, M., Prđun, S., Bubalo, D., & Tlak Gajger, I. (2015). An approach for routine analytical detection of beeswax adulteration using FTIR-ATR spectroscopy. *Journal of Apicultural Science, 59*(2), 37–49. doi:10.1515/JAS-2015-0018

Svečnjak, L., Prđun, S., Baranović, G., Damić, M., & Rogina, J. (2018). Alarming situation on the EU beeswax market: the prevalence of adulterated beeswax material and related safety issues. In Program & Abstracts Book EurBee 8, 8th Congress of Apidology, Ghent, Belgium, 18–20 September 2018, pp. 114–115.

Svečnjak, L., Prđun, S., Bubalo, D., Matošević, M., & Car, J. (2016). Beeswax adulteration issue: aspects of contamination and outcome. In Abstracts Book 6th Apimedica & 5th Apiquality International Symposium – 5th Apiquality, Rome, Italy, 22–25 November 2016, pp. 22–23.

Svečnjak, L., Prđun, S., Bubalo, D., Tlak Gajger, I., & Baranović, G. (2015). Determination of residual paraffin in honeycomb constructed on adulterated foundations. In Proceedings of 44th APIMONDIA International Apicultural Congress, Scientific Program Abstracts, Daejeon, Korea, 15–20 September 2015, p. 436.

Szabo, T. I. (1977). Effect of colony size and ambient temperature on comb building and sugar consumption by honeybees. *Journal of Apicultural Research, 16*(4), 174–183. doi:10.1080/00218839.1977.11099884

Taha, E. A., Manosur, H. M., & Shawer, M. B. (2010). The relationship between comb age and the amounts of mineral elements in honey and wax. *Journal of Apicultural Research, 49*(2), 202–207. doi:10.3896/IBRA.1.49.2.10

Tautz, J. (1996). Honeybee waggle dance: Recruitment success depends on the nature of the dance floor. *Journal of Experimental Biology, 199*, 1375–1381.

Tautz, J., Casas, J., & Sandeman, D. (2001). Phase reversal of vibratory signals in honeycomb may assist dancing honeybees to attract their audience. *The Journal of Experimental Biology, 204*(21), 3737–3746. doi:10.1007/s003590050070

Tautz, J., & Lindauer, M. (1997). Honeybees establish specific sites on the comb for their waggle dances. *Journal of Comparative Physiology A: Sensory, Neural, and Behavioral Physiology, 180*(5), 537–539. doi:10.1007/s003590050070

Thompson, W. D. (1945). *On growth and form* (A new edition). Cambridge/New York: University Press/Macmillan Company, 1118 pp.

Thrasyvoulou, A. T., & Pappas, N. (1988). Contamination of honey and wax with malathion and coumaphos used against the Varroa mite. *Journal of Apicultural Research, 27*(1), 55–61. doi:10.1080/00218839.1988.11100782

Timbers, G. E., & Gochnauer, T. A. (1982). Note on the thermal conductivity of beeswax *Journal of Apicultural Research, 21*(4), 232–235. doi:10.1080/00218839.1982.11100548

Timbers, G. E., Robertson, G. D., & Gochnauer, T. A. (1977). Thermal properties of beeswax and beeswax-paraffin mixtures. *Journal of Apicultural Research. 16*, 49–55. doi:10.1080/00218839.1977.11099860

Tipple, B. J., Chesson, L. A., Erkkila, B. R., Cerling, T. E., & Ehleringer, J. R. (2012). B-HIVE: Beeswax hydrogen isotopes as validation of environment. Part II: Compound-specific hydrogen stable isotope analysis. *Food Chemistry, 134*(1), 494–501. doi:10.1016/j.foodchem.2012.02.106

Titěra, D., & Haklová, M. (2003). Detection method of *Paenibacillus larvae larvae* from beehive winter debris. *Apiacta, 38*, 131–133. Retrieved from https://www.apimondia.com/apiacta/articles/2003/titera_1.pdf

Tobias, H. J., & Brenna, J. T. (1997). On-line pyrolysis as a limitless reduction source for high-precision isotopic analysis of organic-derived hydrogen. *Analytical Chemistry, 69*, 3148–3152. doi:10.1021/ac970332v

Toher, D., Downey, G., & Murphy, T. B. (2007). A comparison of model-based and regression classification techniques applied to near infrared spectroscopic data in food authentication studies. *Chemometrics and Intelligent Laboratory Systems, 89*(2), 102–115. doi:10.1016/j.chemolab.2007.06.005

Toledo, V. A. A. (1991). *Desenvolvimento de colméias hibridas de Apis mellifera e seu comportamento na aceitação e manejo da cera* (Mestredo de Zootecnia Dissertação). Faculdade de Ciências Agrárias e Veterinárias. UNESP., Jaboticabal, SP, Brasil, 196 pp.

Torto, B., Carroll, M. J., Duehl, A., Fombong, A. T., Gozansky, K. T., Nazzi, F., … & Teal, P. E. A. (2013). Standard Methods for chemical ecology research in *Apis mellifera*. In V. Dietemann, J. D. Ellis, & P. Neumann (Eds.), The coloss beebook, Vol. I: Standard methods for *Apis mellifera* research. *Journal of Apicultural Research, 52*(4), 1–34. doi:10.3896/IBRA.1.52.4.06

Tremolada, P., Bernardinelli, I., Colombo, M., Spreafico, M., & Vighi, M. (2004). Coumaphos distribution in the hive ecosystem: Case study for modelling applications. *Ecotoxicology, 13*(6), 589–601. doi:10.1023/B:ECTX.0000037193.28684.05

Tulloch, A. P. (1971). Beeswax: Structure of the esters and their component hydroxy acids and diols. *Chemistry and Physics of Lipids, 6*, 235–265. doi:1016/0009-3084(71)90063-6

Tulloch, A. P. (1973). Factors affecting analytical values of beeswax and detection of adulteration. *Journal of the American Oil Chemists Society, 50*(7), 269–272. doi:10.1007/BF02641800

Tulloch, A. P. (1980). Beeswax – Composition and analysis. *Bee World, 61*(2), 47–62. doi:10.1080/0005772X.1980.11097776

Tulloch, A. P., & Hoffman, L. L. (1972). Canadian beeswax: Analytical values and composition of hydrocarbons, free acids and long chain esters. *Journal of the American Oil Chemists Society, 49*(12), 696–699. doi:10.1007/BF02609202

Tsigouri, A., Menkissoglu-Spiroudi, U., Thrasyvoulou, A. T., & Diamantidis, G. C. (2000). Determination of fluvalinate residues in beeswax by gas chromatography with electron-capture detection. *Journal of AOAC International, 83*(5), 1225–1228.

United States Department of Agriculture (USDA). Animal and Plant Health Inspection Service (APHIS). Retrieved from https://www.aphis.usda.gov/aphis/ourfocus/animalhealth/

export/iregs-for-animal-product-exports/eu-animal-products-not-for-human-consumption

U.S. Food and Drug Administration (FDA). Food additive status list. Retrieved from https://www.fda.gov/food/ingredientspackaginglabeling/foodadditivesingredients/ucm091048.htm#ftnB

Van Buren, N., Marien, J., Velthuis, H., & Oudejans, R. (1992). Residues in beeswax and honey of perizin, an acaricide to combat the mite *Varroa jacobsoni* Oudemans (Acari: *Mesostigmata*). *Environmental Entomology, 21*(4), 860–865. doi:10.1093/ee/21.4.860

Vogt, H. (1911). *Geometrie und Ökonomie der Bienenzelle.* Breslau: Festschrift der Universität, 68 pp.

Walker, A. A., Warden, A. C., Trueman, H. E., Weisman, S., & Sutherland, T. D. (2013). Micellar refolding of coiled-coil honeybee silk proteins. *Journal of Materials Chemistry B, 1*(30), 3644–3651. doi:10.1039/c3tb20611d

Wallner, K. (1992). The residues of P-Dichlorobenzene in wax and honey. *American Bee Journal, 132*(8), 538–541.

Wallner, K. (1993). A method for determination of varroacide residues in bees wax. *Apidologie, 24*, 502–503.

Wallner, K. (1997). The actual beeswax quality in foundations on the market. *Apidologie, 28*, 168–170.

Wallner, K. (1999). Varroacides and their residues in bee products. *Apidologie, 30*(2–3), 235–248. doi:10.1051/apido:19990212

Wallner, K. (2000). Varroacides in beeswax. The actual situation in 10 European countries. *Apidologie, 31*(5), 613–615.

Ward, L., Brown, M., Neumann, P., Wilkins, S., Pettis, J., & Boonham, N. (2007). A DNA method for screening hive debris for the presence of small hive beetle (*Aethina tumida*). *Apidologie, 38*(3), 272–280. doi:10.1051/apido:2007004

Waś, E., Szczęsna, T., & Rybak-Chmielewska, H. (2014a). Determination of beeswax hydrocarbons by gas chromatography with a mass detector (GC-MS) technique. *Journal of Apicultural Science, 58*(1), 145–157. doi:10.2478/JAS-2014-0015

Waś, E., Szczęsna, T., & Rybak-Chmielewska, H. (2014b). Hydrocarbon composition of beeswax (*Apis mellifera*) collected from light and dark coloured combs. *Journal of Apicultural Science, 58*(2), 99–106. doi:10.2478/jas-2014-0026

Waś, E., Szczęsna, T., & Rybak-Chmielewska, H. (2015). Application of gas chromatography with the mass detector (GC-MS) technique for detection of beeswax adulteration with paraffin. *Journal of Apicultural Science, 59*(1), 143–152. doi:10.1515/jas-2015-0015

Waś, E., Szczęsna, T., & Rybak-Chmielewska, H. (2016). Efficiency of GC-MS method in detection of beeswax adulterated with paraffin. *Journal of Apicultural Science, 60*(1), 131–147. doi:10.1515/JAS-2016-0012

Weisman, S., Haritos, V. S., Church, J. S., Huson, M. G., Mudie, S. T., Rodgers, A. J. W., … & Sutherland, T. D. (2010). Honeybee silk: Recombinant protein production, assembly and fiber spinning. *Biomaterials, 31*(9), 2695–2700. doi:10.1016/j.biomaterials.2009.12.021

White, J. W. Jr, Reader, M. K., & Riethof, M. L. (1960). Chromatographic determination of hydrocarbons in beeswax. *Journal of the Association of Official Analytical Chemists, 43*, 778–780.

White, J. W. Jr, Riethof, M. L., & Kushnir, M. (1960). Estimation of microcrystalline wax in beeswax. *Journal of the Association of Official Analytical Chemists, 43*(4), 781–790.

Wiest, L., Bulete, A., Giroud, B., Fratta, C., Amic, S., Lambert, O., … &, Arnaudguilhem, C. (2011). Multi-residue analysis of 80 environmental contaminants in honeys, honeybees and pollens by one extraction procedure followed by liquid and gas chromatography coupled with mass spectrometric detection. *Journal of Chromatography A, 1218*(34), 5743–5756. doi:10.1016/j.chroma.2011.06.079

Wille, K., Claessens, M., Rappe, K., Monteyne, E., Janssen, C. R., Brabander, H. F., & Vanhaecke, L. (2011). Rapid quantification of pharmaceuticals and pesticides in passive samplers using ultra high performance liquid chromatography coupled to high resolution mass spectrometry. *Journal of Chromatography A, 1218*(51), 9162–9173. doi:10.1016/j.chroma.2011.10.039

Williams, G. R., Alaux C., Costa, C., Csáki, T., Doublet, V., Eisenhardt, D., … Brodschneider, R. (2013). Standard methods for maintaining adult Apis mellifera in cages under in vitro laboratory conditions. In V. Dietemann, J. D. Ellis, P Neumann (Eds) The COLOSS BEEBOOK, Volume I: standard methods for Apis mellifera research. *Journal of Apicultural Research, 52*(1). doi:10.3896/IBRA.1.52.1.04

Wilmart, O., Legrève, A., Scippo, M. L., Reybroeck, W., Urbain, B., De Graaf, D. C., … & Saegerman, C. (2016). Residues in beeswax: A health risk for the consumer of honey and beeswax? *Journal of Agricultural and Food Chemistry, 64*(44), 8425–8434. doi:10.1021/acs.jafc.6b02813

Wilson, W. T. (1972). Resistance to American foulbrood in honey bees. XII. Persistence of viable Bacillus larvae spores in the feces of adults permitted flight. *Journal of Invertebrate Pathology, 20*(2), 165–169. doi:10.1016/0022-2011(72)90130-9

Wittmer, C. R., Hu, X., Gauthier, P.-C., Weisman, S., Kaplan, D. L., & Sutherland, T. D. (2011). Production, structure and in vitro degradation of electrospun honeybee silk nanofibers. *Acta Biomaterialia, 7*(10), 3789–3795. doi:10.1016/j.actbio.2011.06.001

Wong, W. W., & Klein, P. D. (1986). A review of techniques for the preparation of biological samples for mass-spectrometric measurements of hydrogen-2/hydrogen-1 and oxygen-18/oxygen-16 isotope ratios. *Mass Spectrometry Reviews, 5*, 313–342. doi:10.1002/mas.1280050304

Wyman, J. (1866). Notes on the cells of the bee. *Proceedings of the American Academy of Arts and Sciences, 7*(1), 1–18.

Yáñez, K. P., Bernal, J. L., Nozal, M. J., Martín, M. T., & Bernal, J. (2013). Determination of seven neonicotinoid insecticides in beeswax by liquid chromatography coupled to electrospray-mass spectrometry using a fused-core column. *Journal of Chromatography A, 1285*, 110–117. doi:10.1016/j.chroma.2013.02.032

Yáñez, K. P., Martín, M. T., Bernal, J. L., Nozal, M. J., & Bernal, J. (2014). Determination of spinosad at trace levels in bee pollen and beeswax with solid-liquid extraction and LC-ESI-MS. *Journal of Separation Science, 37*(3), 204–210. doi:10.1002/jssc.201301069

Yu, J., & Butler, I. S. (2015). Recent applications of infrared and Raman spectroscopy in art forensics: A brief overview. *Applied Spectroscopy Reviews, 50*, 152–157. doi:10.1080/05704928.2014.949733

Zeissloff, E. (2007). Natürliche Zellgröße. *Journal Apicole Luxembourgeois, 3*, 73–78.

Zhang, K., Duan, H., Karihaloo, B. L., & Wang, J. (2010). Hierarchical, multilayered cell walls reinforced by recycled silk cocoons enhance the structural integrity of honeybee combs. *Proceedings of the National Academy of Sciences of the United States of America, 107*(21), 9502–9506. doi:10.1073/pnas.0912066107

Zhang, K., Si, F. W., Duan, H. L., & Wang, J. (2010). Microstructures and mechanical properties of silks of silkworm and honeybee. *Acta Biomaterialia, 6*, 2165–2171. doi:10.1016/j.actbio.2009.12.030

Zimmermann, B., & Kohler, A. (2013). Optimizing Savitzky-Golay parameters for improving spectral resolution and quantification in infrared spectroscopy. *Applied Spectroscopy, 67*, 892–902. doi:10.1366/12-06723

Zimnicka, B., & Hacura, A. (2006). An investigation of molecular structure and dynamics of crude beeswax by vibrational spectroscopy. *Polish Journal of Environmental Studies, 15*(4A), 112–114.

Journal of Apicultural Research, 2019
Vol. 58, No. 2, 1–49, http://dx.doi.org/10.1080/00218839.2016.1222661

REVIEW ARTICLE

# Standard methods for *Apis mellifera* propolis research

Vassya Bankova[a]*, Davide Bertelli[b], Renata Borba[c], Bruno José Conti[d], Ildenize Barbosa da Silva Cunha[e], Carolina Danert[f], Marcos Nogueira Eberlin[g], Soraia I Falcão[h], María Inés Isla[f], María Inés Nieva Moreno[f], Giulia Papotti[b], Milena Popova[a], Karina Basso Santiago[d], Ana Salas[f], Alexandra Christine Helena Frankland Sawaya[e], Nicolas Vilczaki Schwab[g], José Maurício Sforcin[d], Michael Simone-Finstrom[i], Marla Spivak[c], Boryana Trusheva[a], Miguel Vilas-Boas[h], Michael Wilson[c] and Catiana Zampini[f]

[a]*Institute of Organic Chemistry with Centre of Phytochemistry, Bulgarian Academy of Sciences, Acad. G. Bonchev str. bl.9, 1113 Sofia, Bulgaria;* [b]*Dipartimento di Scienze della Vita, Università degli studi di Modena e Reggio Emilia, via Campi 103, 41125 Modena, Italy;* [c]*Department of Entomology, University of Minnesota, St Paul, MN, USA;* [d]*Department of Microbiology and Immunology, Biosciences Institute, UNESP, 18618-970 Botucatu, SP, Brazil;* [e]*Department of Plant Biology, Institute of Biology, State University of Campinas, UNICAMP, Campinas, SP, Brazil;* [f]*Instituto de Química del Noroeste Argentino (INQUINOA), Consejo Nacional de Investigaciones Científica y Técnica (CONICET), Universidad Nacional de Tucumán (UNT), San Lorenzo 1469, San Miguel de Tucumán, Tucumán, Argentina;* [g]*ThoMSon Mass Spectrometry Laboratory, Institute of Chemistry, State University of Campinas, UNICAMP, Campinas, SP, Brazil;* [h]*CIMO/Escola Superior Agrária, Instituto Politécnico de Bragança, Campus de Sta. Apolónia Apartado, 1172, 5301-855 Bragança, Portugal;* [i]*USDA-ARS, Honey Bee Breeding, Genetics and Physiology Research Laboratory, Baton Rouge, LA, USA*

*(Received 25 November 2014; accepted 21 July 2016)*

Propolis is one of the most fascinating honey bee (*Apis mellifera* L.) products. It is a plant derived product that bees produce from resins that they collect from different plant organs and with which they mix beeswax. Propolis is a building material and a protective agent in the bee hive. It also plays an important role in honey bee social immunity, and is widely used by humans as an ingredient of nutraceuticals, over-the-counter preparations and cosmetics. Its chemical composition varies by geographic location, climatic zone and local flora. The understanding of the chemical diversity of propolis is very important in propolis research. In this manuscript, we give an overview of the available methods for studying propolis in different aspects: propolis in the bee colony; chemical composition and plant sources of propolis; biological activity of propolis with respect to bees and humans; and approaches for standardization and quality control for the purposes of industrial application.

**Métodos estándar para investigar el própolis de *Apis mellifera***

El própolis es uno de los productos más fascinante de la abeja de la miel (*Apis mellifera* L.). Es un producto derivado de plantas que las abejas producen a partir de resinas que recogen en diferentes órganos de la planta y que mezclan con la cera de abejas. El própolis es un material de construcción y un agente protector en la colmena de abejas. También juega un papel importante en la inmunidad social de la abeja de la miel, y es ampliamente utilizado por los seres humanos como un ingrediente de nutracéuticos, preparados de venta no regulada y cosméticos. Su composición química varía según la ubicación geográfica, la zona climática y la flora local. La comprensión de la diversidad química del própolis es muy importante en su investigación. En este manuscrito, damos una visión general de los métodos disponibles para el estudio del própolis en diferentes aspectos: própolis en la colonia de abejas; composición química y fuentes vegetales del própolis; actividad biológica del própolis con respecto a las abejas y los seres humanos; y enfoques para la normalización y control de calidad para los fines de aplicación industrial.

蜜蜂蜂胶实验标准方法

摘要

蜂胶是一种重要而特别的蜂产品。蜂胶是通过蜜蜂采集树脂后，混合蜂蜡酿造而成。蜂胶是蜂巢的重要建造材料。蜂胶能够抑制病菌生长，对蜜蜂的社会免疫起到重要作用。不仅如此，蜂胶还广泛用于功能食品和化妆品。不同蜂胶的化学成分也不相同，这主要是和当地的气候与植物群落有关。了解蜂胶的化学成分对于蜂胶研究非常重要。本文将从不同角度概述蜂胶的研究方法：蜜蜂蜂群内的蜂胶分析；蜂胶成分和植物源分析；蜂胶对蜜蜂和人的功能性分析；蜂胶产业化应用和质量标准分析。

**Keywords:** COLOSS; *BEEBOOK*; honey bee; *Apis mellifera*; propolis; chemical composition; plant sources; biological activity; standardization; quality control

*Corresponding author. Email: bankova@orgchm.bas.bg

Please refer to this paper as: Bankova, V; Bertelli, D; Borba, R; Conti, B J; da Silva Cunha, I B; Danert, C; Eberlin, M N; Falcão, S I; Isla, M I; Moreno, M I N; Papotti, G; Popova, M; Santiago, K B; Salas, A; Sawaya, A C H F; Schwab, N V; Sforcin, J M; Simone-Finstrom, M; Spivak, M; Trusheva, B; Vilas-Boas, M; Wilson, M; Zampini, C (2019) Standard methods for *Apis mellifera* propolis research. In V Dietemann, P Neumann, N Carreck and J D Ellis (Eds), *The COLOSS BEEBOOK, Volume III, Part I: standard methods for Apis mellifera hive products research. Journal of Apicultural Research* 58(2): http://dx.doi.org/10.1080/00218839.2016.1222661

## 1. Introduction

Western honey bees (*Apis mellifera* L.) produce propolis (also called bee glue) from resins that they collect from different plant organs and with which they mix beeswax. The term "propolis" is of Greek origin: "pro" meaning "in front of/for" and "polis" meaning "city", that is, in front (or for defense) of the city. Propolis is used by bees as a building material in their hives, for blocking holes and cracks, repairing combs, and strengthening the thin borders of the comb (Ghisalberti, 1979). Feral bees inhabiting tree cavities cover the inside of the cavity with a layer of propolis called the "propolis envelope" (Seeley & Morse, 1976). Propolis plays the role of chemical defense against microorganisms and as an embalmer of larger, dead intruders (insect, small animals) that have died in the hive and are too large to be removed by the bees (Ghisalberti, 1979).

The valuable therapeutic properties of propolis were recognized by human beings millennia ago; historical records suggest the use of propolis dates back to the ancient Egyptians, Romans, and Greeks (Crane, 1999). It is still used as a popular homemade remedy in many countries all over the world, but also as a constituent of food additives, cosmetics and over-the-counter preparations (de Groot, 2013; Sforcin & Bankova, 2011; Suárez, Zayas, & Guisado, 2005).

The biological activity of propolis is due to its chemical composition which, in turn, depends on the source plant(s) from which bees collect the resin. A number of chemical types of propolis have been registered according to their plant source. The understanding of propolis chemical diversity plays a core role in propolis studies.

In this manuscript, an overview is presented of the available methods for studying propolis in different aspects: propolis in the bee colony, chemical composition and plant sources of propolis, biological activity of propolis with respect to bees and humans, and approaches for standardization and quality control for the purposes of industrial application.

## 2. Resin and propolis: sampling and harvesting

Propolis collected from the hive may contain a mixture of resins from various plant sources and beeswax. If individual sources of resin are needed for chemical analysis, it may be necessary to collect the resin from plant tissue or from the hindlegs of returning resin foragers. The procedures described below first describe how to collect resins from plants and individual bees, and then how to collect propolis from within a colony.

### *2.1. Resin sample collection*

#### *2.1.1. Sampling resin from plant tissue*

Identify resinous plants in your area. The most comprehensive guide to resinous plants available is Langenheim (2003), while the most comprehensive guide to resinous plants used by bees is Crane (1990). Also see Bankova, Popova, and Trusheva (2006).

(1) Collect resin from individual plants. If the target resins are foliar, use clean pruning shears to detach 4–6 resinous buds/leaves and place all in a 15 ml screw-top EPA vial. If resins are internal, collect fresh resin from existing or generated wounds.
(2) The number of individual plants sampled will vary by apiary due to availability. Try to collect resin from at least three different individuals per plant species if possible.

#### *2.1.2. Sampling resin from foragers in the field*

(1) Individual resin foragers carrying pure resin can be captured returning to the hive (Figure 1). Block the hive entrance with a mesh screen and observe for 15 min. Capture resin foragers clustering on the hive entrance in wire cages or a suitable screened container and maintain captured bees out of the sun. It is easiest to

Figure 1. Honey bees with resin (on left) and pollen (on right) on hind legs. The resin loads of foragers are semi-translucent and shiny, whilst pollen is opaque and powdery in texture.
Photo: M. Simone-Finstrom.

collect resin foragers from small colonies that are situated on hive stands (see Section 2.2).
(2) Collect samples twice per day (once in the morning and once in the afternoon) as required.
(3) Anesthetize caged bees on ice for 5 min, then remove them from the cage. Remove resin from bee corbiculae using an insect pin. Resin foragers may be marked (see the *BEEBOOK* paper on miscellaneous honey bee research methods by Human et al. (2013)) and released as desired.
(4) Place resin globules from an individual bee inside a small, screw-top glass vial and store on ice while in the field. Place the resin in the freezer (−10 °C) until needed for further use.

### 2.2. Harvesting propolis from hives

#### 2.2.1. Commercial traps

The major commercial beekeeping supply companies sell "propolis traps." These usually are thick sheets of plastic with a series of 1.6 mm grooved slits over the entire surface. This is the width that encourages honey bees to deposit more propolis and less wax to close the opening (Crane, 1990).

(1) Place the propolis trap directly over the top frames of the uppermost box (super) of a colony (Crane, 1990) and cover with a standard colony lid.
(2) Trap success can be improved by increasing air flow and light through the trap (Crane, 1990; Krell, 1996). This can be done easily by placing a wooden rim with holes drilled into its sides over the propolis trap and under the outer cover. Using a migratory cover (a flat cover that does not have an overhang covering the holes in the rim) further supports this process. While this extra step is not necessary, it will increase resin collection (Borba, Simone-Finstrom, Spivak, personal observation).
(3) It is important to note that the amount and quality of propolis collected will vary greatly across colonies based on genetics, environment and colony strength (Butler, 1949; Wilson, Brinkman, Spivak, Gardner, & Cohen, 2015). A strong, high resin-collecting colony can fill a trap full of propolis in a couple of weeks. Other colonies will never close all gaps completely or will use mostly wax to seal the gaps (Borba, Simone-Finstrom, Spivak, personal observation), as there is a genetic component to the level of propolis collection exhibited by bees (e.g. Manrique & Soares, 2002; Nicodemo, Malheiros, De Jong, & Couto, 2014).
(4) To harvest the propolis from the traps, it is best to freeze the traps so that the propolis becomes hard and brittle (Krell, 1996). It then can be knocked or scraped out of the traps.

#### 2.2.2. Non-commercial propolis traps

Many different materials can be utilized to collect propolis (Krell, 1996). The key is making sure that the bees cannot chew away the material and that the gaps are appropriately sized to encourage resin deposition.

(1) One suitable option includes mesh (burlap) bags, like those used for storing corn, potatoes and other crops. These bags doubled-over and placed on top of the colony in the same way as the commercial traps (Section 2.2.1) work particularly well. Landscape cloth also can be used.
(2) Similar to commercial traps (Section 2.2.1), it is best to freeze the cloth prior to harvesting the propolis. Rolling the cloth on a hard surface will release the propolis from the gaps.

#### 2.2.3. Hive scrapings

The most common way for propolis to be harvested in the apicultural setting is simply by scraping propolis from the frame rests, frame edges and from the bottom boards or insides of boxes (Ellis & Hepburn, 2003; Krell, 1996). This is typically done at the end of the season to clean up the boxes for use in the following year and can easily generate a significant amount of propolis. Scrapings may contain propolis from multiple seasons, and it is unknown how age affects propolis quality. More research is needed to determine if the antimicrobial properties of propolis diminish over time.

#### 2.2.4. African-derived bee colonies in Brazil

Honey bees of African origin, such as those found in the tropics of Brazil, deposit large amounts of propolis in tree cavities as well as in commercial bee boxes (Manrique & Soares, 2002). Brazilian beekeepers have developed methods to harvest large quantities of propolis by introducing slats of wood with 4cm gaps to the sides of the hive boxes (Figure 2(a)). The large opening stimulates African-derived bees to fill the slats with propolis. When the gap is completely filled with a thick layer of propolis, the wood slats can be removed and the propolis harvested using a knife to cut out the sheet (Figure 2(b)).

## 3. Propolis chemical analysis

Propolis consists of plant resins and beeswax and the chemical analysis of propolis is directed to the plant derived compounds as they are the components responsible for the bioactivity of propolis. The compounds also indicate the plant(s) that bees have visited for resin collection. The chemical information is important with respect to quality control and standardization purposes. Also, if the propolis type is new and unexplored, it may contain new valuable bioactive compounds.

(a)

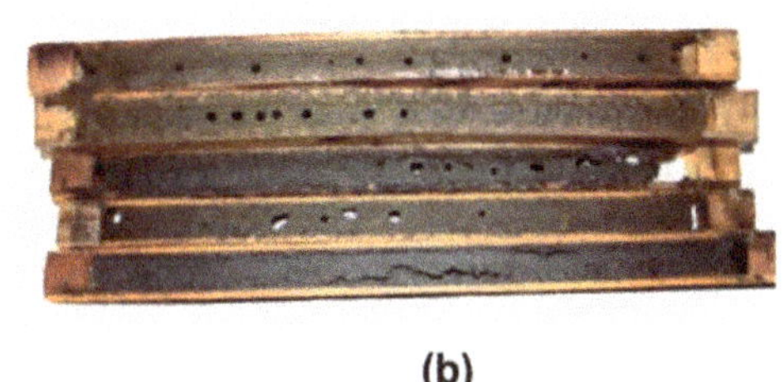

(b)

Figure 2. Brazilian propolis trap. (a) The sides of a hive box are replaced with removal wooden slats, containing 4 cm gaps. (b) The slats are removed for harvesting once they are filled with propolis. The propolis sheet can be cut from the wood with a knife. The bees leave holes in the sheet of propolis naturally.
Photo: R. Borba.

### 3.1. Extraction of propolis

#### 3.1.1. General extraction procedure

The aim of the extraction is to remove the major plant secondary metabolites from any impurities, such as beeswax, for further analysis or for biotests. This is achieved by extraction with 70% ethanol, as noted below.

(1) Keep propolis overnight in a freezer (−20 °C). Powder the frozen propolis using a coffee mill or other similar grinding device to achieve a particle size of about 10–80 μm.
(2) Measure a sample of the powdered propolis, add 70% ethanol (1:30 w:v) and keep it for 24 h at room temperature. Alternatively, sonicate the suspension (propolis in 70% ethanol) for 20 min in an ultrasonic bath at 20 °C.
(3) Filter the resulting suspension at room temperature using a paper filter and repeat the procedure with the part trapped in the filter, extracting the residue again under the same conditions. Experiments have shown that a third extraction under the same conditions is not necessary since the third extract yielded a negligible amount of dry propolis (Popova et al., 2004).
(4) The concentration $C$ of the extract (i.e. the amount of propolis) is determined by evaporating 2 ml of the extract to dryness *in vacuo* to constant weight $g$ and using the formula $C = g/2$ mg/ml (average of three replicates).

The obtained extract can be evaporated to dryness for further use or used as is in further experiments. Alternative extraction procedures might be applied depending on the analysis for which the propolis extract is to be used. For biological tests, a variety of solvents have been used, including methanol, different ethanol-water mixtures (80, 90, and 96%), absolute ethanol, glycerol, water (Park & Ikegaki, 1998; Sforcin & Bankova, 2011), and even DMSO (Netíková, Bogusch, & Heneberg, 2013). It is important to note that water dissolves less than 10% of the weight of propolis.

#### 3.1.2. Extraction of propolis for mass spectrometry fingerprinting

(1) Extract ground propolis by maceration for 7 days in an orbital shaker at a temperature of 30 °C, with 10 ml of absolute ethanol (Merck; Darmstadt, Germany) for every 3 g of crude propolis.
(2) Separate the insoluble portion by filtration; keep the ethanolic solutions in a freezer at −16 °C overnight and filter again at this temperature to reduce the wax content of the extracts.

### 3.2. Extraction of propolis volatiles

Propolis volatile constituents are responsible for the specific pleasant aroma of propolis and contribute to its biological activity, although their amount is seldom greater than 1% of the weight of the sample. They also may play an important role as olfactory cues during resin collection by honey bees (Leonhardt, Zeilhofer, Bluthgen, & Schmitt, 2010). Different methods have been used to extract propolis volatiles: steam distillation, hydrodistillation (Clevenger), distillation-extraction (Likens-Nikerson), solvent extraction (including ultrasound-assisted and microwave-assisted extraction), and static and dynamic head-space, solid-phase microextraction. The method of extraction significantly affects the chemical composition of the volatile constituents of propolis (Bankova, Popova, & Trusheva, 2014). Here, we describe one of the most often used approaches for propolis volatile extraction, distillation-extraction (Bankova, Boudourova-Krasteva, Popov, Sforcin, & Funari, 1998). A review of volatile extraction procedures for hive components in general can be found in Torto et al. (2013).

(1) Keep propolis overnight in a freezer (−20 °C). Powder the frozen propolis using a coffee mill to achieve a particle size of 10–80 μm (Section 3.1.1).
(2) Put 3 g powdered propolis in a 100 mL round-bottom flask and add 80 ml distilled water.
(3) Put 50 ml *n*-pentane - diethyl ether 1:1 (v/v) in another 100 ml round-bottom flask and dip it in an ice bath.
(4) Distill for 4 h in a Likens-Nickerson apparatus (Figure 3, Queiroga, Madruga, Galvão, & Da Costa, (2005)).
(5) After the distillation is over, remove the water layer using a separatory funnel. Keep the organic layer in refrigerator until further processing.
(6) Wash the water layer with 5 ml ice cold *n*-pentane - diethyl ether 1:1 (v/v).
(7) Dry the organic layer over anhydrous $Na_2SO_4$: add 3 g of anhydrous $Na_2SO_4$, shake the flask for 5 min and filter the liquid using a filter paper. Wash the solid on the filter with 1 ml ice cold *n*-pentane - diethyl ether 1:1 (v/v).
(8) Evaporate the solvent under reduced pressure without heating using a rotatory evaporator.

The obtained volatiles can be analyzed further using GC, GC-MS or subjected to biological tests.

### 3.3. Gas chromatography-mass spectrometry analysis of propolis

Gas chromatography-mass spectrometry (GC-MS) is one of the so-called hyphenated analytical techniques extensively used for the chemical analysis of complex mixtures such as propolis. GC-MS combines the features of gas chromatography for compound separation and mass spectrometry to identify different substances. This method is used for chemical profiling of propolis for the needs of comparative analysis, quality control and standardization.

#### 3.3.1. GC-MS analysis of non-volatile propolis constituents

Prior to the GC-MS analysis, derivatization of the propolis extracts is required because propolis contains metabolites that are not volatile enough for gas chromatography (Greenaway, Scaysbrook, & Whatley, 1987). One of the most widely used derivatization reagents is N,O-bis (trimethylsilyl)trifluoroacetamide (BSTFA) (Bankova, Dyulgerov, Popov, & Marekov, 1987; Greenaway & Whatley, 1990). Silyl derivatives (trimethylsilyl ethers) obtained from propolis are less polar and more volatile than their parent compounds and are suitable for analysis by GC-EIMS (gas chromatography – electron impact mass spectrometry).

*3.3.1.1. Sample preparation.* Dry propolis extracts obtained according to Section 3.1.1 are analyzed by GC-MS after derivatization. The derivatization (conversion to trimethylsilyl derivatives) is performed, as follows:

(1) Mix 5 mg of the propolis extract obtained per Section 3.1.1 with 50 μl of dry (water-free) pyridine.
(2) Add 75 μl of bis(trimethylsilyl)-trifluoroacetamide (BSTFA) to the mixture.
(3) Heat the mixture at 80 °C for 20 min.
(4) Subject the silylated extract to GC–MS analysis (see Section 3.3.1.2).

*3.3.1.2. GC-MS analysis.* The GC–MS analysis should be performed with a proper instrument such as a Hewlett–Packard gas chromatograph 5890 series II Plus linked to a Hewlett–Packard 5972 mass spectrometer system (Trusheva et al., 2011).

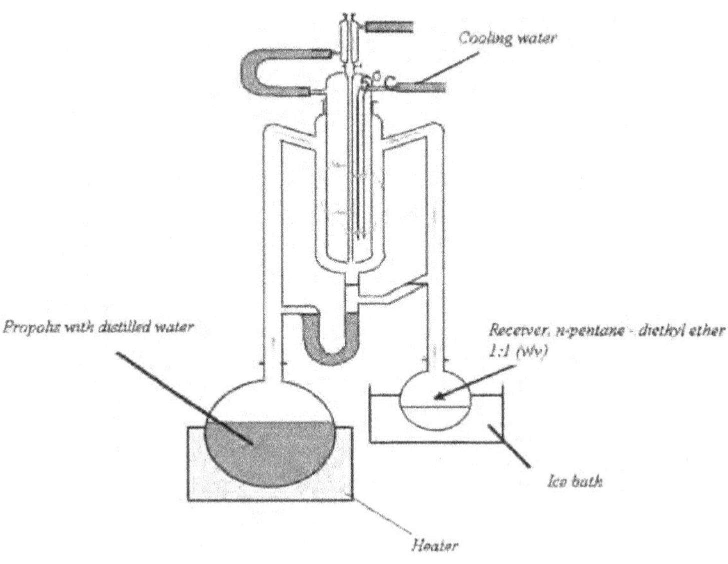


Figure 3. Likens–Nickerson apparatus for distillation-extraction of volatiles.

(1) Use a 30 m long, 0.25 mm ID, and 0.5 μm film thickness HP5-MS capillary column. Other columns with similar characteristics also can be used depending on analytical need.
(2) Program the temperature from 60 to 300 °C at a rate of 5 °C/min, and a 10 min hold at 300 °C.
(3) Helium is used as a carrier gas at a flow rate of 0.8 ml/min.
(4) The split ratio should be 1:10.
(5) The injector temperature should be 280 °C.
(6) The interface temperature should be 300 °C.
(7) The ionization voltage should be 70 eV.
(8) Every extract should be analyzed in duplicate.

The GC conditions can vary depending on the apparatus used and with respect to optimization of chromatographic separation (Isidorov, Szczepaniak, & Bakier, 2014).

### 3.3.2. *GC-MS analysis of propolis volatile constituents*

The GC–MS analysis should be performed with a proper instrument such as a Hewlett–Packard gas chromatograph 5890 series II Plus linked to a Hewlett–Packard 5972 mass spectrometer system (Bankova et al., 1998).

(1) Use a 30 m long, 0.25 mm ID, and 0.25 μm film thickness SPB-1 capillary column. Other columns with similar characteristics can be also used depends on analytical needs.
(2) Program the temperature from 40 to 280 °C at a rate of 6 °C/min.
(3) Helium is used as a carrier gas at a flow rate of 0.8 ml/min.
(4) The split ratio should be 1:10.
(5) The injector temperature should be 280 °C.
(6) The interface temperature should be 300 °C.
(7) The ionization voltage should be 70 eV. Every extract should be analyzed in duplicate.

The GC conditions can vary depending on the apparatus used and with respect to optimization of chromatographic separation (Cheng, Qin, Guo, Hu, & Wu, 2013; Kaškonienė, Kaškonas, Maruška, & Kubilienė, 2014; Nunes & Guerreiro, 2012).

### 3.3.3. *Identification and quantification of compounds*

The identification of individual compounds (such as trimethylsilyl derivatives) can be performed using computer searches on commercial libraries (such as NIST 14, Wiley 10, etc.), comparison with spectra and retention characteristics of authentic samples, and literature data. If no reference spectra are available, identification can be performed based on the characteristic mass-spectral fragmentation, in such cases the compounds are described as "tentative structures".

The quantification of individual constituents is based on internal normalization. This is a general approach used in cases where it is impossible to use other methods such as the internal standard method. The internal normalization method is based on the assumption that all detector response factors are unity, and the following equation should be applied:

$$\%\text{Analyte} = \frac{A_a}{\sum A_i} \times 100$$

where $\Sigma A_i$ is the sum of all the peak areas in the chromatogram. Thus, the percentage of the individual compounds refers to percent of the Total Ion Current (TIC), and the result should not be considered as quantitative in absolute terms (IOFI Working Group on Methods of Analysis, 2011).

## 3.4. *LC-MS chemical profiling of propolis*

### 3.4.1. *Introduction*

The relatively polar nature of propolis constituents (with several hydroxyl groups in their structure), combined with soft ionization techniques compatible with liquid chromatography, make HPLC-DAD and LC-MS the favorite methods for analysis of propolis balsamic content (Sforcin & Bankova, 2011). In the structural identification of new compounds, both mass spectrometry with electrospray ionization (ESI-MS) in the negative (Falcão et al., 2010) or positive ion mode (Piccinelli et al., 2011) studies are satisfactory.

High performance liquid chromatography (HPLC) was and still is the preferred separation technique for the analysis of natural products (Steinmann & Ganzera, 2011). Recent developments of new stationary phases and pumping devices enabling pressures up to 1300 bar are further supporting this trend (Steinmann & Ganzera, 2011). Different detectors can be used, depending on the analytes investigated. The most commonly used detectors for analyzing propolis are DAD and MS detectors.

### 3.4.2. *Separation and analysis of propolis by liquid chromatography-mass spectrometry (LC-MS)*

The use of LC–MS for the qualitative and quantitative analysis of constituents in propolis has increased steadily over the last years.

(1) The extraction of propolis is performed as described in Section 3.1.1.
(2) Dissolve the dry ethanolic extract (10 mg) in 1 ml of 80% of ethanol.
(3) Filter the sample through a 0.2 μm Nylon membrane (Whatman)
(4) Injected 10 μl of the solution into the chromatograph.

The following sub-Section describes in detail the parameters for LC and MS that could be applied for the analysis of propolis.

*3.4.2.1. LC parameters.* HPLC separation is largely dependent on the different affinities between the propolis compounds and the stationary phase. For a particular application, the chemical properties of the packing and physical properties of the column (e.g. particle size and column dimensions) need to be taken into account.

Reversed phase HPLC is doubtlessly the most widely used chromatographic method in propolis analysis (Falcão et al., 2010; Gardana, Scaglianti, Pietta, & Simonetti, 2007; Pellati, Orlandini, Pinetti, & Benvenuti, 2011; Piccinelli et al., 2011; Righi, Negri, & Salatino, 2013; Volpi & Bergonzini, 2006). Most appropriate are octadecylsilane columns (ODS or C18). Nucleosil C18 250 × 4 mm ID, 5μm particle diameter (Falcão et al., 2010); Luna C18 column 150 × 2.0 mm ID, 5 μm (Piccinelli et al., 2011); and CLC-ODS 150 × 6.0 mm ID (Midorikawa et al., 2001) can also give good results. Due to the complex nature of the matrix, a drawback for the use of these columns is the long runs needed, frequently above 50 min per run.

A fast and ultra-fast separation can be achieved with columns packed with sub-2 μm particles operating at ultra-high pressure systems. Ultra-high-performance liquid chromatography (UHPLC) is quite versatile and can be used to increase throughput, particularly suitable for the analysis of complex samples such as plant extracts or their metabolites (Nicoli et al., 2005). Recent work has been performed with propolis in equivalent columns of Waters BEH C18 (50 mm × 2.1 mm ID × 1.7 μm particle size) reducing the time run to 12 min (Novak et al., 2014).

The chromatographic conditions of the HPLC methods include, almost exclusively, the use of UV–Vis diode array detector (DAD) with spectral data for all peaks acquired in the range of 200–600 nm, although 280 nm is the most generic wavelength for phenolic compounds due to the high molar absorptivity of the different phenolic classes at that wavelength.

The eluent is composed of a binary solvent system containing acidified water (solvent A) combined with a polar organic solvent (solvent B). Gradient elution has usually been mandatory in recognition of the complexity of the propolis chemical profile. 0.1% formic or acetic acid can be added to water (as solvent A) and acetonitrile or methanol (as solvent B) are commonly used in propolis analysis. 0.1% formic acid is the most suitable when using a MS detector. The flow rate is dependent on the type of column used, but for the above parameters it is recommended to be 1 ml min$^{-1}$. Temperature control of the column should also be considered to achieve a better peak separation, between 25 and 40 °C, with 30 °C being the most suitable for propolis compound separation. For a flow rate of 1 ml min$^{-1}$, a post-column split of 0.2 ml min$^{-1}$ to MS should be applied (Falcão et al., 2013a).

Table 1 presents the guidelines needed to achieve a good separation and analysis of the phenolic compounds present in propolis.

*3.4.2.2. MS parameters.* Given the unique characteristics of different mass spectrometers, it is critical to choose the suitable MS parameters. Table 1 summarizes the best conditions for the MS analysis of propolis phenolic compounds.

The ion source used should be electron-spray ionization (ESI). ESI is a soft ionization technique for a wide range of compounds (slight fragmentation but adducts are often observed), where ionization is achieved by applying a high electric charge to the sample needle, with voltage between 3 and 5 kV and the capillary temperature between 300 and 350 °C. ESI can be operated in the negative or positive full scan ion mode, although, and concerning the phenolic compounds, a higher sensitivity and better fragmentations can be achieved with the negative ion, thus resulting in more structural information (Cuyckens & Claeys, 2004). A more recent development is atmospheric pressure photoionization (APPI). If the compounds are poorly ionized by ESI and APCI, APPI should be considered as an alternative (Ignat, Volf, & Popa, 2011).

Concerning the mass analyzers, the ion trap is the one most recommended for the profiling of propolis composition since it is specially designed for multiple fragmentation steps (MS$^n$). Regarding target analysis, a tandem-MS detection over a single-stage MS operation is recommended because of the much better selectivity and the wider-ranging information that can be obtained (de Rijke et al., 2006). In linear ion traps, ions are isolated and accumulated due to a special arrangement of hyperbolic and ring shaped electrodes as well as oscillating electric fields. Then the ions can be fragmented by collision-induced decomposition (CID) (Ignat et al., 2011). The MS$^n$ data is simultaneously acquired for the selected precursor ion. The collision induced decomposition (CID)–MS–MS and MS$^n$ experiments should be performed using helium as the collision gas, with collision energy (CE) of 20–40 eV. The CE is dependent on the molecule stability under study. In the negative ion mode, collision energies of 20 eV for phenolic acids and 20–40 eV for flavonoids are suitable (Pellati et al., 2011).

### *3.4.3. Identification of phenolic compounds*

Propolis chemical composition is a rich pool of phenolic compounds. Those, often referred to as polyphenols, embody a class of widely distributed and chemically diverse secondary metabolites synthesized in plants at different developmental stages (Steinmann & Ganzera, 2011). Polyphenols possess at least one aromatic ring with one or more hydroxyl functional groups. Flavonoids,

Table 1. Experimental guidelines for the propolis LC-MS analysis.

| LC parameters | MS parameters |
|---|---|
| *Column*<br>Reversed-phase HPLC octadecylsilane (ODS or C18) with standard measures 250 mm × 4 mm ID, 5 μm particle diameter<br>UHPLC C18 alternative: 50mm × 2.1mm, 1.7 μm particle diameter | *Ionization technique*<br>Electron-spray ionization (ESI) in the negative ion mode<br><br>Capillary voltage: 3–5 kV<br>Capillary temperature: 300–350 °C |
| *Column temperature*<br>30 °C | *Mass analyzer*<br>Ion-trap<br>The collision induced decomposition (CID) –MS–MS and $MS^n$ experiments should be performed using helium as the collision gas, with a collision energy of 20 eV for phenolic acids and between 20–40 eV for flavonoids |
| *Eluents*<br>Mobile phases comprising solvent (A) 0.1% formic acid in water and solvent (B) acetonitrile with 0.1% of formic acid, previously degassed and filtered<br>*Solvent gradient*<br>Start with 80% A and 20% B, reaching 30% B at 10 min, 40% B at 40 min, 60% B at 60 min, 90% B at 80 min, followed by the return to the initial conditions | |
| *Flow*<br>1 ml/min | |
| *Detection*<br>UV–vis DAD detection in the range 200–600 nm, with 280 being the most common wavelength used in the study of phenolic compounds | |

Table 2. Propolis compounds characterized by LC-DAD-MS.

| Compounds | $\lambda_{max}$ (nm) | m/z (ESI polarity) | $MS^2$ (% base peak) | Reference |
|---|---|---|---|---|
| *Phenolic acids* | | | | |
| Quinic acid | | 179 (+) | 143 | (a) |
| Chlorogenic acid | 325 | 353 (−) | 179, 135, 191 | (b) |
| Caffeic acid | 292, 322 | 179 (−) | 135 | (c) |
| Dicaffeoylquinic acid | 325 | 515 (−) | 179, 135, 191 | (b) |
| Ellagic acid | 253, 367 | 301 (−) | 301 (100), 257 (77), 229 (96) | (d) |
| *p*-Coumaric acid | 310 | 163 (−) | 119 | (c) |
| Ferulic acid | 295sh, 322 | 193 (−) | 177 (16), 149 (47), 133 (100) | (c) |
| Dicaffeoylquinic acid | 325 | 515 (−) | 179, 135, 191 | (b) |
| Isoferulic acid | 298, 319 | 193 (−) | 177 (16), 149 (47), 133 (100) | (c) |
| Tricaffeoylquinic acid | 325 | 677 (−) | 179, 135, 191 | (b) |
| Benzoic acid | 229 | 121 (−) | | (d) |
| 3,4-Dimethyl-caffeic acid | 295sh, 322 | 207 (−) | 163 (60), 102 (100) | (c) |
| Cinnamic acid | 277 | 147 (−) | 103 | (c) |
| *p*-Coumaric acid methyl ester | 307 | 177 (−) | 163 (100), 119 (15) | (d) |
| Cinnamylidenacetic acid | 310 | 173 (−) | 129 | (d) |
| Drupanin (3-prenyl-*p*-coumaric acid) | 311 | 232 (−) | 187, 133 | (b) |
| Caffeic acid isoprenyl ester | 298, 325 | 247 (−) | 179 (100), 135 (15) | (d) |
| Caffeic acid isoprenyl ester (isomer) | 298, 325 | 247 (−) | 179 (100), 135 (15) | (d) |
| Caffeic acid benzyl ester | 298, 325 | 269 (−) | 178 (100), 134 (32), 161 (12) | (d) |
| Caffeic acid phenylethyl ester | 295, 325 | 283 (−) | 179 (100), 135 (28) | (d) |
| *p*-Coumaric acid isoprenyl ester | 294, 310 | 231 (−) | 163 (100), 119(12) | (d) |
| *p*-Coumaric acid benzyl ester | 298, 312 | 253 (−) | 162, 145, 118 | (e) |
| *p*-Coumaric acid isoprenyl ester (isomer) | 294, 310 | 231 (−) | 163 (100), 119 (12) | (d) |
| Caffeic acid cinnamyl ester | 295, 324 | 295 (−) | 178 (100), 134 (24) | (d) |
| Caffeic acid cinnamyl ester (isomer) | 295, 324 | 295 (−) | 178 (100), 134 (24) | (d) |
| *p*-Coumaric acid cinnamyl ester | 296, 310 | 279 (−) | 162, 118 | (e) |
| Artepillin C | 311 | 299 (−) | 255, 163, 151, 107 | (b) |
| 3-Prenyl-4-(2-methylpropionyl-oxy)-cinnamic acid | 279.5 | 315 (−) | 271 | (b) |
| 3-(2,2-Dimethyl-3,4-dehydro-8-prenyl-1-benzopyran-6-yl-propenoic acid | 310 | 297 (−) | 253, 149 | (b) |
| 3-Prenyl-4-(dihydrocinnamoyloxi)-cinnamic acid | 279.5 | 363 (−) | 319, 187, 149, 131 | (b) |
| *p*-Methoxi cinnamic acid cinnamyl ester | 279 | 293 (−) | 177, 133 | (b) |
| *p*-Coumaric acid-4-hydroxyphenylethyl ester dimer | 289, 345 | 565 (−) | 455 (10), 417 (36), 283 (100), 269 (43) | (b) |
| *Di-hidroflavonols* | | | | |
| Pinobanksin-5-methyl-ether | 286 | 285 (−) | 267 (100), 252 (13), 239 (27) | (d) |
| Pinobanksin-5-methyl-ether-3-*O*-acetate | 289 | 327 (−) | 285 (100), 267 (18), 239 (31) | (d) |
| Pinobanksin | 292 | 271 (−) | 253 (100), 225 (26), 151 (10) | (d) |
| Pinobanksin-5,7-dimethyl-ether | 292 | 299 (−) | 285, 253, 139 | (b) |
| Pinobanksin-3-*O*-acetate | 292 | 292 (−) | 271 (18), 253 (100) | (d) |
| Pinobanksin-3-*O*-acetate-5-*O*-*p*-hydroxyphenylpropionate | 292 | 292 (−) | 443 (68), 401 (75), 351 (100), 291 (55), 253 (2) | (d) |
| Pinobanksin-3-*O*-propionate | 289 | 289 (−) | 271 (9), 253 (100) | (d) |
| Pinobanksin-5-methyl-ether-3-*O*-pentanoate | 289 | 289 (−) | 285 (53), 267 (65), 239 (100) | (d) |
| Pinobanksin-7-methyl-ether-5-*O*-*p*-hydroxyphenylpropionate | 292 | 292 (−) | 433 (9), 415 (100), 400 (8), 253 (<1) | (d) |
| Pinobanksin-3-*O*-butyrate or isobutyrate | 292 | 292 (−) | 271 (5), 253 (100) | (d) |
| Pinobanksin-3-*O*-pentenoate | 292 | 292 (−) | 271 (7), 253 (100) | (d) |
| Pinobanksin-3-*O*-pentanoate or 2-methylbutyrate | 292 | 292 (−) | 271 (5), 253 (100) | (d) |
| Pinobanksin-*O*-hexenoate | 292 | 292 (−) | 271 (100), 253 (45) | (d) |
| Pinobanksin-3-*O*-phenylpropionate | 292 | 292 (−) | 271 (16), 253 (100) | (d) |
| Pinobanksin-3-*O*-hexanoate | 292 | 292 (−) | 271 (14), 253 (100) | (d) |
| *Flavonols* | | | | |
| Quercetin | 256, 370 | 301 (−) | 179 (100), 151 (60) | (d) |
| Quercetin-3-methyl-ether | 256, 355 | 315 (−) | 300 | (d) |
| Kaempferol | 265, 364 | 285 (−) | 285 (100), 257 (13), 151 (20) | (d) |
| Isorhamnetin | 253, 370 | 315 (−) | 300 | (d) |
| Kaempferol-methyl-ether | 265, 352 | 299 (−) | 284 | (d) |
| Kaempferol-methoxy-methyl-ether | 265, 340 | 329 (−) | 314 | (d) |

(*Continued*)

Table 2. (*Continued*).

| Compounds | $\lambda_{max}$ (nm) | m/z (ESI polarity) | $MS^2$ (% base peak) | Reference |
|---|---|---|---|---|
| Quercetin-dimethyl-ether | 253, 355 | 329 (−) | 314 | (d) |
| Quercetin-tetramethyl-ether | 256, 349 | 359 (−) | 344 | (d) |
| Galangin-5-methyl-ether | 265, 300sh, 352 | 283 (−) | 268 (100), 239 (60), 211 (10) | (d) |
| Rhamnetin | 256, 367 | 315 (−) | 300 (34), 193 (76), 165 (100) | (d) |
| Quercetin-dimethyl-ether | 256, 355 | 329 (−) | 314 | (d) |
| Galangin | 265, 300sh, 358 | 269 (−) | 269 (100), 241 (61), 227 (20), 197 (22), 151 (20) | (d) |
| Kaempferide | 265, 364 | 299 (−) | 284, 151 (<1) | (d) |
| Kaempferol-dimethyl-ether | 265, 346 | 313 (−) | 299 (10), 298 (100) | (d) |
| Myricetin-3,7,4′,5′-tetramethyl-ether | | 375 (+) | 360, 345, 315 | (f) |
| *Flavonol glycosides* | | | | |
| Quercetin-3-*O*-rutinoside | 256, 352 | 609 (−) | 301(100), 300 (87) | (d) |
| Quercetin-3-*O*-glucuronide | 256, 355 | 477 (−) | 301 | (d) |
| Quercetin-3-*O*-glucoside | 256, 355 | 463 (−) | 301(100), 300 (64) | (d) |
| Kaempferol-3-*O*-rutinoside | 265, 349 | 593 (−) | 285 | (d) |
| Isorhamnetin-3-*O*-rutinoside | 253, 355 | 623 (−) | 315 (100), 300 (22) | (d) |
| Isorhamnetin-*O*-pentoside | 253, 346 | 447 (−) | 315 (100), 300 (8) | (d) |
| Quercetin-3-*O*-rhamnoside | 256, 349 | 447 (−) | 301(100), 300 (47) | (d) |
| Isorhamnetin-*O*-glucuronide | 253, 346 | 491 (−) | 315 | (d) |
| Kaempferol-methyl-ether-*O*-glucoside | 265, 343 | 461 (−) | 446 (91), 299 (100), 284 (11) | (d) |
| Isorhamnetin-*O*-acetylrutinoside | 253, 352 | 665 (−) | 623 (18), 315 (100), 300 (14) | (d) |
| Rhamnetin-*O*-glucuronide | 256, 349 | 491 (−) | 315 | (d) |
| Quercetin-dimethyl-ether-*O*-rutinoside | 253, 349 | 637 (−) | 329 (100), 314 (18) | (d) |
| Quercetin-dimethyl-ether-*O*-glucuronide | 253, 349 | 505 (−) | 329 (100), 314 (18) | (d) |
| Kaempferol-*O*-*p*-coumaroylrhamnoside | 265, 322 | 577 (−) | 431 (6), 285 (100) | (d) |
| *Flavones* | | | | |
| Luteolin | 253, 268sh,349 | 285 (−) | 285 (100), 267 (54), 241 (63), 175 (52) | (d) |
| Apigenin | 268, 337 | 269 (−) | 225 (100), 151 (29) | (d) |
| Luteolin-5-methyl-ether | 266, 350 | 299 (−) | 284, 256, 151 | (d) |
| Chrysin-5-methyl-ether | 268, 313 | 267 (−) | 253 (100), 224 (25) | (d) |
| Chrysin | 268, 313 | 268, 313 (−) | 225 (17), 209 (100), 151 (5) | (d) |
| Acacetin | 268, 331 | 268, 331 (−) | 269 | (d) |
| 6-Methoxychrysin | 265, 300sh, 350sh | 283 (−) | 269 | (d) |
| Chrysoeriol-methyl-ether | 250, 268sh, 343 | 313 (−) | 298 | (d) |
| Chrysin-5,7-dimethyl-ether | 265, 311sh | 281 (−) | 267, 165 | (b) |
| *Flavanones* | | | | |
| Pinocembrin-5-methyl-ether | 286 | 269 (−) | 255 (48), 227 (100), 165 (30) | (d) |
| Liquiritigenin | 280, 310 | 257 (+) | 137 (62), 147 (72), 211 (19), 239 (100), 242 (36) | (g) |
| Pinocembrin | 289 | 255 (−) | 213 (100), 211 (32), 151 (48) | (d) |
| Naringenin | 289 | 271 (+) | 153 (100), 149 (100) | (g) |
| Pinocembrin-5-*O*-3-hydroxy-4-methoxyphenylpropionate | 295 | 295 (−) | 415 (3), 401 (31), 323 (15), 309 (100) | (d) |
| 3-Hydroxy-5-methoxyflavanone | 289 | 289 (−) | 254 (100), 251 (54), 165 (22) | (d) |
| *Chalcone* | | | | |
| Isoliquiritigenin | 309, 372sh | 257 (+) | 242 (34), 239 (100), 171 (2), 147 (78), 137 (69) | (g) |
| Dimethylkuraridin | | 425 (+) | 285 | (a) |
| *Isoflavonoids* | | | | |
| Formononetin | 248, 302 | 269 (+) | 254 (100), 237 (39), 213 (35) | (g) |
| Biochanin A | 362, 326sh | 285 (+) | 270 (51), 257 (11), 253 (22), 229 (19) | (g) |
| Vestitol | 280 | 273 (+) | 137 (100), 123 (74) | (g) |
| Neovestitol | 280 | 273 (+) | 137 (100), 123 (70) | (g) |

(*Continued*)

Table 2. (*Continued*).

| Compounds | $\lambda_{max}$ (nm) | *m/z* (ESI polarity) | $MS^2$ (% base peak) | Reference |
|---|---|---|---|---|
| 7-*O*-methylvestitol | | 287 (+) | 163 (10), 137 (100) | (g) |
| Mucronulatol | 280, 340 | 303 (+) | 167 (100), 149 (19), 123 (23) | (g) |
| 7,3′-Dihydroxy-5′-methoxi-isoflavone | 295 | 285 (+) | 270 (100), 253 (55), 225 (18) | (g) |
| Retusapurpurin B | 285, 470 | 523 (+) | 399 (61), 387 (100), 385 (53) | (g) |
| Retusapurpurin A | 285, 480 | 523 (+) | 399 (61), 387 (100), 385 (59) | (g) |
| *Pterocarpans* | | | | |
| Medicarpin | 290 | 271 (+) | 161(44), 137 (100) | (g) |
| Homopterocarpin | | 285 (+) | 137 (100), 161 (51), 137 (100) | (g) |
| Vesticarpan | | 287 (+) | 153 (100), 177 (19) | (g) |
| 3,8-dihydroxy-9-methoxy-pterocarpan | | 287 (+) | 269 (36), 255 (40), 177 (100), 153 (59) | (g) |
| 3,4-dihydroxy-9-methoxy-pterocarpan | | 287 (+) | 161 (23), 139 (100), 137 (55) | (g) |
| 3-dihydroxy-8,9-dimethoxy-pterocarpan | | 301 (+) | 191 (100), 167 (87), 153 (13) | (g) |
| *Polyisoprenylated benzophenones* | | | | |
| Nemorosone | | 501 (−) | 432 | (a) |
| Guttiferone E/xanthochymol | 250, 355 | 603 (+) | 467 (85), 411 (25), 343 (21) | (g) |
| Oblongifolin A | 250, 355 | 603 (+) | 467 (41), 411 (8), 399 (32), 343 (24) | (g) |
| Prenylated benzophenone | | 407 (−) | 338 | (a) |
| *Diterpenes* | | | | |
| Cupressic acid | | $(M-H_2O+H)^+$: 303 (+) | 285, 257, 247 | (f) |
| Isocupressic acid | | $(M-H_2O+H)^+$: 303 (+) | 257, 247, 193 | (f) |
| Imbricatoloic acid | | 323 (+) | 305, 287, 277, 259, 181 | (f) |
| Torulosal | | $(M-H_2O+H)^+$: 287 (+) | 269, 259, 177, 163 | (f) |
| Isogathotal | | $(M-H_2O+H)^+$: 287 (+) | 269, 259, 163, 149 | (f) |
| Torulosol | | $(M-H_2O+H)^+$: 289 (+) | 271, 243, 233, 215, 193, 179 | (f) |
| Agathodiol | | $(M-H_2O+H)^+$: 289 (+) | 271, 243, 231, 215, 193, 179 | (f) |
| Cistadiol | | $(M-H_2O+H)^+$: 291 (+) | 273, 235, 221, 209, 181, 163 | (f) |
| 18-Hydroxy-cis-clerodan-3-ene-15-oic acid | | $(M-H_2O+H)^+$: 305 (+) | 287, 269, 235, 223, 195, 177 | (f) |

Notes: (a) Zhang et al. (2014); (b) Gardana et al.(2007); (c) Falcão et al. (2010); (d) Falcão et al. (2013a); (e) Pellati et al. (2011); (f) Piccinelli et al. (2013); (g) Piccinelli et al. (2011).

whose structures are based on a C6-C3-C6 skeleton, are the most abundant group of phenolic compounds, and are sub-divided into several classes differing in the oxidation state of the central heterocyclic ring (Veitch & Grayer, 2008). These comprise chalcones, flavones, flavonols, flavanones, isoflavonoids, anthocyanidins and flavanols (catechins and tannins). Non-flavonoids comprise simple phenols, phenolic acids, coumarins, xanthones, stilbenes, lignins and lignans. Phenolic acids are further divided into benzoic acid derivatives, based on a C6-C1 skeleton, and cinnamic acid derivatives, which are based on a C6-C3 skeleton (Veitch & Grayer, 2008). The variability of propolis chemical composition contains large numbers of phenolics from different classes including, unexpectedly, glycoside phenolic compounds, clearly highlighting the challenges associated with their analysis.

The structural elucidation of different classes of propolis compounds is achieved by comparing their chromatographic behavior, UV spectra and MS information, to those of reference compounds. When standards are not available, the identity of the compounds can be achieved through comparison of the product ion spectra and retention times with pure compounds isolated from propolis or, alternatively, combining UV data with MS fragmentation patterns previously reported in the literature (Falcão et al., 2013a). Table 2 shows the UV data and MS fragmentation of many compounds described in the literature as propolis constituents. Only compounds with all the information regarding MS fragmentation are present.

Fragmentation patterns are specific for a given compound or class of compounds. For example, for the negative ion mode, phenolic acids demonstrated a common fragmentation pattern, with a loss of the carboxyl group ($CO_2$, −44 Da) (Falcão et al., 2010). In the case of flavonoids, the distinct flavonoids classes differ in their pattern of substitution, which strongly influences the fragment pathway, the interpretation of MS/MS data provides specific structural information about the type of molecules. The $MS^2$ spectrum of many of these flavonoids (Table 2) revealed the fragments at *m/z* 151 or at *m/z* 165, which are resultant from the retro Diels-Alder mechanism (Cuyckens & Claeys, 2004). Also, neutral losses commonly described to occur in these compounds, such as the small molecules CO (−28 Da), $CO_2$ (−44 Da), $C_2H_2O$ (−42 Da), as well as the successive losses of these molecules, were also observed (Cuyckens & Claeys, 2004). In accordance with Cuyckens and Claeys (2004), methylated flavonoids presented a significant $[M\text{-}H\text{-}CH_3]^{-\bullet}$ product ion.

Attention has to be taken to experimental conditions used, such as the type of ion source and mass analyzer, when comparing literature data, since different fragments can be found when different experimental set-up and/or operating conditions are applied. The mass spectra of flavonoids obtained with quadrupole and ion-trap instruments typically are closely similar, even though relative abundances of fragment ions and adducts do show differences. Therefore, direct comparison of spectra obtained with these two instruments is allowed. The main advantage of an ion-trap instrument is the possibility to perform $MS^n$ experiments (Steinmann & Ganzera, 2011).

#### 3.4.4. *Concluding remarks*

LC-MS is a powerful tool that can be used to overcome the difficult task of propolis chemical profiling, due to the high diversity of the resin floral sources collected by honey bees. To enhance the amount of structural information given by the technique, the most important features to be considered in LC-MS propolis chemical profiling are to:

(1) Chose the right LC parameters for the analysis such as a reversed-phase C18 HPLC column, which is the most selective in propolis analysis (Section 3.4.2.1).
(2) Use mobile phases comprising (A) 0.1% formic acid or acetic acid in water and (B) acetonitrile or methanol, (Section 3.4.2.1).
(3) Acquire spectral data with the UV–Vis DAD set at 280 nm, which is the most generic wavelength for phenolic compounds identification (Section 3.4.3).
(4) Use a ESI source and a ion trap mass analyzer, with helium as the collision gas, with CE of 20–40 eV (Section 3.4.2.2).
(5) Compare the UV spectra and MS information to those of reference compounds. If standards are not available, the identity of the compounds can be achieved through comparison of the product ion spectra and retention times with pure compounds isolated from propolis or combining UV data with MS fragmentation patterns previously reported in the literature (Section 3.4.3).

The fast technical evolution of the LC-MS systems, particularly in respect to the mass analyzers, will continue to allow new findings within the chemical composition of propolis.

### 3.5. *Mass spectrometry fingerprinting of propolis*

MS fingerprinting is a qualitative analytical tool used to discern between different types of propolis and to compare the composition of propolis samples to those of plant resins. MS fingerprints are proposed as characteristic of the composition of samples and can be used as a guide for their therapeutic uses. The method used in one study (Sawaya et al., 2004) was only slightly modified in the subsequent applications and can be considered as the standard method for propolis extraction for MS fingerprinting.

#### 3.5.1. *Electrospray ionization mass spectrometry (ESI-MS) fingerprinting of propolis samples*

(1) Extract propolis as described in Section 3.1.1.
(2) Evaporate the solvent (ethanol) on a water bath at a temperature of 50 °C to obtain dry extracts of propolis.
(3) Dissolve these dry extracts in a 70% (v/v) methanol/water solution, containing 50 ng of dry propolis extract per ml of methanolic solution and 5 μl of ammonium hydroxide.
(4) Infuse these solutions directly into the ESI-source of a hybrid high resolution and high-accuracy (5 ppm) Micromass Q-TOF mass spectrometer, via a syringe pump (Harvard Apparatus) at a flow rate of 15 μl/min. The MS conditions should be capillary −3.0 kV, cone 30 V.

Due to the prevalence of acid compounds, the negative ion mode fingerprints result in the clearest discrimination between the groups of propolis samples. This pattern was confirmed by subsequent studies of propolis fingerprinting conducted by Sawaya, da Silva, Cunha, and Marcucci (2011).

A simple chemometric evaluation is applied with Principal Component Analysis (PCA) performed using the 2.60 version of Pirouette software (Infometrix, Woodinville, WA, USA) (see the *BEEBOOK* manuscript on statistical guidelines for more information on using PCA, Pirk et al., 2013). Only the two most characteristic negative ion markers of each sample are selected and expressed as the intensities of these individual ions (variables). The data are preprocessed using auto scale and analyzed using PCA.

Samples are grouped according to their geographic origin (Sawaya et al., 2004). Furthermore, tandem mass

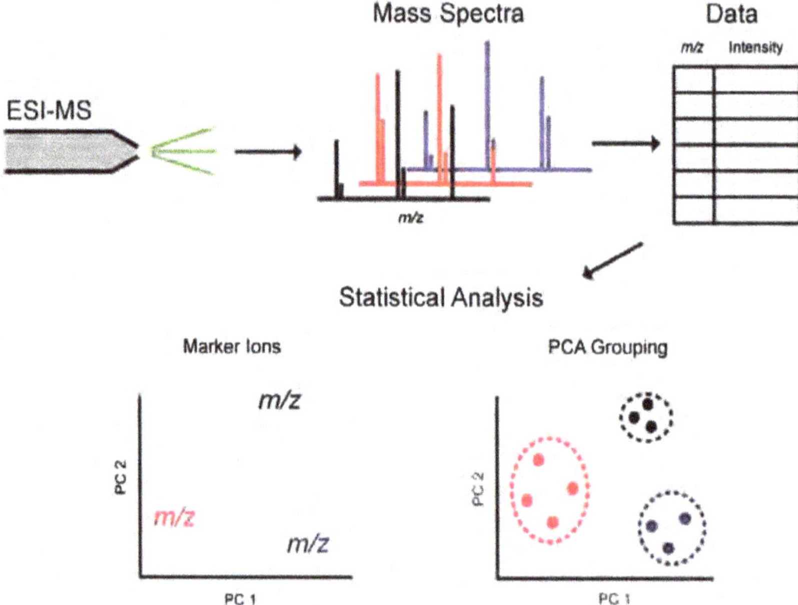


Figure 4. Genaral process used in ESI-MS fingerprinting studies: ionization and anlaysis by ESI-MS, extraction of the *m/z* and intensity of selected ions, statistical analysis of the data via PCA to group samples and indicate the marker ions for each group.

spectrometry with collision induced dissociation (CID) allowed on-line structural identification of certain marker ions such as dicaffeoylquinic acid, 3,5-Diprenyl-4-hydroxycinnamic acid, Pinocembrin, Chrysin, 3-Prenyl-4-hydroxycinnamic acid, 2,2-Dimethyl-6-carboxyethenyl-2H-1-benzopyran and *p*-Coumaric acid (Sawaya et al., 2004). The general flow of these ESI-MS fingerprinting studies is shown in Figure 4.

Using the same extraction and analysis procedures, propolis samples can be compared to the plant sources of their resins. This could allow one to link the resin producing source plant to the propolis from these regions (Marcucci, Sawaya, Custodio, Paulino, & Eberlin, 2008).

#### 3.5.2. *Concluding remarks*

MS fingerprinting may be applied to propolis samples to characterize their composition, identify the plant sources, and indicate their potential therapeutic application. Besides ESI, a new ionization source, named easy ambient sonic ionization (EASI), has been used for this purpose as well (Sawaya et al., 2010). The use of chemometric methods such as PCA to analyze the results is frequently necessary due to the large number of ions observed in each spectrum. The results of the analyses are capable of grouping similar samples, indicating their marker ions and, in some cases, correlating with the biological activity of samples.

### 3.6. *NMR analysis of propolis*

#### 3.6.1. *Introduction*

Since its discovery, the phenomenon of Nuclear Magnetic Resonance (NMR) has been widely exploited as a research tool in analytical laboratories throughout the world. NMR spectroscopy is used to study the structure of molecules (Kwan & Huang, 2008). It also is well known that NMR can be used to analyze complex mixtures such as herbal extracts, foods, biological fluids, etc. (Forseth & Schroeder, 2011). In particular, NMR is used increasingly in the evaluation of food and in the quality assurance of natural products, although all its potential has not been fully exploited. The amount of information available in an NMR spectrum and the ease of sample preparation make this spectroscopic technique very attractive for the assessment of product quality.

One of the main advantages of this technique over that of other methods is its ability to furnish structural and quantitative information on a wide range of chemical species in a single NMR experiment. The mixture analysis by NMR is complex, but potentially very informative (Lin & Shapiro, 1997).

In recent years, the use of much higher magnetic fields and the greater sensitivity and spectral resolution that they bring, have stimulated interest in 1D and 2D NMR spectroscopy as a routine method for the analysis of complex mixtures (Charlton, Farrington, & Brereton, 2002; Fan, 1996).

There are two main strategies for analyzing mixtures via NMR: (a) separate components of the mixture prior to NMR analysis; and (b) analyze the mixture as it is. The first strategy is used when the goal of the work is the characterization of an isolated compound and it is not the subject of this discussion. The second strategy allows one to obtain an overall image of the mixture in question, without any further type of pre-treatment of the sample, except the eventual solubilization in a suitable deuterated solvent. The obtained spectra will be considered as chemical fingerprints of the product under investigation. In this case, the analysis of the spectra, that usually appear very complex, requires tools for the pre-treatment of the signal and for the analysis of the results, normally based on multivariate statistical techniques (Papotti, Bertelli, Plessi, & Rossi, 2010).

### 3.6.2. *Sample preparation*

Since propolis is a solid material, it requires an initial extraction procedure using 70% ethanol (see Section 3.1.1). Obviously, if the extract is analyzed as is, very intense signals related to the solvent will be present in the obtained spectra. To avoid this problem, it is preferable to eliminate the solvents under a light nitrogen stream operating at low temperature. This procedure can be conducted directly in NMR tubes, and by dissolving the solid residue in an appropriate volume of the selected deuterated solvents. The most important thing to remember when choosing the most suitable solvent is that if $D_2O$ is chosen, all the signals relating to alcoholic, phenolic or carboxylic hydroxyls, that are very abundant in propolis, will be lost in the spectrum. If one is interested to observe the signals related to these functional groups, a solvent that does not exchange deuterium with hydroxyls should be used. The most suitable in the case of propolis is the DMSO-$d_6$ (deuterated dimetyl sulfoxyde) (Papotti et al., 2010). There is no ideal ratio of propolis extract and the amount of solvent used; each one must find what works best in each case.

(1) Transfer 1ml of propolis extract (see 3.1.) to an NMR tube and evaporate to dryness at room temperature using a flow of nitrogen gas.
(2) Dissolve the dry residue in 0.5 ml of methyl sulphoxide-d6 (DMSOd6).
(3) Add 20 μl of tetramethylsilane (TMS) as a reference compound.
(4) Use the sample immediately for NMR experiments.

### 3.6.3. *NMR analysis of propolis*

A typical $^1H$ NMR spectrum of a propolis hydroalcoholic extract in DMSO-$d_6$ is reported in Figure 5. There are a high number of signals present in many spectral regions, particularly in the area between 1 and 8 ppm.

For this reason, a simple interpretation of this kind of spectra is rarely possible. Nevertheless, a preliminary assignment of the principal signals is often necessary to permit a correct interpretation of the results. The assignments can be performed using data obtained from one-dimensional NMR experiments and comparing them with literature data or with data obtained from pure standard compounds. The final correct assignment can be obtained using the most informative two-dimensional experiments such as COSY, HSQC and HMBC. In Table 3, the assignments of some well-known propolis components are reported. An example of this application is reported in Bertelli, Papotti, Bortolotti, Marcazzan, and Plessi (2012).

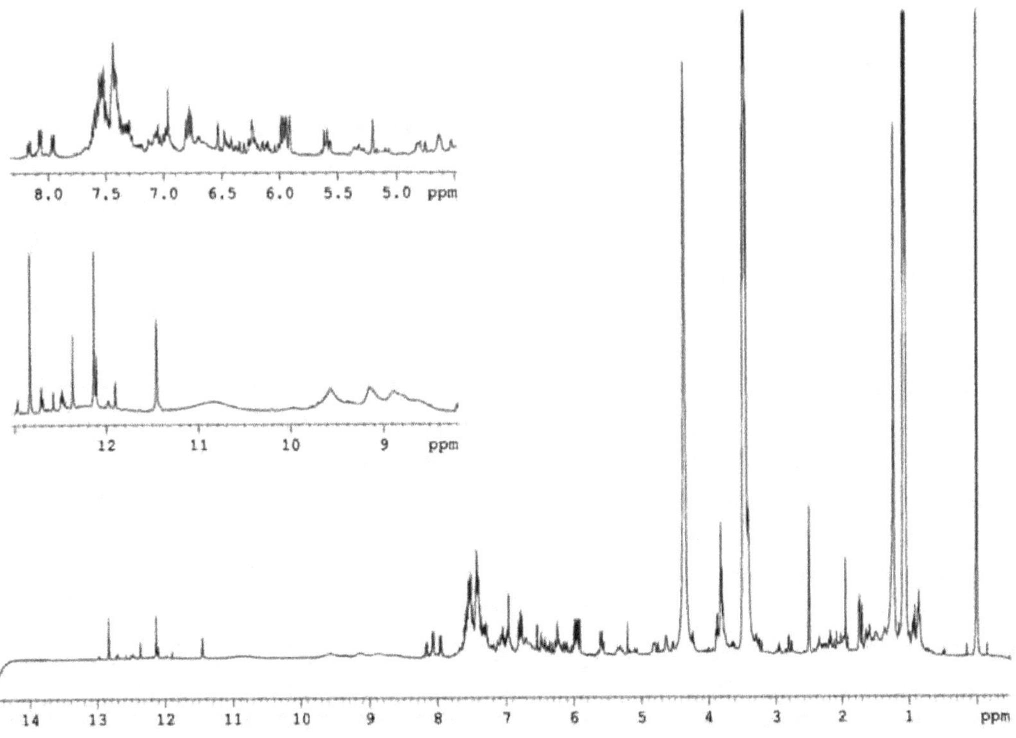


Figure 5. Typical $^1H$ NMR spectrum of propolis extracts in DMSO-d6 (Papotti et al., 2010).

Table 3. $^{1}H$ and $^{13}C$ NMR[a] chemical shifts of some flavonoids and phenolic acids, found in European poplar type propolis.

*Flavones and flavonols*

| Position | Apigenin | | Chrysin | | Galangin | | Kaempferol | | Quercetin | |
|---|---|---|---|---|---|---|---|---|---|---|
| | $\delta_H$[b] (*J* in Hz) | $\delta_C$ | $\delta_H$[b] (*J* in Hz) | $\delta_C$ | $\delta_H$[b] (*J* in Hz) | $\delta_C$ | $\delta_H$[b] (*J* in Hz) | $\delta_C$ | $\delta_H$[b] (*J* in Hz) | $\delta_C$ |
| 2 | – | 164.59 | – | 163.60 | – | 146.11 | – | 147.28 | – | 147.27 |
| 3 | 6.75 (s) | 103.31 | 6.94 (s) | 105.63 | 9.59 (s) *OH* | 137.52 | 9.35 (s) *OH* | 136.11 | 9.18 (s) *OH* | 136.18 |
| 4 | – | 182.19 | – | 182.30 | – | 176.68 | – | 176.36 | – | 176.30 |
| 5 | 12.96 (s) *OH* | 161.62 | 12.82 (s) *OH* | 161.94 | 12.31 (s) *OH* | 161.19 | 12.48 (s) *OH* | 161.18 | 12.48 (s) *OH* | 161.18 |
| 6 | 6.19 (d) (1.5) | 99.30 | 6.22 (d) (1.8) | 99.49 | 6.16 (d) (2.0) | 98.74 | 6.20 (d) (2.2) | 98.67 | 6.19 (d) (1.9) | 98.64 |
| 7 | 10.75 (s) *OH* | 164.19 | 10.90 (s) *OH* | 164.91 | 10.59 (s) *OH* | 164.65 | 10.78 (s) *OH* | 164.35 | 10.75 (s) *OH* | 164.33 |
| 8 | 6.46 (d) (1.5) | 94.42 | 6.51 (d) (1.8) | 94.58 | 6.40 (d) (2.0) | 93.99 | 6.44 (d) (2.2) | 93.94 | 6.41 (d) (1.9) | 93.81 |
| 9 | – | 157.77 | – | 157.91 | – | 156.83 | – | 156.65 | – | 156.60 |
| 10 | – | 104.18 | – | 104.44 | – | 103.65 | – | 103.52 | – | 103.47 |
| 1′ | – | 121.66 | – | 131.19 | – | 131.38 | – | 122.15 | – | 122.42 |
| 2′; 6′ | 7.90 (d) (8.8) | 128.90 | 8.04 (d) (8.8) | 126.84 | 8.08 (d) (8.8) | 127.94 | 8.05 (d) (8.8) | 129.97 | 2′; 7.73 (d) (2.2) 6′; 7.54 (dd) (8.4; 2.2) | 115.53 120.44 |
| 3′; 5′ | 6.92 (d) (8.8) | 116.41 | 3′; 4′; 5′ 7.58 (m) | 129.56 | 3′; 4′; 5′ 7.44 (m) | 128.88 | 6.93 (d) (8.8) | 115.90 | 3′; 9.45 (s) *OH* 5′; 6.89 (d) (8.8) | 145.51 116.06 |
| 4′ | 10.40 (s) *OH* | 161.93 | | 132.42 | | 130.28 | 10.10 (s) *OH* | 159.65 | 9.59 (s) *OH* | 148.16 |

*Flavanones*

| Position | Naringenin | | Pinocembrin | | Pinostrobin | |
|---|---|---|---|---|---|---|
| | $\delta_H$[b] (*J* in Hz) | $\delta_C$ | $\delta_H$[b] (*J* in Hz) | $\delta_C$ | $\delta_H$[b] (*J* in Hz) | $\delta_C$ |
| 2 | 5.46 (dd) (12.8; 2.2) | 78.88 | 5.58 (dd) (12.5; 2.8) | 78.21 | 5.62 (dd) (12.7; 3.0) | 79.03 |
| 3 | 3α; 3.27 (dd) (17.1; 12.8) 3β; 2.67 (dd) (17.1; 2.2) | 42.45 | 3α; 3.23 (dd) (17.1; 12.8) 3β; 2.79 (dd) (17.1; 3.2) | 42.77 | 3α; 3.29 (dd) (17.1; 12.8) 3β; 2.83 (dd) (17.1; 3.0) | 42.63 |
| 4 | – | 196.82 | – | 196.45 | – | 196.92 |
| 5 | 12.16 (s) *OH* | 163.95 | 12.13 (s) *OH* | 164.10 | 12.12 (s) *OH* | 163.71 |
| 6 | 5.89 (d) (2.2) | 96.24 | 5.91 (d) (2.2) | 96.50 | 6.11 (d) (2.2) | 95.24 |
| 7 | 10.79 (s) *OH* | 167.10 | 10.79 (s) *OH* | 167.23 | 3.80 (s) $OCH_3$ | 167.94 |
| 8 | 5.89 (d) (2.2) | 95.42 | 5.95 (d) (2.2) | 95.62 | 6.15 (d) (2.2) | 94.34 |
| 9 | – | 163.40 | – | 163.65 | – | 163.09 |
| 10 | – | 102.23 | – | 102.30 | – | 103.10 |
| 1′ | – | 129.31 | – | 139.64 | – | 138.99 |
| 2′; 6′ | 7.31 (d) (8.8) | 128.77 | 7.52 (d) (7.8) | 127.12 | 7.54 (d) (8.8) | 127.06 |
| 3′; 5′ | 6.79 (d) (8.8) | 115.62 | 3′; 4′; 5′7.41 (m) | 129.13 | 3′; 4′; 5′7.42 (m) | 129.06 |
| 4′ | 9.59 (s) *OH* | 158.19 | | | | |
| $OCH_3$ | – | – | – | – | – | 56.33 |

*Phenolic acids*

| Position | Caffeic acid | | Cinnamic acid | | p-Coumaric acid | | Ferulic acid | |
|---|---|---|---|---|---|---|---|---|
| | $\delta_H$[b] ($J$ in Hz) | $\delta_C$ | $\delta_H$[b] ($J$ in Hz) | $\delta_C$ | $\delta_H$[b] ($J$ in Hz) | $\delta_C$ | $\delta_H$[b] ($J$ in Hz) | $\delta_C$ |
| 1 | 12.10 (s) *COOH* | 168.34 | | | | | | |
| 2 | 6.17 (d) (16.0) | 115.58 | 12.39 (s) *COOH* | 168.00 | 12.09 (s) *COOH* | 168.38 | 12.07 (s) *COOH* | 168.42 |
| 3 | 7.42 (d) (16.0) | 145.04 | 6.53 (d) (16.0) | 119.70 | 6.29 (d) (16.0) | 115.81 | 6.36 (d) (16.0) | 116.09 |
| 1′ | – | 126.16 | 7.59 (d) (16.0) | 144.38 | 7.51 (m) | 144.60 | 7.49 (d) (16.0) | 144.95 |
| 2′ | 7.03 (d) (1.8) | 115.09 | – | 134.70 | – | 130.53 | – | 126.24 |
| 3′ | 9.13 (s) *OH* | 146.02 | 7.68 (m) | 128.65 | 7.51 (m) | 125.74 | 7.28 (d) (2.0) | 111.66 |
| 4′ | 9.52 (s) *OH* | 148.59 | 3′; 4′; 5′ 7.42 (m) | 129.36 | 6.81 (d) (8.8) | 116.20 | 3.82 (s) $OCH_3$ | 148.38 |
| 5′ | 6.76 (d) (8.0) | 116.21 | | 130.67 | 9.97 (s) *OH* | 160.04 | 9.54 (s) *OH* | 149.55 |
| 6′ | 6.96 (dd) (8.0; 2.0) | 121.60 | | 129.36 | 6.81 (d) (8.8) | 116.20 | 6.79 (d) (8.0) | 115.99 |
| $OCH_3$ | – | – | 7.68 (m) | 128.65 | 7.51 (m) | 125.74 | 7.08 (dd) (8.0; 2.0) | 123.27 |
| | | | – | – | – | – | – | 56.17 |

[a]Assignments were from HSQC and HMBC experiments.
[b]Multiplicity in parentheses. (Bertelli et al., 2012).

The application of the NMR technique to propolis samples generates very complicated spectra that need to be processed before spectral calculations and subsequently analyzed by chemometric methods. The NMR signals can be used as intensity or can be integrated.

If the choice is to use spectra as intensity, an ideal preprocessing should include the steps that follow.

(1) Calibrate phased spectra by placing the signal of the standard compound TMS to 0 ppm.
(2) Each spectrum generates a file containing several thousand data points corresponding to the time domain that is the number of points acquired and digitalized by the instrument along the spectral width. Export these files and assemble them in a data-set.
(3) Solve misalignment problems a posteriori using suitable software. A good example of this kind of software is the open source Icoshift program running in Matlab environment (Savorani, Tomasi, & Englesen, 2009). Although the chemical shift of a nucleus is generally assumed to be rather stable, it is necessary to consider that some experimental factors (pH, ionic strength, solvent, field inhomogeneity, temperature) can affect the absolute and the relative position of an NMR signal, producing slight or significant variations in chemical shifts along the spectral width (Bertelli et al., 2012). Unresolved peaks in one spectrum can be resolved or more overlapped in another spectrum. This is particularly important in the analysis of complex mixtures, such as propolis extracts, in which a high number of similar compounds are present.
(4) To reduce the number of data points, do not consider all the spectral regions devoid of signals and the solvent signals, and subsequently remove them. If the number of spectral variables remains very high, reduce it further by lowering the spectral resolution.

### 3.6.4. *Statistical analysis of NMR spectra*

Multivariate chemometric methods can be applied on the data-set containing spectra. There are a number of multivariate techniques that can be used in the analysis of NMR spectra (Brereton, 2013):

(1) Principal Component Analysis (PCA): PCA is an unsupervised technique and allows one to express a large portion of the data's total variance with a smaller number of variables which can be used to represent graphically the population of samples and to identify the most significant original factor(s).
(2) Discriminant Analysis (DA): DA is a supervised technique used to determine whether a given classification of cases into a number of groups is appropriate. DA can be used, for instance, to test whether a particular clustering of cases obtained from a unsupervised method like PCA or Cluster analysis is likely. Also, this analysis can be used to classify unknown samples.
(3) Partial Least Square Discriminant Analysis (PLS-DA): PLS-DA can be described as the regression extension of PCA, giving the maximum covariance between measured data (NMR spectral intensities distribution) and the response variable (represented in this case by the possible classification of samples).

Normally, some statistical pre-treatment should be done before performing one of the above mentioned methods on NMR-generated data. The most useful to improve the results are normalization, mean centering and autoscaling. It is essential that one has a large number of samples in order to cross-validate the obtained models and also to have a test set for external validation. To date, there have not been many published reports where NMR was used to study propolis extracts as mixtures. Meneghelli et al. (2013) used NMR to identify some components of Brazilian propolis using the extracts directly without any kind of isolation and purification steps. They used one- and two dimensional NMR to study the chemical profile of the samples. (Meneghelli et al., 2013).

Two different studies report the use of NMR to compare different types of propolis. Cuesta-Rubio et al. (2007) studied three different varieties of Cuban propolis using $^{1}H$ and $^{13}C$ one-dimensional NMR as chemical fingerprint technique, HPLC-PDA and HPLC-MS. A similar work was published in 2010 by the same authors (Hernandez et al., 2010).

Watson et al. (2006) used NMR and PCA to build a model for the classification of propolis of different geographical origins. In this case, the authors used the bucketing technique. This technique consists of dividing the spectra in different small regions, following which the signals present in each region are integrated and the area results are used as spectral variables. The obtained model was able to classify samples from different areas of the world.

Papotti et al. (2010) published an article regarding the use of NMR to classify propolis samples according to their production procedure. In this work, the authors used not only $^{1}H$ NMR but also $^{1}H$-$^{13}C$ HMBC spectra. In the first case, the spectra were used as intensity and after integration of principal signals. In the latter, the volume of two-dimensional spectra signals were calculated adding together the intensity of the points located in previously manually defined areas surrounding the correlations and all spectra were processed using the same map of regions of interest. On the different obtained data-set, general discriminant analysis (GDA) was used to classify propolis according to their NMR fingerprint (Papotti et al., 2010).

In conclusion, NMR represents a very powerful tool for the study of propolis and the use of NMR coupled with an appropriate data processing procedure and multivariate statistical methods enables the development of sufficiently effective and appropriate models for classifying propolis. It is interesting to note that the best results are normally obtained using the $^{1}H$ NMR which is the simplest and fastest technique.

### 3.7. *Propolis type dereplication*

#### 3.7.1. *Introduction*

Propolis from different locations always demonstrates considerable biological activity even though the chemical composition may vary (Kujumgiev et al., 1999; Seidel, Peyfoon, Watson, & Fearnly, 2008). For this reason, the chemical diversity of different propolis samples also has the potential to provide valuable leads to active components. Thus, the future discovery of new types of propolis from unexplored regions is important with respect to uncovering new biologically active compounds with important pharmacological effects. Investigating propolis from currently unstudied regions is important as it would allow one to determine if the new propolis belongs to an already known propolis type. The rapid identification (dereplication) of known propolis types avoids re-isolation and identification of known propolis constituents and is crucial for fast discovery of new natural/propolis compounds. Dereplication is rapid identification of known bioactive metabolites from chemical profiling of plants and other natural sources.

#### 3.7.2. *GC-MS as a strategy for propolis type dereplication*

GC-EI MS is a powerful analytical platform for dereplication, combining the unprecedented resolving power of capillary GC with the structural information provided by EI mass spectra and supported by rich spectral libraries. In propolis research, GC-MS is one of the most common methods used and thus it is an excellent tool for propolis chemical type dereplication. Propolis ethanol extracts are subjected to GC-MS analysis after silylation (Section 3.3.1).

The first outcome of the GC-MS analysis is the TIC chromatogram. In the case of propolis, this is usually a complicated chromatogram containing several dozen peaks (Figure 6). Although sometimes the practiced eye is able to recognize a characteristic pattern, the analysis of the mass spectra is inevitable. After obtaining the TIC chromatogram, attention is directed towards the most prominent peaks and their mass spectra are analyzed. Let us assume that this analysis has resulted in identification of the major peaks. As soon as the major peaks in the TIC chromatogram are identified, it is necessary to check the characteristic constituents of the known propolis types and determine if these major constituents match one of them. In this case, the dereplication process has been completed.

In this Section 3.7.2, the most important markers for positive identification of the most widespread and well known propolis types are presented. Data about propolis types in Australia, the Middle East, Africa and to some extent North America are scarce and demonstrate diverse chemistry. Thus it is hard to formulate propolis types for these regions.

*3.7.2.1. Poplar type propolis.* Poplar type propolis, originating from *Populus* spp, is characterized by flavonoids, phenolic acids and their esters as bioactive constituents (Ahn et al., 2007; Greenaway, Scaysbrook, & Whatley, 1990; Marcucci, 1995). The most intensive peaks in the TIC chromatogram of a poplar propolis sample typically belong to pinocembrin, chrysin, galangin, pinobanskin 3-acetate and pinobanksin. These compounds are characteristic of propolis originating from the bud exudates of the black poplar *Populus nigra* (Bankova, de Castro, &

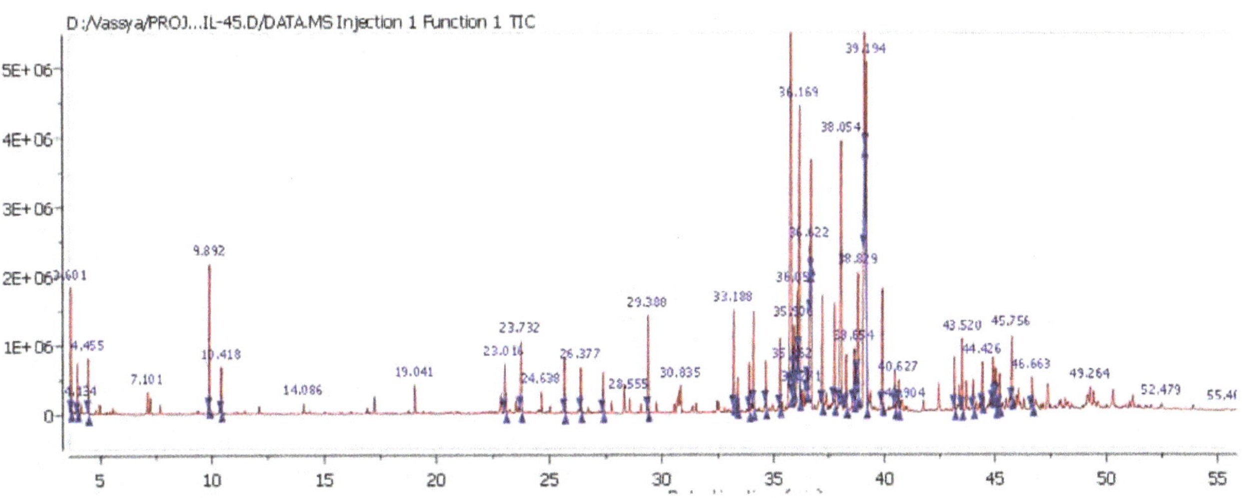


Figure 6. TIC chromatogram of a typical poplar propolis sample (Popova et al., Unpublished data: internal database).

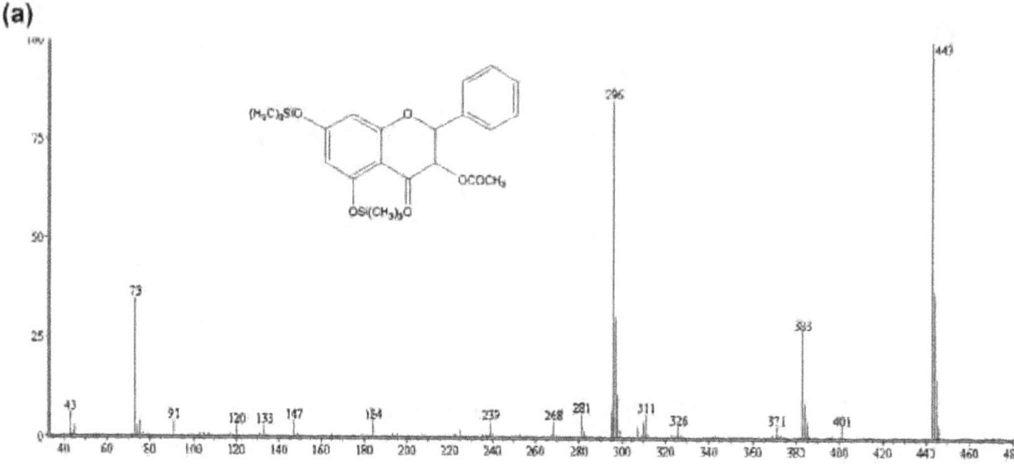


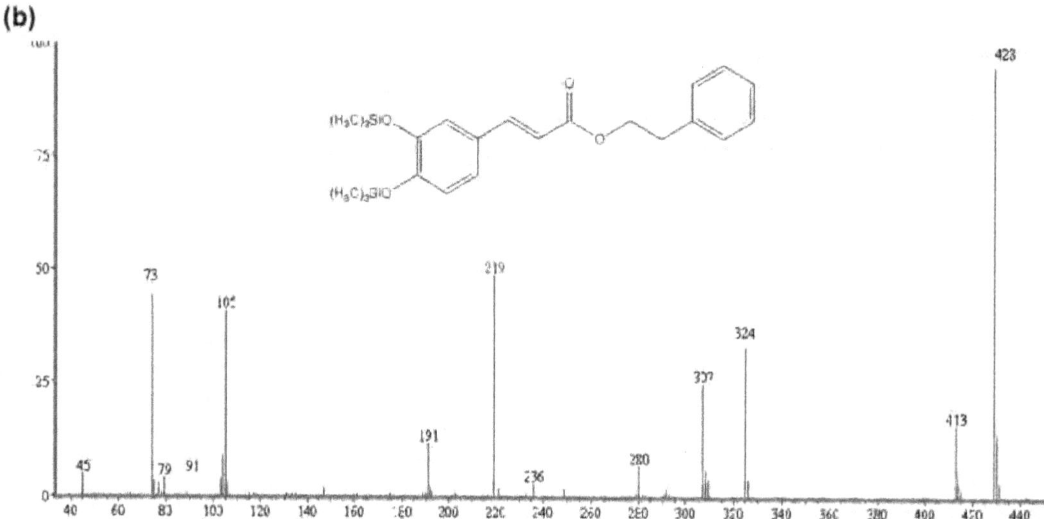


Figure 7. EIMS spectra of the TMS derivatives (a) pinobanksin 3-acetate, $(M)^+$ at *m/z* 458 and (b) phenylethyl caffeate (CAPE), $(M)^+$ at *m/z* 428. (Popova et al., Unpublished data: internal database).

Marcucci, 2000). For positive confirmation of poplar propolis, it is necessary to confirm the presence of the taxonomic markers of the black poplar – esters of substituted cinnamic acids, and especially penteny caffeates and phenylethyl caffeate, as well as pinobanksin 3-acetate. Their mass spectra are presented in Figure 7.

*3.7.2.2. Aspen type propolis.* In northern regions of Europe, the trembling aspen (European aspen) *Populus tremula* is used by bees as a propolis plant source (Bankova, Popova, Bogdanov, & Sabatini, 2002; Isidorov et al., 2014; Popravko, Sokolov, & Torgov, 1982). In the case of aspen propolis, major peaks in the TIC chromatogram belong to *p*-coumaric, ferulic, and benzoic acids, benzyl *p*-coumarate and benzyl ferulate. The minor but discriminant markers of aspen bud exudates are the glycerol esters of substituted cinnamic acids (phenolic glycerides) as 2-acetyl-1,3-di-p-coumaroylglycerol and 1-acetyl-3-feruloyl glycerol (Figure 8).

*3.7.2.3. Brazilian green propolis.* Brazilian green propolis is another well studied propolis type. Its main bioactive constituents include phenolic acids, prenylated phenolic acids and flavonoids which are characteristic for *Baccharis dracunculifolia*, the most important botanical source of Southeastern Brazilian propolis (Bankova et al., 1999; Kumazawa et al., 2003). For this propolis type, the major peaks in TIC chromatogram belong to artepillin C (Figure 9), drupanin, p-coumaric acid and dihydrocinnamic acid. Minor, but important markers are 2,2-dimethyl-6-carboxyethyl prenylbenzopyrane and aromadendrine 4′-methyl ether (Figure 9).

*3.7.2.4. South American red propolis.* The biologically active constituents of red propolis from Cuba and Brazil are isoflavans, isoflavons and pterocarpans (López, Schmidt, Eberlin, & Sawaya, 2014; Lotti et al., 2010; Piccinelli et al., 2011; Trusheva et al., 2006). This type has as major constituents vestitol (Figure 10), medicarpin (Figure 10), neovestitol, 7-O-methylvestitol (isosativan),

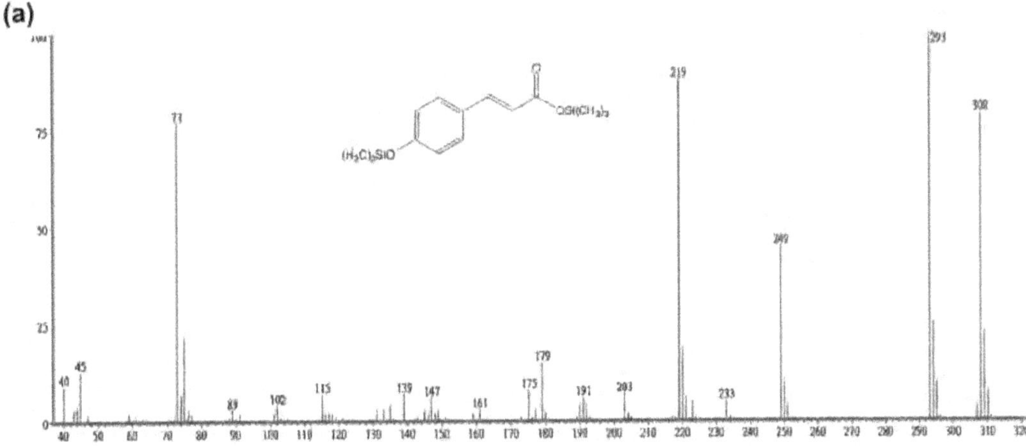


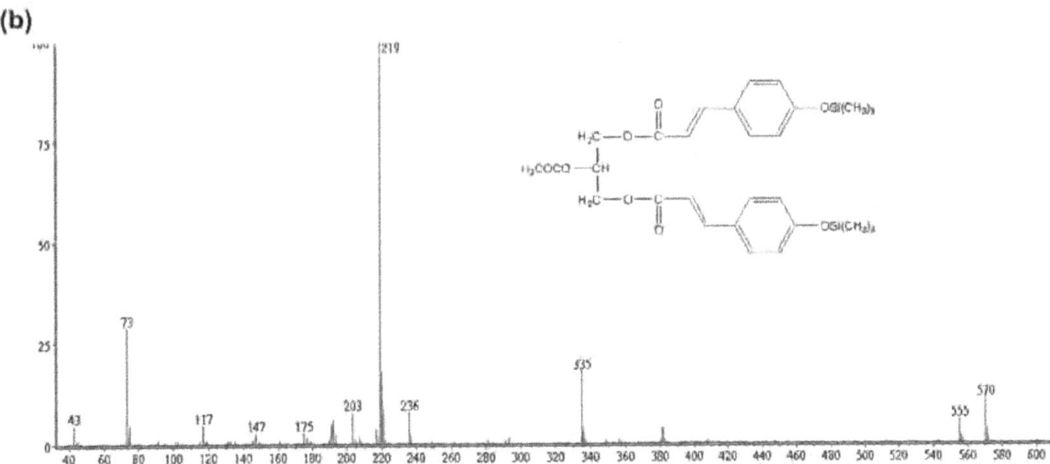


Figure 8. EIMS spectra of the TMS derivatives (a) *p*-coumaric acid, $(M)^+$ at *m/z* 308 and (b) 2-acetyl-1,3-di-p-coumaroylglycerol, $(M)^+$ at *m/z* 570. (Popova et al., Unpublished data: internal database).

and formononetin, all of them taxonomic markers of *Dalbergia ecastophyllum*.

*3.7.2.5. Mediterranean type propolis.* This type is characteristic for propolis samples originating from the Mediterranean region and its major constituents are diterpenes typical for the resin of the cypress tree *Cupressus sempervirens* (Popova, Graikou, Chinou, & Bankova, 2010; Popova et al., 2012). Isocupressic acid (Figure 11), pimaric acid, agathadiol, isoagatholal and totarol (Figure 11) give the major peaks in the TIC chromatogram. The only phenolic compounds in typical cypress propolis are the phenolic diterpenes totarol and totarolone. Cypress propolis usually does not contain flavonoids and phenolic acids.

*3.7.2.6. Pacific type propolis.* This propolis type is characteristic for samples from Pacific islands (Taiwan, Okinawa, Indonesia) (Huang et al., 2007; Kumazawa et al., 2008; Trusheva et al., 2011). Its dereplication includes identification of the prenylated flavanones (propolins) propolin C, propolin D (Figure 12) and propolin F as major peaks in TIC chromatogram. The plant source of these compounds is *Macaranga tanarius*.

*3.7.2.7. Mangifera indica type propolis.* The main bioactive metabolites of this propolis type are a series of phenolic lipids: cardanols, cardols and anacardic acid derivatives – all resin biomarkers of the tree *Mangifera indica* (mango) (Knödler et al., 2008; Trusheva et al., 2011). Among them, heptadecenyl-recorcinol (Figure 13), nonadecenyl-recorcinol, nonadecyl-anacardic acid and heptadecenyl-anacardic acid correspond to the most prominent peaks in TIC chromatogram. Minor, but characteristic constituents are triterpenes from cycloartane type as cycloartenol, mangiferolic acid (Figure 13) and 24-hydroxyisomangiferolic acid.

*3.7.2.8. Mixed propolis types.* In many cases, bees collect resins from two or even three plant sources. In such cases, the characteristic markers of the particular

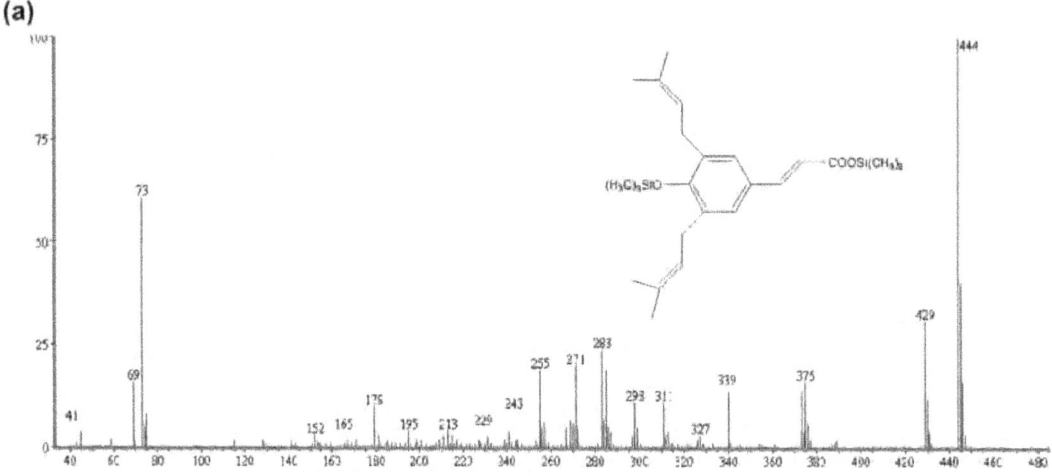


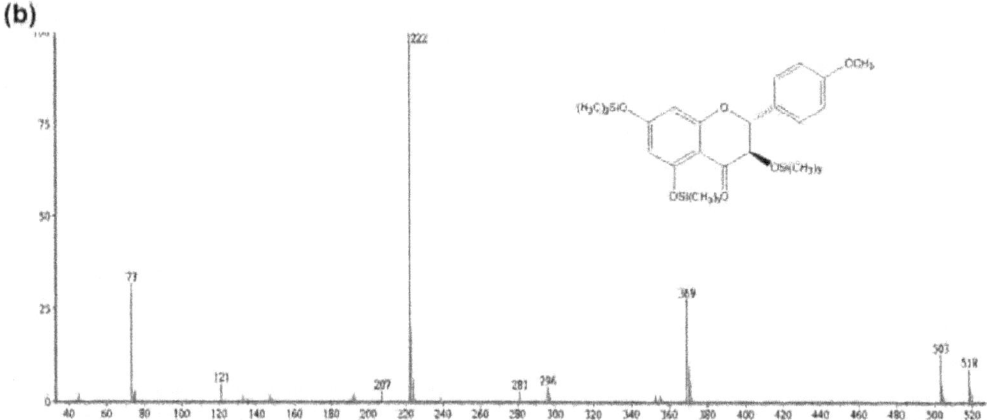


Figure 9. EIMS spectra of the TMS derivatives (a) artepillin C, $(M)^+$ at *m/z* 444 and (b) aromadendrine 4′-methyl ether, $(M)^+$ at *m/z* 518. (Popova et al., Unpublished data: internal database).

source plants can be detected by GC-MS. For this reason, a more detailed analysis of the total ion chromatogram is necessary, in order to consider more than just a limited number of prominent peaks.

Several mixed propolis types have been detected, for example aspen-poplar, *Cupressus*-poplar (Bankova et al., 2002), and Pacific (*Macaranga*)-*Mangifera indicia* propolis (Trusheva et al., 2011).

### *3.7.3. Other possibilities for dereplication*

Other analytical methods also offer the possibility to perform dereplication of the propolis type: LC-MS (Section 3.4), ESI-MS fingerprinting (Section 3.5.1), NMR analysis (Section 3.6), and HPTLC (Morlock, Ristivojevic, & Chernetsova, 2014; Ristivojevic et al., 2014). The important point is to identify the corresponding markers that allow unambiguous positive identification of the source plant(s). If the results of such analyses do not allow the dereplication of propolis type, the metabolomic approach described in Section 3.7.4 should be applied in order to determine the botanical sources of propolis and, respectively, its chemical type based on the chemistry of the source plant. Alternatively, a very recent publication (Jain, Marchioro, Mendonca, Batista, & Araujo, 2014) reports on the application of DNA analysis for determining the botanical origin of red Brazilian propolis.

### *3.7.4. LC-MS-based metabolomic analysis to determine the botanical sources of propolis*

Direct observation of resin forager behavior in the field can be extremely difficult or impossible, as foraging can occur over a large area and in the canopy of trees. This makes analytical analyses an attractive alternative, but one must consider several challenges. First, bees typically have many resinous plants from which to choose in a given environment and these available species may be closely related. For example, six species of *Populus* (a known resin source for honey bees) and numerous hybrids occur in the state of Minnesota, USA, and their resins have some degree of similarity. Second, resins from most species remain uncharacterized and characterization itself is a very labor intensive process. Lastly, further complications occur in the hive where resins from several plant species may be mixed. Therefore, any universal method developed to determine the

(a)

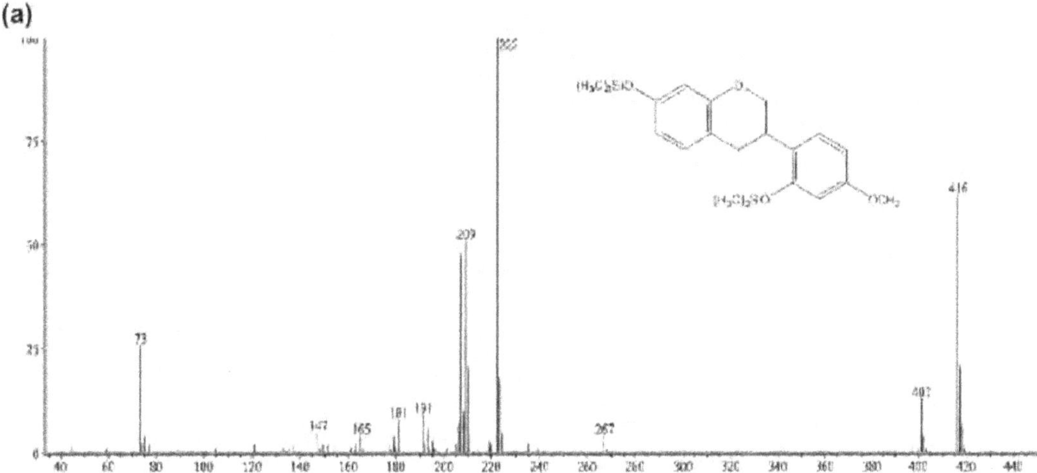

(b)

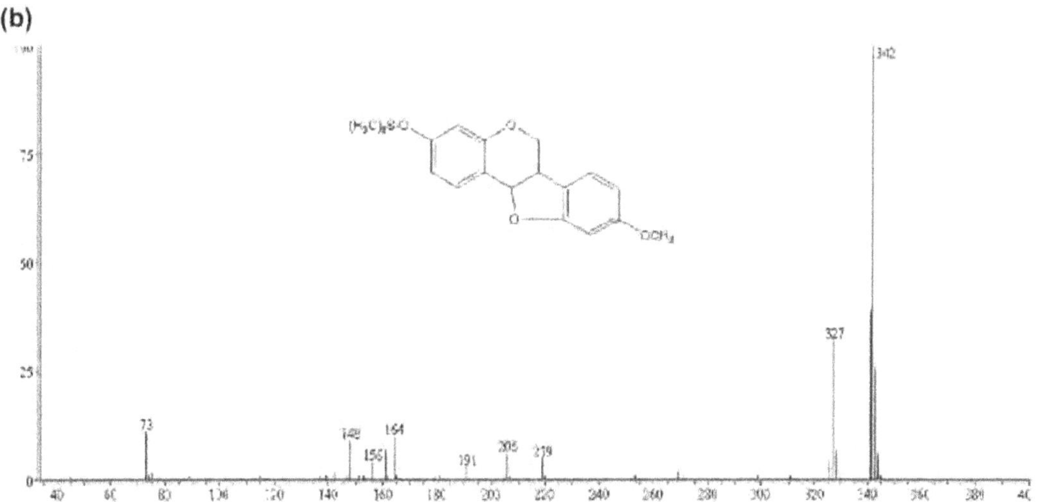

Figure 10. EIMS spectra of the TMS derivatives (a) vestitol, $[M]^+$ at *m/z* 416; (b) medicarpin, $[M]^+$ at *m/z* 342. (Popova et al., Unpublished data: internal database).

botanical sources of propolis must: (1) be powerful enough to discriminate between resins from closely related species; (2) work effectively with uncharacterized resins; and (3) be sensitive enough to sample at the level of individual bees carrying pure resin.

Traditional analytical methods will generally fail to meet our second criteria because comparisons are made regarding specific characterized compounds. Metabolomics is an approach that compares the global pattern of metabolite signals among samples using powerful statistical analyses without regard for the identities of specific compounds. LC-MS based metabolomics analysis fulfils all of our criteria in that: (1) LC-MS can easily generate hundreds of chemical signals that can be used to discriminate between closely related species; (2) metabolomics makes powerful comparisons between sample "fingerprints" without requiring any chemical characterization; and (3) sampling of individual resin foragers can be performed. LC-MS instruments equipped with an auto-sampler have the added capacity to run tens to hundreds of samples easily. Herein, we describe the metabolomics methods used in Wilson, Spivak, Hegeman, Rendahl, and Cohen (2013) to track the resin foraging behavior of individual honey bees.

*3.7.4.1. Sample preparation for LC-MS*

(1) Metabolomics works best with many samples; however, increasing the sample number increases analytical time and difficultly. It is generally reasonable to collect up to 100 samples of resin in total, directly from bees and from plants (Section 2.1).
(2) Weigh resin globules from bees, place in LC-MS vials, and dissolve in HPLC-grade acetonitrile. The final concentration of your samples is highly dependent on your instrumentation; however, we have found that a sample concentration of 1 mg/ml works well for a variety of high and low resolution instruments (Wilson et al., 2013; Wilson, Brinkman, Spivak, Gardner, & Cohen, 2015).

(a)

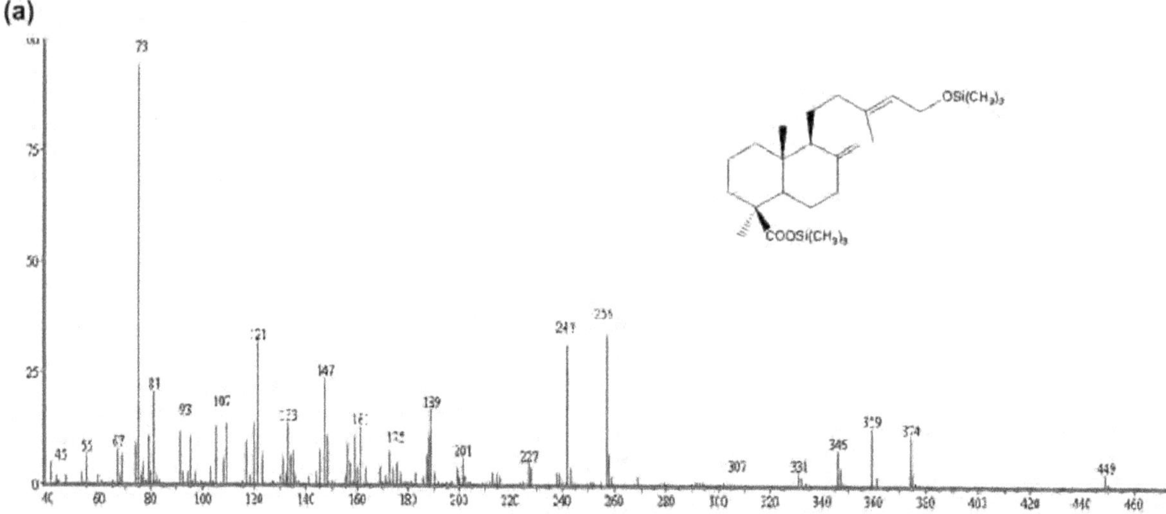

(b)

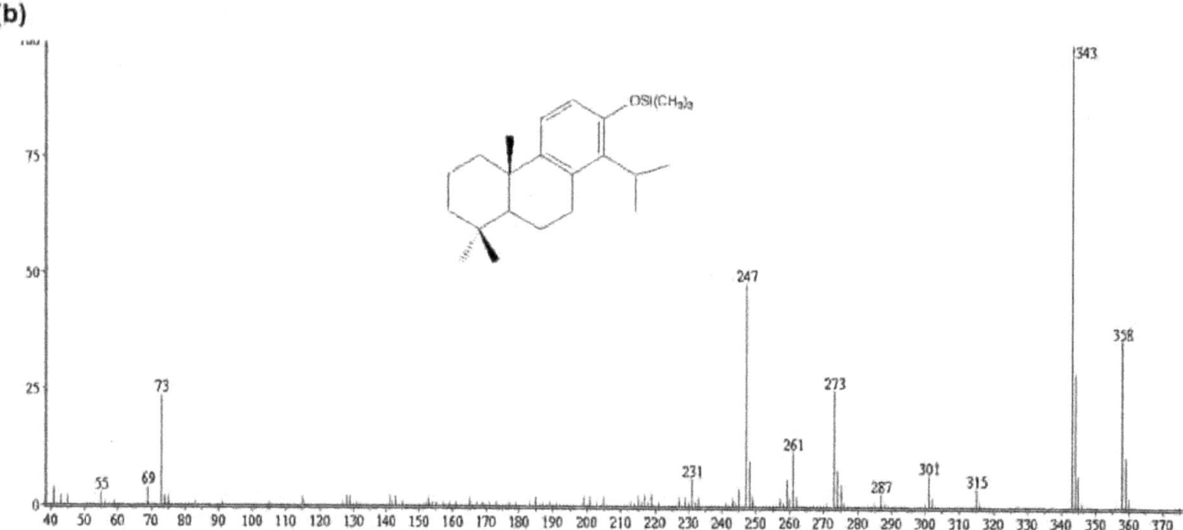

Figure 11. EIMS spectra of the TMS derivatives (a) isocupressic acid, $(M)^+$ at $m/z$ 464 and (b) totarol, $(M)^+$ at $m/z$ 358. (Popova et al., Unpublished data: internal database).

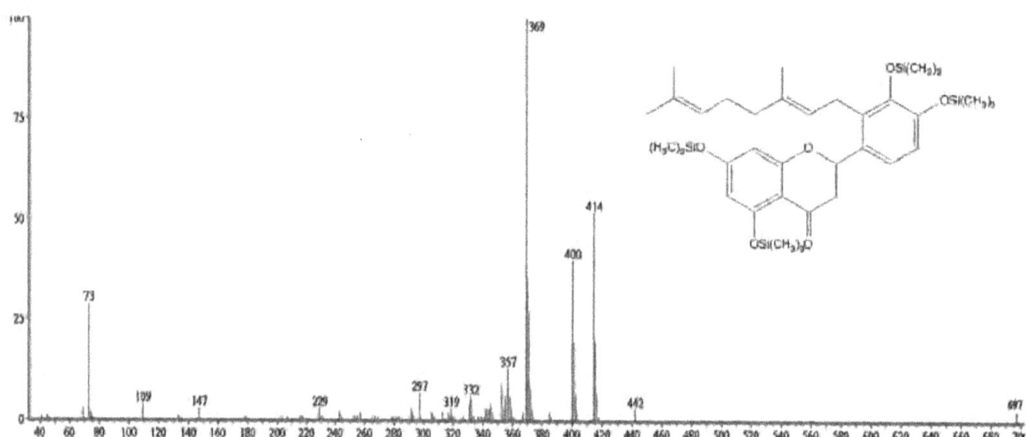

Figure 12. EIMS spectra of the TMS derivative of propolin D, $(M)^+$ at $m/z$ 712. (Popova et al., Unpublished data: internal database).

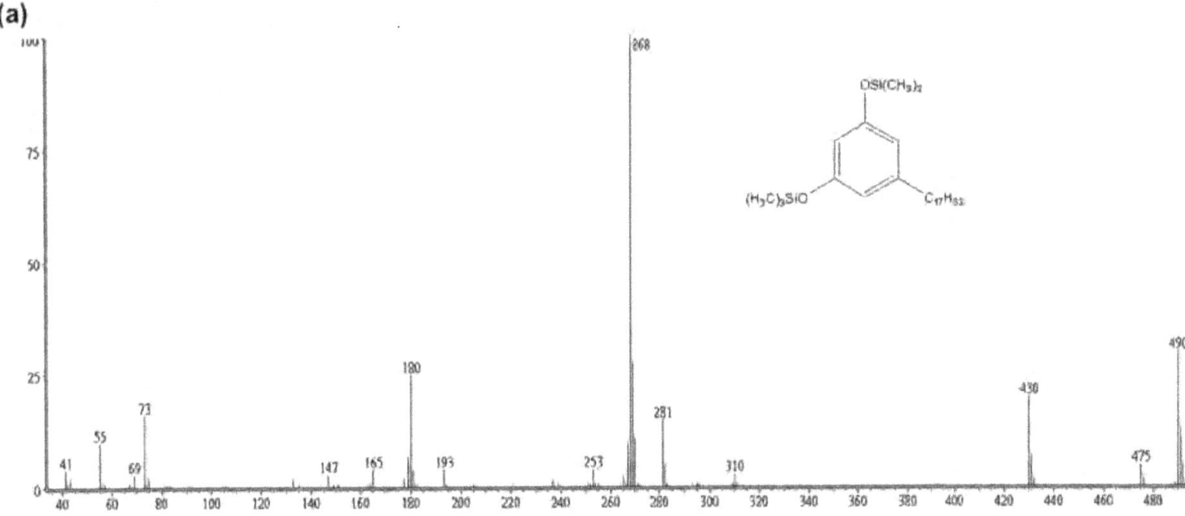


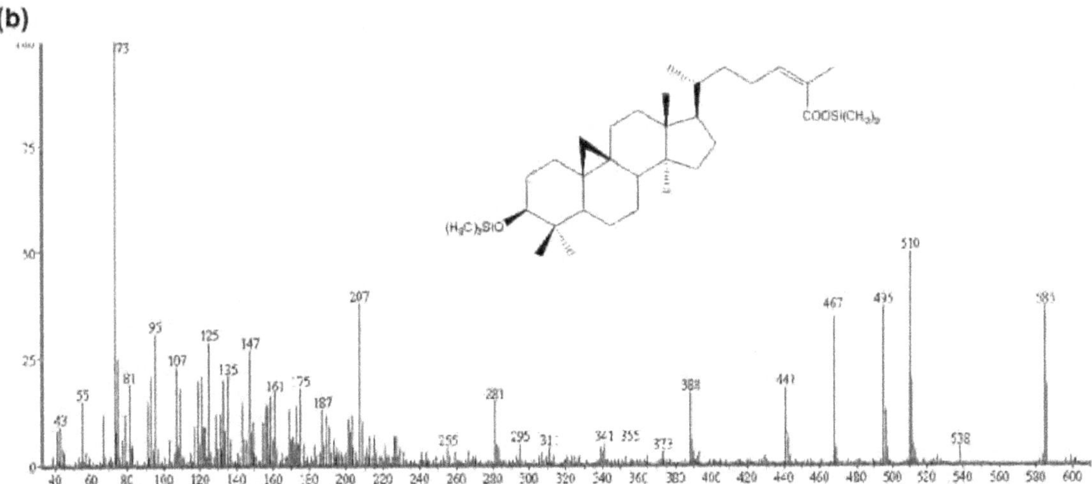


Figure 13. EIMS spectra of the TMS derivatives (a) 5-heptadecenyl-recorcinol, $(M)^+$ at $m/z$ 490 and (b) mangiferolic acid, $(M)^+$ at $m/z$ 600. (Popova et al., Unpublished data: internal database).

Table 4. General LC method for metabolomics analysis.

Column: Agilent Zorbax $C_{18}$, 2.1 × 100 mm, 1.8 μm particle size
Flow rate: 0.45 ml/min

| Time (min) | % A (water w/0.1% formic acid) | % B (acetonitrile w/0.1% formic acid) |
|---|---|---|
| 0 | 90 | 10 |
| 1.5 | 90 | 10 |
| 17.5 | 5 | 95 |
| 19.5 | 5 | 95 |
| 20.5 | 90 | 10 |

(3) Add 5 ml of HPLC-grade acetonitrile to plant tissues or collected resin. Rock gently for 15 min to wash resins off of tissues, then remove tissues using clean forceps. Be careful not to cross-contaminate samples.

(4) Determine the concentration of resin samples from plants using vacuum centrifugation.

(5) Dilute resin samples from plants to 1 mg/ml for analysis.

(6) Create a composite sample for quality control by adding equal volume amounts of each biological sample into a new vial (e.g. If you have 100 total samples from plants and bees, take 10 μl from each and add to a new vial). Since a composite sample

made in this way contains essentially all of the signals that could be produced in all of the biological samples, technical replicates of the composite sample can be used to filter out non-reproducible LC-MS signals during data analysis.

*3.7.4.2. LC-MS data collection.* It is important to recognize that the chemistry of unknown resins cannot be accounted for in the analytical method preemptively. Therefore, we present a general reversed-phase $C_{18}$ approach developed for a Waters Acuity UPLC system connected to either a Waters SQD mass spectrometer (low resolution) or a Waters G2 Synapt mass spectrometer (high resolution) as used in Wilson et al. (2013, 2015) (Table 4). Data can be collected in either negative ion mode, positive ion mode, or both simultaneously, but the composite sample should be run every 5–10 samples, and at least three times during the course of the entire LC-MS run. Remember to utilize best practices for LC-MS analysis. (Viswanathan et al., 2007).

*3.7.4.3. LC-MS data analysis*

(1) Convert data files to CDF format. Waters instruments come with a program called Databridge that will perform this function. This will not be necessary if you plan on using proprietary metabolomics data analysis software.
(2) Smith, Want, O'Maille, Abagyan, and Siuzdak (2006) developed a freely available R script to analyze metabolomics data in CDF format which utilizes XCMS to produce a table of mass/retention time pairs and their intensities by sample for the entire data-set (data matrix). Please refer to Wilson et al. (2013) for a full description. Other metabolomics data analysis software can be used to perform this task, but few can utilize quality control samples in the manner described here, which may result in low quality signals being carried into subsequent analyses.
(3) Perform principle component analysis (PCA) on the data matrix (Pirk et al., 2013). Points representing samples will scatter on the PCA graph based on their LC-MS peak patterns, with samples showing similar peak patterns clustering together (see Wilson et al., 2013). If samples of bee collected resin cluster with samples of plant collected resin, this is a strong indication that bees foraged from this plant.

### 3.8. Spectrophotometric analysis of propolis

Spectrophotometric methods are very useful for fast and easy quantitative determination of phenolic compounds in propolis and for routine control of propolis preparations. There are efficient, precise and reliable spectrophotometric methods that are aimed at the determination of total flavonoids or total phenolics content. Phenolics and flavonoids are major constituents and most important bioactive ingredients of several propolis types and spectrophotometric methods are useful in their rapid characterization.

*3.8.1. Spectrophotometric analysis of poplar type propolis*

The analysis of poplar type propolis consists of the spectrophotometric quantitative determination of the following groups of phenolic compounds: (Popova et al., 2004): flavones and flavonols; flavanones and dihydroflavonols; and total phenolics.

*3.8.1.1. Extraction and sample preparation*

(1) Perform propolis extraction as described in Section 3.1.1. Extract 1 g propolis and make up the volume to 100 ml (volumetric flask). The resulting extract is designated as solution A.
(2) Transfer 1 ml from each of three parallel extracts into a volumetric flask and dilute to 50 ml using methanol. The resulting solution is designated as solution B.
(3) Prepare three parallel extracts for every analyzed sample.

*3.8.1.2. Total flavone and flavonol content.* Total flavone and flavonol content is measured using a spectrophotometric assay based on aluminum chloride complex formation (Bonvehi & Coll, 1994). Methanolic solutions of galangin are used as references to obtain a calibration graph. The analytical procedure for measuring total flavones and flavonols is performed the following way:

(1) To prepare a calibration graph with galangin as the standard, prepare a stock standard solution of galangin 32 μg/ml by dissolving 3.2 mg in methanol in a 100 ml volumetric flask.
(2) Prepare a series of working reference solutions by appropriate dilution of the stock standard solution with methanol (in volumetric flasks) to give a concentration range of 4–32 μg/ml (16.0; 8.0; 6.4; 4.0 μg/ml).
(3) Mix 1 ml of each one of the reference solutions, 10 ml methanol and 0.5 ml 5% $AlCl_3$ in methanol (w/v) in a volumetric flask and make up the volume to 25 ml with methanol.
(4) Let the mixture sit for 30 min and measure the absorbance at 425 nm.
(5) For a blank, use 1ml methanol instead of galangin solution in analogues procedure.
(6) Each reference solution should be analyzed in triplicate.
(7) To obtain the regression, absorbance should be plotted against concentration (International Conference on Harmonization, [ICH], 1996).

(8) For analysis of the propolis sample solution, use B (Section 3.8.1.1), or, if necessary, solution B with additional dilution, and apply the same procedure as described for the reference (steps 3–5).
(9) Perform calculation using the calibration equation for galangin (step 7):

$$c = aA + b$$

where $c$ – concentration, mg/ml; $A$ – absorbance; a – slope of the calibration graph; b – intercept of the calibration graph.

(10) From this value, the percentage of flavones and flavonols in the propolis sample is calculated after the equation:

$$P = \frac{c \times 100 \times 50}{3\bar{M}} \times 100\%$$

where $P$ – percentage in raw propolis; $c$ – concentration, mg/mL (from step 9); $\bar{M}$– mean value of the weight of the three parallel propolis samples, extracted for analysis, mg (Section 3.8.1.1).

(11) In instances when an additional dilution of solution B is provided, it should be reflected in the equation.

*3.8.1.3. Total flavanone and dihydroflavonol content.* For flavanones and dihydroflavonols determination, the colorimetric method from DAB9 was modified for propolis (Nagy & Grancai, 1996; Popova et al., 2004). Methanolic solutions of pinocembrin are used as references to obtain a calibration graph.

(1) To prepare a calibration graph with pinocebmrin as the standard, prepare a stock standard solution of pinocembrin 1.8 mg/ml by dissolving 18.0 mg in methanol in 10 ml volumetric flask.
(2) Prepare a series of working reference solutions by appropriate dilution of the stock standard solution with methanol (in volumetric flasks) to give concentration range of 0.18–1.8 mg/ml (0.9; 0.45; 0.22; 0.18 mg/ml).
(3) Dissolve 1 g of dinytrophenylhydrazine (DNP) in 2 ml 96% sulfuric acid and dilute to 100 ml with methanol (volumetric flask).
(4) Mix 0.5 ml of each one of the reference pinocembrin solutions and 1 ml of the DNP solution.
(5) Heat the mixture at 50 °C for 50 min (water bath).
(6) Cool the mixture to room temperature and dilute it to 5 ml with 10% KOH in methanol (w/v).
(7) Add 0.5 ml of the resulting solution to 10 ml methanol, dilute to 25 ml with methanol (volumetric flasks), and measure absorbance at 486 nm.
(8) As a blank, use 0.5 ml methanol instead of pinocembrin solution in analogues procedure (steps 4–7).
(9) Each reference solution should be analyzed in triplicate.
(10) To obtain the regression, absorbance should be plotted against concentration (International Conference on Harmonization, 1996).
(11) For analysis of the propolis sample, use 0.5 ml of each of the test solutions of each of the three parallel extractions, prepared as described in Section 3.1.1, and apply the same procedure as described for the reference (steps 4–8).
(12) Perform calculation using the regression obtained for pinocembrin. (step 10).

$$c = aA + b$$

where $c$ – concentration, mg/ml; $A$ – absorbance; $a$ – slope of the calibration graph; $b$ – intercept of the calibration graph.

(13) From this value, calculate the percentage of flavanones and dihydroflavonols in the propolis sample using the equation:

$$P = \frac{c \times 100}{\bar{M}} \times 100\%$$

where $P$ – percentage in raw propolis; $c$ – concentration, mg/ml; $\bar{M}$– mean value of the weight of the three parallel samples, extracted for analysis, mg (Section 3.8.1.1).

*3.8.1.4. Total phenolic content.* The Folin–Ciocalteu's method is used for the quantification of total phenolics (Waterman & Mole, 1994) and it is modified for poplar type propolis (Popova et al., 2004). As a reference, methanolic solutions of a mixture of pinocembrin-galangin at a 2:1 ratio (w/w) in the range 25–300 μg/ml are used to obtain a calibration graph.

(1) To prepare a calibration graph with pinocebmrin:galangin 2:1 (w/w) as the standard, prepare a stock standard solution by dissolving of 2.2 mg pinocembrin and 1.1 mg galangin in methanol in a 10 ml volumetric flask. The concentration of the stock solution is 0.33 mg/ml of the mixture pinocebmrin:galangin 2:1.
(2) Prepare a series of working reference solutions by appropriate dilution of the stock standard solution with methanol (in volumetric flasks) to give a concentration range of 33–330 μg/ml (165; 82.5; 41.2; 33 μg/ml) for the mixture pinocebmrin:galangin 2:1.
(3) Transfer 0.5 ml of the reference solution into a 25ml volumetric flask, containing 7.5 ml distilled water.

(4) Add 2 ml of the Folin–Ciocalteu's reagent and 3ml of a 20% sodium carbonate solution in distilled water.
(5) Make up the volume to 25 ml with distilled water and wait for 2 h (±3 min) at room temperature.
(6) Measure the absorbance at 760 nm using a UV–vis spectrophotometer.
(7) As a blank 0.5 ml methanol instead of reference mixture is used following the same procedure (steps 3–6).
(8) Each reference solution should be analyzed in triplicate.
(9) To obtain the regression, the absorbance should be plotted against concentration (International Conference on Harmonization, 1996).
(10) For analysis of the propolis samples, use 0.5 ml of the solution B (Section 3.8.1.1) in analogues procedure (steps 3–6). Every assay is carried out performed in triplicate.
(11) Perform the calculation using the regression obtained for the reference mixture pinocembrin-galangin (2:1, step 9).
(12) Perform calculations using the regression obtained for pinocembrin-galangin (2:1).

$$c = aA + b$$

where $c$ – concentration, mg/ml; $A$ – absorbance; $a$ – slope of the calibration graph; $b$ – intercept of the calibration graph.

(13) From this value, calculate the percentage of total phenolics in the propolis sample using the equation:

$$P = \frac{c \times 100 \times 50}{3M} \times 100\%$$

where $P$ – percentage in raw propolis; $c$ – concentration, mg/ml; $\bar{M}$– mean value of the weight of the three parallel samples, extracted for analysis, mg (Section 3.8.1.1).

*3.8.2. Spectrophotometric analysis of Brazilian green propolis*

The analysis of Brazilian green propolis consists in the spectrophotometric quantitative determination of the following groups of phenolic compounds: flavonoids; and total phenolics.

*3.8.2.1. Extraction of propolis.* The procedure described in Section 3.8.1.1 is used for the extraction of green propolis.

*3.8.2.2. Total flavonoid content.* The procedures described in Section 3.8.1.2 are followed to determine total flavonoid content. Methanolic solutions of quercetin are used for calibration (Woisky & Salatino, 1998).

*3.8.2.3. Total phenolic content.* The procedures described in Section 3.8.1.4 are followed to determine total phenolic content. Methanolic solutions of gallic acid are used for calibration (Woisky & Salatino, 1998).

*3.8.3. Spectrophotometric analysis of Pacific type propolis*

Since the main components and biologically active compounds in the Pacific type propolis are prenylated flavanones, the analysis of this type propolis is made on the basis of their quantification.

*3.8.3.1. Extraction of propolis.* The procedure described in Section 3.8.1.1 is used for the extraction of Pacific type propolis.

*3.8.3.2. Total flavanones content.* The procedures described in Section 3.8.1.3 are followed to determine total flavanone content. However, methanolic solutions of a mixture of propolin C-propolin D 4:1 (wt/wt) are used for calibration (Popova, Chen, Chen, Huang, & Bankova, 2010).

## 4. Quality criteria and standards

Propolis is a bee product of plant origin, so the standardization of propolis is similar to that of medicinal plants: it has to be based on the concentration of biologically active constituents. Different propolis types are characterized by their distinct chemical profiles and obviously there cannot be any uniform chemical criteria for standardization and quality control in this respect. Specific criteria based on the concentration of bioactive secondary metabolites should be formulated for particular propolis chemical types. The International Honey Commission suggests the values for the concentration of biologically active constituents for the two most wide-spread propolis types, European poplar type propolis (Poplar type) and Brazilian green propolis (*Baccharis* type), determined as described in Sections 3.8.1 and 3.8.2. For Brazilian green propolis, the values are determined by Brazilian legislation (Sawaya et al., 2011).

### 4.1. Specific criteria and standard values for particular propolis chemical types

The specific criteria and standard values for the most popular and most commercialized propolis types: poplar and green Brazilian propolis, are summarized in Table 5.

**Important:** Prior to the analysis, the chemical type of propolis should be determined by one of the analytical methods/dereplication strategies listed in Sections 3.3, 3.4, 3.5, 3.6, and 3.7. It is possible to apply by default the specific methodology and criteria for propolis from well-known geographic origins where it has been proved over the years to be of constant plant origin.

Table 5. Specific criteria and standard values for the content of bioactive constituents in propolis.

| Propolis type | | Minimum % by weight in raw propolis | Reference |
|---|---|---|---|
| *Poplar propolis* | Total phenolics | 21 | (Popova et al., 2004) |
| | Total flavones and flavonols | 4 | (Popova et al., 2004) |
| | Total flavanones and dihydroflavonols | 4 | (Popova et al., 2004) |
| *Brazilian green propolis* | Total phenolics | 5 | (Sawaya et al., 2011) |
| | Total flavonoids | 0.5 | (Sawaya et al., 2011) |

In the recent years, the problem of poplar propolis adulteration with poplar extracts emerged, connected mainly to Chinese propolis. An HPLC method was developed, based on detection of catechol as a marker for propolis adulteration (Huang et al., 2014).

### 4.2. *Criteria and standards common for all propolis types*

There are other quality parameters that can be applied to any propolis sample, no matter its plant origin and content of secondary plant metabolites. These include content of matter soluble in 70% ethanol (balsam content), water content, wax content, mechanical impurities, and ash content. The limits of their acceptable values, as suggested by the IHC follow:

Balsam – minimum 45% (Popova et al., 2007; http://www.ihc-platform.net/bankova2008.pdf).

Wax content – Different national standards suggest different values.

Mechanical impurities – maximum 6% (Popova et al., 2007; http://www.ihc-platform.net/bankova2008.pdf).

Water content – maximum 8% (Popova et al., 2007; http://www.ihc-platform.net/bankova2008.pdf).

Ash content – maximum 5% (Falcão, Freire, & Vilas-Boas, 2013b).

For Brazilian green propolis, Brazilian legislation determines a minimum of 35% ethanol extractable substances and a maximum of 25% wax (Sawaya et al., 2011).

#### 4.2.1. *Amount of matter soluble in 70% ethanol (balsam)*

(1) Perform extraction as described in Section 3.1.1.
(2) From each of the three parallel extracts, evaporate 2 mL *in vacuo* to dryness to constant weight g.
(3) Calculate the percentage of balsam *P* in the propolis sample using the following formula.

$$P = \frac{g \times 100}{2M} \times 100\%$$

where g – the weight of the residue after evaporation of 2 ml of propolis 70% ethanol extract; *M* – the weight of the raw propolis sample, g.

#### 4.2.2. *Water content*

Water content is determined according to Woisky and Salatino (1998).

(1) Heat 10 g of powdered raw propolis (see Section 3.1.1, step 1) in an oven at 105 °C for 5 h.
(2) Cool to room temperature and place in a desiccator until constant weight is achieved.
(3) Calculate the percentage of water content *P* in the propolis sample using the following formula.

$$P = \frac{M_0 - M_1}{M_0} \times 100\%$$

where $M_0$ – the weight of the raw propolis sample before heating, g; $M_1$ – the weight of the propolis residue after heating, g.

Figure 14. Determining the wax content of propolis by Soxhlet extraction.
Photo: B. Trusheva.

A mean of the three measurements should be calculated.

#### 4.2.3. *Wax content*

*4.2.3.1. Wax content measurement by extraction.* The wax content is determined according to the procedures described by Woisky and Salatino (1998).

(1) Treat 3 g of the powdered propolis sample (powdered per Section 3.1.1, step 1) with chloroform in a Soxhlet for 6 h (Figure 14), using a weighed cartridge.
(2) Concentrate the extract to dryness under reduced pressure and add 120 ml of hot methanol to the residue.
(3) Boil the mixture until there is a clear solution on top and a small oily residue on the bottom of the flask. The residue should solidify upon cooling.
(4) Filter the methanolic phase through filter paper, taking care to avoid transferring the oily residue. Transfer the methanolic phase, while hot, to a previously weighed 150 ml flask.
(5) Cool the flask containing the methanolic phase to 0 °C and filter the content through a filter paper that has been weighed and the weight recorded.
(6) Wash the flask and the residue with 25 ml cold methanol.
(7) After drying in the air, transfer the flask and the residue to a desiccator until constant weight.
(8) Calculate the percentage of wax content $P_w$ in the propolis sample using the following formula.

$$P_w = \frac{M_w}{M} \times 100\%$$

where $M_w$ – the weight of the wax obtained, g; $M$ – the weight of the propolis sample, g.

(9) The analysis should be performed in duplicate.

*4.2.3.2. Wax content measurement based on differences in specific density.* An alternative procedure for measuring the wax content of propolis has been described by Hogendoorn, Sommeijer, and Vredenbregt (2013).

(1) Add 25 ml de-ionized water to 20 g powdered propolis (powdered per Section 3.1.1, step 1) in a tube with screw-cap. When adding the water to the powdered sample, it is necessary to stir the mixture constantly and carefully to avoid propolis powder floating on the water surface.
(2) Tighten the screw cap loosely to prevent pressure building up while heating and place the tubes vertically in a household microwave apparatus set at medium.
(3) Adjust the time of heating so that the temperature rises to about 100 °C but without the boiling of the water phase (usually about 1 min).
(4) Cool down the sample to room temperature. A three layer system is formed in the tube: the beeswax (upper layer), then water (middle layer), and de-waxed propolis at the bottom.
(5) With a small stainless steel spatula, transfer the beeswax in the upper layer to a weighed paper tissue for the removal of the remaining water.
(6) Weigh the amount of extracted beeswax and calculate the wax content as a percentage of the weight of the original sample.
(7) The analysis should be performed in duplicate.

#### 4.2.4. *Mechanical impurities*

Follow the procedure below in order to determine the amount of mechanical impurities in a propolis sample.

(1) Extract the rest of the propolis sample (i.e. that which remained in the cartridge after the procedure described in Section 4.2.3.1) in the same Soxhlet with ethanol for 4 h (until the extract becomes colorless).
(2) Transfer the weighed cartridge together with the residue (the mechanical impurities), after drying it in the air, to a desiccator until constant weight.
(3) Calculate the percentage of mechanical impurities $P_{mi}$ in the propolis sample using the formula that follows.

$$P_{mi} = \frac{M_{mi}}{M} \times 100\%$$

where $M_{mi}$ – the weight of the residue after extraction, g; $M$ – the weight of the propolis sample, g.

(4) The analysis is performed in duplicate.

#### 4.2.5. *Ash content*

The ash content is determined according to the AOAC method (Association of Official Analytical Chemists, 2000).

(1) Place the crucible and lid in the furnace at 550 °C overnight to ensure that impurities on the surface of the crucible are burnt off.
(2) Cool the crucible in a desiccator for 30 min.
(3) Weigh the crucible and lid to 3 decimal places.
(4) Weigh about 5 g of the powdered propolis sample (Section 3.1.1 step 1) into the crucible. Heat over a low Bunsen flame with the lid half covering the crucible. When fumes are no longer produced, place crucible and lid into the furnace.

(5) Heat at 550 °C overnight. During heating, do not fully cover the crucible with the lid. After heating is complete, fully place the lid over the crucible to prevent the loss of fluffy ash. Cool the crucible down in a desiccator.
(6) Weigh the ash with crucible and lid when the sample turns gray. If the sample does not turn gray, return the crucible and lid to the furnace for the further ashing.
(7) Calculate the ash content using the formula that follows.

$$\text{Ash}(\%) = \frac{\text{Weight\_of\_ash}}{\text{Weight\_of\_sample}} \times 100$$

Figure 15. A cross-section of a feral honey bee hive within a tree cavity found September 2009 in the residential area of Bloomington, Minnesota, USA. The nest interior, where comb is present, is coated with a thin layer of propolis creating a "propolis envelope" around the colony. The upper portion of the cavity had not been lined with propolis, as the colony had not begun to use that space. Mold can be seen growing above the propolis envelope From: Simone-Finstrom and Spivak (2012).

## 5. Health benefits of a propolis envelope to bees

In a natural tree cavity, honey bees line the inside of the cavity with propolis in a contiguous sheet called a propolis "envelope" (Seeley & Morse, 1976). In a tree, the propolis envelope is particularly thick around the entrance and extends from where the combs attach at the top of the nest as far down as the combs are constructed (Simone-Finstrom & Spivak, 2012). Above and below the envelope, molds and fungi can be observed in the tree (Figure 15), which suggests that one purpose of the propolis envelope is to prevent the growth of molds inside the nest. The propolis envelope is an anti-microbial layer surrounding the colony and has quantifiable benefits to the bees' immune systems, and pathogen defense (Simone, Evans, & Spivak, 2009; Simone-Finstrom & Spivak, 2012).

The smooth and solid inner surfaces of standard beekeeping wooden boxes do not elicit resin collection behavior and further construction of a propolis envelope by bees. Instead, the bees deposit propolis in cracks and crevices, such as between boxes and under the frame rests, making it difficult to pry apart boxes and remove frames for beekeeping inspections without use of a hive-tool (Haydak, 1953; Huber, 1814; Ghisalberti, 1979). For this reason, many beekeepers do not like the difficulty that sticky propolis presents in the colony, and over many years, it is likely that queen producers have selected for colonies that do not deposit large quantities of propolis in the nest (Fearnley, 2001). At the same time, some beekeepers have harvested propolis from bee colonies for uses in human medicine (Burdock, 1998; Castaldo & Capasso, 2002; Krell, 1996).

The effects of a propolis envelope on honey bee immunity and on pathogen defense within the colony can be studied in two ways: (1) guide the bees to naturally deposit propolis throughout the nest interior; or (2) apply a propolis extract to the hive walls.

### *5.1. Forming a propolis envelope within standard beekeeping equipment*

#### *5.1.1. A naturally-deposited propolis envelope*

A colony of bees can be encouraged to build a natural propolis envelope within standard beekeeping equipment by modifying the inner walls of bee boxes. If the inside of the bee box is built using unfinished, rough lumber the bees will apply a layer of propolis over the rough surfaces. The inner walls of bee boxes can be scraped with a wire brush; the rougher the surface, the more propolis the bees will deposit on the walls (Simone-Finstrom & Spivak, personal observation). Alternatively, commercial propolis traps, used to harvest propolis, (see Section 2.2.1) can be cut to fit the four inside walls of the hive boxes and stapled with the smooth side of the trap facing the wood and the rough side facing the colony (Borba & Spivak, personal observation; Figure 16). It is recommended to manage

Figure 16. Propolis traps stapled to inside walls of hive to create a propolis envelope.
Photo: R. Borba.

colonies using nine frames instead of ten when using this method in standard 10-frame Langstroth equipment.

#### 5.1.2. *Experimental or artificial propolis envelope*

For experimental purposes when it is necessary to quantify the quantity or concentration of the propolis envelope, a propolis envelope can be painted on the inside surface of the box using an extract of propolis (Simone et al., 2009; Figure 17).

(1) Propolis is harvested using any combination of the methods described below (Section 2.2).
(2) Extraction of propolis – (13% propolis in 70% ethanol, e.g. Simone et al. (2009); see section 3.1 for further details and discussion).
(3) The extracts then can be painted on as a "varnish" for the interior hive walls. Based on the determined concentrations of the extracts ~50 g (for a nucleus colony, 5-frame Langstroth) or ~100 g (for a single deep, 10-frame Langstroth) of propolis should be applied evenly to the 4 side hive walls and the bottom board and cover (Simone et al., 2009; Simone-Finstrom & Spivak, 2012).
(4) In order to apply enough grams of propolis to the hive interior, multiple coats of the propolis extracts may need to be applied to the surfaces if the extract is not sufficiently strong or of high enough concentration for a single coat.
(5) The same volume of solvent used for the propolis extract should be applied to control colonies to account for any effects from the solvent alone.

Figure 17. Example of painting the hive interior with propolis extract to create a propolis envelope. The top box was painted with 70% ethanol, the middle with an extract of Brazilian green propolis and the bottom with MN propolis extract.
Photo: M. Simone-Finstrom.

### 5.2. *Effect of propolis envelope on the immune system of bees*

The honey bee immune response varies with age, so when comparing immune-related gene expression among treatments, it is important to sample bees of the same age. Young bees have greater fat body mass, therefore higher capacity to synthesize antimicrobial peptides, compared to older bees (Wilson-Rich, Spivak, Fefferman, & Starks, 2009). As honey bees age and switch from in-hive tasks to foraging, immune function can be altered both by age and task performance (Schmid, Brockmann, Pirk, Stanley, & Tautz, 2008; Wilson-Rich et al., 2009).

Previous studies on the role of propolis as a social immune trait have focused on younger, in-hive bees (e.g. Simone et al., 2009). However, investigators focusing on environmental effects on immunocompetence should consider collecting samples from other life stages and among behavioral tasks when possible (Human et al., 2013).

Once individuals are collected based on the colony treatments, RNA can be extracted for analysis of gene expression via real-time PCR (Evans et al., 2013; Simone et al., 2009). From current and previous work, gene expression for the antimicrobial peptide hymenoptaecin seems to be affected consistently by exposure to a propolis-enriched environment (e.g. Simone et al., 2009). However, continued work finds other genes involved in cellular immunity and representatives of each of the immune pathways, providing a more robust analysis of immune gene expression.

### 5.3. *Effect of propolis envelope on pathogens and pests in the hive*

In addition to indirect effects of propolis envelope on bee health through the immune system, research is underway to explore if the propolis envelope has direct effects on bee pathogens (e.g. Simone-Finstrom & Spivak, 2012) and pests. Colonies provided with a propolis envelope (either an extract or natural), can be challenged with *Ascosphaera apis*, *Paenibaciullus larvae*, other pathogens, small hive beetles (*Aethina tumida*), varroa (*Varroa destructor*), and other pests as described in *BEEBOOK* Vol II (e.g. De Graaf et al., 2013; Dietemann et al., 2013; Jensen et al., 2013; Neumann et al., 2013). Comparing challenged colonies with unchallenged controls allows quantification of the potential effects of propolis on the pest/pathogen in question.

### 5.4. *Self-medication: monitoring colony-level changes in resin-collection*

Colonies challenged with *A. apis* have been shown to collect significantly more resin after challenge (Simone-Finstrom & Spivak, 2012). Since a resin-enriched environment also reduces overall colony-level infection of this pathogen, resin foragers are self-medicating at the colony level against at least particular pathogens.

High variation across colonies in the number of resin foragers can be an issue when conducting this experiment. The appropriate sample size needs to be calculated carefully. Half of the colonies would be treated or challenged with a pathogen and the other half would remain unchallenged. An experiment to address the question of resin use as self-medication in honey bees combines the methods described above in Sections 2.1 and 5.3.

Statistical analysis of the change in resin foraging after exposure to pathogens can be done following various methods. One method previously used (Simone-Finstrom & Spivak, 2012), determined the change in resin foraging for each colony (total number of resin foragers pre-challenge subtracted from the total number counted post-challenge per colony). The change in resin foraging was then compared across pathogen-challenged and unchallenged colonies. A matched pairs analysis could also be used with treatment (challenged vs. unchallenged) as a factor in the statistical analysis.

The most accurate and direct indicator of increased resin use is by observing foraging rates (Simone-Finstrom & Spivak, 2012). However alternative methods of the assessment of propolis deposition in hives pre- and post-challenge could possibly be used to determine if resin collection rate increases in response to pathogen exposure. Deposition on commercial propolis traps (see Section 2.2.1) could be examined by weight or amount of coverage, although the amount of wax that is incorporated into resins varies highly across colonies and would greatly influence this measure. Similarly, the deposition of propolis on frame edges and in the hive itself, as described in the introduction to Section 5, could be analyzed but this has similar issues in terms of difficultly for accurate quantification (Borba, Simone-Finstrom & Spivak, personal observations).

## 6. Testing the biological activity of propolis *in vitro*

The most studied biological activities of propolis are the antimicrobial and antioxidative ones. Here, tests against both human and bee pathogens will be described.

### 6.1. *Testing the antibacterial activity*

#### 6.1.1. *Activity against human pathogens*

6.1.1.1. *Bacterial strains.* Antibacterial tests have been used to analyze bacterial sensitiveness to propolis. One may compare, for example, its effect on Gram positive and Gram negative bacteria, e.g. *Staphylococcus aureus* and *Escherichia coli* strains. American Type Culture Collection (ATCC) strains should be used in the assays.

6.1.1.2. *Susceptibility tests (macrodilution).* Susceptibility tests are performed by dilution in agar as recommended by the Clinical and Laboratory Standards Institute and minimal inhibitory concentration (MIC) values are determined (Alves et al., 2008; Clinical & Laboratory Standards Institute - CLSI/National Committee for Clinical Laboratory Standards – NCCLS, 2005).

(1) Inoculate bacterial strains in Brain Heart Infusion (BHI – Difco, USA) at 35 °C for 24 h and standardize at 0.5 on the McFarland scale in sterile saline (Sutton, 2011). Perform dilutions of each sample to obtain bacterial suspensions with $1 \times 10^6$ colony-forming units (CFU)/ml.

Figure 18. Steer's multiple inoculator used for bacterial inoculation in the plates.

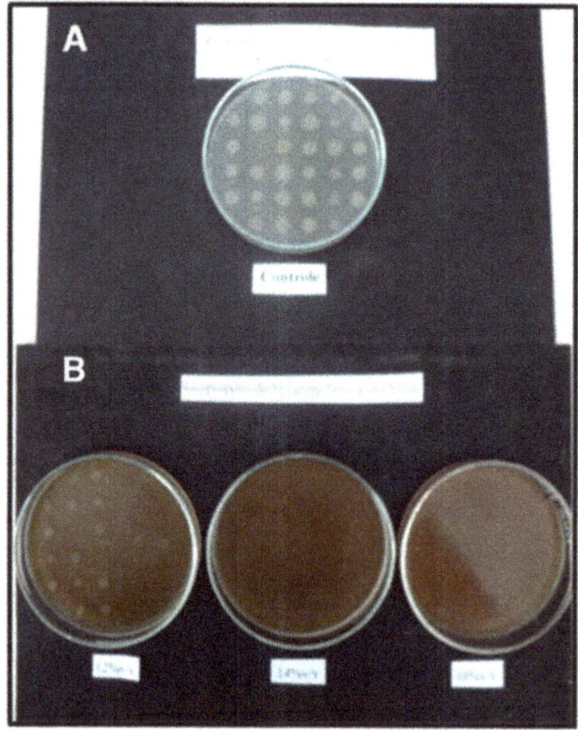


Figure 19. (A) Control plate showing bacterial growth. (B) Plates incubated with propolis showing the partial bacterial growth at left and inhibition of bacterial growth in the plates containing MIC (center and right).

(2) Add propolis to Petri dishes containing Mueller Hinton Agar (MHA) (Difco, USA) at different concentrations, such as: 3, 6, 9, 12, 14, 16, 18 and 20% v/v. Control plates contain only 70% ethanol at the same concentrations found in propolis.
(3) Inoculate bacterial strains in Petri dishes containing different concentrations propolis and 70% ethanol, using a Steer's multiple inoculator (Figure 18), and incubated at 35 °C for 24 h.
(4) $MIC_{90}$ is considered as the lowest concentration of propolis able to inhibit 90% of microorganisms, showing no visible growth or haze on the surface of the culture medium (Figure 19).

*6.1.1.3. Susceptibility tests (microdilution)*

(1) Incubate bacterial strains in BHI at 35 °C for 24 h and standardize at 0.5 on the McFarland scale (Sutton, 2011) in sterile saline. Perform dilutions of each sample to obtain bacterial suspensions with $1 \times 10^6$ CFU/ml.
(2) Add 100 μl of BHI medium containing different concentrations of propolis or ethanol 70% to 96 well plates and then 100 μl of the bacterial suspension. Incubate plates at 35 °C for 24 h (Figure 20).
(3) Read the plates by observing the turbidity of the solution in each well by adding the dye resazurin (50 μl). Record the MIC values of propolis for each strain (Figure 21). Resazurin (7-hydroxy-3H-phenoxazine-3-one-10-oxide) is a redox indicator used to check for the presence of viable cells in microdilution method. It naturally is blue or purple in color. In the presence of viable cells, it oxidizes to resofurin, which is red and promotes the observation of microbial growth (Alves et al., 2008).

*6.1.1.4. Time kill curve.* The time kill curve of bacteria is carried out to observe the bactericidal or bacteriostatic action of propolis over time, using the $MIC_{90}$ values.

(1) Inoculate bacterial suspensions ($1 \times 10^6$ CFU/ml) in tubes or Erlenmeyer flasks (20 ml) containing BHI plus Tween 80% (0.5% v/v) and the $MIC_{90}$ of propolis or 70% ethanol. Bacterial suspensions in BHI plus Tween 80% (0.5% v/v) alone are considered as control.
(2) After 3, 6, 9 and 24 h of incubation at 35 °C, take aliquots (50 μl) of each culture and plate on Plate Count Agar (PCA – Difco; USA) by the pour plate method which is used to count the bacteria. Put 50 μl of each solution in a dish and

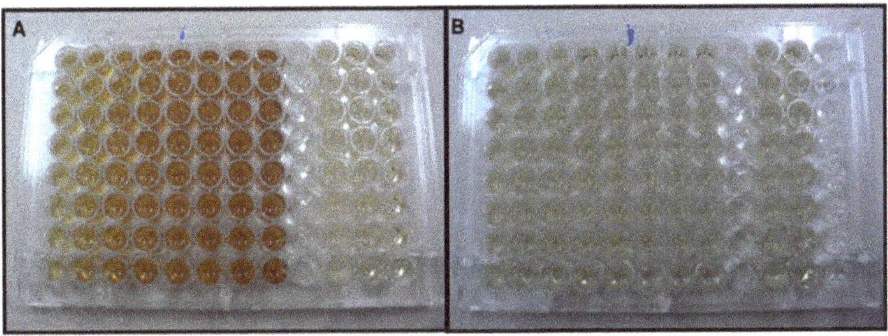


Figure 20. Plates for the microdilution test. In the 8 columns: BHI + propolis in different concentrations (A) or ethanol 70% (B). Column 10 (A and B): positive control (bacteria + BHI) and column 11 (A and B): negative control (BHI alone).

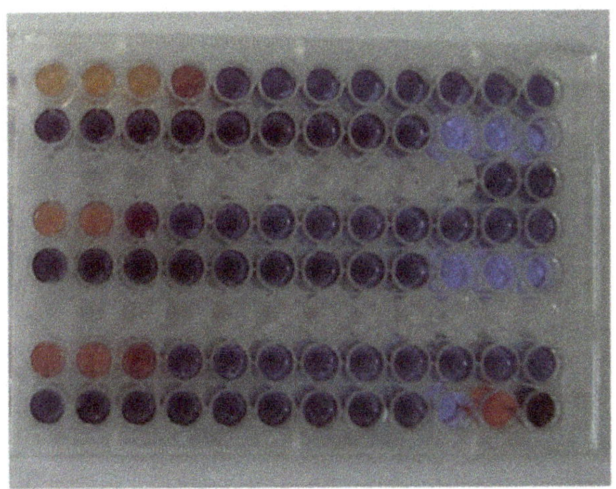

Figure 21. MIC of propolis. Blue color indicates absence of viable cells, while red color indicates the presence of viable ones.

mix with 15 ml of plate count agar (PCA). CFU are counted after incubation at 35 °C for 24 h.

(3) Calculate the survival percentage (Sforcin, Fernandes, Lopes, Bankova, & Funari, 2000) according to the formula:

$$\text{Survival percentage} = \text{CFU sample} \times 100/\text{CFU control}$$

#### 6.1.2. *Testing against bee pathogens: American foulbrood (*Paenibacillus larvae*)*

Described here is a high-throughput susceptibility assay published in Wilson et al. (2015) for testing antimicrobial activity against active *Paenibacillus larvae* cultures in 96-well plate format. Liquid *P. larvae* culturing techniques were adapted from Bastos, Simone, Jorge, Soares, & Spivak, (2008) and De Graaf et al. (2013). This protocol views antimicrobial activity as treated bacterial growth relative to untreated bacterial growth, and includes the equations for making good statistical comparisons of antimicrobial activity between propolis samples.

##### 6.1.2.1. *Culturing* P. larvae

(1) Obtain target strains of *P. larvae*. Many reference strains can be obtained from the USDA Agricultural Research Service culture collection (http://nrrl.ncaur.usda.gov/) and are discussed in De Graaf et al. (2013). Field strains can be isolated from infected larvae according to De Graaf et al. (2013).
(2) Grow stock *P. larvae* cultures in liquid brain/-heart infusion media (BHI) supplemented with 1 mg/l thiamine by shaking and incubating at 37 °C. A 30 ml stock culture started from lyophilized cells or isolated spores needs to be grown for 48 h.
(3) Split the stock culture into three 10 ml aliquots and add 10 ml glycerol to each aliquot and store at −20 °C. These 50% glycerol cultures should last for several months.
(4) Inoculate 29.5 ml of liquid BHI with 0.5 ml of glycerol culture. Shake and incubate at 37 °C for 48 h.

##### 6.1.2.2. *Preparing 96-well plates*

(1) Add propolis extracts (per Section 3.1.1) to flat-bottom 96-well plates in desired dilutions, and then dry extracts to residue under nitrogen. Experiments should include a range of propolis concentrations, with at least 3 replicates per treatment. Negative and positive growth controls should be included in the experiment.
(2) Add 100 µl of liquid BHI media to each propolis-treated well. Cover, shake, and incubate microplates at 37 °C for 15 min to solubilize propolis residue; however, propolis residue is unlikely to be completely soluble if concentrations are too high.
(3) Dilute the 48 h *P. larvae* culture started from glycerol stock 1:50 and add 100 µl of this dilute culture to each well. Measure the initial optical density (OD) at 600 nm with a spectrophotometer, which should be ~0.13 AU in untreated

controls. Cover, shake, and incubate at 37 °C.

(4) Measure final $OD_{600nm}$ at 6 h, which should be ~0.6 AU in untreated controls.

*6.1.2.3. Data analysis*

(1) Subtract the initial $OD_{600nm}$ of each well from the final $OD_{600nm}$ of each well to normalize the growth data.

(2) Bacterial growth can be interpreted relative to untreated controls as a percent:

$$\%\text{Relative growth} = \frac{\text{treated average } OD_{600nm}}{\text{untreated average } OD_{600nm}}$$

Error needs to be propagated between the two means used to calculated relative growth:

$$\%\text{Standard error} = \sqrt{\left(\frac{SE_a}{a}\right)^2 + \left(\frac{SE_b}{b}\right)^b}$$

where '*a*' is the treated average $OD_{600nm}$; '*b*' is the untreated average $OD_{600nm}$; $SE_a$ is the standard error of '*a*'; $SE_b$ is the standard error of '*b*'.

(3) If bacterial growth inhibition is dose-responsive, you should observe a sigmoidal growth curve with less growth at high propolis concentrations and more growth at low propolis concentrations. It is best if experiments are developed so that several of the highest propolis concentrations completely inhibit growth and several of the lowest propolis concentrations allow growth similar to untreated controls.

(4) Sigmoidal growth curves can be fit with a four-parameter logistic equation to calculate $IC_{50}$ values and their standard errors for individual propolis samples,

$$y = \min + \frac{\max - \min}{1 + \left(\frac{x}{IC_{50}}\right)^{-\text{Hillslope}}}$$

This operation can be done by many statistical analysis programs, such as SigmaPlot.

(5) $IC_{50}$ values can be compared pair-wise using confidence intervals:

$$''CI = z \pm \left[1.96\left(''\sqrt{(x^2 + y^2)}\right)\right]$$

where *x* is the standard error of $IC_{50(1)}$; *y* is the standard error of $IC_{50(2)}$; z is the difference between $IC_{50(1)}$ and $IC_{50(2)}$.

If the confidence interval of the difference between $IC_{50(1)}$ and $IC_{50(2)}$ does not include 0, then the difference between the two $IC_{50}$ values can be taken as significant. $\alpha$ = 1.96 in the equation above, which is the value used to test at 95% confidence.

### 6.2. Antifungal activity

*6.2.1. Testing against human pathogens*

*6.2.1.1. Yeasts.* Antifungal tests have been carried out to compare the sensitiveness of yeasts to propolis. As an example, pathogens isolated from human infections such as *Candida albicans*, *Candida guilliermondii* and *Candida tropicalis* may be used (Fernandes, Sugizaki, Fogo, Funari, & Lopes, 1995; Sforcin, Fernandes, Lopes, Bankova, & Funari, 2001). Microorganisms should be identified by current standard microbiological methods and ATCC strains should be used in the assays.

*6.2.1.2. Susceptibility tests (macrodilution).* Susceptibility tests may be performed by dilution in agar as recommended by the Clinical and Laboratory Standards Institute and MIC values are determined (Clinical & Laboratory Standards Institute - CLSI/National Committee for Clinical Laboratory Standards – NCCLS, 2005).

(1) Grow yeast strains in Sabouraud Dextrose Agar (Difco) at 35 °C/24 h. After incubation, suspend five colonies of each strain in 5 mL of sterile phosphate buffer solution (PBS) and dilute 1/100 in PBS to get a final inoculum of approximately $5 \times 10^4$ cells/ml.

(2) Make serial concentrations (% v/v) of propolis from each sample on plates containing Sabouraud Dextrose Agar to achieve 0.4, 0.6, 0.8, 1.0, 1.5, 2.0, 4.0, 6.0, 8.0, 9.0, 10.0, 10.5, 11.0, 11.5, 12.0, 12.5, 13.0, and 14.0%.

(3) Prepare a duplicate set of plates containing culture medium plus ethanol in order to obtain 5.0, 10.0, and 15.0% concentrations of solvent as control.

(4) Perform the inoculation procedures using a multiloop replicator, incubate the plates at 35 °C for 24 h and read MIC endpoints as the lowest propolis concentration that results in no visible growth or haze on the surface of the culture medium. Perform population analyses of data by calculating the MIC for 50 and 90% of the strains of each group of microorganisms.

*6.2.2. Testing against bee pathogens: chalkbrood fungus (Ascophaera apis)*

Described here is a high-throughput susceptibility assay published in Wilson et al. (2015) for testing antimicrobial activity against *Ascophaera apis* spores in 96-well plate format. Liquid culture and propagation techniques are based on those described in Jensen et al. (2013).

This protocol views antimicrobial activity as treated fungal growth relative to untreated fungal growth, and includes the equations for making good statistical comparisons of antimicrobial activity between propolis samples.

#### 6.2.2.1. *Culturing* A. apis

(1) Obtain target strains of *A. apis*. Reference strains can be obtained from the USDA Agricultural Research Service Entopathogenic Fungal Culture Collection (http://www.ars.usda.gov/is/np/systematics/fungibact.htm). USDA #7405 (+ mating type) and USDA #7406 (- mating type) were used in Wilson et al. (2015). Field strains can be isolated from chalkbrood mummies according to Jensen et al. (2013).
(2) Grow and mate strains on solid MY-20 media and then harvest spores into sterile water all according to Jensen et al. (2013). Store spore solution at 4 °C.
(3) Count spores under a microscope with a hemocytometer. There will be a high risk of contamination if spores were isolated from mummies, so proper steps must be taken to ensure that *A. apis* is the organism that grows in assay cultures. For PCR methods to identify *A. apis*, please refer to Jensen et al. (2013).

#### 6.2.2.2. *Preparing 96-well plates*

(1) Add propolis extracts to flat-bottom 96-well plates in desired dilutions, and then dry extracts to residue under nitrogen. Experiments should include a range of propolis concentrations, with at least 5 replicates per treatment. Negative and positive growth controls should be included in the experiment.
(2) Add 180 μl of liquid MY-20 media to each propolis-treated well. Cover, shake, and incubate microplates at 31 °C for 15 min to solubilize propolis residue; however, propolis residue is unlikely to be completely soluble if concentrations are too high.
(3) Add approximately $2.0 \times 10^6$ *A. apis* spores in 20 μl sterile water to each well. Measure initial OD at 600 nm with a spectrophotometer, which should be ~0.13 AU in untreated controls. Cover, shake, and incubate at 31 °C.
(4) Measure final $OD_{600nm}$ at 65 h, which should be ~0.8 AU in untreated controls. It takes ~50 h for spores to germinate, but near maximum growth should be achieved by 72 h.

#### 6.2.2.3. *Data analysis*

(1) Subtract the initial $OD_{600nm}$ of each well from the final $OD_{600nm}$ of each well to normalize the inhibition data.
(2) Bacterial growth can be interpreted relative to untreated controls as a percent:

$$\%\text{Relative growth} = \frac{\text{Treated average } OD_{600nm}}{\text{Untreated average } OD_{600nm}}$$

(3) Error needs to be propagated between the two means used to calculated relative growth:

$$\%\text{Standard error} = \sqrt{\left(\frac{SE_a}{a}\right)^2 + \left(\frac{SE_b}{b}\right)^b}$$

where '*a*' is the treated average $OD_{600nm}$; '*b*' is the untreated average $OD_{600nm}$; $SE_a$ is the standard error of '*a*'; $SE_b$ is the standard error of '*b*'.

(4) If fungal growth inhibition is dose-responsive, you should observe a sigmoidal growth curve with less growth at high propolis concentrations and more growth at low propolis concentrations. It is best if experiments are developed so that several of the highest propolis concentrations completely inhibit growth and several of the lowest propolis concentrations allow growth similar to untreated controls.
(5) Sigmoidal growth curves can be fit with a four-parameter logistic equation to calculate $IC_{50}$ values and their standard errors for individual propolis samples.

$$y = \min + \frac{\max - \min}{1 + \left(\frac{x}{IC_{50}}\right)^{-\text{Hillslope}}}$$

This operation can be done by many statistical analysis programs, such as SigmaPlot.

$$''CI = z \pm \left[1.96\left(''\sqrt{(x^2 + y^2)}\right)\right]$$

(6) $IC_{50}$ values can be compared pair-wise using confidence intervals.

$$''CI = z \pm \left[1.96\left(''\sqrt{(x^2 + y^2)}\right)\right]$$

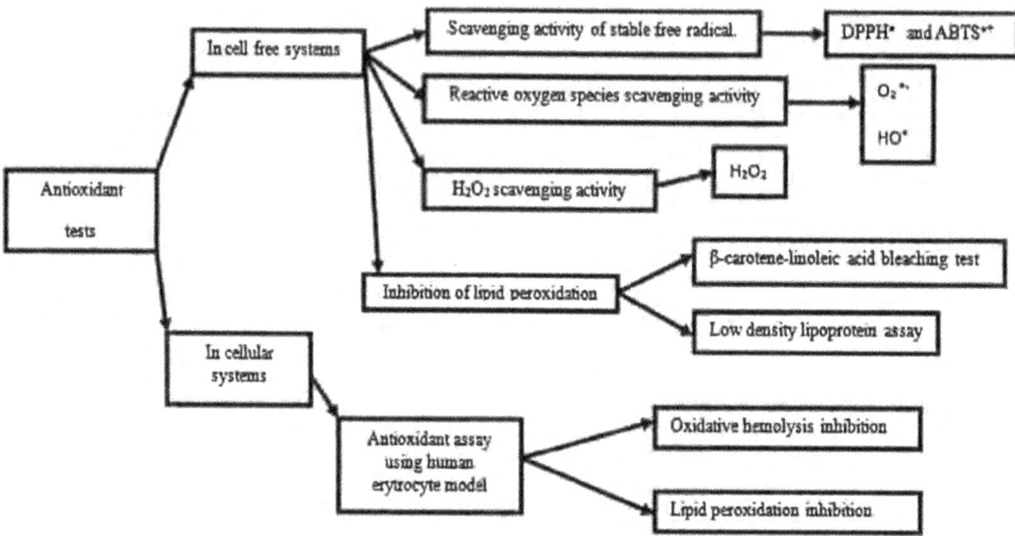


Figure 22. Methods for determination of antioxidant activity of propolis samples.

where $x$ is the standard error of $IC_{50(1)}$; $y$ is the standard error of $IC_{50(2)}$; $z$ is the difference between $IC_{50(1)}$ and $IC_{50(2)}$.

If the confidence interval of the difference between $IC_{50(1)}$ and $IC_{50(2)}$ does not include 0, then the difference between the two $IC_{50}$ values can be taken as significant. $\alpha = 1.96$ in the equation above, which is the value used to test at 95% confidence.

### 6.3. Testing the antioxidant activity of propolis

#### 6.3.1. Introduction

Oxidative stress, originated from an increase in free radical production or from a decrease in the antioxidant network, is characterized by the inability of endogenous antioxidants to counteract the oxidative damage on biological targets. In this context, it has been suggested that the intake of antioxidant is inversely associated with the risk to develop some pathologies like cancer, inflammatory process, cardiovascular diseases, and others (Lobo, Patil, Phatak, & Chandra, 2010; Pisoschi & Pop, 2015; Siti, Kamisah, & Kamsiah, 2015). Thus, attention has been paid to the antioxidant capacity of natural products such as bee products (honey, propolis), medicinal plant extract, and functional food (fruits and vegetable). Different *in vitro* assays have been developed to determine the antioxidant capacity of natural products (Figure 22). However, considering the complexity of *in vivo* antioxidant action mechanisms, several *in vitro* assays have also been used to study the potential antioxidant of natural products.

#### 6.3.2. Evaluation of the antioxidant activity in cell free system

##### 6.3.2.1. Scavenging activity toward stable free radicals ($DPPH^{\bullet}$, $ABTS^{\bullet+}$) by quantitative methods

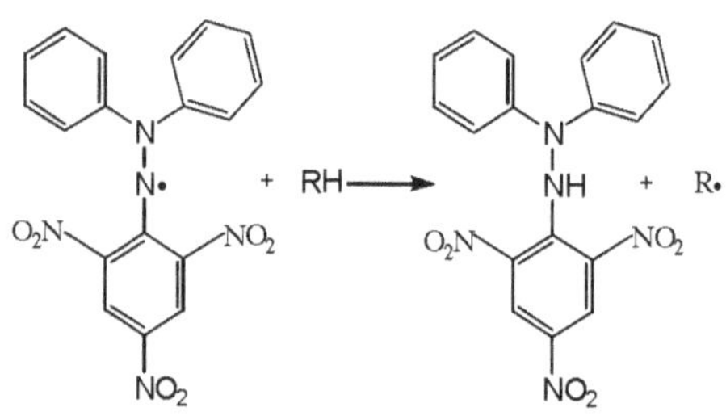

Figure 23. $DPPH^{\bullet}$ radical scavenging process, leading to decoloration which is registered spectrophotometrically in the DPPH assay.

*6.3.2.1.1. DPPH free radical scavenging activity.* The 1,1-diphenyl-2-picrylhydrazine (DPPH) radical scavenging assay is one of the most extensively used antioxidant assays for propolis samples. $DPPH^{\bullet}$ is a stable free radical that reacts with compounds that can donate a hydrogen atom. This method is based on the scavenging of $DPPH^{\bullet}$ through the addition of a radical species or an antioxidant that decolorizes the $DPPH^{\bullet}$ solution (Figure 23). The antioxidant activity is then measured by the decrease in absorption at 515 nm according to Nieva Moreno, Isla, Sampietro, and Vattuone (2000) and Yamaguchi, Takamura, Matoba, and Terao (1998).

*6.3.2.1.1.1. DPPH quantitative analysis using macromethod*

(1) Prepare a solution of $DPPH^{\bullet}$ in 96% ethanol to obtain a 300 μM $DPPH^{\bullet}$ solution.
(2) Add 1.5 ml of this solution to 0.5 ml of different concentrations of dry propolis extract (see Section 3.1.1) dissolved in 96% ethanol.
(3) Maintain during twenty minutes at 25 °C and then measure the absorbance at 517 nm in a

spectrophotometer. A decrease in the absorbance (>20%) of the reaction mixture indicates free radical scavenging activity of the propolis samples.

(4) Calculate the percentage of radical scavenging activity (RSA%) using the following equation:

$$\text{RSA\%} = \left[\frac{A_0 - A_s}{A_0}\right] \times 100$$

where $A_0$ is the absorbance of the control; $A_s$ is the absorbance of the samples at 515 nm.

$SC_{50}$ values denote the μg GAE/ml or μg dry weight of propolis extract/ml required to scavenge 50% DPPH free radicals. Quercetin, an antioxidant natural product or BHT, a synthetic antioxidant, are used as positive controls.

*6.3.2.1.1.2. DPPH quantitative analysis using micro-method.* Reaction mixtures containing different concentrations of propolis extract (0 to 50 μg dry weight of propolis extract (see Section 3.1.1) dissolved in 5 μl DMSO) and 95 μl of DPPH$^{\bullet}$ solution (0.125 mg/ml) in a 96-well microtiter plate are incubated at 25 °C for 30 min. Absorbance is measured at 550 nm in a microplate spectrophotometer. Scavenging activity ($SC_{50}$ values) of different propolis samples is determined by comparison with a DMSO control (Solórzano et al., 2012).

*6.3.2.1.2. ABTS free radical scavenging activity.* Along with the DPPH method (Section 6.3.2.1.1), the ABTS radical cation ($ABTS^{\bullet+}$) scavenging method is one of the most extensively used antioxidant assays for propolis samples. The ABTS radical cation is generated by the oxidation of ABTS with potassium persulfate, and its reduction in the presence of hydrogen-donating antioxidants is measured spectrophotometrically at 734 nm (Re et al., 1999).

$ABTS^{\bullet+}$ is generated by reaction of 7 mM ABTS and 2.45 mM potassium persulfate after incubation at room temperature (23 °C) in the dark for 16 h. The $ABTS^{\bullet+}$ solution is obtained by diluting the stock solution to an absorbance of 0.70 at 734 nm in ethanol, or PBS pH 7.4 according the solvent used to extract preparation.

*6.3.2.1.2.1. ABTS quantitative analysis using macro-method*

(1) Add $ABTS^{\bullet+}$ solution (1 ml) to 0.5 ml propolis extract (see Section 6.3.2.1.1.1 step 2) and mix thoroughly.
(2) The absorbance should be recorded at 734 nm after 6 min.
(3) Calculate the percentage of inhibition using the following formula:

$$\%\text{Inhibition} = \left[\frac{A_0 - A_s}{A_0}\right] \times 100$$

where $A_0$ is the absorbance of the control (blank, without propolis sample); $A_s$ is the absorbance in presence of propolis extract.

$SC_{50}$ values denote the μg GAE/ml required to scavenge 50% ABTS free radicals. This assay measures the total antioxidant capacity in both lipophilic and hydrophilic substances. Trolox, a water-soluble analog of Vitamin E, or quercetin, an antioxidant natural product is used as a positive control.

*6.3.2.1.2.2. ABTS quantitative analysis using micro-method.* Different concentrations of propolis dry extract (see Section 3.1.1) dissolved in ethanol or buffer (20 μl) and 180 μl of $ABTS^{\bullet+}$ are incubated at 25 °C for 6 min. Absorbance is measured at 734 nm in a microplate spectrophotometer. Scavenging activity ($SC_{50}$ values) of different propolis samples is determined by comparison to an ethanol or buffer control.

*6.3.2.1.3. Scavenging activity toward stable free radicals (DPPH$^{\bullet}$, $ABTS^{\bullet+}$) by qualitative methods: autographic assay with DPPH$^{\bullet}$ and $ABTS^{\bullet+}$*

(1) Separate the chemical components of the propolis extract (see Section 3.1.1) by thin layer chromatography (TLC, 4 × 4 cm silica gel plate) using as mobile phase a solvent system such as toluene: chloroform:acetone 4.5:2.5:3.5 v/v/v.
(2) Air-dry the TLC plate.
(3) Distribute 3 ml of medium containing agar 0.9% and 1 ml $ABTS^{\bullet+}$ solution (Figure 23) or DPPH$^{\bullet}$ solution on TLC plates (Vera et al., 2011; Zampini, Ordoñez, & Isla, 2010).

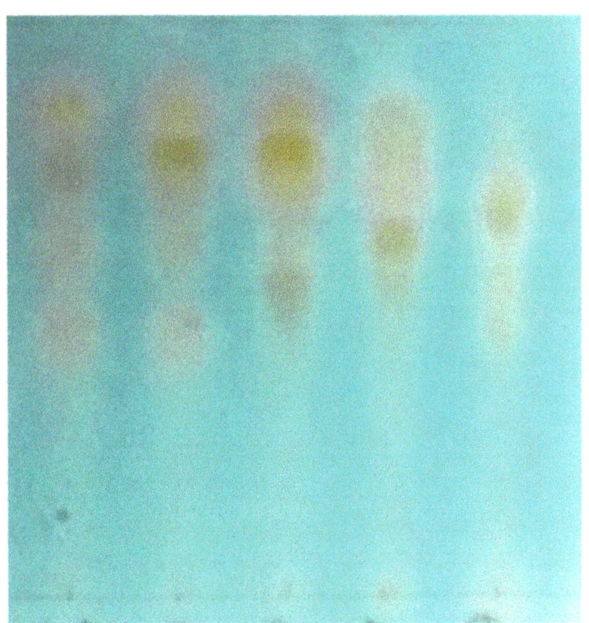

Figure 24. Autographic assay of $ABTS^{\bullet+}$ scavenging activity in propolis samples. The yellow spots on the thin layer chromarography correspond to compounds which scavenge $ABTS^{\bullet+}$ radicals.

(4) Incubate the plate at room temperature for 1 min in the dark.
(5) The antioxidant compounds are visualized as bright areas on a purplish (DPPH) or green blue ($ABTS^{\bullet+}$) background (Figure 24).

#### 6.3.2.2. *Scavenging activity of reactive oxygen species*

*6.3.2.2.1. Superoxide radical scavenging activity-non-enzymatic assay.* Superoxide radicals are generated by the NADH/PMS (phenazine methosulfate) system following a method as described Valentão et al. (2002) and modified by Danert et al. (2014).

(1) Mix a total of 50 μl of the tested propolis extract (see Section 3.1.1) with 40 μl of NADH (2 mM), 20 μl of NBT (nitroblue tetrazolium) (1 mM) and 40 μl of PMS (60 μM).
(2) Dissolve all the reagents in a phosphate buffer (19 mM, pH 7.4).
(3) Dissolve the extracts in DMSO (final concentration of 0.1%).
(4) Incubate the reaction mixture for 30 min at 37 °C and measure the absorbance at 550 nm in a microplate reader. $SC_{50}$ values denote the μg GAE/ml required to scavenge 50% of superoxide free radicals and are obtained from doses-response curves.

*6.3.2.2.2. Hydroxyl radical scavenging activity.* Hydroxyl radical scavenging is carried out by measuring the competition between deoxyribose and each extract for hydroxyl radicals generated from the $Fe^{3+}$/ascorbate/EDTA/$H_2O_2$ system. The attack of the hydroxyl radical on deoxyribose leads to thiobarbituric acid reactive species formation. The reaction is performed according to Chobot (2010) with modifications according to Danert et al. (2014).

(1) Add various concentrations of propolis extract (see Section 3.1.1) to the reaction mixture containing 50 μl of a 10.4 mM 2-deoxy-D-ribose solution, 100 μl of $FeCl_3$ (50 μM) and 100 μl of 52 μM EDTA.
(2) Add 50 μl of 10 mM $H_2O_2$, 50 μl of 1.0 mM ascorbic acid and 50 mM phosphate buffer (pH 7.4) making up a final volume of 0.5 ml, to start the Fenton reaction.
(3) Incubate the reaction mixture at 37 °C for 1 h.
(4) Dissolve 500 μl of 1% 2-thiobarbituric acid in 3% trichloroacetic acid (w/v) and add to each test tube and maintained at 100 °C for 20 min.
(5) To remove the reaction product, add 700 μl of *n*-butanol and vigorously vortex the mixture.
(6) Separate the *n*-butanol layers, each 600 μl, and measure the absorbance at 532 nm.
(7) Assays are performed in triplicate.

Reaction mixtures without the test compound serve as positive controls (100% malodialdehyde). The negative control should contain the full reaction mixture except 2-deoxy-D-ribose. Controls without either EDTA or ascorbic should be performed. $IC_{50}$ values are obtained from dose-response curves.

*6.3.2.2.3. Hydrogen peroxide scavenging activity*

(1) Prepare a solution (4 mM) of hydrogen peroxide in phosphate buffer (PBS, pH 7.4).
(2) Determine hydrogen peroxide concentration spectrophotometrically from absorption at 230 nm using the molar absorptivity 81 $M^{-1}$ $cm^{-1}$.
(3) Add the propolis sample to the hydrogen peroxide solution (0.6 ml).
(4) Measure the absorbance of hydrogen peroxide with and without propolis extract at 230 nm (Aruoma, Grootveld, & Halliwell, 1987; Zampini et al., 2008).

#### 6.3.2.3. *Inhibition of lipid oxidation*

*6.3.2.3.1. β-Carotene-linoleic acid bleaching assay.* The β-carotene linoleic acid bleaching assay is one of the antioxidant assays suitable for propolis samples. In this assay, the antioxidant capacity is determined by the formation of conjugated diene hydroperoxides arising from linoleic acid oxidation, which results in the discolouration of β-carotene. The reaction is carried out according to Velioglu, Mazza, Gao, and Oomah (1998) with slight modifications according to Danert et al. (2014).

(1) Add β-carotene (10 mg) in 50 μl of chloroform to 40 μl of linoleic acid and 400 μl of Tween 40 emulsifier mixture.
(2) After evaporation of the chloroform under vacuum, add 10 ml of distilled water with vigorous shaking.
(3) Add 60 ml of 14 mM $H_2O_2$, transfer 1ml of this mixture into test tubes containing different concentrations of dry propolis extract (see Section 3.1.1) or positive controls (100 μl).
(4) As soon as the emulsion is added to each tube, the zero time point absorbance is measured at 470 nm using a spectrophotometer.
(5) Incubate the emulsion for 2 h at 50 °C. A blank, devoid of $\beta$-carotene is prepared and a control of $\beta$-carotene and propolis. Quercetin, BHT and $\alpha$-tocopherol are used as standards.

*6.3.2.3.2. Inhibition of oxidation of low density lipoprotein.* At present, it is well known that reactive oxygen species (ROS) can play a pivotal role in the initiation, propagation and termination reactions of the low density lipoprotein (LDL) peroxidation processes (Lobo et al., 2010; Pisoschi & Pop, 2015; Siti et al., 2015). *In vitro* assays usually employ cupric sulfate or cupric chloride as initiators of LDL oxidation and the lipid peroxidation processes should be followed with the formation of diene conjugates by UV spectroscopy at 234 nm. The kinetic is characterized by the presence of a lag time associated with the presence of endogenous antioxidants (mainly vitamin E and coenzyme Q) in the LDL particle. After that period, the peroxidation of lipids is evidenced as an increase in the absorbance at 234 nm. In the presence of antioxidants, this lag time is increased. The main advantage of this *in vitro* assay is the use of a biologically relevant target.

(1) Obtain blood by vein puncture of a forearm vein from 12 h fasted individuals.
(2) Receive the blood into tubes without anticoagulant and centrifuge at 1000× *g* for 20 min at 4 °C.
(3) Recover the serum and use it immediately for the assays.
(4) Incubate human serum samples containing 1.23 mg of protein/ml; 0.035 mg of LDL-cholesterol/ml; 0.04 mg of protein of LDL/ml in 10 ml of PBS (10 mM sodium phosphate buffer, pH 7.4, containing 0.15 M NaCl) at 37 °C with or without $CuCl_2$ (final concentration 11.7 mM) and with or without propolis extract (see Section 3.1.1) (final concentration 1–50 µg/ml) for 2 h.
(5) Terminate the oxidation by the addition of 100 µmol of EDTA or 10 µM butylated hydroxytoluene (BHT) and refrigeration at 4 °C (Aviram, 1996; adapted by Isla, Nieva Moreno, Sampietro, & Vattuone, 2001).
(6) The formation of conjugated dienes is followed by the absorbance at 234 nm. Determine the concentration of dienes using the difference in absorbance at zero time and at the end of experiment, using the molar absorption coefficient $\varepsilon_{234} = 29{,}500\ M^{-1}\ cm^{-1}$ for conjugated dienes (Abuja, Murkovic, & Pfannhauser, 1998). Lag times (min) should be determined from the intercept of lines drawn through the linear portions of the lag phase and propagation phase.

#### 6.3.3. *Evaluation of the antioxidant activity of propolis in cellular systems*

Oxidative stress can be induced in whole cell suspension by hydrophilic compounds such as $H_2O_2$ or 2,2′-Azobis-(2-amidinopropane) dihydrochloride (AAPS). $H_2O_2$ that is normally generated *in vivo* mainly by the autoxidation of hemoglobin and dismutation of superoxide gives rise to radicals like hydroxyl ions. AAPH generates peroxyl radicals outside the membrane.

##### 6.3.3.1. *Inhibitory efficiency of propolis extracts on $H_2O_2$-induced lipid peroxidation*

(1) Obtain blood (5–10 ml) from healthy non-smoker adult individuals after informed consent. Isolate human erythrocytes from citrated blood immediately by centrifugation at 1500 rpm for 10 min at 4 °C.
(2) After removal of plasma and buffy coat, wash the erythrocytes three times with phosphate-buffered saline (PBS; pH 7.4) at 4 °C, and, finally, resuspend in PBS to obtain erythrocyte suspensions at 5%.
(3) Dissolve initially the dry propolis extract (collected per Section 3.1.1) in DMSO to obtain stock solutions and further dilute in PBS to obtain different final concentrations of propolis. From these serial dilutions, the DMSO final concentration is never higher than 0.08%.
(4) To study the protective effects of propolis extracts against $H_2O_2$-induced lipid peroxidation, pre-incubate 0.5 ml of an erythrocyte suspension at 5% in PBS suspension, with 10 µl of propolis extract in presence or absence of 4 mM sodium azide, a catalase inhibitor, for 20 min at 37 °C before inducing oxidative stress.
(5) After incubation, centrifuge the mixture, wash with PBS, re-suspend with 0.5 ml of PBS and treat with 0.5 ml of 0.5, 2 and 8 mM of $H_2O_2$ for 4 h at room temperature (Senturk et al., 2001). A negative control (erythrocytes in PBS), a positive control (erythrocytes in PBS with ascorbic acid), and extract controls (erythrocytes in PBS with each extract) are necessary.
(6) Estimate the extension of lipid peroxidation using a modified thiobarbituric acid (TBA) assay. Briefly, take 500 µl of erythrocyte suspensions and incubate at 95 °C for 45 min with 1 ml of TBA–TCA–HCl (0.375% (w/v) TBA, 15% (w/v) TCA, 0.25 M HCl).
(7) Cool at room temperature and centrifuge at 1000× g for 10 min.
(8) Measure the absorbance of the supernatant at 532 nm. Use a standard curve to quantify the amount of MDA.

##### 6.3.3.2. *Protective effect of propolis extracts on $H_2O_2$-induced oxidative hemolysis*

(1) Pre-incubate 0.5 ml of an erythrocyte suspension at 5% in PBS, with 10 µl of propolis extract (collected per Section 3.1.1) in presence or absence of 4 mM sodium azide, a catalase inhibitor, for 20 min at 37° C before inducing oxidative stress.
(2) After incubation, centrifuge the mixture, wash it with PBS, re-suspend with 0.5 ml of PBS and

treat with 0.5 ml of 0.5, 2 and 8 mM of $H_2O_2$ for 4 h at room temperature (Senturk et al., 2001). A negative control (erythrocytes in PBS) and extract controls (erythrocytes in PBS with each extract) are necessary.

(3) Take out aliquots of the reaction mixture at each hour during 4 h of incubation, dilute with saline, and centrifuge at 1000× g for 10 min to separate the erythrocytes.

(4) Determine the percentage of hemolysis by measuring the absorbance of the supernatant (A) at 545 nm and compare with that of complete hemolysis (B) by treating an aliquot with the same volume of the reaction mixture with distilled water.

(5) Calculate the hemolysis percentage using the formula: $A/B \times 100$. $IC_{50}$ values at time 3 h are determined from a concentration–response curve obtained by plotting the percentage of hemolysis inhibition vs. the extract concentration. Use ascorbic acid as the reference antioxidant compound.

### 6.4. *Antiparasitic activity: action against varroa*

The methods are described in Dietemann et al. (2013).

### 6.5. *Other tests, including clinical tests*

Propolis extracts have been tested for many different types of biological and pharmacological activities (Burdock, 1998; Farooqui & Farooqui, 2010; Sforcin & Bankova, 2011), including in clinical trials (Henshaw et al., 2014; Hoheisel, 2001; Paulino, Coutinho, Coutinho, & Scremin, 2014; Soroy, Bagus, Yongkie, & Djoko, 2014; Vaz Coelho et al., 2007). Most significant is the number of clinical trials in dentistry (Anauate-Netto et al., 2014; Pereira et al., 2011; Prabhakar, Karuna, Yavagal, & Deepak, 2015; Purra, Mushtaq, Acharya, & Saraswati, 2014; Torwane et al., 2013). It is impossible to describe standard methods for these numerous and diverse tests here.

However, it is essential to emphasize the importance of using chemically characterized and standardized propolis in any biological and/or clinical test performed with propolis extracts and preparations containing propolis. The fact that propolis chemical composition varies dramatically with the geographic and plant origin makes any pharmacological research done with propolis without chemical characterization irreproducible and completely irrelevant.

## 7. Conclusion

Propolis has been attracting the attention of researchers for over five decades, due to its wide range of valuable pharmacological activity and potential for prevention and treatment of numerous diseases. Only recently have scientists begun to recognize the importance of propolis for honey bees and its significance as a component of their social immunity. Appropriate methods should be developed further for in-depth studies of this aspect of propolis function.

Future studies on propolis should also be directed to the development of procedures for the standardization of propolis types other than poplar type and green Brazilian propolis, and to conduct research on propolis from different geographic regions in order to characterize them chemically and discover their plant source(s). Studies of biological and pharmacological activities of propolis have to be performed only with chemically characterized and standardized propolis in order to get meaningful, reliable and reproducible results. Metabolomics approaches should be applied in combination with biological tests in order to get a holistic picture of the composition-activity relationship.

## Acknowledgements

The COLOSS (Prevention of honey bee COlony LOSSes) Association aims to explain and prevent massive honey bee colony losses. It was funded through the COST Action FA0803. COST (European Cooperation in Science and Technology) is a unique means for European researchers to jointly develop their own ideas and new initiatives across all scientific disciplines through trans-European networking of nationally funded research activities. Based on a pan-European intergovernmental framework for cooperation in science and technology, COST has contributed since its creation more than 40 years ago to closing the gap between science, policy makers and society throughout Europe and beyond. COST is supported by the EU Seventh Framework Program for research, technological development and demonstration activities (Official Journal L 412, 30 December 2006). The European Science Foundation as implementing agent of COST provides the COST Office through an EC Grant Agreement. The Council of the European Union provides the COST Secretariat. The COLOSS network is now supported by the Ricola Foundation – Nature & Culture.

## Disclosure statement

No potential conflict of interest was reported by the authors.

## References

Abuja, P. M., Murkovic, M., & Pfannhauser, W. (1998). Antioxidant and prooxidant activities of Elderberry (*Sambucus nigra*) extract in low-density lipoprotein oxidation. *Journal of Agricultural Food Chemistry, 46*, 4091–4096. doi:10.1021/jf980296g

Ahn, M. R., Kumazawa, S., Usui, Y., Nakamura, J., Matsuka, M., Zhu, F., & Nakayama, T. (2007). Antioxidant activity and constituents of propolis collected in various areas of China. *Food Chemistry, 101*, 1383–1392. doi:10.1016/j.foodchem.2006.03.045

Alves, E. G., Vinholis, A. H. C., Casemiro, L. A., Furtado, N. A. J. C., Silva, M. L. A., Cunha, W. R., & Martins, C. H. G. (2008). Estudo comparativo de técnicas de *screening* para a avaliação da atividade antibacteriana de extratos brutos de espécies vegetais e de substâncias puras [Comparative study of screening techniques for antibacterial activity evaluation of plant crude extracts and pure compounds]. *Química Nova, 31*, 1224–1229. doi:10.1590/S0100-40422008000500052

Anauate-Netto, C., Anido-Anido, A., Leegoy, H. R., Matsumoto, R., Alonso, R. C. B., Marcucci, M. C., … Bretz, W. A. (2014). Randomized, double-blind, placebo-controlled clinical trial on the effects of propolis and chlorhexidine mouthrinses on gingivitis. *Brazilian Dental Science, 17*, 11–15. doi:10.14295/bds.2014.v17i1.947

Aruoma, O., Grootveld, M., & Halliwell, B. (1987). The role of iron in ascorbate dependent deoxyribose degradation. *Journal of Bioorganic Biochemistry, 29*, 289–299. doi:10.1016/0162-0134(87)80035-1

Association of Official Analytical Chemists. (2000). *Official methods of analysis*. Washington, DC: Author.

Aviram, M. (1996). Interaction of oxidized low density lipoprotein (OX-LDL) with macrophage in atherosclerosis and antiatherogenicity of antioxidants. *European Journal of Clinical Chemistry and Biochemistry, 34*, 599–608.

Bankova, V., Dyulgerov, A., Popov, S., & Marekov, N. (1987). A GC/MS study of the propolis phenolic constituents. *Zeitschrift für Naturforschung, C, 42*, 147–151. doi:10.1515/znc-1987-1-224

Bankova, V., Boudourova-Krasteva, G., Popov, S., Sforcin, J. M., & Cunha Funari, S. R. (1998). Seasonal variations in essential oil from Brazilian propolis. *Journal of Essential Oil Research, 10*, 693–696. doi:10.1080/10412905.1998.9701012

Bankova, V., Boudourova-Krasteva, G., Sforcin, J. M., Frete, X., Kujumgiev, A., Maimoni-Rodella, R., & Popov, S. (1999). Phytochemical evidence for the plant origin of Brazilian propolis from Sao Paulo state. *Zeitschrift für Naturforschung C, 54*, 401–405. doi:10.1515/znc-1999-5-616

Bankova, V., de Castro, S. L., & Marcucci, M. C. (2000). Propolis: Recent advances in chemistry and plant origin. *Apidologie, 31*, 3–15. doi:10.1051/apido:2000102

Bankova, V., Popova, M., Bogdanov, S., & Sabatini, A. G. (2002). Chemical composition of European propolis: Expected and unexpected results. *Zeitschrift für Naturforschung, C, 57*, 530–533. doi:10.1515/znc-2002-5-622

Bankova, V., Popova, M., & Trusheva, B. (2006). Plant sources of propolis: An update from a chemist's point of view. *Natural Product Communications, 1*, 1023–1028.

Bankova, V., Popova, M., & Trusheva, B. (2014). Propolis volatile compounds: Chemical diversity and biological activity: A review. *Chemistry Central Journal, 8*, 24. doi:10.1186/1752-153X-8-28

Bastos, E. M. A. F., Simone, M., Jorge, D. M., Soares, A. E. E., & Spivak, M. (2008). *In vitro* study of the antimicrobial activity of Brazilian propolis against *Paenibacillus larvae*. *Journal of Invertebrate Pathology, 97*, 273–281. doi:10.1016/j.jip.2007.10.007

Bertelli, D., Papotti, G., Bortolotti, L., Marcazzan, G. L., & Plessi, M. (2012). $^{1}$H-NMR simultaneous identification of health-relevant compounds in propolis extracts. *Phytochemical Analysis, 23*, 260–266. doi:10.1002/pca.1352

Bonvehi, J. S., & Coll, F. V. (1994). Phenolic composition of propolis from China and from South America. *Zeitschrift für Naturforschung, C, 49*, 712–718. doi:10.1515/znc-1994-11-1204

Brereton, R. G. (2013). *Chemometrics: Data analysis for the laboratory and chemical plant*. Chichester: John Wiley & Sons.

Burdock, G. A. (1998). Review of the biological properties and toxicity of bee propolis (propolis). *Food & Chemical Toxicology, 36*, 347–363. doi:10.1016/S0278-6915(97)00145-2

Butler, C. G. (1949). *The Honeybee: An introduction to her sense-physiology and behaviour*. London: Oxford University Press.

Castaldo, S., & Capasso, F. (2002). Propolis, an old remedy used in modern medicine. *Fitoterapia, 73*, S1–S6. doi:10.1016/S0367-326X(02)00185-5

Charlton, A. J., Farrington, W. H. H., & Brereton, P. (2002). Application of $^{1}$H NMR and multivariate statistics for screening complex mixtures: Quality control and authenticity of instant coffee. *Journal of Agricultural and Food Chemistry, 50*, 3098–3103. doi:10.1021/jf011539z

Cheng, H., Qin, Z. H., Guo, X. F., Hu, X. S., & Wu, J. H. (2013). Geographical origin identification of propolis using GC-MS and electronic nose combined with principal component analysis. *Food Research International, 51*, 813–822. doi:10.1016/j.foodres.2013.01.053

Chobot, V. (2010). Simultaneous detection of pro- and antioxidative effects in the variants of the deoxyribose degradation assay. *Journal of Agricultural and Food Chemistry, 58*, 2088–2094. doi:10.1021/jf902395k

Clinical and Laboratory Standards Institute - CLSI/National Committee for Clinical Laboratory Standards – NCCLS. (2005). *Performance Standards for Antimicrobial Susceptibility Testing* (CLSI document M100-S15). Wayne, PA.

Crane, E. (1990). *Bees and beekeeping: Science, practice, and world resources* (p. 640). Ithaca, NY: Cornell University Press. ISBN: 978-0801424298.

Crane, E. (1999). *The world history of beekeeping and honey hunting* (p. 682). London: Duckworth. ISBN: 978-0715628270.

Cuesta-Rubio, O., Piccinelli, A. L., Campo Fernandez, M., Márquez Hernández, I., Rosado, A., & Rastrelli, L. (2007). Chemical characterization of Cuban propolis by HPLC-PDA, HPLC-MS, and NMR: The brown, red, and yellow Cuban varieties of propolis. *Journal of Agricultural and Food Chemistry, 55*, 7502–7509. doi:10.1021/jf071296w

Cuyckens, F., & Claeys, M. (2004). Mass spectrometry in the structural analysis of flavonoids. *Journal of Mass Spectrometry, 39*, 1–15. doi:10.1002/jms.585

Danert, F. C., Zampini, C., Ordoñez, R., Maldonado, L., Bedascarrasbure, E., & Isla, M. I. (2014). Argentinean propolis as non-conventional functional foods. Nutritional and functional composition. *Natural Product Communications, 9*, 167–170.

De Graaf, D. C., Alippi, A. M., Antúnez, K., Aronstein, K. A., Budge, G., De Koker, D., … Genersch, E. (2013). Standard methods for American foulbrood research. In V. Dietemann, J. D. Ellis, & P. Neumann (Eds.), *The COLOSS BEEBOOK, Volume II: Standard methods for Apis mellifera pest and pathogen research. Journal of Apicultural Research*, 52(1). doi:10.3896/IBRA.1.52.1.11

Dietemann, V., Nazzi, F., Martin, S. J., Anderson, D., Locke, B., Delaplane, K. S., …, Ellis, J. D. (2013). Standard methods for varroa research. In V. Dietemann, J. D. Ellis, & P. Neumann (Eds.), *The COLOSS BEEBOOK, Volume II: Standard methods for Apis mellifera pest and pathogen research. Journal of Apicultural Research, 52*(1). doi: 10.3896/IBRA.1.52.1.09

Ellis, J. D., & Hepburn, H. R. (2003). A note on mapping propolis deposits in Cape honey bee (*Apis mellifera capensis*) colonies. *African Entomology, 11*, 122–124. Retrieved from http://agris.fao.org/aos/records/ZA2003000542

Evans, J. D., Schwarz, R. S., Chen, Y.-P., Budge, G., Cornman, R. S., De La Rua, P., … Pinto, M. A. (2013). Standard methodologies for molecular research in *Apis mellifera*. In V. Dietemann, J. D. Ellis, & P. Neumann (Eds.), *The COLOSS BEEBOOK, Volume I: Standard methods for Apis mellifera research. Journal of Apicultural Research, 52*(4). doi:10.3896/IBRA.1.52.4.11

Falcão, S., Vilas-Boas, M., Estevinho, L. M., Barros, C., Domingues, M. R. M., & Cardoso, S. M. (2010). Phenolic characterization of Northeast Portuguese propolis: Usual and unusual compounds. *Analytical and Bioanalytical Chemistry, 396*, 887–897. doi:10.1007/s00216-009-3232-8

Falcão, S., Vale, N., Gomes, P., Domingues, M. R. M., Freire, C., Cardoso, S. M., & Vilas-Boas, M. (2013a). Phenolic profiling of Portuguese propolis by LC-MS spectrometry: Uncommon propolis rich with flavonoid glycosides. *Phytochemical Analysis, 24*, 309–318. doi:10.1002/pca.2412

Falcão, S., Freire, C., & Vilas-Boas, M. (2013b). A proposal for physicochemical standards and antioxidant activity of Portuguese propolis. *Journal of the American Oil Chemists' Society, 90*, 1729–1741. doi:10.1007/s11746-013-2324-y

Fan, T. W. M. (1996). Metabolite profiling by one- and two-dimensional NMR analysis of complex mixtures. *Progress in Nuclear Magnetic Resonance Spectroscopy, 28*, 161–219. doi:10.1016/0079-6565(95)01017-3

Farooqui, T., & Farooqui, A. A. (2010). Molecular mechanism underlying the therapeutic activities of propolis: A critical review. *Current Nutrition & Food Science, 6*, 186–199. doi:10.2174/157340110792389136

Fearnley, J. (2001). *Bee propolis: Natural healing from the hive*. London: Souvenir Press.

Fernandes, A., Jr, Sugizaki, M. F., Fogo, M. L., Funari, S. R. C., & Lopes, C. A. M. (1995). *In vitro* activity of propolis against bacterial and yeast pathogens isolated from human infections. *Journal of Venomous Animals and Toxins, 1*, 63–65. doi:10.1590/S0104-79301995000200003

Forseth, R., & Schroeder, F. C. (2011). NMR-spectroscopic analysis of mixtures: From structure to function. *Current Opinion in Chemical Biology, 15*, 38–47. doi:10.1016/j.cbpa.2010.10.010

Gardana, C., Scaglianti, M., Pietta, P., & Simonetti, P. (2007). Analysis of the polyphenolic fraction of propolis from different sources by liquid chromatography tandem mass spectrometry. *Journal of Pharmaceutical and Biomedical Analysis, 45*, 390–399. doi:10.1016/j.jpba.2007.06.022

Ghisalberti, E. L. (1979). Propolis: A review. *Bee World, 60*, 59–84. doi:10.1080/0005772X.1979.11097738

Greenaway, W., & Whatley, F. R. (1990). Resolution of complex mixtures of phenolics in poplar bud exudate by analysis of gas chromatography-mass spectrometry data. *Journal of Chromatography A, 519*, 145–158. doi:10.1016/0021-9673(90)85143-J

Greenaway, W., Scaysbrook, T., & Whatley, F. R. (1987). The analysis of bud exudate of *Populus* x *euramericana*, and of propolis, by gas chromatography-mass spectrometry. *Proceedings of the Royal Society B: Biological Sciences, 232*, 249–272. Retrieved from http://www.jstor.org/stable/36240

Greenaway, W., Scaysbrook, T., & Whatley, F. R. (1990). The composition and plant origin of propolis: A report of work at Oxford. *Bee World, 71*, 107–118. doi:10.1080/0005772X.1990.11099047

de Groot, A. C. (2013). Propolis: A review of properties, applications, chemical composition, contact allergy, and other adverse effects.. *Dermatitis, 24*, 263–282. doi:10.1097/DER.0000000000000011

Haydak, M. H. (1953). Propolis, Report Iowa State. Apiarist, 74–87.

Henshaw, F. R., Bolton, T., Nube, V., Hood, A., Veldhoen, D., Pfrunder, L., … Twigg, S. M. (2014). Topical application of the bee hive protectant propolis is well tolerated and improves human diabetic foot ulcer healing in a prospective feasibility study. *Journal of Diabetes and its Complications, 28*, 850–857. doi:10.1016/j.jdiacomp.2014.07.012

Hernandez, I. M., Cuesta-Rubio, O., Campo Fernandez, M., Rosado Perez, A., de Oca, Montes, Porto, R., Piccinelli, A. L., & Rastrelli, L. (2010). Studies on the constituents of yellow cuban propolis: GC-MS determination of triterpenoids and flavonoids. *Journal of Agricultural and Food Chemistry, 58*, 4725–4730. doi:10.1021/jf904527n

Hogendoorn, E. A., Sommeijer, M. J., & Vredenbregt, M. J. (2013). Alternative method for measuring beeswax content in propolis from the Netherlands. *Journal of Apicultural Science, 57*, 81–90. doi:10.2478/jas-2013-0019

Hoheisel, O. (2001). The effects of Herstat (3% propolis ointment) application in cold sores: A double-blind, placebo-controlled clinical trial. *Journal of Clinical Research, 4*, 65–75.

Huang, W. J., Huang, C. H., Wu, C. L., Lin, J. K., Chen, Y. W., Lin, C. L., … Chen, C. N. (2007). Propolin G, a Prenylflavanone, isolated from Taiwanese propolis, induces caspase-dependent apoptosis in brain cancer cells. *Journal of Agricultural and Food Chemistry, 55*, 7366–7376. doi:10.1021/jf0710579

Huang, S., Zhang, C.-P., Li, G. Q., Sun, Y.-Y., Wang, K., & Hu, F.-L. (2014). Identification of catechol as a new marker for detecting propolis adulteration. *Molecules, 19*, 10208–10217. doi:10.3390/molecules190710208

Huber, F. (1814). *New observations upon bees*. (C. P. Dadant, Trans. (1926)). *American Bee Journal*. IL: Hamilton.

Human, H., Brodschneider, R., Dietemann, V., Dively, G., Ellis, J. D., Forsgren, E., … Zheng, H.-Q. (2013). Miscellaneous standard methods for *Apis mellifera* research. In V. Dietemann, J. D. Ellis, & P. Neumann (Eds.), *The COLOSS BEEBOOK, Volume I: Standard methods for Apis mellifera research. Journal of Apicultural Research*, 52(4). doi:10.3896/IBRA.1.52.4.10

Ignat, I., Volf, I., & Popa, V. I. (2011). A critical review of methods for characterisation of polyphenolic compounds in fruits and vegetables. *Food Chemistry, 126*, 1821–1835. doi:10.1016/j.foodchem.2010.12.026

International Conference on Harmonization. (1996). European Agency for the Evaluation of Medical Products. ICH Topic Q2 B Note for guidance on validation of analytical procedures: Methodology CPMP/ICH/281/95. ICH. Retrieved from: https://www.uam.es/personal_pas/txrf/MU5.pdf

IOFI Working Group on Methods of Analysis. (2011). Guidelines for the quantitative gas chromatography of volatile flavoring substances, from the Working Group on Methods of Analysis of the International Organization of the Flavor Industry (IOFI). *Flavor and Fragrance Journal, 26*, 297–299. doi:10.1002/ffj.2061

Isidorov, V. A., Szczepaniak, L., & Bakier, S. (2014). Rapid GC/MS determination of botanical precursors of Eurasian propolis. *Food Chemistry, 142*, 101–106. doi:10.1016/j.foodchem.2013.07.032

Isla, M. I., Nieva Moreno, M. I., Sampietro, A. R., & Vattuone, M. A. (2001). Antioxidant activity of Argentine propolis extracts. *Journal of Ethnopharmacology, 76*, 165–170. doi:10.1016/S0378-8741(01)00231-8

Jain, S., Marchioro, G., Mendonca, L., Batista, M., & Araujo, E. (2014). Botanical origin of the Brazilian red propolis: A new approach using DNA analysis. *Journal of Apicultural Science, 58*, 79–85. doi:10.2478/jas-2014-0024

Jensen, A. B., Aronstein, K., Flores, J. M., Vojvodic, S., Palacio, M. A., & Spivak, M. (2013). Standard methods for fungal brood disease research. In V. Dietemann, J. D. Ellis, & P. Neumann (Eds.), *The COLOSS BEEBOOK: Volume II: Standard methods for Apis mellifera pest and pathogen research. Journal of Apicultural Research, 52*(1) doi:10.3896/IBRA.1.52.1.13

Kaškonienė, V., Kaškonas, P., Maruška, A., & Kubilienė, L. (2014). Chemometric analysis of volatiles of propolis from different regions using static headspace GC-MS. *Central European Journal of Chemistry, 12*, 736–746. doi:10.2478/s11532-014-0521-7

Knödler, M., Conrad, J., Wenzig, E. M., Bauer, R., Lacorn, M., Beifuss, U., …., Schieber, A. (2008). Anti-inflammatory 5-(11′Z-heptadecenyl)- and 5-(8′Z,11′Z-heptadecadienyl)-resorcinols from mango (*Mangifera indica* L.) peels. *Phytochemistry, 69*, 988–993. doi:10.1016/j.phytochem.2007.10.013

Krell, R. (1996). *Value-added products from beekeeping.* FAO Agricultural Services Bulletin, No. 124, Food and Agriculture Organization of the United Nations, Rome, Italy. Retrieved from http://www.fao.org/docrep/w0076e/w0076e00.htm

Kujumgiev, A., Tsvetkova, I., Serkedjieva, Y., Bankova, V., Christov, R., & Popov, S. (1999). Antibacterial, antifungal and antiviral activity of propolis of different geographic origin. *Journal of Ethnopharmacology, 64*, 235–240. doi:10.1016/S0378-8741(98)00131-7

Kumazawa, S., Yoneda, M., Shibata, I., Kanaeda, J., Hamasaka, T., & Nakayama, T. (2003). Direct evidence for the plant origin of Brazilian propolis by the observation of honey bee behavior and phytochemical analysis. *Chemical & Pharmaceutical Bulletin, 51*, 740–742. doi:10.1248/cpb.51.740

Kumazawa, S., Nakamura, J., Murase, M., Miyagawa, M., Ahn, M. R., & Fukumoto, S. (2008). Plant origin of Okinawan propolis: Honey bee behavior observation and phytochemical analysis. *Naturwissenschaften, 95*, 781–786. doi:10.1007/s00114-008-0383-y

Kwan, E. E., & Huang, S. G. (2008). Structural elucidation with NMR spectroscopy: Practical strategies for organic chemists. *European Journal of Organic Chemistry, 2008*, 2671–2688. doi:10.1002/ejoc.200700966.

Langenheim, J. (2003). *Plant resins: Chemistry, evolution, ecology, ethnobotany.* Portland: Timber Press.

Leonhardt, S. D., Zeilhofer, S., Bluthgen, N., & Schmitt, T. (2010). Stingless bees use terpenes as olfactory cues to find resin sources. *Chemical Senses, 35*, 603–611. doi:10.1093/chemse/bjq058

Lin, M., & Shapiro, M. J. (1997). Mixture analysis by NMR spectroscopy. *Analytical Chemistry, 69*, 4731–4733. doi:10.1021/ac970594x

Lobo, V., Patil, A., Phatak, A., & Chandra, N. (2010). Free radicals, antioxidants and functional foods: Impact on human health. *Pharmacognosy Reviews, 4*, 118–126. doi:10.4103/0973-7847.70902

López, B. G. C., Schmidt, E. M., Eberlin, M. N., & Sawaya, A. C. H. F. (2014). Phytochemical markers of different types of red propolis. *Food Chemistry, 146*, 174–180. doi:10.1016/j.foodchem.2013.09.063

Lotti, C., Campo Fernandez, M., Cuesta-Rubio, O., Piccinelli, A. L., Marquez Hernandez, I., & Rastrelli, L. (2010). Chemical constituents of red Mexican propolis. *Journal of Agricultural and Food Chemistry, 58*, 2209–2213. doi:10.1021/jf100070w

Manrique, A. J., & Soares, A. E. E. (2002). Start of Africanized honey bee selection program for increased propolis production and its effect on honey production. *Interciencia, 27*, 312–316. Retrieved from http://www.interciencia.org/v27_06/manrique.pdf

Marcucci, M. C. (1995). Propolis: Chemical composition, biological properties and therapeutic activity. *Apidologie, 26*, 83–99. doi:10.1051/apido:19950202

Marcucci, M., Sawaya, A. C. H. F., Custodio, A. R., Paulino, N., & Eberlin, M. N. (2008). HPLC and ESI-MS typification: New approaches for natural products. In N. Orsolic & I. Basic (Eds.), *Scientific evidence of the use of propolis in ethnomedicine* (Vol. 1, pp. 33–54). Kerala: Transworld Research Network.

Meneghelli, C., Joaquim, L., Félix, G., Somensi, A., Tomazzoli, M., da Silva, D., ... Maraschin, M. (2013). Southern Brazilian autumnal propolis shows anti-angiogenic activity: An *in vitro* and *in vivo* study. *Microvascular Research, 88*, 1–11. doi:10.1016/j.mvr.2013.03.003

Midorikawa, K., Banskota, A. H., Tezuka, Y., Nagaoka, T., Matsushige, K., Messaje, D., ... Kadota, S. (2001). Liquid chromatography-mass spectrometry analysis of propolis. *Phytochemical Analysis, 12*, 366–373. doi:10.1002/pca.605

Morlock, G. E., Ristivojevic, P., & Chernetsova, E. S. (2014). Combined multivariate data analysis of high-performance thin-layer chromatography fingerprints and direct analysis in real time mass spectra for profiling of natural products like propolis. *Journal of Chromatography A, 1328*, 104–112. doi:10.1016/j.chroma.2013.12.053

Nagy, M., & Grancai, D. (1996). Colorimetric determination of flavanones in propolis. *Pharmazie, 51*, 100–101.

Netíková, L., Bogusch, P., & Heneberg, P. (2013). Czech ethanol-free propolis extract displays inhibitory activity against a broad spectrum of bacterial and fungal pathogens. *Journal of Food Science, 78*, M1421–M1429. doi:10.1111/1750-3841.12230

Neumann, P., Evans, J. D., Pettis, J. S., Pirk, C. W. W., Schäfer, M. O., Tanner, G., & Ellis, J. D. (2013). Standard methods for small hive beetle research. In V. Dietemann, J. D. Ellis, & P. Neumann (Eds.), *The COLOSS BEEBOOK: Volume II: Standard methods for Apis mellifera pest and pathogen research. Journal of Apicultural Research, 52*(4). doi:10.3896/IBRA.1.52.4.19

Nicodemo, D., Malheiros, E. B., De Jong, D., & Couto, R. H. N. (2014). Increased brood viability and longer lifespan of honey bees selected for propolis production. *Apidologie, 45*, 269–275. doi:10.1007/s13592-013-0249-y

Nicoli, R., Martel, S., Rudaz, S., Wolfender, J. L., Veuthey, J. L., Carrupt, P. A., & Guillarme, D. (2005). Advances in LC platforms for drug discovery. *Expert Opinion in Drug Discovery, 5*, 475–489. doi:10.1517/17460441003733874

Nieva Moreno, M. I., Isla, M. I., Sampietro, A. R., & Vattuone, M. A. (2000). Comparison of the free radical-scavenging activity of propolis from several regions of Argentina. *Journal of Ethnopharmacology, 71*, 109–114. doi:10.1016/S0378-8741(99)00189-0

Novak, E. M., Silva, M. S. C., Marcucci, M. C., Sawaya, A. C. H. F., López, B. G.-C., Fortes, M. A. H. Z., ... Maria, D. A. (2014). Antitumoural activity of Brazilian red propolis fraction enriched with xanthochymol and formononetin: An *in vitro* and *in vivo* study. *Journal of Functional Foods, 11*, 91–102. doi:10.1016/j.jff.2014.09.008

Nunes, C. A., & Guerreiro, M. C. (2012). Characterization of Brazilian green propolis throughout the seasons by headspace GC/MS andESI-MS. *Journal of the Science of Food and Agriculture, 92*, 433–438. doi:10.1002/jsfa.4596

Papotti, G., Bertelli, D., Plessi, M., & Rossi, M. C. (2010). Use of HR-NMR to classify propolis obtained using different harvesting methods. *International Journal of Food Science and Technology, 45*, 1610–1618. doi:10.1111/j.1365-2621.2010.02310.x

Park, Y. K., & Ikegaki, M. (1998). Preparation of water and ethanolic extracts of propolis and evaluation of the preparations. *Bioscience Biotechnology and Biochemistry, 62*, 2230–2232. doi:10.1271/bbb.62.2230

Paulino, N., Coutinho, J. R., Coutinho, L. A., & Scremin, A. (2014). Clinical evaluation of the anti-inflammatory effect of *Baccharis dracuncuclifolia* propolis gel on cervicits. *Revista Ciência e Estudos Acadêmicos de Medicina, 1*, 31–46. Retrieved from http://periodicos.unemat.br/index.php/revistamedicina/article/view/355

Pellati, F., Orlandini, G., Pinetti, D., & Benvenuti, S. (2011). HPLC-DAD and HPLC-ESI-MS/MS methods for metabolite profiling of propolis extracts. *Journal of Pharmaceutical and Biomedical Analysis, 55*, 5934–5948. doi:10.1016/j.jpba.2011.03.024

Pereira, E. M. R., Duval Cândido da Silva, J. L., Freitas Silva, F., Passos De Luca, M., Ferreira e Ferreira, E., Medeiros Lorentz, T. C., & Rodrigues Santos, V. (2011). Clinical evidence of the efficacy of a mouthwash containing propolis for the control of plaque and gingivitis: A phase II study. *Evidence-Based Complementary and Alternative Medicine, 2011*, Article ID 750249, 7 pages. doi:10.1155/2011/750249

Piccinelli, A. L., Lotti, C., Campone, L., Cuesta-Rubio, O., Campo-Fernandez, M., & Rastrelli, L. (2011). Cuban and Brazilian red propolis: Botanical origin and comparative analysis by high-performance liquid chromatography–photodiode array detection/electrospray ionization tandem mass spectrometry. *Journal of Agricultural and Food Chemistry, 59*, 6484–6491. doi:10.1021/jf201280z

Piccinelli, A. L., Mencherini, T., Celano, R., Mouhoubi, Z., Tamendjari, A., Aquino, R. P., & Rastrelli, L. (2013). Chemical Composition and antioxidant activity of Algerian propolis. *Journal of Agricultural and Food Chemistry, 61*, 5080–5088. doi:10.1021/jf400779w

Pirk, C. W. W., De Miranda, J. R., Fries, I., Kramer, M., Murray, T., Paxton, R., … van Dooremalen, C. (2013). Statistical guidelines for *Apis mellifera* research. In V. Dietemann, J. D. Ellis, & P. Neumann (Eds.), *The COLOSS BEEBOOK, Volume I: Standard methods for Apis mellifera research. Journal of Apicultural Research, 52*(4). doi:10.3896/IBRA.1.52.4.13

Pisoschi, A. M., & Pop, A. (2015). The role of antioxidants in the chemistry of oxidative stress: A review. *European Journal of Medicinal Chemistry, 97*, 55–74. doi:10.1016/j.ejmech.2015.04.040

Popova, M., Bankova, V., Butovska, D., Petkov, V., Nikolova-Damyanova, B., Sabatini, A. G., … Bogdanov, S. (2004). Validated methods for the quantification of biologically active constituents of poplar-type propolis. *Phytochemical Analysis, 15*, 235–240. doi:10.1002/pca.777

Popova, M., Bankova, V., Bogdanov, S., Tsvetkova, I., Naydenski, C., Marcazzan, G. L., & Sabatini, A. G. (2007). Chemical characteristics of poplar type propolis of different geographic origin. *Apidologie, 38*, 306–311. doi:10.1051/apido:2007013

Popova, M., Chen, C.-N., Chen, P.-Y., Huang, C.-Y., & Bankova, V. (2010). A validated spectrophotometric method for quantification of prenylated flavanones in pacific propolis from Taiwan. *Phytochemical Analysis, 21*, 186–191. doi:10.1002/pca.1176

Popova, M. P., Graikou, K., Chinou, I., & Bankova, V. (2010). GC-MS profiling of diterpene compounds in mediterranean propolis from greece. *Journal of Agricultural and Food Chemistry, 58*, 3167–3176. doi:10.1021/jf903841k

Popova, M., Trusheva, B., Cutajar, S., Antonova, D., Mifsud, D., Farrugia, C., & Bankova, V. (2012). Identification of the plant origin of the botanical biomarkers of Mediterranean type propolis. *Natural Products Communication, 7*, 569–570.

Popravko, S. A., Sokolov, I. V., & Torgov, I. V. (1982). New natural phenolic triglycerides. *Chemistry of Natural Compounds, 18*, 153–157. doi:10.1007/BF00577181

Prabhakar, A. R., Karuna, Y. M., Yavagal, C., & Deepak, B. M. (2015). Cavity disinfection in minimally invasive dentistry—Comparative evaluation of *Aloe vera* and propolis: A randomized clinical trial. *Contemporary Clinical Dentistry, 6*, S24–S31. doi:10.4103/0976-237X.152933

Purra, A. R., Mushtaq, M., Acharya, S. R., & Saraswati, V. (2014). A comparative evaluation of propolis and 5.0% potassium nitrate as a dentine desensitizer: A clinical study. *Journal of Indian Society of Periodontology, 18*, 466–471. doi:10.4103/0972-124X.138695

Queiroga, R de C. R. E., Madruga, M. S., Galvão, M. de S., & Da Costa, R. G. (2005). Extraction enhancement of volatile compounds from goat milk using the simultaneous extraction and concentration techniques. *Revista Instituto Adolfo Lutz, 64*, 97–103. Retrieved from http://periodicos.ses.sp.bvs.br/pdf/rial/v64n1/v64n1a15.pdf

Re, R., Pellegrini, N., Proteggente, A., Pannala, A., Yang, M., & Rice-Evans, C. (1999). Antioxidant activity applying an improved ABTS radical cation decolorization assay. *Free Radical Biology and Medicine, 26*, 1231–1237. doi:10.1016/S0891-5849(98)00315-3

Righi, A. A., Negri, G., & Salatino, A. (2013). Comparative chemistry of propolis from eight Brazilian localities. *Evidence-Based Complementary and Alternative Medicine, 2013*, Article ID 267878, 14 pages. doi:10.1155/2013/267878

de Rijke, E., Out, P., Niessen, W. M. A., Ariese, F., Gooijer, C., & Brinkman, U. A. T. (2006). Analytical separation and detection methods for flavonoids. *Journal of Chromatography A, 1112*, 31–63. doi:10.1016/j.chroma.2006.01.019

Ristivojevic, P., Andric, F. L., Trifkovic, J. D., Vovk, I., Stanisavljevic, L. Z., Tesic, Z. L., & Milojkovic-Opsenica, D. M. (2014). Pattern recognition methods and multivariate image analysis in HPTLC fingerprinting of propolis extracts. *Journal of Chemometrics, 28*, 301–310. doi:10.1002/cem.2592

Savorani, F., Tomasi, G., & Englesen, S. B. (2009). Icoshift: A versatile tool for the rapid alignment of 1D NMR spectra. *Journal of Magnetic Resonance, 202*, 190–202. doi:10.1016/j.jmr.2009.11.012

Sawaya, A. C. H. F., Tomazela, D. M., Cunha, I. B. S., Bankova, V., Marcucci, M. C., Custodio, A. R., & Eberlin, M. N. (2004). Electrospray ionization mass spectrometry fingerprinting of propolis. *Analyst, 129*, 739–744. doi:10.1039/B403873H

Sawaya, A. C. H. F., Abdelnur, P. V., Eberlin, M. N., Kumazawa, S., Ahn, M., Bang, K., … Afrouzan, H. (2010). Fingerprinting of propolis by easy ambient sonic-spray ionization mass spectrometry. *Talanta, 81*, 100–108. doi:10.1016/j.talanta.2009.11.043

Sawaya, A. C. H. F., da Silva, Barbosa, Cunha, I., & Marcucci, M. C. (2011). Analytical methods applied to diverse types of Brazilian propolis. *Chemistry Central Journal, 5*, 27. doi:10.1186/1752-153X-5-27

Schmid, M. R., Brockmann, A., Pirk, C. W. W., Stanley, D. W., & Tautz, J. (2008). Adult honey bees (*Apis mellifera* L.) abandon hemocytic, but not phenoloxidase-based immunity. *Journal of Insect Physiology, 54*, 439–444. doi:10.1016/j.jinsphys.2007.11.002

Seeley, T. D., & Morse, R. A. (1976). The nest of the honey bee (*Apis mellifera* L.). *Insectes Sociaux, 23*, 495–512. doi:10.1007/BF02223477

Seidel, V., Peyfoon, E., Watson, D., & Fearnly, J. (2008). Comparative study of the antibacterial activity of propolis from different geographical and climatic zones. *Phytotherapy Research, 22*, 1252–1263. doi:10.1002/ptr.2480

Senturk, U. K., Gunduz, F., Kuru, O., Aktekin, M. R., Kipmen, D., Yalcin, O., … Baskurt, O. K. (2001). Exercise-induced oxidative stress affects erythrocyte in sedentary rats but not exercise-trained rats. *Journal of Applied Physiology, 91*, 1999–2004. Retrieved from http://jap.physiology.org/content/jap/91/5/1999.full.pdf

Sforcin, J. M., & Bankova, V. (2011). Propolis: Is there a potential for the development of new drugs? *Journal of Ethnopharmacology, 133*, 253–260. doi:10.1016/j.jep.2010.10.032

Sforcin, J. M., Fernandes, A., Jr., Lopes, C. A. M., Bankova, V., & Funari, S. R. C. (2000). Seasonal effect on Brazilian propolis antibacterial activity. *Journal of Ethnopharmacology, 73*, 243–249. doi:10.1016/S0378-8741(00)00320-2

Sforcin, J. M., Fernandes, A., Jr., Lopes, C. A. M., Bankova, V., & Funari, S. R. C. (2001). Seasonal effect on Brazilian propolis on *Candida albicans* and *Candida tropicalis*. *Journal of Venomous Animals and Toxins, 7*, 139–144. doi:10.1590/S0104-79302001000100009

Simone, M., Evans, J., & Spivak, M. (2009). Resin collection and social immunity in honey bees. *Evolution, 63*, 3016–3022. doi:10.1111/j.1558-5646.2009.00772.x

Simone-Finstrom, M., & Spivak, M. (2012). Increased resin collection after parasite challenge: A case of self-medication in honey bees? *PLoS ONE, 7*, e34601. doi:10.1371/journal.pone.0034601

Siti, H. N., Kamisah, Y., & Kamsiah, J. (2015). The role of oxidative stress, antioxidants and vascular inflammation in cardiovascular disease (a review). *Vascular Pharmacology, 71*, 40–56. doi:10.1016/j.vph.2015.03.005

Smith, C. A., Want, E. J., O'Maille, G., Abagyan, R., & Siuzdak, G. (2006). XCMS: Processing mass spectrometry data for metabolite profiling using nonlinear peak alignment, matching, and identification. *Analytical Chemistry, 78*, 779–787. doi:10.1021/ac051437y

Solórzano, E., Vera, N., Cuello, S., Ordoñez, R., Zampini, C., Maldonado, L., ... Isla, M. I. (2012). Chalcones in bioactive Argentine propolis collected in arid environments. *Natural Product Communications, 7*, 879–882.

Soroy, L., Bagus, S., Yongkie, I. P., & Djoko, W. (2014). The effect of a unique propolis compound (Propoelix™) on clinical outcomes in patients with dengue hemorrhagic fever. *Infection ND Drug resistence, 7*, 323–329. doi:10.2147/IDR.S71505

Steinmann, D., & Ganzera, M. (2011). Recent advances on HPLC/MS in medicinal plant analysis. *Journal of Pharmaceutical and Biomedical Analysis, 55*, 744–757. doi:10.1016/j.jpba.2010.11.015

Suárez, D., Zayas, D., & Guisado, F. (2005). Propolis: Patents and technology trends for health applications. *Journal of Business Chemistry, 2*, 119–125. Retrieved from http://www.businesschemistry.org/downloads/articles/Issue09-2005_84.pdf

Sutton, S. (2011). Determination of inoculum for microbiological testing. *Microbiology Topics, 15*, 49–53. Retrieved from http://www.microbiol.org/wp-content/uploads/2011/08/Sutton.JGXP_.15_3.pdf

Torto, B., Carroll, M. J., Duehl, A., Fombong, A. T., Gozansky, K. T., Nazzi, F., ..., Teal, P. E. A. (2013). Standard methods for chemical ecology research in *Apis mellifera*. In V. Dietemann, J. D. Ellis & P. Neumann (Eds.), *The COLOSS BEEBOOK, Volume I: Standard methods for* Apis mellifera *research. Journal of Apicultural Research, 52*(4), doi:10.3896/IBRA.1.52.4.06

Torwane, N. A., Hongal, S., Goel, P., Chandrashekar, B. R., Jain, M., & Saxena, E. (2013). A clinical efficacy of 30% ethenolic extract of Indian propolis and Recaldent TM in management of dentinal hypersensitivity: A comparative randomized clinical trial. *European Journal of Dentistry, 7*, 461–468. doi:10.4103/1305-7456.1206

Trusheva, B., Popova, M., Bankova, V., Simova, S., Marcucci, M. C., Miorin, P. L., ... Tsvetkova, I. (2006). Bioactive constituents of Brazilian red propolis. *Evidence-Based Complementary Alternative Medicine, 3*, 249–254. doi:10.1093/ecam/nel006

Trusheva, B., Popova, M., Koendhori, E. B., Tsvetkova, I., Naydenski, C., & Bankova, V. (2011). Indonesian propolis: Chemical composition, biological activity and botanical origin. *Natural Product Research, 25*, 606–613. doi:10.1080/14786419.2010.488235

Valentão, P., Fernandes, E., Carvalho, F., Andrade, P. B., Seabra, R. M., & Bastos, M. (2002). Studies on the antioxidant activity of lippia citriodora infusion: Scavenging effect on superoxide radical, hydroxyl radical and hypochlorous acid. *Biological and Pharmaceutical Bulletin, 25*, 1324–1327. doi:10.1248/bpb.25.1324

Vaz Coelho, L. G., Ferreira Bastos, E. M. A., Carvalho Resende, C., Paula e Silva, C. M., Fernandes Sanches, B. S., de Castro, F. J., ... Trindade, O. R. (2007). Brazilian green propolis on *Helicobacter pylori* infection. *Helicobacter, 12*, 572–574. doi:10.1111/j.1523-5378.2007.00525.x

Viswanathan, C. T., Bansal, S., Booth, B., De Stefano, A. J., Rose, M. J., Sailstad, J., ... Weiner, R. (2007). Quantitative Bioanalytical Methods Validation and Implementation: Best Practices for Chromatographic and Ligand Binding Assays. *Pharmaceutical Research, 24*, 1962–1973. doi:10.1007/s11095-007-9291-7.

Veitch, N. C., & Grayer, R. J. (2008). Flavonoids and their glycosides, including anthocyanins. *Natural Product Reports, 25*, 555–611. doi:10.1039/b718040n

Velioglu, Y., Mazza, G., Gao, L., & Oomah, B. D. (1998). Antioxidant activity and total phenolics in selected fruits, vegetables, and grain products. *Journal of Agricultural and Food Chemistry, 46*, 4113–4117. doi:10.1021/jf9801973

Vera, N., Solorzano, E., Ordoñez, R., Maldonado, L., Bedascarrasbure, E., & Isla, M. I. (2011). Chemical composition of Argentinean propolis collected in extreme regions and its relation with antimicrobial and antioxidant activities. *Natural Product Communications, 6*, 823–827.

Volpi, N., & Bergonzini, G. (2006). Analysis of flavonoids from propolis by on-line HPLC-electrospray mass spectrometry. *Journal of Pharmaceutical and Biomedical Analysis, 42*, 354–361. doi:10.1016/j.jpba.2006.04.017

Waterman, P. G., & Mole, S. (1994). *Analysis of phenolic plant metabolites*. Cambridge, MA: Blackwell Scientific Publications.

Watson, D. G., Peyfoon, E., Zheng, L., Lu, D., Seidel, V., Johnston, B., ... Parkinson, J. (2006). Application of principal components analysis to 1H-NMR data obtained from propolis samples of different geographical origin. *Phytochemical Analysis, 17*, 323–331. doi:10.1002/pca.921

Wilson, M. B., Spivak, M., Hegeman, A. D., Rendahl, A., & Cohen, J. D. (2013). Metabolomics reveals the origins of antimicrobial plant resins collected by honey bees. *PLoS ONE, 8*, e77512. doi:10.1371/journal.pone.0077512

Wilson, M. B., Brinkman, D., Spivak, M., Gardner, G., & Cohen, J. D. (2015). Regional variation in composition and antimicrobial activity of US propolis against *Paenibacillus larvae* and *Ascosphaera apis*. *Journal of Invertebrate Pathology, 124*, 44–50. doi:10.1016/j.jip.2014.10.005

Wilson-Rich, N., Spivak, M., Fefferman, N. H., & Starks, P. T. (2009). Genetic, individual, and group facilitation of disease resistance in insect societies. *Annual Review of Entomology, 54*, 405–423. doi:10.1146/annurev.ento.53.103106.093301.

Woisky, R. G., & Salatino, A. (1998). Analysis of propolis: Some parameters and procedures for chemical quality control. *Journal of Apicultural Research, 37*, 99–105. doi:10.1080/00218839.1998.11100961

Yamaguchi, T., Takamura, H., Matoba, T., & Terao, J. (1998). HPLC method for evaluation of the free radical-scavenging activity of food by using 1,1-diphenyl-2-picrylhydrazil. *Bioscience, Biotechnology and Biochemistry, 62*, 1201–1204. doi:10.1271/bbb.62.1201

Zampini, I. C., Meson, G. J., Ordoñez, R. M., Sayago, J. E., Nieva Moreno, M. I., & Isla, M. I. (2008). Antioxidant and xanthine oxidase inhibitory activities of plant species from the Argentine Puna (Antofagasta, Catamarca). *Recent Progress in Medicinal Plants, 21*, 95–110.

Zampini, I. C., Ordoñez, R. M., & Isla, M. I. (2010). Autographic assay for the rapid detection of antioxidant capacity of liquid and semisolid pharmaceutical formulations using $ABTS^{\cdot+}$immobilized by gel entrapment. *American Association of Pharmaceutical Scientists, 11*, 1159–1163. doi:10.1208/s12249-010-9484-y

Zhang, T., Omar, R., Siheri, W., Mutairi, S. A., Clements, C., Fearnley, J., ... Watson, D. (2014). Chromatographic analysis with different detectors in the chemical characterisation and dereplication of African propolis. *Talanta, 120*, 181–190. doi:10.1016/j.talanta.2013.11.094

Journal of Apicultural Research, 2019
Vol. 58, No. 2, 1–28, http://dx.doi.org/10.1080/00218839.2016.1226606

**REVIEW ARTICLE**

# Standard methods for *Apis mellifera* brood as human food

Annette Bruun Jensen[a*], Josh Evans[b], Adi Jonas-Levi[c], Ofir Benjamin[c,d], Itzhak Martinez[d,e], Bjørn Dahle[f,g], Nanna Roos[h], Antoine Lecocq[a] and Kirsten Foley[a]

[a]*Department of Agriculture and Ecology, University of Copenhagen, Frederiksberg C, Denmark;* [b]*Nordic Food Lab, Department of Food Science, University of Copenhagen, Frederiksberg C, Denmark;* [c]*Food Science Department, Tel Hai College, D.N. Upper Galilee, Israel;* [d]*MIGAL Galilee Research Institute Ltd, Kiryat Shmona, Israel;* [e]*Animal Science Department, Tel Hai College, D.N. Upper Galilee, Israel;* [f]*Norwegian Beekeepers Association, Kløfta, Norway;* [g]*Department of Animal and Aquacultural Sciences, Norwegian University of Life Sciences, Ås, Norway;* [h]*Department of Nutrition, Exercise and Sports, University of Copenhagen, Frederiksberg C, Denmark*

*(Received 25 April 2015; accepted 8 August 2016)*

Insects hold enormous potential to address food and nutritional security issues. The honey bee is a key insect, given its importance for pollination, as well as its products which can be directly consumed, like honey, pollen and brood. Research on edible insects is an emerging field that draws upon methods and techniques from related fields of research. In this paper, we provide recommendations and research protocols centered on production of worker and drone brood for human consumption, on brood harvesting, including hygienic considerations, on nutritional aspects of brood, on sensory analyses of brood and brood products and on the gastronomic applications of honey bee brood; all of which will help elucidate the edible potential of honey bee brood now, and in the future.

**Métodos estándar para el uso de la cría de *Apis mellifera* como alimento humano**

Los insectos tienen un enorme potencial para abordar cuestiones de seguridad alimentaria y nutricional. La abeja de la miel es un insecto clave, dada su importancia para la polinización, así como por sus productos que pueden consumirse directamente, como la miel, el polen y la cría. La investigación sobre insectos comestibles es un campo emergente que se basa en métodos y técnicas de campos de investigación relacionados. En este documento, proporcionamos recomendaciones y protocolos de investigación centrados en la producción de obreras y en la cría de zánganos para el consumo humano, sobre la producción de cría, incluyendo consideraciones sobre la higiene, en los aspectos nutricionales de la cría, en los análisis sensoriales de la cría y los productos de la cría y en las aplicaciones gastronómicas de la cría de la abeja de la miel, todo lo cual ayudará a dilucidar el potencial comestible de la cría de las abejas de la miel, ahora y en el futuro.

将蜜蜂幼虫制备成食物的标准方法

昆虫在保障食物供应和营养安全问题上的作用巨大。蜜蜂不但是重要的授粉昆虫，而且能提供食用蜂产品，比如蜂蜜，花粉和蜜蜂幼虫。研究食用昆虫是一个新兴的领域。食用制备技术与方法也在相关领域的研究中形成。在本文中，我们介绍雄蜂和工蜂幼虫的食用方法，包括幼虫收集、卫生、营养、检测和美食搭配。这些都将有助于阐明目前以及将来蜜蜂幼虫作为可食用昆虫的潜力。

**Keywords:** entomophagy; edible insects; *Apis mellifera*; honey bee; drone brood; worker brood; bee larvae; bee pupae; nutritional analysis; sensory analysis; *Varroa* sp. control; gastronomy

*Corresponding author: Email: abj@plen.ku.dk
Please refer to this paper as: Jensen, A B, Evans, J, Jonas-Levi, A, Benjamin, O, Martinez, I, Dahle, B, Roos, N, Lecocq, A, Foley, K (2019) Standard methods for *Apis mellifera* brood for human food. In V Dietemann, P Neumann, N Carreck and J D Ellis (Eds), *The COLOSS BEE-BOOK, Volume III, Part I: standard methods for Apis mellifera hive products research. Journal of Apicultural Research* 58(2): http://dx.doi.org/10.1080/00218839.2016.1226606

## 1. Introduction

The potential of insects for food and feed has gained global attention in the past few years, and with good reason: many species require fewer resources to rear, have a lower environmental impact than conventional livestock (Oonincx et al., 2010), are highly nutritious (Rumpold & Schlüter, 2013), and are celebrated for their high gastronomic value in much of the world. Approximately 2,000 species are consumed (van Huis et al., 2013) in at least 113 countries (Macevilly, 2000). By no means are insects a "new" food to humanity and their potential for improving global food systems by diversifying our food supplies is considerable high.

Brood of the honey bee (*Apis mellifera*) is a particularly promising edible resource, as honey bees are kept by humans worldwide, and in many cultures eaten as a delicacy. Consumption of honey bee brood is a common cultural practice in regions as diverse as Mexico (Ramos-Elorduy et al., 1997), Ecuador (Onore, 1997), China (Zhi-Yi, 1997), Thailand (Yhoung-Aree, Puwastien, & Attig, 1997), Senegal (Gessain & Kinzler, 1975), Zambia (Mbata, 1995) and Australia (de Foliart, 2002). Honey bee brood is valued for its rich nutritional composition of proteins, fatty acids, vitamins and minerals (Finke, 2005; Hocking & Matsumura, 1960; Rumpold & Schlüter, 2013), as well as its pleasing taste and versatility in culinary preparations (Evans, 2013, 2014).

In certain regions of the world, drone brood removal has become part of regular hive maintenance by beekeepers as a strategy for managing populations of the varroa mite (*Varroa destructor*), widely recognized as the most harmful parasite affecting honey bees worldwide (Dietemann et al., 2013). This practice makes honey bee drone brood a by-product, producing an abundant source of farmed insects with untapped potential.

In 2015, the European Food safety Authority (EFSA) reviewed food safety risks associated with production and consumption of insects (European Food safety Authority (EFSA), 2015). The review covered insect species identified to be relevant for farming in closed farming systems in Europe and elsewhere. Since honey bee production is not a closed production system, the perspective of bee-brood harvested for human consumption was not specifically covered. Also, bee-brood is legislatively an api byproduct. The overall findings for other insect species-such as grasshoppers, crickets and mealworms – were that there were no additional or specific biological safety risks associated with the production and consumption of these insects, when compared to traditional livestock production systems.

Research on honey bee biology and breeding has a long history as compared with other candidates for insect farming and in many parts of the world the link between research and beekeepers are ensured by extension services. Furthermore, compared with other forms of local natural resource development, setting up hives and beekeeping practice require relatively little arable surface and low financial investment. In addition honey bees provide important pollination services; stimulating ecological robustness, agricultural production and related local economies in rural and urban areas. Considering this, we recommend further research into the potential of honey bee brood for human food and food systems. Very limited research has been performed so far, thus scientific protocols have not yet been fully developed on all subjects; hence this chapter will have a mixture of both considerations and more developed protocols.

## 2. Production of worker brood

Honey bees have developed a mechanism to react to changes in the ratio of pollen supply and protein demand of the brood. Under pollen deprivation, brood rearing in a colony is at first limited (Hrassnigg & Crailsheim, 1998; Imdorf, Rickli, Kilchenmann, Bogdanov, & Wille, 1998) later, eggs and young larvae are cannibalized, thereby providing protein to feed other larvae (Boes, 2010). Genetic and environmental factors can influence the bee brood production directly or indirectly by modulating the food flow to the hive. Queen age and colony composition and size are also crucial factors for optimal brood production, a young and healthy queen has higher egg production, while her capacity decreases with age (Gary, 1992), and small colonies of 2,300–9,000 bees produce more bee brood per bee than more populous ones (Harbo, 1986). Colonies without pollen supply maintain brood rearing only for a short time. Thus, research in pollen resources or supplementary diet is essential for a high production of worker brood.

### 2.1. Subspecies and queens

Honey bee subspecies greatly vary genetically, physiologically and behaviorally (Ruttner, 1988), in particularly in response to climatic and environmental factors, such low or high temperature, food utilization and reproductive traits (DeGrandi-Hoffman, Eckholm, & Huang, 2013). For further details on characterization of honey bee subspecies see the *BEEBOOK* chapter of Meixner et al. (2013). The queen quality such as age and mating age are important for her reproduction and thus important to consider before an experimental setup. For further details, see the *BEEBOOK* chapter on standard methods for rearing and selection of *Apis mellifera* queens (Büchler et al., 2013).

(1) Be sure that the same subspecies of honey bee is used for all the treatments in the experiment, including control.
(2) Choose full sister queens in order to minimize the influence of genetic variability.
(3) Choose queens that have been mated by several drones to ensure genetic variation within colony variation.

The reliability of the results of such designs is narrowed to a single genetic population (Deme). Further experiments should be replicated among genetically diverse colonies of the same subspecies to generalize the results. The impact of subspecies on bee brood production should also be studied.

### 2.2. Setting up experimental colonies

It is important to use experimental colonies of uniform strength either by use of the classical objective mode or shook swarm, see the *BEEBOOK* chapter on measuring colony strength parameters (Delaplane et al., 2013). If the classical objective mode is chosen (described in Delaplane et al., 2013), presence/absence of brood at the beginning is an option worth considering. Small nucleus hives containing five frames are called "nucs" by beekeepers) and are ideal for brood production. They may be occupied by ~9,000 workers, thus a size for optimal brood production per worker (Harbo, 1986). However standard hives can also be used in particular in regions with cold weather, in which the colony should be more populous to foster thermoregulation.

Small nucleus hives have the advantages of optimal brood production and they are cheaper, lighter, and portable. Standard hives can be an advantage when conducting food supplement experiments since it will permit sufficient space for food stocking in the control hives (where brood is not removed), thus leaving enough space for the queen to lay new eggs (Martinez, personal observation).

In order to get reliable data from experiments it is important to work with healthy colonies thus check the experimental colonies, or the source of the workers for clinical symptoms of bacterial, fungal and viral diseases (de Graaf et al., 2013; de Miranda et al., 2013; Forsgren, Budge, Charrière, & Hornitzky, 2013; Fries et al., 2013; Jensen et al., 2013). Pesticide residues at sub-lethal doses can have negative effects on brood and adult emergence (Wu, Anelli, & Sheppard, 2011) thus it is also recommended to get wax from combs and foundations analyzed. Pesticide residues can also affect human health (Blair, Ritz, Wesseling, & Freeman, 2014) thus residues in the brood for human consumption should also be tested.

### 2.3. The site of the experimental apiary

If the experiment is based, partly or totally, on natural feeding of the bees – control and treatment colonies have to be placed in the same apiary. If a geographic comparison is part of the experimental design, the hives will be allocated to different apiaries, but each treatment and control should be assigned to each site.

(1) Choose a sunny open site to prevent humidity and to allow sooner activity of the foragers (Ambrose, 1995). At the contrary, in very hot regions, choose a naturally shadowed place, e.g., by high trees. Be aware of the wind regime because it may impact the forager activities, as well as the cooling of the hives.
(2) Locate the experimental apiary on dry and well drained ground, on a stand.
(3) If no natural water source is present around the apiary – supply it at least during the hot season. A barrel full of water, with some pieces of wood or polystyrene on the liquid surface, is enough (Johansson & Johansson, 1978).
(4) In countries with great concern about ants as bee predators, insert each leg of the stand in a cup full of a solution of engine and diesel oil (50:50) to prevent the ants entering the hive It is also possible to fill the cups with water and detergents as detailed by Dainat, Kuhn, Cherix, and Neumann (2011) or grease caps (Seregen, 2004).
(5) The site should be isolated from hazards: insecticides, pollution, robbery, vandalism, predators and bush fire, because it may impact the number of foragers in each colony, and consequently the results of the experiments. Distance from such hazards should be determined following the experience of the professional beekeepers in the region and agriculture extension services for informations (Ambrose, 1995).
(6) Be aware that you can receive valuable data from a meteorological station near the apiary, or equip the apiary with a meteorological field station. This is important because the activity of the foragers is impacted by weather.
(7) Minimize forager drift among colonies by dispersing the hives in the area without any arrangement, not in line, their entrances open to different directions. Paint varying geometric forms of different colors above the entrance that will help the foragers to orientate to their own colony when coming back (see references in Neumann, Moritz, & Mautz, 2000). Drift may impact the results of the experiment because the population of foraging workers may increase in some hives while decreasing in others.
(8) Write directly the code or identity number on both the roofs and on one side of the hives. Use permanent ink.

### 2.4. *Feeding experiments*

Food flow, of both pollen and nectar, has qualitative and quantitative impacts on brood production, mainly via brood food-producing hypo-pharyngeal glands of nurse worker bees (Brodschneider & Crailsheim, 2010; Haydak, 1970; Standifer, Moeller, Kauffeld, Herbert, & Shimanuki, 1977). Supplemental feeding in particular in periods with food shortages can be important for maintaining a higher brood production. Measuring the natural food flow is however important for such studies and methods on various measures e.g., hive weight, food reserves, counting forager workers and pollen traps can be found in the *BEEBOOK* paper on miscellaneous standard methods for *Apis mellifera* research (Human et al., 2013) and *BEEBOOK* paper on measuring colony strength parameters (Delaplane et al., 2013).

(1) For each hive, adopt a particular feeder that can be used uniquely by the foragers of the hive (Table 1).
(2) Check impermeability of the feeders before experiment, using tap water. The ones that are permeable should be treated: pour hot liquefied candle's paraffin or wax on the slots and fissures and repeat impermeability test.
(3) Quantities can be calculated following some criteria: (1) amount of artificial food per adult worker, (2) type of hives and population size in each colony, (3) length of the experiment.
(4) Prepare the food no more than 48 h before feeding the colonies and keep it cold to prevent its deterioration and microbial development.
(5) Artificial feeding should be the last operation of the day in the apiary to avoid robbery. Robbing can be controlled by reducing the entrance of the hive, sealing all cracks or openings, placing straw or grass in front of the entrance. All these actions will help the guarding bees in their mission (Ambrose, 1995; Standifer et al., 1977).

## 3. Production of drone brood

Drone brood production is, like brood in general, dependent on a good pollen supply to meet their protein demand. A drone larva is more costly to produce and maintain than a worker larva and colonies regulate drone production in accordance with the availability of food and season (Hrassnigg & Crailsheim, 1998; Seeley, 2002; Seeley & Mikheyev, 2003). Optimal drone productivity is dependent upon how resources are allocated between the sexes given particular climate, colony status (size and queen ages) and forage availability conditions (Boes, 2010). The amount of drone brood present in the colony also affects drone production by a negative feedback process (Free & Williams, 1975). Therefore, removal of drones from the colony should upregulate drone production when regularly harvested. In temperate regions of the Northern hemisphere the drone production season occurs from May to August (Boes, 2010; Free & Williams, 1975).

Table 1. Types of syrup feeders to be used in feeding experiments for optimal brood production.

| Type of feeder | Place | Size | Advantages | Disadvantages |
|---|---|---|---|---|
| Miller | Internal – on the top of the frames | 3–5 l | Easy to fill. Serve as thermic isolation between the nest and the roof. Does not disturb the nest when filling | Require tight closure against robbers. The sugar solution cools too fast in winter |
| Frame | Internal – in place of 2 frames | 2 l | Easy to fill. Keeping the temperature of the syrup during time | Disturbs the nest |
| Inverted | container | Outside the hive – on the flying board | 1 l | Easy to work with |
| High rates of robbing | | | | |
| Baggie feeder | Internal on the top of the frames, inside an empty super | Following the size to the bag | Great quantities of syrupx | The empty super enlarges the internal sizes of the hive that the bees have to warm |

### 3.1. Production of drone brood and Varroa sp. trapping

Drone production at any one time is limited by the amount of drone comb cells present in the colony (Boes, 2010) particularly in modern beekeeping where frames with imprinted worker cell foundation are used. The tendency for workers to build fresh drone comb when empty frames are placed within the colony during drone production season can be exploited to encourage drone brood production. Current evidence indicates that the practice of drone brood removal is an effective measure of *Varroa* spp. mite control (Calderone, 2005). In Europe, particularly in Nordic countries this technique combined with chemical treatment, is used as part of a *Varroa* spp. IPM strategy.

Typically, an empty frame allocated for drone production is placed in the hive from which capped brood comb sections are later removed containing pupae and trapped *Varroa* sp. mites. Frames without wax foundation either as full frames or frames divided into two to three sections can be used.

(1) Prepare the experimental colonies according to the guidelines given for worker brood production (Sections 2.1, 2.3 and 2.4).)
(2) Use an empty trap frame, full or divided into two or three sections. Alternatively, a trap frame with imprinted drone brood cells can be used.
(3) Placed the frame next to or in the center of the brood nest at the beginning of drone season-the springtime.
(4) The first drone brood is capped approximately eleven days after its introduction so weekly data collection is recommended.
(5) Record the wanted data e.g., worker brood area (capped and uncapped cells), drone brood area (capped/uncapped cells), colony weight, storages (honey and pollen), *Varroa* spp. infested cells. See more details in the *BEEBOOK* chapters: Miscellaneous standard methods for *Apis mellifera* research (Human et al., 2013); Measuring colony strength parameters (Delaplane et al., 2013) and Standard methods for *Varroa* spp. research (Dietemann et al., 2013).
(6) Remove capped drone brood weekly for three section frames or at longer intervals for two section frames (10 days) and full frames (20 days).

*Pros and cons*:
Three section frames (Figure 1) give more uniform brood ages at harvest that could be timed according to the brood ages of interest (Section 4.1), and longer periods for *Varroa* spp. trapping. However it makes weekly removal of drone brood necessary.

### 3.2. Estimations of drone brood from weight of combs

The weight of the edible part of a brood comb can be estimated by the weight of the brood piece or the number of capped drone cells by use of regression equations. Drone pupae lose weight during the pupation (Duay, Jong, & Engels, 2003) thus it is important to make a regression equation based on representative subsamples of the study material.

(1) Weigh out the removed comb pieces on a scale.
(2) Count or estimate the number of capped brood cells Delaplane et al. (2013).

Figure 1. A three section drone brood 12 × 10 frame (306 × 262 cm). The right section is one week, the middle section is two weeks and the left section is three weeks old and ready to be harvested.

(3) Separate drone pupae from the wax and weigh the pupae from each piece.
(4) Plot the comb weight and edible drone pupae weight to generate a regression equation.
(5) Plot the number of capped brood against the edible drone pupae weight to generate a regression equation.
(6) Alternatively, the mean weight of a subsample of individual drone pupae can used together with count of capped drone brood cells.

## 4. Brood harvesting

### 4.1. Time of harvest – age of brood

Honey bee brood can be harvested from the colonies at different stages in the development from larvae to adult bee. To maximize biomass of the harvested brood it should not be harvested before the larvae stop feeding, which is at the time of capping. As pupae grow older, there will be an increasing amount of chitin (from the yellow thorax stage in Figure 2) making the brood less palatable. Just before emergence the brood resembles adults with considerable amounts of indigestible chitinous cuticle and has an aversive taste (Schmidt & Buchmann, 1992). In addition, the pupae will lose weight during the pupation period. Based on these considerations, brood should be harvested before the pupae eyes becomes pink, corresponding to 0–130 h and 0–160 h post-capping for worker and drone brood respectively.

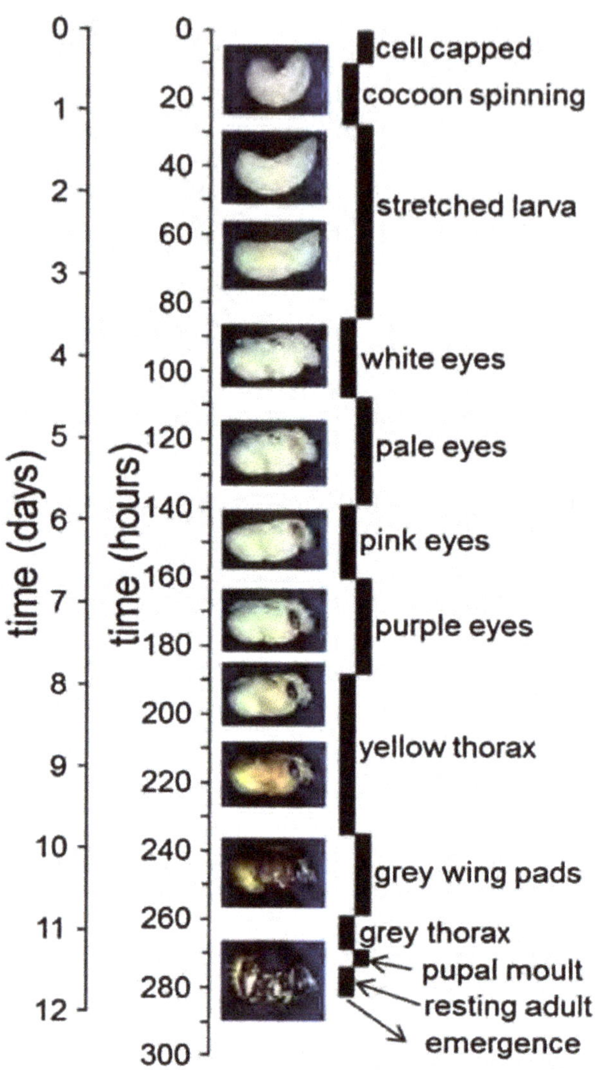


Figure 2. Time and duration of sealed worker brood from capping to emergence including morphological categories. Source: Human et al. (2013).

### 4.2. Hygienic measures

Insects are rich in nutrients and moisture, providing a favorable environment for microbial survival and growth (Klunder, Wolkers-Rooijackers, Korpela, & Nout, 2012). No records of food poisoning from bee brood currently exist. Nevertheless, more research however will be required from a food safety perspective as goes for microbial hazards in brood for human food. In Japan, wasp brood (*Vespula* spp.) is among the main insect food consumed today and also here no records of food poisoning exists (Payne, 2015). The main aspect from a food perspective is not the microflora composition of live animals, but the possibility to safely store and preserve raw material or derived products.

#### 4.2.1. Harvesting in the field and preprocessing storage

(1) To avoid contamination of toxic chemicals brood should not be harvested from colonies treated with toxic substances used for protection against pests and parasites.
(2) To reduce the risk of microbiological contamination harvested brood combs should be kept in clean hive boxes or other suitable containers that allows transportation without physical damages to the brood.
(3) Both open and capped brood will stay alive at room temperature for a few hours allowing for transport from apiaries to facilities where the brood can be processed or stored.
(4) If brood frames are cut out from a frame in the apiary additional care should be given due to the risk of nutritious liquid leaking from damaged brood. Therefore each brood piece should be placed in clean plastic bags or containers and kept cold during transportation to facilities where the brood can be processed or stored.
(5) The brood should be frozen as soon as possible after harvest, at −20 °C ideally within 4–6 h after harvest to ensure the freshness of the product.
(6) Add date and apiary/colony id to the stored batch.

### 4.3. *Separation of the edible brood after freezing*

The brood combs consist of brood and bee wax. Bee wax is used as an additive to food, but is not recommended to ingest in large quantities (European Food safety Authority, 2007), thus, the brood needs to be separated from the wax. Honey bee larvae, prepupae and pupae are very fragile; therefore special consideration is required if the brood is to remain intact. Obtaining larvae from the combs might be easier prior to capping as larvae can then be removed from combs by a stream of water (Schmidt & Buchmann, 1992).

For capped brood combs we recommend freezing the comb down before proceeding with the separation as it preserve the freshness and allows easy breaking up of the comb.

#### 4.3.1. *Intact brood separation at −20 °C*

(1) Work in clean environments with clean hands, gloves and tools according to EU regulations on the hygiene of foodstuffs (European Union, 2004b, Regulation (EC) No 852/2004 (e.g., clean tables with water and soap, warm water, 70% ethanol and warm water). Tools have to be washed in a dishwasher min 80 °C.
(2) Place large (~3 l) and small (0.5–1 l) containers made of material intended to come into contact with food (European Union, 2004a, Regulation (EC) No 1935/2004) into freezer.
(3) Work only with one comb piece at a time and let the rest be in the freezer.
(4) Break the comb piece into small pieces with hands in the large (~3 l) container.
(5) Remove individual brood from the wax with forceps or tweezers into smaller plastic containers.
(6) Every 5 min, before the frozen brood begins to thaw, place the small container of separated brood back into the freezer and pull another small container from the freezer.
(7) Repeat, processing comb from the freezer one piece at a time. Rear small containers once the separated brood is re-frozen and keep in the freezer before using for new separated brood.

*Pros and cons*:
The brood and wax quickly thaw so only small amount of material can be handled at a time, and due to the quick thawing small wax pieces quickly stick to the brood, which still makes it difficult to remove fully. Uncapped brood can be harvested by swing-impact described by Hocking and Matsumura (1960) but will only work if the brood is produced in a stable wood or plastic frame otherwise it the brood comb will break apart due to the swing-impact. If the brood has to be uncapped, the uncapping is difficult without injury of the brood and subsequently the brood will liquefy.

#### 4.3.2. *Intact brood separation using liquid nitrogen*

(1) Work in clean environments with clean hands, gloves and tools (see Section 4.3.1).
(2) Liquid nitrogen should be kept in an appropriate tang (e.g., a Dewar tank) and gloves should be used when handling liquid nitrogen.
(3) Pour liquid nitrogen into an insulated food-grade container.
(4) Drop pieces of brood comb (~100 cm$^2$) into the container.
(5) Let it freeze completely (about 20–30 s).
(6) Use a spoon or a steel strainer to lift the chunk of comb out of the liquid nitrogen and place onto a clean (see Section 4.3.1) plastic tray.
(7) Quickly rub and break up the brood comb using gloved hands to separate all the frozen wax and debris from the brood.
(8) Rub the brood in between a clean (see Section 4.3.1), dry cloth to remove any remaining debris.
(9) Rubbing should be gentle in particular for pupae not to cause breakages between the thorax and abdomen.
(10) Once clean, place the brood in a small container and store in a freezer.

*Pro and cons*:
Access to liquid nitrogen is necessary. The advantages of using liquid nitrogen are the lower temperature of the brood and wax which increase the window of handling time before thawing. In addition the wax becomes more brittle and thus easier to break and remove fully.

#### 4.3.3. *Brood juice by squeeze-method*

(1) Work in clean environments with clean hands, gloves and tools (see 4.3.1).
(2) Use unfrozen or thawed brood combs.
(3) Place a stainless steel sieve (∅ = 18 cm) above a plastic container (∅ = 20 cm).
(4) Squeeze comb pieces above the sieve and let the juices pass the sieve.
(5) The juice needs to be frozen or used quickly.

*Pro and cons*:
The method is much faster and less labor-intensive, however, it has a lower yield and the liquid oxidizes extremely quickly.

### 4.4. *Storage of brood combs and processed brood*

Storage conditions and shelf life information of food products is important (Taoukis & Labuza, 1989), however no research has so far been done with honey bee brood. Due to the high fat content of the larvae and pupae, rancidification, if not properly removed from contact with oxygen, causing a rank, unpleasant smell or taste might be expected to occur. However, brood in the combs has been stored at −20 °C without severe loss or change of taste for up to 6 months and cleaned intact brood in vacuum-sealed bags for up to 10 months (Pers com, Josh Evans).

(1) Place frozen and separated brood in vacuum bags, not too full so they can be sealed properly with space to move.
(2) Seal the brood in vacuum bags, so they are sealed from oxygenated air.
(3) Do not pack brood tightly in the bag but rather with space enough left for them to stay loose from each other.
(4) Keep the sealed bag frozen at −20 °C until required for use.
(5) Once taken from the freezer they are best used immediately.
(6) If only a part of a bag is used, reseal to preserve storability.

## 5. Nutritional value of edible honey bee brood

The high nutritional value of honey bee larvae and pupae that can be compared to beef by its protein quality and quantity (van Huis et al., 2013) is among the main reasons for using it as an alternative food source. The determination of the nutrients is based on standard analytical methods, used for meat in the food science industry and in research (Association of Officiating Analytical Chemists, 1996). Attention is given in this chapter to the unique nutritional ingredients found in insects, particularly honey bees.

### 5.1. Protein analyses

Proteins can be considered as the most valuable food ingredient, as they serve as building blocks of life. We need to obtain the essential amino acids from food, because they cannot be metabolized by ourselves. The amino acids are important for growth, development, and for the proper functioning of organs and systems, such as brain, immune, and endocrine systems, etc., from the cell level to the entire body. Accurate analyses of total proteins and amino acids composition is important when considering bee brood as a nourishing, alternative foodstuff.

#### 5.1.1. Total protein (Kjeldahl method)

Protein quantity can be analyzed by a few standard techniques, most dependent on the amino acid content or covalently bound material. Honey bee brood proteins can be roughly separated into three groups: hemolymph, muscle, cuticular proteins. These groups differ in composition and structure, and the quantitative ratio among them varies according to body part, age, and external factors. The haemolymph proteins are water soluble and can easily be quantified by Bradford assay or bicinchoninic acid assay as described earlier (Hartfelder et al., 2013). The muscle and cuticular proteins are fibrous. They are connected to the insect's exoskeleton. As in other meat and fish food products, Kjeldahl is the preferred method for determining total protein in honey bee brood, as it accurately measures nitrogen content. Nitrogen amount can easily be converted into the amount of protein, after subtracting the nitrogen derived from chitin and divided by the conversion ratio between nitrogen and protein. This method is based on the standard procedures of the Association of Analytical Chemists (Association of Officiating Analytical Chemists, 1996) with minor modification (Jonas-Levi, Benjamin, & Martinez, 2015).

##### 5.1.1.1. Total nitrogen

*Sample preparation*:
Samples can be used either without preparation or after drying (60 °C or freeze-drying until constant weight). While analyzing small samples (<10 g), it is recommended to use intact items in order to prevent testing a nonhomogeneous sample. For analyzing larger samples (>10 g), it is recommended either to homogenize a fresh sample or powder a dry sample before analysis.

*Procedure*:

(1) Weigh approximately 1.0 g (dry weight or equivalent fresh weight) sample and transfer to a Kjeldahl digestion flask without staining the flask's wall.
(2) Add 0.15 g Kjeldahl catalyst, 10 ml concentrated sulfuric acid, and 10 ml 35% hydrogen peroxide.
(3) Boil in a Kjeldahl apparatus at 400 °C for two hours and cool. Add 50 ml distilled water.
(4) Prepare the distillation apparatus: place a 250 ml titration flask containing trapping solution-65 ml boric acid and five drops indicator. Verify that the delivery tube dips into the trapping solution.
(5) Carefully pour down the side of the digestion flask, 60 ml 50% NaOH and connect to the distillation apparatus.
(6) Mix and boil until about 150 ml distillate has been collected.
(7) Titrate with 0.1 N HCl until green color changes to gray and purple.

*Calculation*:
Total nitrogen:

$$Ns = \frac{Vs - Vb \times 0.1 \times 14.007}{10 \times Ws}$$

where Ns is the Total nitrogen in sample (%) of dry bee weight; Vs is the Titration volume for sample (ml); Vb is the Titration volume for blank (ml); Ws is the Sample weight (g).

Total protein is calculated by multiplying the Nitrogen content from the proteins by the conversion factor 6.25 (0.16 g nitrogen per gram of protein). This is very close to reality, especially in young larvae, where the cuticle weight is negligible.

When the precise amount of protein is required, chitin nitrogen should be measured, as described below.

$$\text{Total protein, \%} = \text{Ns} \times 6.25$$

where Ns is the is the Total nitrogen in sample (%) of dry bee weight.

*5.1.1.2. Total protein after diminution of chitins nitrogen.* Extraction of the chitin should be done as described in Section 4.3.3, following nitrogen quantification of the fibrous material by Kjeldahl, as described above (Section 4.1.1). To determine total nitrogen in chitin use the following equation:

$$\text{Nc} = \frac{(\text{Vc} - \text{Vb}) \times 0.1 \times 14.007}{10 \times \text{Ws}}$$

where Nc is the Total nitrogen in chitin (%) of dry bee weight; Vc is the Titration volume for sample (ml); Vb is the Titration volume for blank (ml); Ws is the Sample dry weight (g) (dry bee weight before chitin extraction).

$$\text{Total protein, \%} = (\text{Ns} - \text{Nc}) \times 6.25$$

where Ns is the Total nitrogen in sample (%) of dry bee weight; Nc is the Total nitrogen in chitin (% )of dry bee weight (be aware to calculate Nc as percentage of dry bee weight and not of cuticle weight).

### 5.1.2. Amino acid composition

Amino acids can be analyzed by various methods, such as GC-MS, HPLC with UV and fluorescence detection, NIR spectroscopy, or amino acid analyzer. Sample preparation principles are identical nonetheless and are based on hydrolysis of the proteins to amino acids. Acid hydrolysis is the most common method, but unfortunately affects several amino acids: Tryptophan is destroyed and asparagine and glutamine are deaminated, resulting in aspartic acid and glutamic acid, respectively. Serine, threonin, methionine and cysteine are partly oxidized and use of vacuum or an inert gas in the headspace of the reaction vessel can reduce the level of their oxidative destruction. The peptide bonds between isoleucine and valine and between themselves are not easily or totally cleaved. When HPLC analyses are used, common artificial results can occur due to contamination. For this analysis, therefore, high purity reagents are used and solvents, especially buffers, must be fresh in order to avoid microorganism contamination, which are rich in amino acids. In any case, using an internal standard to monitor losses and variations during amino acid analysis is recommended.

*Sample preparation*:
Use freeze dried and fat free samples.

#### 5.1.2.1. For most amino acids

(1) Weigh 0.2 g into a 15 ml flask, add 8 ml 6.0 N HCl, seal well, and hydrolyze at 110 °C for 24 h.
(2) Withdraw 1 ml and evaporate under vacuum at 45 °C until dry.
(3) Re-dissolve the sample by mixing in 5 ml 0.02 N HCl and centrifuge at 5,000 rpm.

#### 5.1.2.2. For tryptophan

(1) Weigh 0.20 g into a 15 ml flask, add 8 ml 5 N NaOH containing 5% $SnCl_2$ (w/v) for 20 h at 110 °C.
(2) After hydrolysis, neutralize with 6 N HCl and centrifuge at 5,000 rpm.

These samples are ready to use in an amino acid analyzer, in GC-MS, NMR and NIR spectroscopy. For HPLC analysis further derivatization is needed.

*5.1.2.3. Derivatization for analysis by HPLC.* Ion-exchange chromatography followed by postcolumn derivatization with ninhydrin:

Postcolumn derivatization techniques can be used with samples that contain small amounts of buffer components, such as salts and urea, and generally require between 5 and 10 µg of protein sample per analysis. A Na-based cation-exchange system can also be used.

Precolumn derivatization with Phenylisothiocyanate (PITC): Precolumn derivatization techniques are very sensitive and usually require between 0.5 and 1.0 µg of a protein sample per analysis, although may be influenced by buffer salts in the samples. These techniques are typically followed by a reversed-phase HPLC.

*Procedure*:

(1) Place 20 µl amino acid standard solution or hydrolyzed samples in a vial and dry in a vacuum oven at 65 °C for 2 h.
(2) Add 30 µl of methanol–water–TEA (2:2:1), dissolve and vacuum-dry at 65 °C for 10 min.
(3) Add 30 µl of the derivatizing reagent methanol–water–TEA-PITC (7:1:1:1 [v/v]), vortex for 30 s and leave to stand at room temperature for 20 min.
(4) Vacuum-dry the resulting solution at 65 °C for 15 min.
(5) Prior to injection, add 150 µl of diluents, consisting of 5 mM sodium phosphate with
(6) 5% acetonitrile, pH 7.4 and vortex for 15 s.

*Chromatography condition*:
There is a wide variety in chromatography conditions, which can be chosen for amino acid separation, depending on the apparatus, column, and goals.

*Amino acid concentration*:
Identified amino acid concentration is determined according to their peaks area in relation to its specific standard. Protein quality needs to take the amino acid compostion as well as the digestibility of the protein into account to evaluate the nutritional value to humans (FAO, 2013). FAO has recommended replacing the protein digestibility-corrected amino acid score with the digestible indispensable amino acid score.

### 5.2. *Lipid analyses*

Lipids are a diverse group of compounds consisting of tri-, di- and monoacylglycerols, free fatty acids, phospholipids, sterols, and vitamins. From a nutritional standpoint, total lipid analyzes are mostly important because they serve as a major source of fuel energy for the body. Moreover, it is important to specify the concentration of the different types of lipids present in foods due to their effect on several aspects: health, physical and sensory properties, and, also, shelf life stability. In general, all lipid analyses start with lipid extraction following specific analyses. Unsaturated lipids tend to oxidize very quickly. When tests that follow total lipids extraction are requested, it is recommended to grind samples at low temperature, and add an antioxidant (20–50 ppm BHT) to the extracted lipids, to flush containers with nitrogen gas and to avoid dried sample storage or exposure to heat or light.

#### 5.2.1. *Total lipophilic compounds*

Total lipid analyzes are based on separation of the lipophilic compounds from the water soluble compounds in a sample by organic solvent extraction. Extraction efficiency depends on the similarity in the polarity between the lipids and the solvent. Because different lipids have different polarities, it is impossible to find an ideal solvent to extract all the lipid components without affecting all the other components in the sample. It is recommended to use a non-polar solvent with a low boiling point. Acetone, with 0.355 relative polarity and boiling point of 56.2 °C, can be chosen. Hydrophilic compounds, such as glycolipids and lipoproteins, will not be well extracted in the hydrophobic solvent. To overcome this problem, acid hydrolysis is required in order to disconnect the complexes of lipids from proteins and carbohydrates. Moreover, a semi-continuous solvent extract can be used, which increases the extraction efficacy to nearly 100%. The recommended procedure is gravimetric (Association of Officiating Analytical Chemists, 1996).

*Sample preparation*:
The sample should be dried (60 °C or freeze-dried) and grounded (at low temperature). For hydrolysis of lipoproteins, glycolipids and phospholipids, the sample should be heated at a minimum volume of 3 N HCl for 1 h and dried once more.

*Procedure*:

(1) Put 5.0 g dried grounded sample into a cellulose thimble and placed in the extraction chamber of the Soxhlet apparatus.
(2) Add acetone into a flask, weighed in advance, connect it to the lower part of the Soxhlet apparatus, and connect a condenser above it.
(3) Heat the flask for 12 extraction cycles.
(4) Cool the system, and evaporate organic solvent.
(5) Dry the flask and the thimble and weigh them.

*Calculation*:
Calculate the total lipids as percentage of dry sample weight.

$$\text{Total lipids } \% = \frac{\text{sample before extraction} - \text{sample after extraction}}{\text{sample before extraction}} \times 100$$

#### 5.2.2. *Fatty acids composition and content*

Fatty acids composition and content has a large impact on the nutritional value and the functional properties of foods. Alpha-linolenic acid (omega-3) and linoleic acid (omega-6) are known as essential fatty acids. Along with most of the polyunsaturated fatty acids, these acids are beneficial to the body, as they are required for building healthy cell membranes, assisting in development and function of the central and peripheral nervous system, and protecting against cardiovascular diseases. In addition, their composition influences food taste, odor and texture.

Fatty acids analyses consist of three phases: (1) lipid extraction, (2) volatilization of the fatty acids to fatty acid methyl esters (FAME), and (3) separation, identification, and quantification of the FAME by chromatography.

*Sample preparation*:
Lipid extraction: Use the method as described above (Section 4.2.1), for preventing oxidation of the unsaturated fatty acids add 20–50 ppm BHT to the extract. Volatilization of the fatty acids to fatty acid methyl esters (FAME), according to Jonas-Levi et al. (2015).

(1) Put 150 μl lipid extract into an evaporation bulb, add 20 ml methanol 3% $H_2SO_4$, and boil for 1 h.

(2) After cooling, transfer all liquids to a 50 ml volumetric flask, wash the bulb with 3 ml petroleum ether, and add it to the volumetric flask.
(3) Add 10 ml distilled water, mix, and add another10 ml distilled water, and wait until phases separate.
(4) Collect upper phase, which contains the FAME, to a new flask.
(5) Add some sodium sulfate for water residues absorption.
(6) If necessary, store in a sealed glass bottle at −20 °C.

Separation, identification and quantification:
For reliable identification and quantification, use first a FAME standard for analysis in GC-MS or GC-FID chromatography.

*Procedure:*
Inject 1 µl of fatty acid methyl esters standard and 1–2 µl sample to a gas chromatograph with flame ionization detector with a Supleco 2560 100 m × 0.25 mm × 0.2 µm column (Sigma-Aldrich, St Louis, USA). The oven temperature program should be set to 5 min at 140 °C, heating to 240 °C at a rate of 4 °C per min, followed by 20 min constant temperature at 240 °C. Set the detector temperature at 260 °C, $H_2$ 40 ml/min splitless. Use nitrogen as the carrier gas (400 ml/min). Use an external mixed standard of C4-C24 fatty methyl acids to identify the fatty acid peaks and retention times.

*Calculation:*
The identification of a FAME is done according to the retention time of a standard, while the quantification is in relation to the picks area. There are two possibilities for fatty acids amount calculation:

Relative amount of a fatty acid ($C_{FA_x}$) is calculated as the area of the specific fatty acid ($S_{FA_x}$) divided to the total area of all the fatty acids ($\sum S_{FA}$) in the chromatogram multiplied by 100.

$$C_{FA_x} = \frac{S_{FA_x}}{\sum S_{FA}} \times 100$$

Fatty acid concentration is calculated as the area of the fatty acid ($S_{FA_x}$) multiplied by the concentration of a specific standard ($C_{FA_s}$) divided by the area of the standard fatty acid ($S_{FA_S}$).

$$C_{FA_x} = \frac{S_{FA_x} \times C_{FA_s}}{S_{FA_S}}$$

## 5.3. Carbohydrate analyses

Carbohydrates are divided into two main groups, carbohydrates for energy supply and structural polysaccharide – chitin. The energy storing carbohydrates are composed of the immediate metabolic available sugars, glucose and trealose, which are located mainly in the hemolymph, their analysis was described before (Hartfelder et al., 2013) along with the energy storage carbohydrate – glycogen. The simple sugars are responsible for the sweetish taste of honey bee brood.

### 5.3.1. Total carbohydrates

Total carbohydrates include digestible and indigestible sugars. The procedure is based on full hydrolysis of the carbohydrates, turning them into monosaccharides, following Sumner's test for reducing sugars.
*Sample preparation – hydrolysis:*

(1) Weigh 1.000 g sample, add 80 ml 1 N HCl.
(2) Boil for 2 h under reflux, and cool to room temp.
(3) Transfer sample to a 100 ml volumetric flask, adjust pH to 7.5 with NaOH, add distilled water to 100 ml, and mix.

*Glucose standard curve:*
Dilute the glucose to at least six different concentrations from blank up to 0.0125%.

*Procedure:*

(1) Add 2 ml of the tested compounds (glucose, sample and blank) into a glass test tube (≥5 ml, suitable for spectrophotometer).
(2) Add 2 ml Sumner's reagent and mix.
(3) Boil 5 min at 100 °C, and cool on ice.
(4) Evaluate the absorbance at 550 nm after resetting the spectrophotometer on "zero sugar" concentration.

*Calculation:*

(1) Build a standard curve graph – optic absorbance at 550 nm to glucose concentration.
(2) Add a linear trendline with its equation and a $R^2$ value to the graph.
(3) Calculate the sugar concentration (equivalent to glucose) in the samples, according to their optical density and the trendline.

### 5.3.2. Fibrous material

The fibrous material in honey bee brood comes from Chitin, a nondigestible carbohydrate from the insect exoskeleton. It is a long polymer of acetylated glucosamine units linked by beta-1,4 glycosidic bonds. The determination of total fiber is determined by an enzymatic gravimetric procedure, based on the AOAC method (Prosky, Asp, Schweizer, Devries, & Furda, 1988).

*Sample preparation*:
For the procedure use dry fat-free samples as described above (Section 6.2.1).

*Gooch crucible preparation*:

(1) Add 0.5 g Celite filter aid to every gooch, the Celite should cover the entire filter surface.
(2) Dry at 130 °C for 1 h.
(3) Allow cooling in a desiccator and weigh.

*Procedure*:
Enzymatic hydrolysis:

(1) Weigh 1.0 g into a 250 ml beaker and add 50 ml 20 mM phosphate buffer pH-6.
(2) Add 100 μl heat stable alpha-amylase and boil for 15 min. Gently stir once every 5 min.
(3) Cool to 60 °C, adjust the pH to 7.5 by adding 0.5 M NaOH and add 50 μl protease. Cover the beaker and incubate 30 min at 60 °C, stir gently every 5 min.
(4) Adjust pH to 4.5 by 0.5 M HCl and add 100 μl amyloglucosidase. Cover the beaker, incubate 30 min at 60 °C, stir gently every 5 min and cool to RT.

Fiber precipitation and filtration:

(5) Measure the solution volume and add 4 volumes of 95% EtOH. Allow it to precipitate over night at RT.
(6) Stir the solution gently and filter it through the gooch in a vacuum system.
(7) Wash all residues from the beaker through the gooch with 78% EtOH.
(8) Wash with 15 ml 95% EtOH.
(9) Wash with 15 ml acetone.
(10) Dry the gooches at 130 °C for 2 h, cool in a desiccator and weight.

*Calculation*:
Be aware that the sample you used for fiber analysis is lipid free and dry. To calculate the fiber percentage from the dry sample (with the lipids), the complete sample weight before lipid extraction needs to be calculated.

$$W_S = \frac{W_{Lfs} \times 100}{100 - C_L}$$

$$C_F = \frac{W_F}{W_S} \times 100 - C_M$$

where $W_S$ is the Sample weight (g) (complete sample weight before lipids extraction); $W_{Lfs}$ is the Lipid free sample weight (g) (sample weight after lipids extraction); $C_L$ is the Lipids concentration (%); $C_F$ is the Fibers concentration in a dry fat free sample (%); $W_F$ is the Fibers weight (g); $C_M$ is the Mineral concentration (%).

### 5.4. Calories

A calorie (Cal) is the amount of energy needed to raise the temperature of one gram of water by one degree (Celsius or Kelvin) in one atmosphere of pressure. When referring to food and nutrition, it is common to use the value, kcal. Two types of measurements are used in determining food energy: (1) measurement of total energy stored in the chemical bounds of a material, by a bomb calorimeter, and (2) estimation of energy available from digestion of food.

#### 5.4.1. Bomb calorimeter

A bomb calorimeter is used to determine the enthalpy of combustion for organic compounds. It is used to measure the heat created by a sample burned under an oxygen atmosphere in a closed vessel, which is surrounded by water, under controlled conditions. There are different bomb calorimeter versions and sizes, but the principle is the same for all.

*Sample preparation*:
Samples must be dried in advance. It is recommended to compress the sample into a bloc.

*Procedure*:

(1) Weigh ~1.0 g of sample or standard (benzoic acid) into a crucible, and place it inside a stainless steel container.
(2) Attach a nickel or iron fuse from the sample to the ignition wire and seal the container.
(3) Fill it with 30 bar of oxygen and then put the container into a Dewar.
(4) Add a known weight of water ($W$) and close the calorimeter.
(5) Let the temperature stabilize ($T_1$).
(6) Press the ignition unit to generate combustion.
(7) Let the temperature stabilize ($T_2$).

*Calculation*:
Correction factor

$$CF = \frac{6318}{(T_2 - T_1) \times T}$$

where $CF$ is the Correction factor; $T_2$ is the Temperature after combustion ($C$); $T_1$ is the Temperature before combustion ($C$); $W$ is the Water weight (g); 6318 is the Calories in standard, cal/1 g benzoic acid.

Sample calories

$$S_{\text{cal}} = \frac{(T_2 - T_1) \times W \times CF}{S_w}$$

where $S_{\text{cal}}$ is the Calories in sample (cal/g); $T_2$ is the Temperature after combustion ($C$); $T_1$ is the

Temperature before combustion (*C*); *W* is the Water weight (g); *CF* is the Correction factor; $S_w$ is the Sample weight (g); $S_J$ is the Joules in sample (j/g = $S_{cal} \times 4.184$).

*5.4.2. Estimation of energy available to be digested in food*

For estimating the energy available to be digested in food, the information about total proteins, total digestible carbohydrates and total lipids is needed.

$$Cal_d = (L \times 9) + (P \times 4) + (C \times 4)$$

where $Cal_d$ is the Calories available to be digested in the portion (cal); *L* is the Lipids amount in the sample (g); *P* is the Protein amount in the sample (g); *L* is the Carbohydrate amount in the sample (g).

### 5.5. Vitamins

Vitamins are needed for the normal and healthy metabolism of the body. Lack of vitamins can lead to serious diseases and death. Vitamins are divided into two groups according to their chemical characteristic, the fat soluble vitamins (A, D, E, K) and the water soluble (C, B family and some miscellaneous compounds). This separation into groups is important for sample preparations. Many articles and books are written about analytical determination of vitamins, Blake (2007) has reviewed the subject, while Collin and Blum (2005) give specific analytical methods for all vitamins and provitamins.

*5.5.1. Fat soluble vitamins*

Fat soluble vitamins are very easily oxidized and changed, therefore it is necessary to avoid expose to light and heat, and the analysis must be immediately after separation. Internal standard is required in vitamin analyses due to their sensitivity. Sample preparation includes saponification and solvent extraction. The use of HPLC grade solvents is recommended.

*Sample preparation:*

This procedure is according to Salo-Väänänen et al. (2000).

Saponification:

(1) Put 1 g of homogenized sample into a Pyrex tube.
(2) Add 5 ml of 2% ascorbic acid, 10 ml of ethanol and 1 ml of internal standard (0.2 mg of vitamin).
(3) Flush with nitrogen and vortex.
(4) Add 4 ml of KOH.
(5) Flush with nitrogen and seal the tube.
(6) Boil for 20 min. Mix every few min.
(7) Cool the sample on ice for 10 min.
(8) Add 10 ml 10% NaCl.

Extraction:

(9) Add 20 ml extraction solvent, n-hexane – ethyl acetate (8:2) mix for 2 min.
(10) Allow the phases to separate.
(11) Transfer the organic phase to a 100 ml tube.
(12) Repeat this extraction procedure (section 15–17) three times, combine the organic phase.

Washing:

(13) Wash the extract with 20 ml of 5% NaCl,
(14) Transfer the organic phase to a 100 ml bottle,
(15) Add 5 ml ethanol and 5 ml n-hexane, mix and evaporate.
(16) Dissolve sample in n-hexane and transfer to HPLC analysis tube,
(17) Evaporate and dissolve again in 1 ml n-hexane prior to analysis.

*5.5.1.1. Tocopherols, b-carotene, and all-trans-retinol analysis.* HPLC method for the determination of tocopherols, b-carotene, and all-trans-retinol is summarized in Table 2. Condition as according to Salo-Väänänen et al. (2000): step gradient using an mPorasil column (10 µm, 30 cm × 3.9 mm, Waters) with a silica guard column (GuardPak Silica, Waters), adjust temperature of the column oven to 30 °C, injection volume 75 µl.

*5.5.1.2. Vitamin $D_3$ analysis.* Use a HPLC system with a UV detector set at 265 nm, and a Vydac TP 54 column (5 mm, 25 cm, 4.6 mm, The Separations group) with an ODS quard column (Nova-Pak C18, Waters). The temperature of the column oven should be set to 25 °C. The mobile phase: 93% methanol and 7% water. The flow rate of the mobile phase: 1 ml/min, and the injection volume 75 ml. (Salo-Väänänen et al., 2000).

*5.5.2. Water soluble vitamins*

*Sample preparation:*

A suitable sample preparation may be the most important factor for accurate results. The water soluble vitamins are separated into groups for sample preparation according to Leporati et al. (2005).

*5.5.2.1. Vitamins $B_1$, $B_2$, $B_6$, and Niacin (Vitamin $B_3$)*

(1) Grind and homogenate samples.
(2) Transfer 1.0 g of sample into a 50 ml brown glass flask.
(3) Add 25 ml of 0.1 M HCl (for hydrolysis).
(4) Boil in water bath for 30 min (with frequent mixing).
(5) Allow sample to cool, and adjust to 50 ml with 2 M ammonium acetate.

Table 2. HPLC method for determination of tocopherols, b-carotene, and all-trans-retinol: step gradient and detector program (Salo-Väänänen et al., 2000).

| Time, min | n-hexane % | diisopropyl ether, % | Diisopropyl ether-n-hexane (1:9)% | Flow (ml/min) | UV (nm) | Fluorescence (ex/em) |
|---|---|---|---|---|---|---|
| 0 | 97 | 0 | 3 | 1.5 | 450 | 292/325 |
| 4.0 | 97 | 0 | 3 | 1.5 | 325 | 292/325 |
| 4.1 | 20 | 0 | 80 | 1.5 | 325 | 292/325 |
| 24.0 | 20 | 0 | 80 | 1.5 | 325 | 292/325 |
| 24.1 | 70 | 30 | 0 | 1.5 | 325 | 292/325 |
| 45.0 | 70 | 30 | 0 | 1.5 | 325 | 292/325 |
| 45.1 | 97 | 0 | 3 | 3.0 | 325 | 292/325 |

(6) Shake and centrifuge at 12,000 ×g for 10 min.
(7) Filter 1 ml of sample on 0.45 micrometer filter.

5.5.2.2. *Pantothenic acid (Vitamin $B_5$)*

(1) Grind and homogenate 4.0 g of sample and transfer into a 100 ml brown flask.
(2) Add 50 ml of acetate buffer (pH 5.6) and autoclave at 121 °C for 15 min.
(3) Allow cooling.
(4) Adjust to 100 ml with 0.3 M ammonium formate, then mix on a Vortex.
(5) Centrifuge at 12,000 × g for 10 min.
(6) Filter through a 0.45 micrometer membrane.
(7) Perform sample clean-up by solid-phase extraction (Section 5.5.2.4).

5.5.2.3. *Folic acid (Vitamin $B_9$)*

(1) Grind and homogenize 2.5 g of sample and transfer into a 50 ml brown flask.
(2) Add 10 ml of water and 25 ml of 0.05 M sodium phosphate–sodium citrate/ascorbate buffer, pH 8.0 to the sample.
(3) Hydrolyze in a boiling bath for 10 min.
(4) Allow cooling and add 0.2 g of papain and 0.1 g of di-alpha-amylase and incubate at 40 °C for 2 h.
(5) Allow cooling and adjust to 50 ml with 0.3 M ammonium formate.
(6) Mix on a Vortex
(7) Centrifuge at 12,000 × g for 10 min.
(8) Filter through a 0.45 micrometer membrane.
(9) Perform sample clean-up by solid-phase extraction (Section 5.5.2.4).

5.5.2.4. *Sample clean-up by solid-phase extraction (SPE).* 3 ml Supelco LC–SAX cartridge:

(1) Condition with 2 ml of conditioning solution (sodium acetate (20 mM) and sodium sulfate (20 mM), adjusted to pH 5.3 with acetic acid).
(2) Load 1 ml of sample solution.
(3) Wash twice with 1 ml of the first washing solution (sodium sulfate (0.1 M) and sodium chloride (0.5 M), adjusted to pH 5.3 with acetic acid).
(4) Wash with 1 ml of the second washing solution (sodium sulfate (0.5 M) and sodium chloride (1 M), adjusted to pH 5.3 with acetic acid).
(5) Elute with 1 ml of elution solution (sodium sulfate (0.5 M) and sodium chloride (1 M), adjusted to pH 2.5 with trifluoroacetic acid).

5.5.2.5. *Vitamin C.* Vitamin C extraction is according to Behrens and Madère (1987) and Chebrolu, Jayaprakasha, Yoo, Jifon, and Patil (2012).

(1) Homogenize 3.0 g sample in a motor-driven glass homogenizer in 15 ml cold 17% metaphosphoric acid.
(2) Centrifuge at 20,000 g for 10 min.
(3) Transfer clear supernatant and use for analysis.

Determination of ascorbic acids can be carried out according to Scherer, Rybka, Ballus, Meinhart, and Godoy (2012). Use a HP liquid chromatography equipped with degasser, quaternary pump, autosampler adjusted to 20 μl volume injection, RP-C18 column (3 $\mu$m particle size, 150 × 4.6 mm I.D., kept at 25 °C) and a UV–visible diode array detector. The mobile phase should consist of 0.01 mol/l KH2 PO4 buffer solution (pH 2.60 adjusted with o-phosphoric acid), using an isocratic elution procedure with a flow rate of 0.5 ml/min. Detect at 250 nm.

5.5.2.6. *All soluble vitamins (except vitamin C).* All soluble vitamins (except vitamin C) can be analyzed by a HPLC system (Moreno & Salvadó, 2000) using reversed-phase Nova-Pack C18 (150 × 3.9 mm, 4 μm) column. The mobile phase of the HPLC system should consist of a 0.05 M ammonium acetate (solvent A) – methanol (solvent B) gradient as follows: in the range 0–1.5 min, 92.5:7.5; at 1.6 min, 84:16; at 15 min, 70:30. The mobile phase flow-rate should be adjusted to 1 ml/min. Inject 40 μl sample. Monitor column effluents at 270 nm.

### 5.6. *Minerals*

Minerals are needed for the health and function of the body. They comprise macrominerals (Na, Cl, K, Ca, P,

Mg, S) and others, including trace minerals, which are needed in much smaller amounts. Ash content or ash analysis refers to the total inorganic materials in a sample, whereas mineral content or analysis refers to the amount of a specific mineral in a sample.

*5.6.1. Ash analyses*

Ash analysis provides information of the total mineral content in a sample; it may also be used as the first step in preparing samples for analysis of specific minerals. Two main analytical procedures are used to determine ash content of foods: dry ashing and wet ashing.

*5.6.1.1. Dry ashing.* This procedure is carried out in a crucible. Various types of crucibles are available for ashing food samples, including quartz, Pyrex, porcelain, steel and platinum. The most widely used are those made from porcelain; they are relatively inexpensive, can be used up at high temperatures (<1,200 °C), are easy to clean, and are resistant to acids, although not to alkaline. Be cautious, porcelain cracks easily under rapid temperature change.

*Sample preparation*:

The sample should be dried at 105 °C for 1.5 h or at 60 °C overnight or it can be freeze-dried before being grounded. It is generally recommended to extract the fat before dry ashing, but according to previous experience it is not needed for bee brood ash analysis.

*Procedure*:

Put 5.0–10.0 g dried samples into a crucible and ash it in a furnace at 550 °C for 12 h (AOAC, 1990). Take care to dry, burn, and weigh the crucible before weighing the sample in it.

*Calculation*:

$$\% \text{ Ash} = \frac{W_{\text{Ash}}}{W_{\text{Dry}}}$$

where $W_{Ash}$ is the Ash weight (g); $W_{Dry}$ is the Samples dry weight (g).

*5.6.1.2. Wet ashing.* Wet ashing is primarily used in the preparation of samples for subsequent analysis. It completely oxidizes the organic matrix to $CO_2$ and water and leaves the mineral in a clear acidic solution. Wet ashing occurs in strong acids as oxidizing agents and heated about 350 °C. Special care must be taken when working with strong acids. Work with acids needs to be done in a chemical hood, gloves replaced and contact lenses avoided. The acid used for wet ashing depends on the subsequent analysis performed; therefore a specific procedure of sample preparation is described for any analysis.

*5.6.2. Mineral analyses*

There are several options for mineral analysis: (1) ion selective electrodes (ISE), an appealing method due to its simplicity, speed, and ease of use; (2) atomic absorption spectrometry (AAS), the accepted method in food analysis; and (3) the accepted method for phosphorus determination.

All methods for mineral analyses are quite sensitive, thus care must be taken by using pure deionised water, and particularly clean tools, soaked overnight in a 10% nitric acid solution and then rinsed three times with deionised water. Moreover, it is recommended to identify the minerals in a blank sample, which had been exposed to the same treatments and solvents for the other samples, although without the tested compounds.

*5.6.2.1. Ion-selective electrodes.* The principle of the ion-selective electrodes (ISE) method is as pH meters, but with special electrodes to determine the concentration of specific ions. The most common commercial electrodes are: ammonium, barium, calcium, cadmium, copper, lead, mercury, potassium, sodium, silver, bromide, chloride, cyanide, fluoride, iodide, nitrate, nitrite, perchlorate, sulfide and thiocyanate. The explanation of how to perform this method can be summarized easily: two electrodes, a reference electrode and ion selective electrode are inserted into an aqueous extract of the sample. The concentration of a specific mineral is determined from a calibration curve of voltage vs. the logarithm of concentration. There are two possible confounding factors using the ISE method: (1) interference from other types of ions. This problem can be reduced by adjusting pH, complexing or precipitating the interfering ions; and (2) presence of non free-ions, for example when the ions are in complexes with chelating agents or polymers. In most cases this can be prevented by ashing the sample, as described above (Section 5.6.1).

*5.6.2.2. Atomic absorption spectrometry.* Atomic absorption spectrometry (AAS) is one of the most effective ways of determining specific minerals; it can be used to quantify the entire range of minerals in foods in ppm concentrations (Association of Officiating Analytical Chemists, 1996).

*Sample preparation*:

Acidic digestion:

(1) Take 5–10 g of sample, digest with a mixture of concentrated nitric acid, sulphuric acid, and perchloric acid (10:0.5:2, v/v), in a 250 ml Kjeldahl flask.
(2) Dilute the digest to 100 ml and analyze.
(3) Analyze samples directly.
(4) Make sure standards have the same acid concentration as samples.

If calcium is being determined, add 0.5% lanthanum to samples and standards.

Ashing:

(1) Ash 5–10 g of sample as described above (Section 5.6.1).
(2) Dilute the ash to a known volume and analyze.
(3) Ashed samples can be kept in a desiccator until use.

Each mineral has its specific standard condition The PerkinElmer (1996) provides details for all the standard conditions.

*5.6.2.3. Phosphorus analysis by a spectrophotometric method.* P (phosphorus) is generally measured by a spectrophotometric method after its formation to a phosphomolybdate complex at 650 nm (Association of Officiating Analytical Chemists, 1996).

*ANSA reagent solution*:

(1) In a 100-ml volumetric flask dissolve 2.0 g sodium sulfite ($Na_2SO_3$) in approximately 80 ml of reagent water.
(2) Add 0.25 g of 1-amino-2-napthol-4-sulfonic acid.
(3) Dissolve and dilute to the mark.
(4) Prepare a second solution by dissolving 15 g sodium bisulfite ($NaHSO_3$) in 300 ml of reagent water.
(5) Mix the two solutions in a dark plastic container.
(6) Degas with helium. Store refrigerated and discard when color changes and it becomes darker.

*Sample preparation*:

(1) Mineralize 5.0 g sample in 7.0 ml of $HNO_3$ (65%) for 24 h in a Kjeldahl flask.
(2) Add 2.0 ml of $H_2O_2$ (30%).
(3) Heat at 100 °C for 60 min.
(4) Continue heating at 200 °C for 120 min.
(5) If necessary cool, mix and continue heating at 200 °C until the solution is clear.
(6) Cool and dilute to 50.0 ml with water in a 50 ml volumetric flask.

*Procedure:*

(1) Mix 1 ml of sample with 0.6 ml molybdate reagent.
(2) Add 0.2 ml ANSA solution.
(3) Mix and incubate at RT for 10 min.
(4) Measure absorbance at 650 nm.

### 5.7. *Antioxidant activity*

Antioxidants are molecules that inhibit the oxidation of other molecules. They play a significant role in our health, as they decrease oxidative damage in our food as well as in our body. One of the main mechanisms of antioxidants is to be oxidized by themselves, producing a chemically stable oxidized molecule. By way of this mechanism the antioxidants act as a reducing agent, which can be quantified by the Ferric-reducing antioxidant power (FRAP) assay.

#### *5.7.1. Antioxidant power by Ferric-Reducing Antioxidant Power (FRAP) assay*

Anti-oxidant reducing power can be estimated as described earlier (Jonas-Levi et al., 2015).

(1) Add 10 μl samples, standards or blank to a 96 well plate.
(2) Add 200 μl reaction mixture (300 mM acetate buffer pH-3.6, 10 mM 6-tripyridyl-s-triazine and 20 mM $FeCl_3$).
(3) Mix and immediately measure the optical density at 595 nm.
(4) Repeat measurement every 5 min during 1 h.

#### *5.7.2. Antioxidant enzymes*

Antioxidant enzymes are an important part of the natural body defense against reactive oxygen species, in this way they can act in the digestive system before degradation. Enzymes are sensitive macro-molecules that can easily lose their activity. A few simple rules can assure enzyme activity: enzymes should not be heated, they have to be kept on ice until reaction begins, mixing should not be done vigorously, which could produce unnecessary sheer stress; and enzymes should not be exposed to strong acid bases or organic solvents. For comparison between enzyme activities in various samples, it is common to divide the enzyme activity by protein concentration, receiving specific enzyme activity. Protein concentration can be determined by Bradford assay as described by Hartfelder et al. (2013). For reliable results blanks are required. Few different blanks should be done for every test: without enzyme, without substrate, and if possible with a denaturated enzyme.

Samples preparation and enzyme activities for three antioxidant enzymes as described by Jonas-Levi et al. (2015).

(1) Put 200 mg soft tissue (fresh weight) into 1 ml K-phosphate buffer 50 mM, pH 7.0.
(2) Drain liquids and soft tissue.
(3) Mix gently.
(4) Centrifuge (4,000 g) for 5 min and remove firm material from the extraction mixture.

*5.7.2.1. Catalase activity.* Catalase activity can be determined by the decomposition of $H_2O_2$ over time, measured at 240 nm.

(1) In a UV suitable cuvette put 1 ml K-phosphate buffer 50 mM pH 7.0.
(2) Add 10 μl sample extract.
(3) Mix gently.
(4) Reset the spectrophotometer (240 nm) on the reaction mixture.
(5) Add 10 μl 3% $H_2O_2$, mix by pipetting.
(6) Measure absorption immediately.
(7) Continue measuring for 5 min.

*5.7.2.2. Peroxidase activity.* Peroxidase activity can be determined using pyrogallol as substrate and purpurgallin as product.

(1) Add 5 μl samples into a 250 ml polypropylen 96 well microplate suitable for absorbance microplate readers.
(2) Add 200 μl reaction mixtures (10 mM potassium phosphate buffer pH 6.0, 0.03% hydrogen peroxide and 0.5% pyrogallol (w/v)).
(3) Measured optic density at 420 nm during 8 min.

*5.7.2.3. Superoxide dismutase activity.* Superoxide dismutases (SODs) are a class of closely related enzymes that catalyze the breakdown of the superoxide anion into oxygen and hydrogen peroxide. Superoxide dismutase activity can be quantified, based on the competitive inhibition of nitroblue tetrazolium reduction by the superoxide radical.

(1) Into a glass tube suitable for spectrophotometer, add 1.8 ml Phosphate buffer 50 mM pH- 7.8, 2 mM EDTA, 0.1 ml methionin 200 mM, 2 ml nitroblue tetrazolium 55 mM, 50 ml Triton X-100.
(2) Add 25 μl sample.
(3) Mix gently.
(4) Add 38 μl Riboflavin 53 mM, mix.
(5) Shake under strong light for 10 min.
(6) Measure at 560 nm.

Calculation of antioxidant enzyme activities (catalase, peroxidase and SOD)

$$\text{Enzyme activity} = \frac{\Delta OD}{\Delta t}$$

where ΔOD-The remainder between optic density of the blank and the optic density of the sample; Δt-The remainder between end reaction time and beginning reaction time, second or minute.

$$\text{Enzyme specific acitivity} = \frac{\text{Enzyme activity}}{\text{Protein concentration}}$$

### *5.8. Antinutrients*

Antinutrients are compounds that interfere with the absorption of nutrients. They act in two main mechanisms: (1) by interfering with the digestive enzymes *in vivo*, like protease inhibitors, lipase inhibitors and amylase inhibitors. This type of antinutrient is common in plants against insects, but no such inhibitors were reported in insects such as bees, or pollen; (2) by binding nutrients, and not allowing them to be absorbed. Adedunatan (2005) reported presence of small amounts of phytate and tannins in several insects, which are antinutrients in higher concentrations.

## 6. Sensory analyses of edible honey bee brood

In many developing countries, eating insects is a common practice and each type has its own taste, aroma and texture. Honey bee brood has a nutty and slightly sour flavor with a crunchy texture when eaten cooked or dried. There are few studies exploring the flavor and texture characteristics of edible insects. In this section, we describe methods originated from research done on different sorts of edible insects, and practiced on honey bee brood (Bednářová, Borkovcová, Mlček, Rop, & Zeman, 2013; Cai, Koziel, & O'Neal, 2007; Caparros Megido et al., 2014; Kiatbenjakul, Intarapichet, & Cadwallader, 2015). The samples in some cases are raw insects, but in other cases can be processed such as cooked, baked or boiled.

### *6.1. Volatile compound analysis*

There are three main extraction methods to be used for static sampling of volatile organic compounds (VOCs) from the honey bee brood. The methods are: Direct solvent extraction, solvent assisted flavor evaporation (SAFE) and headspace analysis using solid-phase microextraction (SPME). The first two methods deal with liquid extraction using solvent and the third method is gas extraction from the headspace above the sample. The methods should keep three important demands for a good extraction: (1) essential compounds that contribute to the aroma should not be excluded from the extraction, (2) the conditions applied in the extraction procedure should not change the VOC structure and (3) non-volatile compounds that may affect the separation in the column should be removed. In each method, the advantages and disadvantages are shortly discussed.

*Sample preparation*:
Take honey bee larvae, prepupae or pupae from the beehive as a whole and store at −80 °C until the analyses. Optional for longer storages: Dry the material completely under freeze-drier for 48 h, following a crushing step in a fine mixer to get a powder. The advantage in using freeze dry is to maintain the nutritional and sensorial quality of the source material (Jonas-Levi et al., 2015).

#### 6.1.1. *Direct solvent extraction method*

A direct extraction of honey bee pupae with solvent is a simple procedure that requires the knowledge of the solvent properties to be able to achieve maximum yield of volatiles. It is advised to spike the sample with internal pure standard VOCs in order to assess the recovery percentage in the extracted sample to injection. The method is based on non-insect sample, but should work just as well with bees.

(1) Mix 10 gram of pupae sample (fresh or powder) thoroughly with 30 ml solvent (solvent: dichloromethane, diethyl ether, tert-butyl methyl ether or hexane) for an hour.
(2) Filter the solution using Whatman paper #1.
(3) Transfer the mixture solvent-sample to a separating funnel and collect the organic phase.
(4) Collect the residues with 15 ml of the solvent twice, collect the organic phase and add it to the previously collected solution.
(5) Dry the solution with salt $Na_2SO_4$ and concentrate the sample using slow stream of nitrogen gas.

Pros: Simple method, can achieve part of the VOCs exist in the sample.

Cons: A risk that not all the VOCs are recovered depending on the solvent choice and long time for drying using the $N_2$ gas.

#### 6.1.2. *Liquid extraction by SAFE method*

This method involves the use of a distillation unit designed by Engel, Bahr, and Schieberle (1999). The system contains the feeding the funnel of the sample, heating part and condensing part with liquid nitrogen. The unit is connected to high vacuum pressure ($10^{-3}$ Pa). The following procedure is based on a method by Kiatbenjakul et al. (2015) on water-bugs.

(1) Weigh pupae sample (as whole or powder) and blend it with 20 ml deodorized distilled water in a 250 ml Teflon bottle.
(2) Add 30 ml solvent mixture of pentane: diethyl ether (1:1 v/v).
(3) Homogenize the solution at 3,000 rpm for 15 min and collect the organic phase.
(4) Pour the collected solution with the VOCs to the SAFE unit.
(5) Dry the SAFE extract using $Na_2SO_4$ salt and concentrate the sample to 200 μl with nitrogen gas.
(6) Inject 1 μl to the GC-MS for further identification and quantification of the VOCs.

Pros: The method overcomes other liquid extraction methods due to higher extraction yields. In addition the method used vacuum to lower the evaporation temperatures, which results in less chance to form new volatile compounds.

Cons: Time consuming and not all the VOCs from the sample are extracted to the solvent. There is a risk of losing VOCs to the vacuum inlet if the condensation is not quick enough.

#### 6.1.3. *Headspace analysis using SPME*

Solid phase microextraction (SPME) is a recent and simple method to adsorb VOCs to a fiber and thermally desorb it into a GC-MS. More details about the method can be read in the Beebook chapter "Standard methods for chemical ecology research in *Apis mellifera*" (Torto et al., 2013). The method is based on a study by Cai et al. (2007) on aroma compounds from beetles.

*Procedure:*

(1) Weigh insect (fresh or powder) into a 20 ml vial with a polytetrafluoroethylene (PTFE)-lined silicone septum.
(2) Allow the vial to equilibrate at 30 °C for 24 h before sampling with SPME fiber.
(3) Sample the vial by piercing the septum with the SPME needle and exposing the SPME fiber to the gases in the vial.
(4) Desorb the fiber into the GC-MS port at 260 °C for 40 min using an autosampler system.

Pros: The method is not invasive and collects most of the VOCs in the headspace above the sample according to the fiber sensitivity.

Cons: Costly as the fiber needs to be replaced every 100 samples. To obtain a broader spectrum of the VOCs, different types of fibers can be used.

#### 6.1.4. *Volatile compound identification and quantification*

The VOCs are analyzed using gas chromatography mass spectrometry (GC-MS) that is capable to separate the compounds in a column based on polarity and identify them based on mass to charge ratio. More information regarding the GC-MS technique can found in the Beebook chapter "Standard methods for chemical ecology research in *Apis mellifera*" (Torto et al., 2013).

*Analysis operating conditions:*

- The VOCs are separated using either a polar capillary column (e.g., DB-Wax, 30 m, 0.25 mm id) or nonpolar column (e.g., BP5, 30 m, 0.25 mm id).

- The GC oven is programmed to rise from 35 °C to 240 °C at 4 °C $min^{-1}$.
- The MSD conditions are: capillary direct interface temperature, 280 °C; ionization energy, 70 eV; mass range, 35–350 amu; electron multiplier voltage (Autotune + 200 V); scan rate, 5.27 scans/s.

### 6.2. *Taste profile using electronic-tongue*

The electronic-tongue (e-tongue) is based on potentiometric chemical sensors that respond to interactions with many taste substances in specific, non-specific and overlapping signals (Kobayashi et al., 2010). The sensors can analyze both positively and negatively charged taste molecules. There are a couple of different e-tongues systems available today. The e-tongue has been used in many studies with food and pharmaceutical applications (Kobayashi et al., 2010). Multivariate statistical analysis and principle component analysis is mostly used to describe differences in the overall taste profile among different sample groups.

*Sample preparation*:

(1) Weigh whole pieces of honey bee pupae and put them in a tray with an aluminum layer covering the sample. Keep some holes in the layer to allow evaporation to occur.
(2) Put the tray inside a freeze dry machine and run it until there is no change in sample weight.
(3) Mix the dry sample with fine mixer to obtain powder. Dissolve a fixed amount of powder in warm (30 °C) distilled water, filter it with paper and fill the measuring cups up to a volume of 70 ml.

All the procedure steps are done automatically by the e-tongue instrument e.g., SA402B Insent, Japan (Figure 3). The following conditions need to be set in the operating software.

(1) Prepare reference solution (30 mM KCl and 0.3 mM tartaric acid). The sample reading is measured against the reference solution
(2) Measure the sensor output of the samples for 30 s.
(3) Clean the sensors (300 ml ethanol + 100 ml 1 M HCl (for negative) and 10 ml 1 M KOH + 7.46 gr KCl (for positive) + *x* ml distilled water up to 1 l) for 90 s.
(4) Wash the sensors in a reference solution for 90 s twice and then measure a stable reading from a reference solution before continuing to the next sample.
(5) Repeat every sample at least four times and use the last three samples to calculate a mean.

### 6.3. *Texture analyses*

Texture properties of insects can be analyzed on raw material or processed food products. A texture analyzer measures different physical characters such as: firmness, hardness, crunchiness, crispiness, etc.

#### 6.3.1. *Raw material*

(1) Place a fresh piece of pupae in a plastic cup in the center table of the texture analyzer.
(2) Measure the firmness, defined as the maximum positive force and hardness, defined as the total work during the test.
(3) Set the operation conditions as followed: extension rate to about 20 mm/min and stop at 10 mm depth. Use cylinder probe with 5 mm diameter.
(4) Be careful to not crush or squeeze the delicate sample too fast, but to slowly penetrate the sample and measure changes in resistance.

#### 6.3.2. *Processed food product*

(1) Use the raw material, dry it using an oven and mix it into a powder in a fine mixer or freeze the material with liquid nitrogen and crush it into a powder.
(2) Use the bee powder for processed food: bread, snacks, cakes, paste, etc. The insect powder replaces some of the original ingredients and the impact on texture is measured.
(3) In case of bread – take 30 × 30 × 20 mm sample size (Figure 4).
(4) Use a disk probe with 30 mm diameter.
(5) Analyze the texture using a Texture Profile Analysis (TPA) – a double compression test for determining the textural properties of foods. The sample is compressed twice to mimic the feeling of samples when they are chewed in the mouth. Several parameters can be received from the same experiment: firmness, hardness, elasticity (length of base of area 2), cohesiveness (area 2/area 1), chewiness (firmness × elasticity × cohesiveness) and adhesiveness (negative work during the work).
(6) Set the operation conditions: 60 mm/min extension rate, 50% depth compression.

### 6.4. *Sensory analysis of honey bee brood by trained and non-trained panels*

Sensory panel are important tool in the overall sensory analyses of edible insects, including bee brood. Using human senses, it is possible to describe the sensory experience of eating certain insects in terms of aroma, taste, texture and sometimes visual appearance. There are many

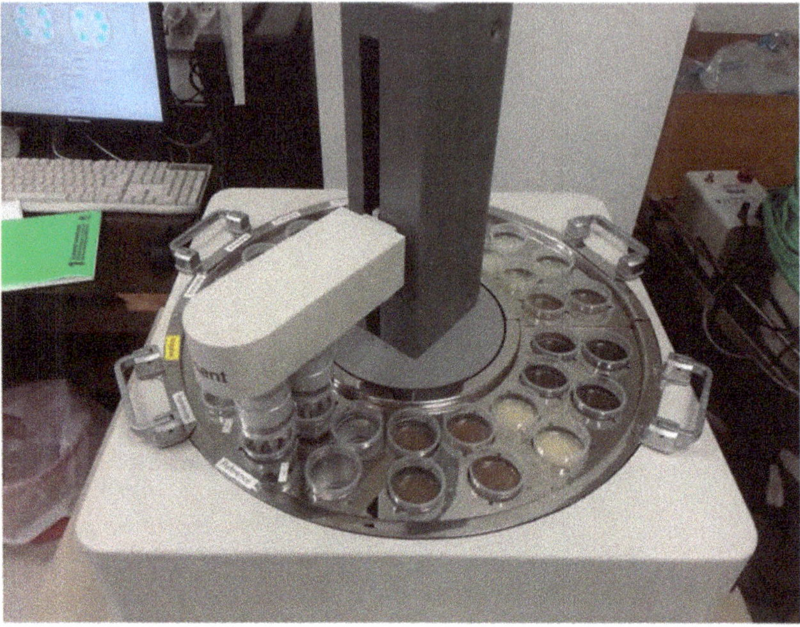

Figure 3. The electronic tongue (Insent SA402B, Japan) analyzing taste profiles of different honey bee pupae samples.

Figure 4. Texture properties on raw or processed food products including bee brood e.g., bread, protein bars, cakes. The texture analyzer measures different physical characters such as: firmness, hardness, crunchiness, crispiness, etc.

different types of sensory tests. Here we describe two kinds: intensity evaluation and hedonic test (Bednářová et al., 2013; Caparros Megido et al., 2014; Kiatbenjakul et al., 2015). The tests are performed in sensory booths under controlled temperature and humidity conditions (usually 20 °C ± 2 °C). Red light can be used to mask the effect of color from the insect samples.

#### 6.4.1. *Trained sensory panel – intensity evaluation*

*Selection and training*:

(1) Recruit a group of panellists that pass a threshold test for basic taste and smell intensity, and that are willing to participate in eating insects. Select about 15 people for the panel, describe the panel (gender and age).
(2) The second part includes basic training to get the panellists familiar with the type of test and common taste, aroma and texture definitions.
(3) Train the panel for the odor descriptive terms arising from honey bee brood samples. The panel has to reach a consensus for a list of attributes for both aroma and taste. For example, nine attributes of giant water bug odors were selected in consensus agreement between eleven panelists. Among the odor attributes were floral, pineapple like, fruity and fishy (Kiatbenjakul et al., 2015).
(4) Assign each attribute with a reference that can be used during the following training meeting. For example, sweetness and saltiness tastes are evaluated using sugar and NaCl solutions at varied concentrations as standards (Benjamin & Gamrasni, 2015).
(5) Lastly, the panel conducts "real" like tests with intensity scales to get used to the type of test and the samples.

*Test procedure:*

(1) Take 1 g of sample in a plastic cup assigned with a three digit random number.
(2) Ask the panelist to evaluate the aroma intensity using an anchored scale between 0 (none) to 15 (extremely strong). A reference can be used during the test to assist with ranking the intensity.
(3) Between samples the participant need to drink water and eat salt-free cracker to avoid carry over between samples.
(4) Analyze the results from triplicate sensory tests using statistics software (Radar charts, analysis of variance analysis ANOVA or principle component analysis PCA).

#### 6.4.2. *Non-trained sensory panel- hedonic test*

Sensory acceptance is very crucial for the success of introducing edible insects to the future food market. Non trained people (more than 100) are used as public judges for "liking" evaluation without previous experience or training with insects. Two types of samples can be used in the hedonic test – fresh samples or processed food products. This method is commonly used in other food products like soft cheese (Gomes et al., 2011).

Before the test starts, each participant needs to be informed about the safety and the potential allergenicity of the samples.

(1) Provide a sample to each participant without any information about the attributes.
(2) Ask the participant to rate the sample in a 10 point scale from 0 (very dislike) to 10 (very like).
(3) Between samples provide the participant with water and salt-free cracker to avoid carry over between samples.

## 7. Gastronomic applications and potential of edible honey bee brood

In many developing countries, eating insects is a common practice and each type has its own taste, aroma and texture. Honey bee brood has some unique organoleptic qualities, which presents many possibilities for gastronomic applications.

### 7.1. *Flavor*

Honey bee brood has a range of flavors which can vary depending on preparation technique. Semi-structured sensory analyses of honey bee larvae have yielded descriptors such as: "raw nuts", "hazelnut", "brazil nut", "avocado", "nutty", "herbal", "vegetal", "liver" for raw and steamed; and "meaty", "chicken-y", "bacon", "mushroom", "liver" for fried or roasted. The honey bee pupae have similar flavor characteristics, but can also develop a more bitter and animal taste as they become more developed, as well as a firmer and, just before becoming imago, a more chewy texture. One descriptive sensory analysis of honey bee drone brood has been conducted, with sensory differences between larvae and pupae demonstrated as the primary axis of variance (Evans et al., 2016).

### 7.2. *Texture*

The textural properties of honey bee brood give it both some of its challenges and its rewards. When not fro-

Figure 5. Kheang Nor Mai Preaw Kai Peung: soup from north-east Thailand with sour fermented bamboo shoots and weaver ant eggs, here also made with (less traditional) honey bee larvae, made by Ranee Udtumthisarn in Copenhagen.

Figure 6. Kai Jiew Mod Dang: Chicken eggs scrambled with garlic, chilli, shallots, scallions, fish sauce, and weaver ant eggs, here also made with (less traditional) honey bee larvae, made by Ranee Udtumthisarn in Copenhagen.

zen, honey bee brood is quite fragile, and can rupture easily, especially if having already been frozen. This softness, however, is also part of their gastronomic potential, and if preserved well it can bring a surprising and unique element to a dish. Raw larvae at room temperature are soft and plump. In the mouth, they can be popped with slight pressure from the tongue, with a mouth-coating liquid within. Raw pupae at room temperature are slightly more firm as a result of their later developmental stage and so resist slightly more pressure than the larvae, though they still contain a similarly viscous interior.

### 7.3. *Culinary transformations*

Honey bee brood is a versatile ingredient in the kitchen. It can be used whole or blended, and can be processed using wet or dry heat, acid or fermentation techniques.

#### 7.3.1. *Wet heat (steaming, simmering, poaching)*

At the appropriate temperature, these methods gently coagulate the proteins in the honey bee brood yielding a plump and soft but solid texture; though at inappropriate temperatures (for example, steaming too quickly or boiling) the brood can become hard and grainy. These

Figure 7. "Peas 'n' bees", a dish developed at the Nordic Food Lab: fresh pea soup, bee larvae (some poached, some fried), blanched peas, lovage and fermented bee pollen ("bee bread").

Figure 8. Honey bee larvae granola, developed at the Nordic Food Lab: oats, seeds of sunflower, pumpkin and flax, coated in a blended mixture of bee larvae and honey and baked until golden brown, with whole dried larvae and pupae added at the end for texture and flavor.

methods tend to bring out the herbal, vegetal and nutty notes, as well as a light but distinct savouriness, and seems to be more common in traditional applications for honey bee brood, for example simmering in soup at the end of cooking, or steaming in egg dishes (Figures 5 and 6).

*7.3.2. Dry heat (frying, roasting, baking, drying)*

These methods often create crispy and crunchy textures and can bring out the stronger "meaty", "bacon-y", "liver" and "mushroom" flavors due to the caramelising Maillard reactions that give toasty flavors and brown colors. These methods are not as common in traditional applications as wet heat methods. They yield fast and satisfying results, though some of their aromatic complexity is lost or transformed (Figures 7 and 8). Heat dried honey bee brood can also be transformed to a fine powder using an electric blender and used directly or mixed with any other flour as an additive in e.g., bread, vegetable dish or soup (Krell, 1996).

Figure 9. Honey bee larvae ceviche, a dish developed at the Nordic Food Lab: bee larvae cured in rhubarb vinegar, garnished with purple oxalis leaves and stems, freeze-dried raspberries, and lemon verbena.

*7.3.3. Acid*

This is a traditional technique used for pickling as well as in the South American technique for ceviche, a dish of seafood cured or "cooked" with citrus juice. Though its application to honey bee brood is not -traditional, the technique firms up the exterior while preserving the liquid interior, depending on the curing time, and create a delicious textural contrast (Figure 9).

*7.3.4. Fermentation*

Fermentation using salt and sometimes grains molded with *Aspergillus oryzae* to produce exogenous enzymes (in Japanese cuisine known as koji) yields a salt-rich umami sauce which can be used as an analog to soy sauce or fish sauce to add a savory element to dishes. In this case the fermented sauce uses honey bee brood as its source of protein (less traditional), generating amino acids when broken down by the enzymes, and thus has the particular qualities of honey bee brood – a light, transparent golden color, with nutty, vegetal and delicate meaty aromas. Another example of bee brood involved in fermentation would be the South African mead "iQhilika".

## 8. Conclusion

From production through to consumption via examples of nutrient analyses, the aim of this chapter was to contribute to the growing field of entomophagy by introducing honey bee brood as a realistic addition to a more sustainable food industry. The methods described are far from exhaustive, but should facilitate more research into e.g., the nutritional value of honey bee, drone and worker, brood. Once again, the near omnipresence of honey bees around the world and its long documented relationship with man could potentially accelerate the acceptance of insects as a foodstuff. Drone brood removal as a part of a sustainable *Varroa* sp. control will add on to the branding of drone brood and could revitalize the industry by providing new sources of revenues to beekeepers.

## Acknowledgements

The COLOSS (Prevention of honey bee COlony LOSSes) Association aims to explain and prevent massive honey bee colony losses. It was funded through the COST Action FA0803. COST (European Cooperation in Science and Technology) is a unique means for European researchers to jointly develop their own ideas and new initiatives across all scientific disciplines through trans-European networking of nationally funded research activities. Based on a pan-European intergovernmental framework for cooperation in science and technology, COST has contributed since its creation more than 40 years ago to closing the gap between science, policy makers and society throughout Europe and beyond. COST is supported by the EU Seventh Framework Program for research, technological development and demonstration activities (Official Journal L 412, 30 December 2006). The European Science Foundation as implementing agent of COST provides the COST Office through an EC Grant Agreement. The Council of the European Union provides the COST Secretariat. The COLOSS network is now supported by the Ricola Foundation - Nature & Culture.

## Disclosure statement

No potential conflict of interest was reported by the authors.

## ORCiD

*Annette Bruun Jensen* http://orcid.org/0000-0002-2044-2274

## References

Adedunatan, S. A. (2005). Nutritional and antinutritional characteristics of some insects foraging in Akure forest reserve Ondo state, Nigeria. *Journal of Food Technology, 3*, 563–567.

Ambrose, J. T. (1992). Management for honey production. In J. M. Graham (Ed.), *The hive and the honey bee.* Hamilton, Illinois: Dadant Publications, Dadant & Sons.

Association of Officiating Analytical Chemists. (1996). *Official methods of analysis* (16th ed.). Washington, DC: Author.

Bednářová, M., Borkovcová, M., Mlček, J., Rop, O., & Zeman, L. (2013). Edible insects-species suitable for entomophagy under condition of Czech Republic. *Acta Universitatis Agriculturae et Silviculturae Mendelianae Brunensis, 61*, 587–593.

Behrens, W. A., & Madère, R. (1987). A highly sensitive high-performance liquid chromatography method for the estimation of ascorbic and dehydroascorbic acid in tissues, biological fluids, and foods. *Analytical Biochemistry, 165*, 102–107.

Benjamin, O., & Gamrasni, D. (2015). Electronic tongue as an objective evaluation method for taste profile of pomegranate juice in comparison with sensory panel and chemical analysis. *Food Analytical Methods., 9*, 1726.

Blair, A., Ritz, B., Wesseling, C., & Freeman, L. B. (2014). Pesticides and human health. *Occupational and Environmental Medicine,.* doi:10.1136/oemed-2014-102454

Blake, C. J. (2007). Analytical procedures for water-soluble vitamins in foods and dietary supplements: A review. *Analytical and Bioanalytical Chemistry., 389*, 63–76.

Boes, K. E. (2010). Honey bee colony drone production and maintenance in accordance with environmental factors: An interplay of queen and worker decisions. *Insectes Sociaux, 57*, 1–9.

Brodschneider, R., & Crailsheim, K. (2010). Nutrition and health in honey bees. *Apidologie, 41*, 278–294.

Büchler, R., Andonov, S., Bienefeld, K., Costa, C., Hatjina, F., Kezic, N., ... Wilde, J. (2013). Standard methods for rearing and selection of *Apis mellifera* queens. In V.Dietemann, J.D.Ellis, & P.Neumann (Eds.), *The COLOSS BEEBOOK, Volume I: standard methods for Apis mellifera research. Journal of Apicultural Research, 52*(1), doi:10.3896/IBRA.1.52.1.07

Cai, L., Koziel, J. A., & O'Neal, M. E. (2007). Determination of characteristic odorants from *Harmonia axyridis* beetles using *in vivo* solid-phase microextraction and multidimensional gas chromatography-mass spectrometry-olfactometry. *Journal of Chromatography A, 1147*, 66–78.

Calderone, N. W. (2005). Evaluation of drone brood removal for management of *Varroa destructor* (Acari: Varroidae) in colonies of *Apis mellifera* (Hymenoptera: Apidae) in the Northeastern United States. *Journal of Economic Entomology, 98*, 645–650.

Caparros Megido, R., Sablon, L., Geuens, M., Brostaux, Y., Alabi, T., Blecker, C., ... Francis, F. (2014). Edible insects acceptance by Belgian consumers: Promising attitude for entomophagy development. *Journal of Sensory Studies, 29*, 14–20.

Chebrolu, K. K., Jayaprakasha, G. K., Yoo, K. S., Jifon, J. L., & Patil, B. S. (2012). An improved sample preparation method for quantification of ascorbic acid and dehydroascorbic acid by HPLC. *LWT–Food Science and Technology, 47*, 443–449.

Collin, K., & Blum, R. (2005). *Method of analysis of vitamins, provitamins and chemically well defined substances having a similar biological effect-1st part "tel quel"*. FEFANA-EU Additives & Premixtures Association Publication, Brussels.

Dainat, B., Kuhn, R., Cherix, D., & Neumann, P. (2011). A scientific note on the ant pitfall for quantitative diagnosis of *Varroa destructor*. *Apidologie, 42*, 740–742.

de Foliart, G.R. (2002) *The human use of insects as a food resource: A bibliographic account in progress* (no. 28, pp. 1–51). Oceania. Retrieved from http://labs.russell.wisc.edu/insectsasfood/the-human-use-of-insects-as-a-food-resource/

de Graaf, D. C., Alippi, A. M., Antúnez, K., Aronstein, K. A., Budge, G., De Koker, D., ... Genersch, E. (2013). Standard methods for American foulbrood research. In V. Dietemann, J. D. Ellis, & P. Neumann (Eds.), *The COLOSS BEEBOOK, Volume II: standard methods for Apis mellifera pest and pathogen research. Journal of Apicultural Research, 52*(1). doi:10.3896/IBRA.1.52.1.11

de Miranda, J. R., Bailey, L., Ball, B. V., Blanchard, P., Budge, G., Chejanovsky, N., ... van der Steen, J. J. M. (2013). Standard methods for virus research in *Apis mellifera*. In V. Dietemann, J. D. Ellis, & P. Neumann (Eds.), *The COLOSS BEEBOOK, Volume II: standard methods for Apis mellifera pest and pathogen research. Journal of Apicultural Research, 52*(4). doi:10.3896/IBRA.1.52.4.22

DeGrandi-Hoffman, G., Eckholm, B. J., & Huang, M. H. (2013). A comparison of bee bread made by Africanized and European honey bees (*Apis mellifera*) and its effects on hemolymph protein titers. *Apidologie, 44*, 52–63.

Delaplane, K. S., Dag, A., Danka, R. G., Freitas, B. M., Garibaldi, L. A., Goodwin, R. M., & Hormaza, J. I. (2013). Standard methods for pollination research with *Apis mellifera*. In V. Dietemann, J. D. Ellis, & P. Neumann (Eds.), *The COLOSS BEEBOOK, Volume I: standard methods for Apis mellifera research. Journal of Apicultural Research, 52*(4) doi:10.3896/IBRA.1.52.4.12

Dietemann, V., Nazzi, F., Martin, S. J., Anderson, D., Locke, B., Delaplane, K. S., ... Ellis, J. D. (2013). Standard methods for varroa research. In V. Dietemann, J. D. Ellis, & P. Neumann (Eds.), *The COLOSS BEEBOOK, Volume II: standard methods for Apis mellifera pest and pathogen research. Journal of Apicultural Research, 52*(1). doi:10.3896/IBRA.1.52.1.09

Duay, P., Jong, D., & Engels, W. (2003). Weight loss in drone pupae (*Apis mellifera*) multiply infested by Varroa destructor mites. *Apidologie, 34*, 61–65.

Engel, W., Bahr, W., & Schieberle, P. (1999). Solvent assisted flavour evaporation-a new and versatile technique for the careful and direct isolation of aroma compounds from complex food matrices. *European Food Research and Technology, 209*, 237–241.

European Food safety Authority. (2007). Scientific opinion of the Panel on Food additives, Flavorings, Processing aids and Materials in Contact with Food (AFC) on a request from the Commission on the safety in use of bees wax. *The EFSA Journal, 615*, 1–28.

European Food safety Authority. (2015). Scientific committee: Risk profile related to production and consumption of insects as food and feed. *The EFSA Journal, 13*, 4257–4372. doi:10.2903/j.efsa.2015.4257

European Union. (2004a). *Regulation (EC) No 1935/2004 of the European Parliament and of the Council of 27 October 2004 on materials and articles intended to come into contact with food and repealing Directives 80/590/EEC and 89/109/EEC.*

European Union. (2004b). *Regulation (EC) No 852/2004 of the European parliament and of the council of 29 April 2004 on the hygiene of foodstuffs.*

Evans, J. (2013). *Insect gastronomy. Cereal Magazine., 3*, 62–71.

Evans, J. (2014). *Observations from the frontier of deliciousness MAD Dispatches: what is cooking?* (pp. 80–91). Copenhagen: MAD.

Evans, J., Müller, A., Jensen, A. B., Dahle, B., Flore, R., Eilenberg, J., Frøst, M. B. (2016, In press). An exploratory descriptive sensory analysis of honey bee drone brood from Denmark and Norway. *Journal for Food and Feed.*

FAO. (2013). *Dietary protein quality evaluation in human nutrition* (Report of an FAO Expert Consultation, FAO Food and Nutrition Paper n.92). Rome: FAO. Retrieved from http://www.fao.org/ag/humannutrition/35978-02317b979a686a57aa4593304ffc17f06.pdf

Finke, M. D. (2005). Nutrient composition of bee brood and its potential as human food. *Ecology of Food and Nutrition, 44*, 257–270.

Forsgren, E., Budge, G. E., Charrière, J.-D., & Hornitzky, M. A. Z. (2013). Standard methods for European foulbrood research. In V. Dietemann, J. D. Ellis, & P. Neumann (Eds.), *The COLOSS BEEBOOK: Volume II: Standard methods for Apis mellifera pest and pathogen research. Journal of Apicultural Research, 52*(1). doi:10.3896/IBRA.1.52.1.12

Free, J. B., & Williams, I. H. (1975). Factors determining the rearing and rejection of drones by the honey bee colony. *Animal Behaviour, 23*, 650–675.

Fries, I., Chauzat, M.-P., Chen, Y.-P., Doublet, V., Genersch, E., Gisder, S., ... Williams, G. R. (2013). Standard methods for nosema research. In V. Dietemann, J. D. Ellis, & P. Neumann (Eds.), *The COLOSS BEEBOOK: Volume II: Standard methods for Apis mellifera pest and pathogen research. Journal of Apicultural Research, 52*(1). doi:10.3896/IBRA.1.52.1.14

Gary, N. E. (1992). Activities and behavior of honey bees. In J. M. Graham (Ed.), *The hive and the honey bee*. Hamilton, IL: Dadant Publications, Dadant & Sons.

Gessain, M., & Kinzler, T. (1975). Miel et insectes à miel *chez* les Bassari et dáutres populations du Sénégal Oriental [Honey and honey making insects in the Bassari and other populations of Eastern Senegal]. In R. Pujol (Ed.), *L'homme et l'annimal* (pp. 247–254). Paris: Premier Colloque d'Éthnozoologie.

Gomes, A. P., Cruz, A. G., Cadena, R. S., Celeghini, R. M., Faria, J. A., Bolini, H. M., Pollonio, M. A., & Granato, D. (2011). Manufacture of low-sodium Minas fresh cheese: Effect of the partial replacement of sodium chloride with potassium chloride. *Journal of Dairy Science, 94*, 2701–2706.

Harbo, J. R. (1986). Effect of population size on brood production, worker survival and honey gain in colonies of honey bee. *Journal of Apicultural Research, 25*, 22–29. doi:10.1080/00218839.1986.11100687

Hartfelder, K., Gentile Bitondi, M. M., Brent, C., Guidugli-Lazzarini, K. R., Simões, Z. L. P., Stabentheiner, A., ... Wang, Y. (2013). Standard methods for physiology and biochemistry research in *Apis mellifera*. In V. Dietemann, J. D. Ellis, & P. Neumann (Eds.), *The COLOSS BEEBOOK, Volume I: standard methods for Apis mellifera research. Journal of Apicultural Research, 52*(1). doi:10.3896/IBRA.1.52.1.06

Haydak, M. H. (1970). Honey bee nutrition. *Annual Review of Entomology, 15*, 143–156.

Hocking, B., & Matsumura, F. (1960). *Bee World, 41*, 113–120. doi:10.1080/0005772X.1960.11096777

Hrassnigg, N., & Crailsheim, K. (1998). The influence of brood on the pollen consumption of worker bees (*Apis mellifera* L.). *Journal of Insect Physiology, 44*, 393–404.

Human, H., Brodschneider, R., Dietemann, V., Dively, G., Ellis, J. D., Forsgren, E., ... Zheng, H.-Q. (2013). Miscellaneous standard methods for *Apis mellifera* research. In V. Dietemann, J. D. Ellis, & P. Neumann (Eds.), *The COLOSS BEEBOOK, Volume I: standard methods for Apis mellifera research. Journal of Apicultural Research, 52*(4). doi:10.3896/IBRA.1.52.4.10

Imdorf, A., Rickli, M., Kilchenmann, V., Bogdanov, S., & Wille, H. (1998). Nitrogen and mineral constituents of honey bee worker brood during pollen shortage. *Apidologie, 29*, 315–325.

Jensen, A. B., Aronstein, K., Flores, J. M., Vojvodic, S., Palacio, M. A., & Spivak, M. (2013). Standard methods for fungal brood disease research. In V. Dietemann, J. D. Ellis, & P. Neumann (Eds.), *The COLOSS BEEBOOK: Volume II: Standard methods for Apis mellifera pest and pathogen research. Journal of Apicultural Research, 52*(1). doi:10.3896/IBRA.1.52.1.13

Johansson, T. S. K., & Johansson, M. P. (1978). Providing honey bees with water. *Bee World, 59*, 54–64. doi:10.1080/0005772X.1978.11097693

Jonas-Levi, A., Benjamin, O., & Martinez, J. I. (2015). Does a parasite infestation change the nutritional value of an insect? Varroa mites on honey bees as a model. *Journal of Insects as Food and Feed, 1*, 141–147.

Kiatbenjakul, P., Intarapichet, K. O., & Cadwallader, K. R. (2015). Characterization of potent odorants in male giant water bug (*Lethocerus indicus* Lep. and Serv.), an important edible insect of Southeast Asia. *Food Chemistry, 168*, 639–647.

Klunder, H. C., Wolkers-Rooijackers, J., Korpela, J. M., & Nout, M. J. R. (2012). Microbiological aspects of processing and storage of edible insects. *Food Control, 26*, 628–631. doi:10.1016/j.foodcont.2012.02.013

Kobayashi, Y., Habara, M., Ikezazki, H., Chen, R., Naito, Y., & Toko, K. (2010). Advanced taste sensors based on artificial lipids with global selectivity to basic taste qualities and high correlation to sensory scores. *Sensors, 10*, 3411–3443.

Krell, R. (1996) *Value-added products from beekeeping. FAO Agricultural Services Bulletin No. 124*. Rome: Food and Agriculture Organization of the United Nations.

Leporati, A., Catellani, D., Suman, M., Andreoli, R., Manini, P., & Niessen, W. M. A. (2005). Application of a liquid chromatography tandem mass spectrometry method to the analysis of water-soluble vitamins in Italian pasta. *Analytica Chimica Acta, 531*, 87–95.

Macevilly, C. (2000). Bugs in the system. *Nutrition Bulletin, 25*, 267–268.

Mbata, K. J. (1995). Traditional use of arthropods in Zambia. I. The food insects. *Food Insects Newsletter, 8*, 5–7.

Meixner, M. D., Pinto, M. A., Bouga, M., Kryger, P., Ivanova, E., & Fuchs, S. (2013). Standard methods for characterising subspecies and ecotypes of *Apis mellifera*. In V. Dietemann, J. D. Ellis, & P. Neumann (Eds.), *The COLOSS BEEBOOK, Volume I: standard methods for Apis mellifera research. Journal of Apicultural Research, 52*(4). doi:10.3896/IBRA.1.52.4.05

Moreno, P., & Salvadó, V. (2000). Determination of eight water- and fat-soluble vitamins in multi-vitamin pharmaceutical formulations by high-performance liquid chromatography. *Journal of Chromatography A, 870*, 207–215.

Neumann, P., Moritz, R., & Mautz, D. (2000). Colony evaluation is not affected by drifting of drone and worker honey bee (*Apis mellifera* L.) at a performance testing apiary. *Apidologie, 31*, 67–79.

Onore, G. (1997). A brief note on edible insects in Ecuador. *Ecology of Food and Nutrition, 36*, 277–285.

Oonincx, D. G. A. B., Van Itterbeeck, J., Heetkamp, M. J. W., Van Den Brand, H., Van Loon, J. J. A., & Van Huis, A. (2010). An exploration on greenhouse gas and ammonia production by insect species suitable for animal or human consumption. *PLoS ONE, 5*(12), 1–7. e14445.

Payne, C. L. (2015). Wild harvesting declines as pesticides and imports rise: The collection and consumption of insects in contemporary rural Japan. *Journal of Insects as Food and Feed, 1*, 57–65.

PerkinElmer. (1996). *Analytical methods for atomic absorption spectroscopy*. Retrieved from http://www.lasalle.edu/~prushan/Intrumental%20Analysis_files/AA-Perkin%20Elmer%20guide%20to%20all!.pdf

Prosky, L., Asp, N. G., Schweizer, T. F., Devries, J. W., & Furda, I. (1988). Determination of insoluble, soluble, and total dietary fiber in foods and food products: Interlaboratory study. *Journal of the Association of Official Analytical Chemists, 71*, 1017–1023.

Ramos-Elorduy, J., Moreno, J. M. P., Prado, E. E., Perez, M. A., Otero, J. L., & de Guevara, O. L. (1997). Nutritional value of edible insects from the state of Oaxaca, Mexico. *Journal of Food Composition and Analysis, 10*, 142–157.

Rumpold, B., & Schlüter, O. (2013). Nutritional composition and safety aspects of edible insects. *Molecular Nutrition and Food Research, 57*, 802–823.

Ruttner, F. (1988). *Biogeography and taxonomy of honey bees* (284 pp.). Berlin: Springer.

Salo-Väänänen, P. S., Ollilainen, V., Mattila, P., Lehikoinen, K., Salmela-Mölsä, E., & Piironen, V. (2000). Simultaneous HPLC analysis of fat-soluble vitamins in selected animal products after small-scale extraction. *Food Chemistry, 71*, 535–543.

Scherer, R., Rybka, A. C. P., Ballus, C. A., Meinhart, A. J. T., & Godoy, H. T. (2012). Validation of a HPLC method for simultaneous determination of main organic acids in fruits and juices. *Food Chemistry, 135*, 150–154.

Schmidt, J. O., & Buchmann, S. L. (1992). Other products of the hive. In J. M. Graham (Ed.), *The hive and the honey bee* (pp. 927–988). Hamilton, IL: Dadant & Sons.

Seeley, T. D. (2002). The effect of drone comb on a honey bee colony's production of honey. *Apidologie, 33*, 75–86.

Seeley, T. D., & Mikheyev, A. S. (2003). Reproductive decisions by honey bee colonies: Tuning investment in male production in relation to success in energy acquisition. *Insectes Sociaux, 50*, 134–138.

Segeren, P. (2004). *Beekeeping in the tropics. Agrodok 32* (90 pp.). Wageningen: Agromisa Foundation.

Standifer, L. N., Moeller, F. R., Kauffeld, N. M., Herbert, Jr., E. W., & H.Shimanuki (1977) *Supplemental feeding of honey bee colonies. Agriculture Information Bulletin 43* (pp. 8). United States Department of Agriculture, Washington, DC.

Taoukis, P. S., & Labuza, T. P. (1989). Applicability of time-temperature indicators as shelf life monitors of food products. *Journal of Food Science, 54*, 783–788.

Torto, B., Carroll, M. J., Duehl, A., Fombong, A. T., Gozansky, K. T., Nazzi, F., ... Teal, P. E. A. (2013). Standard methods for chemical ecology research in *Apis mellifera*. In V. Dietemann, J. D. Ellis, & P. Neumann (Eds.), *The COLOSS BEEBOOK, Volume I: standard methods for Apis mellifera research. Journal of Apicultural Research, 52*(4). doi:10.3896/IBRA.1.52.4.06

van Huis, A., Van Itterbeeck, J., Klunder, H., Mertens, E., Halloran, A., Muir, G., & Vantomme, P. (2013) *Edible insects: future prospects for food and feed security* (p. 201). Rome: FAO. Retrieved from http://www.fao.org/docrep/018/i3253e/i3253e.pdf

Wu, J. Y., Anelli, C. M., & Sheppard, W. S. (2011). Sub-lethal effects of pesticide residues in brood comb on worker honey bee (*Apis mellifera*) development and longevity. *PLoS ONE, 6*, e14720.

Yhoung-Aree, J., Puwastien, P., & Attig, G. A. (1997). Edible insects in Thailand: An unconventional protein source? *Ecology of Food and Nutrition, 36*, 133–149.

Zhi-Yi, L. (1997). Insects as food in China. *Ecology of Food and Nutrition, 36*, 201–207.

*Journal of Apicultural Research*, 2020
Vol. 59, No. 3, 1–62, https://doi.org/10.1080/00218839.2020.1738135

REVIEW ARTICLE

# Standard methods for *Apis mellifera* honey research

Ligia Bicudo de Almeida-Muradian[a]*, Ortrud Monika Barth[b], Vincent Dietemann[c], Michael Eyer[c], Alex da Silva de Freitas[d], Anne-Claire Martel[e], Gian Luigi Marcazzan[f], Carla Marina Marchese[g], Carla Mucignat-Caretta[h], Ana Pascual-Maté[i,j], Wim Reybroeck[k], M. Teresa Sancho[i] and José Augusto Gasparotto Sattler[a]

[a]*Pharmaceutical Science School, University of São Paulo, São Paulo, Brazil;* [b]*Instituto Oswaldo Cruz, Fiocruz, Rio de Janeiro, Brazil;* [c]*Agroscope, Swiss Bee Research Centre, Bern, Switzerland;* [d]*Instituto de Geociências, Universidade Federal Fluminense, Niterói, Brazil;* [e]*Sophia Antipolis Laboratory, Honeybee Pathology Unit, ANSES, Sophia, France;* [f]*CREA – Council for Agricultural Research and Economics, Research Centre for Agriculture and Environment, Bologna, Italy;* [g]*Red Bee Honey, American Honey Tasting Society, Weston, FL, USA;* [h]*Department of Molecular Medicine, University of Padova, Padua, Italy;* [i]*Departamento de Biotecnología y Ciencia de los Alimentos, Nutrición y Bromatología, Universidad de Burgos, Burgos, Spain;* [j]*Facultad de Ciencias de la Salud, Universidad Isabel I, Spain;* [k]*Research Institute for Agriculture, Fisheries and Food; Technology and Food Science Unit, Melle, Belgium*

*(Received 3 May 2018; accepted 27 February 2020)*

Honey is an important food for man and has been used as a natural sweetener since ancient times. It is a viscous and aromatic product made by honey bees using the nectar of flowers or honeydew. Honey is composed of a complex mixture of carbohydrates and other substances such as organic acids, amino acids, proteins, minerals, vitamins, lipids, aroma compounds, flavonoids, pigments, waxes, pollen grains, several enzymes, and other phytochemicals. This chapter presents some properties of *Apis mellifera* honey as well as the main methods of honey analysis. All methods are based on specialized literature, including the Codex Alimentarius, AOAC, and publications of the International Honey Commission. Herein, we describe methods related to honey authenticity, botanical origin, geographical origin, physicochemical analysis, radioentomology, pesticide and antibiotic contamination, chemotherapeutics, and sensory analysis. All methods are described in a step-by-step model in order to facilitate their use.

**Métodos estándar para la investigación de la miel de *Apis mellifera***

La miel es un alimento importante para el hombre y se ha utilizado como edulcorante natural desde la antigüedad. Es un producto viscoso y aromático hecho por las abejas de la miel usando el néctar de las flores o el rocío de miel. La miel está compuesta por una mezcla compleja de carbohidratos y otras sustancias como ácidos orgánicos, aminoácidos, proteínas, minerales, vitaminas, lípidos, compuestos aromáticos, flavonoides, pigmentos, ceras, granos de polen, varias enzimas y otros fitoquímicos. En este capítulo se presentan algunas propiedades de la miel de *Apis mellifera*, así como los principales métodos de análisis de la miel. Todos los métodos se basan en literatura especializada, incluido el Codex Alimentarius, AOAC y publicaciones de la Comisión Internacional de la Miel. A continuación, describimos métodos relacionados con la autenticidad de la miel, el origen botánico, el origen geográfico, el análisis fisicoquímico, la radioentomología, la contaminación por pesticidas y antibióticos, la quimioterapia y el análisis sensorial. Todos los métodos se describen en un modelo paso a paso para facilitar su uso.

**西方蜜蜂蜂蜜研究标准方法**

蜂蜜是人类重要的食品，自古以来就被用作天然甜味剂。它是蜜蜂使用花蜜或蜜露酿造而成的粘性芳香产品。蜂蜜由碳水化合物和其他物质（例如有机酸、氨基酸、蛋白质、矿物质、维生素、脂质、香气化合物、类黄酮、色素、蜡、花粉粒、几种酶和其他植物化学物质）组成的复杂混合物。本章介绍西方蜜蜂蜂蜜的一些特性以及主要分析方法。所有方法均基于专业文献，包括食品法典，AOAC和国际蜂蜜委员会的出版物。在此，我们描述与蜂蜜真实性、蜜源植物、地理来源、理化分析、放射昆虫学、农药和抗生素污染、化学治疗和感官分析有关的方法。为了便于使用，所有方法都采用分步描述的方式加以介绍

**Keywords:** *Apis mellifera*; honey; physicochemical analysis; honey authenticity; botanical and geographical origin; diagnostic radioentomology; pesticides; sensory; antibiotics and chemotherapeutics

*Corresponding author. Email: ligiabi@usp.br
Please refer to this paper as: de Almeida-Muradian, L. B., Barth, O. M., Dietemann, V., Eyer, M., da Silva de Freitas, A., Martel, A.-C., Marcazzan, G. L., Marchese, C. M., Mucignat-Caretta, C., Pascual-Matè, A., Reybroeck, W., Sancho, M. Teresa, & Sattler, J. A. G. (2020) Standard methods for *Apis mellifera* honey research. In V. Dietemann, P. Neumann, N. Carreck, & J. D. Ellis (Eds.), *The COLOSS BEEBOOK – Volume III, Part 2: Standard methods for* Apis mellifera *product research. Journal of Apicultural Research* 59(3), 1–62. https://doi.org/10.1080/00218839.2020.1738135.

## 1. Introduction

The honey chapter is divided into seven parts describing the main properties of honey as well as their methods of analysis. Each method is described step-by-step. The authors come from many countries such as Belgium, Brazil, France, Italy, Spain, and Switzerland, with many of the authors being members of the International Honey Commission. Honey is the most important product from the apiary and is consumed globally, making it economically important (Almeida-Muradian et al., 2012). As the only natural sweetener, honey historically was an important food for man. Honey was mentioned in many ancient cultures (the holy books of ancient India (the Vedas), book of songs Shi Jing (written in the sixth century BC in ancient China), the Holy Bible, just to name a few), and depicted in many wall drawings in ancient Egypt, Rome (mentioned by the writers Vergil, Varro, and Plinius) and also in medieval high cultures. In ancient Greece, the honey bee, a sacred symbol of Artemis, was an important design on Ephesian coins for almost six centuries (Bogdanov, 2011).

Honey is a viscous and aromatic product, widely appreciated by man and prepared by honey bees (Hymenoptera: Apidae, *Apis* spp.) from the nectar of flowers or honeydew, the latter being produced through an intermediary, generally an aphid (Ouchemoukh et al., 2007; Almeida-Muradian et al., 2013; Ferreira et al., 2009) . This natural product is generally composed of a complex mixture of carbohydrates and other less frequent substances, such as organic acids, amino acids, proteins, minerals, vitamins, lipids, aroma compounds, flavonoids, vitamins, pigments, waxes, pollen grains, several enzymes, and other phytochemicals (Almeida-Muradian et al., 2013; Almeida-Muradian & Penteado, 2015; Gomes et al., 2010; Lazarevic et al., 2010).

All methods related to honey research cited herein are based on the specialized literature, and the authors recommend reading the primary references such as Codex Alimentarius, Association of Official Agricultural Chemists (AOAC), and International Honey Commission publications. Herein, we include the main methods related to research with honey, from the hive to the final product consumed.

## 2. Honey authenticity—botanical, geographical, and entomological origin

Pollen analysis is the initial technique used to determine the botanical origin of the honey produced by bees. This knowledge enables one to characterize the honey producing area by means of pollen grain identification. *Apis* and *Meliponini* bees show different preferences when foraging plant products.

The management of honey harvesting must follow quality procedures, aiming not only at its efficient collection, but mainly at maintaining its original characteristics and the quality of the final product. The beekeeper, when handling the harvest, must be wear appropriate clothing for beekeeping. It is recommended not to collect honey on rainy days or with high relative humidity, as it could lead to an increase in the moisture content of the product. Harvesting should take place selectively during the opening of the beehive. The beekeeper must inspect each frame, giving priority to removing only the frames that have at least 90% of their operculated storage cells, this being indicative of the maturity of the honey in relation to the percentage of moisture. In order to guarantee the quality of the final product, during the extraction of honey, all the equipment and utensils used in the various stages of handling must be specific to this activity, and there is no room for adaptation. In the case of equipment and utensils that will have direct contact with the product, all must be stainless steel, specific for food products.

Pollen analysis is the initial technique used to determine the botanical origin of the honey produced by bees. This knowledge enables one to characterize the honey producing area by means of pollen grain identification. *Apis* and *Meliponini* bees show different preferences when foraging plant products. For palynological analysis, honey samples of *Apis* have to be received in glass vials and kept at room temperature. In case of stingless bees, the honey has to be kept in a refrigerator at about 8 °C to avoid fermentation.

For more details in sampling, please see "Sampling" sections.

### 2.1. *Botanical origin*

Pollen grains and other structured elements that compose honey sediments are indicative of the botanical origin (Barth, 1989; Vit et al., 2018). A complete palynological analysis reveals the real provenance of the raw material. Pollen grains are introduced into honey by different means (Barth, 1989). Besides the nectariferous plants, that secrete nectar in addition to producing pollen, the polliniferous plants mainly produce pollen grains and have low concentrations of nectar (Barth, 1989).

#### *2.1.1. Pollen grains*

The first universal technique aiming at the recognition and evaluation of the pollen in honey was elaborated by the French, Swiss, and Dutch specialists on honey analysis. This started with Zander's work in 1935, which was published by Louveaux et al. (1970a, 1970b) and updated by Louveaux et al. (1978). The latter includes many standard methods, including the acetolysis method. A more recent publication on pollen analysis was published in von der Ohe et al. (2004).

These methodologies are used today. For the first technique, the pollen grain content is not removed, making it difficult to observe some of the structures. However, more of the elements occurring in honey sediments are preserved. The acetolysis technique enables a better visualization of pollen grain structures when only the outer pollen grain wall (exine) is preserved, but there is a significant loss of no resistant elements such as oils, fungi hyphae, yeast, insect fragments, and organic matter (Haidamus et al. 2019). A detailed analysis of the technique established by Louveaux et al. (1978) was presented also by von der Ohe et al. (2004) considering quantitative and qualitative pollen counting.

#### 2.1.2. Other structured elements

Honey sediments may contain several types of particles besides pollen grains ( Barth, 1989; Zander & Maurizio, 1975). These structured elements, not nectar indicative, show what happens during honey sampling and storage, and reflect honey quality. Some adulterants include a variety of fungal spores, yeasts, colorless molds and air dispersed spores, green algae, bacteria, insect bristles and scales, and pathogenic mites, as well as starch and synthetic fibers.

### 2.2. Methodology to prepare honey sediment following Louveaux et al. (1970a)

(1) Weigh 10 g of honey into a beaker.
(2) Add 20 mL of distilled water and dissolve well by stirring (10–15 min).
(3) If the honey is crystallized, heat the mixture for a few seconds in a microwave oven to complete the dissolution process of the sugar crystals.
(4) Divide the solution eventually between two conical centrifuge tubes of 15 mL capacity.
(5) Centrifuge at least 252 RCF (relative centrifugal force or G) that, according to a centrifuge model, may comprise on average 1500 rpm (revolutions per minute), for 15 min.
(6) Discard the supernatant and rapidly invert the tubes for approximately 30 s before returning to normal position.
(7) Add 10 mL of distilled water to the pellet.
(8) Centrifuge at 1500 rpm for 15 min.
(9) Discard the supernatant and rapidly invert the centrifuge tubes for approximately 30 s before returning to normal position.
(10) Add 5 mL of glycerin water (solution 1: 1 with distilled water and glycerin) to the pellet.
(11) Wait for 30 min and then centrifuge at least 1500 rpm for 15 min.
(12) Discard the supernatant and rapidly invert the tube onto absorbent paper (never return to the normal position) to allow all the liquid to drain.
(13) Prepare two microscope slides.
(14) Put a little piece of glycerin jelly (0.5 $mm^3$) onto a needle tip and collect part of the sediment into the centrifuge tube.
(15) Place the sediment on the center of the microscope slide, heat gently on a hot plate, only to dissolve the glycerin jelly.
(16) Cover with a coverslip, and seal with paraffin.
(17) The pollen grain identification is based on the reference pollen slides library from regional/country vegetation and with the aid of reference catalogs (e.g., Roubik & Moreno, 1991).
(18) After the identification and palynological analysis, the prepared slides were stored in a pollen library.

### 2.3. The acetolysis technique

Acetolysis (Erdtman, 1952) is an artificial fossilization method to prepare pollen grains in order to get the best transparency of the grain envelope.

(1) Weigh 10 g of honey into a beaker.
(2) Add 20 mL of distilled water and dissolve well by stirring (10–15 min).
(3) If the honey is crystallized, heat the mixture for a few seconds in a microwave oven to complete the dissolution process of the sugar crystals.
(4) Divide the solution evenly between two centrifuge tubes of 15 mL capacity.
(5) Centrifuge at least 1500 rpm for 15 min.
(6) Discard the supernatant and rapidly invert the tubes for approximately 30 s before returning to normal position.
(7) Add 5 mL of acetic acid (32%), centrifuge at least 1500 rpm for 15 min and discard the supernatant
(8) Add the acetolysis mixture (4.5 mL of acetic anhydride (98%) and 0.5 mL of sulfuric acid (99%)).
(9) Heat in a water bath just to 80 °C for 3 min.
(10) Centrifuge at least 1500 rpm for 15 min.
(11) Discard the supernatant and add 20 mL of distilled water.
(12) Wash twice.
(13) Discard the supernatant and rapidly invert the tubes for approximately 30 s before returning to normal position.
(14) Add 5 mL of glycerin water (solution 1: 1 with distilled water and glycerin) to the pellet. Wait for 30 min.
(15) Centrifuge at least 1500 rpm for 15 min.
(16) Discard the supernatant and rapidly invert the tube onto absorbent paper (never return to the normal position) to allow all the liquid to drain.
(17) Prepare two microscope slides.
(18) Put a little piece of glycerin jelly (0.5 $mm^3$) onto a needle tip and collect part of the sediment in the centrifuge tube.
(19) Place the sediment on the center of the microscope slide, heat gently on a hot plate, only to dissolve the glycerin jelly.
(20) Cover with a coverslip, and seal with paraffin.

### 2.4. Additional techniques

A number of other techniques to prepare honey samples were purposed and published. In general, they are complex to execute and require a broad assortment of equipment.

Table 1. Evaluation of a *Croton* monofloral honey sample, a strong nectariferous and subrepresented plant taxon.

| | | |
|---|---|---|
| Total number of pollen grains counted | 962 | = 100% of the pollen |
| Total of anemophylous pollen counted | 842 | = 87.5% of the pollen |
| Total of polliniferous pollen counted | 52 | = 5.4% of the pollen |
| Total of nectariferous pollen counted (=100%) | 68 | = 7.1% of the pollen |
| *Croton* | 18 | = 26.5% of nectariferous pollen, a subrepresented taxon |
| *Solanum* | 18 | = 26.5% of nectariferous pollen |
| *Myrcia* | 10 | = 1.0% of nectariferous pollen |
| Final evaluation: Monofloral honey of *Croton* (Euphorbiaceae). | | |

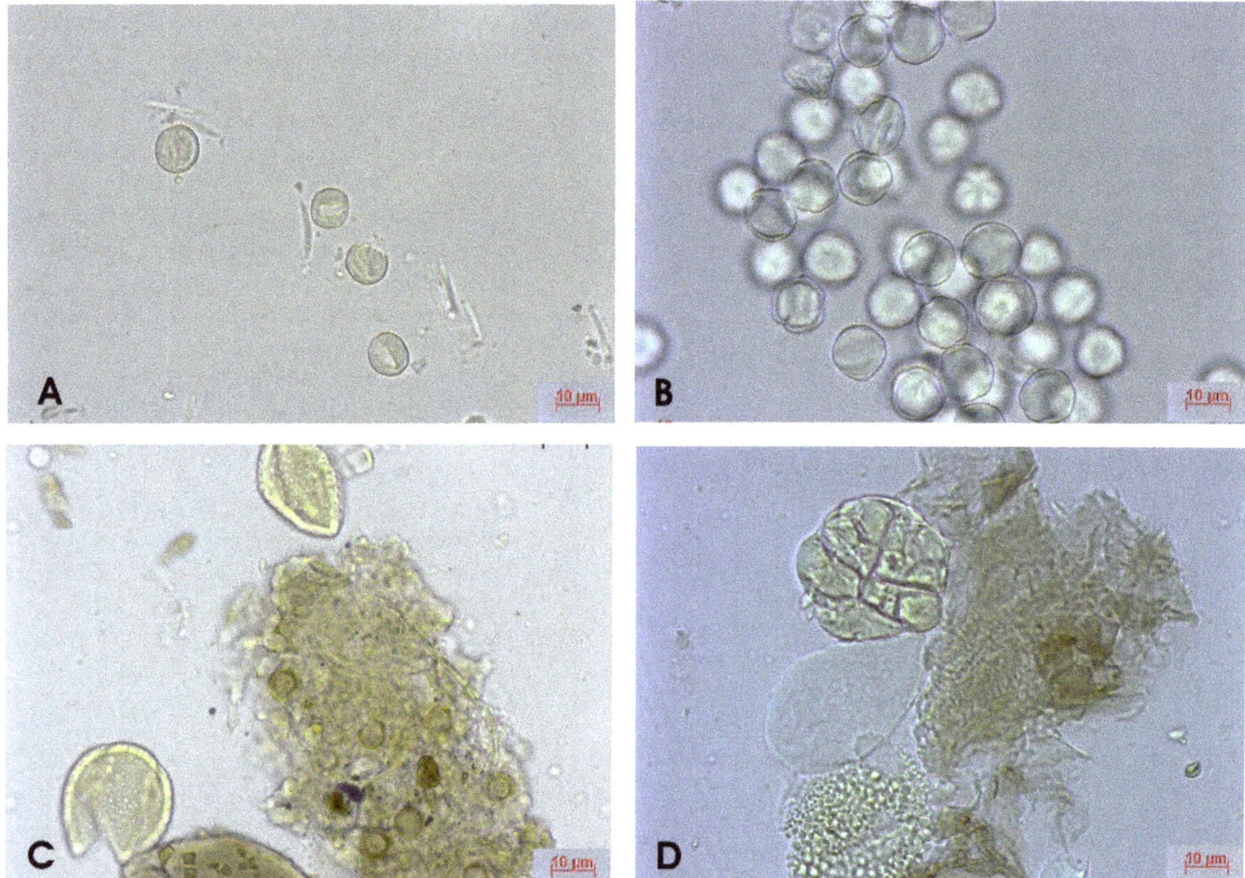


Figure 1. (A) Anemophylous pollen of *Piper*; (B) Polliniferous pollen of Melastomataceae; (C) *Citrus* monofloral honey with honeydew contribution (organic material and spores); (D) Group of algae cells and a broken pollen grain of *Croton* of a honeydew honey.

Filtration (Lutier & Vaissière, 1993), ethanol dilution (Bryant & Jones, 2001), mid-infrared spectroscopy (Ruoff et al., 2006), multivariate analysis Principal Component Analysis (PCA) (Aronne & de Micco, 2010), ultraviolet spectroscopy (Roshan et al., 2013), isotopes (Wu et al., 2015), and pollen grain DNA metabarcoding (Hawkins et al., 2015) have all been used to determine honey origin.

### 2.5. Honey sample evaluation

Since the standard work of Louveaux et al. (1978), the qualitative and quantitative evaluation of honey samples (Maurizio, 1977) has been a subject of many investigations until the data assembly in von der Ohe et al. (2004). The reliability of pollen analysis in honey samples was discussed in detail by Behm et al. (1996). Pollen coefficients were discussed by the *R*-values of honey pollen coefficients by Bryant and Jones (2001).

#### 2.5.1. Super and subrepresentation of pollen types

For each region of honey production, there are plants that are significant nectar or/and pollen producers. The pollen grains present in the honey sediments can reveal the bee foraging behavior. In general, the number of pollen grains present on one microscope slide provides information to use to assign a frequency class, indicating if the pollen types observed in the honey samples are predominant (PP), accessory (AP), important minor pollen (IMP), or minor pollen (MP) (Louveaux et al., 1970a, 1970b).

Few pollen grains found in a honey sample suggests that the bees visited plant species of a low pollen production, but that have a high nectar disponibility like *Citrus* spp. and *Croton* sp. (Table 1, Figure 1). A high quantity of pollen grains from polliniferous plants (e.g., Poaceae and Melastomataceae) can indicate a super

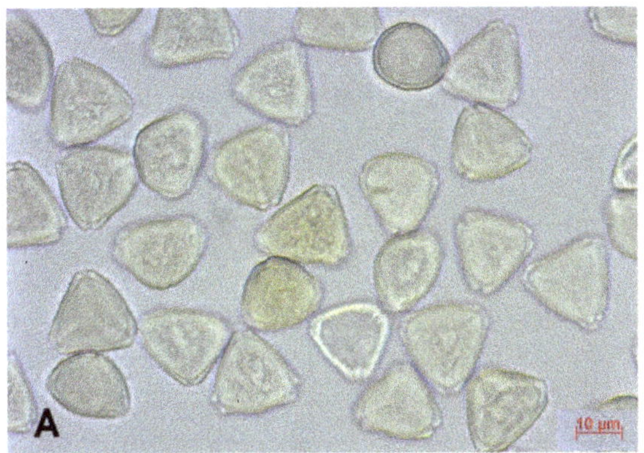


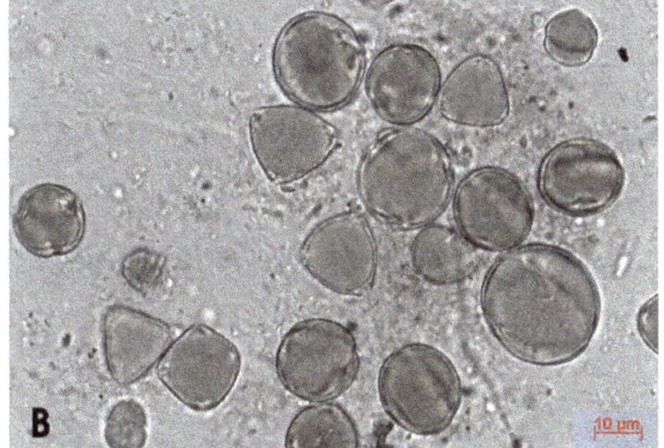


Figure 2. (A) Monofloral *Myrcia* honey. (B) Heterofloral honey.

representation since these species produce a lot of pollen but little or no nectar (Louveaux et al., 1978).

### *2.5.2. Monofloral and heterofloral honeys*

(1) Monofloral/unifloral honeys contain predominantly pollen grains from an unique plant species (≥45% of all nectariferous pollen grains counted), denominated a ("predominant pollen type," PP)

(2) Bifloral honeys contain pollen grains from two plant species with a frequency of 15–45% per nectariferous species ("accessory pollen types," AP)

(3) Heterofloral/plurifloral honeys contain pollen grains from three or more nectariferous plant species with frequencies in the 3–15% ("important minor pollen types," IMP) or <3% ("minor pollen types," MP) (Barth, 1989; Louveaux et al., 1978) (Figure 2).

### *2.5.3. Honeydew*

Bees can use the excreta of aphids (Aphidae) that feed on plant xylem to create honeydew. This bee product can be difficult to classify and presents a high electrical conductivity and mineral salt content. Generally, pollen grain occurrence is low and a moderate number of fungal spores can be observed (Barth, 1971; Maurizio, 1959) (Figure 1).

## *2.6. Geographical origin*

Pollen grain analysis allows one to determine the nectar origin of honey. The assembly of plant species identified by pollen analysis provides information about honey's local or regional origin. A reference pollen slide collection (pollen library) must be available in order to compare the morphology of pollen grains obtained directly from flowers with the ones obtained in the honey sediment collection. The scope of this collection depends upon the study area and the radius of action of the bees. Plant assemblies recognized in the honey sediment by means of pollen grain morphology inform about the nectariferous and non-nectariferous, local and regional potential of honey production (Aronne & de Micco, 2010; Consonni & Cagliani, 2008; Persano Oddo & Bogdanov, 2004; Salgado-Laurenti et al., 2017).

Unfortunately, some honeys do not contain much pollen. In these cases, the honey origin is difficult to determine. It may derive from strong nectar secreting plant species such as those from the Asclepiadaceae. Sometimes, certain flowers secrete abundant nectar that will drip from the flower inflorescences, making it unlikely to contain pollen. This happens with *Dombeya* (Sterculiaceae, a shrub) and with some plants that grow in very humid environments.

## *2.7. Entomological origin*

For purposes of this manuscript, we are discussing honey produced by bees of the family Apidae, genus *Apis*, species *A. mellifera*. There are several subspecies/races of *A. mellifera* occurring in different parts of the world (Carreck et al., 2013; Meixner et al., 2013). Humans have cultivated this bee since antiquity given that it has provided a source of sugar. There are other types of honey producing bees. Collectively called stingless bees (also Apidae), these bees occur mainly in the tribes Meliponini and Trigonini. Bees from both tribes produce a different type of honey than that produced by *A. mellifera.*

Stingless bee honey contains less sugar and more water than the honey produced by *Apis* spp. These bees are very important pollinators in their native habitats, being responsible for maintenance of the local vegetation, mainly the forests, and for seed production. There are circa 500–800 species occurring exclusively in the tropical and sub-tropical Neotropical regions. *Apis* spp. and *Meliponini* spp. forage on some of the same plants, but many different plants as well. Species specific preferences for various forage resources depend upon the bee species' affinity, as well as the flowering vegetation strata and environmental conditions (Barth et al., 2013).

## 3. Standard honey physicochemical analysis for nutritional composition, chemical composition, and antioxidant activity

In this section, we present the main physicochemical analyses for nutritional composition, chemical composition, and antioxidant activity of honey in step-by-step format. Review articles on the composition and properties of honey from *A. mellifera* colonies are present in the literature (De-Melo et al., 2017; Pascual-Maté et al., 2018).

We recommend reading of the primary reference on this topic: Codex Standard for Honey (2001).

### 3.1. Water content determination (moisture)

Water content of honey is related to the botanical and geographical origin of the nectar, edaphic and climatic conditions, season of harvesting, manipulation by beekeepers, processing/storage conditions, etc. (De-Melo et al., 2017). It is an important physicochemical parameter for honey shelf life (Bogdanov, 2011) and it normally ranges between 13% and 25% (Uran et al., 2017), with the average being about 17% (Fallico et al., 2009). Honeys with high moisture content (>18%) have a higher probability of fermenting upon storage (Bogdanov, 2011) while those with a low moisture content (<15%) are likely to granulate.

The Council Directive 2001/110/EC (European Commission, 2001) as well as the Brazilian legislation (Ministério da Agricultura Pecuária e Abastecimento, 2000) recommend the limit of 20% for moisture in honey samples (Thrasyvoulou et al., 2018).

The standard method for honey moisture determination, indicated by Codex Standard for Honey (2001), uses the refractometric method which is simple and reproducible. It is possible to use an Abbé refractometer or digital refractometers for this purpose (AOAC, 1992; Bogdanov, 2009, 2011). The refractometric method is described here as a reference method (AOAC, 1990; Bogdanov, 2009).

#### 3.1.1. Principle

Water content or moisture determined by the refractometric method is based on the fact that the refractive index increases with solid content. In honey, the refraction index can be converted in moisture content using the Chataway Table (Bogdanov et al., 1997).

#### 3.1.2. Apparatus and reagents

- Water bath or incubator
- Honey sample
- Refractometer

#### 3.1.3. Procedure

##### 3.1.3.1. Samples and standards preparation and determination

(1) If honey is granulated, place it in an airtight closed flask and dissolve crystals in a water bath or incubator at 50 °C.
(2) Cool the solution to room temperature and stir.
(3) Ensure that the prism of the refractometer is clean and dry.
(4) Cover the surface of the prism from the refractometer with the sample and read the refractive index.
(5) Read the corresponding moisture content from the Chataway Table (Table 2), making the temperature correction, if necessary.

Other methods can be used to determine water content of honey including Karl Fischer method and nonstandardized method Fourier transform infrared spectroscopy (FTIR) (Almeida-Muradian et al., 2013, 2014; Almeida-Muradian & Gallmann, 2012; Bogdanov, 2009).

### 3.2. Sugar content determination (HPLC method)

Honey is mainly composed of sugars (60 – 85% of the total volume). Monosaccharides (fructose and glucose) represent about 75% of the sugars found in honey. Disaccharides (sucrose, maltose, turanose, isomaltose, maltulose, trehalose, nigerose, and kojibiose) compose about 10–15% of the sugars. The rest is composed of a small amount of other sugars, including trisaccharides (mainly maltotriose and melezitose).

Due to the wide variety of botanical sources, the sugar profile presents great variation; however, it has been demonstrated that among the same species of plants, the profile of sugars and other physicochemical components can be used as fingerprint for the identification of the botanical origin of monofloral honeys (El Sohaimy et al., 2015; Ruoff et al., 2005, 2006).

According to the publication of the Harmonized Methods of the International Honey Commission (2009), the methodologies for sugar determination are grouped as specific and nonspecific. Liquid chromatography - refractive index detector (specific methodology) and a methodology proposed by Lane-Eynon (nonspecific methodology) are among the most used methodologies for the quantification of sugars presented in honeys (Bogdanov, 2009).

There are other methods described in the literature for the determination of sugars. These include (1) enzymatic determination (Gómez-Díaz et al., 2012; Huidobro & Simal, 1984), (2) high performance liquid chromatography with pulsed amperometric detector and anion exchange column (HPLC-PAD) (Bogdanov, 2009; Cano et al., 2006; Nascimento et al., 2018; Ouchemoukh et al., 2010), (3) ultra-performance LC with and evaporative light scattering detector (UPLC-ELSD) (Zhou et al., 2014), (4) gas chromatography (GC) (Bogdanov, 2009; Ruiz-Matute et al., 2007), (5) capillary electrophoresis (CE) (Dominguez et al., 2016), (6) electrochemical determination (EL Alami EL Hassani et al., 2018), (7) FTIR (Almeida-Muradian et al., 2012; Almeida-Muradian, Sousa, et al., 2014; Almeida-Muradian, Stramm, et al., 2014; Anjos et al., 2015), and (8) Raman spectroscopy (RAMAN) (Özbalci et al., 2013). High-Performance Liquid Chromatography with Refractive Index Detection (HPLC-RI) (Bogdanov, 2009) is described herein.

Table 2. Chataway table—refractive index and moisture content (%) 20 °C. (source: International Honey Commission, 2009).

| Water content (g/100 g) | Refractive index (20 °C) | Water content (g/100 g) | Refractive index (20 °C) |
|---|---|---|---|
| 13.0 | 1.5044 | 19.0 | 1.4890 |
| 13.2 | 1.5038 | 19.2 | 1.4885 |
| 13.4 | 1.5033 | 19.4 | 1.4880 |
| 13.6 | 1.5028 | 19.6 | 1.4875 |
| 13.8 | 1.5023 | 19.8 | 1.4870 |
| 14.0 | 1.5018 | 20.0 | 1.4865 |
| 14.2 | 1.5012 | 20.2 | 1.4860 |
| 14.4 | 1.5007 | 20.4 | 1.4855 |
| 14.6 | 1.5002 | 20.6 | 1.4850 |
| 14.8 | 1.4997 | 20.8 | 1.4845 |
| 15.0 | 1.4992 | 21.0 | 1.4840 |
| 15.2 | 1.4987 | 21.2 | 1.4835 |
| 15.4 | 1.4982 | 21.4 | 1.4830 |
| 15.6 | 1.4976 | 21.6 | 1.4825 |
| 15.8 | 1.4971 | 21.8 | 1.4820 |
| 16.0 | 1.4966 | 22.0 | 1.4815 |
| 16.2 | 1.4961 | 22.2 | 1.4810 |
| 16.4 | 1.4956 | 22.4 | 1.4805 |
| 16.6 | 1.4951 | 22.6 | 1.4800 |
| 16.8 | 1.4946 | 22.8 | 1.4795 |
| 17.0 | 1.4940 | 23.0 | 1.4790 |
| 17.2 | 1.4935 | 23.2 | 1.4785 |
| 17.4 | 1.4930 | 23.4 | 1.4780 |
| 17.6 | 1.4925 | 23.6 | 1.4775 |
| 17.8 | 1.4920 | 23.8 | 1.4770 |
| 18.0 | 1.4915 | 24.0 | 1.4765 |
| 18.2 | 1.4910 | 24.2 | 1.4760 |
| 18.4 | 1.4905 | 24.4 | 1.4755 |
| 18.6 | 1.4900 | 24.6 | 1.4750 |
| 18.8 | 1.4895 | 24.8 | 1.4745 |
| | | 25.0 | 1.4740 |

*3.2.1. Apparatus and reagents*

Apparatus:

- HPLC system consisting of pump, sample applicator, temperature-regulated RI-detector (30 °C), and temperature-regulated column oven (30 °C).
- Ultrasonic bath
- Micro-membrane 0.45 µm
- Volumetric flask (100 mL)
- Syringe (1 mL)
- Beaker (50 and 100 mL)
- Pipette (25 mL)
- Sample vials (1.5 mL)

Reagents:

- Water (HPLC grade)
- Methanol (HPLC grade)
- Acetonitrile (HPLC grade)
- Analytical stainless-steel column, containing amine-modified silica gel with 5 µm particle size, 250 mm in length; 4.6 mm in diameter.

*3.2.2. Procedure*

***3.2.2.1. Sample preparation and determination***

(1) Weigh 5 g of honey into a beaker and dissolve in 40 mL distilled water.
(2) Pipette 25 mL of methanol into a 100 mL volumetric flask and transfer the honey solution quantitatively to the flask.
(3) Complete the volumetric flask with water until 100 mL.
(4) Filter the solution through a 0.45 µm membrane and collect the filtered solution into sample vials.
(5) Store for four weeks in the refrigerator (4 °C) or for six months at freezer (−18 °C).

***3.2.2.2. Chromatographic conditions.***

- Flow rate: 1.3 mL/min.
- Mobile phase: acetonitrile: water (80:20, v/v).
- Column and detector temperature: 30 °C.
- Sample volume: 10 µL.

***3.2.2.3. Calculation and expression of results.*** Honey sugars can be identified and quantified by comparison of their retention times and the peak areas with those of the standard sugars (external standard method).

The mass percentage of the sugars (W) is calculated according to the following formula:

$$W = A_1 \times V_1 \times m_1 \times 100 / A_2 \times V_2 \times m_0$$

$W$ = Mass percentage of sugars (g /100 g).

$A_1$ = Peak areas or peak heights of the given sugar compound in the sample solution (expressed as units of area, length, or integration).

$A_2$ = Peak areas or peak heights of the given sugar compound in the standard solution (expressed as units of area, length, or integration).

$V_1$ = Total volume of the sample solution (mL).

$V_2$ = Total volume of the standard solution (mL).

$m_1$ = Mass amount of the sugar in grams in the total volume of the standard ($V_2$).

$m_0$ = sample weight (g)

### 3.3. Reducing sugars

Apparent reducing sugars and apparent sucrose content of honey are most commonly determined for honey quality control purposes. The Lane and Eynon method (Codex Alimentarius Standard, 1969) is used for the determination of sugars. Apparent reducing sugars are defined as those sugars that reduce a Fehling's reagent under the conditions specified (Bogdanov et al., 1997; Bogdanov & Martin, 2002; Bogdanov, 2009; Granato & Nunes, 2016).

#### 3.3.1. Apparatus and reagents

Apparatus:

- Burette (25 mL)
- Volumetric pipette (2 mL)
- Volumetric flasks (25 and 100 mL)
- Erlenmeyer flasks (250 mL)
- Beakers (50 and 100 mL)
- Pipette (25 mL)
- Heating plate

Reagents:

- Copper (II) sulfate pentahydrate ($CuSO_4$. $5H_2O$)
- Potassium sodium tartrate tetrahydrate ($C_4H_4KNaO_6$ . $4H_2O$)
- Sodium hydroxide (NaOH)
- Glucose

#### 3.3.2. Procedure

**3.3.2.1. Solutions preparation.**

Fehling's solution A:

(1) Weigh 69.28 g of copper sulfate pentahydrate ($CuSO_4.5H_2O$) in a beaker and solubilize by stirring with distilled water.
(2) Transfer to a 1 L volumetric flask.

Fehling's solution B:

(1) Weigh 346 g of sodium and potassium double tartrate ($C_4H_4KNaO_6.4H_2O$) in 300 mL of distilled water.
(2) Separately, dissolve 100 g of sodium hydroxide (NaOH) into 200 mL of distilled water in a plastic beaker and ice bath in the hood.
(3) Combine the solutions of sodium and potassium tartrate and sodium hydroxide in a 1 L volumetric flask.
(4) Homogenize by stirring.
(5) Raise the volume of the flask to full with distilled water.

**3.3.2.2. Standardization of Fehling solutions.** It is necessary to standardize Fehling solutions to obtain the correction factor.

(1) Prepare a 0.5% (w/v) glucose solution.
(2) Weigh 0.5 g of glucose in a beaker, add distilled water, dissolve, and transfer to a 100 mL flask.
(3) Fill the burette with the glucose solution and holder as described (in section 3.3.2.4).
(4) For each standardization, make a minimum of four titrations. Apply the average of these values in the formula to obtain the correction factor (F) of the solutions of Fehling:

$$F = (\%\text{ of glucose}) \times (\text{average of volumes spent}) \times 0.01$$

The correction factor corresponds to the amount of sugar required to reduce 10 mL of the Fehling solutions.

**3.3.2.3. Preparation of the sample.**

(1) Dissolve 5.0 g of honey in distilled water up to 25 mL in a volumetric flask (Solution 1:5).
(2) From this homogenized solution, transfer 2 mL to a 100 mL volumetric flask.
(3) Make up to volume with distilled water and homogenize (final dilution 1:250). The initial diluted solution 1:5 should be reserved for the analysis of apparent sucrose.

**3.3.2.4. Titration.** The titration is done with a 25 mL burette containing the diluted honey (solution 1:5) and an Erlenmeyer flask with 5 mL of Fehling A solution, 5 mL of Fehling B, and 40 mL of distilled water.

(1) Heat the Erlenmeyer to a hot plate, using a tripod and a screen.
(2) Heat the solution with the heating plate until boil, then the titration is initiated, releasing in one go 5 mL of the sugar solution of the burette. With the restart of the boiling, the Fehling solution becomes reddish, but still with a lot of blue color ($Cu^{2+}$ ions).
(3) The titration should be restarted, this time dropwise, under stirring and the color modification being observed. The reaction is over when the solution, against a fluorescent light, does not present any shade or blue reflection, being colored an intense red. Titration should not exceed 3 min.

**3.3.2.5. Calculation.**

The calculation of the percentage of reducing sugars is given by the following formula:

$$\%AR = \frac{(\text{dilution of honey solution} = 250) \times F \times 100}{\text{average of the volumes spent in the titration}}$$

AR = reducing sugars

$F$ = correction factor obtained from the standardization of Fehling's solutions.

### 3.4. Apparent sucrose

Apparent sucrose content is determined indirectly by calculating difference in total reducing sugar before and after inversion of sugars in honey. The determination of apparent sucrose requires the use of acid hydrolysis to break the glycoside bonds of the disaccharides, thus releasing reducing sugars such as glucose and fructose (Bogdanov et al., 1997; Granato & Nunes, 2016).

#### 3.4.1. Apparatus and reagents

Apparatus:

- 100 mL glass beaker
- 100 and 500 mL volumetric flask
- 1 and 2 mL volumetric pipettes
- 25 mL burette
- 250 mL Erlenmeyer flask
- Hot plate

Reagents:

- Concentrated hydrochloric acid (HCl)
- Fehling A solution
- Fehling B solution
- Sodium hydroxide (NaOH)

#### 3.4.2. Procedure

##### 3.4.2.1. Sodium hydroxide solution (5 mol/L)

(1) Weigh 100 g of NaOH and dissolve in distilled water.
(2) Increase the volume in a volumetric flask to 500 mL with distilled water.
(3) Condition the solution in a plastic container.

##### 3.4.2.2. Sample preparation

(1) From the 1:5 diluted honey solution prepared in the reducing sugars analysis, transfer 2 mL to a 100 mL glass beaker and add 40 mL of distilled water and 1 mL of concentrated HCl.
(2) This solution must be boiled, cooled, neutralized by adding 5 mol/L NaOH until a pH of 7 ± 0.2.
(3) Add distilled water up to 100 mL in a volumetric flask. and make the titration according to the same procedure as "Reducing sugars" (3.3.2.4.). The result is expressed as percentage of total sugars (% AT):

$$\%AT = \frac{(\text{dilution of honey solution} = 250) \times F \times 100}{\text{average of the volumes spent in the titration}}$$

AT = total sugars

$F$ = correction factor obtained from the standardization of Fehling's solutions

Heating to boiling under strongly acid conditions ensures that the sucrose is hydrolyzed; having as products the separate molecules of glucose and fructose. The molecular mass of sucrose (MM = 342 mol/L) is 95% of the molecular mass of glucose and fructose together (MM = 360 mol/L), taking into account that 5% of the mass refers to water (MM = 18 mol/L) which was used in the hydrolysis. Thus, the percentage of apparent sucrose in a sample is calculated using the following formula:

$$\%\,\text{apparent sucrose} = (\%\,\text{total sugars} - \%\,\text{reducing sugars}) \times 0.95$$

### 3.5. Proteins (proline)

The presence of protein in honey is low and varies depending on the species of honey bee (*A. mellifera* 0.2–1.6% and *A. cerana* 0.1–3.3%) (Lee et al., 1998) and the contact of these with flower pollen, honeydew, and nectar (Davies, 1975).

In the last decades, the use of more sensitive and robust methodologies has enabled the identification of a larger quantity of proteins and amino acids in honey samples from several countries (De-Melo et al., 2017). Among the currently used methodologies, the most preferred is HPLC (Hermosín et al., 2003) associated with identification by mass spectrometry (Chua et al., 2013).

The classical method to determine protein uses the total nitrogen content that is usually calculated by multiplying the total nitrogen content from the Kjeldahl method by a factor of 6.25 (AOAC, 1990). However, about 40–80% of the total honey's nitrogen comes from the protein fraction and most of the remainder resides in the free amino acids (Chua et al., 2013).

According to the published update "Harmonized Methods of the International Honey Commission" in 2009, the proline content is recommended for quality control (being considered a criterion of honey ripeness) and it can be an indicator of sugar adulteration (Bogdanov, 2009). Since most of the amino acids are present in trace quantities, only proline content is included for the quality parameters of honey. No other methods for protein quantification are included in the methods referenced above or in the Codex Standard for Honey (2001).

#### 3.5.1. Principle

Proline and ninhydrin form a colored complex. After adding 2-propanol, color extinction of the sample solution and a reference solution at a wavelength maximum is determined. The proline content is determined from the ratio of both and it is expressed in mg/kg. The method is based on the original method of Ough (1969).

*3.5.2. Apparatus and reagents*

Apparatus:

- Spectrophotometer measuring in the range of 500–520 nm
- Cells 1 cm
- 20 mL tubes with screw cap or stopper
- 100 mL measuring flask and 100 mL beaker
- Water bath

Reagents:

- Distilled Water
- Formic acid (H.COOH), 98–100%
- Solution of 3% ninhydrin in ethylene glycol monomethylether (methyl-cellosolve)
- Proline reference solution: Prepare an aqueous proline stock solution containing 40 mg proline to 50 mL distilled water (volumetric flask). Dilute 1 mL to 25 mL (volumetric flask) with distilled water (solution containing 0.8 mg/25 mL)
- 2-propanol solution, 50% by volume in water

Proline reference solution.

(1) Prepare an aqueous proline stock solution containing 40 mg proline to 50 mL distilled water.
(2) Dilute 1 mL to 25 mL with distilled water (solution containing 0.8 mg/25 mL).
(3) Prepare a 2-propanol solution, 50% by volume in water.

*3.5.3. Procedure*

***3.5.3.1. Preparation of the sample solution.***

(1) Weigh to the nearest milligram about 5 g honey into a beaker and dissolve in 50 mL distilled water.
(2) After dissolving, transfer all the solution to a 100 mL volumetric flask.
(3) Complete the volume with distilled water and shake well.

***3.5.3.2. Determination.*** Note that the coefficient of extinction is not constant. Therefore, for each series of measurements, the average of the extinction coefficient of the proline standard solution must be determined at least in triplicate.

(1) Pipette 0.5 mL of the sample solution in one tube, 0.5 mL of distilled water (blank test) into a second tube, and 0.5 mL of proline standard solution into each of three additional tubes.
(2) Add 1 mL of formic acid and 1 mL of ninhydrin solution to each tube.
(3) Cap the tubes carefully and shake vigorously using a vortex machine for 15 min.
(4) Place tubes in a boiling water bath for 15 min, immersing the tubes below the level of the solution.
(5) Transfer to a water bath at 70 °C for 10 min.
(6) Add 5 mL of the 2-propanol-water-solution to each tube and cap immediately.
(7) Leave to cool and determine the absorbance using a spectrophotometer for 45 min after removing from the 70 °C water bath at 510 nm, using 1 cm cells.

Note: Adherence to the above times is critical.

***3.5.3.3. Calculation and expression of results.*** Proline in mg/kg honey at one decimal place is calculated according to following equation:

$$\text{Proline (mg/kg)} = \frac{E_s}{E_a} \times \frac{E_1}{E_2} \times 80$$

$E_s$ = Absorbance of the sample solution

$E_a$ = Absorbance of the proline standard solution (average of two readings)

$E_1$ = mg proline taken for the standard solution

$E_2$ = Weight of honey in grams

80 = Dilution factor

In Germany, a honey with < 180 mg proline/kg honey is considered either unripe or adulterated (Bogdanov, 2009).

### 3.6. Vitamins (HPLC)

The presence of vitamins in honey is quite variable and is related to the botanical origin of the honey. Water soluble vitamins (vitamins C and B-complex) occur in higher levels in honey than do lipid-soluble vitamins (A, D, E, and K) due to the low presence of lipids in honey. In general, this foodstuff cannot be considered as an important source of vitamins due to the fact that they are in very low quantities. As mentioned by León-Ruiz et al. (2013), the identification of vitamins in different honeys allows their characterization to botanical type and is a valuable approach of honey quality control, but no official methods have been described (De-Melo et al., 2017).

There are few studies that identified and quantified vitamins in honey. Some authors such as Ciulu et al. (2011) and León-Ruiz et al. (2013), published works aiming to investigate the presence of vitamins in honey using HPLC. The method outlined by León-Ruiz et al. (2013) is described herein.

*3.6.1. Apparatus and reagents*

Apparatus:

- HPLC system consisting of pump, sample applicator, UV-vis variable wavelength detector
- Micro-membrane 0.45 μm
- Volumetric flask (25 mL)
- Syringe (1 mL)

Table 3. Elution program in the RP-HPLC gradient elution of water-soluble vitamins in honey.

| Time (min) | Solvent A (TFA aqueous solution, 0.025%, v/v) (%) | Solvent B (acetonitrile) (%) |
|---|---|---|
| 0 [a] | 100 | 0 |
| 5 [a] | 100 | 0 |
| 11 [a] | 75 | 25 |
| 11 [b] | 75 | 25 |
| 19 [b] | 55 | 45 |
| 20 [b] | 60 | 40 |
| 22 [b] | 100 | 0 |

*Note*: flow rate: 1.0 mL/min; a—Operative wavelength: 254 nm; b—Operative wavelength: 210 nm.

- Beaker (50 and 100 mL)
- Pipette (25 mL)
- Sample vials (1.5 mL)

Reagents:

- Trifluoroacetic acid (0.025%, v/v) (HPLC grade)
- Acetonitrile (HPLC grade)
- Analytical column (C18 column, 250 mm × 4.6 mm, 5 µm particle size).

*3.6.2. Procedure*

***3.6.2.1. Sample preparation.***

(1) Weigh 10 g of honey into a 50 mL beaker and dissolve in 10 mL of ultra-pure water.
(2) Add 1 mL of NaOH 2M to complete solubilization.
(3) Add 12.5 mL of phosphate buffer 1M (pH = 5.5).
(4) Raise the volume to 25 mL using ultra-pure water.
(5) Filter the solution through a 0.45 µm membrane and collect in sample vials.
(6) Store in the refrigerator if it is necessary.

***3.6.2.2. Standards preparation.*** The stock standard solution is prepared by weighing in a 100 mL volumetric flask: 10.0 mg of vitamin B2; 25.0 mg of vitamin B5; and 10.0 mg of vitamin B9.

(1) Add 40 mL of ultra-pure water.
(2) Add 4 mL of NaOH to complete dissolution.
(3) Add 50 mL of phosphate buffer 1M (pH = 5.5); 10.0 mg of vitamin B3; then 10.0 mg of vitamin C.
(4) Fill the flask to 100 mL using distilled water.
(5) Store the standard solution in the dark at 4 °C.
(6) Prepare the solution fresh daily as needed.

***3.6.2.3. Chromatographic analysis.*** The vitamins are identified and quantified by comparison of the retention times and the peak area of the pure standards. The elution program in the RP-HPLC gradient elution of water-soluble vitamins in honey is described in Table 3.

### *3.7. Minerals*

The mineral content of honey samples has been receiving significant attention from many scientists globally. In recent years, the mineral profile of honey became an important indicator of quality control and environmental contamination by agrochemicals (Almeida-Silva et al., 2011; Kacaniová et al., 2009). In the same way, some authors proposed that mineral profile can be used as an authentication analysis for botanical identification or designation of geographical origin (Anklam, 1998; Louppis et al., 2017).

Many factors can contribute to the mineral composition of honey, including the botanical origin of nectar, the harvesting treatment and material of storage, seasonal climatic variations, and geographical origin (Almeida-Silva et al., 2011; Anklam, 1998). Other minerals in honey may derive from those present in the air, water, and soil. Furthermore, bees can be exposed to minerals while visiting flowers, contacting branches and leaves, drinking water from pools, and while flying (Kacaniová et al., 2009).

The mineral content in honey ranges from 0.04% in light honeys to 0.2% in dark honeys (Alqarni et al., 2014). Potassium is the most abundant mineral in honey, composing generally up to 1/3 of the total mineral content (Alqarni et al., 2014; Yücel & Sultanoglu, 2013). In smaller quantities, honey also contains sodium, iron, copper, silicon, manganese, calcium, and magnesium. Macro elements (such as potassium, calcium, and sodium) and trace minerals (such as iron, copper, zinc, and manganese) perform a fundamental function in biological systems: maintaining normal physiological responses, inducing the overall metabolism, influencing the circulatory system and reproduction, and acting as catalysts in various biochemical reactions (Alqarni et al., 2014; da Silva et al., 2016).

Honey should be free from heavy metals in amounts that may represent a hazard to human health. However, maximum residue levels of these potentially toxic elements in honey have not been established. The World Health Organization (WHO, 1996), has proposed acceptable levels of 15 µg/kg for arsenic, 25 µg/kg for lead, 5 µg/kg for mercury, and 7 µg/kg for cadmium.

As described by Bogdanov et al. (2007), the characterization of trace elements by inductively coupled sector-field plasma mass spectrometry (ICP-SFMS) is significantly improved by virtue of enhanced sensitivity and separation of polyatomic interferences in a high resolution mode. In contrast, the option of inductively coupled plasma with mass spectroscopy detector (ICP-

Table 4. Instrumental working parameters for mineral analysis of honey.

| Torch injector | Quartz |
|---|---|
| Spray chamber | Peltier cooled cyclonic |
| Sample uptake | 0.3 rps (rounds per second) |
| Nebulizer type | MicroMist |
| Interface | Pt-cones |
| RF power | 1550W |
| Ar gas flow rate (L/min) | Plasma 15; auxiliary 0.9 |
| Nebulizer pump | 0.1 rps |
| He gas flow rate | 0.03 mL/min |
| Ion lenses model | X-lens |
| Lens voltage | 10.7 V |
| Omega bias | −90V |
| Omega lens | 10.2 V |
| Acquisition mode | Spectrum |
| Peak Pattern | 1 point |
| Integration time | 2000 ms |
| Replicate | 3 |
| Sweeps/replicate | 100 |
| Tune mode<br>(Stabilization time; Integ. Time/mass) | No gas: 0 s; 0.1 s<br>He: 5 s; 0.5 s HEHe: 5 s; 1 s |
| ICP-MS (standard mode) | No gas: $Be^{9}$, $Na^{23}$, $Mg^{24}$, $Al^{29}$, $K^{39}$ $As^{75}$,$Mo^{95}$,$Ag^{107}$,$Sb^{121}$,$Ba^{137}$, $Tl^{205}$,$Pb^{208}$,$U^{23}$<br>He mode: $Ca^{43}$,$V^{51}$,$Cr^{52}$,$Mn^{55}$, $Co^{59}$,$Ni^{60}$,$Cu^{63}$,$Zn^{66}$,$Cd^{111}$ HEHe: $Se^{78}$,$Fe^{56}$ |
| Internal standards | $^{209}Bi$, $^{115}In$, $^{45}Sc$ |

MS) is widely used by researchers and yields accurate results. Bilandžić et al. (2017) published a work with mineral composition of 24 honey samples using an ICP-MS system with a great percentual recovery in the determinations. This work shows detailed parameters of analysis validation and we chose this method as the standard analysis for determining mineral content.

#### 3.7.1. Apparatus and reagents

Apparatus:

- High-pressure microwave oven
- Inductively coupled plasma instrument with mass detector
- Teflon dishes
- Volumetric flasks (10 and 50 mL)

Reagents:

- Nitric acid ($HNO_3$)
- Certified standards consisting of Ag, Al, As, Ba, Be, Ca, Cd, Co, Cr, Cu, Fe, K, Mg, Mn, Mo, Na, Ni, Pb, Se, Sb, Th, U, V, and Zn
- Internal Standard Multi-Element Mix consisting of Li, Sc, Ga, Y, In, Tb, and Bi
- Ultrapure water

#### 3.7.2. Procedure

##### 3.7.2.1. Sample extraction

(1) Weigh 0.5 g of honey samples into a Teflon dish.
(2) Add 3 mL ultra-pure water.
(3) Add 2.5 mL $HNO_3$ (65%).
(4) Perform a wet digestion of the solution using a microwave oven set at a digestion program consisting of three potency steps: first step at 500 W for 2.5 min, second step at 1000 W for 20 min, and the third step at 1200 W for 30 min.
(5) Following the wet digestion, cool the samples to room temperature.
(6) Transfer the digested clear solution to a 50 mL volumetric flask and fill it using ultra-pure water.
(7) All solutions should be spiked with the internal standard to a final concentration of 10 µg/L.

##### 3.7.2.2. Calculation and expression of results.

(1) Perform a quantitative analysis using the calibration curve method.
(2) Calibration curves should be constructed with a minimum of five concentrations of standards per element.
(3) The limits of detection (LODs) should be calculated as three times the standard deviation of 10 consecutive measurements of the reagent blank, multiplied by the dilution factor used for sample preparation (Table 4).

### 3.8. Calories

Honey calories can be determined indirectly using calculation criteria according to the Brazilian Food Composition Table (Food Research Center, 2017) and (WHO/FAO, 2002) which uses the Atwater general factors system.

The calculation is done by multiplying the content of proteins and available carbohydrates by the factors described as follows:

- Proteins: 17.0 kJ/g or 4.0 kcal/g
- Available carbohydrates: 16.0 kJ/g or 4.0 kcal/g

*Observation:* In the case of lipids, dietary fiber, and alcohol, it is zero for honey.

### *3.9. Hydroxymethylfurfural*

Hydroxymethylfurfural (HMF) is a compound produced by sugar degradation (dehydration of hexoses in an acidic medium) (Bogdanov, 2011). HMF is a freshness parameter of honey; and in fresh honeys, it is absent or occurs in trace amounts. High values of HMF are naturally present in honeys from areas with warm climates (Sodré et al., 2011).

HMF is enhanced with honey processing and heat treatment. Adulteration with commercial sugars and long storage can also enhance HMF (Bogdanov, 2011).

There is more than one method found in literature to determine HMF content (HPLC and spectrophotometric method). We have chosen to summarize the spectrophotometric method of Bogdanov et al. (1997).

#### *3.9.1. Principle*

The spectrophotometric method is based on the principle that HMF absorbs at 284 nm.

#### *3.9.2. Apparatus and reagents*

Apparatus:

- Spectrophotometer able to measure 284 and 336 nm
- Quartz cells 1 cm (two cells)
- Vortex mixer
- Beaker (50 mL)
- Volumetric flask (50 mL)
- Filter paper
- Pipettes (0.5 and 5 mL)
- Test tubes (18 x 150 mm)
- Carrez solution I
- Carrez solution II
- Sodium bisulphite solution 0.2%
- Sample solution (initial honey solution = 5.0 mL)

Reagents:

- Distilled water.
- Carrez solution I
- Carrez solution II
- Sodium bisulphite solution 0.2%
- Sample solution (initial honey solution = 5.0 mL)

#### *3.9.3. Procedure*

The HMF spectrophotometric procedure can be found in AOAC (1990) and Bogdanov et al. (1997).

##### *3.9.3.1. Samples, standards preparation and determination*

*Carrez I Solution*—Dissolve 15 g of ferrocyanide in water and fill to 100 mL in volumetric flask.

*Carrez II Solution*—Dissolve 30 g of zinc acetate in water and fill to 100 mL in volumetric flask.

*Sodium Bisulfite Solution 0.2%*—Dissolve 0.20 g of sodium bisulfite in water and dilute to 100 mL in volumetric flask.

(1) Weigh approximately 5 g of honey into a 50 mL beaker.
(2) Dissolve the sample in approximately 25 mL of distilled water and transfer quantitatively into a 50 mL volumetric flask.
(3) Add 0.5 mL of Carrez solution I to the 50 mL volumetric flask and mix with a vortex mixer.
(4) Add 0.5 mL of Carrez solution II to the 50 mL volumetric flask, mix with a vortex mixer, and fill to 50 mL total volume with distilled water.
(5) Filter the liquid through filter paper (rejecting the first 10 mL of the filtrate).
(6) Pipette 5.0 mL of the filtered liquid into each of two test tubes (18 × 150 mm).
(7) Add 5.0 mL of distilled water to one of the test tubes and mix with the vortex (sample solution).
(8) Add 5.0 mL of sodium bisulphite solution 0.2% to the second test tube and mix with a vortex (reference solution).
(9) To the sample solution (initial honey solution = 5.0 mL), add 5.0 mL of distilled water.
(10) To the reference solution (initial honey solution = 5.0 mL), add 5.0 mL of 0.2% sodium bisulphite solution.
(11) Determine the absorbance of the sample solution against the reference solution at 284 and 336 nm in 10 mm quartz cells within 1 h.
(12) If the absorbance at 284 nm exceeds a value of about 0.6, dilute the sample solution with distilled water and the reference solution with sodium bisulphite solution (as described before) in order to obtain a sample absorbance low enough for accuracy.

##### *3.9.3.2. Calculations.*

Results are calculated using the formula:

$$\text{HMF in mg/kg} = (A_{284} - A_{336}) \times 149.7 \times 5 \times D/W$$

Where:

$A_{284}$ = absorbance at 284 nm;
$A_{336}$ = absorbance at 336 nm;

$$149.7 = \frac{1.26 \times 1000 \times 1000}{16830 \times 10 \times 5} = \text{Constant}$$

126 = molecular weight of HMF
16,830 = molar absorptivity $\varepsilon$ of HMF at $\lambda = 284$ nm
1000 = conversion g into mg
10 = conversion 5 into 50 mL
1000 = conversion g of honey into kg
5 = theoretical nominal sample weight

$D$ = dilution factor (in case dilution is necessary)
$W$ = Weight in g of the honey sample
Express the results in mg/kg.

### 3.10. Ashes

Ashes as a quality parameter is reported for Brazilian regulation, but this method will probably be replaced by the conductivity measurement. Ash content is related to honey origin (e.g., blossom honeys have lower ash content compared with honeydew honey) (Bogdanov, 2009).The procedure for ash determination is described in Bogdanov (2009).

#### 3.10.1. Principle

Samples are ashed using a temperature < 660 °C and the residue is weighted.

#### 3.10.2. Apparatus

- Crucible
- Electric furnace (can run a temperature of 600 °C)
- Desiccator with dry agent
- Analytical balance

#### 3.10.3. Procedure

**3.10.3.1. Samples, standards preparation and determination**

(1) Prepare the ash dish by heating it in the electrical furnace (ashing temperature 600 °C at least 1 h).
(2) Cool the ash dish in a desiccator to room temperature and weigh to 0.001g ($m_2$).
(3) Weigh 5–10g of the honey sample to the nearest 0.001 g into an ash dish that has been previously prepared in (1).
(4) Cool the ash dish in the desiccator (2 h) and weigh.
(5) Continue the ashing procedure (steps 4 and 5) until a constant weight is reached ($m_1$).

**3.10.3.2. Calculations.** Make the calculation using the following formula:

$$W_A = \frac{(m_1 - m_2)100}{m_0}$$

Where:
$W_A$ is expressed in g/100 g
$m_0$ = weight of honey
$m_1$ = weight of dish + ash
$m_2$ = weight of dish

### 3.11. Free acidity

Free acidity is determined using the method of recommended by the AOAC (1990). The International Honey Commission has proposed 50 milliequivalents as the maximum permitted acidity in honey (Bogdanov, 2009). Furthermore, the Council Directive 2001/110/EC (European Commission, 2001) mentions free acidity as a measured honey quality. Finally, the Brazilian regulation established the maximum of 50 milliequivalents/kg of honey.

#### 3.11.1. Principle

A honey sample is dissolved in distilled water and free acidity is measured by titration with 0.1 M sodium hydroxide solution at pH 8.0.

#### 3.11.2. Apparatus and reagents

Apparatus:

- pH meter
- Magnetic stirrer
- Burette or automatic titrator
- Beaker (250 mL)

Reagents:

- Distilled water (carbon dioide free)
- Buffer solutions (pH 4.0 and 9.0)
- 0.1M sodium hydroxide solution

#### 3.11.3. Procedure

The procedure for free acidity is found in Bogdanov (2009).

**3.11.3.1. Samples and standards preparation.**

(1) The pH meter should be calibrated.
(2) Dissolve 10g of the honey sample in 75 mL of carbon dioxide-free water in a 250 mL beaker.
(3) Stir with the magnetic stirrer.
(4) Immerse the pH electrodes in the solution and record the pH.
(5) Titrate with 0.1M NaOH to a pH of 8.30 (complete the titration within 2 min).
(6) Record the reading to the nearest 0.2 mL when using a 10 mL burette and to 0.01 mL if the automatic titrator has enough precision.

**3.11.3.2. Calculations.** Free acidity is calculated using the formula:

$$\text{Free acidity (milliequivalents or millimoles acid/kg honey)} = \text{mL of 0.1M NaOH} \times 10$$

### 3.12. Insoluble solids

Honey insoluble solids include pollen, honey-comb debris, bees and filth particles, and is considered a criterion of honey cleanness. It is possible to determine insoluble solid content by filtering honey diluted with distilled water at 80 °C in porous plate crucibles (Bogdanov, 2009; Bogdanov et al., 1997; Codex, 2001).

#### 3.12.1 Apparatus and reagents

- Drying oven at 105 °C
- Drying oven

- Pump vacuum
- Thermometer
- Desiccator with silica gel
- Metal spatula
- Beakers (50 and 250 mL)
- Porous crucible # 3
- 1000 mL kitassato flask

*3.12.2 Procedure*

(1) Weigh about 20 g of honey into a 250 mL Beaker.
(2) Dilute with hot distilled water (80 °C) until the sample is dissolved.
(3) Transfer to a porous crucible # 3 coupled in the kitassato flask (the crucible must be pre-dried at 105 °C for 12 h, cooled in desiccator and weighed).
(4) Filter under vacuum and wash the honey sample in the crucible with distilled water at 80 °C until the volume of each filtrate reaches 1 L.
(5) Dry the crucible again at 105 °C for 12 h.
(6) Cool the crucible in a desiccator and weigh it.

*3.12.3 Calculation*

The percentage of insoluble solids is calculated by the following ratio:

$$\text{Original sample mass} = 100\%\ \text{Mass of solids}\ (\text{filtered and dried crucible} - \text{dried crucible}) = x\%$$

### 3.13. Diastase activity

Diastases (α- and β-amylases) are enzymes naturally present in honey. Diastase content depends on the floral and geographical origins of the honey. Their function is to digest the starch molecule in a mixture of maltose (disaccharide) and maltotriose (trisaccharide). Diastase is sensitive to heat (thermolabile) and consequently are able to indicate overheating of the product and the degree of preservation (Ahmed, 2013; da Silva et al., 2016; Granato & Nunes, 2016).

For this analysis, the method recommended by The International Honey Commission (2009) and Codex Alimentarius (2001) is based on the "diastase activity" that corresponds to the activity of the enzyme present in 1 g of honey, which can hydrolyze 0.01 g of starch in 1 h at 40 °C, expressed as the diastase number in Göthe units (Bogdanov et al., 1997). In this step-by-step method, the modifications proposed by Santos et al. (2003) were included to give improvements in the solutions preparation and spectrophotometer procedures.

*3.13.1. Apparatus and reagents*

Apparatus:

- UV/VIS Spectrophotometer
- Bucket 1 cm
- Analytical balance
- pH meter
- Thermometer
- Stopwatch
- Water bath
- 250 mL Erlenmeyer flask
- Volumetric flasks of 50, 100, and 500 mL
- 1 L dark volumetric flask
- 50 mL Beaker
- 100 mL test tube
- 20 mL volumetric pipettes
- Micropipette with a variable volume of 100–1000 μL

Reagents:

- Sodium acetate ($CH_3COONa$)
- Glacial acetic acid ($CH_3COOH$)
- Sodium hydroxide (NaOH)
- Soluble starch
- Anhydrous sodium chloride
- Twice sublimated Iodine
- Potassium iodide (KI)
- Destilated water

*3.13.2. Standard preparation*

Acetate buffer solution 0.1 mol/L pH 5.3:

(1) Pipette 0.57 mL acetic acid into a 50 mL volumetric flask and fill with distilled water (Solution A).
(2) Weigh 1.64 g of sodium acetate and dissolve it into 80 mL of distilled water. Adjust the volume to 100 mL in a volumetric flask (Solution B).
(3) Mix 10.5 mL of Solution A with 39.5 mL of Solution B.
(4) Check the pH value using the pH meter and adjust, if necessary, to 5.3 with sodium acetate if the pH value is less than 5.3 or with acetic acid if the pH value is above 5.3.
(5) Transfer the solution to a 100 mL volumetric flask and fill with distilled water. Keep refrigerated.

Sodium chloride solution 0.1 mol/L:

(1) Weigh 0.585 g of sodium chloride into a 50 mL Beaker.
(2) Dissolve in distilled water.
(3) Transfer to a 100 mL volumetric flask and fill up to volume with distilled water.

Iodine solution 0.02 mol/L:

(1) Weigh 4 g of KI into a 100 mL test tube.
(2) Transfer with about 40 mL of distilled water to a dark volumetric flask, glass stopper, with a capacity of 1 L.
(3) Weigh 2.54 g of twice sublimated iodine and transfer to the KI-containing flask. Fill up to volume with distilled water.

(4) Shake until complete dissolution of iodine.
(5) Transfer to a 500 mL amber milled flask.
(6) Store the solution at room temperature with no light.

Starch solution 1% (m/v):

(1) Weigh 1.0 g of soluble starch into a 250 mL Erlenmeyer flask.
(2) Mix with 70 mL of distilled water.
(3) Heat to boiling under constant stirring and keep for 3 min. Transfer the final volume of 100 mL in a volumetric flask and cool down to room temperature in running water. The solution should be prepared at the time of use, do not store to avoid contamination.

Sodium hydroxide solution 0.1 mol/L:

(1) Weigh 2.0 g of NaOH and dissolve in distilled water by filling the volume in a volumetric flask to 500 mL.
(2) Store the solution in a plastic container.

#### *3.13.3. Experimental procedure*

(1) Weigh 5.0 g of honey into flask.
(2) Add 20 mL of distilled water to the flask.
(3) Correct the pH of this solution to a value of 5.3 by adding 0.1 mol/L NaOH until the correct pH is obtained.
(4) Fill the flask to 50 mL with distilled water.
(5) Reaction system: Add 5 mL of the honey solution to a test tube.
(6) Add 500 µL of the acetate buffer 0.1 mol/L pH 5.3 to the tube and mix. It is essential that the honey solution is buffered before contact with sodium chloride, because at pH below 4, diastase activity is inhibited.
(7) Add 500 µL of the sodium chloride solution 0.1 mol/L to the tube.
(8) Add 150 mL of the solution of 0.02 mol/L iodine and 9.6 mL distilled water to the tube and mix.
(9) Keep the tube and its contents in a water bath at 40 °C ± 1 °C.
(10) Add 250 µL of the 1% starch solution (m/v).
(11) Start the stopwatch, shaking the solution until complete homogenization occurs.
(12) Transfer a part of the volume of the reaction system to complete a 1 cm cell and measure the absorbance of the solution in the spectrophotometer at 660 nm against a water blank. This first reading is the initial absorbance value ($Ab_{si}$). The tube with the solution should remain in the water bath at 40 °C.
(13) Make periodic readings of absorbance, always returning the tube to the water bath when not being read, until a value between 0.24 and 0.20 is reached.
(14) When this value is reached, stop the timer and record the elapsed time value.
(15) The last recorded absorbance value is considered the final absorbance ($Abs_f$).
(16) Calculate the diastase index using the following formula:

$$\text{Diastase index} = \frac{(\text{Abs}_i - \text{Abs}_f) \times 0.3}{T(\text{h}) \times V \times 0.016}$$

0.3 = absorbance constant = 0.3 $mg^{-1}$ (previously determined by honey-free assay, given by method)

$T$ (h) = time (in hours) between $Abs_i$ and $Abs_f$ measurements

$V$ = volume of the 10% honey solution in the test tube (mL)

0.016 = total volume in liters of the solution in the test tube (16 mL)

### *3.14. Electrical conductivity*

This method is easy, quick, and involves inexpensive instrumentation. Electrical conductivity determination depends on the ash and acid contents of honey (e.g., the higher ash and acids, the higher the result for electrical conductivity). This parameter has been used as criterion of the botanical origin of honey (Bogdanov, 2009).

According to the European regulation (European Commission, 2001), electrical conductivity of blossom honey must be $< 0.8$ mS/cm of EC, while electrical conductivity of honeydew honey and chestnut honey must be $> 0.8$ mS/cm. Exceptions are honeys from *Arbutus, Banksia, Erica, Leptospermum, Melaleuca, Eucalyptus, Tilia*, and blends (De-Melo et al., 2017).

#### *3.14.1. Principle*

Electrical conductivity uses the measurement of the electrical resistance.

#### *3.14.2. Apparatus and reagents*

- Conductimeter lower range $10^{-7}$ S.
- Conductivity cell (platinized double electrode—immersion electrode)
- Thermometer
- Water bath
- Volumetric flasks (100 mL and 1000 mL)
- Beakers
- Distilled water
- Potassium chloride

#### *3.14.3. Procedure*

The standard method for measuring honey electrical conductivity was outlined by Bogdanov (2009).

##### ***3.14.3.1. Samples and standards preparation.***

(1) If the cell constant of the conductivity cell is not known, transfer 40 mL of the potassium chloride solution (0.1 M) to a beaker.

(2) *Potassium chloride solution (0.1 M) preparation:* dissolve 7.4557 g of potassium chloride (KCl), dried at 130 °C, in freshly distilled water and fill up to 1000 mL volumetric flask. Prepare fresh on the day of use.
(3) Connect the conductivity cell to the conductivity meter.
(4) Rinse the cell thoroughly with the potassium chloride solution and immerse the cell in the solution, together with a thermometer.
(5) Read the electrical conductance of this solution in mS after the temperature has equilibrated to 20 °C.
(6) Calculate the cell constant $K$, using the formula: $K = 11.691 \times 1/G$, where: $K$ = cell constant $cm^{-1}$; $G$ = Electrical conductance (mS), measured with the conductivity cell; 11.691 = the sum of the mean value of the electrical conductivity of freshly distilled water in $mS.cm^{-1}$ and the electrical conductivity of a 0.1 M potassium chloride solution, at 20 °C).
(7) Rinse the electrode with distilled water after the determination of the cell constant.
(8) Dissolve an amount of honey, equivalent to 20 g of dry matter of honey, in distilled water.

#### 3.14.3.2. Determination.

(1) Transfer 20 mL of the solution created in section "Samples and standards preparation" to a 100 mL volumetric flask and fill up to 100 mL with distilled water.
(2) Pour 40 mL of the sample solution into a beaker and place the beaker in a thermostated water bath at 20 °C.
(3) Rinse the conductivity cell thoroughly with the remaining part of the sample solution (Step 1).
(4) Immerse the conductivity cell in the sample solution and read the conductance in mS after temperature equilibrium has been reached.
(5) If the determination is carried out at a different temperature, because of a lack of thermostated cell, a correction factor can be used for calculating the value at 20 °C. For temperatures above 20 °C, subtract 3.2% of the value per °C. For temperatures below 20 °C, add 3.2% of the value per °C.

**3.14.3.3. Calculations.** Calculate the electrical conductivity of the honey solution, using the following formula:

$$S_H = K \times G$$

Where: $S_H$ = electrical conductivity of the honey solution in $mS.cm^{-1}$; $K$ = cell constant in $cm^{-1}$; $G$ = conductance in mS. Express the result to the nearest 0.01 $mS.cm^{-1}$.

Also, FTIR could be used in order to determine honey electrical conductivity (Almeida-Muradian et al., 2012, 2013).

## 3.15. Trolox equivalent antioxidant capacity (TEAC)

Determining the scavenging ability of honey to the radical cation of ABTS [2,2′-azinobis-(3-ethyl-benzothiazoline-6-sulfonic acid)], using trolox, a water-soluble vitamin E analog, as the standard for the calibration curve (Miller et al., 1993; Re et al., 1999; Sancho et al., 2016) is the method we propose for measuring antioxidant capacity. The described method is a modification of Re et al. (1999) procedure, optimized by Sancho et al. (2016):

### 3.15.1. Apparatus and reagents

- Spectrophotometer (Visible), to measure absorbance (A) at 734 nm
- Glass or Plastic cells 1 cm
- Ultrasonic bath
- Shaker
- Stopwatch
- 100 mL volumetric flask
- 100 mL beakers
- Pipettes
- Distilled water
- ABTS
- $K_2S_2O_8$
- Trolox

### 3.15.2. Solutions

#### 3.15.2.1. Preparing the solutions.

(1) 7mM ABTS aqueous solution: Dissolve 0.3841 g 2,20 – azino-bis (3-ethylbenzothiazoline)-6-sulfonic acid diammonium salt (ABTS) in distilled $H_2O$, stir and dilute to 100 mL with distilled $H_2O$.
(2) 2.45 mM $K_2S_2O_8$ aqueous solution: Dissolve 0.0662 g potassium persulfate (di-potassium peroxidisulfate) in distilled $H_2O$, stir, and dilute to 100 mL with distilled $H_2O$.
(3) 5 mM trolox solution: Dissolve 0.1251 g (±)-6-hydroxy-2,5,7,8-tetramethylchromane-2–carboxylic acid (trolox) into a mixture of absolute ethanol-distilled $H_2O$ (1:1). The use of an ultrasonic bath for 5 min aids dissolution. Dilute to 100 mL with distilled $H_2O$.

#### 3.15.2.2. Generation and dilution of ABTS radical cation ($ABTS^{\bullet+}$)

(1) Undiluted $ABTS^{\bullet+}$ solution: mix 7mM ABTS aqueous solution and 2.45 mM $K_2S_2O_8$ aqueous solution in equal amounts (1:1, v/v) and allow them to react in the dark for 16 h at room temperature. Undiluted $ABTS^{\bullet+}$ solution is stable two days, stored in the dark at room temperature.

(2) Working $ABTS^{•+}$ solution: Dilute the undiluted $ABTS^{•+}$ solution with distilled water to obtain an absorbance between 0.70 and 0.80 at 734 nm. The working $ABTS^{•+}$ solution must be prepared fresh daily.

***3.15.2.3. Samples and standards preparation.***
Sample preparation (H-sample):

(1) Weigh 10 g honey into a beaker.
(2) Dissolve in distilled water.
(3) Transfer to a 100 mL volumetric flask.
(4) Further dilute to 100 mL with distilled water.

$$\text{TEAC}\ (\mu\text{mol trolox/g honey}) = \frac{(\%\ \text{inhibition H-samples} - b)}{a} \times \frac{100}{\text{honey weight (g)}}$$

Trolox calibration curve (standard):

(1) In six volumetric flasks, dilute the 5 mM trolox solution with distilled water to obtain the following concentrations: 0.25 mM, 0.50 mM, 0.75 mM, 1.00 mM, 1.25 mM, and 1.50 mM.

***3.15.2.4. Determination.***

(1) Using a pipette, transfer 990 µL of the diluted $ABTS^{•+}$ working solution (see "Generation and dilution of ABTS radical cation ($ABTS^{•+}$)" section) into each cell. One cell corresponds to distilled water (no antioxidant), six cells correspond to the calibration curve (with trolox standard solutions of section "Samples and standards preparation"), and another cell corresponds to the sample (of section "Samples and standards preparation").
(2) Start the reaction by adding 10 µL of distilled water (DW), trolox standard solution (STANDARD), or sample solution (H-sample) to the corresponding cell.
(3) Mix immediately by turning the cell upside down. Determine the absorbance ($A_0$).
(4) Determine the absorbance ($A$) after 6 min ($A_6$) at 734 nm in the cell, against a blank of distilled water.

This procedure is performed in triplicate.

***3.15.2.5. Calculations.***
Percentage of inhibition:

$$\%\ \text{inhibition} = \frac{(A_0 - A_6) \times 100}{A_0}$$

$A_0$ is the absorbance at initial time of distilled water (DW), standard solution (STANDARD), or sample solution (H-sample), referred to section "Determination" (3).

$A_6$ is the absorbance measured at 6 min of distilled water (DW), standard solution (STANDARD), or sample solution (H-sample), referred to section "Determination" (4).

Linear regression: a standard trolox calibration curve is drawn by representing trolox concentrations (mM) on the *x*-axis, and % inhibition on the *y*-axis.

% inhibition $= ax + b$

$x$ = concentration trolox (mM)

$a$ = slope

$b$ = intercept

TEAC value of the honey sample (µmol trolox/g honey): TEAC antioxidant activity of honey is calculated as follows:

$a$ = slope

$b$ = intercept

Where 100 (of the formula) = 1000 (µmol/mmol) * 100 (mL honey solution)/1000 (mL/L).

### 3.16. Color of honey by the CIELab system

The colorimeter measures the reflected color of the honey samples, both liquid and crystallized, of the whole visible spectrum in a wavelength interval between 380 and 740 nm. Other methods for measuring a honey sample's color require honey dilution before measuring its color by spectrophotometry at a given wavelength. Or, it is necessary to dissolve any crystals in the honey to obtain a completely liquid honey that is transparent. Both procedures modify the honey color. The advantage of the colorimeter method is that it is the only instrument that gives the samples' color the way the human eye does, whether the sample is liquid or crystallized. Moreover, honey remains suitable for use in other analysis (Sancho et al., 2016).

*3.16.1. Apparatus*

- $L^*a^*b^*$ Colorimeter glass
- Ultrasonic bath

*3.16.2. Procedure*

(1) Fill the colorimeter glass completely with honey. Bubbles in the honey can interfere with honey color. They can be removed from the honey if it is placed in an ultrasonic bath at room temperature for a few minutes before color determination (avoiding honey heating).
(2) Determine the color parameters $L^*$ (lightness, 100 for white and 0 for black), and the chromaticity coordinates $a^*$ (positive values for redness and

negative values for greenness) and $b^*$ (positive values for yellowness and negative values for blueness), using illuminant D65, 10° observation angle and 45°/0° geometry illumination.

(3) Measure the color in triplicate (Sancho et al., 2016).

(4) Other interesting values for honey comparison or characterization, calculated from $a^*$ and $b^*$, are the coordinates $C^*$ (chroma, saturation, vividness, or purity of a color) and $h^*$ (hue angle or tone) (Tuberoso et al., 2014).

$$C^* = \sqrt{a^{*2} + b^{*2}}$$
$$h^* = \mathrm{arctg}\, b^*/a^*$$

## 4. Investigating honey production with diagnostic radioentomology

Diagnostic radioentomology is a technique based on the measurement of density from component materials using X-rays. The density measured by computer tomography (CT) scanning reflects the degree to which the energy of the X-ray beam is reduced when penetrating a certain material. This technique allows for the quantification and visualization of differences in density with minimal disturbance of the colony since the hive does not need to be opened to introduce light into the dark nest and since no sample need be collected. Thus, it becomes possible to investigate, noninvasively, phenomena that are affected by other common research methods that require destructive sampling.

Honey originates from nectar and honeydew and undergoes a process of concentration (i.e., dehydration). The increase in sugar concentration affects honey density. This can be tracked using radioentomology.

Despite the importance of honey for beekeeping, the processes that lead to its production within the honey bee nest have rarely been investigated (Eyer, Greco, et al., 2016; Eyer, Neumann, et al., 2016). Diagnostic radioentomology was recently used to shed light on nectar storage strategies of bees by allowing for the measure of sugar concentration of the content of a large number of storage cells (Eyer et al., 2016; Greco et al., 2013). By following the evolution of the sugar concentration in individual cells over time, it is also possible to monitor several stages of honey production by the workers, from initial deposition of nectar or honeydew in cells to the capping of mature honey (Eyer et al., 2016).

This technique could be used further to determine the influence of environmental (e.g., high or low nectar flow, diverse or monotonous nectar flow, varying weather conditions) and internal factors (e.g., hive design, air circulation, colony demography) on honey ripening and storage. A better knowledge of the factors affecting honey production processes by honey bee workers could not only lead to a better knowledge of how the honey bee superorganism manages its food stock but could also lead to the improvement of beekeeping management practices by favoring a rapid ripening of honey or by increasing productivity. A further application of this technique is to track the deposition and spread of substances applied into hives (Rademacher et al., 2013) that might contaminate honey stores. The concentration of crop content of live, yet motionless honey bees can also be measured with this technique. Rapidly moving individuals will appear blurred in the scan and thus hinder precise measurements. In the following sections, we describe how to perform measurements of sugar concentration of the content of carbohydrate reserves stored in wax combs with a CT scanner.

### *4.1. Experimental conditions and material required to conduct diagnostic radioentomology trials*

#### *4.1.1. Hive to scanner size ratio*

The most commonly available CT-scanners are those used for human or veterinary medicine. Their dimensions allow for scanning hives of most standard sizes without honey supers. Larger scanners exist for veterinary applications and can be used to study hives with supers. When high resolution is required, it can be obtained by reducing the size of the experimental hive (e.g., using Miniplus (R) hives, 30 × 30 × 34 cm). The focalization of the X-ray beam on a smaller area increases scan resolution. Micro-CT scanners are also available to study small items (e.g., individual motionless workers, a few cells) with high resolution.

#### *4.1.2. Material to be scanned*

Combs built from foundation sheets in moveable frames have a regular and planar shape, which facilitates scanning and data analysis. However, diagnostic radioentomology can be used to study honey production in combs of any shape (e.g., in wild nests in logs) thanks to 3D reconstruction tools. Depending on the research question, combs already containing stores can be used as starting material. When the starting point of the study is an empty cell, a drawn empty comb can be inserted in the test colony. Single combs without workers, workers alone or entire hives with workers can be scanned. Single combs can be gathered in an empty hive or super box and all scanned at once to decrease the number of scans required (Figure 3).

When hives containing honey bees are to be scanned, safety of scanning facility staff and of bystanders needs to be considered. Hive entrances should be closed early on the day of scanning before foragers start their activity and tight hives should be used to prevent honey bees from escaping. Ventilation (e.g., through a screened bottom) should nevertheless be possible for the workers to be able to adjust hive temperature. Combs and hives have to be labeled in a way that they can be recognized on the scan images. Notches cut in the wood or thin metallic letters or numbers can be

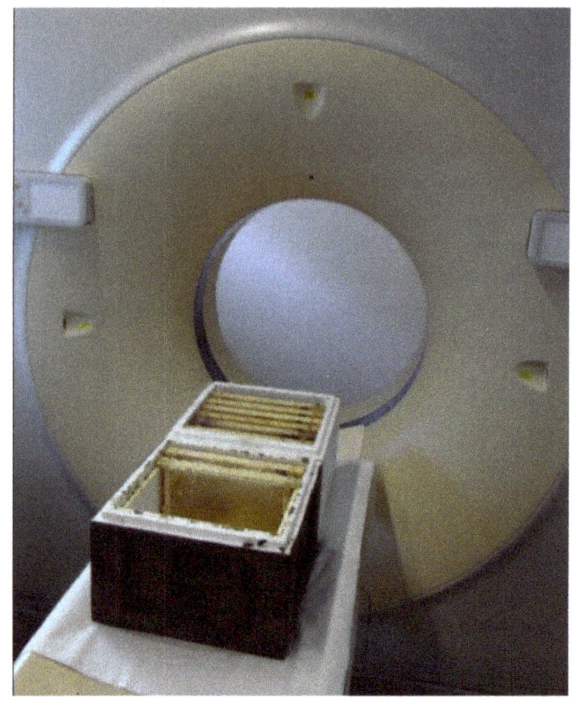

Figure 3. Experimental combs are placed in hive bodies for a CT scan.

used. Metallic parts like nails and wires deflect the X-ray beam and produce glares that make data analysis difficult. It is thus necessary to remove all metallic parts from the material to be scanned or to use plastic or polystyrol hive material.

#### *4.1.3. Monitoring the storage of artificial diets*

Nectar and honeydew collected by foragers are highly diverse in their composition. Consequently, it can be useful to use artificial diets to simplify the study of honey production. The storing of artificial diets (e.g., commercial feeding solutions) can also be the purpose of the experiment. Commercial feeding solutions or a solution of controlled sugar concentration can be provided to the colonies in feeders within the hive or outside of the hive at a close distance, which reflects a more natural foraging situation.

To prevent the foragers from collecting other sources of carbohydrates than those experimentally provided, test colonies can be placed in tunnel cages (see Medrzycki et al. (2013), section "Scanning settings"). When tunnel cages are not available, field colonies can be used during a period without natural nectar flow. To confirm absence of incoming nectar, the concentration of returning foragers' crop content can be measured at regular interval using the following method:

(1) Collect returning foragers that have no pollen baskets on their legs at the hive entrance with the help of forceps.
(2) Gently squeeze their abdomens between your fingers to force them to regurgitate a small droplet of crop content.
(3) Place this droplet onto a refractometer adapted to the range of expected concentrations (see Human et al. (2013), section 4.7.3.1).
(4) Read the sugar concentration through the refractometer.
(5) Repeat this procedure daily for 10–50 foragers per test-colony.

If the concentration of crop content does not correspond to that of the diet provided, the foragers are collecting nectar from other sources. If this input cannot be accounted for in the analysis, the experiment should be performed when naturally available resources do not interfere with the aim of the study.

#### *4.1.4. Studying the storage of natural diets*

When the aim of the experiment is to study natural diets, field colonies with freely flying honey bees can be used. Choose a period with the nectar or honeydew flow of interest for your research question. The range of sugar concentrations collected by foragers can be verified by squeezing the abdomen of returning foragers (see section "Monitoring the storage of artificial diets").

#### *4.1.5. Labeling of the diet to increase contrast*

When the storage of qualitatively different solutions of similar densities needs to be studied, it is possible to discriminate between these solutions by labeling one of them with a contrast agent (Eyer et al., 2016). One example of a suitable contrast agent is Visipaque, which is harmless to honey bee workers at a concentration of 10%. Visipaque (iodixanol Injection) is a contrast media solution, with multiple indications in human medicine (aortography, venography, urography, etc.) for use in X-ray scans. Other contrast agents may be used but their toxicity for honey bees need to be determined (Medrzycki et al. 2013).

If a contrast agent is used, whether it mixes with the solution or sediments in the cell needs to be determined prior to the experiment (Eyer et al., 2016). This verification is required to exclude possible bias from the contrast agent when monitoring density patterns or measuring density of cell or crop content.

#### *4.1.6. Calibration*

The density of cell content is measured in Hounsfield units. This unit is used as a proxy for sugar concentration. A regression of sugar concentrations on density of the artificial carbohydrate solutions or natural sources is thus required to convert Hounsfield units in % sugar concentration (Eyer et al., 2016). To

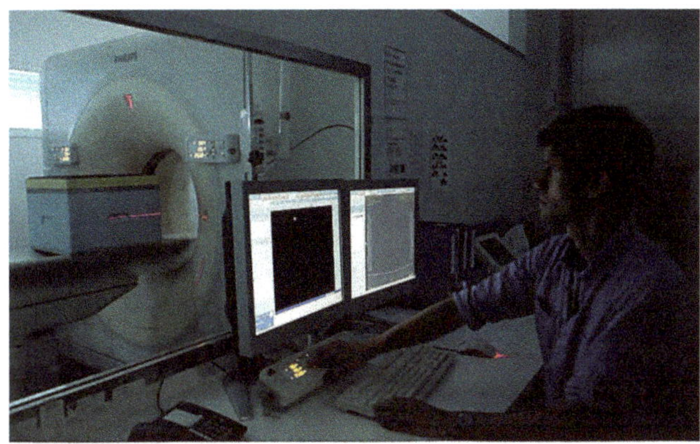

Figure 4. Movement of the scanner bed is programed and the hive box is positioned using the scanner's laser beams.

generate a calibration curve, the following steps are conducted.

#### 4.1.6.1. Artificial diet.

(1) Prepare vials (20–50 mL) with carbohydrate solutions (mixing granulated sugar with water) covering the range of concentrations required for the experiment (e.g., 30%, 50%, 60%, 70%, and 80% sucrose).
(2) Vortex to homogenize solution.
(3) Scan the vials (see section "Computer tomography scanning" below).
(4) Measure density of 10 randomly selected points in each vial (using the point measure tool of eFilm, for example) to consider putative density variations within the vial.
(5) Average these measures.
(6) Plot the averages for each vial with density measured on the $x$-axis and sugar concentration on the $y$-axis.
(7) Add a linear trend line to the chart.
(8) Obtain the equation describing the trend line in order to determine the slope ($m$) and the constant ($t$) of the calibration line.
(9) Calculate the sugar concentration ($y$) by multiplying the density measured by the slope value and by adding the constant ($t$) ($y = (m^*x) + t$). Repeat this calculation for each density value measured.

#### 4.1.6.2. Natural diet.

(1) Sample aliquot of nectar, cell, or crop content.
(2) Place aliquot on a refractometer adapted to the expected sugar concentration (see Human et al. (2013), section 4.7.3.1).
(3) Read the sugar concentration through the refractometer.
(4) Repeat this procedure until a wide enough range of concentrations is obtained to perform calibration.
(5) Perform steps (5–9) as indicated in "Artificial diet" section.

**4.1.6.3. *Labeled diet.*** A specific calibration curve has to be generated for the calculations of the sugar concentrations in labeled diets. Repeat steps described in Section "Monitoring the storage of artificial diets" with labeled solutions of various sugar concentrations. Densities of labeled and nonlabeled diets should not overlap at any point of the concentration range.

### 4.2. Computer tomography scanning

#### 4.2.1. Handling and transporting of the combs

Ideally, a portable CT-scanner should be used to avoid interfering with the biological processes in the colony. However, until the use of portable devices in the field becomes more practical and affordable and since the most frequently accessible devices are fixed, it is recommended to minimize transport of the test hives or combs by conducting the field component of the experiments near the CT-facility. In cases when transportation cannot be avoided, its effects can be minimized by careful handling, avoidance of shaking and shocking the hives or combs to be tested. If combs are scanned outside of the hive, not using smoke during comb collection will limit methodological bias by not provoking cell content take up by workers. Workers can be removed from the combs by gently brushing them off, taking care that the bristles do not contact the cell content. Shaking the combs to get rid of workers is not recommended due to possible effects on cell content, especially if storage of nectar of low sugar concentration is investigated. In the absence of honey bees, further biases such as cell content evaporation can be minimized by scanning the comb immediately after collection.

#### 4.2.2. Scanning settings

The scans can be performed with a Philips Brilliance CT 16-slice apparatus (e.g., Philips Healthcare, 5680 DA Best, The Netherlands) using 120.0 kVp and 183.0 mA as settings. For other models, detailed device setting should be discussed with scanner operator according to the output required.

#### 4.2.3. Performing a scan

Constant conditions during scanning should be maintained, holding the temperature in the scanning room at

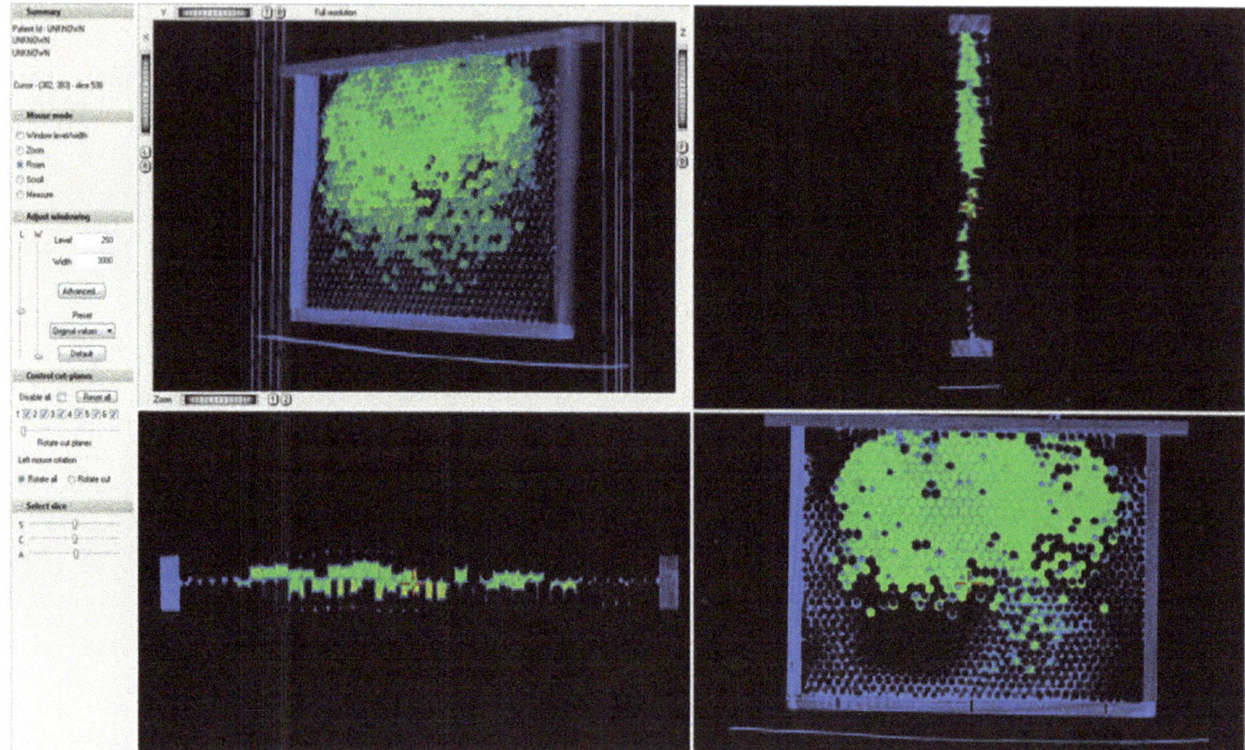

Figure 5. Screenshot of a 3D imaging software with 3D reconstruction panel (top left) and three section planes (sagittal: top right, transverse: bottom left, coronal bottom right).

18–20 °C. If required by the study question, scans can be repeated at defined time intervals. Adhere to safety measures as instructed by scanner operator to avoid the danger of exposure to X-rays.

(1) Place the hive or combs to be scanned on the scanner bed.
(2) Adjust bed movement range and position of material to be scanned using the device's positioning laser beams (Figure 4).
(3) Place yourself in the designated, protected area; personal other than operators leave the room.
(4) Press scan and survey the scan-procedure on the computer.
(5) Wait until scan is complete (usually indicated by the extinction of an acoustic signal) before leaving the designated area.
(6) Save the scan output to a specific location on a server or a hard drive.

### 4.3. Data analysis

#### 4.3.1. Analysis of density patterns in individual cell

For the visualization of cell content density patterns, CT-images can be analyzed with specific 3D rendering software that permits the visualization of differences in density with a color gradient (Figure 5). This feature enables visual monitoring of density measured following this procedure:

(1) Load file.
(2) Choose parameter settings. Detailed device settings for the windowing feature should be discussed with the scanner operator according to the output required. A dark color represents low density content, whereas light color represents high density content. Image analysis software also has several coloration presets (e.g., Figure 5) that can be tried to render the best image output for the analysis. There is no empirical manner to determine the best color settings. The appearance of the images generated will determine which is the most informative.
(3) Apply and record the chosen settings.
(4) Inspect cells from sagittal, transverse, and coronal perspectives for specific cell content patterns (Figures 5 and 6).

#### 4.3.2. Measuring density of cell content

Density, quantified in Hounsfield Units (HU), can be measured using the software eFilm, for example (30-day test-version is available under https://estore.merge.com/na/index.aspx). Density of individual cell content can be measured with the ellipse tool as described below. If required, more precise measurements of cell content can be obtained with the point measurement tool.

(1) Load image file.
(2) Navigate/Scroll vision plane to the cells of interest.

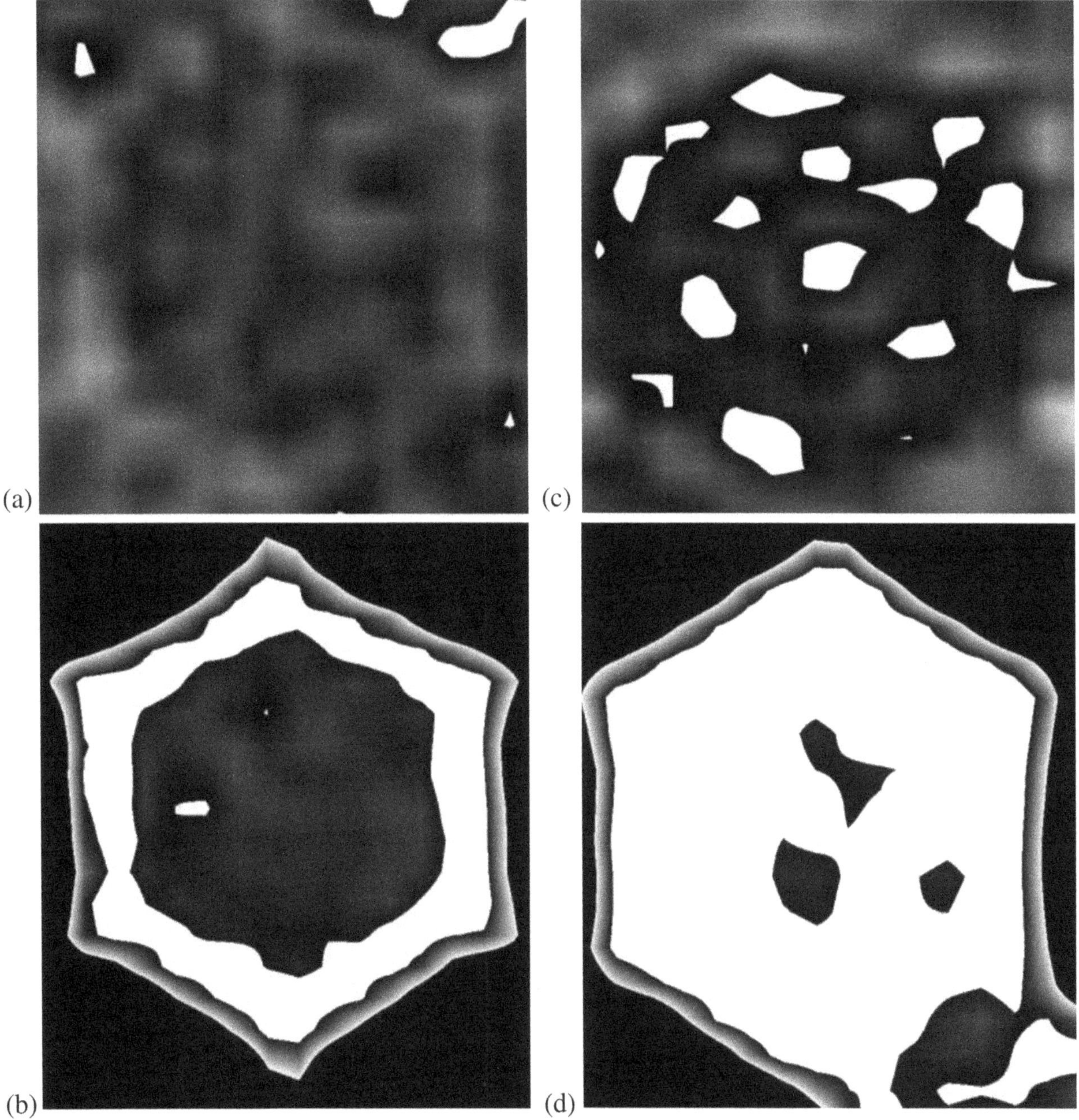


Figure 6: Density patterns in individual cells: (a) low density homogeneous pattern; (b) ring of high density content along the cell walls; (c) inhomogeneous pattern with high density speckles; (d) homogeneous high density cell content. Modified from images published in Eyer et al., 2016 under CC BY 4.0.

(3) Use the ellipse tool to draw a circle over the content of the cell, three scanning frames away from its bottom end.
(4) Repeat the measurement of the same cell, three frames below the surface of the cell content.

Steps 3 and 4 are necessary to take into account within cells variation of density (see Eyer et al., 2016). Scanning a few frames away from cell bottoms and openings helps avoiding the inclusion of air volumes or wax in the circles, which would strongly bias the measurement.

(5) Average the values of these measurements.

*4.3.3. Measuring cell filling status*

The filling status of the cells can be estimated as the number of scanning frames for which the content fills the whole cell diameter.

(1) Load image file.
(2) Place vision plane at cell bottom.
(3) Scroll through each frame with full content while counting.

*4.3.4. Analyzing content of individual cells over time*

(1) Take a picture of each test comb or load the image file with eFilm.

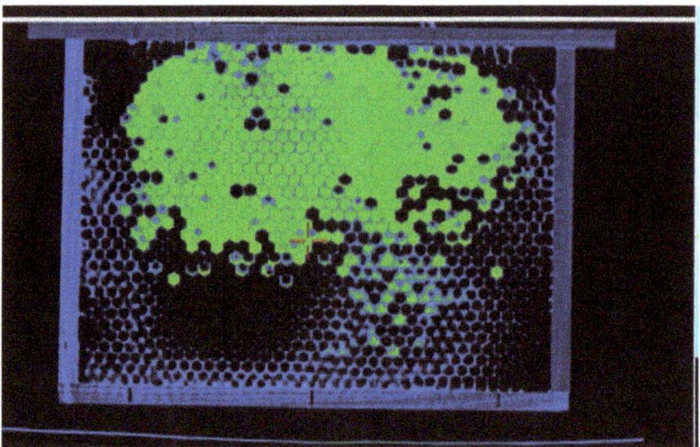

Figure 7. A coronal scan through a partly filled comb. Colors reflect differences in density, with black for low density to yellow for high density. Cell content appear in green-yellow shades, wax and wood in blue, air in black.

(2) Select and mark 10 individual cells (on picture, screenshots (Figure 7), or printouts) for each test comb.
(3) Define and use specific landmarks (e.g., particular shapes on the wax combs or marks on the frames that are not likely to change over time) on the image files/pictures to easily identify individual cells in subsequent scans.
(4) Analyze content of individual cells with eFilm or other dedicated software on subsequent scan times. This allows the investigation of their filling and ripening dynamics over time.

In case the whole process of honey production is to be monitored but not all cells are capped at the end of the experiment, the selection of cells measured can be done *a posteriori*, once cells have been capped.

#### 4.3.5. Spatial analyses at comb level

The following method can be used to investigate spatial patterns in nectar processing (Eyer et al., 2016).

(1) Select combs that contain a sufficient number of filled cells (e.g., Figure 7).
(2) Select a single vertical scan frame perpendicular to the cells' long axes and parallel to the comb midrib for each comb. Choose this frame in order to maximize the number of cells showing content.
(3) Measure the density of content of a predefined number of cell (e.g., using the ellipse or point tool of eFilm, see section "Measuring density of cell content").
(4) Determine the projected *X* and *Y* coordinates of each cell (e.g., with the help of the probe tool of eFilm).
(5) Enter coordinates in a spreadsheet in which the density values will be recorded.
(6) Analyze spatial patterns (e.g., using the spatial autocorrelation Moran's I test statistic in Arc GIS 10.2).
(7) Enter feature locations (projected coordinates) and attribute values (sugar concentration) to calculate the spatial autocorrelation (I index).
(8) Chose the fixed distance model with the mean distance between the centers of two neighboring cells. This distance can be measured with the line tool of eFilm.
(9) Run the analysis and consider the I-index obtained: I-indices close to zero indicate random pattern, whereas positive indices indicate a tendency toward clustering. Negative indices indicate a tendency toward uniformity (Eyer et al., 2016). *P*-values indicate whether the distribution patterns are significant.

### 4.4. Pros and cons

Pros: The use of CT-scanning permits nondestructive observations and measurements within the dark hive. Repeating the observations and measurements in time is thus possible without disrupting the phenomenon under scrutiny. Three-dimension imaging is possible.

Cons: Different materials of similar density cannot be distinguished based on their Hounsfield value. Unless a portable device is available, the material to be scanned has to be transported to the scanning facility. Only snapshots and hence low frequency time lapse images can be captured, filming is not possible to observe the behavior of workers for example. Metallic parts create glare and need to be removed from the scanned material.

### 4.5. Perspectives

Diagnostic radioentomology is a powerful tool to monitor otherwise difficult to observe processes. However, this recently developed method could be further developed to increase its usefulness. Scans performed immediately after behavioral observations in hives with transparent sides can help relate density patterns with the behavior that generated them. Better labeling techniques (e.g., avoiding sedimentation of contrast agent) of cell content would provide further opportunities to monitor processes involved in nectar storage and honey production with an even higher resolution. Scanning of storage combs should also be performed more frequently as done to date, aiming at investigating the processes of honey production with higher time resolution. Further, generating a database of densities of nectar and honey of specific origins (with various sugar compositions) or at different ripening stages could improve the sets of tools available and might relieve the need for calibration before each experiment. Scanning honey bee workers, to investigate crop content, for example, is limited by the relatively low scanning speed. This is especially the case with older devices and when workers are moving. However, the motors of new generations of scanners are spinning the X-ray and detection units at high speeds and will allow for the freezing of a greater proportion of the worker's movement. Other imaging techniques (e.g., radioactive labeling and laser scanning microscopy) could also be employed for studying honey production by workers.

## 5. Pesticide residues in honey

The term "pesticides" represents many different substances used in various crop protection products to treat plants against pests and in veterinary drugs used against animal pests/parasites (including those used in honey bee colonies to control bee pests/parasites), in/around structures to protect against structural pests, etc.

Pesticides applied on crops can contaminate plants, soil, water, and air, and honey bees may be exposed to them via contact with these matrices. The bees collect and transport contaminated products (nectar, pollen, and water) into the hive. There is also a risk of finding pesticide residues in bee products following treatment of the hive.

Honey samples can be screened for pesticide residues for food safety purposes ("Commission Regulation (EU) No 37/2010", European Union Commission, 2009; "Regulation (EU) No 396/2005", 2005). The European Union (EU) requires that honey be tested for pesticide residues within the framework of the monitoring program covered by the Council Directive 96/23/EC (European Communities, 1996). Maximum residue limits (MRLs) are defined for pesticides in honey (EU Pesticides Database) which mostly are between 0.01 and 0.05 mg/kg. For acaricides used in beekeeping, two MRLs were defined in the Commission Regulation (EU) No 37/2010 (2009) for amitraz (and its metabolite, 2,4-dimethylaniline) and coumaphos: 0.2 and 0.1 mg/kg, respectively.

### 5.1. Chemical families of pesticides

The most common use of pesticides is as plant protection products in agriculture. Pesticides are classified in three main groups according to the nature of the "pest" to be controlled: herbicides, fungicides, and insecticides (including acaricides). In each group, there are different chemical families of pesticides.

#### *5.1.1. Herbicides*

Herbicides are used to eliminate weeds which disturb the growth of the crops. They are widely used in agriculture to kill plants or to inhibit their growth or development. Residues are found in air, water, and on various plant parts. There are multiple chemical classes of herbicides: carbamates, triazines, triazoles, and ureas to name a few.

#### *5.1.2. Fungicides*

Fungicides are used to kill fungi in plants, stored products, or soil, or to inhibit their development. Fungicides can either be contact, translaminar, or systemic. The main chemical families of fungicides used are benzimidazoles, dicarboximides, triazoles, chloronitriles, and carbamates.

#### *5.1.3. Insecticides and acaricides*

Insecticides and acaricides are used to kill or disrupt the growth/development of insects or mites. Insecticides are applied on crops to protect them against pests. They can be classified into two groups. (i) Systemic insecticides are incorporated into the tissues of treated plants. Insects ingest the insecticide while feeding on the plants. (ii) Contact insecticides are toxic by contact to insects. Acaricides are also used against honey bee parasites such as *Varroa* in hives (Dietemann et al., 2013). The main chemical insecticide families are organochlorines, organophosphorus, pyrethroids, neonicotinoids, carbamates, and phenylpyrazoles.

### 5.2. Analytical methods

For protecting the health of consumers, the analytical challenge is to achieve limits of quantification at or below the MRL specified for pesticides under EU or other similar legislation. The laboratories conducting the residue analyses usually develop and validate a multiresidue method. Within the framework of official controls on pesticide residues, laboratories follow the requirements specified in the guidance document on analytical quality control and validation procedures for pesticide residues analysis in food and feed ("Document SANTE/11813/2017", European Commission, 2018).

The development of residue analysis methods depends on the properties of both the matrices (honey, wax, pollen, etc.) and pesticides. Residue analysis involves several steps. There are many extraction and clean-up procedures used by different authors to determine the amount of pesticides in honey (Bargańska & Namieśnik, 2010). Solvent extraction and solid-phase extraction (SPE) are the techniques most commonly used for the extraction of pesticides from honey.

There are two general analytical approaches used to determine residues in food and environmental samples: (i) specific methods where a single pesticide and its metabolites are quantitatively determined in the sample, and (ii)

Table 5. Analysis of pesticide residues in honey using a liquid–liquid extraction.

| Pesticide analyzed | Extraction protocol | References |
|---|---|---|
| Organochlorines (α-, β-, and γ-hexachlorocyclohexane (HCH), hexachlorobenzene (HCB), aldrin, p,p'-DDE, p,p'-DDD, o,p'-DDT and p,p'-DDT) | 1. Dissolve 5 g of honey with 50 mL 4% aqueous solution of sodium sulfate in a centrifuge tube.<br>2. Add 20 mL of ethyl acetate to the sample. Repeat the liquid-liquid extraction two times with 15 mL of ethyl acetate.<br>3. Centrifuge at 3000 rpm for 10 min if emulsion is formed.<br>4. Filtrate the organic phase through anhydrous sodium sulfate.<br>5. Evaporate the organic phase under a stream of nitrogen to 2.5 mL for analysis in graduated centrifuge tube.<br>6. Analyze by GC-ECD, and Gas Chromatography/Mass Spectrometry (GC-MS) for confirmatory analysis following authors' protocols. | Blasco, Lino et al. (2004) |
| Coumaphos, bromopropylate, amitraz and tau-fluvalinate | 1. Mix 20 g of honey in an Ultra-Turrax blender with a mixture of *n*-hexane (60 mL), propanol-2 (30 mL) and 0.28% of ammonia. The pH of this mixture is 8.<br>2. Filter the solution through a filter paper.<br>3. Repeat steps 1 and 2 with the mixture of *n*-hexane (60 mL), propanol-2 (30 mL) and 0.28% of ammonia.<br>4. Rinse the Ultra-Turrax with 40 mL of *n*-hexane and filter this washing solution on the same filter paper as in step 2.<br>5. Transfer the combined extracts from steps 2, 3 and 4 to a separating funnel (500 mL).<br>6. Add 50 mL of distilled water and 0.28% of ammonia (pH 10).<br>7. Shake the separating funnel vigorously.<br>8. Allow the filtrate to separate into two phases.<br>9. Discard the aqueous phase (lower).<br>10. Repeat steps 6, 7, 8 and 9 twice.<br>11. Filter the *n*-hexane phase through a layer of anhydrous sodium sulfate placed in a funnel plugged with a filter paper.<br>12. Concentrate the extract by evaporation to dryness under reduced pressure in rotary evaporator using a 35–40 °C water bath.<br>13. Recover the residue obtained with 1 mL of acetone before analysis.<br>14. Analyze by HPLC-DAD following authors' protocol. | Martel and Zeggane (2002) |

multiresidue methods which are analytical methodologies for the simultaneous analysis of trace amounts of a large number of analytes. The number of pesticides tested in the sample can be limited in order to get reliable results, higher recoveries, and lower quantification limits.

Several analytical methods have been used to separate and detect pesticides in honey. Gas chromatography (GC) and liquid chromatograpy (LC) are used for the detection and quantification of pesticide residues (Souza Tette et al., 2016). The choice of the separation technique depends mostly on the characteristics of the pesticides of interest. The volatile, semi-volatile, and thermally stable compounds can be determined by GC, whereas nonvolatile and/or thermally unstable ones should be determined by LC.

At least three steps are required for the analysis of pesticides, among them extraction, separation, and detection. Each one of these steps will be described in the sections that follow.

#### 5.2.1. Sampling

The samples to be analyzed must be representative of the entire honey batch in question. The different steps for the sampling of honey are as follow:

(1) If the honey sample contains impurities (e.g., wax), the sample should be filtered through a stainless-steel sieve. If needed, the honey can be gentle pressed through the sieve with a spatula.
(2) If a honeycomb is sampled, the honeycomb is drained through a 0.5 mm sieve without heating in order to separate honey from the comb.
(3) For crystallized honey, the sample is homogenized with a spatula and an analytical test portion is collected by coring of the honey.
(4) Honey must be homogenized before analysis.
(5) According to the protocol of sample preparation applied for extraction of pesticide residues in honey, collect 1–20 g honey into a centrifuge tube or a beaker for sample preparation.

#### 5.2.2. General requirements for pesticide residue analyses

(1) All analyses should include negative control honey (honey free from pesticide residues—the "blank") and matrix calibration standards prepared by adding pesticides to blank honey before pretreatment of the samples.
(2) The sample spiked with pesticides at the level corresponding to the limit of quantification (LOQ) for

Table 6. Analysis of pesticide residues in honey using a solid-phase extraction (SPE).

| Pesticide analyzed | Extraction protocol | References |
|---|---|---|
| 450 pesticides | 1. Dilute 15 g of the test sample in a 250 mL glass jar with 30 mL of water.<br>2. Shake for 15 min at 40 °C in a shaking water bath.<br>3. Add 10 mL of acetone to the jar.<br>4. Transfer the jar contents to a 250 mL separating funnel.<br>5. Rinse the jar with 40 mL of dichloromethane and transfer this rinse to the separating funnel for partitioning.<br>6. Shake the funnel eight times and pass the bottom layer through a funnel containing anhydrous sodium sulfate into a 200 mL pear-shaped flask.<br>7. Add 5 mL of acetone and 40 mL of dichloromethane into the separating funnel.<br>8. Repeat steps 6 and 7 twice.<br>9. Evaporate the organic phase to about 1 mL with a rotary evaporator at 40 °C for clean-up.<br>10. Add sodium sulfate into a graphitized carbon black cartridge to about 2 cm.<br>11. Connect the cartridge to the top of the aminopropyl cartridge in series.<br>12. Condition the cartridges with 4 mL acetonitrile/toluene 3:1 (v/v).<br>13. Add 1 mL of the sample.<br>14. Rinse the pear-shaped flask with 3 × 2 mL acetonitrile/toluene 3:1 (v/v) and decant it into the cartridges.<br>15. Elute the pesticides with 25 mL acetonitrile/toluene 3:1 (v/v).<br>16. Evaporate the eluate to about 0.5 mL using a rotary evaporator at 40 °C.<br>17. Analyze by GC-MS or LC-MS/MS following authors' protocols according to the group of pesticides. | Pang et al. (2006) |
| Organochlorines (α-, β-, and ɣ-hexachlorocyclohexane (HCH), alachlor, heptachlor, aldrin, endosulfan II, 4,4′-DDE, dieldrin, endrin and 4,4′-DDD) and primiphos-ethyl (internal standard) | 1. Add 10 g of honey in a jar and heat the honey at 35 °C for 15 min.<br>2. Add 50 mL of distilled water to dissolve honey.<br>3. Extract with 3 × 30 mL portions of a binary mixture of petroleum ether/ethyl acetate 80:20 (v/v) in a separating funnel.<br>4. Dry the combined organic extract over anhydrous sodium sulfate.<br>5. Evaporate to 2 mL with rotary evaporator and transfer into a 5 mL glass tube concentrator.<br>6. Evaporate to dryness under a stream of nitrogen.<br>7. Dissolve the dried residue with 0.4 mL of *n*-hexane.<br>8. Condition the florisil cartridge with 10 mL of *n*-hexane.<br>9. Load the concentrated extract obtained in step 7 onto the cartridge.<br>10. Elute the pesticides with 25 mL of 20% (v/v) of diethyl ether in *n*-hexane.<br>11. Evaporate the eluate to dryness under a stream of nitrogen.<br>12. Dissolve the dried residue with 0.4 mL of *n*-hexane containing 1 μg/mL of internal standard.<br>13. Analyze by GC-MS/MS following authors' protocol. | Tahboub et al. (2006) |
| 15 organophosphorus (OP), 17 organochlorines (OC), 8 pyrethroids (PYR), 12 N-methyl-carbamate (NMC), bromopropylate and the internal standards: 4-bromo-3,5-dimethylphenyl-Nmethylcarbamate (4-Br-NMC), triphenylphosphate (TPP) and polychlorinated biphenyl (PCB) 209 | 1. Dissolve 10 g of honey in a 60 mL glass tube with 10 mL of water and 10 mL of acetone.<br>2. Homogenize the mixture with Ultra Turrax for 2 min.<br>3. Wash the Ultra Turrax with about 2 mL of acetone and collect the washings into the glass tube.<br>4. Load the solution obtained in step 3 into an EXtrelut®NT 20 column.<br>5. Allow to drain for 10 min to obtain an even distribution into the filling material.<br>6. Elute pesticides with 5 × 20 mL of dichloromethane using the first aliquot to wash the glass tube.<br>7. Collect the eluate into a 150 mL Erlenmeyer flask.<br>8. Concentrate to nearly 1 mL with a rotary evaporator at 40 °C (reduced pressure) and by drying manually by rotating the flask.<br>9. Dissolve the residue in 2 mL of *n*-hexane.<br>10. Divide the sample solution obtained in step 9 in two portions.<br>11. Transfer one portion of the sample (1 mL) into a 25 mL Erlenmeyer flask and dry by manually rotating the flask.<br>12. Dissolve the residue in 0.5 mL water/acetonitrile 50:50 (v/v).<br>13. Add the appropriate amount of 4-Br-NMC as an internal standard.<br>14. Inject the sample into the LC/DD/FI (liquid chromatography-double derivatization coupled with spectrofluorimetric detector) for the determination of NMC pesticides following author's protocol.<br>15. Add the appropriate internal standards (PCB 209 and TPP) in the second portion of the original sample left in the Erlenmeyer flask (1 mL).<br>16. Inject this portion in GC-MSD and in GC-FPD following authors' protocols in order to analyze bromopropylate, OC, PYR and OP residues respectively. | Amendola et al. (2010) |

each target analyte must be prepared with other samples to control the sensitivity of the equipment.

(3) A supplementary test (named "test sample") is conducted to measure the recovery of all target analytes.

(4) For each analytical sequence, the extracts are injected into the analytical instrument in the following order: blank solvent, negative control honey (blank sample), samples spiked with pesticides for calibration (the "matrix calibration standards"), blank sample, unknown samples (samples to quantify), test sample (to calculate the recovery for each pesticide) and, again, a spiked sample from the calibration to control any variation during the sequence and blank sample.

(5) At the end of the sequence, the sample spiked at level 1 (corresponding to the LOQ) is injected to verify the ability of the equipment to detect the LOQ. Validation of the analytical method should be performed according to the Guidance document on analytical quality control and method validation procedures for pesticides residues analysis in food and feed (Document SANTE/11813/2017, 2018).

### 5.2.3. Sample preparation (extraction and clean-up)

The purpose of the extraction is to extract the substance from the sample with minimum co-extractives matrix interferences. The choice of the solvent is important. Acetone, acetonitrile, ethyl acetate, and hexane are commonly used in this step. The analytes are extracted from the matrix and then, through a clean-up process, the co-extractives are removed.

The most frequent technique used for sample extraction is via the homogenization of the honey sample with an organic solvent or water followed by clean-up with a liquid–liquid extraction (LLE) or a solid phase extraction (SPE) on column. Relatively new techniques like the dispersive solid phase extraction (dSPE) and solid phase microextraction (SPME) are used. Main parameters of methodologies mentioned in the literature for the extraction of pesticides from honey are reported by Rial-Otero et al. (2007) and Souza Tette et al. (2016). Different techniques are presented below with corresponding protocols. The main criterion is to find one method that gives acceptable recoveries for all analytes with one protocol.

***5.2.3.1. Liquid-liquid extraction (LLE).*** In multiresidue methods, an important step is the extraction procedure, especially for complex matrices such as honey, which contains high sugar content. LLE is the most common extraction and purification technique used in the determination of pesticides in honey (Souza Tette et al., 2016). However, LLE usually employs large sample sizes and toxic organic solvents. It is also characterized by the use of multiple sample handling steps, which makes it susceptible to error and contamination. Furthermore, it usually enables the extraction of analytes belonging to only one chemical class. Despite the disadvantages described above, LLE continues to be used in the analysis of pesticides in honey. Bargańska and Namieśnik (2010) reviewed in their paper the techniques already used to extract pesticides from samples of honey. The general protocol involves dissolving honey in water and applying a LLE with various solvents (such as petroleum ether, hexane, dichloromethane, ethyl acetate or mixtures of petroleum ether/ethyl acetate (Tahboub et al., 2006), acetonitrile/acetone/ethyl acetate/dichloromethane (Rissato et al., 2007), acetone/dichloromethane (Pang et al., 2006), hexane/propanol-2 (Martel & Zeggane, 2002)) to extract pesticides according to the pesticide's polarity. The experimental procedures are given in Table 5.

The extracts obtained before analysis by chromatography are clean but these methods use a considerable amount of solvent. Furthermore, these methods are limited to the extraction of a few pesticides due to their different solubility.

***5.2.3.2. Solid phase extraction (SPE) and dispersive solid phase extraction (dSPE).*** Sometimes, after a LLE, a clean-up step on a SPE cartridge with different adsorbents may be necessary before quantification of pesticide residues in honey. Protocols using SPE are given in Table 6 for the main classes of pesticides.

The SPE technique (Bargańska & Namieśnik, 2010; Rial-Otero et al., 2007; Singh et al., 2014; Souza Tette et al., 2016) combines in a single step the extraction and the clean-up procedures based on the separation of LC, where the solubility and functional group interactions of sample, solvent, and sorbent are optimized to affect retention and elution. In SPE, the sample is passed through a cartridge or a packed column filled with a solid sorbent where the pesticides are adsorbed and then eluted with an organic solvent. Moderately polar to polar analytes are extracted from nonpolar solvents on polar sorbents. Nonpolar to moderately polar analytes are extracted from polar solvents on nonpolar sorbents. Most multiresidue methods include a clean-up step using adsorption columns on polar sorbents (florisil, alumina, silica gel) for normal-phase SPE and on nonpolar sorbent like C18 for reversed-phase SPE. The reversed-phase C18 cartridge is by far the most common choice used by researchers for the extraction of insecticides, acaricides, fungicides, herbicides, organochlorines, and organophosphorus pesticides from honey (Rial-Otero et al., 2007). As florisil retains some lipids preferentially (25 g florisil with 3% water retains 1 g of fat), it is particularly well suited for the clean-up of fatty foods. When a florisil column is eluted with solvent mixtures of low polarity, nonpolar residues are recovered almost quantitatively (Singh et al., 2014). Florisil sorbent has been used for pyrethroids, organochlorines and organophosphorus pesticides. Lores et al. (1987) used a silica gel clean-up method for organophosphorus pesticide analysis.

Other SPE clean-up approaches include the combination of GCB (graphitized carbon black) and PSA (primary secondary amine) columns, the combination of C18, GCB and aminopropyl, and the combination of GCB, PSA, and SAX (strong anion-exchange sorbent) columns. GCB is such a sorbent, being nonspecific and generally of hydrophobic nature. Contrary to sorbents based on $SiO_2$, these may be used without the pH of the treated solutions being considered. Because of difficulties with elution of certain planar or aromatic pesticides from GCB, only PSA is used for very efficient clean-up of acetonitrile extracts. The dSPE with PSA is effectively used to remove many polar matrix components, such as organic acids, certain polar pigments, and sugars. Thus, the PSA clean-up method is selected as the most efficient for cleaning honey samples.

The choice of one sorbent or another depends on the analyte polarity and on the possible co-extracted interferences. Sample pH can be critical to obtain high yields of pesticide retention in the sorbent. Thus, in some cases, sample pH modification can be necessary in order to stabilize the pesticides and increase their absorption in the solid phase. Once the pesticides have been retained in the SPE cartridges, they are then eluted with an organic solvent such as acetone, dichloromethane, ethyl acetate, hexane, methanol, tetrahydrofurane or mixtures of hexane/ethyl acetate, hexane/dichloromethane, and methanol/water or methanol/ethyl acetate/dichloromethane.

There is also a method based on SPE, the on-column liquid-liquid extraction (OCLLE) or liquid-liquid extraction on a solid support (SLE), a technique based on classical LLE principle, but assisted by inert solid support (Pirard et al., 2007). This inert matrix consists of diatomaceous earth, well-known for its high porosity, its high dispersing capacities and its high capacity for aqueous adsorption. Pesticides in honey were studied by Amendola et al. (2010) using EXtrelut®NT 20 columns packed with a specially processed wide pore kieselguhr with a high pore volume as support for the repartition process. Honey is dissolved in a mixture of water and acetone and is loaded into an EXtrelut®NT 20. The majority of the co-extractive compounds are retained on the adsorbent material of the column, while the pesticides are eluted by dichloromethane (Amendola et al., 2010). Other columns can be used such as ChemElut 5 mL cartridges (Kujawski et al., 2014).

***5.2.3.3. Quick Easy Cheap Effective Rugged and Safe (QuEChERS).*** The QuEChERS method consists in salting-out LLE using acetonitrile, $MgSO_4$, and NaCl salts and a dSPE step based on primary and secondary amine bonded silica (PSA) to remove co-extractive impurities (U.S. EPA, 2013; Bargańska & Namieśnik, 2010). The QuEChERS method is particularly applied for the determination of polar, middle polar, and nonpolar pesticide residues in various matrices. This method is combined with sensitive analytical techniques such as LC-MS/MS and GC-MS/MS. Tomasini et al. (2012) demonstrated that the matrix effect depends on the floral origin of honey samples and that quantification by the standard addition method in blank matrix is needed.

With this approach, the sample should be >75% water. Then, an initial dissolution of the honey sample is required. Acetonitrile is used as the water-miscible extraction solvent and the separation phase is achieved by the addition of $MgSO_4$. The heat produced by the water binding process promotes extraction to acetonitrile. The addition of NaCl also increases the extraction efficiency (Kujawski et al., 2014). The supernatant is further extracted and cleaned using a dSPE technique. The dSPE centrifuge tube format (available in 2 mL and 15 mL sizes) contains magnesium sulfate (to remove residual water) and PSA sorbent (to remove sugars and fatty acids). These tubes are available with or without GCB (to remove pigments and sterols) and/or C18-EC (endcapped) packing (to remove nonpolar interferences such as lipids). Blasco et al. (2011) showed that the QuEChERS method presented the highest recoveries (mean recovery 91.67%) followed by the SPE (mean recovery 90.25%) whereas the solid-phase microextraction (SPME) showed the lowest recovery (mean recovery of 49.75%) for the pesticides studied. Thus, the QuEChERS method was the most adapted method with around 58% of recoveries >90%.

The QuEChERS multiresidue procedure replaces many complicated analytical steps commonly employed in traditional methods with easier ones. The QuEChERS method has been the most commonly used method for the analysis of pesticides in honey. However, one limitation is that the sample should be >75% water; thus, an initial dissolution of the honey sample is required, which leads to lower concentration of the sample compared to other sample preparation techniques. In order to overcome this limitation, Wiest et al. (2011) added a sample concentration step by evaporation which was satisfactory for extraction of organohalogens, organophosphorus, pyrethroids, and insect growth regulators in honey. According to these authors, evaporation may be necessary when the MRL is lower than the method LOQ. A general protocol is given in the Table 7.

***5.2.3.4. Solid phase microextraction (SPME).*** SPME is a rapid and simple procedure of extraction that can be easily automated and does not need an organic solvent. This technique consists of two separate steps: an extraction step and a desorption step. Both steps must be optimized for the procedure to be successful (Singh et al., 2014).

Table 7. Analysis of pesticide residues in honey using the QuEChERS methodology.

| Pesticide analyzed | Extraction protocol | References |
|---|---|---|
| 12 organophosphorus and carbamates insecticides (bromophos-ethyl, chlorpyrifos-methyl, chlorpyrifos-ethyl, diazinon, fenoxycarb, fonofos, phenthoate, phosalone, pirimiphos-methyl, profenofos, pyrazophos and temephos) | 1. Weigh 1.5 g of honey into a 50 mL polypropylene centrifuge tube.<br>2. Add 3 mL of hot water and vortex until dissolution.<br>3. Add 3 mL of acetonitrile to the sample and shake the tube vigorously by hand for 30 s.<br>4. Pour the sample and extract into the appropriate tube containing 6 g of $MgSO_4$ and 1.5 g of NaCl.<br>5. Shake the tube vigorously by hand for 1 min (avoiding formation of oversized $MgSO_4$ agglomerates).<br>6. Centrifuge the tube at 3000 rpm for 2 min.<br>7. Transfer 1 mL of acetonitrile extract (upper layer) to the dispersive-SPE tube containing 150 mg anhydrous of $MgSO_4$ and 50 mg of PSA.<br>8. Vortex the dSPE tube for 30 s and centrifuge at 3000 rpm for 2 min.<br>9. Transfer 0.5 mL of the final extract into the labeled autosampler vial.<br>10. Analyze by LC-MS/MS following authors' protocol. | Blasco et al. (2011) |
| 80 environmental contaminants | 1. Weigh 5 g of honey in a 50 mL centrifuge tube.<br>2. Add 10 mL of water in the centrifuge tube.<br>3. Shake the tube to dissolve honey.<br>4. When the mixture is homogeneous, add 10 mL of acetonitrile (ACN), 4 g of anhydrous $MgSO_4$, 1 g of sodium chloride, 1 g of sodium citrate dihydrate and 500 mg of disodium citrate sesquihydrate.<br>5. Shake the tube immediately by hand.<br>6. Vortex one minute and then centrifuge for 2 min at 5000 rpm.<br>7. Transfer 6 mL of supernatant in a pre-prepared 15 mL PSA tube (900 mg of anhydrous $MgSO_4$, 150 mg of PSA bonded silica).<br>8. Then, shake the tube immediately by hand.<br>9. Vortex 10 s and centrifuge for 2 min at 5000 rpm.<br>10. Evaporate 4 mL of the extract in a 10 mL glass cone-ended centrifuge tube until 50 μL are left.<br>11. Kept the remaining extract at −18 °C until analysis by LC-MS/MS or GC-MS following authors' protocols according to the group of pesticides. | Wiest et al. (2011) |
| Chlorothalonil, heptachlor, captan, $\alpha$-endosulfan, $\beta$-endosulfan, endosulfan sulfate, and dieldrin | 1. Place 10 g of honey sample into a polypropylene tube (50 mL) of conical base.<br>2. Homogenize with 10 mL of high purity water.<br>3. Add 15 mL of 1% acetic acid in ethyl acetate extraction solvent, also containing 6 g of $MgSO_4$ and 1.5 g of $CH_3COONa$ anhydrous to the tube.<br>4. Shake by hand vigorously for 1 min.<br>5. Centrifuge at 5000 rpm for 5 min.<br>6. Transfer an aliquot of 1 mL of the supernatant to a 2 mL polypropylene tube containing 50 mg of PSA and 150 mg of $MgSO_4$.<br>7. Shake by hand vigorously for 30 s and centrifuge at 5000 rpm for 5 min.<br>8. Put 500 μL of the extract obtained in step 7 into the 1.5 mL vial and complete with 500 μL of ethyl acetate. | Vilca et al. (2012) |

(*Continued*)

Table 7. (*Continued*).

| Pesticide analyzed | Extraction protocol | References |
|---|---|---|
| | 9. Analyze by GC-μECD following authors' protocol. | |
| 200 pesticides | 1. Weigh 5 g of the honey sample into a 50 mL polypropylene centrifuge tube.<br>2. Add 10 mL of deionized water.<br>3. Vortex and incubate in a water bath at 40 °C until complete homogeneity is obtained.<br>4. Add 10 mL of acetonitrile into the tube.<br>5. Shake the content for 1 min using a mechanical shaker.<br>6. Add the QuEChERS salt kit (containing 4 g of anhydrous magnesium sulfate, 1 g of sodium chloride, 1 g of sodium citrate and 0.50 g of sodium hydrogen citrate sesquihydrate) and immediately shake for further 1 min.<br>7. Centrifuge at 15,000 rpm at 4–8 °C for 5 min.<br>8. Transfer the whole acetonitrile fraction into a 15 mL dSPE polypropylene tube (containing 150 mg of anhydrous magnesium sulfate and 25 mg of PSA).<br>9. Shake the tube for 1 min and centrifuge for 2 min at 15,000 rpm using a cooling centrifuge.<br>10. Transfer 2 mL of the supernatant into 50 mL round bottom glass flask.<br>11. Evaporate under vacuum at 40 °C till complete dryness.<br>12. Reconstitute the residue into 2 mL of hexane/acetone 9:1 (v/v).<br>13. Ultra-sonicate and filter the sample through a disposable 0.45 μm PTFE membrane filter into an amber glass vial.<br>14. Analyze by GC-MS/MS following the authors' protocol. | Shendy et al. (2016) |

A fused silica fiber coated with a polymeric film is immersed into an aqueous sample for a given amount of time. The analyte enrichment is by partitioning between the polymer and the aqueous phase according to their distribution constant. Factors influencing the extraction step include fiber type, extraction time, ionic strength, sample pH, extraction temperature, and sample agitation. The pesticides are adsorbed into the stationary phase and later thermically desorbed into the injection port of a gas chromatograph (Rial-Otero et al., 2007). Variables affecting the desorption step include temperature, desorption time, focusing oven temperature, the solvent used, and its volume. Most pesticides have been extracted with polydimethylsiloxane (PDMS, 100 μm) fiber.

The PDMS has the following advantages: (i) enhanced reproducibility, (ii) lower detection limits, (iii) extended linearity, (iv) improved correlation coefficients, (v) low extraction time, and (vi) better chromatograms. The use of a sol-gel CROWN ETHER® fiber (40 μm) was also proposed by Yu et al. (2004) to remove 11 organophosphorus pesticides from honey with good relative recoveries (74–105%) and low detection limits (<0.001 mg/kg). Blasco et al. (2011) employed silica fibers coating with 50 μm carbowax/template resins (CW/TPR) to analyze fenoxycarb, penthoate, temephos, fonofos, diazinon, pyrazophos, phosalone, profenofos, pirimiphos-ethyl, bromophos-ethyl, chlorpyrifos-methyl, and chlorpyrifos-ethyl in honey. The SPME is accurate as a monitoring method for the extraction of the selected pesticides from honey but cannot be implemented as currently applied as a quantification method due to its low recovery for pyrazophos, chlorpyrifos-methyl, temephos, and bromophos-ethyl. The application of the internal standard should be considered.

***5.2.3.5. Stir bar sorptive extraction (SBSE).*** SBSE is a technique theoretically similar to SPME. It has been used with success for the extraction of organic compounds from aqueous food, biological, and environmental samples (Rial-Otero et al., 2007).

For SBSE, a stir bar is coated with a sorbent and immersed in the sample to extract the analyte from solution. The sample is stirred for a given time until the analyte reaches equilibrium between the polymer and the aqueous phase according to their distribution constant. Then, the analytes are desorbed by high temperatures into the injector port of the GC or by liquid removal for liquid chromatography-tandem mass spectrometry (LC-MS) analysis. Blasco et al. (2004) have

Table 8. Analysis of pesticide residues in honey using the SBSE protocol (Blasco et al., 2004).

| Pesticide analyzed | Extraction protocol |
|---|---|
| 6 organophosphorus insecticides (chlorpyrifos-methyl, diazinon, fonofos, phenthoate, phosalone, and pirimiphos-ethyl) | 1. Place 2.5 g of honey into a 50 mL glass beaker.<br>2. Dilute 1/10 ratio with water and homogenize over 15 min using a magnetic stirring bar coated with PDMS.<br>3. Carry out the sorption for 120 min while stirring at 900 rpm.<br>4. Remove the stir bar from the aqueous sample with tweezers.<br>5. Perform the desorption of the analytes into 2 mL vial filled with 1 mL of methanol.<br>6. Perform the desorption of the pesticides by agitating for 15 min.<br>7. Inject 5 μL of this extract into the LC-MS system following the authors' protocol. |

applied this technique for the extraction of six organophosphorus pesticides from honey. These authors also compared the use of SBSE with SPME and concluded that although linearity and precision obtained by both techniques are similar, SBSE is more accurate and sensitive and the effect of honey matrix in the quantification is lower than that of SPME.

The most important advantages of SBSE are the same as those for SPME; however, higher recoveries are obtained with SBSE (Blasco et al., 2004) because of the thicker polydimethylsiloxane (PDMS) coating. The procedure is presented in Table 8. Nevertheless, recoveries are less than those obtained using other techniques (LLE, SPE, QuEChERS).

***5.2.3.6. Dispersive liquid-liquid microextraction (DLLME).*** In 2006, the use of DLLME was developed in the field of separation science for the preconcentration of organic and inorganic analytes from aqueous matrices. The basic principle of DLLME is the dispersion of an extraction solvent (immiscible with water) and a disperser solvent (miscible in water and extraction solvent) in an aqueous solution that provides a large contact area between the aqueous phase and the extraction solvent. In DLLME, extraction and dispersive solvents are simultaneously and rapidly injected into the aqueous sample using a syringe. The main advantages of DLLME over conventional techniques are simplicity of operation, rapidity, low cost, easy handling, low consumption of organic solvents, high recovery, high factor enrichment, and compatibility with chromatographic techniques such as liquid chromatography (LC) and gas chromatography (GC). However, the QuEChERS method is demonstrated to be more robust and more suitable for the determination of pesticides in complex samples (Tomasini et al., 2011). A simple DLLME protocol for the determination of 15 organochlorine pesticides residues in honey is proposed by Zacharis et al. (2012) and is described in Table 9. The final DLLME protocol involves the addition of 750 μL acetonitrile (disperser) and 50 μL chloroform (extraction solvent) into a 5 mL aqueous honey solution followed by centrifugation. The sedimented organic phase (chloroform) is analyzed directly by GC-IT/MS (gas chromatography-ion trap mass spectrometry) or evaporated and reconstituted in acetonitrile prior to the GC-ECD analysis.

#### *5.2.4. Chromatographic detection*

The quantification is performed by gas or liquid chromatography (Bargańska & Namieśnik, 2010; Rial-Otero et al., 2007) according to the characteristic of the analyte: (i) Gas chromatography (GC) with different detectors: electron capture (ECD), nitrogen-phosphorus (NPD) and mass spectrometry (GC-MS or GC-MS/MS); (ii) liquid chromatography coupled with mass spectrometry (LC-MS/MS).

Traditionally nonpolar and middle polar pesticides are analyzed with GC. This is the case for organochlorines, pyrethroids, and organophosphorus pesticides. With the emergence of new pesticides (e.g., neonicotinoids) and due to their physicochemical properties, the LC is used.

For the multiresidue analysis of pesticides, the most convenient detector would be a mass spectrometer (MS/MS, MS/TOF) coupled with either GC or LC, depending on the type of pesticides of interest. Volatile, semi-volatile, and thermally stable ones can be determined by GC, whereas nonvolatile and/or thermally unstable ones should be determined by LC. When there is a positive result, it has to be confirmed using mass spectrometry coupled with chromatography (GC-MS/MS and LC-MS/MS). Mass spectrometry coupled to a chromatographic separation method is a very powerful combination for identification of an analyte in the sample extract.

The choice of the GC column is a very important task in pesticide analysis. The stationary phase should be selected as a function of the polarity of the pesticides. Nonpolar columns (5% phenyl, 95% dimethylpolysiloxane) are the most commonly used for pesticide analysis in honey (e.g., DB-5, HP-5MS, and DB-XLB). Furthermore, other column parameters such as length, inner diameter, or film thickness can be optimized as a function of the number of pesticides to be determined simultaneously. Usually, the parameters of the column are: 30 m × 0.25 mm × 0.25 μm (length × inner diameter × film thickness). In LC, a C18 column (e.g., 4.6 and 2.1 mm i.d.) is almost consensus for the separation of pesticides.

Table 9. Analysis of pesticide residues in honey using the DLLME protocol.

| Pesticide analyzed | Extraction protocol | References |
|---|---|---|
| 15 pesticides (etridiazole, chloroneb, propachlor, trifluralin, hexachlorobenzene, chlorothalonil, cyanazine, chlorpyrifos, DCPA, trans-chlordane, cis-chlordane, trans-nonachlor, chlorobenzilate, cis-permethrin, trans-permethrin) | 1. Prepare a solution of honey at 50 g/L (dissolve 10 g of honey in a flask with 200 mL of water).<br>2. Leave the sample to equilibrate for at least for 15 min prior to performing the DLLME extraction.<br>3. Transfer an aliquot of 5 mL of the diluted sample into a 10 mL screw cap glass tube with conical bottom.<br>4. Inject rapidly into the sample solution a mixture of 750 µL acetonitrile (disperser) and 50 µL chloroform.<br>5. Gently shake the mixture by hand for 1 min.<br>6. Extract the pesticides and the ISTD (1-bromo decahexane) from the aqueous matrix/phase into the fine chloroform microdroplets.<br>7. Centrifuge the mixture for 3 min at 2500 rpm for phase separation.<br>8. Remove the sedimented chloroform volume using a microsyringe.<br>9. Transfer the extract into an autosampler vial with 50 µL insert and 2 µL of the organic solvent; directly inject this into the GC-IT/MS (GC with ion-trap mass spectrometer detector) following the authors' protocol.<br>10. For GC-ECD analysis, evaporate 20 µL of chloroform extract (obtained in step 8) to dryness by a gentle stream of nitrogen.<br>11. Reconstitute the sample by the same volume (20 µL) of acetonitrile. | Zacharis et al. (2012) |
| Aldrin, endrin, lindane, 2,4′-DDT, 2,4′-DDD, 2,4′-DDE, 4,4′-DDT, 4,4′-DDE and α-endosulfan | 1. Dissolve in a centrifuge tube 0.5 g of a homogenized honey sample with 3 mL of ultrapure water.<br>2. Prepare and inject rapidly a mixture of 450 µL acetone (disperser solvent) and 100 µL chloroform (extract) into the sample to obtain an emulsion.<br>3. After 20 s (including 5 s of shaking), centrifuge the sample (5 min, 4000 rpm). A two-phase solution is obtained.<br>4. The resulting volume of sediment phase is 80 µL. During the extraction, a precipitate formed between chloroform and aqueous phase.<br>5. Collect the chloroform phase at the bottom of the conical vial with a microlitre syringe.<br>6. Analyze by GC-MS (2 µL of sample are injected) following the authors' protocol. | Kujawski et al. (2012) |

## 6. Antibiotics and chemotherapeutics in honey

Honey is generally considered as a natural, healthy, and residue-free product. However, in the early 2000s, some imported honeys were often contaminated with antimicrobial residues, even with residues of chloramphenicol, a forbidden substance (Reybroeck, 2018). Importers of honey started with regular screening of honey for residues and food authorities increased the number of honey samples at border inspection posts and in national monitoring plans. In the beginning, the monitoring was focused on streptomycin, chloramphenicol, tetracyclines, and sulfonamides. Later, the scope was enlarged to include other compounds. Presently, honey is monitored for a large list of antibiotics and chemotherapeutics of interest in apiculture: tetracyclines (oxytetracycline), aminoglycosides (streptomycin), sulfonamides (sulfamethazine, sulfathiazole, sulfadiazine, sulfamethoxazole, sulfamerazine, sulfadimethoxine), macrolides (tylosin, erythromycin), lincosamides (lincomycin), amphenicols (chloramphenicol), nitrofurans (furazolidone, furaltadone, nitrofurazone), nitroimidazoles (metronidazole), fluoroquinolones (enrofloxacin (ciprofloxacin), norfloxacin), and fumagillin (Reybroeck et al., 2012).

In general, residues of antimicrobials in honey originate from apicultural use since bee diseases caused by microorganisms such as American (*Paenibacillus larvae*, de Graaf et al. 2013) and European foulbrood (*Melissococcus plutonius*, Forsgren et al. 2013) can be cured by anti-infectious agents. Also nosemosis, caused by spores of the fungi or fungi-related *Nosema apis* or *N. ceranae* (Fischer & Palmer, 2005; Fries et al. 2013) is sometimes treated with antimicrobials like sulfonamides.

High levels (mg/kg or ppm) of antimicrobial residues of veterinary drugs applied to hives can be found in honey, especially the first week after dosing (Reybroeck et al., 2012). Afterward, the concentration of residues in honey decreases via a dilution effect by incoming nectar and consumption of contaminated honey by the bees. Some compounds (oxytetracycline, tylosin, furazolidone) are also degraded by metabolization. In contrast to other food producing animals, honey bees do not metabolize the drugs so residues could remain in the honey for more than a year (Reybroeck et al., 2012), see Table 10.

There are some cases reported of residues of antimicrobials in the honey due to agricultural practices, beekeeping practices, and environmental or fraud issues (Reybroeck, 2014): contaminated nectar from fruit trees treated with streptomycin against fireblight, natural production of streptomycin by certain Streptomyces bacteria, robbery by bees of contaminated honey, feeding bees contaminated honey, mixing clean honey with honey containing residues, the collection by bees of medicated drinking water from farms or surface water from fields where antibiotic-containing manure has been spread, contamination of nectar with sulfanilamide as a degradation product of the herbicide asulam, migration of residues from polluted wax foundation, semicarbazide formed from azodicarbonamide (ADC: a blowing agent used in the manufacturing of plastic gaskets in metal lids), and finally semicarbazide in heather honey formed from elevated arginine levels.

For residue analysis in honey, it is worth noting that some pharmacologically active compounds metabolize or degrade in honey. Thus, it is important to look for the suitable marker residue. For example, honey should be screened for both tylosin A and desmycosin (tylosin B) (Thompson et al., 2007). An overview of the most suitable marker residues for some antimicrobials of interest in beekeeping is given in Table 11. Reybroeck (2018) published a review about residues of antibiotics and chemotherapeutics in honey. This review can be a basis to decide which compounds to analyze in honey.

### *6.1. Legislation regarding residues of veterinary drugs in honey*

MRLs were established for residues of veterinary medicinal products in foodstuffs of animal origin in order to protect public health. No MRLs have been established so far for antibiotics and sulfonamides in honey ("Commission Regulation (EU) No 37/2010 of 22 December 2009," 2009). This leads to the interpretation that the use of antibiotics in beekeeping is not permitted in the EU. However, based on the "cascade" system which is open to all animal species (including honey bees), antibiotics can be used for the treatment of bee diseases (Anonymous, 2007b) on condition that the active substance concerned is registered as an allowed substance in Table 1 in the Annex of Commission Regulation (EU) No 37/2010, and a withholding period has been specified by the prescribing veterinarian. In this period, no honey could be harvested for human consumption. In reality, a very long withdrawal period needs to be considered.

Some EU Member States (Belgium, France) applied action limits, recommended target concentrations, nonconformity, or tolerance levels for antimicrobial residues in honey (Reybroeck et al., 2012). Presently, in application of the Commission Implementing Regulation (EU) 2018/470 for honey produced within the EU, the MRL to be considered for control purpose shall be the lowest of all the MRLs established for other target tissues in any animal species. However, some Member States do not allow treatment with antibiotics and chemotherapeutics under "cascade" in beekeeping which is making international honey trade more complex.

For certain prohibited or unauthorized analytes in food of animal origin, the regulatory limit is the Minimum Required Performance Limit (MRPL) or the Reference Point for Action (RPA). MRPLs were fixed to harmonize the level of control of those substances and to ensure the same level of consumer protection in the Community. So far in honey, a MRPL of 0.3 µg/kg was set for chloramphenicol (European Communities Commission, 2003), while the MRPL of 1 µg/kg for nitrofuran (furazolidone, furaltadone, nitrofurantoin, and nitrofurazone) metabolites in poultry meat and aquaculture products generally is considered as also applicable in honey (SANCO, 2004).

For the importation of products of animal origin from third countries, the MRPLs should be employed where they exist as RPAs to ensure a harmonized implementation of Council Directive 97/78/EC (Commission, 2005).

In a guidance paper of the Community Reference Laboratories (CRLs), recommended concentrations for testing were suggested for the harmonization of the performance of analytical methods for national residue control plans for substances without MRLs (Anonymous, 2007a). An overview of European regulatory limits and recommended concentrations for testing in honey for residues of antibiotics and chemotherapeutics of interest for use in beekeeping is given in Table 11.

In the USA, there are no authorized residue limits for antibiotics in honey despite the authorized use of certain antibiotic drugs (oxytetracycline, tylosin, and lincomycin) in beekeeping (Administration, 2017).

In Europe, for national residue monitoring plans, groups of substances that need to be monitored in honey are indicated ("Council Directive 96/23/EC of 29 April 1996," 1996). For honey, it concerns group B1 (antibacterial substances, including sulfonamides and quinolones), B2c (other veterinary drugs - carbamates and pyrethroids), B3a and B3b (organochlorine and organophosphorus compounds), and B3c (chemical elements).

Table 10. Classification of some veterinary drugs based on their stability in honey.

| Compounds that do not degrade in honey | Compounds metabolizing or degrading in honey |
|---|---|
| Streptomycin | Tetracyclines: epimerization to 4-epimers |
| Sulfa drugs | Tylosin: degradation to desmycosin (tylosin B) |
| Lincomycin | Furazolidone: metabolization to AOZ* |
| Chloramphenicol | |

**Note*: AOZ: 3-amino-2-oxazolidone.

Despite not being indicated, honey should also be monitored ("essential") for prohibited substances (group A6) such as chloramphenicol, nitrofurans, and nitroimidazoles (Anonymous, 2017).

Sampling rules are given in Annex IV of the Council Directive 96/23/EC. The same Directive specifies the frequencies and level of sampling while the Commission Decision 97/747/EC (European Communities Commission, 1997) provides levels and frequencies of sampling. The number of honey samples to be taken each year must equal 10 per 300 t of the annual production for the first 3,000 t of production, and one sample for each additional 300 t. Rules for official sampling procedures and the official treatment of samples until they reach the laboratory responsible for analysis are given in Commission Decision 98/179/EC (Commission, 1998).

### 6.2. Determination of residues of antibiotics and chemotherapeutics in honey

#### *6.2.1. Sampling*

The analyzed honey samples should be representative of the honey lot.

(1) Crystallized honey should be homogenized before starting the analysis. This can be done by placing the sample in a water bath at maximum 40 °C until the honey is fully liquified and all of the sugar crystals are dissolved.
(2) The honey should be homogenized before starting the analysis by stirring thoroughly (at least 2 min).
(3) In the event the honey contains extraneous matter, the honey should be strained through a stainless-steel sieve with a mesh diameter of 0.5 mm.
(4) If needed, the honey could be gently pressed through the sieve with a spatula.
(5) Comb honey needs to be uncapped and drained through a 0.5 mm sieve without heating in order to separate honey from the comb.

#### *6.2.2. Sample pretreatment*

Some honeys require a special sample pretreatment. In honey, sulfonamides tend to bind sugars via the formation of N-glycosidic bonds through their aniline group (Sheth et al., 1990). Therefore, the sulfonamides need to be released from the sugar concentrates by hydrolysis using strong acids (Schwaiger & Schuch, 2000) prior to analysis. Otherwise, it is possible that the sulfonamides are missed or underestimated.

#### *6.2.3. General remarks for honey testing on antimicrobial residues*

It is necessary to integrate at least one negative and one positive control sample in each test run. The negative honey is honey free from antimicrobial residues and, by preference, a mixture of the different types of honey (origin, color, and texture) all tested as negative in prior analysis. The positive control sample is prepared by contaminating blank honey with standard material. If commercial kits are used, follow the kit instructions set by the kit manufacturer and validate the method internally before using the method in routine testing. In Europe, validation should be performed according to Commission Decision 2002/657/EC (European Union Commission, 2002). For screening methods, the guidelines published by the Community Reference Laboratories for Residues could also be taken into account (Anonymous, 2010).

#### *6.2.4. Microbiological screening tests*

Microbiological screening is not often used for screening honey for antibiotic residues, this due to the very low detection capabilities (CC$\beta$) that need to be reached because of the issue of zero-tolerance for antimicrobial residues in honey applied in many countries. The high sugar content in this special matrix makes the use of microbial inhibitor tests also less evident. A broad-spectrum detection of antimicrobials in honey by Eclipse 50 (ZEULAB, Zaragosa, Spain) and PremiTest (R-Biopharm AG, Darmstadt, Germany) was suggested by Gaudin et al. (2013). In their validation, a high false positive rate of 5 and 14%, respectively, was observed. This study also showed that the detection capabilities are not in line with the action or reporting limits or recommended concentrations for testing for different compounds (e.g., streptomycin) applied in some European countries. The extraction and test procedure for the Eclipse 50 and PremiTest for honey testing are given in Tables 12 and 13, respectively.

Noncommercial microbiological methods, two for the detection of tetracyclines and one for the detection of tylosin, were published for honey based on the use of *Bacillus cereus* (Gordon, 1989), *Bacillus subtilis*

Table 11. Overview of European regulatory limits in honey and recommended concentrations for testing for residues of anti-infectious agents of interest for use in beekeeping.

| Group | Substance | Marker residue | MRL (in μg/kg) | Recommended concentration for testing (in μg/kg) |
|---|---|---|---|---|
| Amphenicols | chloramphenicol | chloramphenicol | —[a] | 0.3, MRPL |
| Nitrofurans | group (furazolidone, …) | AOZ, AHD, SEM and AMOZ | —[a] | 1, MRPL |
| Nitro-imidazoles | ronidazole | hydroxy-metabolites | —[a] | 3 |
| | dimetridazole | hydroxy-metabolites | —[a] | 3 |
| | metronidazole | hydroxy-metabolites | —[a] | 3 |
| Tetracyclines | tetracyclines | sum of parent drug and its 4-epimer | —[b], 100[c] | 20 |
| Sulfonamides (all substances belonging to the sulfonamide group) | sulfonamides | the combined total residues of all substances within the sulfonamide group | —[b], 100[c] | 50 |
| Aminoglycosides | streptomycin | streptomycin | —[b], 200[c] | 40 |
| Macrolides | erythromycin | erythromycin A | —[b], 40[c] | 20 |
| | tylosin | tylosin A | —[b], 50[c] | 20 |

Note: MRL: Maximum Residue Limit, Regulation (EC) No 470/2009 of the European Parliament and of the Council (The European Parliament & the Council of European Union, 2009) and Commission Regulation (EU) No 37/2010 and amendments as of August 1, 2019 (Commission, 2010); MRPL: Minimum Required Performance Limit, Commission Decision 2003/181/EC (European Communities Commission, 2003); Recommended concentration for testing: CRL guidance paper (Anonymous, 2007a); AOZ: 3-amino-2-oxazolidone; AHD: 1-aminohydantoin; SEM: semicarbazide; AMOZ: 3-amino-5-morpholinomethyl-2-oxazolidone; a: prohibited substance, Table 2 in Annex of Commission Regulation (EU) No 37/2010 (Commission, 2010); b: no MRL fixed in honey; c: MRL based on application of the Commission Implementing Regulation (EU) 2018/470 for honey produced within the EU.

Table 12. Extraction and test protocol for the Eclipse 50 for honey (Gaudin et al., 2013).

**Extraction protocol**

1. Weigh 2 g of honey in a 15 mL centrifuge tube.
2. Add 5 mL of acetonitrile/acetone (v/v, 70/30).
3. Mix for 30–40 s using a vortex.
4. Incubate at 62.5 ± 2.5 °C for 5 min ± 30 s to dissolve honey completely.
5. Centrifuge at 4000 rpm for 15 min.
6. Transfer the supernatant to a clean tube.
7. Evaporate under a nitrogen stream at 55 °C for 15–20 min.
8. Dissolve the residue in 250 μL of UHT consumption milk using a vortex for 1 min.

**Test protocol**

1. Transfer 50 μL of extract into an Eclipse 50 microplate well and seal the wells.
2. Incubate at 65 °C until the negative control changes color or when the absorbances for the negative control are between 0.2 and 0.4 (absorbance = difference of measurement at 590 nm (filter 1) and 650 nm (reference filter)).
3. Interpret the end color:
   - visual reading: yellow is negative; purple is positive; intermediate color is doubtful.
   - optical reading (measurement of absorbance = difference of measurement at 590 nm (filter 1) and 650 nm (reference filter)): absorbance > absorbance for negative control + 0.2: positive.

Table 13. Extraction and test protocol for the Premitest for honey (Gaudin et al., 2013).

**Extraction protocol**

1. Weigh 2 g of honey in a 15 mL centrifuge tube.
2. Add 5 mL of acetonitrile/acetone (70/30).
3. Mix for 30–40 s (vortex).
4. Sonicate for 5 min.
5. Vortex for 30–40 s.
6. Centrifuge at 4500 rpm for 10 min at 4 °C.
7. Remove the supernatant.
8. Evaporate under a nitrogen stream at 40–45 °C.
9. Resuspend the residue in Lab Lemco broth 8 g per L (Oxoid Cat No. CM0015) and mix well.

**Test protocol**

1. Transfer 100 μL of this mixture into the PremiTest ampoule.
2. Incubate at 64 °C until the negative control changes color.
3. Interpret the end color:
   - visual reading: yellow is negative; purple is positive; intermediate color is doubtful.
   - instrumental reading with flatbed scanner and Premiscan software: z-value <0: negative; z-value ≥0: positive.

Table 14. Extraction and test protocol for the Chloramphenicol Residue Rapid Test Device.

**Extraction procedure**
1. Weigh 4 g of honey into a 15 mL centrifuge tube.
2. Add 1 mL PBS A and 1 mL PBS B orderly, shake to dissolve fully.
3. Add 8 mL ethyl acetate, cap and mix by inversion for approx. 8 min.
4. Keep still for clear separation, transfer 5 mL of supernatant into a 5 mL centrifuge tube.
5. Dry it at 65 °C with a stream of air evaporator or nitrogen evaporator.
6. Dissolve the dried residue in 160 μL of CAP PBST buffer.

**Test protocol**
1. Use a dropper to collect at least 3 drops (100 μL) of prepared sample, hold the dropper vertically and transfer 3 full drops (around 100 μL) of solution to the specimen well (S) of the test kit.
2. Start the timer.
3. Wait for red bands to appear.
4. Interpret the result; result should be read in approximately 3–5 min:
   - visual reading – the test line (T) is the same as or darker than the control line (C): negative; the test line (T) is lighter than the control line (C) or there is no test line, it is positive; reference line fails to appear: invalid.

ATCC6633 (Khismatoullin et al., 2003), and *Micrococcus luteus* ATCC 9341 (Khismatoullin et al., 2004), respectively. However, the preparation of the test medium and the testing procedures themselves are not fully and clearly described.

### *6.3. Immunological and receptor assays*

#### *6.3.1. Lateral flow devices*

For use at the apiary, on-site honey tests in the format of lateral flow devices based on the technology of Colloidal Gold Immune Chromatographic Strip Assay (GICA) are commercially available for the detection of residues of several veterinary drugs in honey from Nankai Biotech Co., Ltd. (Binjiang District, Hangzhou, Zhejiang, China). These test kits are individually packed and contain all the reagents needed. Different test devices are available for Chloramphenicol, Streptomycin, Tetracyclines, Sulfadiazine, Sulfaguanidine, Sulfamethazine, Sulfathiazole, Sulfonamides, Furazolidone Metabolite (AOZ), Furaltadone Metabolite (AMOZ), Furacilin Metabolite (AHD), Nitrofurans Metabolite, Fluoroquinolones, Quinolones, Gentamicin, Kanamycin, and Tylosin Residue Rapid Test Device. The following SmarK!T kits are available from the same company: Gentamicin, Kanamycin, Penicillin, Streptomycin, Sulfadiazine, Sulfamethazine, Sulfathiazole, Sulfonamides, Furaltadone Metabolite (AMOZ), Nitrofurazone (SEM), Furantoin (AHD), Furazolidone (AOZ), Nitrofurans 4-in-1, and Fluoroquinolones Rapid Test Kit. Another Chinese company (Shenzhen Bioeasy Biotechnology Co., Ltd., Shenzhen, Guangdong, China) is producing the Chloramphenicol and Tetracycline Rapid Test Kit for Honey.

The tests are utilizing gold conjugated antibodies as signal reagents and a drug protein conjugate as a solid phase capture reagent. As the sample flows through the absorbent sample pad, the liquid reconstitutes the dried monoclonal gold conjugate. The drug in the sample will bind to the conjugate antibody and will migrate further up the membrane to the test line. If there is no drug in the sample, the free antibody conjugate will bind to the test line giving a negative result. In case the sample contains drug residues, the antibody conjugate will not bind to the test line giving a positive result. There is a short sample pretreatment for honey but without the need of instrumentation for most of the tests. With all these test devices, a result is obtained within 3–5 min and visual interpretation of the result is possible. As an example, the test protocol of the Chloramphenicol Residue Rapid Test Device (Nankai Biotech Co., Ltd.) is given in Table 14.

The TetraSensor Honey KIT008 (25 tests)/KIT009 (100 tests) (Unisensor s.a., Liège, Belgium) sensitively (<10 μg/kg) detects the four most important tetracyclines in honey in 30 min, without any special equipment, making analysis at the production site possible (Alfredsson et al., 2005; Reybroeck et al., 2007; Gaudin et al., 2013) . The test procedure of the TetraSensor Honey is given in Table 15. During this first incubation period, tetracyclines possibly present in the honey bind with the specific receptor forming a stable complex. Afterward, during the second incubation, the liquid is absorbed by the dipstick; and while flowing over the dipstick, the liquid passes through the green capture lines. In case the honey is free from tetracycline residues, the first line captures the remaining active receptor and a strong red line will appear. The second line, serving as a control line, takes a certain amount of the excess of reagent that passed through the first line. A red control line should always become visible; otherwise, the test is invalid. Results can be read visually or by means of a ReadSensor, comparing the color intensity of both capture lines.

During this first incubation period, tetracyclines possibly present in the honey bind with the specific receptor. Afterward, the dipstick is dipped into the vial and a second incubation at room temperature occurs for 15 min. When the liquid passes through the green

Table 15. Extraction and test protocol for the TetraSensor Honey KIT008/009.

**Sample pretreatment**

1. Fill the lid of the plastic 'Honey Dilution Tube' with honey (around 600 mg).
2. Reclose the plastic 'Honey Dilution Tube'.
3. Shake well until all honey is dissolved in the buffer.

**Test protocol**

1. Add 200 µL of this mixture to the lyophilised receptor in the glass vial.
2. Mix well.
3. Incubate at room temperature (20 ± 5 °C) for 15 min.
4. Bring the dipstick into the glass vial.
5. Continue incubating at room temperature (20 ± 5 °C) for 15 min.
6. Interpret the result:
   - visual reading: intensity of the test line is stronger compared to the reference line: negative result; intensity of the test line is equal or weaker compared to the reference line: positive result.
   - instrumental reading (ReadSensor): the instrument is calculating the ratio test line/reference line. Ratio ≤ 1.10 (cut-off): negative; 0.90 ≥ratio< 1.10: low positive; ratio <0.90: positive; reference line fails to appear: invalid.

Table 16. Extraction and test protocol for the Sulfasensor Honey KIT033.

**Extraction protocol**

1. Fill the lid of the empty plastic 'Honey Tube' with honey (around 650 mg).
2. Add 600 µL of ACID Buffer in the empty tube.
3. Reclose the tube.
4. Shake well until complete honey dissolution.
5. Incubate the tubes 5 min in boiling water (90–100 °C).
6. Add 600 µL of NEUTRALIZING Buffer, close the tube and vortex.
7. Add 1800 µL of HONEY-II Buffer.

**Test protocol**

1. Place the reagent microwell in the Heatsensor at 40 °C.
2. Add 200 µL of sample and mix 5–10 times until complete reagent dissolution.
3. Incubate for 5 min at 40 °C.
4. Bring the dipstick into the microwell.
5. Continue incubating for 15 min at 40 °C.
6. Remove the dipstick, remove the filter pad from the dipstick and interpret the result:
   - visual reading: intensity of the test line is stronger compared to the reference line: negative result; intensity of the test line is equal or weaker compared to the reference line: positive result.
   - instrumental reading (ReadSensor): the instrument is calculating the ratio test line/reference line. Ratio ≤ 1.20 (cut-off): negative; 1.00 ≥ratio< 1.20: low positive; ratio <1.00: positive; reference line fails to appear: invalid.

capture lines, a red color appears. The first line captures the remaining active receptor and the second line takes a certain amount of the excess reagent that passed through the first line. The second line serves as a control line and always has to become visible; otherwise the test is invalid. Results can be read visually or by means of a ReadSensor, comparing the color intensity of both capture lines.

Only limited laboratory equipment is required to run the Sulfasensor Honey KIT033 (Unisensor s.a.), a generic monoclonal antibody test, for the detection of sulfonamides in honey in 20 min. A sample pretreatment (acid hydrolysis) by heating a mixture of honey sample and buffer at 95 °C for 5 min is needed to release the sulfonamides that are chemically bound to the sugars (Chabottaux et al., 2010; Gaudin et al., 2012; Reybroeck & Ooghe, 2010). The kit manufacturer claims a detection of several sulfa drugs at 25 µg/kg in honey. The extraction procedure and test protocol are shown in Table 16.

With the competitive multiplex dipstick Bee4Sensor KIT059 (Unisensor s.a.), the screening for tylosin, (fluoro)quinolones, sulfonamides, and chloramphenicol in honey is possible (Heinrich et al., 2013). The test could be used in two different ways: as a field test or as a lab test with better detection capabilities. The sample extraction and test protocol of the field test is described in Table 17. To run the lab test, one aliquot (A) is dissolved using acid hydrolysis, whereas the other aliquot (B) is dissolved in water. After liquid/liquid partitioning of both aliquots with ethyl-acetate, the organic layers are evaporated until dry under nitrogen. After reconstitution in a buffer, aliquots A and B are combined and applied to the well of a Bee4Sensor test kit for 5 min at 40 °C, as one sample extract. Afterward, a dipstick is then incubated in this prepared well for 15 min at 40 °C. The dipstick can be assessed visually or instrumentally via the ReadSensor.

Table 17. Extraction and test protocol (field test) for the Bee4Sensor KIT059.

**Extraction protocol**

1a. Fill the lid of an empty plastic 'Honey Tube' (A) with honey (around 650 mg).
2a. Add 300 µL of ACID Buffer (Solution 1) in the empty tube A.
3a. Reclose the tube.
4a. Shake well until complete honey dissolution.
5a. Incubate the tube 5 min in boiling water (90–100 °C).
6a. Add 300 µL of NEUTRALIZING Buffer (Solution 3)(preheated), close the tube and mix by inversion.
7a. Let the tube cool down.
8a. Add Dipstick Buffer (Solution 2) to a final extract volume of 3.5 mL and mix well by inversion.
1b. Fill the lid of an empty plastic 'Honey Tube' (B) with honey (around 650 mg).
2b. Add 2 mL of Dipstick Buffer (Solution 2) in the empty tube B.
3b. Reclose the tube.
4b. Shake well until complete honey dissolution.
5b. Incubate the tube 5 min in boiling water (90–100 °C).
6b. Let the tube cool down.
7b. Add Dipstick Buffer (Solution 2) to a final extract volume of 3.5 mL and mix well by inversion.

**Test protocol**

1. Mix A& B in 1:1 ratio.
2. Take a reagent microwell and open the microwell.
3. Add 200 µL of mixture of sample extract (A&B) and mix 5–10 times until complete reagent dissolution.
4. Incubate for 5 min at room temperature.
5. Bring the dipstick into the microwell.
6. Continue incubating for 20 min at 40 °C.
7. Remove the dipstick, remove the filter pad from the dipstick and interpret the result visually.

#### *6.3.2. Enzyme-linked immunosorbent assays (ELISA)*

There are many commercial and noncommercial ELISA's developed for the detection of antibiotic residues in different matrices, including honey. Table 18 summarizes existing ELISA kits (the list is nonexhaustive). The sample preparation for honey ranges from very simple, just a dilution followed by a filtration, to more complex, including acidic hydrolysis, derivatization, solvent extraction, or purification on a SPE column. The technical brochures of the different kit manufacturers should be followed. Validation studies of ELISA kits have been published regarding the detection of chloramphenicol (Scortichini et al., 2005), tylosin and tilmicosin (Peng et al., 2012), nitrofuran metabolites (Elizabeta et al., 2012), and tylosin en streptomycin (Gaudin et al., 2013) (Figure 8).

Competitive enzyme immunoassays are used in most cases for the detection of small molecular weight components such as antimicrobial residues. Such assays use either an enzyme-linked antibody or an enzyme-linked analyte to detect a particular antigen (drug). As an example, the sample preparation and test procedure for the detection of chloramphenicol in honey by means of the Chloramphenicol ELISA of EuroProxima (Arnhem, The Netherlands) is shown in Table 19. In this kit, the walls of the wells of the microtiterplate are precoated with sheep antibodies to rabbit IgG. A specific antibody (rabbit anti-CAP), enzyme labeled CAP (enzyme conjugate) and CAP standard or sample are added to the precoated wells followed by a single incubation step. The specific antibodies are bound by the immobilized antibodies. At the same time, free CAP (present in the standard solution or sample) and enzyme conjugated CAP compete for the CAP antibody binding sites. After an incubation time of 1 h, the nonbound (enzyme labeled) reagents are removed in a washing step. The amount of CAP enzyme conjugate is visualized by the addition of a chromogen substrate. The bound enzyme conjugate elicits a chromogenic signal. The substrate reaction is stopped to prevent eventual saturation of the signal. The color intensity is measured photometrically at 450 nm. The optical density is inversely proportional to the CAP concentration in the sample.

In some laboratories, noncommercial ELISA methods are used to monitor honey for the presence of antimicrobial residues (Heering et al., 1998; Jeon & Rhee Paeng, 2008). A microplate and magneto iELISA was developed for the detection of sulfonamides in honey using magnetic beads to reduce nonspecific matrix interferences (Muriano et al., 2015).

#### *6.3.3. Enzyme-linked aptamer assays (ELAA)*

Assays were developed using DNA or RNA aptamers to get specific binding to streptomycin (Zhou et al., 2013) or tetracyclines (Wang et al., 2014) in honey.

#### *6.3.4. Radio-labeled receptor/antibody techniques (Charm II tests)*

The Charm II (Charm Sciences Inc., Lawrence, MA) is a scintillation-based detection system for chemical families of drug residues utilizing class-specific receptors or an antibody in immune-binding assay format. The sample preparation is mostly just a dilution of the honey with an extraction buffer supplied as part of

Table 18. Commercial and non-commercial ELISA kits for the detection of antibiotic residues in honey (the list is non-exhaustive).

| Drug (family) | ELISA Kit | Kit manufacturer or reference |
|---|---|---|
| AHD (1-aminohydantoin) | Nitrofuran (AHD) ELISA Test Kit | Unibiotest Co.,Ltd (Wuhan, China) |
| | MaxSignal Nitrofurantoin (AHD) ELISA Kit | Bioo Scientific (Austin, TX) |
| AMOZ (3-amino-5-morpholinomethyl-2-oxazolidone) | Furaltadone (AMOZ) ELISA | Abraxis (Warminster, PA) |
| | I'screen AMOZ v2 | Tecna s.r.l. (Trieste, IT) |
| | MaxSignal Furaltadone (AMOZ) ELISA Test Kit | Bioo Scientific (Austin, TX) |
| | Nitrofuran ( AMOZ) ELISA Kit | Unibiotest Co.,Ltd (Wuhan, China) |
| AOZ (3-amino-2-oxazolidone) | Furazolidone (AOZ) ELISA | Abraxis (Warminster, PA) |
| | Nitrofuran ( AOZ ) ELISA Kit | Unibiotest Co.,Ltd (Wuhan, China) |
| Chloramphenicol | Chloramphenicol (CAP) ELISA Kit | Unibiotest Co.,Ltd (Wuhan, China) |
| | Chloramphenicol ELISA | Abraxis (Warminster, PA) |
| | Chloramphenicol ELISA | EuroProxima bv (Arnhem, NL) |
| | Chloramphenicol ELISA | Randox (Crumlin, UK) |
| | I'screen CAP | Tecna s.r.l. (Trieste, IT) |
| | I'screen CAP v2 | Tecna s.r.l. (Trieste, IT) |
| | MaxSignal Chloramphenicol (CAP) ELISA Test Kit | Bioo Scientific (Austin, TX) |
| | RIDASCREEN Chloramphenicol | r-Biopharm (Darmstadt, DE) |
| Doxycycline | MaxSignal Doxycycline ELISA Kit | Bioo Scientific (Austin, TX) |
| Enrofloxacin | Enrofloxacin ELISA | EuroProxima bv (Arnhem, NL) |
| | MaxSignal Enrofloxacin ELISA Test Kit | Bioo Scientific (Austin, TX) |
| Erythromycin | Erythromycin ELISA | EuroProxima bv (Arnhem, NL) |
| | Erythromycin ELISA Test Kit | Unibiotest Co.,Ltd (Wuhan, China) |
| Flumequine | Flumequine ELISA | Randox (Crumlin, UK) |
| Fluoroquinolones | Fluoroquinolones (Generic) ELISA | EuroProxima bv (Arnhem, NL) |
| | Fluoroquinolones II ELISA | EuroProxima bv (Arnhem, NL) |
| | MaxSignal Fluoroquinolone ELISA Kit | Bioo Scientific (Austin, TX) |
| | QUINOLONES ELISA KIT | Tecna s.r.l. (Trieste, IT) |
| | Quinolones (QNS) ELISA Kit | Unibiotest Co.,Ltd (Wuhan, China) |
| Lincomycin | MaxSignal Lincomycin ELISA Test Kit | Bioo Scientific (Austin, TX) |
| Norfloxacin | MaxSignal Norfloxacin ELISA Test Kit | Bioo Scientific (Austin, TX) |
| Oxytetracycline | MaxSignal Oxytetracycline ELISA Kit | Bioo Scientific (Austin, TX) |
| | Oxytetracycline ELISA | Randox (Crumlin, UK) |
| SEM (semicarbazide) | Nitrofuran (SEM) ELISA Test Kit | Unibiotest Co.,Ltd (Wuhan, China) |
| Streptomycin | I'screen STREPTO | Tecna s.r.l. (Trieste, IT) |
| | MaxSignal Streptomycin ELISA Test Kit | Bioo Scientific (Austin, TX) |
| | MaxSignal Streptomycin ELISA Test Kit For Honey Samples | Bioo Scientific (Austin, TX) |
| | RIDASCREEN Streptomycin | r-Biopharm (Darmstadt, DE) |
| | Streptomycin ELISA | EuroProxima bv (Arnhem, NL) |
| | Streptomycin ELISA | Randox (Crumlin, UK) |
| Sulfadiazine | MaxSignal Sulfadiazine ELISA Test Kit | Bioo Scientific (Austin, TX) |
| Sulfamethazine | MaxSignal Sulfamethazine ELISA Test Kit | Bioo Scientific (Austin, TX) |
| Sulfamethoxazole | MaxSignal Sulfamethoxazole ELISA Test Kit | Bioo Scientific (Austin, TX) |
| Sulfaquinoxaline | MaxSignal Sulfaquinoxaline (SQX) ELISA Test Kit | Bioo Scientific (Austin, TX) |
| Sulfonamides | B ZERO SULFA | Tecna s.r.l. (Trieste, IT) |
| | I'screen SULFA | Tecna s.r.l. (Trieste, IT) |
| | I'screen SULFA QL | Tecna s.r.l. (Trieste, IT) |
| | MaxSignal Sulfonamide ELISA Test Kit | Bioo Scientific (Austin, TX) |
| | Total Sulfonamides ELISA kit | Unibiotest Co.,Ltd (Wuhan, China) |
| Tetracyclines | B ZERO TETRA HS | Tecna s.r.l. (Trieste, IT) |
| | MaxSignal Tetracycline (TET) ELISA Test Kit | Bioo Scientific (Austin, TX) |
| | RIDASCREEN Tetracycline | r-Biopharm (Darmstadt, DE) |
| | SuperScreen TETRA | Tecna s.r.l. (Trieste, IT) |
| | SuperScreen Tetra HS | Tecna s.r.l. (Trieste, IT) |
| | Tetracycline ELISA | EuroProxima bv (Arnhem, NL) |
| | Tetracyclines ELISA | Abraxis (Warminster, PA) |
| | Tetracyclines ELISA | Randox (Crumlin, UK) |
| | Tetracyclines(TCs) ELISA Test Kit | Unibiotest Co.,Ltd (Wuhan, China) |
| Tilmicosin | MaxSignal Tilmicosin ELISA Test Kit | Bioo Scientific (Austin, TX) |
| Tylosin | I'screen TYLOSIN | Tecna s.r.l. (Trieste, IT) |
| | MaxSignal Tylosin ELISA Test Kit | Bioo Scientific (Austin, TX) |
| | Tylosin Plate Kit | Abraxis (Warminster, PA) |

*Note*: AOZ: 3-amino-2-oxazolidone; AHD: 1-aminohydantoin; SEM: semicarbazide; AMOZ: 3-amino-5-morpholinomethyl-2-oxazolidone.

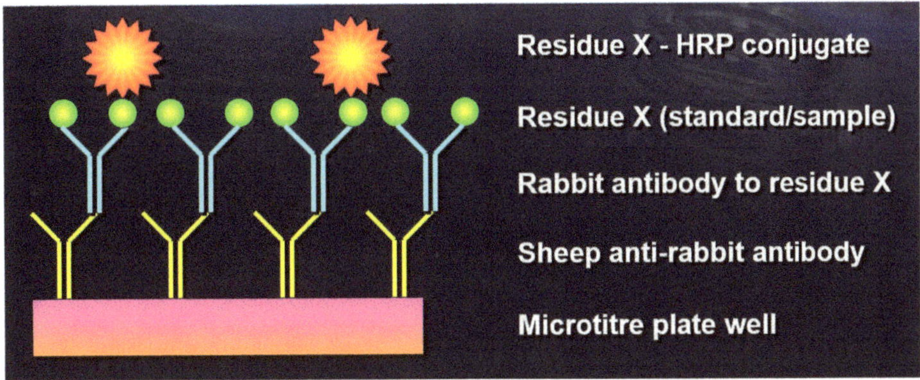


Figure 8. Format of a competitive ELISA. Residue X = target molecule; HRP = horseradish peroxidase enzyme.

the kit. The assay itself takes < 30 min to complete. Charm II kits are available for the detection of sulfonamides, tetracyclines, macrolides (and lincosamides), aminoglycosides ((dihydro)streptomycin), amphenicols (chloramphenicol), nitrofuran AOZ metabolite, and beta-lactams. Some Charm II assays were improved or adapted by the integration of solid phase extraction as extract clean-up to limit false positive results due to matrix quenching effects (McMullen et al., 2004). As an example, the procedure of the Charm II Macrolide Test for Honey is given in Table 20. This screening test detects macrolides and lincosamides in honey.

#### 6.3.5. Biochip-based methods

Some biochip-based methods such as Biacore (GE Healthcare Europe GmbH, Freiburg, DE) and Anti Microbial Arrays (Randox Laboratories Limited, Crumlin, UK) allow the detection of multiple drug residues in honey (McAleer et al., 2010).

The Biacore biosensor system is based on surface plasmon resonance. Analytical Qflex Kits are offered for use with the Biacore Q-instrument for screening for the antibiotics chloramphenicol, streptomycin, sulfonamides, and tylosin (Caldow et al., 2005). The sample preparation includes dissolving the honey in an aqueous buffer and filtering. Analysis for chloramphenicol requires an extra LLE. For sulfonamides, a hydrolysis step is needed to release the sulfonamides bound to sugars. With this system, a high throughput and a rapid (around 5 min) multi-analyte screening in honey is possible (Weigel et al., 2005). However, the instrument costs are high.

Randox Laboratories Limited offers a multi-analyte quantitative testing platform, the Evidence Investigator using Biochip Array Technology for the monitoring of honey for antimicrobials (Daniela et al., 2012; Gaudin et al., 2014; O'Mahony et al., 2010; Popa et al., 2012). An overview of available reagents for the monitoring of honey on antimicrobials is given in Table 21. The sample preparation typically takes about 20 min for dilution. This does not include the Antimicrobial Array III which requires 4 h due to the derivation step or 50 min if CAP only due to extraction step. The incubation and assay times take 2 h for all arrays. Arrays for the detection of nitroimidazoles in honey are in development.

#### 6.3.6. Methods using MIPs and other immunotechniques

Molecularly imprinted polymers (MIP) with specific recognitions sites are sometimes used as a preconcentration step as for the determination of chloramphenicol in honey (Thongchai et al., 2010). MIP sensors based on electropolymerized oligothioaniline-functionalized gold nanoparticles are prepared and applied to the detection of tetracycline in honey (Bougrini et al., 2016). Furthermore, there are publications on the development of a immunosensor for the detection of residues of sulfonamides in honey (Muriano et al., 2013; Valera et al., 2013).

### 6.4. Physico-chemical methods (chromatographic techniques)

Methods for the detection of residues of veterinary drugs in honey based on chromatographic techniques have been described in several publications (Benetti et al., 2004, 2006; Carrasco-Pancorbo et al., 2008; Dubreil-Chéneau et al., 2014; Edder et al., 1998; Kaufmann et al., 2002; Kivrak et al., 2016; Maudens et al., 2004; Nozal et al., 2006; Sporns et al., 1986; Van Bruijnsvoort et al., 2004; Verzegnassi et al., 2002) and in a review article (Bargańska et al., 2011). In the 1990s, the use of HPLC was popular; but today, confirmation of antibiotic residues in honey is performed by LC-MS, mainly LC-MS$^2$ (tandem mass spectrometry) (Blasco et al., 2007). When mass fragments are measured using techniques other than full scan, the system of identification points (IP) is applied. Confirmatory methods must fulfil the criteria listed in Commission Decision 2002/657/EC (European Union Commission, 2002) and must be based on molecular spectrometry providing direct information concerning the molecular structure of the

Table 19. Extraction and test protocol (laboratory test) for the Chloramphenicol ELISA (EuroProxima B.V.).

**Extraction protocol**

1. Bring 10 ± 0.1 g of honey into a centrifugation tube of 50 mL; fill also a centrifugation tube with the same amount of blank honey, honey doped with 0.3 μg/kg of chloramphenicol, and honey doped with 0.1 μg/kg of chloramphenicol, respectively.
2. Add to each sample 5 mL of distilled water.
3. Vortex the samples until all honey is dissolved.
4. Add 5 mL of ethyl acetate.
5. Mix (head over head) for 15 min.
6. After centrifugation (5 min at 2,600 g), pipette 2 mL of the upper layer (ethyl acetate) into a glass test tube.
7. Evaporate at 45 ± 5 °C under a mild stream of nitrogen.
8. Dissolve the residue in 2 mL of *n*-hexane for defatting purposes and vortex.
9. Add 1 mL of reconstitution buffer and vortex for 20 s.
10. After centrifugation (10 min at 2600 g), pipette the layer underneath (reconstitution buffer, ± 800 μL) into a short glass test tube.
11. Add again 1 mL of n-hexane and vortex for 20 s.
12. Add 1 mL of reconstitution buffer and vortex for 20 s.
13. After centrifugation (10 min at 2,600 g), pipette the layer underneath (reconstitution buffer, ± 800 μL) into a short glass test tube.
14. Use the extract in the ELISA (extract could be stored for one day at 4 ± 2 °C).

**Test protocol**

1. Before starting the test, the reagents should be brought up to ambient temperature by taking them out of the refrigerator ~ 20 min before use. Keep the substrate away from light. After analysis, store the remaining reagents as soon as possible in the refrigerator.
2. Identify the wells of the microtiterstrip upon the plate configuration sheet.
3. Pipette 100 μL of reconstitution/zero standard buffer into well A1 (blank).
4. Pipette 50 μL of reconstitution/zero standard buffer into well A2 (zero standard).
5. Pipette 50 μL of chloramphenicol-free extract buffer into wells B1 and B2 (blank control sample).
6. Pipette 50 μL of extract of samples doped at 0.1 μg/kg into wells C1 and C2 (positive control sample 0.1 ppb).
7. Pipette 50 μL of extract of samples doped at 0.3 μg/kg into wells C1 and C2 (positive control sample 0.3 ppb).
8. Pipette 50 μL of each sample extract in duplicate into the remaining wells of the microtiter plate.
9. Add 25 of conjugate (CAP-HRPO) into all wells except in well A1.
10. Add 25 of antibody solution into all wells except in well A1.
11. Cover the plate with aluminum foil and shake the plate for 1 min.
12. Incubate for 1 h in the dark (refrigerator) at 2–8 °C.
13. Discard the solution from the microtiter plate and wash three times with rinsing buffer. Fill all the wells each time with rinsing buffer to the rim. Place the inverted plate on absorbent paper and tap the plate firmly to remove residual washing solution. Take care that none of the wells dry out before the next reagent is dispensed.
14. Shake the substrate solution before use. Pipette 100 μL of substrate solution into each well. Cover the plate and shake the plate slowly.
15. Incubate for 30 min in the dark at room temperature (20–25 °C).
16. Add 10 μL of stop solution into each well.
17. Read the absorbance values (OD) immediately at 450 nm in a spectrophotometer.

**Interpretation of results**

1. Subtract the optical density (OD) value of the blank well (A1) from the individual OD of the other wells.
2. Calculate the Cut-off by adding 3xSD of repeatability for the positive control sample spiked at 0.3 ppb (value calculated out of the validation data) to the mean of both corrected OD values for the two positive control samples spiked at 0.3 ppb.
3. Interpret the control samples: the negative control sample should test negative and the positive control sample spiked at 0.1 ppb should give an OD below the OD of the negative control and higher than the OD of the positive control sample at 0.3 ppb. The run is invalid if the control samples are not giving correct results.
4. Compare the corrected value for each sample to the cut-off value. If the corrected OD of the sample is equal to or below the cut-off: the sample is considered as suspect for the presence of chloramphenicol at 0.3 μg/kg; if the corrected OD of the sample is higher than the cut-off: the sample is free from residues of chloramphenicol at 0.3 μg/kg.

analyte under examination, such as GC-MS and LC-MS (De Brabander et al., 2009).

Some compounds such as nitroimidazoles (Polzer et al., 2010), fumagillin (Daeseleire & Reybroeck, 2012; Kanda et al., 2011; Tarbin et al., 2010), and nitrofuran residues (Khong et al., 2004; Lopez et al., 2007; Tribalat et al., 2006) are mostly directly screened in honey using LC-MS/MS detection, since no or only few immunochemical methods for the detection in honey have been developed. By the development of multiresidue methods, LC-MS is used more and more for multiclass screening for antimicrobial residues in honey (Azzouz & Ballesteros, 2015; Galarini et al., 2015; Hammel et al., 2008; Lopez et al., 2008).

Table 20. Extraction and test protocol for Charm II Macrolide Honey Test (Charm Sciences Inc.).

**Honey sample preparation**

1. Label a 50 mL conical centrifuge tube for each sample.
2. Add 20 grams of honey to an appropriately labeled centrifuge tube. Also, prepare two negative control samples by using blank honey and one positive control sample of blank honey spiked with 20 μg/kg of erythromycin A.
3. Add 30 mL MSU Extraction Buffer to each tube. Mix well until honey is completely dissolved by putting the samples for 30 min on a shaker or by vortexing the tubes.
4. Add 9–10 drops of M2-buffer. Check pH with pH indicator strips; pH of extract should be equivalent to 7.5 (–8.0) on pH strip.
5. If the pH is still too low, add M2 Buffer dropwise, mix, and retest pH until the desired pH is reached.
   Note: If pH is high, add 0.3 mL (300 μL) 0.1 M HCl, mix, and retest. If pH is still high, add 0.1 M HCl drop-wise, mix, and retest.
6. The extract solutions can be kept for 1 day at 4 ± 2 °C.

**Sample filtration**

1. Using the syringe assembly, filter the entire sample through a glass fiber filter (e.g., Millipore-Millex AP prefilter of 25 mm).
2. Collect the entire sample into a clean container (50 mL conical tube or beaker).
3. After the sample has been pushed through, detach the filter holder. Rinse syringe and bivalve with 10 mL deionized water.

**Clean-up over C18 cartridge**

1. Prepare C18 cartridge by attaching the C18 cartridge to adapter, attaching the adapter to the bivalve in the syringe assembly.
2. Activate the C18 cartridge by pushing through 5.0 mL of methanol. The cartridge should be used within 10 min of activation.
3. Wash cartridge with 5.0 mL of deionized or distilled water.
4. Repeat the washing step with 5.0 mL of water.
5. Perform the extraction by adding the filtered honey solution to the syringe. Push the solution slowly through the preactivated C18 cartridge one drop at a time. The sample may be thick and difficult to push through the cartridge. Hold cartridge with two hands to prevent cartridge from popping off due to backpressure. Discard liquid that flows through the cartridge.
6. Wash the cartridge with 5.0 mL distilled water and discard the flow through.
7. Remove the C18 cartridge from the assembly and add 3.0 mL methanol directly into the cartridge. Attach cartridge to the assembly and bring labeled test tube in position.
8. Slowly push methanol one drop at a time through the cartridge and collect the eluate in a labeled test tube.
9. Dry eluate for each sample. Dry under nitrogen or air in a 40–45 °C heat block or water bath and remove from the heat block or water bath when methanol is completely evaporated.
10. Once the methanol is completely evaporated, reconstitute the dried eluate with 5 mL of Zero Control Standard (ZCS) and vortex extensively.
11. Cool the diluted samples on ice for 10 min prior to running the Charm II Macrolide Test Procedure.

**Charm II Macrolide test procedure**

1. Label test tubes and scintillation vials. Let Charm II reagents reach room temperature.
2. Add the white tablet to the empty test tube.
3. Add 300 ± 100 μL water. Mix 10 s to break up the tablet. Take additional time if required to be sure the tablet is broken up.
4. Add 5 ± 0.25 mL diluted sample or control. Use a new tip for each sample.
   Immediately mix by swirling sample up and down 10 times for 10 s.
5. Incubate at 65 ± 2 °C for 2 min.
6. Add green tablet (< 0.19kBq $C^{14}$ labeled erythromycin). Immediately mix by swirling the sample up and down 10 times for 15 s. The tablet addition and mixing of all samples should be completed within 40 s.
7. Incubate at 65 ± 2 °C for 2 min.
8. Centrifuge for 5 min at 1750 G.
9. Immediately pour off liquid completely. While draining, remove fat ring and wipe dry with swabs. Do not disturb the pellet.
10. Add 300 ± 100 μL water. Mix thoroughly to break up the pellet. The pellet must be suspended in water before adding scintillation fluid.
11. Add 3.0 ± 0.5 mL of scintillation fluid. Cap and invert (or shake) until mixture has a uniform cloudy appearance.
12. Count in liquid scintillation counter for 60 s. Read cpm (counts per minute) on [14C] channel. Count within 10 min of adding of scintillation fluid.
13. Calculate the Control Point by taking the highest negative control CPM –35%.
14. Recount if greater than and within 50 cpm of the Control Point.

**Interpretation of results**

1. The cpm for the positive control sample should be below the Control Point, otherwise the run is not valid.
2. If the sample cpm is greater than the Control Point, the sample is negative. Report as "Not Found". If the sample cpm is less than or equal to the Control Point, the sample is positive. The presence of macrolides/lincosamides could be confirmed by a confirmatory method (LC-MS/MS).

Table 21. Overview of antimicrobial arrays (Randox Laboratories Limited) for the monitoring for antimicrobials in honey.

**Antimicrobial Array I Plus**

| Assay | Compound(s) |
|---|---|
| Sulfadimethoxine | Sulfadimethoxine |
| Sulfadiazine | Sulfadiazine |
| Sulfadoxine | Sulfadoxine |
| Sulfamethizole | Sulfamethizole<br>Sulfachlorpyridazine |
| Sulfachlorpyridazine | Sulfachlorpyridazine |
| Sulfamethoxypyridazine | Sulfamethoxypyridazine<br>Sulfaethoxypyridazine |
| Sulfamerazine | Sulfamerazine |
| Sulphisoxazole | Sulphisoxazole |
| Sulfathiazole | Sulfathiazole<br>Sulfadiazine |
| Sulfamethazine | Sulfamethazine |
| Sulfaquinoxaline | Sulfaquinoxaline |
| Sulfapyridine | Sulfapyridine<br>Sulfasalazine |
| Sulfamethoxazole | Sulfamethoxazole<br>Sulfamethizole<br>Sulfachlorpyridazine |
| Sulfamonomethoxine | Sulfamonomethoxine |
| Trimethoprim | Trimethoprim |

**Antimicrobial Array II**

| Assay | Compound(s) |
|---|---|
| Quinolones | Norfloxacin<br>Pefloxacin<br>Enrofloxacin<br>Ciprofloxacin<br>Ofloxacin<br>Pipemidic Acid<br>Fleroxacin<br>Levfloxacin<br>Nadifloxacin<br>Orbifloxacin<br>Danofloxacin<br>Marbofloxacin<br>Oxolinic Acid<br>Difloxacin<br>Pazufloxacin<br>Sarafloxacin<br>Enoxacin |
| Ceftiofur | Ceftiofur<br>Desfuroyceftiofur |
| Thiamphenicol | Florphenicol<br>Thiamphenicol |
| Streptomycin | Streptomycin<br>Dihydrostreptomycin |
| Tylosin | Tylosin<br>Tilmicosin |
| Tetracyclines | Tetracycline<br>4-epitetracycline<br>Rolitetracycline<br>4-epioxytetracylcine<br>Oxytetracycline<br>Chlortetracycline<br>Demeclocycline<br>Doxycycline<br>4-epichlortetracycline<br>Methacycline |

**Antimicrobial Array III**

| Assay | Compound |
|---|---|
| AOZ | 4-NP-AOZ<br>Furazolidone |
| AMOZ | 4-NP-AMOZ<br>Furaltadone |
| AHD | 4-NP-AHD<br>Nitrofurantoin |
| SEM | 4-NP-SEM<br>5-Nitro-2-furaldehyde<br>Semicarbazone |
| Chloramphenicol | Chloramphenicol<br>Chloramphenicol glucuronide |

**Antimicrobial Array III (CAP only)**

| Assay | Compound |
|---|---|
| Chloramphenicol | Chloramphenicol<br>Chloramphenicol<br>Glucuronide |

**Antimicrobial Array IV (continued)**

| Assay | Compound |
|---|---|
| Spiramycin/Josamycin | Spiramycin<br>Kitasamycin<br>Spiramycin I<br>Acetylspiramycin<br>Josamycin |

**Antimicrobial Array IV (continued)**

| Assay | Compound |
|---|---|
| Spectinomycin | Spectinomycin |

(Continued)

Table 21. (*Continued*).

| Antimicrobial Array IV (continued) | | Antimicrobial Array IV (continued) | |
|---|---|---|---|
| Apramycin | Apramycin | Amikacin/Kanamycin | Amikacin<br>Kanamycin A<br>Kanamycin B |
| Bacitracin | Bacitracin | Lincosamides | Lincomycin<br>Clindamycin<br>Pirlimycin |
| Neomycin/Paromomycin | Neomycin<br>Paromomycin | Erythromycin | Erythromycin<br>Clarithromycin<br>Roxithromycin<br>N-Demethyl Erythromycin A |
| Tobramycin | Tobramycin<br>Kanamycin B | Streptomycin/ Dihydrostreptomycin | Streptomycin<br>Dihydrostreptomycin |
| Tylosin B/Tilmicosin | Tylosin B<br>Tilmicosin<br>Tylvalosin<br>Tylosin A | Virginiamycin | Virginiamycin |

*Note*: AOZ: 3-amino-2-oxazolidone; AHD: 1-aminohydantoin; SEM: semicarbazide; AMOZ: 3-amino-5-morpholinomethyl-2-oxazolidone.

## 7. Standard method for honey sensory analysis

Sensory analysis of honey describes the organoleptic profile and may be used to assess the quality, as well as the botanical and geographical origin of a honey sample. Furthermore, it may be used to assess the pleasantness to the end-consumers and to monitor the quality of honey during harvesting, packaging, and storage.

In order to evaluate uniflorality, "honey may be designated according to floral or plant source if it comes wholly or mainly from that particular source and has the organoleptic, physicochemical and microscopic properties corresponding with that origin" (Codex 2001). It should be noted that some pollen is either under or over-represented in a honey sample and the percentage of one specific pollen does not necessarily correspond to the amount of nectar from that specific species. Moreover, the pollen entering the nectar during bee collection (primary pollen spectrum, strictly related to botanical origin) can be significantly modified from secondary pollen contamination, that is, pollen grains entering in honey in the hive, during transformation from nectar to honey by bees, or as a consequence of beekeeper operations. Using only pollen analysis for determining unniflorality may be misleading and a unique generalized threshold may be too high for under-represented and too low for over-represented pollens (see, e.g., the descriptive sheets Person Oddo & Piro, 2004). Thus, sensory evaluation, in addition to physico-chemical and melissopalinological analyses, is essential for determining unniflorality, also because organoleptic characteristics are the only ones that consumers can identify and evaluate.

Traditional sensory analysis was first applied to honey by Michel Gonnet who trained specialists to evaluate honey on the basis of their sensory experience (Gonnet et al., 1985; Gonnet & Vache, 1979, 1992). Modern techniques relying on a panel of assessors and controlled protocols are applied more regularly now (Piana et al., 2004). One selected technique for the sensory analysis of honey is presented here, while a wider guide is published in Marcazzan et al. (2018).

The following proposed standard method will be developed by referring to specific articles (Marcazzan et al., 2014; Mucignat-Caretta et al., 2012; Piana et al., 2004; Sabatini et al., 2007) or to ISO standards (ISO 5492, 2008; ISO 6658, 2017). For detailed information regarding testing rooms and the selection and training of assessors or technicians, one can refer to the specific ISO standards (ISO 8586, 2012; ISO 8589, 2007). For a complete list of ISO standards on sensory analysis, see the ISO catalog (http://www.iso.org).

### 7.1. Methods

Many methods for the sensory analysis of honey are available. The general requirements and tasting procedures are common to all methods. The descriptive semiquantitative method will be described in detail, as it requires less training and has a wider use; the other methods are described in Marcazzan et al. (2018), and references herein.

### 7.2. General requirements

#### 7.2.1. Test room

Evaluations should be conducted under known and controlled conditions with minimal distractions. Individual testing booths should be used (Figure 9) and located in rooms with controlled lighting and temperature, without interfering noises and odors. The general guidance for the design of test rooms is reported in ISO 8589 (2007).

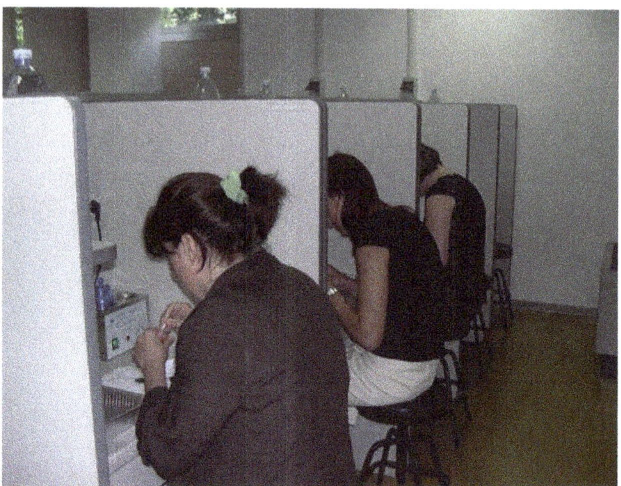

Figure 9. Assessors in testing booths.

Figure 10. Balloon stemmed glasses proposed for sensory honey evaluation.

*7.2.2. Selection and training of assessors and choice of panel*

The guidelines for the selection, training, and monitoring of assessors are reported in ISO 8586 (2012). Panel sessions involve at least five specialized expert assessors working as part of a panel managed by a panel leader (ISO 13300, 2006a, 2006b) who oversees the selection, training, monitoring, and performance of the assessors. The panel leader is responsible for organizing and coordinating the tests as well as collecting and processing the results.

*7.2.3. Product storage*

Recommended temperature for honey sample storage should be <20 °C to maintain the stability of honey. Avoid high temperatures (>35 °C) and exposure to sunlight.

*7.2.4. Preparation and presentation of the samples*

The sensory analysis is performed in containers as detailed below. It is useful to separate the visual from the olfactory-gustatory assessment: general aspect and certain defects should be assessed in the original container if possible. In any case, the container should not have any sort of identification except a code. Otherwise, honey samples must be presented in the containers mentioned below.

***7.2.4.1. Sampling glasses.***

The same type of glasses or containers should be used within each tasting session. The glasses/containers should meet the following requirements:

- Containers should be odorless.
- Present samples in a homogeneous and anonymous way (identical containers with no distinguishing marks apart from the identification code).
- Cover samples to prevent the contamination and dispersion of the honey odors.

Table 22. Guidelines for assessors.

| |
|---|
| 1. In the day of assessment, the assessors should: |
| • avoid perfumed toiletries, and |
| • avoid anything else that may introduce odors into the tasting room. |
| 2. For 1 h before the evaluation, the assessors should: |
| • avoid smoking, and |
| • avoid eating. |
| 3. During the session, the assessors should: |
| • work alone, |
| • follow the instructions of the panel leader, and |
| • abstain from expressing any comment. |

- Use a 150/200 mL capacity balloon stemmed wine glass for samples (Figure 10) or other types of glasses if they satisfy the requirements mentioned above, maintaining the sample/volume ratio near 1/5.
- Use colored or opaque containers if it is necessary to mask the color of the honey. For example, oil sampling beakers defined in the COI/T.20/Doc. No. 5 norm (International Olive Oil Council, 2007) and ISO standard (ISO 16657, 2006) meet these requirements. The color of honey may be masked by using a red light.

***7.2.4.2. Preparation of the samples.***

(1) Assign each sample a random three-digit code.
(2) Pour at least 30 g of the sample (suggested for a 150 mL glass, one for each assessor) into the container.
(3) Cover samples with a suitable lid (e.g., petri dish, aluminum foil, cling film).
(4) Use minimal manipulation during transferring of samples to ensure that the subsamples look the same.
(5) To guarantee anonymity, the technician preparing the samples should not perform the testing.
(6) Maintain the sample at room temperature, preferably between 20 and 25 °C.
(7) The evaluation should ideally occur within 24 h after sample preparation.

### *7.3. Tasting procedures*

*7.3.1. Guidelines*

Each assessor will be given instructions and materials, including evaluation forms and computers, necessary to conduct the sensory analysis. The assessors must have at their disposal neutral spoons (without any odor and/or taste, metallic material should not be used) to stir the honey during the olfactory assessment and to take the product to be tasted. Water and a juicy, slightly acidic apple are suggested to cleanse the mouth between tastings. Also rose-hip tea or low-salt bread may be used. The assessors must follow simple but important rules (ISO 6658, 2017; Piana et al., 2004), see Table 22.

***7.3.1.1. Session protocol.***

(1) Limit the number of samples according to the type of honey to prevent oversaturation and fatigue of the senses.
(2) The evaluation should occur when sensory sensitivity is at its optimum, at least 2 h after a meal.
(3) Rest at least 30 min between sessions.
(4) Randomly present the order of the samples to each assessor.

*7.3.2. Evaluation of the visual characteristics*

(1) The evaluation is carried out under day light in a well illuminated room or using an artificial illumination as lights with a correlated color temperature of 6 500 °K.
(2) Evaluate the visual cleanliness and quality of the crystallization of a honey in its original container.
(3) Evaluate all other characteristics by assessing the sample in a separate glass.

*7.3.3. Evaluation of the olfactory characteristics*

Evaluate olfactory characteristics before tasting. Avoid receptor desensitization by taking short sniffs of the sample.

(1) Spread the honey around the inside of the glass using the spoon. This increases the surface area and releases the volatiles.
(2) Take first sniff to evaluate odor immediately.
(3) Wait 15–20 s before taking a second sniff, so that the full intensity of the odor is perceived.
(4) Leave part of crystallized honey untouched for tactile evaluation.

*7.3.4. Evaluation of the olfactory-gustatory and tactile characteristics*

(1) Bring to the mouth 1/2 g of honey with a spoon.
(2) Allow the honey sample to dissolve slowly in the mouth before swallowing.
(3) Evaluate olfactory-gustatory characteristics immediately.
(4) Evaluate the taste (sweet, salty, sour, and bitter).
(5) Evaluate the retro-nasal odor: aroma intensity and quality, persistence, after-taste.
(6) Evaluate other mouth sensations (viscosity, texture, type of crystals).
(7) Concentrate on the chemical aspects of the mouth sensations apart from the tactile characteristics.
(8) Wait 1 or 2 min before taking a second taste, longer in the case of persistent honeys. Eat a slice of apple or bread or cleanse the mouth with water or rose-hip tea if necessary.
(9) Evaluate tactile characteristics simultaneously. We encourage doing so in a second taste. Press the sample between tongue and palate in order to perceive and assess the viscosity, texture, and crystals characteristics.
(10) Wait a few minutes between samples, eat a slice of apple or bread or cleanse the mouth with water or rose-hip tea.

### *7.4. Descriptive semiquantitative analysis*

All numbered lists correspond to a scale, while bulleted lists correspond to a nonexhaustive list of descriptors. An example of descriptive card is presented in Figure 11.

*7.4.1. Appearance and visual descriptors*

The principal visual descriptors for honey are:

(1) Physical state: liquid or crystallized.
(2) Color intensity: 1. very light; 2. light; 3. medium; 4. dark; 5. very dark.
(3) Color tone: normal honey color; bright yellow; grayish; reddish; with orange tone; fluorescent green; dull; bright.
(4) Presentation: depending on the aim of the evaluation (e.g., competition or checking botanical declaration etc.), the importance of the assessment of some or all of these elements may vary. The following characteristics can be assessed: clarity (for liquid honey); impurities; air bubbles or foam.
(5) Defects can be assessed as marbling or white spots, incomplete crystallization, phase separation.

Viscosity, cohesion and type of crystals are observed in both visual and tactile evaluation but assessed only in the latter (see Section "Texture and other tactile descriptors" for description).

*7.4.2. Olfactory descriptors*

(1) The intensity of the odor may be defined as: 0—absent; 1—weak; 2—medium; 3—strong.
(2) The description of the odors may be classified in families and subfamilies. These are shown in Table 23. The examples used in the table to describe the families should be considered useful but not exhaustive.

## HONEY DESCRIPTION CARD

**SAMPLE CODE:** ..........................................

**VISUAL ASSESSMENT**

Physical aspect ...................................... Colour intensity 1 ☐ 2 ☐ 3 ☐ 4 ☐ 5 ☐

Colour tone ......................................

**OLFACTORY ASSESSMENT:**

**Intensity of odour:**

absent ☐ weak ☐ medium ☐ strong ☐

| Family | | Sub-family | |
|---|---|---|---|
| floral | ☐ | subtle | ☐ |
| | | heady | ☐ |
| fruity | ☐ | fresh fruit | ☐ |
| | | tropical fruit | ☐ |
| | | syrup | ☐ |
| | | transformed fruit | ☐ |
| | | vine | ☐ |
| warm | ☐ | subtle | ☐ |
| | | lactic | ☐ |
| | | caramelized | ☐ |
| | | roasted | ☐ |
| | | malted | ☐ |
| | | burned | ☐ |
| aromatic | ☐ | spicy | ☐ |
| | | resinous | ☐ |
| | | woody | ☐ |
| | | camphor | ☐ |
| | | citrus peel | ☐ |
| | | bitter almond | ☐ |
| chemical | ☐ | phenolic | ☐ |
| | | medicine | ☐ |
| | | household soap | ☐ |
| | | petrochemical | ☐ |
| | | smoky | ☐ |
| | | vinegar | ☐ |
| | | ammonia | ☐ |
| vegetal | ☐ | green | ☐ |
| | | damp | ☐ |
| | | dry | ☐ |
| animal | ☐ | sulphur | ☐ |
| | | protein | ☐ |
| | | valerianic | ☐ |
| | | cat's urine | ☐ |

OTHER: ..........................................

**AROMA ASSESSMENT:**

**Intensity of aroma:**

absent ☐ weak ☐ medium ☐ strong ☐

| Family | | Sub-family | |
|---|---|---|---|
| floral | ☐ | subtle | ☐ |
| | | heady | ☐ |
| fruity | ☐ | fresh fruit | ☐ |
| | | tropical fruit | ☐ |
| | | syrup | ☐ |
| | | transformed fruit | ☐ |
| | | vine | ☐ |
| warm | ☐ | subtle | ☐ |
| | | lactic | ☐ |
| | | caramelized | ☐ |
| | | roasted | ☐ |
| | | malted | ☐ |
| | | burned | ☐ |
| aromatic | ☐ | spicy | ☐ |
| | | resinous | ☐ |
| | | woody | ☐ |
| | | camphor | ☐ |
| | | citrus peel | ☐ |
| | | bitter almond | ☐ |
| chemical | ☐ | phenolic | ☐ |
| | | medicine | ☐ |
| | | household soap | ☐ |
| | | petrochemical | ☐ |
| | | smoky | ☐ |
| | | vinegar | ☐ |
| | | ammonia | ☐ |
| vegetal | ☐ | green | ☐ |
| | | damp | ☐ |
| | | dry | ☐ |
| animal | ☐ | sulphur | ☐ |
| | | protein | ☐ |
| | | valerianic | ☐ |
| | | cat's urine | ☐ |

OTHER: ..........................................

**FLAVOUR ASSESSMENT:**

**TASTE:**

**Sweetness:**

absent ☐ weak ☐ medium ☐ strong ☐

**Sourness:**

absent ☐ weak ☐ medium ☐ strong ☐

**Saltiness:**

absent ☐ weak ☐ medium ☐ strong ☐

Bitterness:

absent ☐ weak ☐ medium ☐ strong ☐

**Aroma/aftertaste persistence:**

absent ☐ short ☐ medium ☐ long ☐

**Aftertaste:** absent ☐ present ☐

**Aftertaste description** :..........................................

**Astringency:** absent ☐ present ☐

**Piquant:** absent ☐ present ☐

**Refreshing:** absent ☐ present ☐

**TEXTURE ASSESSMENT**

Consistency/viscosity: ..........................................

..........................................

Crystals: ..........................................

**NOTES (and possible defects)**

..........................................

..........................................

Figure 11. Example of descriptive card.

Table 23. Attributes for honey olfactory description.

| | Attribute | Definition | Reference |
|---|---|---|---|
| 1 | Floral | Odor reminiscent of flowers | orange flower (distilled) water, $\beta$-ionone (subfamily "*fine*")<br>rose, ylang ylang, phenylacetaldehyde (subfamily "*heady*") |
| 2 | Fruity | Odor reminiscent of fruit | fruit aroma (strawberry, pear, apple, raspberry) (subfamily "*fresh fruit*")<br>isoamyl acetate, $\gamma$-decalactone, $\delta$-decalactone (subfamily "*tropical fruit*")<br>grape sugar (concentrated rectified must) (subfamily "*syrup*")<br>apricot fruit juice, peeled tomatoes (subfamily "*processed fruit*")<br>white wine (subfamily "*fermented fruit*") |
| 3 | Warm | Odor reminiscent of food characterized by sweet taste | vanillin, benzoin, beeswax (subfamily"*fine*")<br>concentrated sugared milk, diacetyl (subfamily "*lactic*")<br>cane sugar, furanone (subfamily "*candied*")<br>toasted hazel nuts, toasted almonds, toasted peanuts (subfamily "*toasted*")<br>barley malt syrup, rice syrup (subfamily "*malted*")<br>toasted barley, toasted bread (charred part) (subfamily "*burnt*") |
| 4 | Aromatic | Odor reminiscent of something fresh, balsamic, cosmetic, not necessarily related to food products | cloves, nutmeg (subfamily "*spicy*")<br>mugo pine, incense (subfamily "*resinous*")<br>oak musk (subfamily "*woody*")<br>peppermint, anise, eucalyptus (subfamily "*balsamic*")<br>lemon, sweet orange (subfamily "*citrus fruit*")<br>coumarin (subfamily "*bitter almond*") |
| 5 | Chemical | Odor generally not related to food products, characterized by a certain degree of aggressiveness | phenol, vegetal tar (subfamily "*phenolic*")<br>castile soap (not perfumed) (subfamily "*soap*")<br>jute smoke, cigarette butt extract (subfamily "*smoky*")<br>white wine vinegar (subfamily "*vinegar*")<br>ammonia (subfamily "*ammonia*") |
| 6 | Vegetal | Odor reminiscent of non-aromatic plants | fresh raw beans, cut grass, fresh ivy leaves (subfamily "*green*")<br>raw champignon mushrooms, lightly boiled spinach (subfamily "*damp*")<br>green tea, hay (subfamily "*dry*") |
| 7 | Animal | Odor reminiscent of animal world or decomposition | hardboiled egg, dimethyl sulphide, methionine (subfamily "*sulfurized*")<br>dried ceps mushrooms, stock cube (subfamily "*proteic*")<br>isovalerianic acid, butyric acid, hexanoic acid (subfamily "*valerianic*")<br>cassis (subfamily "*cassis*") |

Table 24. Attributes for honey taste description.

| | Attribute | Definition | Reference |
|---|---|---|---|
| 1 | Sweetness | Basic taste produced by dilute aqueous solutions of natural or artificial substances such as sucrose or aspartame | Sucrose in water 12 g/L |
| 2 | Sourness | Basic taste produced by dilute aqueous solutions of mostly acid substances (e.g., citric acid and tartaric acid) | Citric acid in water 0.6 g/L |
| 3 | Saltiness | Basic taste produced by dilute aqueous solutions of various substances such as sodium chloride | Sodium chloride in water 2 g/L |
| 4 | Bitterness | Basic taste produced by dilute aqueous solutions of various substances such as quinine or caffeine | Caffeine in water 0.27 g/L |

#### *7.4.3. Flavor descriptors*

Flavor is the "complex combination of the olfactory, gustatory and trigeminal sensations perceived during tasting. It may be influenced by tactile, thermal, painful and/or kinaesthesic effects" (ISO 5492, 2008).

(1) The intensity of the taste sensation sweetness, sourness, salty, and bitterness (Table 24) is quantified according to the scale: 0. absent; 1. weak; 2. medium; 3. strong. The taste of umami is not considered; if perceived, it is reported as present.

(2) The odor perceived via retro-nasal pathways (aroma) is described as reported in olfactory description.

(3) Other sensations we can perceive in the mouth include: (i) piquant: an aggressive and burning sensation perceived by the mucous membranes and/or in the throat; (ii) metallic: a feeling similar to a slight tingling sensation due to the presence of traces of iron (or perhaps other metals) in ionic form in some honeys. In honey, it indicates a defect.

(4) Persistence can be evaluated. Persistence refers to the duration of the aroma sensations after swallowing. Less than 30 s is considered "short" persistence, and longer than 5 min, "long" persistence. Persistence is evaluated according to the scale: 0. absent; 1. short; 2. medium; 3. long.

(5) Aftertaste indicates the presence of aroma sensations in the mouth that remain after swallowing the product and are different from those initially perceived. The quality of the sensations can be described with the same terminology used to explain the olfactory sensations (Table 23).

*7.4.4. Texture and other tactile descriptors*

Tactile characteristics refer to those perceived in the mouth, although the visual inspection and handling of the sample contribute to its assessment.

(1) Viscosity is evaluated in liquid honey according to the scale: 1. very fluid; 2. fluid; 3. normal viscosity; 4. viscous; 5. very viscous.
(2) Texture of the mass in crystallized honeys is evaluated and can be described as: 1. deliquescent; 2. soft; 3. pasty; 4. firm; 5. compact.
(3) Crystals can be differentiated according to their size: 0. not perceptible; 1. fine; 2. medium; 3. large.
(4) Crystals can be differentiated depending on the shape and can differ according to solubility, that is, to say the tendency to melt in the mouth more or less easily. The crystals can be described as: sharp, gritty, or roundish.

A further tactile descriptor for honey is astringency, a complex sensation due to substances such as the tannins of persimmon or unripe fruit, which precipitate the proteins in the saliva responsible for lubricating the mucous membranes in the mouth. The astringency is perceived as a lack of lubrication and dryness of mucous membranes; it is often associated with a bitter taste.

In finely crystallized honey, it is common to experience a refreshing sensation. This is not due to the temperature of the product, but it is linked to the solubility of some type of crystals (glucose).

## Acknowledgments

The following authors were responsible for the respective manuscript sections:

Ligia Bicudo de Almeida-Muradian, (Introduction); Ortrud Monika Barth and Alex da Silva de Freitas (Honey authenticity—botanical, geographical, and entomological origin); Ligia Bicudo de Almeida-Muradian, Maria Teresa Sancho, Ana Pascual-Maté, José Augusto Gasparotto Sattler (Honey physicochemical analysis for nutritional, chemical composition, and antioxidant activity); Michael Eyer and Vincent Dietemann (Investigating honey production with diagnostic radioentomology); Anne-Claire Martel (Pesticide residues in honey); Wim Reybroeck (Antibiotics and chemotherapeutics in honey); Gian Luigi Marcazzan, Carla Mucignat, Carla Marina Marchese (Sensory analysis). The authors thanks Lucia Piana for participation in the Sensory Analysis section. Ana Pascual-Maté and Maria Teresa Sancho-Ortiz are very grateful to Cipriano Ramos and Mónica Pascual for their valuable help. Ligia Bicudo de Almeida-Muradian thanks Stefan Bogdanov and Peter Gallmann for their technical support and also Adriane Alexandre Machado de Melo. We further thank Leila Aparecida Bonadio for helping with the reference management.

The COLOSS (Prevention of honey bee COlony LOSSes) Association aims to explain and prevent massive honey bee colony losses. It was funded through the COST Action FA0803. COST (European Cooperation in Science and Technology) is a unique means for European researchers to jointly develop their own ideas and new initiatives across all scientific disciplines through trans-European networking of nationally funded research activities. Based on a pan-European intergovernmental framework for cooperation in science and technology, COST has contributed since its creation more than 40 years ago to closing the gap between science, policy makers and society throughout Europe and beyond. COST is supported by the EU Seventh Framework Program for research, technological development, and demonstration activities (Official Journal L 412, 30 December 2006). The European Science Foundation as implementing agent of COST provides the COST Office through an EC Grant Agreement. The Council of the European Union provides the COST Secretariat. The COLOSS network is now supported by the Ricola Foundation—Nature & Culture.

We would also like to express our gratitude to Professor Jamie Ellis for his kind support all the way, and for providing helpful and constructive suggestions.

## Disclosure statement

No potential conflict of interest was reported by the author(s).

## ORCID

*Ligia Bicudo de Almeida-Muradian* http://orcid.org/0000-0001-5322-1358
*Ortrud Monika Barth* http://orcid.org/0000-0002-2418-8605
*Vincent Dietemann* http://orcid.org/0000-0002-0791-9939
*Michael Eyer* http://orcid.org/0000-0003-1047-1714
*Alex da Silva de Freitas* http://orcid.org/0000-0002-8665-7649
*Anne-Claire Martel* http://orcid.org/0000-0002-1179-4286
*Gian Luigi Marcazzan* http://orcid.org/0000-0002-1695-9713
*Carla Marina Marchese* http://orcid.org/0000-0002-7495-0688
*Carla Mucignat-Caretta* http://orcid.org/0000-0001-5307-0414
*Wim Reybroeck* http://orcid.org/0000-0002-2269-6534
*M. Teresa Sancho* http://orcid.org/0000-0002-9128-9422
*José Augusto Gasparotto Sattler* http://orcid.org/0000-0002-7760-0501

## References

Administration, U. S. F. & D. (2017). *Title 21 of the U.S. Code of Federal Regulations Part 556, Subpart B. Specific tolerances for residues of new animal drugs. Section 556.360 Lincomycin; Section 556.500 Oxytetracycline; Section 556.740 Tylosin.* http://www.ecfr.gov/cgi-bin/text-idx?SID=635d1951lf68344e66597151f2534107&mc=true&node=pt21.6.556&rgn=div5#sp21.6.556.b

Ahmed, M. (2013). The relationship between bioactive compounds with diastase activity and antibacterial synergy of honey and potato starch combinations against *Klebsiella pneumonia. Clinical Microbiology: Open Access*, 2(3), 2–5. doi:10.4172/2327-5073.1000131

Alfredsson, G., Branzell, C., Granelli, K., & Lundström, Å. (2005). Simple and rapid screening and confirmation of tetracyclines in honey and egg by a dipstick test and LC–MS/MS. *Analytica Chimica Acta*, *529*(1–2), 47–51. doi:10.1016/j.aca.2004.08.050

Almeida-Muradian, L. B., & Gallmann, P. (2012). Generalizability of PLS calibrations with FT-IR ATR spectrometry for the prediction of some physicochemical measurands of honey. *ALP Science*, *541*, 1–20.

Almeida-Muradian, L. B., Luginbühl, W., Badertscher, R., & Gallmann, P. (2012). Generalizability of PLS calibrations with FT-IR ATR spectrometry for the prediction of some physicochemical measurands of honey. *ALP Science, 441*, 1–20.

Almeida-Muradian, L. B., & Penteado, M. V. (2015). *Vigilância Sanitária. Tópicos sobre Legislação e Análise de Alimentos* (2nd ed.). Guanabara Koogan.

Almeida-Muradian, L. B., Sousa, R. J., Barth, O. M., & Gallmann, P. (2014). Preliminary data on brazilian monofloral honey from the northeast region using ft-ir atr spectroscopic, palynological, and color analysis. *Quimica Nova, 37*(4), 716–719. 10.5935/0100-4042.20140115

Almeida-Muradian, L. B., Stramm, K. M., & Estevinho, L. M. (2014). Efficiency of the FT-IR ATR spectrometry for the prediction of the physicochemical characteristics of Melipona subnitida honey and study of the temperature's effect on those properties. *International Journal of Food Science & Technology, 49*(1), 188–195. doi:10.1111/ijfs.12297

Almeida-Muradian, L. B., Stramm, K. M., Horita, A., Barth, O. M., Freitas, A. S., & Estevinho, L. M. (2013). Comparative study of the physicochemical and palynological characteristics of honey from Melipona subnitida and *Apis mellifera*. *International Journal of Food Science & Technology, 48*(8), 1698–1706. doi:10.1111/ijfs.12140

Almeida-Silva, M., Canha, N., Galinha, C., Dung, H. M., Freitas, M. C., & Sitoe, T. (2011). Trace elements in wild and orchard honeys. *Applied Radiation and Isotopes, 69*(11), 1592–1595. doi:10.1016/j.apradiso.2011.01.013

Alqarni, A. S., Owayss, A. A., Mahmoud, A. A., & Hannan, M. A. (2014). Mineral content and physical properties of local and imported honeys in Saudi Arabia. *Journal of Saudi Chemical Society, 18*(5), 618–625. doi:10.1016/j.jscs.2012.11.009

Amendola, G., Pelosi, P., & Dommarco, R. (2010). Solid-phase extraction for multi-residue analysis of pesticides in honey. *Journal of Environmental Science and Health, Part B, 46*(1), 24–34. doi:10.1080/03601234.2010.515170

Anjos, O., Campos, M. G., Ruiz, P. C., & Antunes, P. (2015). Application of FTIR-ATR spectroscopy to the quantification of sugar in honey. *Food Chemistry, 169*, 218–223. doi:10.1016/j.foodchem.2014.07.138

Anklam, E. (1998). A review of the analytical methods to determine the geographical and botanical origin of honey. *Food Chemistry, 63*(4), 549–562.(98)00057-0 doi:10.1016/S0308-8146(98)00057-0

Anonymous. (2007a). *CRLs view on state of the art analytical methods for national residue control plans.* CRL Guidance Paper of 7th December. https://ec.europa.eu/food/sites/food/files/safety/docs/cs_vet-med-residues_guideline_validation_screening_en.pdf

Anonymous. (2007b). Note of the Commission dated 24 July 2007 to the members and observers of the Veterinary Pharmaceutical Committee (GG). In *Summary record of the Standing Committee on the Food Chain and Animal Health held on 18 September in Brussels.* https://ec.europa.eu/food/sites/food/files/safety/docs/reg-com_biosec_sum_18092007_en.pdf

Anonymous. (2010). *Guidelines for the validation of screening methods for residues of veterinary medicines (initial validation and transfer).* https://ec.europa.eu/food/sites/food/files/safety/docs/cs_vet-med-residues_guideline_validation_screening_en.pdf

Anonymous. (2017). *Table 2 Substances or Group of substances (1) to be monitored for in the relevant commodity. E = "essential" HD = "highly desirable".* https://ec.europa.eu/food/sites/food/files/safety/docs/cs_vet-med-residues_animal-imports-non-eu_residue-monitoring-progs_required-table2.pdf

AOAC (Association of Official Agricultural Chemists). (1990). *AOAC official methods of analysis* (15th ed., Vol. 1, pp. 136–138).

AOAC. (1992). AOAC 969.38B, MAFF validated method V21 for moisture in honey. *Journal of the Association of Public Analysts, 28*, 183–187.

Aronne, G., & DE Micco, V. (2010). Traditional melissopalynology integrated by multivariate analysis and sampling methods to improve botanical and geographical characterisation of honeys. *Plant Biosystems - an International Journal Dealing with All Aspects of Plant Biology, 144*(4), 833–840. doi:10.1080/11263504.2010.514125

Azzouz, A., & Ballesteros, E. (2015). Multiresidue method for the determination of pharmacologically active substances in egg and honey using a continuous solid-phase extraction system and gas chromatography-mass spectrometry. *Food Chemistry, 178*, 63–69. doi:10.1016/j.foodchem.2015.01.044

Bargańska, Z., & Namieśnik, J. (2010). Pesticide analysis of bee and bee product samples. *Critical Reviews in Analytical Chemistry, 40*(3), 159–171. doi:10.1080/10408347.2010.490484

Bargańska, Z., Namieśnik, J., & Ślebioda, M. (2011). Determination of antibiotic residues in honey. *TrAC Trends in Analytical Chemistry, 30*(7), 1035–1041. doi:10.1016/j.trac.2011.02.014

Barth, O. M. (1989). *O pólen no mel brasileiro.* Online. http://scholar.google.com/scholar?hl=en&btnG=Search&q=intitle:O+Pólen+no+Mel+Brasileiro#0

Barth, O. M. (1971). Mikroskopische Bestandteile brasilianischer Honigtau-Honigen. *Apidologie, 2*(2), 157–167. doi:10.1051/apido:19710202

Barth, O. M., Freitas, A. S., Sousa, G. L., & Almeida-Muradian, L. B. (2013). Pollen and physicochemical analysis of Apis and Tetragonisca (Apidae) honey. *Interciencia, 38*(4), 280–285. http://www.redalyc.org/html/339/33926985012/

Behm, F., von der Ohe, K., & Henrich, W. (1996). Zuverlässigkeit der Pollenanalyse von Honig. *Deutsche Lebensmittel-Rundschau, 192*, 183–188.

Benetti, C., Dainese, N., Biancotto, G., Piro, R., & Mutinelli, F. (2004). Unauthorised antibiotic treatments in beekeeping: Development and validation of a method to quantify and confirm tylosin residues in honey using liquid chromatography-tandem mass spectrometric detection. *Analytica Chimica Acta, 520*(1–2), 87–92. doi:10.1016/j.aca.2004.04.070

Benetti, C., Piro, R., Binato, G., Angeletti, R., & Biancotto, G. (2006). Simultaneous determination of lincomycin and five macrolide antibiotic residues in honey by liquid chromatography coupled to electrospray ionization mass spectrometry (LC-MS/MS). *Food Additives and Contaminants, 23*(11), 1099–1108. doi:10.1080/02652030600699338

Bilandžić, N., Tlak Gajger, I., Kosanović, M., Čalopek, B., Sedak, M., Solomun Kolanović, B., Varenina, I., Luburić, Đ. B., Varga, I., & Đokić, M. (2017). Essential and toxic element concentrations in monofloral honeys from southern Croatia. *Food Chemistry, 234*, 245–253. doi:10.1016/j.foodchem.2017.04.180

Blasco, C., Fernández, M., Picó, Y., & Font, G. (2004). Comparison of solid-phase microextraction and stir bar sorptive extraction for determining six organophosphorus insecticides in honey by liquid chromatography-mass spectrometry. *Journal of Chromatography A, 1030*(1–2), 77–85. doi:10.1016/j.chroma.2003.11.037

Blasco, C., Lino, C. M., Picó, Y., Pena, A., Font, G., & Silveira, M. I. N. (2004). Determination of organochlorine pesticide residues in honey from the central zone of Portugal and the Valencian community of Spain. *Journal of Chromatography A, 1049*(1–2), 155–160. doi:10.1016/j.chroma.2004.07.049

Blasco, C., Picó, Y., & Torres, C. M. (2007). Progress in analysis of residual antibacterials in food. *TrAC Trends in Analytical Chemistry, 26*(9), 895–913. doi:10.1016/j.trac.2007.08.001

Blasco, C., Vazquez-Roig, P., Onghena, M., Masia, A., & Picó, Y. (2011). Analysis of insecticides in honey by liquid chromatography-ion trap-mass spectrometry: Comparison of different extraction procedures. *Journal of Chromatography A, 1218*(30), 4892–4901. doi:10.1016/j.chroma.2011.02.045

Bogdanov, S. (2009). Harmonised methods of the international honey commission. International Honey Commission. Retrieved from http://www.ihc-platform.net/ihcmethods2009.pdf

Bogdanov, S. (2011). The honey book. http://www.bee-hexagon.net/honey/

Bogdanov, S., & Martin, P. (2002). Honey authenticity: A review. *Mitteilungen Aus Dem Gebiete Der Lebensmitteluntersuchung Und Hygiene*, *6*, 1–20. http://www.bee-hexagon.net/files/file/fileE/Honey/AuthenticityRevue_Internet.pdf

Bogdanov, S., Haldimann, M., Luginbühl, W., & Gallmann, P. (2007). Minerals in honey: Environmental, geographical and botanical aspects. *Journal of Apicultural Research*, *46*(4), 269–275. doi:10.1080/00218839.2007.11101407

Bogdanov, S., Martin, P., & Lüllmann, C. (1997). Harmonized methods of the European Honey Commission. *Apidologie Extra Issue*, 1–59.

Bougrini, M., Florea, A., Cristea, C., Sandulescu, R., Vocanson, F., Errachid, A., Bouchikhi, B., El Bari, N., & Jaffrezic-Renault, N. (2016). Development of a novel sensitive molecularly imprinted polymer sensor based on electropolymerization of a microporous-metal-organic framework for tetracycline detection in honey. *Food Control*, *59*, 424–429. https://doi.org/https://doi.org/10.1016/j.foodcont.2015.06.002 doi:10.1016/j.foodcont.2015.06.002

Bryant, V. M., & Jones, G. D. (2001). The R-values of honey: Pollen coefficients. *Palynology*, *25*(1), 11–28. doi:10.1080/01916122.2001.9989554

Caldow, M., Stead, S. L., Day, J., Sharman, M., Situ, C., & Elliott, C. (2005). Development and validation of an optical SPR biosensor assay for tylosin residues in honey. *Journal of Agricultural and Food Chemistry*, *53*(19), 7367–7370. doi:10.1021/jf050725s

Cano, C. B., Felsner, M. L., Bruns, R. E., Matos, J. R., & Almeida-Muradian, L. B. (2006). Optimization of mobile phase for separation of carbohydrates in honey by high performance liquid chromatography using a mixture design. *Journal of the Brazilian Chemical Society*, *17*(3), 588–593. doi:10.1590/S0103-50532006000300024

Carrasco-Pancorbo, A., Casado-Terrones, S., Segura-Carretero, A., & Fernández-Gutiérrez, A. (2008). Reversed-phase high-performance liquid chromatography coupled to ultraviolet and electrospray time-of-flight mass spectrometry on-line detection for the separation of eight tetracyclines in honey samples. *Journal of Chromatography A*, *1195*(1–2), 107–116. doi:10.1016/j.chroma.2008.05.003

Carreck, N. L., Andree, M., Brent, C. S., Cox-Foster, D., Dade, H. A., Ellis, J. D., Hatjina, F., & vanEngelsdorp, D. (2013). Standard methods for *Apis mellifera* anatomy and dissection. *Journal of Apicultural Research*, 52(4), 1–39.

Chabottaux, V., Bonhomme, C., Muriano, A., Stead, S., Diserens, J. M., Marco, M. P., & Granier, B. (2010). Effect of acidic hydrolysis on sulfamides detection in honey with new generic antibody-based dipstick assay. In Abstract book of the 6th International Symposium on Hormone and Veterinary Drug Residue Analysis (Vol. 63, pp. 1–4). http://www.conffidence.eu/wp-content/uploads/2014/08/enews_2_VDRA_abstract_CHABOTTAUX_SULFASENSOR.pdf

Chua, L. S., Lee, J. Y., & Chan, G. F. (2013). Honey protein extraction and determination by mass spectrometry. *Analytical and Bioanalytical Chemistry*, *405*(10), 3063–3074. doi:10.1007/s00216-012-6630-2

Ciulu, M., Solinas, S., Floris, I., Panzanelli, A., Pilo, M. I., Piu, P. C., Spano, N., & Sanna, G. (2011). RP-HPLC determination of water-soluble vitamins in honey. *Talanta*, *83*(3), 924–929. doi:10.1016/j.talanta.2010.10.059

Codex. (2001). Codex Alimentarius standard for honey 12-1981. Revised Codex standard for honey. Standards and standard methods (Vol. 11). Retrieved April, 2020, from http://www.codexalimentarius.net

Codex Alimentarius Commission. (1969). Recommended European regional standard for honey (CAC/RS 12-1969).

Codex Standard for Honey. (2001). *CODEX STAN 12-1981* (pp. 1–8).

Commission, E. (1998). Commission Decision 98/179/EC of 23 February 1998 laying down detailed rules on official sampling for the monitoring of certain substances and residues thereof in live animals and animal products. *Official Journal of the European Communities*, *L65*, 31–34. http://eur-lex.europa.eu/legal-content/EN/TXT/PDF/?uri=CELEX:31998D0179&from=EN

Commission, E. (2005). Commission Decision 2005/34/EC of 11 January 2005 laying down harmonized standards for the testing for certain residues in products of animal origin imported from third countries. *Official Journal of the European Union*, *L16*, 61–63. http://eur-lex.europa.eu/legal-content/EN/TXT/PDF/?uri=CELEX:32005D0034&from=EN

Commission, E. (2010). Commission Regulation (EU) No 37/2010 of 22 December 2009 on pharmacologically active substances and their classification regarding maximum residue limits in foodstuffs of animal origin. *Official Journal of the European Union*, *L15*, 1–72. https://ec.europa.eu/health/sites/health/files/files/eudralex/vol-5/reg_2010_37/reg_2010_37_en.pdf

Commission, E. (2018). Commission Implementing Regulation (EU) 2018/470 of 21 March 2018 on detailed rules on the maximum residue limit to be considered for control purposes for foodstuffs derived from animals which have been treated in the EU under Article 11 of Directive 2001/82/EC. *Official Journal of the European Union*, *L79*, 16–18. https://eur-lex.europa.eu/legal-content/EN/TXT/?qid=1561736258606&uri=CELEX:32018R0470.

Consonni, R., & Cagliani, L. R. (2008). Geographical characterization of polyfloral and Acacia honeys by nuclear magnetic resonance and chemometrics. *Journal of Agricultural and Food Chemistry*, *56*(16), 6873–6880. doi:10.1021/jf801332r

Council Directive. (1996). Council Directive 96/23/EC of 29 April 1996 on measures to monitor certain substances and residues thereof in live animals and animal products and repealing Directives 85/358/EEC and 86/469/EEC and Decisions 89/187/EEC and 91/664/EEC. *Official Journal of the European Communities*, *L125*, 10–32. http://eur-lex.europa.eu/legal-content/EN/TXT/PDF/?uri=CELEX:31996L0023&from=EN

da Silva, P. M., Gauche, C., Gonzaga, L. V., Costa, A. C. O., & Fett, R. (2016). Honey: Chemical composition, stability and authenticity. *Food Chemistry*, *196*, 309–323. doi:10.1016/j.foodchem.2015.09.051

Daeseleire, E., & Reybroeck, W. (2012). Development and validation of a liquid chromatographic-tandem mass spectrometric method for the detection of fumagillin in honey: Use in a stability study. In *I International Symposium on Bee Products* (pp. 9–12). Bragança, Portugal.

Daniela, I., Morariu, P., Schiriac, E., Matiut, S., & Cuciureanu, R. (2012). Method validation for simultaneous determination of 12 sulfonamides in honey using biochip array technology. *Farmacia*, *60*, 143–154. https://pdfs.semanticscholar.org/96c0/944ef59d8e1a03fc7c966532a528bb232c43.pdf

Davies, A. M. C. (1975). Amino acid analysis of honeys from eleven countries. *Journal of Apicultural Research*, *14*(1), 29–39. doi:10.1080/00218839.1975.11099798

De Brabander, H. F., Noppe, H., Verheyden, K., Vanden Bussche, J., Wille, K., Okerman, L., Vanhaecke, L., Reybroeck, W., Ooghe, S., & Croubels, S. (2009). Residue analysis: Future trends from a historical perspective. *Journal of Chromatography A*, *1216*(46), 7964–7976. doi:10.1016/j.chroma.2009.02.027

de Graaf D., Alippi A., Antúnez K., Aronstein K., Budge G., De Koker D., De Smet L., Dingman D., Evans J., Foster L., Fünfhaus A., Garcia-Gonzalez E., Gregore A., Human H., Murray K., Nguyen B., Poppinga L., Spivak M., van

Engelsdorp D., Wilkins S. & Genersch E. (2013). Standard methods for American foulbrood research, *Journal of Apicultural Research*, 52(1), 1–28. https://doi.org/10.3896/IBRA.1.52.1.11

De-Melo, A. A. M., Almeida-Muradian, L. B., Sancho, M. T., & Pascual-Maté, A. (2017). Composition and properties of *Apis mellifera* honey: A review. *Journal of Apicultural Research, 57*(1), 5–37. doi:10.1080/00218839.2017.1338444

De-Melo, A. A. M., Almeida-Muradian, L. B., Sancho, M. T., & Pascual-Maté, A. (2017). Composition and properties of *Apis mellifera* honey: A review. *Journal of Apicultural Research, 8839*(June), 1–33. doi:10.1080/00218839.2017.1338444

Dietemann, V., Nazzi, F. J., Martin, S., L., Anderson, D., Locke, B., S., Delaplane, K., Wauquiez, Q., Tannahill, C., Frey, E., Ziegelmann, B., Rosenkranz, P., D., & Ellis, J. (2013). Standard methods for varroa research. In *V. Dietemann, J. D. Ellis & P. Neumann (Eds), The COLOSS* BEEBOOK, *Volume II: standard methods for* Apis mellifera *pest and pathogen research. Journal of Apicultural Research, 52(1).* doi:10.3896/IBRA.1.52.1.09

DOCUMENT SANTE/11813/2017. (2018). *Guidance document on analytical quality control and method validation procedures for pesticides residues analysis in food and feed European Commission.*

DOCUMENT SANTE/11945/2015. (2015). *Guidance document on analytical quality control and method validation procedures for pesticides residues analysis in food and feed European Commission.*

Dominguez, M. A., Jacksén, J., Emmer, Å., & Centurión, M. E. (2016). Capillary electrophoresis method for the simultaneous determination of carbohydrates and proline in honey samples. *Microchemical Journal, 129*, 1–4. doi:10.1016/j.microc.2016.05.017

Dubreil-Chéneau, E., Pirotais, Y., Verdon, E., & Hurtaud-Pessel, D. (2014). Confirmation of 13 sulfonamides in honey by liquid chromatography-tandem mass spectrometry for monitoring plans: Validation according to European Union Decision 2002/657/EC. *Journal of Chromatography A, 1339*(37), 128–136. doi:10.1016/j.chroma.2014.03.003

Edder, P., Cominoli, A., & Corvi, C. (1998). Dosage de résidus de streptomycine dans le miel par HPLC avec postdérivatisation et détection fluorimétrique. *Mitteilungen Aus Dem Gebiete Der Lebensmitteluntersuchung Und Hygiene, 89*(3), 369–382.

EL Alami EL Hassani, N., Tahri, K., Llobet, E., Bouchikhi, B., Errachid, A., Zine, N., & EL Bari, N. (2018). Emerging approach for analytical characterization and geographical classification of Moroccan and French honeys by means of a voltammetric electronic tongue. *Food Chemistry, 243*, 36–42. doi:10.1016/j.foodchem.2017.09.067

EL Sohaimy, S. A., Masry, S. H. D., & Shehata, M. G. (2015). Physicochemical characteristics of honey from different origins. *Annals of Agricultural Sciences, 60*(2), 279–287. doi:10.1016/j.aoas.2015.10.015

Elizabeta, D. S., Gordana, A., Zehra, H. M., Biljana, S. D., Risto, U., Aleksandra, T., & Goran, S. (2012). In-house validation and quality control of commercial enzyme-linked immunosorbnet assays for screening of nitrofuran metabolites in food of animal origin. *Macedonian Veterinary Review, 35*(1), 13–21. https://www.researchgate.net/profile/Risto_Uzunov/publication/260184219_In-house_validation_and_quality_control_of_commercial_enzyme-linked_immunosorbnet_assays_for_screening_of_nitrofuran_metabolites_in_food_of_animal_origin/links/0a85e52fea989dc8bc00000

Erdtman, G. (1952). *Pollen morphology and plant taxonomy - Angiosperms.* Almqvist&Wiksell.

EU PESTICIDES DATABASE. http://ec.europa.eu/food/plant/pesticides/eu-pesticides-database/public/?event=product.resultat&language=EN&selectedID=375

European Commission. (2001). Council Directive 2001/110/EC of 20 December 2001 relating to honey. Retrieved from http://www.ihc-platform.net/honeydirective2001.pdf

European Communities Commission. (1997). Commission Decision 97/747/EC of 27 October 1997 fixing the levels and frequencies of sampling provided for by Council Directive 96/23/EC for the monitoring of certain substances and residues thereof in certain animal products. *Official Journal of the European Communities, L303*, 12–15. http://eur-lex.europa.eu/legal-content/EN/TXT/PDF/?uri=CELEX:31997D0747&from=EN

European Communities Commission. (2003). Commission Decision 2003/181/EC of 13 March 2003 as regards the setting of minimum required performance limits (MRPLs) for certain residues in food of animal origin. *Official Journal of the European Union, L71*, 17–18. http://eur-lex.europa.eu/legal-content/EN/TXT/PDF/?uri=CELEX:32003D0181&from=EN

European Union Commission. (2002). Commission Decision 2002/657/EC of 12 August 2002 implementing Council Directive 96/23/EC concerning the performance of analytical methods and the interpretation of results. *Official Journal of the European Communities*, L221, 8–36. http://eur-lex.europa.eu/legal-content/EN/TXT/PDF/?uri=CELEX:32002D0657&from=EN

Eyer, M., Greco, M. K., Lang, J., Neumann, P., & Dietemann, V. (2016). No spatial patterns for early nectar storage in honey bee colonies. *Insectes Sociaux, 63*(1), 51–59. doi:10.1007/s00040-015-0432-4

Eyer, M., Neumann, P., & Dietemann, V. (2016). A look into the cell: Honey storage in honey bees, *Apis mellifera. PLoS One, 11*(8), e0161059–20. doi:10.1371/journal.pone.0161059

Fallico, B., Arena, E., & Zappala, M. (2009). Prediction of honey shelf life. *Journal of Food Quality, 32*(3), 352–368. doi:10.1111/j.1745-4557.2009.00253.x

Ferreira, I. C. F. R., Aires, E., Barreira, J. C. M., & Estevinho, L. M. (2009). Antioxidant activity of Portuguese honey samples: Different contributions of the entire honey and phenolic extrac. *Food Chemistry, 114*(4), 1438–1443. doi:10.1016/j.foodchem.2008.11.028

Fischer, W. M., & Palmer, J. D. (2005). Evidence from small-subunit ribosomal RNA sequences for a fungal origin of Microsporidia. *Molecular Phylogenetics and Evolution, 36*(3), 606–622. doi:10.1016/j.ympev.2005.03.031

Food Research Center. (2017). *Tabela Brasileira de Composição de Alimentos (TBCA).*

Forsgren, E., Budge, G., Charrière, J.D., & Hornitzky, M. (2013). Standard methods for European foulbrood research, *Journal of Apicultural Research*, 52(1), 1–14. https://doi.org/10.3896/IBRA.1.52.1.12

Fries, I., Chauzat, M.P., Chen, Y.P., Doublet, V., Genersch, E., Gisder, S., Higes, M., McMahon, D., Martín-Hernández, R., Natsopoulou, M., Paxton, R., Tanner, G., Webster, T., & Williams G. (2013). Standard methods for Nosema research, *Journal of Apicultural Research*, 52(1), 1–28, https://doi.org/10.3896/IBRA.1.52.1.14l

Galarini, R., Saluti, G., Giusepponi, D., Rossi, R., & Moretti, S. (2015). Multiclass determination of 27 antibiotics in honey. *Food Control, 48*, 12–24. doi:10.1016/j.foodcont.2014.03.048

Gaudin, V., DE Courville, A., Hedou, C., Rault, A., Diomandé, S. E., Creff-Froger, C., & Verdon, E. (2013). Evaluation and validation of two microbiological tests for screening antibiotic residues in honey according to the European guideline for the validation of screening methods. *Food Additives & Contaminants: Part A, 30*(2), 234–243. doi:10.1080/19440049.2012.738367

Gaudin, V., Hedou, C., & Verdon, E. (2013). Validation of two ELISA kits for the screening of tylosin and streptomycin in honey according to the European decision 2002/657/EC. *Food Additives & Contaminants: Part A, 30*(1), 93–109. doi:10.1080/19440049.2012.722696

Gaudin, V., Hedou, C., Soumet, C., & Verdon, E. (2014). Evaluation and validation of biochip multi-array technology for the screening of six families of antibiotics in honey according to the European guideline for the validation of screening methods for residues of veterinary medicines. *Food Additives & Contaminants: Part A*, *31*(10), 1699–1711. doi:10.1080/19440049.2014.952784

Gaudin, V., Rault, A., & Verdon, E. (2012). Validation of a commercial receptor kit Sulfasensor® Honey for the screening of sulfonamides in honey according to Commission Decision 2002/657/EC. *Food Additives & Contaminants: Part A*, *29*(6), 942–950. doi:10.1080/19440049.2012.668718

Gaudin, V., Rault, A., & Verdon, E. (2013). Validation of a commercial receptor kit for tetracycline residues in honey according to the European guideline for screening methods. *Food and Agricultural Immunology*, *24*(1), 111–128. doi:10.1080/09540105.2011.651446

Gomes, S., Dias, L. G., Moreira, L. L., Rodrigues, P., & Estevinho, L. (2010). Physicochemical, microbiological and antimicrobial properties of commercial honeys from Portugal. *Food and Chemical Toxicology*, *48*(2), 544–548. doi:10.1016/j.fct.2009.11.029

Gómez-Díaz, D., Navaza, J. M., & Quintáns-Riveiro, L. C. (2012). Physicochemical characterization of Galician honeys. *International Journal of Food Properties*, *15*(2), 292–300. doi:10.1080/10942912.2010.483616

Gonnet, M., & Vache, G. (1979). Technique de dégustation des miels et recherche d'un système de notation et de classification objectif pour apprécier leur qualité par l'analyse sensorielle. In *27th Apimondia International Apicultural Congress* (pp. 499–506). Athènes.

Gonnet, M., & Vache, G. (1992). *The taste of honey*. Apimondia.

Gonnet, M., Vache, G., & Jean-Prost, P. (1985). *Le goût du miel*. UNAF.

Gordon, L. (1989). Quantitative determination of oxytetracycline in honey by cylinder plate microbioassay. *Australian Journal of Agricultural Research*, *40*(4), 933–940. doi:10.1071/AR9890933

Granato, D., & Nunes, D. S. (2016). *Análises químicas, propriedades funcionais e controle de qualidade de alimentos e bebidas: uma abordagem teórico-prática*. Elsevier Brasil.

Greco, M. K., Lang, J., Gallmann, P., Priest, N., Feil, E., & Crailsheim, K. (2013). Sugar concentration influences decision making in *Apis mellifera* L. workers during early-stage honey storage behaviour. *Open Journal of Animal Sciences*, *3*(3), 210–218. doi:10.4236/ojas.2013.33031

Haidamus, S.L., Lorenzon, M.C.A., & Barth, O.M. (2019). Biological elements and residues in Brazilian honeys. *Greener Journal of Biological Sciences*, 9(1), 8–14. https://doi.org/10.15580/GJBS2019.1.022119038.

Hammel, Y. A., Mohamed, R., Gremaud, E., Lebreton, M. H., & Guy, P. A. (2008). Multi-screening approach to monitor and quantify 42 antibiotic residues in honey by liquid chromatography–tandem mass spectrometry. *Journal of Chromatography A*, *1177*(1), 58–76. http://re.indiaenvironmentportal.org.in/files/Multi screening approach to monitor.pdf

Hawkins, J., de Vere, N., Griffith, A., Ford, C. R., Allainguillaume, J., Hegarty, M. J., Baillie, L., & Adams-Groom, B. (2015). Using DNA metabarcoding to identify the floral composition of honey: A new tool for investigating honey bee foraging preferences. *PLoS One*, *10*(8), e0134735–20. doi:10.1371/journal.pone.0134735

Heering, W., Usleber, E., Dietrich, R., & Märtlbauer, E. (1998). Immunochemical screening for antimicrobial drug residues in commercial honey. *The Analyst*, *123*(12), 2759–2762. doi:10.1039/a805047c

Heinrich, K., Macarthur, R., VON Holst, C., & Sharman, M. (2013). An inter-laboratory validation of a multiplex dipstick assay for four classes of antibiotics in honey Rapid Detection in Food and Feed. *Analytical and Bioanalytical Chemistry*, *405*(24), 7875–7884. doi:10.1007/s00216-013-7070-3

Hermosín, I., Chicón, R. M., &., & Dolores Cabezudo, M. (2003). Free amino acid composition and botanical origin of honey. *Food Chemistry*, *83*(2), 263–268. doi:10.1016/S0308-8146(03)00089-X

Huidobro, J. F., & Simal, J. (1984). Contribution to the sugars determination on honey. *Anales de Bromatología*, *36*, 247–264.

Human, H., Brodschneider, R., Dietemann, V., Dively, G., Ellis, J. D., Forsgren, E., Fries, I., Hatjina, F., Hu, F.-L., Jaffé, R., Jensen, A. B., Köhler, A., Magyar, J. P., Özkýrým, A., Pirk, C. W. W., Rose, R., Strauss, U., Tanner, G., Tarpy, D. R., … Zheng, H.-Q. (2013). Miscellaneous standard methods for *Apis mellifera* research. In *V. Dietemann, J. D. Ellis, & P. Neumann (Eds), The COLOSS* BEEBOOK, *Volume I: standard methods for* Apis mellifera *research. Journal of Apicultural Research*, *52*(4), 1–53. doi:10.3896/IBRA.1.52.4.10

International Olive Oil Council. (2007). *Sensory analysis of olive oil. Standard. Glass for oil tasting*. COI/T.20/Doc. No 5/Rev.1. Retrieved from https://www.internationaloliveoil.org

ISO 16657. (2006). Sensory analysis: Apparatus – Olive oil tasting glass. International Organization for Standardization. Retrieved from https:// www.iso.org

ISO 13300. (2006a). *Sensory analysis: General guidance for the staff of a sensory evaluation laboratory - Part 1: Staff responsibilities*. International Organization for Standardization. https://www.iso.org

ISO 13300. (2006b). *Sensory analysis: General guidance for the staff of a sensory evaluation laboratory - Part 2: Recruitment and training of panel leaders*. International Organization for Standardization. https://www.iso.org

ISO 5492. (2008). *Sensory analysis: Vocabulary*. International Organization for Standardization. https://www.iso.org

ISO 6658. (2017). *Sensory analysis: Methodology - general guidance*. International Organization for Standardization. https://www.iso.org

ISO 8586. (2012). *Sensory analysis: General guidelines for the selection, training and monitoring of selected assessors and expert sensory assessors*. International Organization for Standardization. https://www.iso.org

ISO 8589. (2007). *Sensory analysis: General guidance for the design of test rooms*. International Organization for Standardization. https://www.iso.org

Jeon, M., & Rhee Paeng, I. (2008). Quantitative detection of tetracycline residues in honey by a simple sensitive immunoassay. *Analytica Chimica Acta*, *626*(2), 180–185. doi:10.1016/j.aca.2008.08.003

Kacaniová, M., Knazovicka, V., Melich, M., Fikselova, M., Massanyi, P., Stawarz, R., & Putała, A. (2009). Environmental concentration of selected elements and relation to physicochemical parameters in honey. *Journal of Environmental Science and Health - Part A Toxic/Hazardous Substances and Environmental Engineering*, *44*(4), 414–422. doi:10.1080/10934520802659802

Kamal, M. A., & Klein, P. (2011). Determination of sugars in honey by liquid chromatography. *Saudi Journal of Biological Sciences*, *18*(1), 17–21. doi:10.1016/j.sjbs.2010.09.003

Kanda, M., Sasamoto, T., Takeba, K., Hayashi, H., Kusano, T., Matsushima, Y., Nakajima, T., Kanai, S., & Takano, I. (2011). Rapid determination of fumagillin residues in honey by liquid chromatography–tandem mass spectrometry using the QuEChERS method. *Journal of AOAC International*, *94*(3), 878–899. http://www.ingentaconnect.com/content/aoac/

jaoac/2011/00000094/00000003/art00023 doi:10.1093/jaoac/94.3.878

Kaufmann, A., Roth, S., Ryser, B., Widmer, M., & Guggisberg, D. (2002). Quantitative LC/MS-MS determination of sulfonamides and some other antibiotics in honey. *Journal of AOAC International*, *85*(4), 853–860. http://www.ingentaconnect.com/contentone/aoac/jaoac/2002/00000085/00000004/art00009?-crawler=true&mimetype=application/pdf&casa_token=cl5pUDQmtlQAAAAA:A3Syql7alZndvjCZA0bR2luaZyBleUKUQ-U1YQT5p8t6vqxxk30JLPlxd2xbeYzlwMQb9-lQl

Khismatoullin, R. G., Lyapunov, Y. E., Kuzyaev, R. Z., Yelovikova, E. A., Legotkina, G. I., & Zakharov, A. V. (2004). Bacteriological detection of tylosin residues in honey [Paper presentation]. *Proceedings of the First European Conference of Apidology* (pp. 19–23).

Khismatoullin, R., Kuzyaev, R., Lyapunov, Y., & Elovikova, E. (2003). Modification of microbiological detection method of tetracycline in honey. *Apiacta*, *38*, 246–248. https://www.apimondia.com/apiacta/articles/2003/khismatoullin_1.pdf

Khong, S. P., Gremaud, E., Richoz, J., Delatour, T., Guy, P. A., Stadler, R. H., & Mottier, P. (2004). Analysis of matrix-bound nitrofuran residues in worldwide-originated honeys by isotope dilution high-performance liquid chromatography-tandem mass spectrometry. *Journal of Agricultural and Food Chemistry*, *52*(17), 5309–5315. doi:10.1021/jf0401118

Kivrak, İ., Kivrak, Ş., & Harmandar, M. (2016). Development of a rapid method for the determination of antibiotic residues in honey using UPLC-ESI-MS/MS. *Food Science and Technology*, *36*(1), 90–96. doi:10.1590/1678-457X.0037

Kujawski, M. W., Bargańska, Z., Marciniak, K., Miedzianowska, E., Kujawski, J. K., Ślebioda, M., & Namieśnik, J. (2014). Determining pesticide contamination in honey by LC-ESI-MS/MS - Comparison of pesticide recoveries of two liquid-liquid extraction based approaches. *LWT - Food Science and Technology*, *56*(2), 517–523. doi:10.1016/j.lwt.2013.11.024

Kujawski, M. W., Pinteaux, E., & Namieśnik, J. (2012). Application of dispersive liquid-liquid microextraction for the determination of selected organochlorine pesticides in honey by gas chromatography-mass spectrometry. *European* Food Research and Technology, *234*(2), 223–230. doi:10.1007/s00217-011-1635-1

Lazarevic, J. S., Palic, R. M., Radulovic, N. S., Ristic, N. R., & Stojanovic, G. S. (2010). Chemical composition and screening of the antimicrobial and antioxidative activity of extracts of Stachys species. *Journal of the Serbian* Chemical Society, *75*, 1347–1359. doi:10.2298/JSC100601117L

Lee, D.-C., Lee, S.-Y., Cha, S.-H., Choi, Y.-S., & Rhee, H.-I. (1998). Discrimination of native bee-honey and foreign bee-honey by SDS-PAGE. *Korean Journal of Food Science and Technology*, *30*(1), 1–5.

León-Ruiz, V., Vera, S., González-Porto, A. V., & San Andrés, M. P. (2013). Analysis of water-soluble vitamins in honey by isocratic RP-HPLC. *Food Analytical Methods*, *6*(2), 488–496. doi:10.1007/s12161-012-9477-4

Lopez, M. I., Feldlaufer, M., Williams, A., & Chu, P.-S. (2007). Determination and confirmation of nitrofuran residues in honey using LC-MS / MS. *Journal* of Agricultural and Food Chemistry, *55*(4), 1103–1108. doi:10.1021/jf0625712

Lopez, M. I., Pettis, J. S., Smith, I. B., & Chu, P.-S. (2008). Multiclass determination and confirmation of antibiotic residues in honey using LC-MS/MS. *Journal of Agricultural and Food Chemistry*, *56*(5), 1553–1559. doi:10.1021/jf073236w

Lores, E. M., Moore, J. C., & Moody, P. (1987). Improved silica gel cleanup method for organophosphorus pesticides. *Chemosphere*, *16*(5), 1065–1069. doi:10.1016/0045-6535(87)90043-9

Louppis, A. P., Karabagias, I. K., Kontakos, S., Kontominas, M. G., & Papastephanou, C. (2017). Botanical discrimination of Greek unifloral honeys based on mineral content in combination with physicochemical parameter analysis, using a validated chemometric approach. *Microchemical Journal*, *135*, 180–189. doi:10.1016/j.microc.2017.09.004

Louveaux, J., Maurizio, A., & Vorwohl, G. (1970a). Internationale Kommission Für Bienenbotanik Der I. U. B. S. Methodik Der Melissopalynologie. *Apidologie*, *1*(2), 193–209. doi:10.1051/apido:19700205

Louveaux, J., Maurizio, A., & Vorwohl, G. (1970b). Methods of melissopalynology. *Bee World*, *51*(3), 125–138. doi:10.1080/0005772X.1970.11097312

Louveaux, J., Maurizio, A., & Vorwohl, G. (1978). Methods of melissopalynology. *Bee World*, *59*(4), 139–157. doi:10.1080/0005772X.1978.11097714

Lutier, P. M., & Vaissière, B. E. (1993). An improved method for pollen analysis of honey. *Review of Palaeobotany and Palynology*, *78*(1–2), 129–144.(93)90019-Q doi:10.1016/0034-6667(93)90019-Q

Marcazzan, G. L., Magli, M., Piana, L., Savino, A., & Stefano, M. A. (2014). Sensory profile research on the main Italian typologies of monofloral honey: Possible developments and applications. *Journal of Apicultural Research*, *53*(4), 426–437. doi:10.3896/IBRA.1.53.4.09

Marcazzan, G. L., Mucignat-Caretta, C., Marchese, C. M., & Piana, M. L. (2018). A review of methods for honey sensory analysis. *Journal of Apicultural Research*, *57*(1), 75–87. doi:10.1080/00218839.2017.1357940

Martel, A. C., & Zeggane, S. (2002). Determination of acaricides in honey by high-performance liquid chromatography with photodiode array detection. *Journal of Chromatography A*, *954*(1–2), 173–180.(02)00126-7 doi:10.1016/S0021-9673(02)00126-7

Maudens, K. E., Zhang, G. F., & Lambert, W. E. (2004). Quantitative analysis of twelve sulfonamides in honey after acidic hydrolysis by high-performance liquid chromatography with post-column derivatization and fluorescence detection. *Journal of Chromatography A*, *1047*(1), 85–92. doi:10.1016/j.chroma.2004.07.007

Maurizio, A. (1959). Zur Frage der Mikroskopie von Honigtau-Honigen. *Annales de L'abeille*, 2, 145–157. doi:10.1051/apido:19590203

Maurizio, A. (1977). Honig-Mikroskopie. In H. Freud (Ed.), *Handbuch der Mikroskopie in der Technik* (vol. VIII, pp. 627–652). Umschau Verlag.

McAleer, D., Mcconnell, R. I., Tohill, A., & Fitzgerald, S. P. (2010). Biochip arrays for multi-analytical screening of antibiotics in honey [Paper presentation]. *Abstract Book of the 6th International Symposium on Hormone and Veterinary Drug Residue Analysis* (Vol. 129, pp. 1–4). Ghent.

McMullen, S. E., Lansden, J. A., & Schenck, F. J. (2004). Modifications and adaptations of the Charm II rapid antibody assay for chloramphenicol in honey. *Journal of Food Protection*, *67*(7), 1533–1536. doi:10.4315/0362-028X-67.7.1533

Medrzycki, P., Giffard, H., Aupinel, P., Belzunces, L. P., Chauzat, M.-P., Claßen, C., Colin, M. E., Dupont, T., Girolami, V., Johnson, R., Le Conte, Y., Lückmann, J., Marzaro, M., Pistorius, J., Porrini, C., Schur, A., Sgolastra, F., Delso, N. S., van der Steen, J. J. M., … Vidau, C. (2013). Standard methods for toxicology research in *Apis mellifera*. In *V. Dietemann, J. D. Ellis, & P. Neumann (Eds), The COLOSS* BEEBOOK, *Volume I: standard methods for* Apis mellifera *research*. *Journal of Apicultural Research*, *52*(4), 1–60. doi:10.3896/IBRA.1.52.4.14

Meixner, M.D., Pinto, M.A., Bouga, M., Kryger, P., Ivanova E., & Stefan Fuchs, S. (2013). Standard methods for characterising subspecies and ecotypes of *Apis mellifera*. *Journal of Apicultural Research*, 52(4), 1–27.

Miller, N. J., Rice-Evans, C., Davies, M. J., Gopinathan, V., & Milner, A. (1993). A novel method for measuring antioxidant capacity and its application to monitoring the antioxidant status in premature neonates. *Clinical Science, 84*(4), 407–412. doi:10.1042/cs0840407

MINISTÉRIO DA AGRICULTURA PECUÁRIA E ABASTECIMENTO, B (2000). *Instrução Normativa n° 11, de 20 de outubro 2000.*

Mucignat-Caretta, C., Marcazzan, G. L., & Piana, M. L. (2012). Il Miele. In *Atlante sensoriale dei prodotti alimentari* (pp. 286–295). Techniche Nuove.

Muriano, A., Chabottaux, V., Diserens, J. M., Granier, B., Sanchez-Baeza, F., & Marco, M. P. (2015). Rapid immunochemical analysis of the sulfonamide-sugar conjugated fraction of antibiotic contaminated honey samples. *Food Chemistry, 178*, 156–163. doi:10.1016/j.foodchem.2015.01.037

Muriano, A., Pinacho, D.-G., Chabottaux, V., Diserens, J.-M., Granier, B., Stead, S., Sanchez Baeza, F., Pividori, M. I., & Marco, M.-P. (2013). A portable electrochemical magnetoimmunosensor for detection of sulfonamide antimicrobials in honey Rapid Detection in Food and Feed. *Analytical and Bioanalytical Chemistry, 405*(24), 7885–7895. doi:10.1007/s00216-013-7219-0

Nascimento, K. S. D., Gasparotto Sattler, J. A., Lauer Macedo, L. F., Serna González, C. V., Pereira de Melo, I. L., da Silva Araújo, E., Granato, D., Sattler, A., & de Almeida-Muradian, L. B. (2018, January). Phenolic compounds, antioxidant capacity and physicochemical properties of Brazilian *Apis mellifera* honeys. *LWT, 91*, 85–94. doi:10.1016/j.lwt.2018.01.016

Nozal, M. J., Bernal, J. L., Gómez, M. T., Jiménez, J. J., Bernal, J., & Higes, M. (2006). Trace analysis of antibacterial tylosin A, B, C and D in honey by liquid chromatography-electrospray ionization-mass spectrometry. *Journal of Separation Science, 29*(3), 405–413. 10.1002/jssc.200500423

O'Mahony, J., Moloney, M., Mcconnell, I., Benchikh, E., Lowry, P., & Danaher, M. (2010). Biochip array detection of nitrofuran metabolites in honey [Paper presentation]. *Abstract Book of the 6th International Symposium on Hormone and Veterinary Drug Residue Analysis* (Vol. 176, pp. 1–4). Ghent.

Ouchemoukh, S., Louaileche, H., & Schweitzer, P. (2007). Physicochemical characteristics and pollen spectrum of some Algerian honeys. *Food Control, 18*(1), 52–58. doi:10.1016/j.foodcont.2005.08.007

Ouchemoukh, S., Schweitzer, P., Bachir Bey, M., Djoudad-Kadji, H., & Louaileche, H. (2010). HPLC sugar profiles of Algerian honeys. *Food Chemistry, 121*(2), 561–568. doi:10.1016/j.foodchem.2009.12.047

Ough, C. (1969). Rapid determination of proline in grapes and wines. *Journal of Food Science, 34*(3), 228–230. doi:10.1111/j.1365-2621.1969.tb10327.x

Özbalci, B., Boyaci, I. H., Topcu, A., Kadilar, C., & Tamer, U. (2013). Rapid analysis of sugars in honey by processing Raman spectrum using chemometric methods and artificial neural networks. *Food Chemistry, 136*(3–4), 1444–1452. doi:10.1016/j.foodchem.2012.09.064

Pang, G.-F., Fan, C.-L., Liu, Y.-M., Cao, Y.-Z., Zhang, J.-J., Fu, B.-L., Li, X.-M., Li, Z.-Y., & Wu, Y.-P. (2006). Multi-residue method for the determination of 450 pesticide residues in honey, fruit juice and wine by double-cartridge solid-phase extraction/ gas chromatography-mass spectrometry and liquid chromatography-tandem mass spectrometry. *Food Additives and Contaminants, 23*(8), 777–810. doi:10.1080/02652030600657997

Pascual-Maté, A., Osés, S. M., Fernández-Muiño, M. A., & Sancho, M. T. (2018). Methods of analysis of honey. *Journal of Apicultural Research, 57*(1), 38–74. doi:10.1080/00218839.2017.1411178

Peng, D., Ye, S., Wang, Y., Chen, D., Tao, Y., Huang, L., Liu, Z., Dai, M., Wang, X., & Yuan, Z. (2012). Development and validation of an indirect competitive enzyme-linked immunosorbent assay for the screening of tylosin and tilmicosin in muscle, liver, milk, honey and eggs. *Journal of Agricultural and Food Chemistry, 60*(1), 44–51. doi:10.1021/jf2037449

Persano Oddo, L., & Bogdanov, S. (2004). Determination of honey botanical origin: Problems and issues. *Apidologie, 35*(Suppl. 1), S2–S3.:2004044 doi:10.1051/apido:2004044

Piana, M. L., Persano Oddo, L., Bentabol, A., Bruneau, E., Bogdanov, S., & Guyot Declerck, C. (2004). Sensory analysis applied to honey: State of the art. *Apidologie, 35*(Suppl. 1), S26–S37.:2004048 doi:10.1051/apido:2004048

Pirard, C., Widart, J., Nguyen, B. K., Deleuze, C., Heudt, L., Haubruge, E., Haubruge, E., DE Pauw, E., & Focant, J. F. (2007). Development and validation of a multi-residue method for pesticide determination in honey using on-column liquid – liquid extraction and liquid chromatography – tandem mass spectrometry. *Journal of Chromatography A, 1152*(1–2), 116–123. doi:10.1016/j.chroma.2007.03.035

Polzer, J., Kindt, K., Radeck, W., & Gowik, P. (2010 Determination of nitro-imidazoles in honey: Method development, validation and relevance in practice [Paper presentation]. *Abstract Book of the 6th International Symposium on Hormone and Veterinary Drug Residue Analysis* (Vol. 185, pp. 1–4). Ghent.

Popa, I. D., Schiriac, E. C., & Cuciureanu, R. (2012). Multi-analytic detection of antibiotic residues in honey using a multiplexing biochip assay. *Revista Medico-Chirurgicala a Societatii de Medici Si Naturalisti Din Iasi, 116*(1), 324–329.

Quintana-Rizzo, J., Salter, R. S., & Saul, S. J. (2005). Solid phase extraction procedures for validation of Charm II sulfonamide, streptomycin and chloramphenicol positives and detection of tylosin in honey. *Apiacta, 40*, 55–62. http://www.apimondia.com/apiacta/articles/2005/quintana-rizzo_1.pdf

Rademacher, E., Fahlberg, A., Raddatz, M., Schneider, S., & Voigt, K. (2013). Galenics: Studies of the toxicity and distribution of sugar substitutes on *Apis mellifera*. *Apidologie, 44*(2), 222–233. doi:10.1007/s13592-012-0174-5

Re, R., Pellegrini, N., Proteggente, A., Pannala, A., Yang, M., & Rice-Evans, C. (1999). Antioxidant activity applying an improved ABTS radical cation decolorization assay. *Free Radical Biology and Medicine, 26*(9–10), 1231–1237.(98)00315-3 doi:10.1016/S0891-5849(98)00315-3

Regulation, E. (2005). Regulation (EU) No 396/2005 on maximum residue levels of pesticides in or on food and feed of plant and animal origin, and amending Council Directive 91/414/EEC. *Official Journal of the European Union*, L70, 1–16. https://eur-lex.europa.eu/legal-content/EN/TXT/PDF/?uri=CELEX:32005R0396&from=en

Reybroeck, W. (2014). Quality control of honey and bee products. In *Beekeeping for poverty alleviation and livelihood security* (pp. 481–206). Springe. 10.1007/978-94-017-9199-1.

Reybroeck, W. (2018). Residues of antibiotics and chemotherapeutics in honey. *Journal of Apicultural Research, 57*(1), 97–112. doi:10.1080/00218839.2017.1338129

Reybroeck, W., & Ooghe, S. (2010). Detection of sulfa drugs in honey by Sulfa-Sensor Honey and Charm II Sulfa Test Honey: A comparison [Paper presentation]. *Proceedings of the 4th European Conference of Apidology* (p. 148). Kence M.

Reybroeck, W., Daeseleire, E., de Brabander, H. F., & Herman, L. (2012). Antimicrobials in beekeeping. *Veterinary Microbiology, 158*(1–2), 1–11. doi:10.1016/j.vetmic.2012.01.012

Reybroeck, W., Ooghe, S., DE Brabander, H., & Daeseleire, E. (2007). Validation of the tetrasensor honey test kit for the screening of tetracyclines in honey. *Journal of Agricultural and Food Chemistry, 55*(21), 8359–8366. doi:10.1021/jf071922p

Rial-Otero, R., Gaspar, E. M., Moura, I., & Capelo, J. L. (2007). Chromatographic-based methods for pesticide determination in honey: An overview. *Talanta, 71*(2), 503–514. doi:10.1016/j.talanta.2006.05.033

Rissato, S. R., Galhiane, M. S., DE Almeida, M. V., Gerenutti, M., & Apon, B. M. (2007). Multiresidue determination of pesticides in honey samples by gas chromatography-mass spectrometry and application in environmental contamination. *Food Chemistry*, *101*(4), 1719–1726. doi:10.1016/j.foodchem.2005.10.034

Roshan, A. R. A., Gad, H. A., EL-Ahmady, S. H., Khanbash, M. S., Abou-Shoer, M. I., & Al-Azizi, M. M. (2013). Authentication of monofloral yemeni sidr honey using ultraviolet spectroscopy and chemometric analysis. *Journal of Agricultural and Food Chemistry*, *61*(32), 7722–7729. doi:10.1021/jf402280y

Roubik, D.W., & Moreno, J.E. (1991). *Pollen and spores of Barro Colorado Island*. Missouri Botanical Garden (pp. 268). (Monographs in *Systematic Botany*, n.36).

Ruiz-Matute, A. I., Sanz, M. L., & Martinez-Castro, I. (2007). Use of gas chromatography-mass spectrometry for identification of a new disaccharide in honey. *Journal of Chromatography A*, *1157*(1-2), 480–483. doi:10.1016/j.chroma.2007.05.056

Ruoff, K., Karoui, R., Dufour, E., Luginbühl, W., Bosset, J.-O., Bogdanov, S., & Amadò, R. (2005). Authentication of the botanical origin of honey by front-face fluorescence spectroscopy. A preliminary study. *Journal of Agricultural and Food Chemistry*, *53*(5), 1343–1347. doi:10.1021/jf048384q

Ruoff, K., Luginbühl, W., Künzli, R., Iglesias, M. T., Bogdanov, S., Bosset, J. O., von der Ohe, K., von der Ohe, W., & Amadò, R. (2006). Authentication of the botanical and geographical origin of honey by mid-infrared spectroscopy. *Journal of Agricultural and Food Chemistry*, *54*(18), 6873–6880. doi:10.1021/jf060838r

Sabatini, A. G., Bortolotti, L., & Marcazzan, G. L. (Eds.). (2007). *Conoscere il miele*. Avenue Media.

Salgado-Laurenti, C. R., Tellería, M. C., & Coronel, J. M. (2017). Botanical and geographical origin of honey from the dry and humid Chaco ecoregions (Argentina). *Grana*, *56*(6), 450–461. doi:10.1080/00173134.2016.1276619

Sancho, M. T., Pascual-Maté, A., Rodríguez-Morales, E. G., Osés, S. M., Escriche, I., Periche, Á., & Fernández-Muiño, M. A. (2016). Critical assessment of antioxidant-related parameters of honey. *International Journal of Food Science & Technology*, *51*(1), 30–36. doi:10.1111/ijfs.12988

SANCO. (2004). Summary record of the Standing Committee on the Food Chain and Animal Health held in Brussels on 21 September 2004 (Section Biological Safety of the Food Chain) (Section Controls and Import Conditions). Article 16, Statement of the Commission. SANCO -E.2(04)D/521927. https://ec.europa.eu/food/sites/food/files/safety/docs/reg-com_cic_summary35_en.pdf

Santos, K. S., Malaspina, O., & Palma, M. S. (2003, March). Cinética da diástase em méis de diferentes origens florais. Um novo protocolo experimental. *Mensagem Doce*, *70*, 2–4.

Schwaiger, I., & Schuch, R. (2000). Bound sulfathiazole residues in honey – need of a hydrolysis step for the analytical determination of total sulfathiazole content in honey. *Deutsche Lebensmittel-Rundschauche Lebensmittel-Rundschau*, *96*(3), 93–98.

Scortichini, G., Annunziata, L., Haouet, M. N., Benedetti, F., Krusteva, I., & Galarini, R. (2005). ELISA qualitative screening of chloramphenicol in muscle, eggs, honey and milk: Method validation according to the Commission Decision 2002/657/EC criteria. *Analytica Chimica Acta*, *535*(1–2), 43–48. doi:10.1016/j.aca.2004.12.004

Shendy, A. H., Al-Ghobashy, M. A., Mohammed, M. N., Gad Alla, S. A., & Lotfy, H. M. (2016). Simultaneous determination of 200 pesticide residues in honey using gas chromatography-tandem mass spectrometry in conjunction with streamlined quantification approach. *Journal of Chromatography A*, *1427*, 142–160. doi:10.1016/j.chroma.2015.11.068

Sheth, H. B., Stiles, M. E., Sporns, P., Yaylayan, V. A., & Low, N. H. (1990). Reaction of reducing sugars with sulfathiazole and importance of this reaction to sulfonamide residue analysis using chromatographic, colorimetric, microbiological, or ELISA methods. *Journal of Agricultural and Food Chemistry*, *38*(4), 1125–1130. doi:10.1021/jf00094a047

Singh, S. K., Dwevedi, R., Deopa, P., & Krishna, V. (2014). A gradual advancement of the extraction and cleanup techniques for residue analysis of pesticides in food matrices and liquid products : A review. *Iosr Journal of Environmental Science, Toxicology and Food Technology*, *8*(1), 113–118. https://pdfs.semanticscholar.org/2b36/bf5f5e0e60d0c972ceb53b71b830ce87779f.pdf doi:10.9790/2402-0815113118

Sodré, G. D. S., Marchini, L. C., Moreti, A. C. D. C. C., Otsuk, I. P., & Carvalho, C. A. L. D. (2011). Physico-chemical characteristics of honey produced by *Apis mellifera* in the Picos region, state of Piauí, Brazil. *Revista Brasileira de Zootecnia*, *40*(8), 1837–1843. doi:10.1590/S1516-35982011000800030

Souza Tette, P. A., Guidi, L. R., DE Abreu Glória, M. B., & Fernandes, C. (2016). Pesticides in honey: A review on chromatographic analytical methods. *Talanta*, *149*, 124–141. doi:10.1016/j.talanta.2015.11.045

Sporns, P., Kwan, S., & Roth, L. A. (1986). HPLC analysis of oxytetracycline residues in honey. *Journal of Food Protection*, *49*(5), 383–388. doi:10.4315/0362-028X-49.5.383

Tahboub, Y. R., Zaater, M. F., & Barri, T. A. (2006). Simultaneous identification and quantitation of selected organochlorine pesticide residues in honey by full-scan gas chromatography-mass spectrometry. *Analytica Chimica Acta*, *558*(1–2), 62–68. doi:10.1016/j.aca.2005.11.004

Tarbin, J. A., Read, W., & Sharman, M. (2010 Development and validation of a rapid method for the determination of fumagillin in honey [Paper presentation]. *Abstract Book of the 6th International Symposium on Hormone and Veterinary Drug Residue Analysis* (Vol. 217, pp. 1–4). Ghent.

The European Parliament and the Council of European Union. (2009). Regulation (EC) No 470/2009 of the European Parliament and of the Council of 6 May 2009 laying down Community procedures for the establishment of residue limits of pharmacologically active substances in foodstuffs of animal origin, repealing Council Regulation (EEC) No 2377/90 and amending Directive 2001/82/EC of the European Parliament and of the Council and Regulation (EC) No 726/2004 of the European Parliament and of the Council laying down a Community procedure for the establishment of maximum residue limits of veterinary medicinal products in foodstuffs of animal origin. Official Journal of the European Union, L152, 11–22.

Thompson, T. S., Pernal, S. F., Noot, D. K., Melathopoulos, A. P., & VAN DEN Heever, J. P. (2007). Degradation of incurred tylosin to desmycosin-Implications for residue analysis of honey. *Analytica Chimica Acta*, *586*(1–2), 304–311. doi:10.1016/j.aca.2006.09.043

Thongchai, W., Liawruangath, B., Liawruangrath, S., & Greenway, G. M. (2010). A microflow chemiluminescence system for determination of chloramphenicol in honey with preconcentration using a molecularly imprinted polymer. *Talanta*, *82*(2), 560–566. doi:10.1016/j.talanta.2010.05.007

Thrasyvoulou, A., Tananaki, C., Goras, G., Karazafiris, E., Dimou, M., Liolios, V., Kanelis, D., & Gounari, S. (2018). Legislation of honey criteria and standards. *Journal of Apicultural Research*, *57*(1), 88–96. doi:10.1080/00218839.2017.1411181

Tomasini, D., Sampaio, M. R. F., Caldas, S. S., Buffon, J. G., Duarte, F. A., & Primel, E. G. (2012). Simultaneous determination of pesticides and 5-hydroxymethylfurfural in honey by the modified QuEChERS method and liquid chromatography coupled to tandem mass spectrometry. *Talanta, 99*, 380–386. doi:10.1016/j.talanta.2012.05.068

Tomasini, D., Sampaio, M. R. F., Cardoso, L. V., Caldas, S. S., & Primel, E. G. (2011). Comparison of dispersive liquid–liquid microextraction and the modified QuEChERS method for the determination of fipronil in honey by high performance liquid chromatography with diode-array detection. *Analytical Methods, 3*(8), 1893. doi:10.1039/c1ay05221g

Tribalat, L., Paisse, O., Dessalces, G., & Grenier-Loustalot, M. F. (2006). Advantages of LC-MS-MS compared to LC-MS for the determination of nitrofuran residues in honey. *Analytical and Bioanalytical Chemistry, 386*(7–8), 2161–2168. doi:10.1007/s00216-006-0878-3

Tuberoso, C. I. G., Jerković, I., Sarais, G., Congiu, F., Marijanović, Z., & Kuś, P. M. (2014). Color evaluation of seventeen European unifloral honey types by means of spectrophotometrically determined CIEL â -Cab â - Hab â chromaticity coordinates. *Food Chemistry, 145*, 284–291. doi:10.1016/j.foodchem.2013.08.032

U.S. Environmental Protection Agency (EPA). (2013, June). *A summary of analytical methods used for the analyses of pesticide residues in bees and hive matrices*. U. S. Environmental Protection Agency Region 5, Land and Chemicals Division, Chemicals Management Branch, pp. 1–10. https://www.epa.gov/sites/production/files/2013-09/documents/bee-analytical-methods.pdf

Uran, H., Aksu, F., & Altiner, D. D. (2017). A research on the chemical and microbiological qualities of honeys sold in Istanbul. *Food Science and Technology, 37*(suppl 1), 30–33. (doi:10.1590/1678-457x.32016

Valera, E., Muriano, A., Pividori, I., Sánchez-Baeza, F., & Marco, M. P. (2013). Development of a Coulombimetric immunosensor based on specific antibodies labeled with CdS nanoparticles for sulfonamide antibiotic residues analysis and its application to honey samples. *Biosensors and Bioelectronics, 43*(1), 211–217. doi:10.1016/j.bios.2012.12.017

VAN Bruijnsvoort, M., Ottink, S. J. M., Jonker, K. M., & DE Boer, E. (2004). Determination of streptomycin and dihydrostreptomycin in milk and honey by liquid chromatography with tandem mass spectrometry. *Journal of Chromatography A, 1058*(1–2), 137–142. doi:10.1016/j.chroma.2004.07.101

Verzegnassi, L., Savoy-Perroud, M. C., & Stadler, R. H. (2002). Application of liquid chromatography-electrospray ionization tandem mass spectrometry to the detection of 10 sulfonamides in honey. *Journal of Chromatography A, 977*(1), 77–87. doi:10.1016/S0021-9673(02)01341-9

Vilca, F. Z., Corrreia-Oliveira, M. E., Leal, R. M. P., Monteiro, S. H., Zanardi, O. Z., Marchini, L. C., & Tornisielo, V. L., (2012). Quechers approach for the determination of seven pesticide residues in Brazilian honey samples using GC-μECD. *Journal of Food Science and Engineering*, 2, 163–169.

Vit, P., Pedro, S. R., & Roubik, D. (2018). *Pot-pollen in stingless bee melittology*. Springer, 481p.

von der Ohe, W., Persano Oddo, L., Piana, M. L., Morlot, M., & Martin, P. (2004). Harmonized methods of melissopalynology. *Apidologie, 35*(Suppl. 1), S18–S25. 10.1051/apido:2004050

Wang, S., Yong, W., Liu, J., Zhang, L., Chen, Q., & Dong, Y. (2014). Development of an indirect competitive assay-based aptasensor for highly sensitive detection of tetracycline residue in honey. *Biosensors and Bioelectronics, 57*, 192–198. doi:10.1016/j.bios.2014.02.032

Weigel, S., Gatermann, R., & Harder, W. (2005). Screening of honey for residues of antibiotics by an optical biosensor. *Apiacta, 40*, 63–69. http://www.apimondia.com/apiacta/articles/2005/weigel_1.pdf

WHO (World Health Organization). (1996). *Trace elements in human nutrition and health World Health Organization*, pp. 1–360.

WHO/FAO. (2002). *Food energy – methods of analysis and conversion. Fao Food and Nutrition Paper*. https://doi.org/0254-4725

Wiest, L., Buleté, A., Giroud, B., Fratta, C., Amic, S., Lambert, O., Pouliquen, H., & Arnaudguilhem, C. (2011). Multi-residue analysis of 80 environmental contaminants in honeys, honey bees and pollens by one extraction procedure followed by liquid and gas chromatography coupled with mass spectrometric detection. *Journal of Chromatography A, 1218*(34), 5743–5756. doi:10.1016/j.chroma.2011.06.079

Wu, Z., Chen, L., Wu, L., Xue, X., Zhao, J., Li, Y., Ye, Z., & Lin, G. (2015). Classification of Chinese honeys according to their floral origins using elemental and stable isotopic compositions. *Journal of Agricultural and Food Chemistry, 63*(22), 5388–5394. doi:10.1021/acs.jafc.5b01576

Yu, J., Wu, C., & Xing, J. (2004). Development of new solid-phase microextraction fibers by sol-gel technology for the determination of organophosphorus pesticide multiresidues in food. *Journal of Chromatography A, 1036*(2), 101–111. doi:10.1016/j.chroma.2004.02.081

Yücel, Y., & Sultanoglu, P. (2013). Characterization of honeys from Hatay Region by their physicochemical properties combined with chemometrics. *Food Bioscience, 1*, 16–25. doi:10.1016/j.fbio.2013.02.001

Zacharis, C. K., Rotsias, I., Zachariadis, P. G., & Zotos, A. (2012). Dispersive liquid-liquid microextraction for the determination of organochlorine pesticides residues in honey by gas chromatography-electron capture and ion trap mass spectrometric detection. *Food Chemistry, 134*(3), 1665–1672. doi:10.1016/j.foodchem.2012.03.073

Zander, E., & Maurizio, A. (1975). Der Honig. In A. Maurizio (Ed.), *Handbuch der Bienenkunde* (2nd ed., pp. 212). Verlag Eugen Ulmer, Stuttgart.

Zhou, J., Qi, Y., Ritho, J., Duan, L., Wu, L., Diao, Q., Li, Y., & Zhao, J. (2014, February). Analysis of maltooligosaccharides in honey samples by ultra-performance liquid chromatography coupled with evaporative light scattering detection. *Food Research International, 56*, 260–265. doi:10.1016/j.foodres.2014.01.014

Zhou, N., Wang, J., Zhang, J., Li, C., Tian, Y., & Wang, J. (2013). Selection and identification of streptomycin-specific single-stranded DNA aptamers and the application in the detection of streptomycin in honey. *Talanta, 108*, 109–116. doi:10.1016/j.talanta.2013.01.064

*Journal of Apicultural Research*, 2021
Vol. 60, No. 4, 1–31, https://doi.org/10.1080/00218839.2020.1801073

REVIEW ARTICLE

# Standard methods for *Apis mellifera* venom research

Dirk C. de Graaf[a†*], Márcia Regina Brochetto Braga[b†], Rusleyd Maria Magalhães de Abreu[c], Simon Blank[d], Chris H. Bridts[e], Luc S. De Clerck[e], Bart Devreese[f], Didier G. Ebo[e], Timothy J. Ferris[g], Margo M. Hagendorens[e], Débora Laís Justo Jacomini[h], Iliya Kanchev[i], Zenon J. Kokot[j], Jan Matysiak[k], Christel Mertens[e], Vito Sabato[e], Athina L. Van Gasse[e] and Matthias Van Vaerenbergh[a]

[a]*Laboratory of Molecular Entomology and Bee Pathology (L-MEB), Department of Biochemistry and Microbiology, Ghent University, Ghent, Belgium;* [b]*Department of Biology, Biosciences Institute, Universidade Estadual Paulista (UNESP), Rio Claro, SP, Brazil;* [c]*Center for Biological Sciences and Nature (CCBN), Federal University of Acre (UFAC), Rio Branco, AC, Brazil;* [d]*Center of Allergy and Environment (ZAUM), Technical University and Helmholtz Center Munich, Munich, Germany;* [e]*Department of Immunology, Allergology, and Rheumatology and the Infla-Med Centre of Excellence, University of Antwerp, Antwerp, Belgium;* [f]*Laboratory of Microbiology, Department of Biochemistry and Microbiology, Ghent University, Ghent, Belgium;* [g]*Extract and Box Limited, Haruru, New Zealand;* [h]*Faculdades Claretianas, Rio Claro, SP, Brazil;* [i]*IGK Electronics Limited, Vladislavovo, Varna, Bulgaria;* [j]*The President Stanislaw, Wojciechowski State University of Applied Sciences, Kalisz, Poland;* [k]*Department of Inorganic and Analytical Chemistry, Poznan University of Medical Science, Poznan, Poland*

*(Received 8 July 2020; accepted 22 July 2020)*

Honey bees have a sting which allows them to inject venomous substances into the body of an opponent or attacker. As the sting originates from a modified ovipositor, it only occurs in the female insect, and this is a defining feature of the bee species that belong to a subclade of the Hymenoptera called Aculeata. There is considerable interest in bee venom research, primarily because of an important subset of the human population who will develop a sometimes life threatening allergic response after a bee sting. However, the use of honey bee venom goes much further, with alleged healing properties in ancient therapies and recent research. The present paper aims to standardize selected methods for honey bee venom research. It covers different methods of venom collection, characterization and storage. Much attention was also addressed to the determination of the biological activity of the venom and its use in the context of biomedical research, more specifically venom allergy. Finally, the procedure for the assignment of new venom allergens has been presented.

**Métodos estándar para la investigación del veneno de *Apis mellifera***

Las abejas melíferas tienen un aguijón que les permite inyectar sustancias venenosas en el cuerpo de un oponente o atacante. El aguijón es un ovipositor modificado que solo se manifiesta en el insecto hembra, siendo este una característica que define a las especies de abejas que pertenecen al subclado de himenópteros llamada Aculeata. Hay un interés considerable en la investigación del veneno de abeja, principalmente debido a que un porcentaje importante de la población humana desarrollará una respuesta alérgica - a veces mortal - a la picadura de abeja. Sin embargo, el uso del veneno de la abeja melífera abarca mucho más, con presuntas propiedades curativas en terapias antiguas e investigaciones recientes. El presente trabajo tiene como objetivo estandarizar métodos seleccionados para la investigación del veneno de las abejas melíferas. Cubre diferentes métodos de recolección, caracterización y almacenamiento de veneno. También se prestó mucha atención a la determinación de la actividad biológica del veneno y su uso en el contexto de la investigación biomédica, más específicamente la alergia al veneno. Finalmente, se ha presentado el procedimiento para la asignación de nuevos alérgenos de veneno.

西方蜜蜂蜂毒研究标准方法

蜜蜂可以使用螫针将毒液注射到对手或攻击者的体内。螫针起源于改良的产卵器，因此只存在于雌性昆虫中，也是膜翅目针尾组昆虫的决定性特征。人们对蜂毒研究有很大的兴趣，这主要是因为相当一部分人在被蜜蜂螫到后会产生过敏反应，严重时会危及生命。然而，蜂毒的使用不止于此，在古代疗法和最新的研究中据称具有治愈作用。本文旨在对蜂毒研究中选定的方法进行规范，涵盖了蜂毒收集、表征和储存的不同方法。蜂毒的生物学活性的测定及其在生物医学研究（更具体讲是蜂毒过敏）中的用途也引起了人们的极大关注。最后，介绍了给新的蜂毒过敏原定名的流程。

**Keywords:** acid phosphatase; *Apis mellifera*; esterase; honey bee; hyaluronidase; phospholipase A2; sting allergy; venom characterization; venom collection; venom composition; venom storage; COLOSS; *BEEBOOK*

*Corresponding author. Email: Dirk.deGraaf@UGent.be
†These authors contributed equally to this work.

Please refer to this paper as: de Graaf, D. C., Brochetto Braga, M. R. B., Magalhães de Abreu, R.M., Blank, S., Bridts, C. H., De Clerck, L. S., Devreese, B., Ebo, D.G., Ferris, T. J., Hagendorens, M. M., Justo Jacomini, D. L. J., Kanchev, I., Kokot, Z. K., Matysiak, J., Mertens, C., Sabato, V., Van Gasse, A. L., Van Vaerenbergh, M. (2020) Standard methods for *Apis mellifera* venom research. In V. Dietemann, P. Neumann, N. L. Carreck, & J. D. Ellis (Eds.), The *COLOSS BEEBOOK* – Volume III, Part 2: Standard methods for *Apis mellifera* product research. *Journal of Apicultural Research?*. https://doi.org/10.1080/00218839.2020.1801073.

## 1. Introduction

Honey bees belong to a subclade of Hymenopteran insects called Aculeata, with the defining feature that the ovipositor has been modified into a sting (Moreau, 2013). Thus, the apparatus that originally served for egg laying has been modified into a structure that delivers venom. While some of the aculeate species have a predatory lifestyle and use the venom for prey capture purposes, bees shifted to a diet of pollen and nectar and only used their sting for individual defense or defense of the colony. This can be accomplished by killing other insects or by inflicting pain in higher organisms, including humans. Only species that belong to the families Vespidae (wasps), Apidae (bees) and Formicidae (ants) sting humans frequently (de Graaf et al., 2009; King & Spangfort, 2000). Venom is produced in epidermal glands (venom glands) that are derived from the female accessory reproductive glands and is stored in a venom sac/reservoir until release (Bridges & Owen, 1984). The earliest efforts to unravel the composition of honey bee venom date back to the work of Langer in 1897, who found that it consists of active and hemolytic basic components (Zalat et al., 1999). Today honey bee worker bee venom is one of the best characterized among the Hymenopteran insects with more than 100 identified proteins and peptides (Ali, 2014; Benton & Morse, 1968; Dotimas & Hider, 1987; van Vaerenbergh et al., 2014) and some biogenic amines.

While the toxic activity of Hymenoptera venoms is only of medical importance in case of massive sting events, their allergenic properties are of more important concern for human health. In man, early exposure to bee venom evokes IgG1, IgG2 and to a lesser extent IgG4 antibody responses, whereas long-term exposure often found in beekeepers drives the immunity to an IgG4 type of humoral response (Larché et al., 2006; Muller, 2005). However, some people develop a venom allergy, which is an IgE-mediated type I hypersensitivity of non-atopic origin (de Graaf et al., 2009). The treatment of allergic symptoms depends on the severity of the allergic reaction. Large local reactions are treated with topical/systemic corticosteroids and anti-histamines, and by cooling the swollen area. In case of systemic reactions, also auto-injectable adrenaline should be used as emergency medication (Ruëff et al., 2011). To provide protection from future stings, venom immunotherapy (VIT) is the only effective and potentially life-saving treatment for patients with the most severe grades of allergy (Müller, 1966; von Moos et al., 2013). The VIT procedure consists of subcutaneous injections of venom extract in two phases: the incremental and the maintenance dose phase.

Many research papers on honey bee venom are devoted to the diagnosis and therapy of sting allergy. Bee venom has been used as a therapeutic agent in Eastern Asia since at least the second century BC (Yin & Koh, 1998). Although still controversial, its use in the so-called meridian therapy receives some attention, for instance in the context of bee venom acupuncture, a kind of herbal acupuncture where extracted and processed bee venom is administered on acupoints for stimulation (Cha et al., 2004). In addition, recent studies reported the beneficial role of bee venom in cancer therapy among others (for review see Oršolić, 2012). Although the authors of this article wish to distance themselves from any alleged properties of bee venom that are insufficiently substantiated scientifically, we believe that any bee venom research benefits from an improving standardization of methods. The present paper describes standard methods for bee venom collection, determination of its composition and biological activity, the purification of its compounds and its use in the context of venom allergy research.

## 2. Worker bee venom collection

### 2.1. Electric stimulation

The main method for bee venom collection is by stimulating bees with an electrical current as first described by Markovic and Mollnar (1954) and thereafter by Palmer (1961) and others (Benton et al., 1963; Gunnison, 1966; Morse & Benton, 1964a; Nobre, 1990). All modern and effective devices for collection of bee venom employ this principle. The main benefit of this method is that after the release of bee venom, bees do not lose their sting and stay alive, and are able to operate normally (Morse & Benton, 1964b; Palmer, 1961). Therefore, this is considered the safest bee venom collection method. The best results in collecting bee venom are obtained in the peak season: summer-autumn. The highest amount of bee venom can be produced during summer – bees' active season. The next best season for collection is autumn, and the lowest amounts are produced during the spring period.

Devices for bee venom collection have continuously been improved, and according to Bogdanov (2017) they mostly consist of four parts:

(1) Battery of 12–15V and 2 Amp; AC 25V; 1200 Hz powered by, or directly plugged into the power grid;
(2) Electrical impulse generation with frequency from 50 to 1000 Hz, duration of 2–3 sec and pauses of 3–6 sec;
(3) Electrical stimulator – surface that consists of stretched uninsulated wires, at a distance of 3–4 mm from each other;
(4) Glass slide on which the bee venom is secreted.

A

B

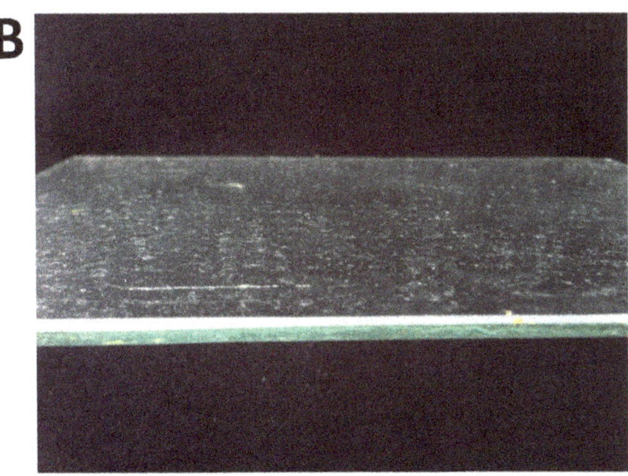

Figure 1. Bee venom collector Beewhisper 5.0 is based on the electric stimulation protocol (http://www.beewhisper.com). In (A) it is shown how the equipment is put in front of the hive entrance. After completion of the collection cycle, the venom that has been deposited on the glass slide (B) can be scraped down.

The electrical stimulator should be designed in such a way that the distance between the active conductors and the glass slide does not allow bees to be stuck or trapped between them. The active wires must be cleaned after the end of each working day – dried bee venom is not electro-conductive and interferes with the device. The efficiency of the device can decrease down to 10% if the wires are not cleaned and maintained. Especially important concerning electrical stimulation are the parameters of the electrical impulses that work on the bees: maximum voltage, pulse width, maximum current that can pass through a bee.

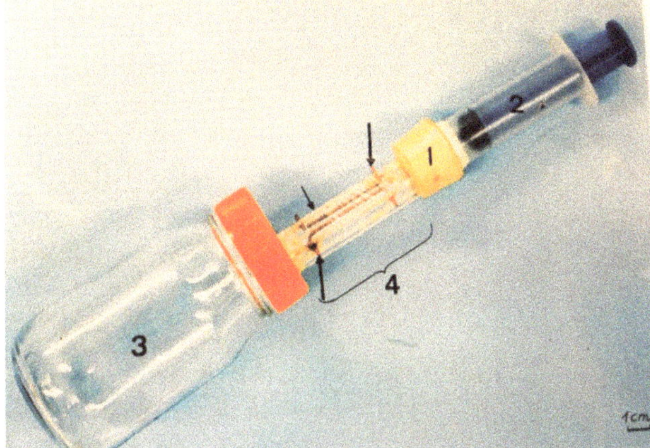


Figure 2. Manufactured device used to extract venom from individual honey bees; 1 = cork; 2 = piston; 3 = glass container; 4 = acrylic tube; smaller arrow = plastic; larger arrows = copper wire.

In order to best limit potential damage to bees, the power must be gauged continuously, and limited to a harmless level. The maximum voltage of the pulse needs to be kept in proportion to ensure that it is harmless, and should be maintained at this level through the full device work cycle. The electrical impulses that stimulate bees must be extremely short to avoid irreversible bee health deterioration. It is best if the bee venom collector is able to work autonomously, using a battery power source. Any unnecessary external wiring will needlessly irritate bees.

There are two ways to collect bee venom – external and internal, based on where the device is located. When the device is placed inside the hive, the whole colony becomes stressed, work is interrupted, and the temperature inside the hive rises to dangerous levels. This could be very harmful during summer. The less invasive and suggested way of bee venom collection is placing the device outside the hive, by the entrance, which avoids most negative impacts (Figure 1).

Bee venom collection has no impact on bees' survivability during winter. Honey production might be lower by about 10% if bee venom collection is done daily and more than three to four hours per day. Weekly, a break of two to three days should be introduced. The bee venom collected for a period of a few hours of uninterrupted stimulation is less than the amount collected for the same time, but with breaks included. The optimal time for electro-stimulation is 30–60 min, while the optimal break time is 45–90 min. Tests have been conducted during which electro-stimulation has been introduced for a four-hour period, with short one–minute breaks every 30 min, required to collect the venom. The result is that bee venom secretion drops exponentially with time, with the peak being in the first 30 min. In the last 30 min, the amount falls to 10% of the peaks collected. Some brands have an apparatus with a function to automatically power down after, for instance, 45 min.

An apparatus for venom collection have been constructed by Abreu et al. (2000) to collect venom from

A

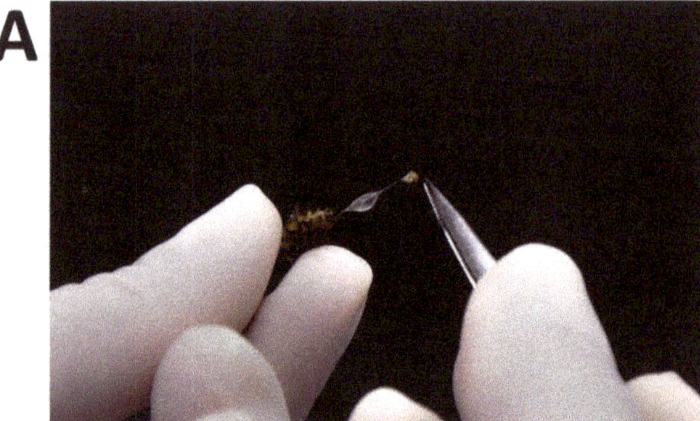

B

Figure 3. Steps of dissection and separation of the venom reservoirs from the sting apparatus. (A) Pull venom reservoir together with sting apparatus out of the bee's body using a fine-tipped tweezers; (B) with the two structures completely removed from the bee's body, but still stuck together, use two tweezers (or one micro scissors and one tweezers) to separate one from another.

single specimen (Figure 2). It consists of an acrylic tube with a section area equal to 0.66 $cm^2$. A plastic PVC film replaces its base, in which are arranged two parallel copper wires, leaving between them a space of 0.5 cm. The system is powered by a 12-volt battery. The worker bee is induced to penetrate the acrylic tube and due to lack of space, can just walk through it without flying. At the time that its legs touch simultaneously the two copper wires, it receives an intermittent shock for 10 sec, corresponding to an electrical current of 29 mA. At this current intensity, bees respond by stinging the plastic bottom, without losing their sting apparatus.

Here we present, by way of example, the operational procedure guidelines of Extract and Box Limited, New Zealand, which commercializes devices for venom collection at the hive entrance.

*2.1.1. Equipment sanitization*

(1) Wash glass slides (= collection plate) with degreaser dish soap, air dry.

(2) Wipe glass slides and electro-stimulation wires with isopropyl alcohol to sterilize.

*2.1.2. Precipitation and temperature check*

Do not collect bee venom in the presence of dew or with anticipated rain. Only collect bee venom using the hive entrance placement technique when the ambient temperature is above 12 °C. These parameters are based around the natural tendencies of the hive colony and are established to reduce any harm caused to the colony.

*2.1.3. Collection*

(1) Place collector at entrance of hive.

(2) Activate collector when guard bees have advanced onto the electro-stimulation wires.

(3) Collect venom by scraping the dried venom (powder) from the glass slide using a sharp scraper.

During the off-season collection months, it is possible to enhance the electro-stimulation by fanning the released isopentyl acetate and other components of the alarm pheromone blend, with a sterilized smoker bellows toward the hive entrance. On completion of the collection cycle, remove the glass slide and lay another slide on top of it (venom exposed surfaces against each other) to reduce oxygen exposure. Reduce atmospheric moisture by encasing both glass slides in plastic wrap. By placing the slides in a dark container you can actively reduce catalytic light exposure during transport to the processing facility.

Remove the plastic wrap from the plates, scrape down the dried venom from the glass using a sharp scraper and weigh the collected venom powder. With the dispatch packaging selection of a 15 ml amber glass bottle you can further reduce light exposure of the venom sample. Within the transfer tank system you can reduce the atmospheric oxygen by displacing the air and create a nitrogen blanket. It is also possible to add further moisture controls by including the placement of a food grade silica sachet in the dispatch bottle while in the transfer packaging process. Remove the bottle from the tank and label it for dispatch. Record in processing, log the date, weight, sites collected from and issue a batch number. Depending on the equipment dimensions, the intensity of the swarm and season, it is possible to collect an average of 1 g per hive in 40 min (e.g., strong swarms, in summer) (Benton et al., 1963).

### 2.2. Reservoir disruption

The reservoir disruption protocol is designed for venom collection at a laboratory scale (Schmidt, 1995). It is one of the methods that avoid unwanted processes to occur between venom deposition on the glass plate and the collection as seen in the electric stimulation protocol (drying, oxidation, degradation).

*2.2.1. Reservoir dissection*

(1) After collection, anesthetize the insects at low temperature (–20 °C) (see the corresponding *BEEBOOK*

chapter; Human et al., 2013) . This will induce bees to go dormant (state of torpor, hibernation-like). According to Wieser (1973) cold temperatures can temporarily immobilize adult honey bees by reducing the amplitude of neuron action potentials.

(2) Keep the insects in petri dishes on ice.

(3) Remove the sting apparatus along with the venom reservoirs by pulling them out of bee bodies with two fine-tipped tweezers as shown in Figure 3.

(4) Use the same tweezers to carefully separate the venom reservoirs (or sacs) from the sting apparatus. For a detailed description of the honey bee anatomy please refer to Carreck et al. (2013).

(5) Rinse the reservoirs, still attached with the tweezers, soon after dissection very carefully and quickly in 10 ml to 50 ml of distilled and deionized water at 4 °C to avoid contamination with hemolymph and/or other secretions.

(6) Sequentially dry each one gently (just by a light touch) on a filter paper. At this point, the reservoirs can be lyophilized in an appropriate tube or suspended in distilled and deionized water at 4 °C as described below in section 2.2.2.

Note: Depending on the purpose of the study, after washing and very gentle drying of venom reservoirs in water, they can be suspended in an aqueous solution at 4 °C, containing a cocktail of protease inhibitors (2 mM AEBSF, 0.3 μM aprotinin, 130 μM bestatin, 1 mM EDTA, 14 μM E-64 and 1 μM leupeptin, Sigma-Aldrich) in a proportion of 1:1 (reservoir: solution).

(7) Store the suspended reservoirs (in water or aqueous solution) at −20 °C for short term (up to six months) or lyophilized at −80 °C for extended storage (more than one year).

#### 2.2.2. *Reservoir disruption and extract preparation*

Using fresh reservoirs:

(1) Suspend the freshly dissected reservoirs (20-60) in a small volume of distilled and deionized water at 4 °C in an Eppendorf tube (1.5 ml).

(2) Macerate them using a micropestle specially designed for microtubes. The ratio of 1:1 (reservoir: solvent), whether or not containing a cocktail of protease inhibitors, is recommended.

(3) Centrifuge the suspensions at 10,000 g for 15 min at 4 °C.

(4) Discard precipitates and transfer supernatants (crude venom extract) to a new microcentrifuge tube. These extracts can be used as source of enzymes, proteins and other venom components.

Using lyophilized reservoirs:

(1) Hydrate lyophilized and frozen reservoirs (at −20 or −80 °C) in the proportion described above (ratio of 1:1; reservoir: solvent).

(2) Macerate with slight pressure using a micropestle to cause disruption.

(3) Thaw with subsequent release of the venom to the suspension.

(4) Proceed with the suspensions in the same way as described for fresh reservoirs.

### *2.3. Manual milking*

Since the introduction of mass spectrometric analysis of bee venom, another sampling technique was introduced: the so-called manual milking (Peiren et al., 2005). Because of the extreme sensitivity of the mass spectrometry for determining the composition of the venom (for instance: Fourier transform ion cyclotron resonance mass spectrometry), the researchers wanted to avoid that the wound or the hemolymph proteins or hemocytes that were attached to the external side of the reservoir would contaminate the venom sample (van Vaerenbergh et al., 2014).

The procedure is very similar to the reservoir disruption, and is as follows:

(1) Remove the sting apparatus along with the venom reservoirs by pulling them out of bee bodies with two fine-tipped tweezers.

(2) Keep the sting submerged in sample buffer.

(3) Simultaneously, exert pressure on the reservoir by a pinching movement between two fingers (using examination gloves) to push out the venom and to release it through the sting.

The venom is stabilized immediately upon release in sample buffer. The procedure works only with freshly dissected reservoirs and sting apparatus.

### *2.4. Controlling de novo venom production: reservoir emptying and replenishment*

Abreu et al. (2000) conducted a study on the resynthesis process of *Apis mellifera* venom, during three consecutive summers and two winters. To this study, worker bees at ages of 7, 14, 21, 28, 35 and 40 days, after having received electrical shocks of 29 mA (using the device for venom collection of single specimen; see section 2.1) were confined in varying numbers by age for a period of 0, 24, 48 and 96 hours after shock. Then, their venom glands were dissected for analysis of the venom protein content as well as by morpho-histological techniques. Results showed that compared to control group no change in the histology of the secretory tubules was observed over the 96 hours after the extraction of venom, neither at all ages studied nor during the summer and winter. It was demonstrated that the venom replacement that occurs over a period of 96 hours after extraction is insignificant and independent of the age of the workers as well as of the season, and besides has no association with histological changes of glandular secreting cells.

The procedure of subsequent venom samplings from the same bee specimens is as follows:

(1) Keep brood combs containing workers about to emerge in an incubator at 33 °C for 24 h.
(2) Remove about 600 bees from different colonies for each experiment/season.
(3) Mark the newly emerged bees on the thorax with synthetic enamel Suvinil (see the corresponding *BEEBOOK* chapter; Human et al., 2013) .
(4) Return the marked bees back to the colony.
(5) Collect the bees at their age of interest. The highest protein content in venom of worker bees is found at the age of 28 or 21 days during summer or winter, respectively (Abreu et al., 2000).
(6) Proceed the venom extraction using an extracting apparatus for single specimen at a time (see section 2.1).
(7) Collect bees soon after having received electrical shocks in a glass container that is connected to the acrylic tube (see Figure 2).
(8) Transfer the bees from the glass container to a laboratory cage (see the corresponding *BEEBOOK* chapter; Williams et al., 2013) .
(9) Gradually collect bees in varied number, by age, at the selected time (for instance: 24 and 96 hours) post-shocks from the laboratory cages.
(10) Perform another round of venom collection starting at step 6.

## 3. Honey bee queen venom collection

Qualitative and quantitative differences have been demonstrated through several studies in venoms of queens, winter and summer workers honey bees (de Abreu et al., 2010; Nocelli et al., 2007; Owen, 1979; Owen & Bridges, 1976; Schmidt, 1995; van Vaerenbergh et al., 2013; Vlasak & Kreil, 1984). However, there is only a limited number of papers that describe collection of venom from honey bee queens. The most recent study (Danneels et al., 2015) used the same manual milking method as described in section 2.3. as was designed to perform the extreme sensitive nano-liquid chromatography FT-ICR MS/MS approach. For queen venom collection, both reservoir disruption (see section 2.2) and manual milking (see section 2.3) are the methods of choice as they allow individual application.

## 4. Venom characteristics and composition

Efforts to characterize bee venom or to determine its composition can in itself be a research objective, for example when searching for new sting allergens or components with a biomedical application. It can also be part of a quality control in the production process that strives for a characteristic composition of the product.

The integrity of whole bee venom and of its components is highly dependent on the chosen methodology, care and handling the extract preparation and these in turn are related to the type of research work to be accomplished. Particular care should be taken to prevent possible events of oxidation, proteolysis and precipitation (by denaturation) of the components from venom extract, mainly if more refined analyses as high performance liquid chromatography (HPLC), capillary zone electrophoresis (CZE), mass spectrometry (MS) and microscopic examination will be developed. In all of these cases it is recommended to prepare the venom extract with a cocktail of protease inhibitors (see section 2.2.1) or at least using only one protease inhibitor (e.g., 1 mM PMSF). From our experience (Brochetto-Braga et al., 2005) 10% glycerol has proved to be an interesting protective to maintain the enzymatic activities during and after ultrafiltration chromatography gel processes.

Total protein content determination is a primary and essential part of any work involving proteinous secretion, including bee venom. However, the most common methodologies for this, like the Bradford assay and the bicinchoninic acid (BCA) assay, were already well explained in another chapter of the *BEEBOOK*, to which we would like to refer (Hartfelder et al., 2013). The same applies for the methods of proteins separation and detection in polyacrylamide gels using Coomassie Brilliant Blue and silver salts staining (Hartfelder et al., 2013). In this section we will focus on two common techniques used to examine the composition of bee venom: high performance liquid chromatography and mass spectrometry. The former is a common separation technique for profiling the venom composition with both analytical and preparative applications. The latter can also be used for profiling but has the advantage of identifying the venom compounds in very low concentrations, even the ones that remained undiscovered.

### *4.1. High performance liquid chromatography*

High performance liquid chromatography (HPLC) is commonly used for separation and identification of most components of honey bee venom. The most important compounds from protein/peptide fraction are: melittin, apamine, MCDP, tertiapin and some enzymes (phospholipase A2 (PLA2), hyaluronidase). Due to the lack of reference materials with certified values of bee venom components it is very difficult to confirm the results obtained. Applying an internal standard is an absolute necessity in order to eliminate the matrix effects and to improve accuracy and precision.

The method is as follows:

(1) Prepare 25 µg/ml cytochrome c solution (internal standard) in deionized water and use it as diluent of all samples.

(2) Prepare the honey bee venom solutions by diluting 3 mg of the product in 10 ml of internal standard solution (see step 1).
(3) Prepare the standard solutions of apamine, mast cell degranulating peptide, PLA2 and melittin by dissolving them in internal standard solution. The solutions will be used for calibration curves construction. The recommended concentrations of standard solutions are as follows (at least six dilutions): apamine – from 2 to 20 μg, mast cell degranulating peptide – from 5 to 30 μg, PLA2 – from 10 to 100 μg, melittin – from 30 to 300 μg. Sonicate all prepared solutions for 5 min and then filter through 0.45-μm membrane filters.
(4) Perform HPLC analysis of prepared standard solutions and honey bee venom solutions.
    - Use the SynChropack C8 6.5 μm, 4.6 x 100 mm column or the column with the same parameters.
    - Separation conditions: linear gradient 5% B – 80% B at 30 min (eluent A – 0.1% TFA in water, eluent B – 0.1% TFA in acetonitrile: water (80:20)); flow rate = 1 ml/min, injection volume = 40 μl, separation temperature = 25 °C, λ = 220 nm.
(5) Identify apamine, mast cell degranulating peptide, PLA2 and melittin using their retention times.
(6) Construct the calibration curves for apamine, mast cell degranulating peptide, PLA2 and melittin using corresponding relative peak areas (peak area of an analyte divided by peak area of internal standard).
(7) Calculate the concentrations of analyzed honey bee venom from the standard calibration curves equations.

### *4.2. Mass spectrometry*

For a comprehensive qualitative analysis of the protein content of bee venom, liquid chromatography-mass spectrometry (LC-MS) is the method of choice (de Graaf et al., 2009). However, the method is compromised by the large differences in concentration of individual proteins in the venom extracts. In particular, peptides originating from mellittin and the different isoforms of phospholipase C are overly abundant in tryptic digest mixtures performed on bee venom samples. To extend the dynamic range of proteomic analysis, several methods are described to pre-fractionate proteins prior to analysis. Typically, cocktails of antibodies or other affinity reagents are used to remove highly abundant proteins. We here outline the use of a commercially available combinatorial peptide library, which act as an "equalizer". This peptide library consists of hexapeptides coupled to beads. Saturation results in the removal of surplus abundant protein during the washing steps. Proteins recovered from the beads are separated by SDS-PAGE, cut into peptides and then analyzed via LC-MS. We describe here the use of an FT-ICR-MS mass spectrometer, though the peptide analysis can be performed on any high-resolution mass spectrometer designed for use in proteomic analysis (Orbitrap, Q-TOF). It should be emphasized that by using the peptide library approach, quantitative information on differences in protein abundancy are lost.

#### *4.2.1. Venom sample preparation*

Venom of 100–150 honey bee workers is collected as described above (see section 2). The venom is pooled, yielding a protein concentration of +/– 60 mg/ml.

#### *4.2.2. Protein enrichment*

For peptide library enrichment, the small-capacity protein enrichment kit (Proteominer, Bio-Rad Laboratories, catalog# 163-3006 and 163-3008) is used. The protocol is based on the manufacturer's instructions. The kit contains spin columns, collection tubes and washing and collection buffers. We refer to the manufacturer's instructions for buffer content and safety measures.

(1) Place the spin column in a capless collection tube.
(2) Centrifuge at 1000 g during 60 sec to remove the storage solution.
(3) Discard the collected material, which is the flow-through (storage) solution.
(4) Replace the bottom cap.
(5) Add 200 μl washing buffer.
(6) Replace the top cap.
(7) Manually rotate the column end-to-end for five minutes.
(8) Remove the bottom cap.
(9) Place the column in a capless collection tube.
(10) Centrifuge at 1000 g during 60 sec to remove buffer.
(11) Discard collected material.
(12) Repeat steps 4–11.
(13) Replace the bottom cap on the spin column
(14) Centrifuge the venom extract.
(15) Remove any precipitate.
(16) Pipette 200 μl of venom extract on the column.
(17) Replace the top cap.
(18) Rotate the column on a platform or rotational shaker.
(19) Incubate at RT during two hours.
(20) Remove the bottom cap.
(21) Place the column in a capless collection tube.
(22) Centrifuge at 1000 g for 60 sec.
(23) Recover collected material: this contains unbound protein and might be used for other experiments.
(24) Replace the bottom cap.
(25) Add 200 μl of washing buffer (150 mM NaCl, 10 mM $NaH_2PO_4$, pH7) to the column.
(26) Replace the top cap.
(27) Rotate from end to end during five minutes.
(28) Remove the bottom cap.

(29) Place column in a capless collection tube.
(30) Centrifuge at 1000 *g* for 30 sec.
(31) Discard collected material.
(32) Repeat steps 24–31 three more times.
(33) After all wash buffer has been removed, replace the bottom cap.
(34) Add 200 μl deionized water.
(35) Attach the top cap.
(36) Rotated end to end for one minute.
(37) Remove the cap.
(38) Place column in a capless collection tub.
(39) Centrifuge at 1000 *g* for 30 sec.
(40) Discard collected material.
(41) Attach the bottom cap to the column.
(42) Add 20 μl of rehydrated elution reagent (8M urea, 2% CHAPS, 5% acetic acid) to the column.
(43) Replace the top cap.
(44) Lightly vortex for five seconds.
(45) Incubate column at RT.
(46) Lightly vortex three times over a period of 15 min.
(47) Remove the cap.
(48) Place in a clean collection tube.
(49) Centrifuge at 1000 *g* for 60 sec. This elution contains the proteins. Do not discard.
(50) Repeat steps 41–49 two times.
(51) Pool the elutions.
(52) Quantify the protein content using conventional protein concentration analysis methods (Bradford or BCA method as explained in another *BEEBOOK* chapter; Hartfelder et al., 2013).

*4.2.3. SDS-page*

In this protocol, we describe the use of SDS-PAGE on a 10% SDS-PAGE. When interested in the small proteins $<$ 10 kDa), a Tris-tricine-SDS-PAGE (Schägger & VON Jagow, 1987) is preferred.

***4.2.3.1. Casting the SDS-PAGE gel.*** To cast an SDS-PAGE, a commercial casting stand (mini gel format) and frame can be used. Assemble the casting stand and place two glass plates separated by a spacer in the holder. Check for leaks using deionized water.

(1) Prepare the separating gel solution by mixing 3.8 ml deionized water with 3.4 ml acrylamide/bisacrylamide (30%/0.8%w/v), 2.6 ml 1.5 M Tris (pH 8) and 0.1 ml 10% SDS. Mix this solution.
(2) Add 100 μl 10% (w/v) ammoniumpersulfate.
(3) Shake the solution gently.
(4) Add 10 μl TEMED.
(5) Shake gently.
(6) Pipette the solution between the glass plates to approximately 2 cm below the edge of the front plate.
(7) Pipette isopropanol into gap until overflow.
(8) Allow the gel to polymerize 20–30 min.
(9) Meanwhile prepare the stacking gel solution by mixing 3.975 ml deionized water with 0.67 ml acrylamide/bisacrylamide (30%/0.8%w/v), 1.6 ml 0.5 M Tris-HCl (pH 6.8) and 0.05 ml 10% SDS.
(10) Discard the isopropanol.
(11) Add 0.05 ml 10% ammoniumpersulfate to the stacking gel solution.
(12) Shake gently.
(13) Add 0.005 ml TEMED.
(14) Shake gently.
(15) Pipette the solution above the separation gel.
(16) Insert the well-forming comb without trapping air under the teeth.
(17) Wait at least for 20–30 min for gelation.
(18) Store the gel for 2–3 h on RT to allow polymerization to complete. Alternatively, store the gel overnight at 4 °C.

***4.2.3.2. Preparation for gel electrophoresis.***

(1) Take out the comb.
(2) Take the glass plates with the gel out of the casting frame and set them in the cell buffer tank.
(3) Pour running buffer (25 mM Tris-HCl, 200 mM glycine, 0.1% SDS) into the inner chamber and continue after overflow until buffer reaches the required level in the outer chamber.
(4) Prepare 1-ml sample buffer solution containing 10% (w/v) SDS, 10 mM dithiothreitol, 20% (w/v) glycerol, 0.2 M Tris-HCl, pH6.8 and 0.05% (w/v) bromophenolblue. This is a 5x concentrated buffer that should be diluted 5x before use.
(5) Mix 50 μg of eluted proteins from the protein enrichment step with appropriate amount of sample buffer to a total volume of 10 μl.

***4.2.3.3. Running the gel electrophoresis.***

(1) Load this sample into a well, preferably the third or fourth lane. Make sure not to overflow.
(2) Add marker protein solution in the first lane.
(3) Cover the top and connect electrodes.
(4) Set the voltage to 100 V. Continue electrophoresis until the blue bromophenol front reaches the bottom of the gels.
(5) Disassemble the gel cassette and place the gel in a box for washing steps.

*4.2.4. Staining and protein modification*

(1) Prepare the staining solution: dissolve 70 mg of Coommassie Brilliant Blue G-250 in 1 l of deionized water.
(2) Stir for 2–4 h. Add 3 ml of concentrated HCl (concentrated HCl should be used in a fume hood, wear protective gloves). This solution can be stored in the dark for several weeks.
(3) Wash the gel by adding 100 ml deionized water.
(4) Heat for a few minutes at 50 °C (or 30 sec when using microwave oven).

(5) Place the box with the gel on a shaker for 3–5 min.
(6) Replace the water twice and repeat shaking.
(7) Add Coomassie Brilliant Blue G250 solution to completely cover the gel.
(8) Heat the box in an oven (or in a microwave oven) for 10 sec without boiling.
(9) Place the box on a shaker until visualization of the bands (typically after 15–30 min).
(10) Pour off the staining solution.
(11) Add 50–100 ml double distilled water to destain the background of the gel.
(12) Place the box on a shaker. If necessary, replace with fresh water. Alternatively, 30% methanol solution can be used.
(13) Scan or photograph the gel.
(14) For mass spectrometry analysis, cysteine bridges are preferably reduced and alkylated. Therefore, replace the destaining solution by reducing solution (25mM dithiothreitol/25 mM $NH_4HCO_3$).
(15) Place 45 min at 56 °C.
(16) Replace reducing solution by alkylating solution (55 mM iodoacetamide/25 mM $NH_4HCO_3$).
(17) Place at RT for 45 min.
(18) Remove the alkylating solution.
(19) Replace by 25 mM $NH_4HCO_3$.

*4.2.5. Protein digestion*

(1) Cut the gel in slices and place them in 0.5 ml Eppendorf tubes. Also gel parts without any visible protein bands are excised.
(2) Wash the residual Coomassie staining with 150 µl of 200 mM $NH_4HCO_3$/50% acetonitrile for 30 min at 37 °C.
(3) Dry the gel pieces in a Speedvac.
(4) Add 12 µl of trypsin solution (0.002 µg/µl in 50 mM $NH_4HCO_3$). Use sequencing grade trypsin (e.g., modified trypsin from Promega).
(5) Allow the trypsin solution to be absorbed by gel.
(6) Add 25 µl of 50 mM $NH_4HCO_3$. Confirm that gel pieces are completely immersed.
(7) Incubate overnight at 37 °C.
(8) Collect the supernatant of each gel piece in a separate Eppendorf tube. This contains mainly hydrophilic peptides.
(9) Extract hydrophobic peptides by two sequential incubation steps (15 min at 30 °C) with, respectively, 60 and 40 µl of 0.1% formic acid/60% acetonitrile.
(10) Pool the extract with the corresponding hydrophilic peptide extract.
(11) Dry the combined extracts in a speedvac.

*4.2.6. Liquid chromatography-mass spectrometry*

(1) Dissolve the dried peptide extract in 15 µl 2% acetonitrile/0.1% formic acid.
(2) Vortex.
(3) Pipette 5 µl in a HPLC sample vial.
(4) Analyze the sample using a 1–3h gradient using LC-MS. Most laboratories will access such a system via a core facility or specialized laboratory. Interested readers are referred to specialized literature.

*4.2.7. Data analyses*

The raw LC-MS data file can be processed by the freely available software tool Maxquant for database analysis. Background on the principles underlying such database searching has been covered elsewhere (Cox & Mann, 2008). Database searching should be performed against the NCBI protein database selecting *A. mellifera* as the organism.

## 5. Storage of venom

According to Krell (1996) – since bee venom is not considered as an official drug or as a food – there are no official quality standards regarding its composition. The quantitative analyses of some of the bee venom components (e.g., melittin dopamine, histamine among others) have been an indirect way of measuring its purity and quality, and these in turn can be strongly affected by the venom storage method. Degradation of the venom components (e.g., by autolysis due to presence of proteases in the whole bee venom) can be avoided by using a drying method such as the low-temperature vacuum freeze-drying. This method can be used in both the venom obtained by electric stimulation (see section 2.1) and the freshly dissected reservoirs (see section 2.2) soon after being collected.

### *5.1. Short-term storage*

Dried venom or venom extracts, obtained from reservoir disruption (see section 2.2) or electric stimulation (see section 2.1), followed by water/buffer resuspension, can be kept refrigerated in a well-sealed dark bottle for a few weeks (Krell, 1996). The addition of 10% glycerol proved to be a good strategy to protect enzymatic activities (Brochetto-Braga et al., 2005). However, this stability can be improved by the additional use of a cocktail of protease inhibitors as referred to earlier (see section 2.2.1).

### *5.2. Long-term storage*

To extend the storage period (several months), both the dry venom and venom extracts should also be kept in very well sealed amber bottles, but frozen at −80 °C. The integrity of bee venom can be also achieved by lyophilization at low temperatures of the freshly dissected venom reservoirs and their subsequent freezing by up to six months (–20 °C) or years (–80 °C).

Table 1. 'Pros and cons' to help selecting the appropriate method for the determination of hyaluronidase activity.

| Method | Pros | Cons |
|---|---|---|
| Capillary Zone Electrophoresis | - Based on the determination of three of the HA degradation products (disaccharide, tetrasaccharide and hexasaccharide).<br>- Very good precision, accuracy and linearity in the HYASE activity range from 16.4 to 124.2 U.<br>- It allows evaluating the HYASE activity originating from different sources.<br>- Rapid test, small sample volume, small amount of buffers used and relative low cost of analysis.<br>- Recommended for specific studies on HYASE activity involving HA degradation products.<br>- Fully compatible with the turbidimetric and viscosimetric methods.<br>- No HYASE purification procedure required. | - Requires specific equipment.<br>- The capillary should be thoroughly washed after each analysis.<br>- Sample dilution is required when HYASE concentration exceeds the operating range.<br>- More complex method compared to the turbidimetric and viscosimetric methods, especially when larger numbers of samples should be analyzed. |
| Turbidimetry | - Very simple, rapid and reproducible method for the determination of HYASE activity and of 10 to 200 μg of isolated acid mucopolysaccharides.<br>- Based on the determining of the intact, non-digested HA molecule.<br>- Works with common and inexpensive reagents.<br>- Very useful for analysis of numerous samples (for instance: chromatography fractions).<br>- It can be down-scaled to smaller volumes (microplate level).<br>- Requires no specific equipment (only spectrophotometer). | - If the acid mucopolysaccharide is dissolved in distilled water instead of in salt solution, much less turbidity develops on addition of CTAB.<br>- Low specificity. |
| Viscosimetry | - High sensitivity and reproducibility.<br>- Based on the decrease in viscosity of the substrate HA by the HYASE activity.<br>- Preferred method for studying the kinetics HA degradation and screening of HYASE inhibitors. | - Requires specific equipment.<br>- Not recommended for analysis of large number of samples. |

## 6. Determination of the biological activity of venom

### 6.1. Hyaluronidase activity

Honey bee venom hyaluronidase (Api m2) belongs to the family of glycosyl hydrolases (EC 3.2.1.35). Hyaluronidase (HYASE) is an enzyme capable of the hydrolysis of hyaluronic acid (HA), which is a natural adhesive of interstitial tissue and thus makes it possible to maintain the adhesive properties of cells in a compact structure. Digestion of HA causes increase of the permeability of the tissue and vascular walls, which leads to free diffusion of other allergens and toxic components of the venom. Api m 2 is the second major allergen of honey bee venom (Vetter et al., 1999). Approximately half of the population allergic to bee venom has IgE antibodies to hyaluronidase. Due to the similarity of the structure of bee and wasp hyaluronidase, the allergen is a major cause of cross-reactivity of IgE antibodies directed against the venoms of these insects. Because of the high importance of honey bee venom HYASE in allergy practice (venom allergy diagnosis and immunotherapy) and their high prevalence in biological fluids, tissues, venoms and toxins, it is necessary to use standard validated methods for determination of hyaluronidase activity. Several methods for determination of the enzyme activity were developed, based on a variety of techniques, including biological, physicochemical (turbidimetric (Pukrittayakamee et al., 1988; Queiroz et al., 2008; Morey et al., 2006; Magalhaes et al., 2008 ) and viscosimetric methods (Vercruysse et al., 1995)), chemical (colorimetric method based on Morgan-Elson reaction (Muckenschnabel et al., 1998)), fluorescence (Krupa et al., 2003; Nakamura et al., 1990; Zhang & Mummert, 2008), immunoenzymatic (ELISA) (Delpech et al., 1987; Stern & Stern, 1992), zymographic (Steiner & Cruce, 1992), SDS-PAGE (Ikegami-Kawai & Takahashi, 2002) and capillary zone electrophoresis (CZE) methods (Matysiak et al., 2013; Pattanaargson & Roboz, 1996). The literature suggests that turbidimetric and viscosimetric methods used for determining the HYASE activity are well correlated with the pharmacological activity of the enzymes when applied as a spreading factor (Humphrey & Jaques, 1953), which is one of the main roles of honey bee venom hyaluronidase. Both turbidimetric and viscosimetric assays are pharmacopoeial methods. In turn, the method based on CZE assay allows not only quantifying HYASE activity but also the monitoring of digestion of HA by HYASE in a wide range of time (from several minutes to several hours or days of digestion). Therefore, only these three methods are described in detail in this paper. Table 1 helps to select the appropriate method for the determination of hyaluronidase activity.

*6.1.1. Determination of HYASE activity by CZE*

The determination of hyaluronidase (HYASE) activity is based on the capillary zone electrophoresis analysis of degradation products of the hyaluronic acid (HA), formed as a result of the activity of HYASE present in an analyzed sample after half an hour of digestion at 37 °C (Matysiak et al., 2013). According to the literature (Hofinger et al., 2008; Koketsu & Linhardt, 2000) the degradation products of HA appear immediately after mixing of HA with hyaluronidase. In order to determine the HYASE activity, it is necessary to build a multiple regression analysis model.

(1) Prepare at least 10 solutions of HYASE of increasing concentrations in acetate buffer (pH 4; 2 mM).
(2) Place the resulting solution in an ultrasonic bath for 10 min.
(3) Filter through 0.45-μm membrane filters.
(4) Prepare a solution of HA (200 μg/g) in the acetate buffer (pH 4; 2 mM).
(5) Mix thoroughly.
(6) Stored at 5 °C HA solution is stable for two weeks.
(7) Pre-incubate solutions of HYASE and solution of HA for 15 min at 37 °C.
(8) Mix HYASE solutions with HA solution in the ratio 1:1.
(9) Incubate half an hour at 37 °C.
(10) Perform electrophoretic analysis of prepared mixtures.
   - Use 64.5 cm total length, 56 cm effective length, 75 μm ID and 360 μm OD uncoated fused-silica capillary.
   - Separation conditions: phosphate buffer pH = 8.10; voltage 20 kV; injection 7 sec; $\lambda$ = 200,00 nm, temperature = 25 °C.
   - New capillary must be conditioned by rinsing with NaOH (1 M) for 20 min, then with MeOH and NaOH (0.1 M) for 10 min, respectively.
   - Before every analysis rinse the capillary with NaOH (1 M), MeOH, deionized water and running buffer (50 mM phosphate buffer; pH 8.1) for one minute.
   - After every analysis wash the capillary with HCl (0.1 M), MeOH and deionized water.
   - Use peak areas of three main HA degradation products (terta-, hexa- and octasaccharides of HA) as explanatory variables in the multiple regression analysis model.
(11) Analyze solutions of the honey bee venom samples of unknown HYASE activity in the same way as in the analysis of solutions of this enzyme needed to build the multiple regression analysis model.
(12) The regression analysis results in creating an equation of the HYASE activity of the general formula:

$$\text{HYASE activity [U]} = aA + bB + cC + d$$

where: A, B, C – peak areas of HA degradation products: tetra-, hexa- and octasaccharides of HA, respectively; a, b, c – constants of proportionality calculated from the model; d – realization of the random component.

*6.1.2. Determination of HYASE activity by turbidimetry*

The method is based on turbidimetric determination of hyaluronic acid (HA) left intact by hyaluronidase (HYASE). The hexadecyltrimethylammonium bromide (CTAB), a cation surface active reagent, used in this assay stops the enzymatic digestion of HA by HYASE and produces turbidity. The degree of turbidity, developed when CTAB forms insoluble complexes with HA solution, is proportional to the amount of undigested HA in the system.

(1) Prepare a calibration curve performing series of mixtures consisting of 0.5 ml of 200 μg/g HA solution and 0.5 ml solution of bovine testicular HYASE of various activity.
(2) Thoroughly mix the obtained solutions.
(3) Incubate at 20 °C for 10 min.
(4) Add to all solutions 2 ml of 2.5% (w/v) solution of CTAB in 2% (w/v) NaOH solution.
(5) Incubate the mixtures at 20 °C for 15 min.
(6) Measure optical density (OD) at the wavelength of 410 nm against blank (1 ml acetate buffer, pH 4 + 2 ml of CTAB).
(7) Analyze solutions of the honey bee venom samples of unknown HYASE activity in the same way as the solutions needed for calibration curve.
(8) Repeat all the measurements three more times.
(9) Calculate the average activity of the analyzed samples with the standard deviation.

*6.1.3. Determination of HYASE activity by viscosimetry*

The method compares the hydrolysis rate of hyaluronic acid (HA) with the rate obtained for the reference product (a comparator) using the slope ratio test. The study is performed using a Ubbelohde capillary viscometer, capillary II, with the constant viscometer value of approximately 0.1 $mm^2/s^2$.

(1) Prepare the substrate solution dissolving 0.10 g of HA in 20 ml of distilled water.
(2) Mix with phosphate buffer of pH 6.4.
(3) Prepare the reference product solution dissolving an appropriate amount of HYASE in the in distilled water, so as to obtain a solution containing 0.6 IU of hyaluronidase per ml.
(4) Prepare the test solution dissolving an appropriate amount of honey bee venom in distilled water, so as to obtain a solution containing 0.6 ± 0.3 IU hyaluronidase per ml (Hoechstetter, 2005).
(5) Incubate at 37 °C.

Table 2. 'Pros and cons' to help selecting the appropriate method for the determination of PLA2 activity.

| Method | Pros | Cons |
|---|---|---|
| Colorimetry – method 1 | - Works with common and inexpensive reagents.<br>- Simple, fast and sensitive test (30–50 ng of enzyme).<br>- Very useful for analysis of numerous samples (for instance: chromatography fractions, where PLA2 is the major component).<br>- Requires no specific equipment (only spectrophotometer). | - Low specificity (other enzymes which catalyze hydrolysis of phospholipids and produce acids will also be detected).<br>- For quantitative results in absolute units it must be considered that the pH indicator may inhibit some phospholipases.<br>- The pH of dye solution may sometimes need to be readjusted (when a change in absorbance is observed). |
| Colorimetry – method 2 | - Simple, fast and sensitive test that is done at physiological pH.<br>- Works with common reagents, though a little more expensive than those from method 1.<br>- Recommended for diluted samples (1- 5 μg/ml) in 96 well microplates assays. | - Alternative to method 1, with the same disadvantages.<br>- Requires a microplate reader. |
| Spectrophotometry | - Rapid test that works common and inexpensive reagents.<br>- Recommended to use with purified PLA2 samples.<br>- Useful for kinetic studies of PLA2 inhibitors (for instance in the context of inflammatory processes). | - Requires N2 atmosphere to prepare aqueous phosphatidylcholine substrate.<br>- Lipoxygenase from soybean is used as the coupling enzyme.<br>- A clear assay medium is required, because any turbidity due to the phospholipase, lipoxygenase, or phospholipid substrate can interfere with measurements. |
| Fluorimetry | - Rapid, simple and very sensitive NanoDrop assay (10 ng of PLA2 render optimal enzyme activity).<br>- It uses a non-fluorescence molecule having fluorescence characteristics by virtue of its chemical nature that do not interfere with the reducing enzymatic activity of PLA2.<br>- It is specific for PLA2 activity.<br>- This assay can be employed for screening PLA2 activity from various sources in either purified or crude form and for analysis of numerous samples.<br>- Also useful for screening of library of compounds targeting inhibition of PLA2. | - Requires specific equipment (NanoDrop fluorospectrometer).<br>- More expensive than the other methods described here. |

(6) Mix the test solution with the substrate solution at the time of $t_l = 0$.

(7) Record the flow time $t_x$ of the test solution and substrate solution mixture several times during a period of ~20 min (Hoechstetter, 2005; Stern & Jedrzejas, 2006; Zhong et al., 1994).

(8) Calculate the viscosity ratio from the following formula:

$$\eta_r = \frac{k * t_X}{0.6915}$$

where: $k$ – viscometer constant in $mm^2/s^2$, $t_x$ – solution flow time in seconds, 0.6915– kinematic viscosity. Since the enzymatic reaction developed during measurement of the flow time, the actual response time is equal to $(t_l + t_x)/2$. Crossed out is $1/(\ln \eta_r)$ as a function of reaction time $(t_l + t_x)/2$ in seconds. It results in a linear interrelationship. Calculate the specific activity of HYASE in international units per milligram using the following formula:

$$A_b = \frac{b_t}{b_r} * \frac{E_r}{E_t} * A$$

where: $A_b$, A – the specific activity of HYASE in International Units per milligram for honey bee venom (test solution) and reference product, respectively; $b_t$, $b_r$ – the slope coefficients for honey bee venom (test solution) and the reference product, respectively.

(9) Perform the measurements three times.

(10) Calculate the average activity of honey bee venom HYASE with the standard deviation.

### 6.2. Phospholipase A2 activity

Phospholipase A2 (PLA2) (EC 3.1.1.4; phosphatide 2-acylhydrolase) was first identified by Neumann and Habermann (1954) in honey bee venom. Besides being an important (12% by dry weight) component in this venom (Habermann, 1972) it is also its most important allergen (Sobotka et al., 1976). This enzyme catalyzes the hydrolysis of membrane phospholipids at the sn-2 position yielding lysophospholipids and arachidonic acid (Habermann, 1972). There are several methods for measurement of this enzyme activity, some of which are commercialized. The latter will not be described here. Table 2 helps select the appropriate method for the determination of PLA2 activity.

#### *6.2.1. Determination of PLA2 activity by colorimetry*

***6.2.1.1. Pla2 colorimetric assay - method 1.*** The following protocol is based on measuring the pH change due to the fatty acids liberation by PLA2 activity. A pH indicator such as phenol red is used to detect these changes (Araújo & Radvanyi, 1987).

(1) Set up PLA2 assays as 2.5 ml (including the sample volume) of a mixture containing the following reagents at final concentrations:
   - 4 mM Tris-HCl pH 7.9.
   - 100 mM NaCl.
   - 5 mM $CaCl_2.2H_2O$.
   - 0.2 mg/ml phosphatidylcholine dissolved in 50 mM Triton X-100.
   - 0.088 mM phenol red as indicator.
(2) Start reaction from each sample by adding 40 to 80 μg (0.1–0.4 ml) of total venom protein.
(3) Make reaction blanks with no sample.
(4) Incubate all reactions for 5 min at 37 °C.
(5) Read absorbance at 558 nm, against blanks. One unit of activity is defined as the amount of enzyme necessary to hydrolyze 1 μmol of fatty acid/mg of total protein, based on a standard curve of known amounts of phosphatidylcholine versus absorbance at 558 nm performed under the assay conditions.

***6.2.1.2. Pla2 colorimetric assay - method 2.*** A second method was described by Price (2007) and works with inexpensive commercially available substrate for 96-well microplates. To this protocol, excepting the substrate, all reaction reagents can be mixed and stored at 4 °C. The substrate can be prepared in large amounts, aliquoted and stored at −70 °C. A single reaction mixture (total volume: 200 μl) comprises the final concentration of the following:

- 5 mM Triton X-100.
- 5 mM phosphatidylcholine (Sigma P5394 phosphatidylcholine (">60%"), Sigma, St. Louis MO). The stock solution of phosphatidylchloline is prepared by dissolution of 160 mg/ml in methanol, stored at −70 °C, and resuspended at 45 °C immediately before using. The methanol final concentration in the assay mixture is at 5% (v/v).
- 2 mM HEPES pH 7.5
- 10 mM calcium chloride.
- 0.124% (w/v) bromothymol blue dye dissolved in water

Protocol outline:

(1) Dilute samples in cold saline containing 2 mM HEPES at pH 7.5 just before starting assays (performed in triplicate).
(2) Apply to a 96-well microplate 20 μl of each sample or buffer (2 mM HEPES at pH 7.5) as negative control.
(3) Adjust pH and temperature to 7.5 and 37 °C, respectively.
(4) Add 180 μl of the assay mixture.
(5) Analyze plates immediately at 620 nm with a microplate reader apparatus, by reading at 1-min intervals.
(6) Verify the reaction linearity by a linear regression curve using a simple program like MSExcel 2000.
(7) Adjust the reaction rates – absorbance change per minute (averaged of assays performed in triplicate) – in relation to reaction blank.
(8) Report the enzyme activity as: absorbance change per minute x dilution factor x 1000/mg of total protein.

#### *6.2.2. Determination of PLA2 activity by spectrophotometry*

This method, described by Jiménez et al. (2003), involves a coupled assay using dilinoleoyl phosphatidylcholine (DL-PC) as substrate and lipoxygenase as the coupling enzyme. It is based on the oxidation by lipoxygenase of the linoleic acid released by the action of PLA2 on the dilinoleoyl phosphatidylcholine. According to Egmond et al. (1976), the oxidation of linoleic acid by lipoxygenase produces the corresponding hydroperoxide derivative ($\varepsilon_{234} = 25{,}000\ M^{-1}\ cm^{-1}$) that can be measured under a continuous and spectrophotometric form by the increasing of absorbance at 234 nm. The limit of sensitivity is approximately 0.4 nmol/ml of the reaction product. The protocol includes the following steps:

(1) Prepare substrate aqueous phosphatidylcholine:
   - Prepare a stock solution of dilinoleoyl phosphatidylcholine in chloroform at 12.5 mg/ml.
   - Dry aliquots of this solution under a $N_2$ flow.
   - Disperse the obtained film to a final concentration of 1.3 mM, in 10 mM deoxycholate dissolved in 50 mM Tris buffer, pH 8.5.
   - Incubate the substrate solution for 10 min at 25 °C (Reynolds et al., 1994).
(2) Assemble, to each assay performed in triplicate, a reaction mixture (1 ml) of 50 mM Tris buffer, pH 8.5, containing the following components at final concentration:
   - 3 mM deoxycholate.
   - 65 μM DL-PC.

- 0.23 µg/ml lipoxygenase.

(3) Start reaction with addition of samples.
Note: The linearity of this assay showed to be between 0 and 0.4 µg/ml to the concentration of the purified hog pancreatic phospholipase A (EC 3.1.1.4; 500 U/mg; Fluka, Spain) (Jiménez et al., 2003).

(4) Make negative controls (blanks) without one of the two components in the assay: lipoxygenase or sample containing PLA2.

(5) Incubate at 25 °C.
Note: Since this is a continuous spectrophotometric method, the reaction time can be established for the better response in terms of the absorbance range at 234 nm. For instance, in Jiménez et al. (2003) the spectra analysis obtained for the oxidation by lipoxygenase of the linoleic acid released by the action of PLA2 on dilinoleoyl phosphatidylcholine was performed in the period of 4–20 min.

(6) Measure the increase in absorbance at 234 nm in the assays against reaction blanks.

(7) Report the enzyme activity as the average of the absorbance change for triplicate assays per minute/µg g of total protein.

*6.2.3. Determination of PLA2 activity by fluorimetry*

This improved assay described by Vivek et al. (2014) is based on fluorescence due to binding of 8- anilino-1-naphthalenesulfonic acid (ANS) to a hydrophobic core (the cationic group of choline head) of substrate 1,2-dimyristoyl-sn-glycero-3-phosphocholine (DMPC). DMPC is a non-fluorogenic and non-chromogenic phospholipid substrate and ANS, a strong anion, is used as an interfacial hydrophobic probe. When ANS binds to substrate, it acts as a fluorescent molecule (Matulis & Lovrien, 1998). The ANS-based fluorescence assay follows PLA2 hydrolysis by direct substrate consumption and change to a hydrophilic environment, causing a decrease in relative fluorescence units (RFU).

Protocol steps:

(1) Prepare DMPC substrate (working solution):
- 1 mM DMPC in methanol containing 2 mM Triton X-100 in Milli-Q water.
- Spin substrate solution at 1000 *g* for 5 min to form uniform mixed micelles.

(2) Prepare quench solution to stop reaction:
- 50 µM ANS.
- 2 mM NaN3.
- 50 mM ethylene glycol tetra acetic acid (EGTA).

(3) Mix for a single reaction (1 ml final volume):
- 50 mM Tris/HCl buffer pH 7.5.
- 10 mM CaCl2.
- 10 µl of substrate (1 mM DMPC and 2 mM Triton X-100).

(4) Pre-incubate this reaction mixture for 5 min at 37 °C.

(5) Start reaction by adding sample ranging from 0 to 50 ng of protein.

(6) Incubate assay for 30 min at 37 °C.

(7) After this time, add 50 µl of quenching solution.

(8) Vortex for 30 sec.

(9) Incubate for 5 min at RT.

(10) Pipette 2 µl of this mixture in a NanoDrop equipment to measure RFU using excitation with ultraviolet (UV) LED (370 ± 10 nm) and emission at 480 nm in dark conditions.

(11) Calculate enzyme activity with the following equation:

$$\mathrm{DRFU} = \mathrm{RFU}_{(c)} - \mathrm{RFU}_{(t)}$$

where ΔRFU is the average of changes in RFU of triplicate assays with respect to control. The resultant RFU is compared with the standard curve of LPC to determine the PLA2 activity. $\mathrm{RFU}_{(t)}$ and $\mathrm{RFU}_{(c)}$ correspond to the RFU of DMPC with enzyme and without enzyme, respectively.

(12) Construct LPC standard curve:
- Use different LPC concentrations (0–300 µM).
- Set up a reaction mixture of 0.1 ml (final volume) containing 50 mM Tris/HCl buffer pH 7.5, 10 mM $CaCl_2$ and different concentrations of LPC.
- Add Triton X-100 in the ratio of 1:2.
- Incubate for 5 min at RT.
- Add 50 µl of quenching solution.
- Vortex for 30 sec.
- Incubate for 5 min at RT.
- Use 2 µl of this reaction mixture to measure RFU as described above.

Note: All reagents used in this assay are stable for up to six months at 4 °C.

### 6.3. Acid phosphatase activity

Acid phosphatases (AcP) (EC 3.1.3.2.) are a family of enzymes that non-specifically catalyze the hydrolysis of monoesters and anhydrides of phosphoric acid to produce inorganic phosphate at acidic pHs. These enzymes have been found and studied in venoms of ants (Schmidt & Blum, 1978), wasps (da Silva et al., 2004) ; Hoffman, 1977; and bees (Barboni et al., 1987; Georgieva et al., 2009; Grunwald et al., 2006; Hoffman, 1977). In honey bee (*Apis mellifera*) venom, the AcP is denominated Api m 3 and is considered as one of the three most potent venom allergens, together with the PLA2 and hyaluronidase.

The colorimetric assay is the most common, easy and cheapest to measure AcP activity. It has been described by Brochetto-Braga et al. (2005) and runs as follows:

(1) Set up reactions (2 ml final volume each) in triplicate by mixing:
- 1.9 ml of 5 mM (209 mg/100 ml) *p*-nitrophenyl sodium phosphate in 10 mM sodium acetate buffer, pH 5.5

Table 3. 'Pros and cons' to help selecting the appropriate method for the determination of protease activity.

| Method | Pros | Cons |
|---|---|---|
| **Non-specific proteolytic activity assay – method 1** | - Works with common and inexpensive reagents.<br>- Recommended for initial protease characterization studies in venom samples.<br>- The assays can be performed in the presence of various specific protease inhibitors (serino-, aspartic, metallo- and so on) giving clues about the classification of proteases existing in the sample.<br>- Very useful for analysis of numerous samples (for instance: chromatography fractions).<br>- Requires no specific equipment (only spectrophotometer). | - Low sensitivity<br>- Care should be taken in step 7 of this method (removal of the supernatant), thus avoiding contamination with the precipitate, which can significantly interfere with the reproducibility of the assays. |
| **Non-specific proteolytic activity assay – method 2** | - Alternative to method 1, but more sensitive due to chromogenic nature of the substrate Azocoll.<br>- Based on Azocoll hydrolysis yielding soluble, colored peptides in proportion to enzyme concentrations at fixed incubation times.<br>- As it is hydrolyzed by a diversity of proteases, is useful for initial protease characterization studies in venom samples.<br>- Very useful for analysis of numerous samples when performed in microplates. | - Care should be taken with the rigorous Azocoll prewash to eliminate any inhibitory material present in commercial preparations which may interfere with the linearity of the reaction over time.<br>- Due to the non-homogeneity of the substrate solution, the tubes or microplates containing the reaction mixtures should be shaken during all the incubation time to ensure maximum enzyme contact with the substrate. |
| **Serine protease activity** | - Simple, fast and specific assay of high sensitivity to detect serine proteases.<br>- Based on that cleavage of a single amide bond of the substrate converts the non-fluorescent bisamide substrate into a highly fluorescent monoamide product.<br>- Very useful in the study of proteolytic enzymes and for identifying inhibitors from different sources<br>- It works with small amounts of enzyme, in range of ng.<br>- Very useful for analysis of numerous samples (for instance: protease purification procedures). | - Requires specific equipment (fluorescence spectrophotometer equipped with a universal digital readout).<br>- More expensive than the other methods here described. |

- 0.1. ml of venom samples (7–20 µg of total protein) or only buffer to reaction blanks.

(2) Incubate for one hour at 37 °C.

(3) Stop reactions by adding 1 ml of 1 M NaOH.

(4) Read absorbance at 405 nm against reaction blanks (without enzyme). The reaction product p-nitrophenol is yellow at alkaline pH and its concentration is determined with base on the Lambert-Beer's law, considering its extinction coefficient ($\varepsilon_{405}$ = 16,900 $M^{-1}.cm^{-1}$).

(5) One unit of activity is defined as the amount of enzyme needed to liberate 1 µmol of p-nitrophenol/mg of total protein, under assay conditions.

### *6.4. Esterase activity*

Esterases (EST) (EC 3.1.1.1) are enzymes capable of hydrolyzing esters into an acid and an alcohol. They constitute a broad group of enzymes that exhibits many differences in relation to their substrate specificity, protein structure, and biological function. In insects, esterase acts extensively on several types of substrates (Turunen & Chippendale, 1977) and shows high polymorphism (Matthiensen et al., 1993). In Hymenoptera venoms the role of esterase is not known, but according to Schmidt et al. (1986) they may play a role in algogenicity, but not in lethality. Benton (1967) reported for the first time the presence of esterase activity in honey bee venom using α-naphthyl as substrate.

The colorimetric assay using α-naphthylacetate as substrate and FAST Garnet GBC for measuring esterase, provides a very common, easy and cheap method for determining esterase activity in honey bee venom.

(1) Prepare mixture reaction:
- 1.6 mM α-naphthylacetate in 300 mM phosphate/Na buffer, pH 7.0.

(2) Prepare stop solution:
- 0.1 mg/ml FAST Garnet GBC dissolved in distilled deionized water.

(3) Set up triplicate reactions for final volume of 2.2 ml using 3-ml disposable plastic cuvettes. Otherwise, the final volume of assays as well as of samples can be reduced to half by using 1.5-ml cuvettes.
(4) Begin pipetting samples (0.2 ml, ~ 10 μg of total protein) or buffer to blanks, keeping cuvettes on ice.
(5) Start reactions by adding 2 ml of mixture reaction to each cuvette.
(6) Close lids or cover with a piece of plastic film.
(7) Turn cuvettes up and down very gently to homogenize this solution.
(8) Incubate assays immediately at 37 °C for one hour.
(9) Stop reactions by adding 1 ml of 0.1 mg/ml FAST Garnet GBC solution.
(10) Incubate 10 min at RT.
(11) Read absorbance at 490 nm against reaction blanks.

Note: One unit of enzymatic activity is defined as the amount of enzyme needed to hydrolyze 1 μmol of α-naphthylacetate/μg of total protein, under the assay conditions. The amount of the product *p*-nitrophenol generated (yellow color) is determined by comparison to a standard curve performed with known quantities of this compound. The conversion of read absorbance into μmols of *p*-nitrophenol generated can be also accomplished considering that each 100 μmol of this compound gives a value of 0.100 at Abs 490 nm, under conditions above established.

### 6.5. Protease activity

Proteases or proteolytic enzymes are one of the wider and important groups of enzymes. Proteases are a subclass of the class of hydrolases and are able to hydrolyze peptide bonds from proteins and peptides. They constitute a large family divided into exopeptidases and endopeptidases or proteinases according to the peptide bond position to be cleaved into the polypeptide chain. The serine proteases include two distinct families: those like mammalian serine proteases (chymotrypsin, trypsin and elastase) and that like bacterial subtilisin protease. Although they have an active site and a common enzyme mechanism, they differ in amino acid sequence and three-dimensional structure. In Hymenoptera venom these enzymes have been increasingly studied and characterized, due to their important effects as allergen (de Lima et al., 2000;; Hoffman & Jacobson, 1996; Winningham et al., 2004) and/or on the victim's hemostatic system by acting on a variety of components of the coagulation cascade (Han et al., 2008). Table 3 helps selecting the appropriate method for the determination of protease activity.

#### *6.5.1. Determination of non-specific proteolytic activity*

***6.5.1.1. Non-specific proteolytic activity assay – method 1.*** This method described by McDonald and Chen (1965; with few modifications) uses casein.

Reagents:

- Substrate solution: 2% casein in 0.1 M glycine-NaOH pH 9.5 buffer. Keep it on ice.
- Stop solution: 50% (w/v) trichloroacetic acid (TCA) in distilled and deionized water.
- Lowry solutions freshly prepared soon before experiment: mix 2% (w/v) $Na_2CO_3$, 10% (w/v) NaOH, Na and K tartrate 2.5% (w/v) and 1% (w/v) $CuSO_4$ in the proportion of 100:10:1:1 (v:v:v:v), respectively.
- Folin reagent diluted 1: 1 (v:v) with water.

Procedure:

(1) Assemble triplicate assays and blanks for a final volume of 1 ml, so that they can be made in 1.5-ml microcentrifuge tubes.
(2) Add to each tube 0.5 ml of substrate solution.
(3) Start reactions by adding samples of venom extracts containing about 20 μg of total protein, or buffer for reaction blanks.
(4) Incubate at 37 °C for two hours.
(5) Stop reactions with 100 μl of 50% TCA for each tube. For this step it is possible to let tubes stand overnight at 4 °C to ensure a better precipitation of unreacted sample.
(6) Centrifuge tubes for 10 min at 5000 *g*.
(7) Remove carefully 900 μl of the supernatant using a micropipette.
(8) Bring supernatant in a new tube (5–10 ml).
(9) Add 4.5 ml of freshly prepared Lowry solution.
(10) Add 0.45 ml of the diluted (1:1) Folin reagent.
(11) Mix well.
(12) Let stand for one hour at RT.
(13) Read absorbance at 700 nm against reactions blank in order to quantify the content of free amino acid and released peptides by proteinase action.

One unit of proteolytic activity is defined as the quantity of liberated amino acids per microgram of total protein under the assay conditions using a standard tryptophane curve as reference. Otherwise, one unit of protease activity can be defined as the amount of enzyme required to produce an increase of 0.1 in optical density at 700 nm under the defined conditions.

***6.5.1.2. Non-specific proteolytic activity assay – method 2.*** This is a non-specific proteolytic activity assayed with a chromogenic substrate as described by Schmidt et al. (1986) for Hymenoptera venoms. Azocoll, the Azo dye-impregnated collagen (Sigma) is used as a chromogenic non-specific substrate for protease activity. Upon proteolysis, soluble peptide fragments which are purple in color due to Azo dye impregnation, are released and can be detected by absorbance at 550 nm (Jiang et al., 2007).

For assaying Hymenoptera venom proteinases we recommend the method described by Schmidt et al. (1986) that runs as follows:

(1) Triplicate reactions are made in 2.5 ml final volume consisting of:
- Venom samples (50 μg maximum).
- 12 mg azocoll (Calbiochem) dissolved in 2.5 ml of 0.1 M buffer (4.8 mg substrate/ml of buffer). As the pH-optimum of the venom proteases may differ, it is recommended to run the test in different pH ranges by using these different buffers: 0.1 M phosphate/Na pH 7.0; 0.1 N glycine/NaOH pH 9.2 and 0.1 N acetate/NaOH pH 4.0.
  Note: In order to perform assays in 2.5 ml final volume it is better to prepare a more concentrated azocoll stock solution (e.g., 5 mg/ml). Thus, less volume of this substrate solution and sample can be used (2.4 ml and 0.1 ml, respectively, in this example).

(2) Incubate reactions (in triplicate) and blanks (without enzyme) at 37 °C for two hours.
Note: Due to non-homogeneity of the substrate solutions, the tubes or microplates containing the reaction mixtures should be shaken throughout the incubation time to ensure maximum enzyme contact with the substrate.

(3) Then filter each solution rapidly.

(4) Read absorbance read at 520 nm against reactions blank.

Considering the use of the enzyme Trypsin (EC 3.4.21.4; Sigma) as reference, 1 U of protease activity is determined as units of trypsin activity by comparison with the reference.

### *6.5.2. Determination of serine protease activity*

This method uses fluorogenic substrate for serine proteinases, bis(N-benzyloxycarbonyl-L-argininamido) rhodamine [$(\text{Cbz-Arg-NH})_2$-rhodamine] as described by Leytus et al. (1983):

(1) Perform protease assays with (Cbz-Arg-NH)2-Rhodamine and 7-(N-Cbz-L-argininamido)-4-methylcoumarin at 22 °C in 10 mM Hepes buffer, pH 7.5, containing 10% (v/v) dimethyl sulfoxide.

(2) Dilute stock solutions of substrates and of enzymes into the 10 mM Hepes buffer, pH 7.5, before the assay.
Note: For all assays, the enzyme concentration (or sample total protein) should be chosen so that less than 5% of the substrate will be hydrolyzed.

(3) Mix 0.01 ml of enzyme solution with 0.04 ml of substrate solution at the bottom of a disposable plastic cuvette.

(4) Leave to react for 5 min.

(5) Add 0.95 ml of 10-mM Hepes buffer, pH 7.5, containing 15% (v/v) ethanol.

(6) Record the fluorescence immediately with a fluorescence spectrophotometer equipped with a universal digital readout.
Note: The excitation and emission wavelengths for (Cbz-Arg-NH)2-Rhodamine are 492 nm and 523 nm, respectively, both set with a bandwidth of 4 nm (Leytus et al., 1983). According to these authors the fluorescence spectrophotometer should be standardized with a polymethacrylate block in which are embedded Rhodamine B to ensure that the relative fluorescence is comparable in different experiments. The excitation and emission wavelengths for 7-(N-Cbz-L-argininamido)-4-methylcoumarin are 380 nm and 460 nm, respectively (Zimmerman et al., 1976; 1977), both set with a bandwidth of 4 nm.

(7) Convert relative fluorescence units into molar concentrations of Cbz-Arg-NH-Rhodamine or 7-amino-4-methylcoumarin produced by using standard curves correlating fluorescence with molar concentrations of either Cbz- Arg-NH-Rhodamine or 7-amino-4-methylcoumarin in 10 mM Hepes buffer, pH 7.5, containing 15% (v/v) ethanol.

## 7. Biomedical research – venom allergy

For most people Hymenopteran stings result in a transient and cumbersome local inflammatory response characterized by a painful, sometimes itchy, local wheal rarely exceeding 2 cm in diameter, surrounded by a swelling of the subcutaneous tissue. However, mass envenomation can be life-threatening and fatal. Besides these toxic reactions, Hymenoptera venom may also trigger hypersensitivity (allergic) reactions.

The routine diagnosis and treatment of venom allergy fall out of the scope of this *BEEBOOK* chapter. However, thanks to new mass spectrometric approaches (see section 4.2) our understanding of the bee venom composition has improved significantly in the past decades, which in turn has given a new dynamic to the biomedical research on bee venom allergens. As the identification of novel allergens among the newly discovered bee venom compounds demands a standardization of the methodology used, it is also covered in this *BEEBOOK* chapter. We have selected two antibody detection tests and two procedures of the basophil activation technique that are commonly used in this type of biomedical research. In addition, the procedure for the assignment of new allergens will be explained (section 7.3).

Since working with patients and human blood samples is subject to specific legislation and regulations, in particular with respect to ethics, privacy and bio-safety, the standard methods presented here below are intended only for appropriately trained researchers working in licensed institutions. We recommend that investigators who wish to enter this type of research inquire in advance with the competent authorities about the specific circumstances applicable in their country.

### *7.1. Antibody detection test*

Several tests for venom-specific antibody detection exist. Most commonly used are the measurement of venom-specific IgE antibodies (by enzyme-linked immunosorbent assay or Western blotting) and the intradermal venom skin test; the latter will not be covered in this *BEEBOOK* chapter as it is an *in vivo* test procedure. We will focus only on the enzyme-linked immunosorbent assay (ELISA) and the Western blotting, which are often used in the context of allergen research. The ELISA is a plate-based assay technique designed for detecting and quantifying immunogenic substances, whereas the Western blotting is an analytical technique in which the antibody binding is preceded by a separation of the proteins by gel electrophoresis.

#### *7.1.1. Enzyme-linked immunosorbent assay*

(1) Coat Nunc MaxiSorp® flat bottom 96-well plates with 150 µl of honey bee venom (2 µg/ml, Sigma-Aldrich) in coating buffer (100 mM bicarbonate/carbonate buffer, pH 9.6) at 4 °C overnight.
(2) Wash three times with Phosphate Buffered Saline, pH 7.4 with 0.05% Tween-20 (PBST).
(3) Block with 50 mg/ml skimmed milk powder in PBS at RT for two hours.
(4) Wash three times.
(5) Perform for each serum sample a two-fold dilution series from 1:40 to 1:20480 using blocking buffer.
(6) Incubate plates for 45 min at 37 °C.
(7) Wash three times.
(8) Incubate plates for 1 h at 37 °C with 150 µl of HRP-conjugated mouse anti-human antibodies (IgE or IgG4; depending on the purpose; Southern Biotech) at the dilution prescribed by the manufacturer.
(9) Wash three times.
(10) Add 200 µl of substrate solution (SIGMAGAST OPD, Sigma-Aldrich) to each well.
(11) After 30 min, stop the reaction with 100 µl of stop solution (3M HCl).
(12) Read the plates at 490 nm.
(13) Define antibody titer as the highest dilution with a reading above the mean of the negative controls plus 1.96 SDs.

#### *7.1.2. Western blotting*

(1) Load 12-µg proteins on a 15% SDS-PAGE gel under reducing and denaturing conditions in a discontinuous system, along with molecular weight markers (Bio-Rad).
(2) Finish the run when the dye-front reaches the end of the gel.
(3) Assemble the transfer sandwich and make sure no air bubbles are trapped in the sandwich. The blot membrane (polyvilylidene difluoride membrane) should be on the cathode and the gel on the anode.
(4) Transfer the protein bands to the blot membrane in a cold-room at a constant current of 10 mA (wet transfer) or 1 A (semi-dry).
(5) Stain the blot membrane with Ponceau S solution by simple immersion.
(6) Wash until the protein bands are visible (if no bands appear the blotting failed).
(7) Cut the membrane into strips.
(8) Block with 50 mg/ml skimmed milk powder in PBS at RT for two hours.
(9) Add 2 ml of diluted serum (1/16 diluted in blocking solution).
(10) Incubate overnight at 4 °C.
(11) Wash three times with blocking solution.
(12) Incubate in 2 ml of HRP-conjugated mouse anti-human antibodies (IgE or IgG4; depending on the purpose; Southern Biotech) at the dilution prescribed by the manufacturer.
(13) Wash three times with PBST.
(14) Wash once with PBS.
(15) Dissolve 3,3'-diaminobenzidine tetrahydrochloride (DAB) in 50 mM Tris, pH 7.2 at 1 mg/ml.
(16) Immediately before use, add an equal volume of 0.02% hydrogen peroxide.
(17) Immerse the blot membrane in the DAB staining solution until the desired development is achieved. Typical incubations are from 5 to 15 min.

### *7.2. Basophil activation test*

Upon challenge with antigen (bee venom) of peripheral basophils of bee venom allergic patients, cross-linking of bee venom-specific IgE antibodies bound to IgE receptors (FcεRI) will activate the cells. As a consequence, expression of specific membrane markers will up-regulate and histamine (and other mediators) will be released. These membrane changes and release of histamine can easily be measured on individual basophils by flow cytometry (Bridts et al., 2014).

Basophils represent about 1% of the white blood cells. Immunophenotyping can be done with different markers including FcεRI, IL-3 receptor (CD123) in combination with HLA-DR, CD203c and staining of IgE bound to FcεRI (Eberlein et al., 2014). CD203c is the only basophil lineage marker and in combination with IgE these are the best markers to define basophils as CD203c+/IgE+. However, the CD203c density on the cell surface can be too low to differentiate cells.

Once activated, CD63 is the most useful activation marker besides CD107a and CD203c, which also up-regulate after activation. This activation cannot only be measured with membrane markers, but also with intracellular markers like phosphorylated proteins (Verweij et al., 2010; 2012) or quantification of the intracellular histamine content (Sabato et al., 2015).

Basophils of about 5–10% of the patients will not respond to a positive control like anti-IgE or anti-FcεRI. One can decide to stimulate with the allergen alone, which will cause basophil activation only if the patient is bee venom allergic. Nevertheless, if the allergen also gives a negative response, the procedure has to be stopped as it will yield a false negative response and results will be inconclusive.

### 7.2.1. *Reagents*

Wear gloves, lab coat, and safety glasses while handling all human blood or chemical products. Dispose of all pipettes, etc. into bagged waste collection bins. Wipe work surfaces with disinfectant before and after running tests.

***7.2.1.1. Washing buffer (must be prepared sterile before use).*** Prepare 20 ml washing buffer with 17 ml pure water Milli Q, 1.9 ml 10x concentrated Hank's balanced salt solution with $Ca^{2+}$ $Mg^{2+}$ (HBSS: ThermoFisher Scientific: Invitrogen), 0.6 ml 7.5% $NaHCO_3$ and 0.4 ml Hepes (1 M) (ThermoFisher Scientific: Invitrogen). Set to pH 7.4.

***7.2.1.2. Lysing/fixing solution.*** Prepare a lysing and fixing solution by diluting 5x concentrated Lyse/Fix buffer (Becton-Dickinson cat. 558049) with pure water. Store at RT.

(1) BD Phosflow Lyse/Fix buffer (5x concentrate: BD Biosciences catalogue #558049) is used to lyse and fix whole blood for use with intracellular staining techniques. Store at RT.
(2) Dilute 1/5 just before use in $H_2O$.

***7.2.1.3. Phosphate buffered saline.*** Prepare phosphate buffered saline (PBS) by diluting 10x concentrated PBS without $Ca^{2+}$ or $Mg^{2+}$ (ThermoFisher Scientific: InVitrogen) with pure water. Measure pH = 7.4.

PBS is the basis of different other phosphate buffered saline solutions:

- PBS-EDTA: Add 10 mM $Na_2EDTA$ to PBS (see above), pH 7.4
- PBS-$NaN_3$: Add 0.5 g/l $NaN_3$ to PBS, pH = 7.4 (store at RT)
- PBS-0.1%BSA: Mix 50 ml PBS with 0.666 ml BSA (Albumin Bovine Fraction V 7.5%; Sigma 84112), pH 7.4.
- PBS-TX100: Add 0.1 ml pure Triton X-100 (Sigma-Aldrich catalogue # T8787) to 100 ml PBS, pH 7.4 (Store at RT). This solution is needed for permeabilization of the cells.

***7.2.1.4. Antibodies.*** The following antibodies (labelled or not) can be used to activate cells or to detect membrane or intracellular markers depending on the technique. The concentrations are indicative and must be titrated before use (McLaughlin et al., 2008). This can be done by using different dilutions of the antibody and, using the correct flow cytometric settings, looking for the most optimal resolution in comparison with an FMO (Fluorescence Minus One) measurement as a control (Roederer, 2002).

- CD63 is a 53 kDa Type III lysosomal protein and is expressed on the cell membrane of activated basophils. PE or FITC mouse anti-human CD63 (BD Biosciences) 20 µl per test, store at 4 °C
- Mouse anti-human IgE (clone G7-18: BD Biosciences Pharmingen 35171D) is used to stimulate basophils as a positive control. Stock solution at 0.5 mg/ml. Dilute to a working solution of 10 µg/ml in HBSS with $Ca^{++}$ and $Mg^{++}$. Make other dilutions to construct a dose response curve.
- CD203c is a type II transmembrane glycoprotein of the E-NPP family of ectoenzymes and is expressed on basophils and mast cells. CD203c is used as a basophil marker and as basophil activation marker in flow cytometric techniques. Five µl CD203c-APC (clone NP4D6: BD Biosciences) is added to 100 µl peripheral blood.
- Anti-human IgE Alexa Fluor488 (aIgE-AF488) or mouse monoclonal anti-human IgE (clone GE-1) is used as a marker for IgE bound to the FcεRI and hence as a basophil marker together with CD203c. This clone is only available as a non-labelled antibody and must be conjugated with a fluorochrome such as Alexa Fluor. The labelling procedure is as follows:

(1) Prepare the following solutions:
    - A working solution of mouse monoclonal anti-human IgE (clone GE-1,Sigma I 6510) of 5 mg/ml in $NaHCO_3$ (0.1 M), pH= 8.3.
    - Prepare AlexaFluor 488 succinimidyl ester (Alexa Fluor 488 NHS: ThermoFisher Scientific, Molecular Probes A20000) according to manufacturer's procedure. Note that stock solutions can be stored at –20 °C for maximum three months. Once reconstituted in DMSO, the solution becomes unstable and must be used immediately.
    - The working solution contains 1 mg in 0.1 ml dimethylsulfoxide (DMSO) and must be freshly prepared.
(2) Mix 2 ml anti-IgE (working solution) slowly with 0.1 ml AlexaFluor 488 NHS (working solution) during one hour at RT in the dark.
(3) Mix regularly.
(4) Dialyse overnight with PBS at 4 °C or use a gel filtration column like Sephadex G-25 to separate conjugated antibody from free dye.
(5) Add 0.05% $NaN_3$.
(6) Store at 2–8 °C.

Note: labels of antibodies can be changed, depending on the technique or flow cytometer. New labels/antibodies must be titrated and or validated in the technique used.

***7.2.1.5. Diamine oxidase.*** Diamine oxidase (DAO) is an enzyme involved in the metabolisation of histamine (histaminase). DAO can be labelled with a fluorochrome to detect histamine in fixed and permeabilized cells like basophils and mast cells in a so-called enzyme affinity technique (Ebo et al., 2012). DAO-PE or DAO-V500 (BD Biosciences) must be diluted in PBS-EDTA. Every new solution must be titrated.

(a) Diamine oxidase or histaminase (Sigma-Aldrich catalogue # D7876) has been conjugated with phycoerythrine (PE) or Horizon V500 at the laboratories of BD Biosciences.
(b) A working solution 1:1 (DAO-PE) or 1:20 (DAO-V500) was made in PBS/EDTA. Every new batch should be titrated and measured with a flow cytometer in order to find the optimal staining concentration for intracellular histamine.

***7.2.1.6. Calibration particles.***

- Rainbow Calibration Particles 6 peaks (BD Biosciences catalogue # 556286) or 8 peaks (# 559123).
- Store undiluted at 4 °C in the dark.
- Mix well before use.
- Dilute 3–5 drops in 1 ml buffer.
- Use immediately.
- Analyze according to the manufacturer's procedure.

***7.2.1.7. Allergens.*** Allergens can be protein extracts, recombinant proteins, chemicals or drugs (Peiren et al., 2006). They must be water soluble and used in non-toxic concentrations for basophils. To test toxicity, incubate basophils with different concentrations of the allergen or mouse monoclonal anti-human IgE (positive control) during 20 min at 37 °C and measure viability using propidium iodide or Live/Dead© viability/Cytotoxicity Kit for mammalian cells (ThermoFisher Scientific: Molecular Probes).

Do not use the available preparations for skin testing. These solutions contain preservatives which can inhibit basophil activation *in vitro*. The optimal stimulation concentration of an allergen to activate basophils must be determined in a dose response curve with well-defined patients positive to the antigen and stung controls negative without any reaction to bee allergen. Because such a response can result in a bell shape curve, in routine practice at least two concentrations should be used to define a positive result.

Bee venom can be obtained by ALK-Abello, Denmark as a lyophilized allergenic venom extract.

*7.2.2. Equipment*

A flow cytometer capable of measuring the available fluorochromes (installation and setup is beyond the scope of this chapter) and a warm water bath (WWB) at 37 °C are needed.

*7.2.3. Sample*

Although different BAT techniques are described using EDTA or even ACD or citrated blood, the best results are obtained with heparinized blood as basophil activation is calcium dependent and chelators or calcium binding products can have deleterious effect on test results.

Anti-histamines have no influence on basophil activation tests in contrast to corticoids, immune-suppressive drugs or statins.

As a sample, about 10 ml heparinized blood must be available. To obtain reproducible results, basophil activation must be executed within three to four hours after sample collection.

*7.2.4. Procedure*

***7.2.4.1. Basophil activation technique as measured by membrane markers – method 1.*** All procedures must be carried out at RT and in a laminar air flow cabinet, to avoid contamination of aerogenic allergens, unless otherwise specified.

To exclude so-called non-responders, i.e., samples non-reactive to anti-IgE or allergen stimulation, available basophils are stimulated to control their reactivity.

(a) Incubate 100 µl blood with 100 µl HBSS buffer as a negative control and anti-IgE as a positive control during 20 min at 37 °C.
(b) Stain with aIgE-AF488 and CD203c-APC using the procedure described below and read on a flow cytometer.
(c) When basophils do not react one can decide to stimulate with two concentrations of the allergen to exclude non-reactivity to anti-IgE.

Allergen procedure

(1) Pre-warm all blood samples, buffers and allergens before mixing during 15 min at 37 °C in WWB.
Notes: Basophil activation results in an immediate response (within 1 min) and lasts about 20–30 min at maximum response. Prewarming all solutions at 37 °C will ensure the highest reproducibility.
(2) Add 100 µl allergen solution to the allergen incubation tubes and buffer (HBSS) or positive control (anti-IgE) to control tubes. When needed make dilutions of the allergen to construct doses-response curves.

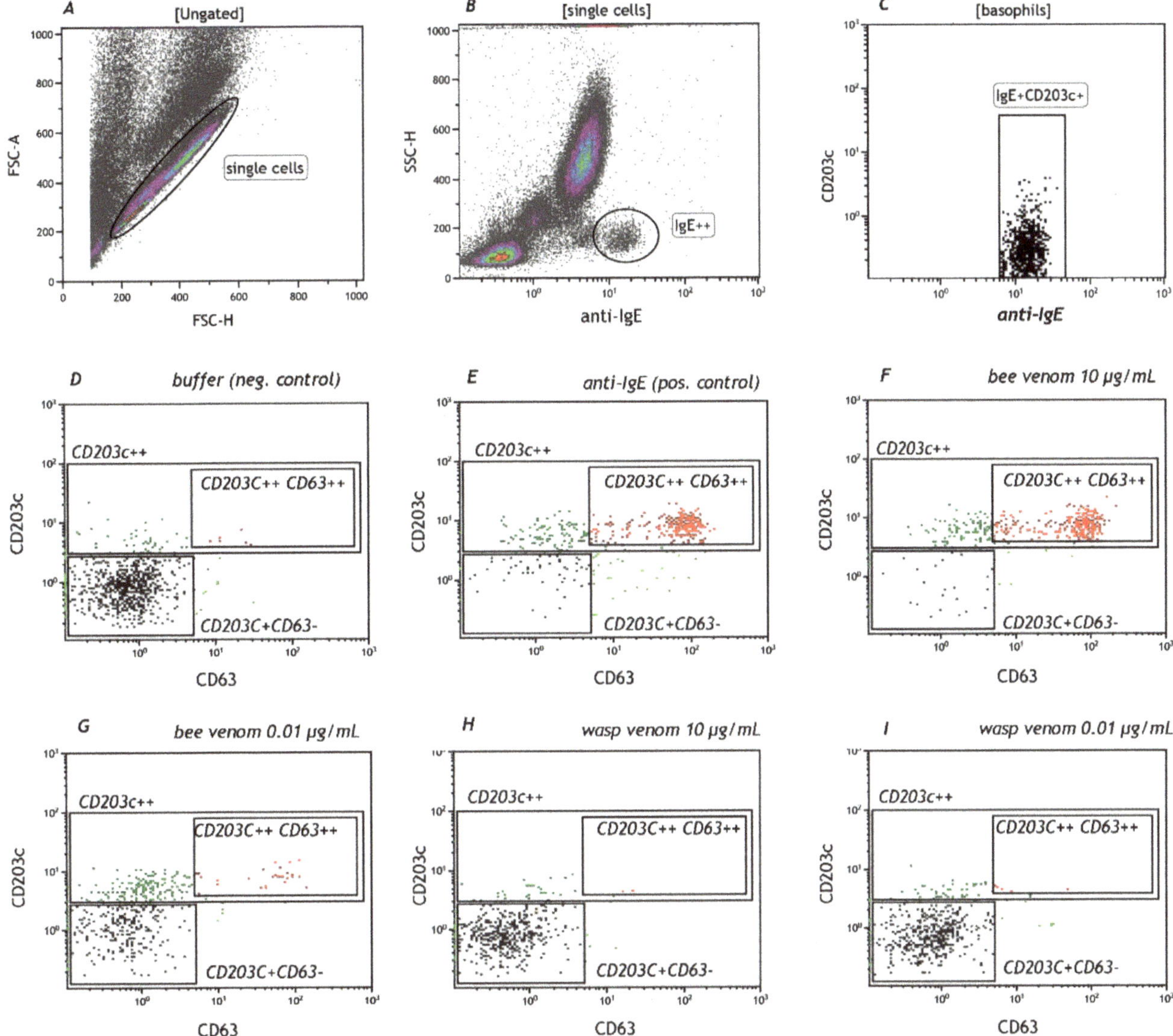


Figure 4. An example of a basophil activation test in a bee venom allergic patient. In panel **A** single cells are isolated from cell aggregates in a forward scatter plot FCS-H (height) versus FCS-A (area). In **B**, High IgE positive basophils are selected and analyzed in panel **C** for their positivity in CD203c (basophil specific). Cells stimulated with buffer (**D**) and anti-IgE (**F**) are analyzed as a negative and positive control, respectively. CD203c++ denotes activated basophils, CD203c++CD63+ denotes degranulating cells. In panels **F-I** cells stimulated with bee venom (F and G) or wasp venom (H and I) are analyzed. It shows clearly a positive dose dependent reaction to bee venom and not wasp venom as expressed by the higher proportion of CD203C++CD63+ degranulating cells.

(3) Add to all tubes 100 µl pre-warmed whole blood and incubate during 20 min at 37 °C in WWB.
(4) Stop reaction by adding 1 ml PBS-EDTA.
(5) Spin at 200 *g* for 10 min at 4 °C, set the brake of the centrifuge off.
(6) Remove supernatant.
(7) Add 20 µL aIgE-AF488, 10 µl anti-CD63-PE.
(8) Add 5 µl anti-CD203c-APC.
(9) Incubate on ice or at 2–8 °C for 20 min in the dark.
(10) Lyse the red blood cells by adding 2 ml lysis solution.
(11) Incubate for 20 min at RT.
(12) Spin for 10 min at 250 *g*, RT.
(13) Remove supernatant.
(14) Wash by adding 2 ml PBS-EDTA.
(15) Spin again for 10 min 250 *g* at RT.
(16) Remove supernatant.
(17) Add 300 µl PBS-$NaN_3$
(18) Count at least 500 basophils by flow cytometry which are defined or gated out as IgE + and CD203c+.
(19) Store all list data before analysis with available flow cytometric software.
(20) Define CD203c+CD63-, CD203c++CD63- and CD203c++CD63+ populations (Figure 4) by following the instructions of the flow cytometric software package.

***7.2.4.2. Basophil activation with intracellular staining of histamine (HistaFlow®) – method 2.*** All procedures are carried out at RT, in a laminar air flow cabinet when necessary, or unless otherwise specified. The procedure is almost identical to that in method 1.

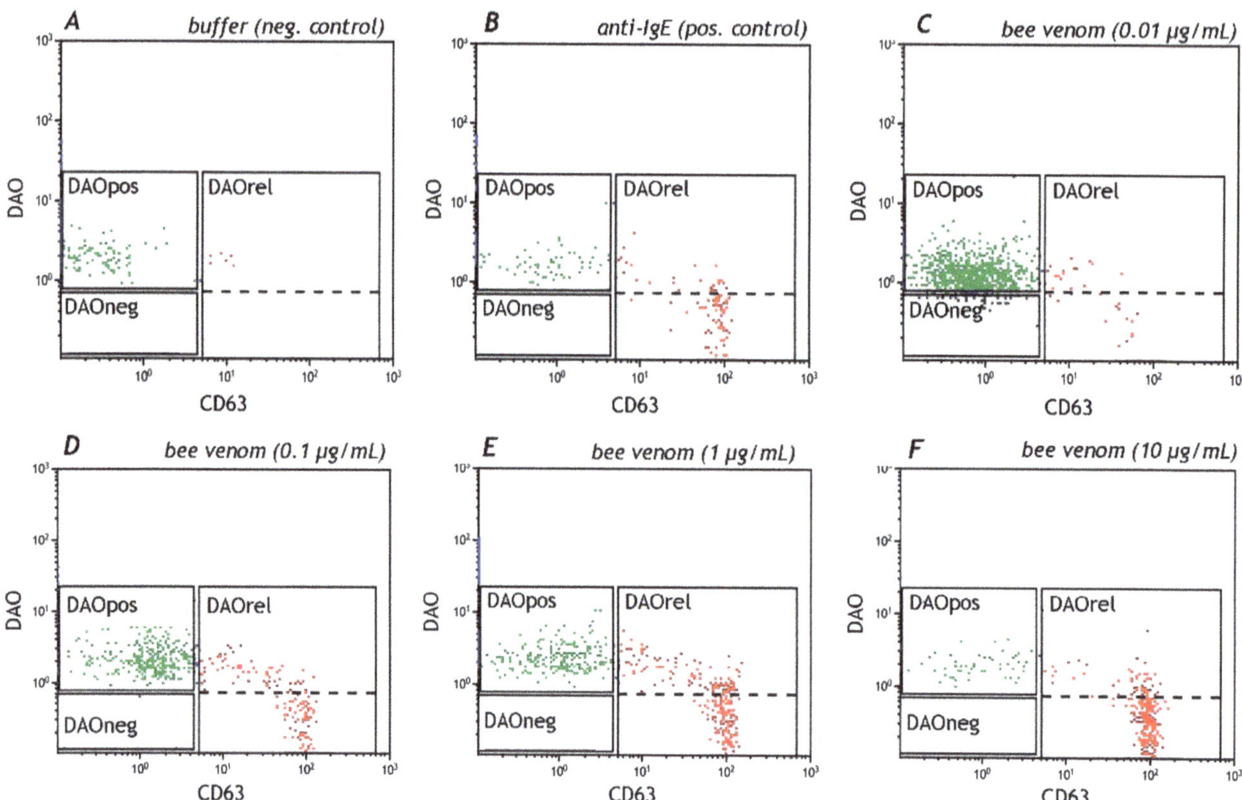


Figure 5. Basophils selected as in panels A-C in Figure 6, were analyzed for their histamine content or DAO positivity. **A** and **B** are the necessary negative and positive controls. C-F is an example of a dose response curve in a bee venom allergic patient showing clearly that DAO positive cells (DAOpos) becoming CD63+ (x-axis) and start to release histamine (DAOrel). The dotted line is the difference between histamine positive and histamine releasing cells.

Allergen procedure

(1) Pre-warm all blood samples, buffers and allergens before mixing during 15 min at 37 °C in WWB.
(2) Add 100 µl allergen solution to the allergen incubation tubes and buffer (HBSS) or positive control (anti-IgE) to control tubes. When needed make dilutions of the allergen to construct dose response curves.
(3) Add to all tubes 100 µl pre-warmed whole blood and incubate during 20 min at 37 °C in WWB.
(4) Stop reaction by adding 1 ml PBS-EDTA.
(5) Spin at 200 *g* for 10 min at 4 °C, set the brake of the centrifuge off.
(6) Remove supernatant.
(7) Add 20 µl aIgE-AF488, 10 µl anti-CD63-PE.
(8) Add 5 µl aCD203c-APC.
(9) Incubate on ice or at 2–8 °C for 20 min in the dark.
(10) Add 2 ml Phosflow Lyse/Fix work solution.
(11) Incubate at 37 °C in WWB for 20 min.
(12) Spin for 10 min at 500 *g* on RT.
(13) Remove supernatant.
(14) Add 2 ml 0.1% PBS-TX100.
(15) Spin again for 10 min at 500 *g* at RT.
(16) Remove supernatant.
(17) Add 100 µl 0.1% PBS-TX100 and 10 to 40 µl (depending on titration) DAO-V500.
(18) Stain for 45 min at 37 °C in WWB. Mix several times during stain procedure.
(19) Add 2 ml PBS.
(20) Spin at 500 *g* for 10 min at RT.
(21) Remove supernatant.
(22) Add 0.5 ml PBS-$NaN_3$
(23) Count at least 1000 basophils by flow cytometry which are defined as IgE+and CD203c+.
(24) Store data in a FCS formatted file, which can be analyzed with flow cytometric software.
(25) Prepare incubation tubes.
(26) Add 200 µl allergen solution to the allergen tubes. Dilutions depend on the allergen used.
(27) Use BAT buffer as a negative control or for FMO evaluation. Define $CD203c^{hi+}CD63^{-}$ basophils and $CD203c^{hi+}CD63^{hi+}$ as activate basophil populations.
(28) Define histamine content using the compensation setting described below. This setting must be unique for every sample. Figure 5 is an example of HistaFlow® analysis of basophils stimulated with bee venom in a bee venom allergic patient.

Compensation procedure

A functional assay like the HistaFlow® technique (Ebo et al., 2012) needs an optimal setting for measurement of histamine release. Also, we must be aware

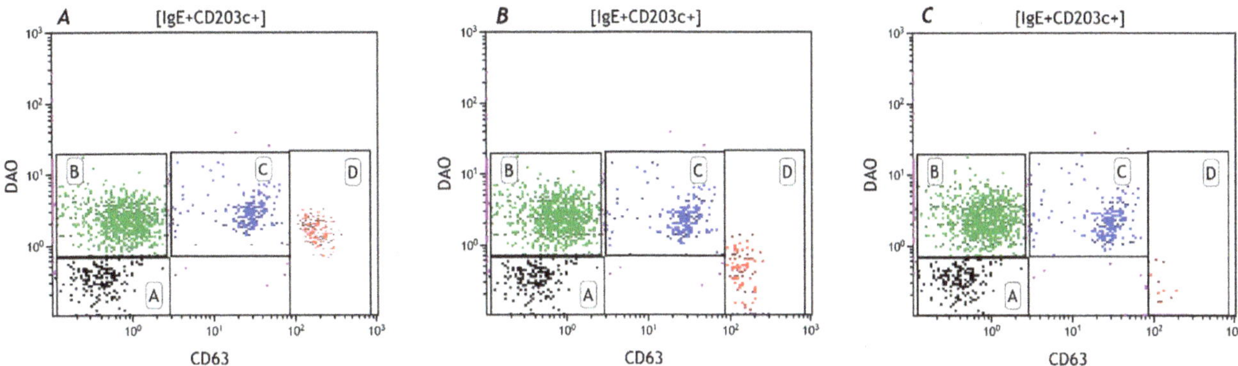


Figure 6. This figure shows the correct compensation setting as describe in Table 4. **A** is undercompensated, **C** is overcompensated and **B** shows the correct compensation setting.

Table 4. Compensation method.

| Tube<br>Basophil type | 1<br>CD63-DAO- | 2<br>CD63-DAO+ | 3<br>CD63++DAO+ | 4<br>CD63++DAO- |
|---|---|---|---|---|
| Explanation | Negative cells | DAO + cells without activation | DAO + and intracellular CD63+ staining without stimulation | DAO and CD63 staining with stimulation |
| Incubation / activation procedure (see method 2) | | | | |
| Whole blood | 200 µl | 200 µl | 200 µl | 200 µl |
| BAT buffer | 200 µl | 200 µl | 200 µl | – |
| Anti-IgE (pos. ctl.) | – | – | – | 200 µl |
| Incubation (see method 2) | | | | |
| Membrane staining procedure | | | | |
| CD63 | None | None | None | none |
| IgE/CD203c | 20 µl/10 µl | 20 µl/10 µl | 20 µl/10 µl | 20 µl/10 µl |
| Lyse / fix (see method 2) | | | | |
| Wash in PBA with 0.1% Triton X | | | | |
| Intracellular staining procedure (see method 2) | | | | |
| PBS-TX100 | 100 µl | 100 µl | 100 µl | 100 µl |
| CD63 | – | – | 10 µl | 10 µl |
| DAO | – | 10 µl | 10 µl | 10 µl |
| Population by correct compensation (see Figure 6) | Grey cell in gate A (no staining for DAO and CD63) | Green cells in gate B (DAO + staining alone) | Blue cells in gate C (all basophils are positive for DAO and CD63, note that there is no CD203c up-regulation) | Red cells in gate D (histamine releasing cells) |

that not all basophils contain/release the same amount of histamine, hence different cell populations can be observed. By mixing the data of the four tubes in the compensation setting (Roederer, 2002) of the HistaFlow® technique (see Table 4), a correct setting can be obtained. The first tube is without staining for DAO/CD63. These are the histamine negative basophils. The second tube is DAO staining alone. These are the DAO (or histamine) positive basophils. The third tube is special: DAO as well as CD63 staining AFTER fixation and WITHOUT stimulation: this will give us the optimal staining for all populations. The fourth tube is DAO and CD63 after activation (just like a positive control): This tube gives us the possibility to optimize the compensation setting (under an optimal PMT setting). By fine tuning we observe the different histamine releasing basophil populations. After running the staining procedure for each tube, we can mix them and find out the optimal setting or we measure every single tube and combine the data in one data-file using appropriate software. After analysis tubes of the same patient are evaluated with this setting. Be aware to repeat this for every patient, because the histamine content is different in every patient.

### *7.3. Assignment of new allergens*

The IUIS Allergen Nomenclature Sub-Committee, under the auspices of the World Health Organization (WHO) and the International Union of Immunological Societies (IUIS), has devised a unified nomenclature system for purified allergens (King et al., 1994; Marsh et al., 1986; Radauer et al., 2014). They are phenotypically designated by the first three letters of the genus followed by a space, the first letter of the species, another space, and finally an Arabic number. Accordingly, allergens of the honey bee *Apis mellifera* are all named Api m x (x stands for the consecutive number).

To assign a new allergen:

(1) Go to the homepage of the IUIS Allergen Nomenclature Sub-Committee at http://www.allergen.org and download the New Allergen Submission Form (MS Word format).

(2) Enter the personal data of the submitter of the new allergen.

(3) Enter the scientific name (genus and species), the family and the order of the allergen source. Thereby refer to the taxonomic system used within the UniProt (www.uniprot.org/taxonomy/) and NCBI (www.ncbi.nlm.nih.gov/sites/entrez?db=taxonomy) databases. Additionally, the common name of the allergen source can be included. For the honey bee *Apis mellifera* the taxonomic data should be entered as follows: *Apis mellifera, Apidae, Hymenoptera*.

(4) Propose an allergen name consisting of three letters of the genus (Api for *Apis*), one letter of the species (e.g., m for *mellifera* or c for *cerana*) and an allergen number. Justifications for numbering can be the next available number or the existence of a homologous allergen with the same number in another species. Additionally, propose an isoallergen and a variant number. If the allergen is a new allergen in the source the isoallergen number (two digits) is 01, if the allergen is the first isoallergen to an existing allergen in the source the isoallergen number is 02 and so on. Isoallergens show a similar molecular size, identical biological function (if known) and an amino acid sequence identity above 67% (this value is only a guideline). Each isoallergen may have multiple forms of closely similar sequences, which are designated as variants.

(5) Enter the tissue or organ of expression of the natural allergen (e.g., venom gland) and, if known, its biochemical name. Additionally, state if a recombinant protein was produced and, if so, by using which expression system. Specify at least one molecular weight data of the mature protein and state which method was used for determination (e.g., deduced from the sequence, mass spectrometry or SDS-PAGE). Additionally, if known, indicate post-translational modifications of the allergen such as glycosylation.

(6) Provide sequence and, if available, structure (PDB; Protein Data Bank) accession numbers of the allergen. For cloned gene products indicate a nucleotide accession number (GenBank, EMBL or DDBJ) and, if available, also the accession number of the protein sequence (GenPept or Uniprot). Protein sequence accession numbers are obligatory for natural proteins that were identified by Edman degradation or mass spectrometry.

(7) Provide the complete protein sequence in one-letter code and, if available, also the nucleotide sequence. Specify how the N-terminus of the mature protein was determined and, if available, add sequence features such as the position of a signal sequence, a propeptide sequence and of the mature sequence.

(8) State the level of sequence confirmation (e.g., sequence coverage by MS/MS). If the sequence was obtained by a PCR strategy indicate the position of used primers by clicking to the predetermined answers. Moreover, specify a sequence reference (unpublished, a publication accessible via PubMed or a congress abstract or publication not accessible via PubMed).

(9) Specify the allergenicity of the new allergen. Allergens are only incorporated into the official list of allergens if binding of IgE antibodies from at least five sera of patients allergic to the respective allergen source were shown or if binding of IgE from at least 5% of all tested sera of patients allergic to the respective allergen source was demonstrated. For this purpose, state if the allergic reaction of the patients from the study population to the allergen source was documented by case history or allergen challenge and if the presence of allergen-specific IgE was demonstrated by an *in vitro* IgE test, skin test or cellular tests. Additionally, indicate the route of exposure to the allergen (e.g., sting or inhalation), describe the experimental evidence of allergenicity by indicating the test method, the total number of tested patients and the number of patients showing a positive test with the putative new allergen. If the natural allergen was tested specify the methods used to confirm the identity and purity of the protein. Moreover, if the allergen is glycosylated, indicate the experiments that were performed to exclude that IgE exclusively binds the glycan moieties.

(10) Submit the completed New Allergen Submission Form by email to the chairman of the IUIS Allergen Nomenclature Sub-Committee (currently Dr. Richard E. Goodman; rgoodman2@unl.edu).

(11) Two or more members of the executive committee will review the submission and assess whether the allergen fulfills the immunological and molecular requirements for inclusion into the nomenclature. The review process will take approximately one month.

(12) The chairman of the executive committee will notify the submitter about the status of the proposed new allergen. If the allergen does not fulfill

the requirements the reviewers may ask for more data or for further clarification. If the allergen fulfills the requirements the data will be included in the Allergen Nomenclature database.

## Acknowledgements

We thank the Editors Peter Neumann and Vincent Dietemann for their insights and advice. The COLOSS (Prevention of honey bee COlony LOSSes) Association aims to explain and prevent massive honey bee colony losses. It was funded through the COST Action FA0803. COST (European Cooperation in Science and Technology) is a unique means for European researchers to jointly develop their own ideas and new initiatives across all scientific disciplines through trans-European networking of nationally funded research activities. Based on a pan-European intergovernmental framework for cooperation in science and technology, COST has contributed since its creation more than 40 years ago to closing the gap between science, policy makers and society throughout Europe and beyond. COST is supported by the EU Seventh Framework Program for research, technological development, and demonstration activities (Official Journal L 412, 30 December 2006). The European Science Foundation as implementing agent of COST provides the COST Office through an EC Grant Agreement. The Council of the European Union provides the COST Secretariat. The COLOSS network is now supported by the Ricola Foundation - Nature & Culture.

## Disclosure statement

No potential conflict of interest was reported by the authors.

## ORCID

*Dirk C. de Graaf* http://orcid.org/0000-0001-8817-0781
*Márcia Regina Brochetto Braga* http://orcid.org/0000-0002-7982-0743
*Rusleyd Maria Magalhães de Abreu* http://orcid.org/0000-0003-1603-1808
*Simon Blank* http://orcid.org/0000-0002-5649-1078
*Chris H. Bridts* http://orcid.org/0000-0002-3324-7320
*Luc S. De Clerck* http://orcid.org/0000-0002-4622-6289
*Bart Devreese* http://orcid.org/0000-0002-9764-2581
*Didier G. Ebo* http://orcid.org/0000-0003-0672-7529
*Margo M. Hagendorens* http://orcid.org/0000-0001-6361-9503
*Zenon J. Kokot* http://orcid.org/0000-0003-4950-9759
*Jan Matysiak* http://orcid.org/0000-0002-9993-1504
*Christel Mertens* http://orcid.org/0000-0003-2359-0771
*Vito Sabato* http://orcid.org/0000-0002-1321-314X
*Athina L. Van Gasse* http://orcid.org/0000-0002-3434-4333
*Matthias Van Vaerenbergh* http://orcid.org/0000-0002-4676-7464

## References

Abreu, R. M. M., Moraes, R. L. M., Silva de., & Malaspina, O. (2000). Histological aspects and protein content of *Apis mellifera* L. worker venom glands: The effect of electrical shocks in summer and winter. *Journal of Venomous Animals and Toxins*, *6*(1), 87–98. https://doi.org/10.1590/S0104-79302000000100006

Ali, E. M. (2014). Contributions of some biological activities of honey bee venom. *Journal of Apicultural Research*, *53*(4), 441–451. https://doi.org/10.3896/IBRA.1.53.4.13

Araújo, A. L., & Radvanyi, F. (1987). Determination of phospholipase A2 activity by a colorimetric assay using a pH indicator. *Toxicon*, *25*(11), 1181–1188. https://doi.org/10.1016/0041-0101(87)90136-X

Barboni, E., Kemeny, D. M., Campos, S., & Vernon, C. A. (1987). The purification of acid phosphatase from honey bee venom (*Apis mellifera*). *Toxicon*, *25*(10), 1097–1103. https://doi.org/10.1016/0041-0101(87)90266-2

Benton, A. W. (1967). Esterases and phosphatases in honey bee venom. *Journal of Apicultural Research*, *6*(2), 91–94. https://doi.org/10.1080/00218839.1967.11100164

Benton, A. W., & Morse, R. A. (1968). Venom toxicity and proteins of the genus Apis. *Journal of Apicultural Research*, *7*(3), 113–118. https://doi.org/10.1080/00218839.1968.11100200

Benton, A. W., Morse, R. A., & Stewart, J. D. (1963). Venom collection from honey bees. *Science (New York, NY)*, *142*(3589), 228–230. https://doi.org/10.1126/science.142.3589.228

Bogdanov, S. (2017). Bee venom: Production, composition, quality. In: *The bee venom book*, Chapter 1, Bee product science. Retrieved from May 2017. http://www.bee-hexagon.net/venom/production-compostion-quality/

Bridges, A. R., & Owen, M. D. (1984). The morphology of the honey bee (*Apis mellifera* L.) venom gland and reservoir. *Journal of Morphology*, *181*(1), 69–86. https://doi.org/10.1002/jmor.1051810107

Bridts, C. H., Sabato, V., Mertens, C., Hagendorens, M. M., DE Clerck, L. S., & Ebo, D. G. (2014). Flow cytometric allergy diagnosis: Basophil activation techniques. *Methods in Molecular Biology (Clifton, NJ)*, *1192*, 147–159. https://doi.org/10.1007/978-1-4939-1173-8_11

Brochetto-Braga, M. R. ; D., Lima, P. R. M., Chaud-Netto, J., Arab, A., DA Silva, G. P., & Cursino-Santos, J. R. (2005). Enzymatic variability among venoms from different subspecies of *Apis mellifera* (Hymenoptera: Apidae). *Sociobiology*, *45*(3), 797–809.

Carreck, N. L., Andree, M., Brent, C. S., Cox-Foster, D., Dade, H. A., Ellis, J. D., Hatjina, F., & Vanengelsdorp, D. (2013). Standard methods for Apis mellifera anatomy and dissection. In V. Dietemann; J. D. Ellis, & P. Neumann (Eds.), The COLOSS BEEBOOK, Volume I: Standard methods for Apis mellifera research. *Journal of Apicultural Research*, 52(4), 1–40. https://doi.org/10.3896/IBRA.1.52.4.03

Cha, J. D., Jeong, S. M., Kim, G. O., Kim, G. S., & Kim, N. O. (2004). The comparison of effectiveness between acupuncture and its cotreatment with bee venom acua-acupuncture therapy on the treatment of herniation of nucleus pulpous. *Journal of the Korean Acupunture and Moxibustion Society*, *21*, 149–158.

Cox, J., & Mann, M. (2008). MaxQuant enables high peptide identification rates, individualized p.p.b.-range mass accuracies and proteome-wide protein quantification. *Nature Biotechnology*, *26*(12), 1367–1372. https://doi.org/10.1038/nbt.1511

da Silva, G. P., Brochetto-Braga, M. R., Ruberti, M., Ternero, M. L., & Gobbi, N. (2004). A comparative study of protein and enzymatic activity in venoms of some common wasps (Hymenoptera:Vespidae) from São Paulo State. *Sociobiology*, *44*(2), 271–282.

Danneels, E. L., van Vaerenbergh, M., Debyser, G., Devreese, B. ; D., & Graaf, D. C. (2015). Honeybee venom proteome profile of queens and winter bees as determined by a mass spectrometric approach. *Toxins*, *7*(11), 4468–4483. https://doi.org/10.3390/toxins7114468

de Abreu, R. M., Silva DE Moraes, R. L., & Camargo-Mathias, M. I. (2010). Biochemical and cytochemical studies of the enzymatic activity of the venom glands of workers of honey bee *Apis mellifera* L. (Hymenoptera, Apidae). *Micron (Oxford,*

*England: 1993)*, *41*(2), 172–175. https://doi.org/10.1016/j.micron.2009.09.003

de Graaf, D. C., Aerts, M., Danneels, E., & Devreese, B. (2009). Bee, wasp and ant venomics pave the way for a component-resolved diagnosis of sting allergy. *Journal of Proteomics*, *72*(2), 145–154. https://doi.org/10.1016/j.jprot.2009.01.017

de Lima, P. R. M., Brochetto-Braga, M. R., & Chaud-Netto, J. (2000). Proteolytic activity of Africanized honey bee (*Apis mellifera*: Hymenoptera, Apidae) venom. *Journal of Venomous Animals and Toxins*, *6*(1), 64–76. https://doi.org/10.1590/S0104-79302000000100004

Delpech, B., Bertrand, P., & Chauzy, C. (1987). An indirect enzymoimmunological assay for hyaluronidase. *Journal of Immunological Methods*, *104*(1–2), 223–229. https://doi.org/10.1016/0022-1759(87)90508-4

Dotimas, E. M., & Hider, R. C. (1987). Honey bee venom. *Bee World*, *68*(2), 51–70. https://doi.org/10.1080/0005772X.1987.11098915

Eberlein, B., Hann, R., Eyerich, S., Pennino, D., Ring, J., Schmidt-Weber, C. B., & Buters, J. (2014). Optimizing of the basophil activation test: Comparison of different basophil identification markers. *Cytometry Part B: Clinical Cytometry*, *88*(3), 183–189. https://doi.org/10.1002/cytob.21203

Ebo, D. G., Bridts, C. H., Mertens, C. H., Hagendorens, M. M., Stevens, W. J., & DE Clerck, L. S. (2012). Analyzing histamine release by flow cytometry (HistaFlow): A novel instrument to study the degranulation patterns of basophils. *Journal of Immunological Methods*, *375*(1–2), 30–38. https://doi.org/10.1016/j.jim.2011.09.003

Egmond, M. R., Brunori, M., & Fasella, P. M. (1976). The steady-state kinetics of the oxygenation of linoleic acid catalysed by soybean lipoxygenase. *European Journal of Biochemistry*, *61*(1), 93–100. https://doi.org/10.1111/j.1432-1033.1976.tb10001.x

Georgieva, D., Greunke, K., Genov, N., & Betzel, C. (2009). 3-D Model of the bee venom acid phosphatase: Insights into allergenicity. *Biochemical and Biophysical Research Communications*, *378*(4), 711–715. https://doi.org/10.1016/j.bbrc.2008.11.101

Grunwald, T., Bockisch, B., Spillner, E., Ring, J., Bredehorst, R., & Ollert, M. W. (2006). Molecular cloning and expression in insect cells of honeybee venom allergen acid phosphatase (Api m 3)). *The Journal of Allergy and Clinical Immunology*, *117*(4), 848–854. https://doi.org/10.1016/j.jaci.2005.12.1331

Gunnison, A. F. (1966). An improved method for collecting the liquid fraction of bee venom. *Journal of Apicultural Research*, *5*(1), 33–36. https://doi.org/10.1080/00218839.1966.11100129

Habermann, E. (1972). Bee and wasp venoms. *Science (New York, NY)*, *177*(4046), 314–322. https://doi.org/10.1126/science.177.4046.314

Han, J., You, D., Xu, X., Han, W., Lu, Y., Lai, R., & Meng, Q. (2008). An anticoagulant serine protease from the wasp venom of *Vespa magnifica*. *Toxicon: Official Journal of the International Society on Toxinology*, *51*(5), 914–922. https://doi.org/10.1016/j.toxicon.2008.01.002

Hartfelder, K., Bitondi, M. M. G., Brent, C. S., Guidugli-Lazzarini, K. R., Simoes, Z. L. P., Stabentheiner, A., Tanaka, E. D., & Wang, Y. (2013). Standard methods for physiology and biochemistry research in Apis mellifera. In V Dietemann; J D Ellis; P Neumann (Eds) The COLOSS BEEBOOK, Volume I: standard methods for Apis mellifera research. *Journal of Apicultural Research*, *52*(1), 1–48. https://doi.org/10.3896/IBRA.1.52.1.06

Hoechstetter, J. (2005). *Characterization of bovine testicular hyaluronidase and a hyaluronate lyase from* Streptococcus agalactiae*: Investigations on the effect of pH on hyaluronan degradation and preclinical studies on the adjuvant administration of the enzymes in cancer therapy* [PhD-dissertation]. University of Regensburg.

Hoffman, D. R. (1977). Allergens in bee venom III. Identification of allergen B of bee venom as an acid phosphatase. *The Journal of Allergy and Clinical Immunology*, *59*(5), 364–366. https://doi.org/10.1016/0091-6749(77)90019-7

Hoffman, D. R., & Jacobson, R. S. (1996). Allergens in Hymenoptera venom XXVII: Bumblebee venom allergy and allergens. *The Journal of Allergy and Clinical Immunology*, *97*(3), 812–821. https://doi.org/10.1016/S0091-6749(96)80159-X

Hofinger, E. S., Hoechstetter, J., Oettl, M., Bernhardt, G., & Buschauer, A. (2008). Isoenzyme-specific differences in the degradation of hyaluronic acid by mammalian-type hyaluronidases. *Glycoconjugate Journal*, *25*(2), 101–109. https://doi.org/10.1007/s10719-007-9058-8

Human, H., Brodschneider, R., Dietemann, V., Dively, G., Ellis, J., Forsgren, E., Fries, I., Hatjina, F., Hu, F.-L., Jaffé, R., Jensen, A. B., Köhler, A., Magyar, J., Özkýrým, A., Pirk, C. W. W., Rose, R., Strauss, U., Tanner, G., Tarpy, D. R., … Zheng, H.-Q. (2013). Miscellaneous standard methods for Apis mellifera research. In V Dietemann; J D Ellis; P Neumann (Eds) The COLOSS BEEBOOK, Volume I: standard methods for Apis mellifera research. *Journal of Apicultural Research*, *52*(4), 1–53. https://doi.org/10.3896/IBRA.1.52.4.10

Humphrey, J. H., & Jaques, R. (1953). Hyaluronidase: Correlation between biological assay and other methods of assay. *The Biochemical Journal*, *53*(1), 59–62. https://doi.org/10.1042/bj0530059

Ikegami-Kawai, M., & Takahashi, T. (2002). Microanalysis of hyaluronan oligosaccharides by polyacrylamide gel electrophoresis and its application to assay of hyaluronidase activity. *Analytical Biochemistry*, *311*(2), 157–165. https://doi.org/10.1016/S0003-2697(02)00425-6

Jiang, N., Tan, N. S., Ho, B., & Ding, J. L. (2007). Azocoll protease activity assay. *Protocol Exchange*, *8*(10), 1114–1122. https://doi.org/10.1038/nprot.2007.484

Jiménez, M., Cabanes, J., Gandía-Herrero, F., Escribano, J., García-Carmona, F., & Pérez-Gilabert, M., (2003). A continuous spectrophotometric assay for phospholipase A2 activity. *Analytical Biochemistry*, *319*(1), 131–137. https://doi.org/10.1016/S0003-2697(03)00331-2

King, T. P., Hoffman, D., Lowenstein, H., Marsh, D. G., Platts-Mills, T. A., & Thomas, W. (1994). Allergen nomenclature. WHO/IUIS Allergen Nomenclature Subcommittee. *International Archives of Allergy and Immunology*, *105*(3), 224–233. https://doi.org/10.1159/000236761

King, T. P., & Spangfort, M. D. (2000). Structure and biology of stinging insect venom allergens. *International Archives of Allergy and Immunology*, *123*(2), 99–106. https://doi.org/10.1159/000024440

Koketsu, M., & Linhardt, R. J. (2000). Electrophoresis for the analysis of acidic oligosaccharides. *Analytical Biochemistry*, *283*(2), 136–145. https://doi.org/10.1006/abio.2000.4649

Krell, R. (1996). Venom. In: *Value-added products from beekeeping. FAO Agricultural Services Bulletin* no. 124, chap. 7. Food and Agriculture Organization of the United Nations Rome 1996. The Chief Editor, FAO, Rome, Italy. http://www.fao.org/3/a-w0076e/w0076e18.htm#7.9.

Krupa, J. C., Butler, M. A., & Mort, J. S. (2003). Quantitative bead assay for hyaluronidase and heparinase I. *Analytical Biochemistry*, *319*(2), 280–286. https://doi.org/10.1016/S0003-2697(03)00295-1

Larché, M., Akdis, C. A., & Valenta, R. (2006). Immunological mechanisms of allergen-specific immunotherapy. *Nature Reviews. Immunology*, *6*(10), 761–771. https://doi.org/10.1038/nri1934

Leytus, S. P., Melhado, L. L., & Mangel, W. F. (1983). Rhodamine-based compounds as fluorogenic substrates for serine proteinases. *The Biochemical Journal*, *209*(2), 299–307. https://doi.org/10.1042/bj2090299

Magalhaes, M. R., da Silva, N. J., Jr., & Ulhoa, C. J. (2008). A hyaluronidase from *Potamotrygon motoro* (freshwater stingrays) venom: Isolation and characterization. *Toxicon: Official Journal of the International Society on Toxinology*, *51*(6), 1060–1067. https://doi.org/10.1016/j.toxicon.2008.01.008

Markovic, O., & Mollnar, L. (1954). Isolation of and determination of bee venom. *Chemicke Zvesti*, *8*, 80–90.

Marsh, D. G., Goodfriend, L., King, T. P., Lowenstein, H., & Platts-Mills, T. A. (1986). Allergen nomenclature. *Bulletin of the World Health Organization*, *64*(5), 767–774.

Matthiensen, A. E., Tellechea, E., & Levy, J. A. (1993). Biochemical characterization for the genetic interpretation of esterase isozymes in *Micropogonias furnieri* (Pisces, Sciaenidae) in the south of Brazil. *Comparative Biochemistry and Physiology Part B: Comparative Biochemistry*, *104*(2), 349–352. https://doi.org/10.1016/0305-0491(93)90378-I

Matulis, D., & Lovrien, R. (1998). 1-Anilino-8-naphthalene sulfonate anion-protein binding depends primarily on ion pair formation. *Biophysical Journal*, *74*(1), 422–429. https://doi.org/10.1016/S0006-3495(98)77799-9

Matysiak, J., Dereziński, P., Urbaniak, B., Klupczyńska, A., Zalewska, A., & Kokot, Z. J. (2013). A new method for determination of hyaluronidase activity in biological samples using capillary zone electrophoresis. *Biomedical Chromatography: BMC*, 27(8), 1070–1078. https://doi.org/10.1002/bmc.2909

McDonald, C. E., & Chen, L. L. (1965). Lowry modification of the Folin reagent for determination of proteinase activity. *Analytical Biochemistry*, *10*, 175–177. https://doi.org/10.1016/0003-2697(65)90255-1

McLaughlin, B. E., Baumgarth, N., Bigos, M., Roederer, M. D., Rosa, S. C., Altman, J. D., Nixon, D. F., Ottinger, J., Oxford, C., Evans, T. G., & Asmuth, D. M. (2008). Nine-color flow cytometry for accurate measurement of T cell subsets and cytokine responses. Part I: Panel design by an empiric approach. *Cytometry. Part A: The Journal of the International Society for Analytical Cytology*, *73*(5), 400–410. https://doi.org/10.1002/cyto.a.20555

Moreau, S. J. M. (2013). "It stings a bit but it cleans well": Venoms of Hymenoptera and their antimicrobial potential. *Journal of Insect Physiology*, *59*(2), 186–204. https://doi.org/10.1016/j.jinsphys.2012.10.005

Morey, S. S., Kiran, K. M., & Gadag, J. R. (2006). Purification and properties of hyaluronidase from *Palamneus gravimanus* (Indian black scorpion) venom. *Toxicon: Official Journal of the International Society on Toxinology*, 47(2), 188–195. https://doi.org/10.1016/j.toxicon.2005.10.014

Morse, R. A., & Benton, A. W. (1964a). Mass collection of bee venom. *Gleaning Bee Culture*, *92*(1), 42–45. 54.

Morse, R. A., & Benton, A. W. (1964b). Notes on venom collection from honey bees. *Bee World*, *45*(4), 141–143. https://doi.org/10.1080/0005772X.1964.11097072

Muckenschnabel, I., Bernhardt, G., Spruss, T., Dietl, B., & Buschauer, A. (1998). Quantitation of hyaluronidases by the Morgan-Elson reaction: comparison of the enzyme activities in the plasma of tumor patients and healthy volunteers. *Cancer Letters*, *131*(1), 13–20. https://doi.org/10.1016/S0304-3835(98)00196-7

Müller, H. L. (1966). Diagnosis and treatment of insect sensitivity. *Journal of Asthma Research*, *3*(4), 331–333.

Muller, U. R. (2005). Bee venom allergy in beekeepers and their family members. *Current Opinion in Allergy and Clinical Immunology*, *5*(4), 343–347.

Nakamura, T., Majima, M., Kubo, K., Takagaki, K., Tamura, S., & Endo, M. (1990). Hyaluronidase assay using fluorogenic hyaluronate as a substrate. *Analytical Biochemistry*, *191*(1), 21–24. https://doi.org/10.1016/0003-2697(90)90380-R

Neumann, W., & Habermann, E. (1954). Über die Phospholipase A des Bienengiftes. *Hoppe-Seyler's Zeitschrift Für Physiologische Chemie*, *296*(Jahresband), 166–179. https://doi.org/10.1515/bchm2.1954.296.1.166

Nobre, A. A. B. (1990). Innovations: A device to provoke venom release from honey bees. *Bee World*, *71*(4), 151–152. https://doi.org/10.1080/0005772X.1990.11099057

Nocelli, R. C. F., Roat, T. C., & Cruz-Landim, C. (2007). Alterations induced by the juvenile hormone in glandular cells of the *Apis mellifera* venom gland: Applications on newly emerged workers (Hymenoptera, Apidae). *Micron (Oxford, England: 1993)*, *38*(1), 74–80. https://doi.org/10.1016/j.micron.2006.03.009

Oršolić, N. (2012). Bee venom in cancer therapy. *Cancer Metastasis Review*, *31*, 173–194.

Owen, M. D. (1979). Relationship between age and hyaluronidase activity in the venom of queen and worker honey bees (*Apis mellifera* L.). *Toxicon: Official Journal of the International Society on Toxinology*, *17*(1), 94–98. https://doi.org/10.1016/0041-0101(79)90260-5

Owen, M. D., & Bridges, A. R. (1976). Aging in the venom glands of queen and worker honey bees (*Apis mellifera* L.): Some morphological and chemical observations. *Toxicon: Official Journal of the International Society on Toxinology*, *14*(1), 1–5. https://doi.org/10.1016/0041-0101(76)90113-6

Palmer, D. J. (1961). Extraction of bee venom for research. *Bee World*, 42(9), 225–226. https://doi.org/10.1080/0005772X.1961.11096884

Pattanaargson, S., & Roboz, J. (1996). Determination of hyaluronidase activity in venoms using capillary electrophoresis. *Toxicon: Official Journal of the International Society on Toxinology*, *34*(10), 1107–1117. https://doi.org/10.1016/0041-0101(96)00083-9

Peiren, N. D., de Graaf, D. C., Brunain, M., Bridts, C. H., Ebo, D. G., Stevens, W. J., & Jacobs, F. J. (2006). Molecular cloning and expression of icarapin, a novel IgE-binding bee venom protein. *FEBS Letters*, *580*(20), 4895–4899. https://doi.org/10.1016/j.febslet.2006.08.005

Peiren, N., Vanrobaeys, F. D., Graaf, D. C., Devreese, B., van Beeumen, J., & Jacobs, F. J. (2005). The protein composition of honeybee venom reconsidered by a proteomic approach. *Biochimica et Biophysica Acta*, *1752*(1), 1–5. https://doi.org/10.1016/j.bbapap.2005.07.017

Price, J. A. (2007). A colorimetric assay for measuring phospholipase A2 degradation of phosphatidylcholine at physiological pH. *Journal of Biochemical and Biophysical Methods*, *70*(3), 441–444. https://doi.org/10.1016/j.jbbm.2006.10.008

Pukrittayakamee, S., Warrell, D. A., Desakorn, V., McMichael, A. J., White, N. J., & Bunnag, D. (1988). The hyaluronidase activities of some Southeast Asian snake venoms. *Toxicon: Official Journal of the International Society on Toxinology*, *26*(7), 629–637. https://doi.org/10.1016/0041-0101(88)90245-0

Queiroz, G. P., Pessoa, L. A., Portaro, F. C. V., Furtado, M. d F. D., & Tambourgi, D. V. (2008). Interspecific variation in venom composition and toxicity of Brazilian snakes from Bothrops genus. *Toxicon: Official Journal of the International Society on Toxinology*, *52*(8), 842–851. https://doi.org/10.1016/j.toxicon.2008.10.002

Radauer, C., Nandy, A., Ferreira, F., Goodman, R. E., Larsen, J. N., Lidholm, J., Pomes, A., Raulf-Heimsoth, M., Rozynek, P., Thomas, W. R., & Breiteneder, H. (2014). Update of the WHO/IUIS allergen nomenclature database based on analysis of allergen sequences. *Allergy*, *69*(4), 413–419. https://doi.org/10.1111/all.12348

Reynolds, L. J., Hughes, L. L., Yu, L., & Dennis, E. A. (1994). 1-Hexadecyl-2-arachidonoylthio-2-deoxy-sn-glycero-3-

phosphorylcholine as a substrate for the microtiterplate assay of human cytosolic phospholipase A2. *Analytical Biochemistry*, *217*(1), 25–32. https://doi.org/10.1006/abio.1994.1079

Roederer, M. (2002). Compensation in flow cytometry. Current Protocols in Cytometry, 22(1), 1.14.1–1.14.20. (Editorial Board, ROBINSON, J P; Managing Editor, Chapter 1, Unit 1 14). https://doi.org/10.1002/0471142956.cy0114s22

Ruëff, F., Chatelain, R., & Przybilla, B. (2011). Management of occupational Hymenoptera allergy. *Current Opinion in Allergy and Clinical Immunology*, *11*(2), 69–74. https://doi.org/10.1097/ACI.0b013e3283445772

Sabato, V., Faber, M., Bridts, C. H., DE Clerck, L., & Ebo, D. G. (2015). Measuring histamine content and release at single-cell level during venom allergy immunotherapy. *The Journal of Allergy and Clinical Immunology*, *135*(4), 1089. https://doi.org/10.1016/j.jaci.2014.12.1946

Schägger, H., & VON Jagow, G. (1987). Tricine-sodium dodecyl sulfate-polyacrylamide gel electrophoresis for the separation of proteins in the range from 1 to 100 kDa. *Analytical Biochemistry*, *166*(2), 368–379. https://doi.org/10.1016/0003-2697(87)90587-2

Schmidt, J. O. (1995). Toxinology of venoms from the honey bee genus *Apis*. *Toxicon*, *33*(7), 917–927. https://doi.org/10.1016/0041-0101(95)00011-A

Schmidt, J. O., & Blum, M. S. (1978). The biochemical constituents of the venom of the harvest ant *Pogonomyrmex badius*. *Comparative Biochemistry and Physiology Part C: Comparative Pharmacology*, *61*(1), 239–247. https://doi.org/10.1016/0306-4492(78)90137-5

Schmidt, J. O., Blum, M. S., & Overal, W. (1986). Comparative enzymology of venoms from stinging Hymenoptera. *Toxicon: Official Journal of the International Society on Toxinology*, *24*(9), 907–921. https://doi.org/10.1016/0041-0101(86)90091-7

Sobotka, A. K., Franklin, R. M., Adkinson, N. F., Jr., Valentine, M., Baer, H., & Lichtenstein, L. M. (1976). Allergy to insect stings. II. Phospholipase A: The major allergen in honey bee venom. *Journal of Allergy and Clinical Immunology*, *57*(1), 29–40. https://doi.org/10.1016/0091-6749(76)90076-2

Steiner, B., & Cruce, D. (1992). A zymographic assay for detection of hyaluronidase activity on polyacrylamide gels and its application to enzymatic activity found in bacteria. *Analytical Biochemistry*, *200*(2), 405–410. https://doi.org/10.1016/0003-2697(92)90487-R

Stern, R., & Jedrzejas, M. J. (2006). Hyaluronidases: Their genomics, structures, and mechanisms of action. *Chemical Reviews*, *106*(3), 818–836. https://doi.org/10.1021/cr050247k

Stern, M., & Stern, R. (1992). An ELISA-like assay for hyaluronidase and hyaluronidase inhibitors. *Matrix (Stuttgart, Germany)*, *12*(5), 397–403. https://doi.org/10.1016/S0934-8832(11)80036-3

Turunen, S., & Chippendale, G. M. (1977). Esterase and lipase activity in the midgut of *Diatraea grandiosella*: Digestive functions and distribution. *Insect Biochemistry*, 7(1), 67–71. https://doi.org/10.1016/0020-1790(77)90058-0

van Vaerenbergh, M., Cardoen, D., Formesyn, E. M., Brunain, M., van Driessche, G., Blank, S., Spillner, E., Verleyen, P., Wenseleers, T., Schoofs, L., Devreese, B. ; D., & Graaf, D. C. (2013). Extending the honey bee venome with the antimicrobial peptide apidaecin and a protein resembling wasp antigen 5. *Insect Molecular Biology*, 22(2), 199–210. https://doi.org/10.1111/imb.12013

van Vaerenbergh, M., Debyser, G., Devreese, B., & de Graaf, D. C. (2014). Exploring the hidden honeybee (*Apis mellifera*) venom proteome by integrating a combinatorial peptide ligand library approach with FTMS. *Journal of Proteomics*, *99*, 169–178. https://doi.org/10.1016/j.jprot.2013.04.039

Vercruysse, K. P., Lauwers, A. R., & Demeester, J. R. (1995). Absolute and empirical determination of the enzymatic activity and kinetic investigation of the action of hyaluronidase on hyaluronan using viscosimetry. *Biochemical Journal*, *306*(1), 153–160. https://doi.org/10.1042/bj3060153

Verweij, M. M., DE Knop, K. J., Bridts, C. H., DE Clerck, L. S., Stevens, W. J., & Ebo, D. G. (2010). P38 mitogen-activated protein kinase signal transduction in the diagnosis and follow up of immunotherapy of wasp venom allergy. *Cytometry: Part B, Clinical Cytometry*, *78*(5), 302–307. https://doi.org/10.1002/cyto.b.20531

Verweij, M. M., Sabato, V., Nullens, S., Bridts, C. H., DE Clerck, L. S., Stevens, L. S., & Ebo, D. G. (2012). STAT5 in human basophils: IL-3 is required for its FcεRI-mediated phosphorylation. *Cytometry: Part B, Clinical Cytometry*, *82*(2), 101–106. https://doi.org/10.1002/cyto.b.20629

Vetter, R. S., Visscher, P. K., & Camazine, S. (1999). Mass envenomations by honey bees and wasps. *Western Journal of Medicine*, *170*(4), 223–227.

Vivek, H. K., Swamy, S. G., Priya, B. S., Sethi, G., Rangappa, K. S., & Nanjunda Swamy, S. (2014). A facile assay to monitor secretory phospholipase A2 using 8-anilino-1-naphthalenesulfonic acid. *Analytical Biochemistry*, *461*, 27–35. https://doi.org/10.1016/j.ab.2014.05.024

Vlasak, R., & Kreil, G. (1984). Nucleotide sequence of cloned cDNAs coding for preprosecapin, a major product of queen-bee venom glands. *European Journal of Biochemistry*, *145*(2), 279–282. https://doi.org/10.1111/j.1432-1033.1984.tb08549.x

von Moos, S., Graf, N., Johansen, P., Mullner, G., Kundig, T. M., & Senti, G. (2013). Risk assessment of Hymenoptera re-sting frequency: Implications for decision-making in venom immunotherapy. *International Archives of Allergy and Immunology*, *160*(1), 86–92. https://doi.org/10.1159/000338942

Wieser, W. (1973). Effects of temperature on ectothermic organisms. In W. Wieser (Ed.), *Temperature relations of ectotherms. A speculative review*. New York: Springer Verlag; pp. 1–23.

Williams, G. R., Alaux, C., Costa, C., Csaki, T., Doublet, V., Eisenhardt, D., Fries, I., Kuhn, R., McMahon, D. P., Medrzycki, P., Murray, T. E., Natsopoulou, M. E., Neumann, P., Oliver, R., Paxton, R. J., Pernal, S. F., Shutler, D., Tanner, G., van der Steen, J. J. M., & Brodschneider, R. (2013). Standard methods for maintaining adult Apis mellifera in cages under in vitro laboratory conditions. In V Dietemann; J D Ellis; P Neumann (Eds) The COLOSS BEEBOOK, Volume I: standard methods for Apis mellifera research. *Journal of Apicultural Research*, 52(1), 1–36. https://doi.org/10.3896/IBRA.1.52.1.04

Winningham, K. M., Fitch, C. D., Schmidt, M., & Hoffman, D. R. (2004). Hymenoptera venom protease allergens. *The Journal of Allergy and Clinical Immunology*, *114*(4), 928–933. https://doi.org/10.1016/j.jaci.2004.07.043

Yin, C. S., & Koh, H. G. (1998). The first documental record on bee venom therapy in Oriental medicine: 2 prescriptions of bee venom in the ancient Mawangdui books of Oriental medicine. *Journal of the Korean Acupunture and Moxibustion Society*, *15*, 143–147.

Zalat, S., Nabil, Z., Hussein, A., & Rahka, M. (1999). Biochemical and haematological studies of some solitary and social bee venoms. *Egyptian Journal of Biology*, *1*, 57–71.

Zhang, L. S., & Mummert, M. E. (2008). Development of a fluorescent substrate to measure hyaluronidase activity. *Analytical Biochemistry*, *379*(1), 80–85. https://doi.org/10.1016/j.ab.2008.04.040

Zhong, S. P., Campoccia, D., Doherty, P. J., Williams, R. L., Benedetti, L., & Williams, D. F. (1994). Biodegradation of hyaluronic acid derivatives by hyaluronidase. *Biomaterials*, *15*(5), 359–365. https://doi.org/10.1016/0142-9612(94)90248-8

Zimmerman, M., Ashe, B., Yurewicz, E. C., & Patel, G. (1977). Sensitive assays for trypsin, elastase, and chymotrypsin using new fluorogenic substrates. *Analytical Biochemistry*, *78*(1), 47–51. https://doi.org/10.1016/0003-2697(77)90006-9

Zimmerman, M., Yurewicz, E., & Patel, G. (1976). A new fluorogenic substrate for chymotrypsin. *Analytical Biochemistry*, *70*(1), 258–262. https://doi.org/10.1016/S0003-2697(76)80066-8

*Journal of Apicultural Research*, 2021
Vol. 60, No. 4, 1–109, https://doi.org/10.1080/00218839.2021.1948240

**REVIEW ARTICLE**

# Standard methods for pollen research

Maria G. Campos[a*], Ofélia Anjos[b,c], Manuel Chica[d], Pascual Campoy[e], Janka Nozkova[f], Norma Almaraz-Abarca[g], Lidia M. R. C. Barreto[h], João Carlos Nordi[i], Leticia M. Estevinho[i], Ananias Pascoal[i], Vanessa Branco Paula[i], Altino Chopina[i], Luis G. Dias[i], Živoslav L. j. Tešić[j], Mirjana D. Mosić[j], Aleksandar Ž. Kostić[k], Mirjana B. Pešić[k], Dušanka M. Milojković-Opsenica[j], Wiebke Sickel[l], Markus J. Ankenbrand[m], Gudrun Grimmer[n], Ingolf Steffan-Dewenter[m], Alexander Keller[o], Frank Förster[p], Chrysoula H. Tananaki[q], Vasilios Liolios[p], Dimitrios Kanelis[q], Maria-Anna Rodopoulou[q], Andreas Thrasyvoulou[q], Luísa Paulo[q], Christina Kast[s], Matteo A. Lucchetti[s], Gaëtan Glauser[t], Olena Lokutova[u], Ligia Bicudo de Almeida-Muradian[v], Teresa Szczęsna[w] and Norman L. Carreck[x,y]

[a]*Coimbra Chemistry Centre (CQC, FCT Unit 313) (FCTUC) & Observatory of Drug-herb Interactions, University of Coimbra, Coimbra, Portugal;* [b]*IPCB– Instituto Politécnico de Castelo Branco, Castelo Branco, Portugal;* [c]*CEF, Instituto Superior de Agronomia, Universidade de Lisboa, Tapada da Ajuda, 1349-017, Lisboa, Portugal;* [d]*Andalusian Research Institute DaSCI "Data Science and Computational Intelligence", University of Granada, Granada, Spain;* [e]*Automatics and Robotics Center, Universidad Politécnica de Madrid, Madrid, Spain;* [f]*Centro Interdisciplinario de Investigación para el Desarrollo Integral Regional unidad Durango, Instituto Politécnico, Nacional, México;* [g]*Interdisciplinary Research Center for the Integral Regional Development-Durango, National Polytechnic Institute, Mexico City, Mexico;* [h]*Apicultural Center of Research of Taubaté University, UNITAU - Universidade de Taubaté-SP, Taubaté, Brazil;* [i]*Centro de Investigação de Montanha (CIMO), Instituto Politécnico de Bragança, Bragança, Portugal;* [j]*Faculty of Chemistry, Department of Analytical Chemistry, University of Belgrade, Belgrade, Serbia;* [k]*Faculty of Agriculture, Department of Chemistry and Biochemistry, University of Belgrade, Belgrade, Serbia;* [l]*Thünen Institute, Institute of Biodiversity, Braunschweig, Germany;* [m]*DZHI, Universitätsklinikum Würzburg, Würzburg, Germany;* [n]*Department of Animal Ecology and Tropical Biology, Biocenter, University of Würzburg, Würzburg, Germany;* [o]*Center for Computational and Theoretical Biology & Department of Bioinformatics, Biocenter, University of Würzburg, Würzburg, Germany;* [p]*Bioinformatics & Systems Biology, Heinrich-Buff-Ring 58, Justus-Liebig-Universität, Gießen, Germany;* [q]*Laboratory of Apiculture-Sericulture, School of Agriculture, Aristotle University of Thessaloniki, Thessaloniki, Greece;* [r]*CATAA- Associação Centro de Apoio Tecnológico Agro-Alimentar de Castelo Branco, Castelo Branco, Portugal;* [s]*Agroscope, Swiss Bee Research Centre, Bern, Switzerland;* [t]*Neuchâtel Platform of Analytical Chemistry, University of Neuchâtel, Neuchâtel, Switzerland;* [u]*The National Institute of Horticultural Research Konstytucji 3 Maja 1/3, 96 - 100, Skierniewice, Poland;* [v]*Pharmaceutical Science School, University of São Paulo, São Paulo, Brazil;* [w]*The National Institute of Horticultural Research Konstytucji 3 Maja 1/3, 96 - 100, Skierniewice, Poland;* [x]*Carreck Consultancy Ltd. Woodside Cottage, Shipley, West Sussex, UK;* [y]*Laboratory of Apiculture and Social Insects, School of Life Sciences, University of Sussex, Brighton, East Sussex, UK*

*(Received 20 March 2020; accepted 22 June 2021)*

"Bee pollen" is pollen collected from flowers by honey bees. It is used by the bees to nourish themselves, mainly by providing royal jelly and brood food, but it is also used for human nutrition. For the latter purpose, it is collected at the hive entrance as pellets that the bees bring to the hive. Bee pollen has diverse bioactivities, and thus has been used as a health food, and even as medication in some countries. In this paper, we provide standard methods for carrying out research on bee pollen. First, we introduce a method for the production and storage of bee pollen which assures quality of the product. Routine methods are then provided for the identification of the pollen's floral sources, and determination of the more important quality criteria such as water content and content of proteins, carbohydrates, fatty acids, vitamins, alkaloids, phenolic and polyphenolic compounds. Finally, methods are described for the determination of some important bioactivities of bee pollen such as its antioxidant, anti-inflammatory, antimicrobial and antimutagenic properties.

**Métodos estándar Para la investigación del polen**

El "polen de abeja" es el polen recogido de las flores por las abejas melíferas. El polen de abeja es utilizado para nutrir a las propias abejas, principalmente para proporcionar jalea real y alimento para las crías, pero también se utiliza para la nutrición humana. Para este último fin, se recoge en la entrada de la colmena en forma de gránulos que las abejas llevan a la colmena. El polen de abeja tiene diversas bioactividades, por lo que se hautilizado como alimento para la salud, e incluso como medicamento en algunos países. En este artículo, proporcionamos métodos estándar para llevar a cabo investigaciones sobre el polen de abeja. En primer lugar, presentamos un método de producción y almacenamiento de polen de abeja que garantiza la calidad del producto. A continuación, se ofrecen métodos de rutina para la identificación de las fuentes florales del polen y la determinación de los criterios de calidad más importantes, como el contenido de

*Corresponding author. Email: mgcampos@ff.uc.pt and ofelia@ipcb.pt

*Please refer to this paper as: Campos, M.G., Anjos, O., Chica, M., Campoy, P., Nozkova, J., Almaraz-Abarca, N., Barreto, L.M.R.C., Nordi, J.C., Estevinho, L.M., Pascoal, A., Paula, V.B., Chopina, A., Dias, L., Tešić, Z.L., Mosić, M.D., Kostić, A.Z., Pešić, M.B., Milojković-Opsenica, D.M., Sickel, W., Ankenbrand, M.J., Grimmer, G., Steffan-Dewenter, I., Keller, A., Förster, F., Tananaki, C.H., Liolios, V., Kanelis, D., Rodopoulou, M.A., Thrasyvoulou, A., Paulo, L., Kast, C., Lucchetti, M.A., Glauser, G., Lokutova, O., Bicudo de Almeida-Muradian, L., Szczęsna, T., Carreck, N.L. (2021) Standard methods for pollen research. In V. Dietemann, P. Neumann, N. L. Carreck, & J. D. Ellis (Eds.), The COLOSS *BEEBOOK* - Volume III: Standard methods for *Apis mellifera* hive product research. *Journal of Apicultural Research*, 60(4). https://doi.org/10.1080/00218839.2021.1948240

agua y de proteínas, carbohidratos, ácidos grasos, vitaminas, alcaloides y compuestos fenólicos y polifenólicos. Por último, se describen métodos para la determinación de algunas bioactividades importantes del polen de abeja, como sus propiedades antioxidantes, antiinflamatorias, antimicrobianas y antimutagénicas.

花粉研究的标准方法

"蜂花粉"是蜜蜂从花中采集的花粉。首先, 它用于蜜蜂本身的营养, 主要是提供蜂王浆和哺育幼虫的食物, 其次也用于人类营养。对于后一目的, 花粉作为蜜蜂带到蜂巢的颗粒, 在蜂巢入口处被收集。蜂花粉具有多种生物活性, 因此被用作健康食品, 甚至在一些国家被用作药物。本文提供了研究蜂花粉的标准方法。首先, 我们介绍了一种生产和储存蜂花粉的方法, 以保证产品的质量。然后提供常规鉴定方法来鉴定花粉源及更重要的质量标准, 如水含量和蛋白质、碳水化合物、脂肪酸、维生素、生物碱、酚类和多酚类化合物的含量等。最后, 描述了测定蜂花粉的一些重要生物活性的方法, 如其抗氧化、抗炎、抗菌和抗诱变特性。

**Keywords:** pollen; harvest; storage; multi-classification; computer; vision; corbicula; proteins; lipids; vitamins; antioxidant; minerals; toxicity; alkaloids; UHPLC-HRMS; phenolic; flavonoids; ABTS; FTIR-ATR; antioxidant; free radical scavenging activity; ORAC; DPPH*; antimicrobial; antimutagenic; COLOSS; *BEEBOOK*

## 1. Introduction

Pollen production on an industrial scale can be considered as a recent development, so there is a need for an international directive for quality control. In this chapter, the authors propose optimized methodologies based on the major skills developed in the field which can be used in the near future to fill that gap. This section aims to aid the understanding of actual production and the development of sanitary standards in the light of food security, standardized beekeeping management actions, collection of the product in the bee hives, transportation, sanitary treatment, processing and storage by agribusiness, resulting in a product with flavor, very delightful aroma and texture of excellent quality for food or nutraceutical purposes. To achieve the above characteristics, we also discuss corbicular pollen itself in terms of weight, size, color, shape and surface texture and, unified systems (descriptors) for the evaluation of corbicula morphological parameters of monofloral pollen.

Pollen is considered by some researchers to be one of the naturally available resources that may be regarded as a perfect food, since it is rich in essential amino acids, trace elements, enzymes, B-complex and some C and E vitamins. The pollen grains present a huge variation of morphological characteristics, established by genetic heredity and do not vary depending on environmental events or changes. In this context, the pollinic analysis of bee pollen is the best method to determine the plants used by honey bees, making it possible, among other inferences, to establish the plant *taxa* in the regions used for the pollen forage. The goal is to provide global information regarding the pollinic analysis of this matrix and the analytical procedures that are most frequently applied, from the more laborious and time consuming to the quicker and more advanced. Among the last, flavonoid / phenolic acid profiles using a gradient HPLC/DAD method is described. Given the *specie-specific* feature of those profiles obtained with different *taxa* collected by bees, they represent an important tool for quality control concerning the botanical origin of both monofloral and complex mixtures of bee pollen.

The first step in the determination of phenolic compounds is their appropriate extraction. The complexity of the structure of pollen grains requires multiple-stage extraction of polyphenols. Here there will be explained different procedures for the preparation of pollen extracts containing phenolic compounds and proposed a general procedure. Also presented are the applications of SPE for pre-concentration of the obtained phenolic fractions which facilitate further analysis. One of the sections describes LC and LC/MS techniques used for the identification and quantification of phenolic compounds with a detailed procedure for their determination by the UHPLC- HESI-MS/MS technique.

Another important question to be discussed is the fact that traditional pollen analysis via light microscopy has limitations in sample throughput as well as taxonomic resolution. Recently, pollen meta-barcoding methods have been developed as alternative approaches, where plant species identification of pollen grains works via DNA sequencing. However, these currently utilize different genetic markers and sequencing platforms lessening study comparability. We here describe a detailed protocol of the latest development in this field as a standard method for pollen meta-barcoding. It is highly cost-efficient, requires no palynological knowledge, is performable in standard laboratories and profits from a well-established reference database.

For the main macro and micronutrients standard methods are proposed. Bee pollen has been considered by some researchers as Mother Naturés perfect food by containing nearly all nutrients required by humans as proteins, sugars, amino acids, lipids, vitamins, minerals, nucleic acids, enzymes, phenolics and many more compounds. Pollen can also be characterized as a supplementary food with varied enhancing effects in human health due to its nutritional properties. Once pollen was exclusively a source of protein for bees' diet, but recently, it has gained further attention as a potential food source for human consumption too. The Kjeldahl method is widely used for protein determination. In this section, we examine the methodology concerning the volumes of the reagents, as well as the distillation time and the quantity of the sample, in order to elucidate the most suitable procedure and a method been validated using bee pollen. There is increasing knowledge of the nutritional value of pollen as a supplementary food or medicine. One species of pollen is different from another, and not one has the same standard chemical composition, nor contains all the characteristics attributed to "pollen" in general. The lipid fraction is one of the main pollen constituents, and its determination is highly important. In this chapter, we describe a method for the determination of lipid content of pollen investigated and optimized, by minimizing the reagent consumption and the sample quantity.

Suggested methods to analyze micronutrients are also presented. Pollen vitamins as well as the literature found about them are included in the text. Methods for B-complex ($B_1$, $B_2$, $B_6$ and PP, with vitamers) and antioxidant vitamins (vitamin E, C and carotenes as provitamin A) are explained and involve HPLC, open column and titrimetric determinations. The composition of minerals in pollen collected by honey bees is dependent on the plant species visited. The variety of the samples collected is closely associated with the geographical apiary locality given by the floral sources *(taxa)* and the environmental conditions in the area such as soil composition, climatic conditions and agricultural crops grown close to the apiary. Because plants can cause bioremediation of some pollutants and are part of some other anthropogenic processes, they should also be considered as an additional source of minerals in pollen (e.g., Cd, Cr, Cu, Fe, Ni, Pb and Zn). Monitoring these minerals in this matrix, as well in other food products is very important as a quality parameter. Despite this, the publications in the field are scarce. Here we propose, from the research published on the mineral composition on pollen, different methodologies that can be carried out.

Regarding toxic compounds, research has been limited. However, as a starting point, attention has focused on pyrrolizidinic alkaloids (PA). Many plant species produce pollen containing PAs that honey bees collect and bring as a protein source into the bee hive. Methods for identification of PA in pollen, together with the extraction and analysis of PAs, are essential tools for plant and food research. Mainly two approaches are commonly used: In the first approach, the total PAs (*N*-oxides and free-bases) are analyzed as the sum of the "necine bases" (the basic structures of PAs) by gas chromatography coupled with mass spectrometry (GC-MS). The second approach is based on liquid chromatography coupled with mass spectrometric detection (LC-MS). This method can detect both the free alkaloids and their *N*-oxides simultaneously, and has become the method of choice in recent years. Due to the lack of standards on the market and the multitude of PAs discovered so far, the detection, identification and quantification of PAs still represent a continuous challenge. In the following chapter, methods for the collection, extraction and profiling of PA plant pollen from *Echium vulgare* L. and *Eupatorium cannabinum* L. are presented.

In the advanced methods that allow multiple analysis with a small amount of sample, it will be shown the potential of new integrated methods as the Fourier Transform Infrared (FTIR) spectroscopic method with Attenuated Total Reflectance (ATR) used, for example, to predict the total content of phenolic and polyphenolic compounds. Determination of Free radical scavenging ($EC_{50}$) and antioxidative capacity (ABTS) of bee pollen is also included. FTIR-ATR can be a useful technique for the quantification of total phenolic compounds, total flavonoids and antioxidant capacity of bee pollen that could improve the speed of laboratory analyses of this product.

Concerning the bioactivity of pollen, several studies have been developed in order to elucidate the usefulness of pollen, and therapeutic effects have been reported among others in the treatment of benign prostatitis and for oral desensitization of children who have an allergy, effects as anti-inflammatory, antimicrobial, anti-fungicidal, anti-mutagenic and immunomodulatory. Some studies reported that bee pollen accelerates mitotic rate, promotes tissue repair, enhances greater toxic elimination and reduces excessive cholesterol levels. Pollen has also been used to alleviate or cure conditions such as colds, flu, ulcers, premature aging, anemia, colitis and today is increasingly used as a health food supplement.

For these reasons, bee pollen has attracted especial attention by researchers throughout the world, and many scientific publications have been produced. In the last 17 years, at least 42 reports have been published about the antioxidant properties from different *taxa*, evaluating these properties by various methods. The available information corresponds to bee pollen of 21 countries including Brazil, Mexico, New Zealand and Portugal, the countries where most attention has been dedicated to this topic. The reports cited in the current work give information about the antioxidant capabilities of bee pollen coming from around 75 well identified plant species belonging to approximately 52 families. The 2,2-diphenyl-1-picrilhydrazyl (DPPH*) assay is the most popular method to evaluate the free radical scavenging properties of bee pollen. It is based on the single electron transfer mechanism, is rapid, easy, accurate, reproducible, and presents good correlations with other assays like TEAC and $\beta$-carotene bleaching. The oxygen radical absorbance capacity (ORAC) assay is based on the hydrogen atom transfer mechanism, which reflects biological conditions, and has been proposed for evaluating antioxidant properties in foods. For all these reasons, DPPH* and ORAC assays are proposed as standard methods to assess the antioxidant properties of bee pollen. The steps to carry out both assays are described. Due to related properties for human health, some information about standard methods commonly used for the determination of antimicrobial and antimutagenic activities in bee pollen are provided.

In conclusion, a significant quantity of standard methods for pollen research is now validated, and will be helpful to be included in the future International Directive for Quality Control of bee pollen in the near future.

## 2. Producing bee pollen from harvest to storage

To provide a good quality product it is necessary to get results in the conception of a professional, technical and competitive apiculture, mainly dealing with the opening trade between the nations, MERCOSUR, European Union, and others (Barreto et al., 2006; Panetta, 1998). The beekeeping segment in the industry context, fits into the conception, "in search of excellence and quality," concerning the significant role that the products represent to humanity (Bassi, 2000; Germano et al., 2000).

Bee pollen harvesting requires special skills and guidelines. In this work we will represent the best methods for a better quality of product in line with the good practices in the field.

The main guidelines to obtain the best raw material are

1. Production plan – Such planning should aim at the highest yield in both quantity and quality. Inside an organization, technical support is crucial, and can be achieved by courses, training, internships and even from specialized consultancy, according to the needs of the production (Barreto et al., 2006);
2. Directing the raw material – It is of an unquestionable importance to direct the raw material for processing and acquisition in an ideal standard with the following sequence: production, harvest, transport, processing and storage.

- Production: Standardized handling results in the harvest schedule.
- Harvest: Harvest is the first stage from raw material operations to processing; this process should undergo rigorous contamination control.
- Transport: Incorrect transport can negatively affect the quality of raw material. The following parameters should be considered (Barreto et al., 2006):
    - i. Adaptability of the vehicle to the type of raw material;
    - ii. Suitable packaging for the quality and type of the collected material;
    - iii. Necessity of traveling long distances (weather conditions, temperature, sun exposure, dust, rainfall, etc.);
    - iv. Internal environment of the transport (odor-free).
- Processing: In this phase, the facilities should be planned according to the flowchart that is demonstrated, as well as the necessary equipment to process the product such as freezers, dehydrators, sieves, cold chambers and others (Barreto et al., 2006).
- Storage: In this phase, consider the following requirements:
    - a. Type of packing;
    - b. Product identification, label;
    - c. Temperature, humidity, luminosity;
    - d. Time of travel to the facilities should be the lowest possible (Barreto et al., 2006).
- During the next steps, other situations can occur, for example: Deterioration of raw material: It usually occurs by the action of living organisms (insects, microorganisms, etc.), by physical processes (freezing, heating, pressurizing, etc.) and biological (fermentation, rancidity, etc.). Follow all the previous recommendations to avoid this situation, otherwise do not use the product (Franco, 2002; Silva, 2000).
- Contamination of raw material: Avoid the following causes of contamination (Gilliam et al., 1989; Silva, 1999):
    - a. Lack of cleanliness in rooms and personnel;
    - b. Exposure of raw material to fresh air;
    - c. Exposure of raw material to environment with dirt, garbage, animals with injuries and infectious illness and poor and badly kept premises.

### *2.1. Production and handling of bee pollen before processing*

Bee pollen is the result of specific conditions of the bee colonies, normally with regular population density of a bee hive in development, where the bees are stimulated to intensify pollen collection. Since it is a protein material production, this raw material will be susceptible to a more rigorous standard in the hygiene requirements, from the production up to its processing and storage (Barreto et al., 2006).

#### *2.1.1. Quality control in the apiary*

The quality control of bee products begins in the apiary because these products do not get their quality by handling, but by nature's attribute. The regulation of good practices in the elaboration and food production is a broad subject that is applicable in most of the production. It is known that the microflora in the bee pollen can have two different origins: (1) by the normal microflora, originated from pollen, being yeast, mold and bacteria; (2) by outside microflora, originated from practices of handling, elaboration and product storage.

To achieve a better-quality product some steps should be done follow this check list:

a. Systematized handling of bee colonies;
b. Feeding the bees with high quality and reliable products when forage is scarce;
c. Daily collection of bee pollen;
d. Disposal of bee pollen collected on rainy days;
e. Sanitation of equipment in contact with the pollen at the harvest period (pollen press, hive tool and brush);
f. Sanitized bee gloves and beekeeper's clothing;
g. Packing of harvest and transport previously sanitized and dried;
h. Enough harvest time and transport to keep the integrity level of the product until the processing facility.

***2.1.1.1. Smoke.*** Smoke is a safety procedure used during the handling of bee colonies. However, the fuel materials, sawdust, should be sieved to avoid producing unwanted dust that can affect bee pollen pellets. The smoke should have a pleasant odor, and pine, cypress and eucalyptus leaves and lemongrass can be used as additives to the mixture. Animal manure must not be used. The smoke should not be aimed at the combs with honey (avoiding that the product acquires unpleasant flavor) or brood. The smoke should be applied horizontally in the bee hive when there is no exposed frame. It should never be aimed directly at the raw material in production (Barreto et al., 2006).

***2.1.1.2. bee hives.*** Nowadays bee hives are often already impermeable, which avoids the previous practice of painting then, with the risk of heavy metal contaminated paints (Barreto et al., 2006)

***2.1.1.3. Clothing.*** The overalls or coats used in the apiary should be cleaned after each handling session to avoid contamination in the bee hive by moth eggs and microorganisms that may adhere to clothes (Barreto et al., 2006).

***2.1.1.4. Tools.*** Hive tool, clean rubber gloves. After each handling, in case any bee hive is found contaminated, these utensils should be substituted so as not to contaminate the next hive (Barreto et al., 2006).

***2.1.1.5. Inputs.*** The main input used in the bee hive is the bees wax. The beekeeper should acquire this material in stores or suitable processing centers. Whenever buying bees wax, make sure it is packed with PVC film, whether there is moth or not (crumbled aspect in the plastic foil). Any wax that is directly packed with newspaper should be avoided as printing ink is a powerful source of heavy metals (Barreto et al., 2006).

***2.1.1.6. Handling.*** The choice of technology to be used in production handling is crucial. The material applied in the production has to be standardized, and it needs to be easily handled to avoid higher bee losses, as well as allow perfect sanitation of the product. For example, some hive boxes made of wood are now being replaced by nontoxic plastic boxes that permit efficient sanitation. Another point is the handling standard; it refers to the conduction of the production in a way that the harvest can be realized in all bee hives in a timely fashion (Petersen et al., 2011).

*2.1.2. Types of pollen traps*

Among frontal, floor, intermediate or bottom pollen traps (Figure 1), each one has its advantages and disadvantages. The ideal pollen trap is one that can receive the pollen in a safe and hygienic manner. The frontal pollen trap is practical and hygienic but because it is attached to the main floor, it receives a significant amount of dirt resulting from the internal bee hive cleaning. The bees continuously remove particles from inside to outside of the hive, making the pollen trap dirty. In the intermediate internal pollen trap, it is opened a secondary drawer, the main one being closed with a ventilation screen. It is noticed that the dirt level decreases and even the pollen has a lower humidity level than the previously mentioned. However, it can cause an increase in the bees' defensibility due to the obstruction of the main entrance. It is recommended to choose a pollen trap that can be easily placed and removed. In addition, a pollen trap that makes the handling easier, allowing better ventilation and sanitation, allowing the flight of drones and protected against weather changes. In tropical countries such as Brazil, beekeepers have been using the "Tropical Africanized Type" (Figure 1) with good results (Barreto, 2004).

Figure 1. A tropical Africanized pollen trap (TTA).

Figure 2. Pollen collection by beekeepers.

The collection of pollen (Figure 2) should be done daily, there are some devices that prescribe the weekly collection of the product, but there is no consensus among scientists about such procedure.

Figure 3. Conditions to be avoided in the apiary and surrounding area.

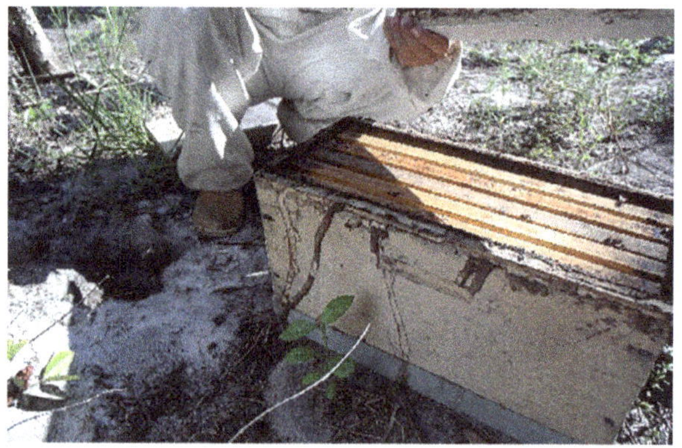

Figure 4. Bad conditions of the hives and packaging containers; hives on the ground are subject to rot and termite attack.

*2.1.3. Chosen area of apiary production*

Choice of apiary location can be problematic. Today, the choice should be careful, taking into account the use of agrochemicals, both in crops and in animals. Pollution is not exclusive to urban areas, and unfortunately some rural areas can show contamination in air, water and or soil. Rubbish in the apiary attracts undesirable insects, exposing the raw material to the risk of unwanted contamination (Figures 3 and 4).

*2.1.4. Quality control in the collection*

Damp compressed pollen should be collected separately from the raw materials intended for processing, and should only be used for bee feed (Figure 5). The collecting box should not be exposed to contamination from the ground.

Figure 5. Damp compressed pollen.

*2.1.5. Quality control in the transport of raw material*

At harvest, the beekeeper should be aware of the type of packaging used to transport the product from the apiary to the processing sector, thus avoiding:

a. crushing of raw material;
b. contamination with dust;
c. transfer of odors and humidity as well as high temperatures, which may interfere with the quality of the product to be processed.

### 2.2. *Handling of bee pollen in the processing sector*

This sector should be provided with a reception room for production (raw material) and processing rooms, according to the Technical Regulation on Hygiene and Sanitary Conditions and Good Preparation Practices for Developers of Establishments/Industrialized Food (Example for Brazil – Ordinance 368 of 4 September 1997 (Brasil, 1997) obtained in www.agricultura.gov.br/ada/dipoa/). Other countries have similar regulations.

In these areas, verify the rules:

a. machinery that will be in contact with the raw material should be in stainless steel (AISI 304);
b. manipulation, the process of cleaning and sanitizing should be adopted integrally;
c. use of heat, dehumidifiers, chemical agents, must have precise control;
d. water used for cleaning the room and machinery should be of known origin with recognized purity.

*2.2.1. Flowchart from production to processing of pollen*

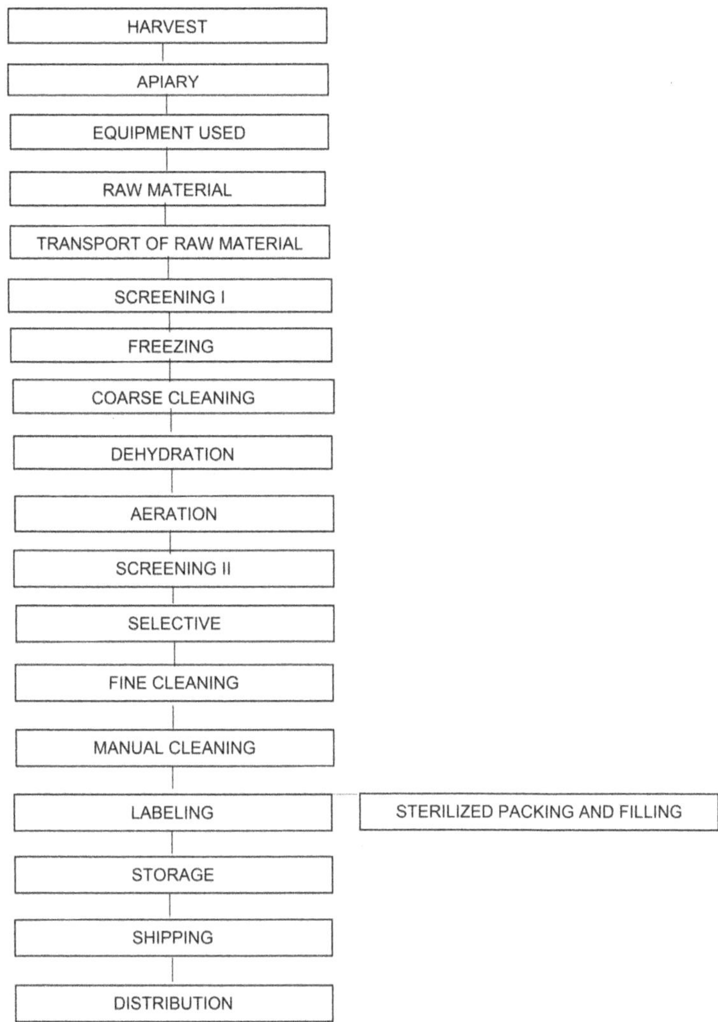


*2.2.2. Processing bee pollen*

In the processing area, the handler must necessarily follow the sequence below (Barreto et al., 2006):

a. Reception of raw material ("In Natura" pollen);
b. Primary labeling, date and place of harvest;
c. Removal of coarse dirt, dead bees, chalk brood mummies;
d. Screening;
e. Freezing to −18 °C for 48 hours minimum and maximum 15 days;
f. Dehydration step I at a maximum temperature of 42 °C for 8 hours;
g. Cleaning by ventilation;
h. Manual cleaning;
i. Dehydration step II maximum temperature of 42 °C for 4 hours;
j. Checking moisture content of dehydrated bee pollen;
k. Final filling of the dehydrated product;
l. Final product labeling valid up to 12 months;
m. Storage;
n. Shipping.

### 2.3. Materials and equipment involved in the process

The collection tray often used is a model made of non-toxic, broad and shallow plastic, suitable for food. Fresh pollen contains high humidity content (20 to 30%), consequently, fragile to large overlap, becoming brittle and causing an increase of pollen powder, representing the most significant loss at the time of processing. These trays must be provided with a cover which will prevent dirt from becoming incorporated into the product during transport (Almeida-Muradian et al., 2012; Brasil, 1997, 2001).

In Figures 6 and 7 can be seen the steps involved from the reception to the delivery of bee pollen. When the product arrives at the facilities, it has to be:

a. Transferred to lots;
b. Pass through wet sieving (usually with maximum capacity of 5 kg);
c. Freeze. Pollen should be in a sterile polyethylene container and immediately placed in a freezer. Bee pollen will remain there for a minimum of 48 hours to destroy

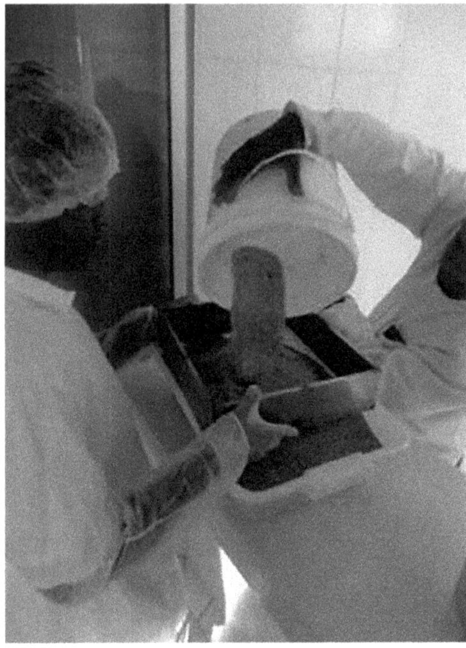

Figure 6. Processing "fresh" bee pollen; first screening.

possible mites, eggs or larvae of the wax moth (*Galleria* spp.), and other insects. The freezing will also have the property of stabilizing and controlling the development of microorganism related to normal microflora contained in pollen (Ibrahim & Spivak, 2006). The collected pollen can be stored in the freezer for a period of a few weeks. After a month, it begins to show change in taste, texture and color;

d. Dehydrated. For this step is necessary to have a greenhouse with adjustable thermostat, keeping the temperature between 40 and 42 °C. Internally, the equipment consists of stainless steel with shelves that receive trays with screens in stainless steel AISI 304 that will allow efficient cleaning and sanitizing. The dehydrating aims at the water removal until it reaches the maximum humidity of 4–8% (Campos et al., 2008), but for tropical countries 4% is often recommended (Thakur & Nanda, 2020).

Frozen pollen is usually placed in a greenhouse, which facilitates its uniform distribution in the screened trays, but it should not be submitted to an immediate temperature of 42 °C. It is essential to respect a thawing period of the material to prevent damaging the product due to thermal shocks. It is suggested that after removing the pollen from the freezer, it should be transferred to a refrigerator for approximately 2 hours for defrosting. The distribution of pollen in the tray should be done in thin layers. Every instrument in contact with the pollen must be stainless steel to prevent contamination. The pollen must remain in the greenhouse from 8 to 12 hours. After dehydration, the packing of bee pollen will involve three steps:

a. Transferring dehydration, sealed loads;
b. Cleaning – First undergo the pollen in sieve particle size to have the larger lumps undone and separate the commercial pollen powder after applying dry air to remove dirt such as bee fragments that are possibly present. The final phase of cleaning is grooming materials not previously removed, for example, any propolis that is still mixed with bee pollen;
c. Packing – Should be done after cleaning, reducing the rehydration possibilities of the processed material. For this procedure, you must use a nontoxic plastic bag, suitable for food packaging as primary, which remains protected by secondary packaging of nontoxic plastic barrels of first use or food cardboard. Before closing it, air must be removed from the bag with help of a vacuum pump. Storage should be in dry conditions, with mild temperature and away from light. For fractional packing of bee pollen, it is suggested placing silica gel pads for foods – that will keep humidity level of the product stable (Brasil, 1997).

## 3. Inexpensive pollen load authentication by means of computer vision and classification algorithms

Bee-keepers, bee-keeping associations, and laboratories are interested in detecting fraud in pollen, and require tools to standardize and authenticate bee pollen origin to guarantee their nutritive and health benefits. Microscopic analysis of pollen grains, which form bee pollen loads, is a precise method for identifying pollen origin. However, this process requires the laboratory work of melissopalynology experts, and is thus time consuming and costly. There have been many attempts to automate pollen grain identification by computer algorithms but there is no inexpensive, complete, and automated process (Allen, 2006; Boucher et al., 2002; Rodríguez-Damián et al., 2006). Image processing techniques and one-class classification algorithms can be used to identify unknown pollen types. Concretely, the mean shift algorithm (Comaniciu & Meer, 2002) filters the pollen image and is used to homogenize pollen load color information. Then, one-class classification algorithms (Chandola et al., 2009; Moya et al., 1993) identify each local pollen type. The method makes use of a multi-classifier algorithm, designed to aggregate one-class classifier outputs, given a unique response with a confidence measure. Finally, an ambiguity discovery algorithm is also included to detect identical pollen type and reduce misclassification.

The chapter provides general details of the algorithms involved in the methodology, describes the necessary hardware for the standard method, and presents the results of applying the image processing algorithms and one-class classifiers. In addition, the chapter presents the overall process to create a software dictionary with known local pollen types, how to use the graphical user interface, and a guide to separate and analyze the pollen loads of the images by type. The validation of the proposed standard methodology is carried out for the authentication of four pollen types, *Cistus ladanifer*, *Rubus*, *Echium*, and *Quercus ilex*, which serve as an example of the methodology (Otero & Losada, 2002). Totally, a dataset of around 2000 instances has been used to validate the trained system.

(a)

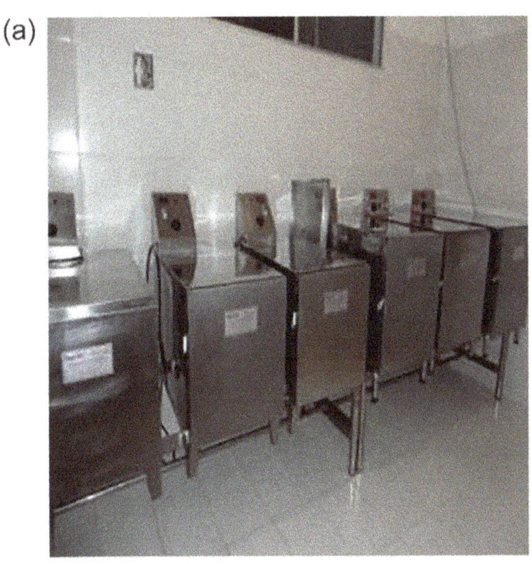

(b)

(c)

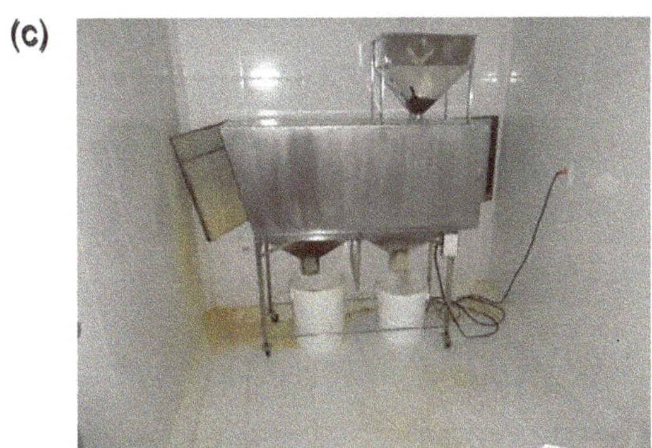

Figure 7. Example of a warehouse for bee pollen production in Brazil. This is one of the largest in Latin America, located in the state of Bahia, northeast Brazil. The main source of pollen is the coconut palm, achieving high standards of excellence and quality of the product: (a) Dehydration room 40 °C; (b) distribution trays; (c) aeration of the dehydrated product room.

### 3.1. General overview of the method

Figure 8 shows the outline of the method. In a nutshell, it starts with preparation of a pollen load sample to acquire the image by the computer vision system. Second, the pollen loads of the image are segmented from background (see Section 3.2.1) and filtered (see Section 3.2.2). Third, the color instances of the processed pollen loads are used to train a multi-classifier model based on one-class classifiers (one for each local pollen type). Finally, the multi-classifier outputs the authentication of each color instance, classifying them as a known local or non-local (outlier) pollen type. Section 3.2.3 provides more details about the classification algorithms.

### 3.2. Description of the computational techniques

#### 3.2.1. Segmentation algorithm

The Otsu segmentation algorithm (Otsu, 1979) is the one applied to the gray-scale image to extract the pollen loads from the background. Later, a morphological opening operation is applied to the thresholded binary image (Gonzalez & Woods, 2008). The goal of this computer vision operation is to remove those small objects having less than 50 connected pixels in an 8-connected neighborhood.

Those pixels extracted in the latter phase are analyzed by the remaining processing algorithms. Their color information can be represented in several ways. The most common is the RGB space where colors are represented by their red, green, and blue components in an orthogonal Cartesian space. However, the RGB space does not represent the higher levels processes which allow human color perception. Color is better represented in terms of hue, saturation, and intensity, as HSI or HSV spaces do (Lucchese & Mitra, 2001). However, the latter color spaces are not perceptually uniform. The CIE and $L^*u^*v^*$ and $L^*a^*b^*$ are ideal for color recognition because of the following three properties: a) separation of achromatic information from chromatic information, b) uniform color space, and c) similarity to human visual perception. In these color spaces, for instance, the Euclidean distance between two color points can be easily calculated. This property will ease the work of the classification algorithms.

#### 3.2.2. Homogenizing the pollen loads of the image by mean shift filtering

Each extracted pollen load from the image has many different $L^*u^*v^*$ color values, possibly as many as pixels contained in the load. This has a negative impact on pollen load color authentication since human experts identify each pollen load as a unique color. Therefore, we

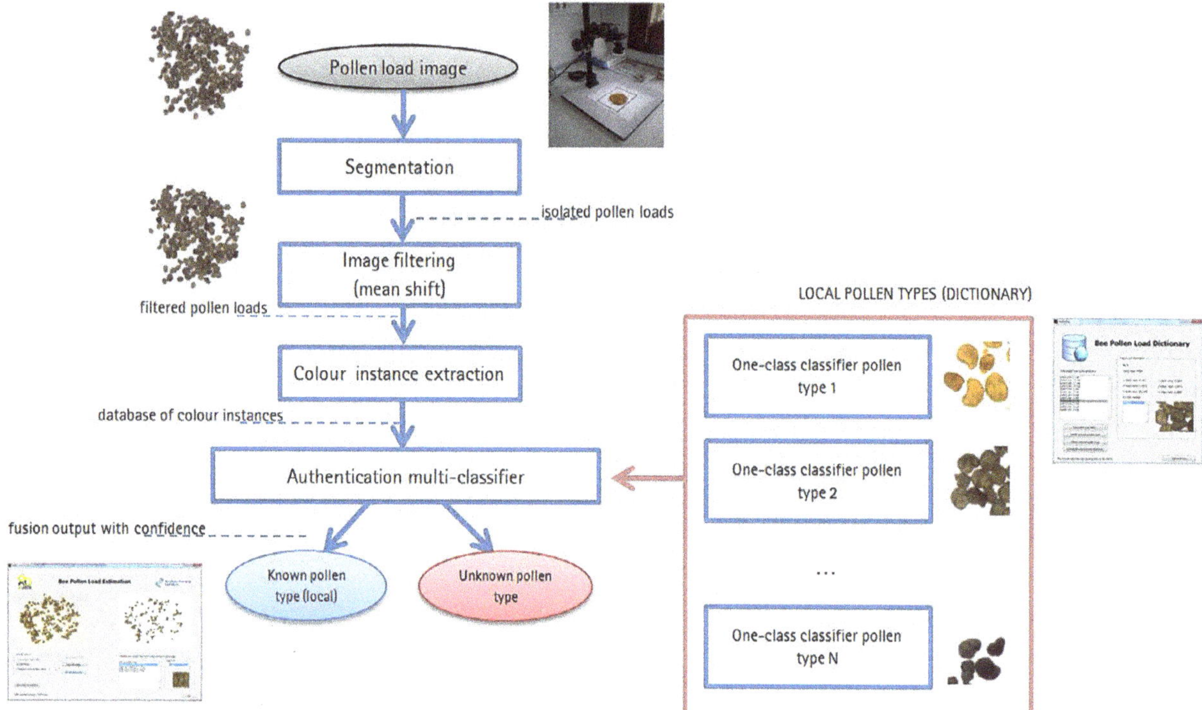


Figure 8. The proposed standard method for detecting unknown pollen loads. The processing chain finishes with an authentication output.

need a procedure of homogenizing the images of pollen loads, which is divided into three different steps:

1. First, we apply an image processing algorithm before using the classification methods. The goal is to homogenize and smooth the large quantity of different color points of a pollen load in just a few unique and representative color instances. One of the best methods for discontinuity-preserving smoothening in image processing is the mean shift algorithm, proposed in Comaniciu and Meer (2002), and in line with the feature space analysis. The main strengths of the mean shift algorithm are: (a) it is an application independent tool, (b) it is suitable for real data analysis, (c) it does not assume any prior shape on data clusters, (d) it can handle arbitrary feature spaces and (e) it has only one parameter, the bandwidth selection window size. The mean shift procedure, originally presented by Fukunaga and Hostetler (1975), is a procedure for locating the maxima of a density function given discrete data sampled from that function. It is also an iterative method, starting with an initial estimate or points $x_i$ and a $G$ kernel function. This kernel function determines the weight of nearby points for the re-estimation of the mean. Denote by $y_j$ with $j = 1, 2, \dots$ the sequence of successive locations of the kernel $G$:

$$y_{j+1} = \frac{\sum_{i=1}^{n} x_i G(\| \frac{x - x_i}{h} \|)^2}{\sum_{i=1}^{n} G(\| \frac{x - x_i}{h} \|)^2}, \tag{1}$$

is the weighted mean at $y_j$ computed with the kernel $G$ and $y_1$ is the center of the initial position of the kernel. Taking this theory as a base, Comaniciu and Meer (2002) extended the mean shift procedure for filtering and segmenting images. A color image is typically represented as a two-dimensional lattice of 3-dimensional vectors (pixels). The space of the lattice is known as the spatial domain, while the color is the range domain. For both, the mean shift algorithm will work with the Euclidean metric defined in the previous section for the color space.

2. When both domains are concatenated, the dimensions of the joint spatial range domain are compensated by a proper normalization. Thus, the multivariate kernel is defined as the product of two radially symmetric kernels, and the Euclidean metric allows a single bandwidth parameter for each domain:

$$K_{h_s,h_r}(x) = \frac{C}{h_s^2 h_r^p} k\left(\| \frac{x^s}{h_s} \|^2\right) k\left(\| \frac{x^r}{h_r} \|^2\right), \tag{2}$$

where $x^s$ is the spatial part, $x^r$ is the range part of a feature vector, $k(x)$ the common profile used in both domains, $h_s$ and $h_r$ the kernel bandwidths, and $C$ the normalization constant. A kernel process usually provides satisfactory results and then the user just has to provide one parameter $h = (h_s, h_r)$ which controls the size of the kernel, and thus the smoothening resolution. Nevertheless, replacing the pixel in the center of the window by the average of the pixels in the window blurs the image. Discontinuity-preserving smoothening techniques, on the other hand, reduce the amount of smoothening near abrupt changes. The mean shift algorithm uses a bilateral filtering which works in the joint spatial-range domain. The data are independently weighted in the two domains and the centered pixel is computed as the weighted average of the window. The kernel in the mean shift procedure moves toward the maximum increase in joint density gradient, while bilateral filtering uses a fixed static window.

3. The last step is the application of the mean shift filtering algorithm. The algorithm works as follows. Let $x_i$ and $z_i$, with $i = 1, 2, \ldots, n$ be the (p + 2)-dimensional input and filtered image pixels in the joint spatial-range domain. Being the superscripts $s$ and $r$ the spatial and range components of a vector, respectively, and $c$ the point of convergence, for each pixel:

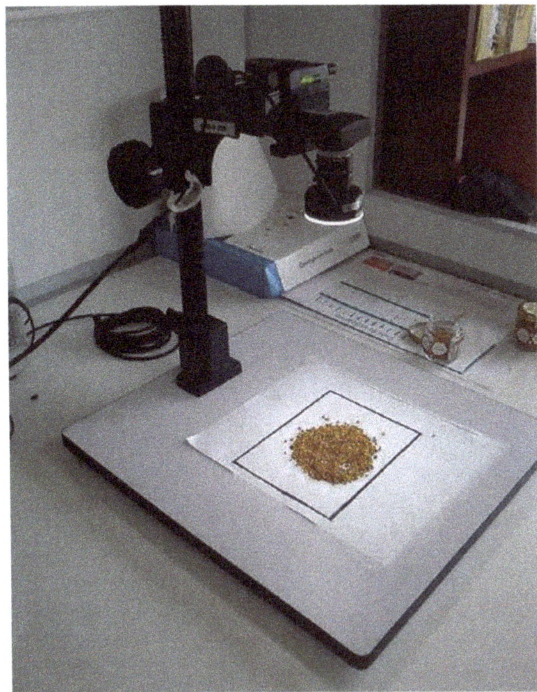

Figure 9. The computer vision system needs inexpensive hardware such as a camera and a stand to control illumination.

a. Initialize $j = 1$ and $y_{i,1} = x_i$;
b. Compute $y_{i,j+1}$ according to Equation (2.1) until convergence, $y = y_{i,c}$;
c. Assign $z_i = \left(x_i^s, y_{i,c}^r\right)$.

The spatial bandwidth has a distinct effect on the output when compared to the range (color) bandwidth. Only features with large spatial support are represented in the filtered image when $h_s$ increases. On the other hand, only features with high color contrast remain when $h_s$ is large.

#### 3.2.3. One-class multi-classification algorithm based on distance

In Chica and Campoy (2012) authors compared four different one-class classification approaches for the problem: Gaussian, Parzen classifier, SVDD and *kNN*. The *kNN* algorithm with $k = 1$ showed the best performance. Then, this is the proposed algorithm for the pollen recognition system.

*kNN*, originally provided by Dasarathy (1991), is a distance-based one-class classifier based on the assumption that normal data instances occur in dense neighborhoods while anomalies occur far from their closest neighbors. The basics of the algorithm for one-class classification is that the anomaly

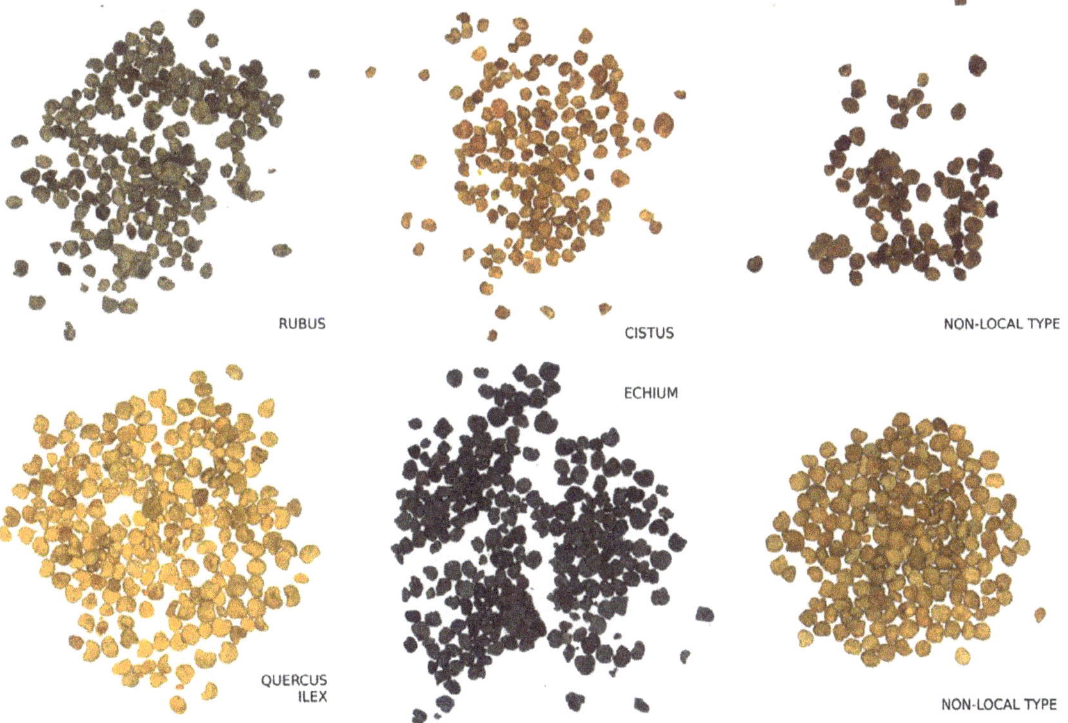


Figure 10. Different images of pollen load samples which were taken with our vision system. Left and central images belong to known pollen types (*Rubus*, *Cistus ladanifer*, *Quercus ilex*, and *Echium*, respectively). Right images are non-local samples and must be rejected by the system.

score of a data instance is defined as the distance with its $k^{th}$ nearest neighbor in each dataset.

Nearest neighbor classifiers always require the definition of distance or similarity measures between two data instances. For continuous features, the Euclidean distance is the most popular choice. In the case of choosing $k = 1$ as the parameter of the algorithm, each new instance z will be considered as target or outlier depending on the classification of its closest neighbor in the training data.

In multi-class anomaly detection (our pollen authentication problem) training data contains labeled instances belonging to multiple normal classes but it does not contain anomalous instances. A test instance is considered anomalous if it is not classified as *normal* by any of the classifiers. To do this, a confidence score obtained from the prediction output of the classifier is normally provided. If none of the *kNN* classifiers are confident in classifying the test instance, the instance is then labeled as anomalous (Chandola et al., 2009).

We have followed the latter approach, modeled as follows: $|C|$ being a set of known local bee pollen load types, the training data will contain instances belonging to $|C|$ classes. In order to use one-class *kNN* classifiers and be able to reject unknown pollen load types, we have to decompose the classification system in $|C|$ binary sub-problems. Thus, we will train $|C|$ different *kNN* classifiers: $f_1, f_2, \ldots f_{|C|}$. An ensemble scheme is used to fuze all of them in a multi-class authentication output.

Therefore, for each pollen color instance $x$ we first map each one-class classifier output $f_i(x)$ to a posterior probability $P(y = c|x)$. These probabilities are also normalized in the range $[0, 1]$. The posterior probability of each classifier's target can be considered as the confidence $CF_{oc}(y = c|x)$ for one instance $x$ to belong to class $c$. In order to classify an incoming pollen load sample as one of the $|C|$ possible pollen types we build a multi-classifier. The multi-classifier will compare the confidence $CF_{oc}(y|x)$ of all the one-class classifiers and will provide a global prediction from the most reliable one-class classifier. The multi-classifier prediction $\omega$ is given by:

$$\omega = \max_{1 \leq c \leq |C|} CF_{oc}(c|x). \quad (3)$$

### 3.2.4. Multi-classification confidence and ambiguity discovery process

It is also necessary to estimate the confidence of the multi-classifier prediction. To do this we first introduce two parameters, $T_{oc}$ and $T_m$, as done in Goh et al. (2005) in Equations (4) and (5):

$$T_{oc} = CF_{oc}(\omega x); \quad (4)$$

$$T_m = T_{oc} - \max_{1 \leq c \leq |C|, c \neq \omega} CF_{oc}(c|x). \quad (5)$$

$T_{oc}$ is the highest confidence factor from the $|C|$ binary one-class classifiers and determines the multi-classifier prediction class $\omega$. However, $T_{oc}$ might not be enough to estimate the global confidence of the multi-classifier prediction. This is the reason for introducing the second parameter, the multi-class margin $T_m$. Wrong predictions could have high $T_{oc}$ but small $T_m$. Then, correct predictions must have high multi-class margin values $T_m$.

Goh et al. (2001) showed that there is a better separation of correct from erroneous predictions if the multi-class margin variable is used. After a preliminary experimentation, we set parameters $T_{oc}$ and $T_m$ to 0.5 and 0.01, respectively, to be used in the final decision stage of the multi-classifier, as shown in the rule of Equation (6).

$$\begin{cases} \omega \textit{ is accepted, if } T_{oc} \geq 0.5 \textit{ and } T_m \geq 0.001, \\ \textit{outlier,} \quad \textit{otherwise.} \end{cases} \quad (6)$$

Sometimes, one or more bee pollen types could have exactly the same color description as another. In that case, the *kNN* multi-classification system must be able to detect, during the training phase, that one incoming local pollen type is identical to one already existing (i.e., it is already included in the dictionary). The goal for the software prototype is to also warm the user not to introduce duplicated entries in the dictionary. This mechanism is called ambiguity discovery. Being $\varepsilon_{m_1}$ a sensitivity-specificity error of a multi-classifier before the inclusion of the new pollen type, and $\varepsilon_{m_2}$ the error of a multi-classifier after the inclusion of the classifier of the new pollen type, we define $\Delta_\varepsilon = (\varepsilon_{m_2} - \varepsilon_{m_1})$ as the difference between them. In our case we have used the F-measure (van Rijsbergen, 1979) as the $\varepsilon_{m_1}$ error measure.

The ambiguity discovery process is launched every time the $\Delta_\varepsilon$ parameter is higher than a fixed value. Then, if $\Delta_\varepsilon$ exceeds a threshold value, ambiguity discovery is triggered, and the process works as follows:

1. The confusion matrix of the new multi-classifier for the testing data is computed;
2. The maximum value of $(\Lambda_i, L_j)$ is calculated with $i \neq j$, $\Lambda$ being the vector of real classes and $L$ the vector of predicted classes;
3. The user is asked about merging conflicting classes $c_i$ and $c_j$ into a unique class $c_i$;
4. The multi-classifier is trained according to the response of the user in the third step.

## 3.3. Experimentation

Section 3.3.1 explains the data used for the experimentation. Section 3.3.2 explains the image processing and filtering results. Section 3.3.3 presents the results to analyze the behavior of the one-class classifiers and multi-classification systems. Finally, Section 3.3.4 shows the validation of both the software prototype and proposed standardization process.

### 3.3.1. Experimental setup: equipment and data

The main equipment used in the experiments is composed of:

- Camera and optics device to obtain images of the pollen load samples. The resolution does not need to be extremely high since only color and edge information will be processed. (Example: color camera uEye UI-1485LE-C (www.ids-imaging.com) with an Aptina CMOS sensor in 5 M Pixels resolution (2560 × 1920 pixels). The light-weight housing of the UI-1485LE features a C/CS lens mount with adjustable flange back distance. A focal lens obtained from Goyo Optical (www.goyooptical.com) is also used.
- The lighting conditions are decisive and must be controlled. An external lighting generator, which points out the pollen load sample is employed to ensure that the color of the pollen loads do not change for the different runs. The external lighting consists of a 50 LED-based ring light from CCS Direct Lighting (www.ccsgrp.com). This lighting device can be mounted on a CS mount.
- A support device is needed to allow a fixed position for the camera, lens, and lighting. The sample of pollen loads is placed on the base of this supporting device in order to get a stable and low-cost system to acquire images of the sample for a posterior processing and analysis (see Figure 9).
- Finally, a computer must be connected to the camera via USB to install the software. Proprietary software can be used to take snapshots of the pollen load samples. Then, the proposed computer vision and classification system will be controlled by the software prototype to ease the labor of the expert.

Different samples from Spanish bee pollen loads were obtained from beekeepers to build the authentication models and validate them against nonlocal samples. Samples belonging to four Spanish local pollen types (*Rubus*, *Echium*, *Cistus ladanifer* and *Quercus ilex*) and non-local samples were identified, labeled, and grouped by experts. Figure 10 shows these pollen loads samples. In this figure we can see how, even for experts, the separation and classification of the pollen loads by color is difficult and subjective. One can also see in the figure how non-local samples can be misleading (bottom right loads of Figure 10).

### 3.3.2. *Image processing results*

The first step of the method is processing the images of the pollen samples (taken in TIFF format and resolution of 1024 × 768 pixels). The software will apply the filtering algorithm after segmenting pollen loads from background. As explained in Section 3.2.2, the selection of an appropriate bandwidth parameter for the mean shift algorithm is not trivial. This selection depends on the type of images. The implemented mean shift algorithm receives the spatial bandwidth $h_s$, the range bandwidth $h_s$, and the minimum segment area in number of pixels. The last parameter is fixed to a high value (20 pixels) not to segment a pollen load into different parts.

The selection of the other two parameters is more difficult. High spatial bandwidth values merge different bee pollen loads because they are normally close to each other. Low range bandwidths do not effectively aggregate the entire color information of the pollen loads. We conducted a preliminary experiment with different values. In order to illustrate the importance of this parameter selection, we show in Figure 11 an original *Rubus* data sample together with three output images after applying the filtering algorithm with different values of $h_s$ : 7, 15, and 30. Additionally, it is important to remark that higher bandwidth values mean higher computational time. Thus, we set the bandwidth values to $h = (h_s, h_r) = (15, 20)$ for the experimentation, which seems one of the best parameter combination for this pollen load authentication problem.

We generate, from the processed and filtered images, a set of 3146 color instances. Each color instance has three input features, corresponding to the $L^*u^*v^*$ color space values, and its class (one of the four pollen types or an outlier class). From this set of color instances, 400 of them are used for training the four one-class classifiers (100 for each pollen type), 800 to test the one-class classifiers (400 belonging to the four know pollen types and 400 were outliers). The rest of the 1946 instances are used to validate the multi-classification system (Section 3.3.4). Table 1 summarizes these groups of experimental data.

Table 1. Description of the three independent datasets used to train, test the classifiers, and validate the one-class multi-classifier.

| Dataset name | Total size | Pollen types instances | Outlier instances |
|---|---|---|---|
| Training | 400 | 400 | 0 |
| Test | 800 | 400 | 400 |
| Validation | 1946 | 1016 | 930 |

Table 2. Evaluation measures obtained by the one-class *kNN* classifiers for each of the four pollen types.

| *Rubus* | | | *Echium* | | |
|---|---|---|---|---|---|
| FN rate | FP rate | F-measure | FN rate | FP rate | F-measure |
| 0.05 | 0 | 0.9744 | 0.09 | 0 | 0.9529 |
| *Cistus ladanifer* | | | *Quercus Ilex* | | |
| FN rate | FP rate | F-measure | FN rate | FP-rate | F-measure |
| 0.06 | 0 | 0.9691 | 0.01 | 0.0028 | 0.9851 |

Lower values mean better performance.

Table 3. Evaluation measures of the final multi-classifier.

| *kNN* multi-classifier model | Accuracy (%) | FN rate | FP rate |
|---|---|---|---|
| 0% rejected during training | 92.5488 | 0.0108 | 0.1161 |
| 10% rejected during training | 94.6043 | 0.0935 | 0.0167 |

Two cases are listed; considering 0%, and considering 10% of data as outliers when training the one-class classifiers. FN and FP rates are calculated taking all the known pollen types as the positive class and the outliers as the negative.

Table 4. Confusion matrix of the multi-classification system formed by the four one-class kNN classifiers (one for each pollen type).

| | | Predicted pollen type | | | | | |
|---|---|---|---|---|---|---|---|
| | | Rubus | Echium | Cistus | Quercus | Outlier | Total |
| Real pollen type | Rubus | 248 | 0 | 0 | 0 | 33 | 281 |
| | Echium | 0 | 357 | 0 | 0 | 43 | 400 |
| | Cistus | 0 | 0 | 275 | 1 | 10 | 286 |
| | Quercus | 0 | 0 | 0 | 48 | 1 | 49 |
| | Outlier | 4 | 0 | 5 | 8 | 913 | 930 |
| | Total | 252 | 357 | 280 | 57 | 1000 | 1946 |

10% of the training data were rejected as outliers during the training phase of the one-class classifier.

### 3.3.3. *Numerical performance of the multi-classification algorithms*

In this section we employ the classification accuracy, false negative and positive rates, F-measure, and confusion matrix to numerically validate the multi-classifier performance. A false negative (FN) occurs when the outcome of the classifier is incorrectly predicting as outlier when it is actually a target. A false positive (FP), on the other hand, occurs when the outcome is incorrectly predicting as target when it is actually an outlier. The FN rate measures the number of FNs out of the total number of negatives or outliers, and the FP rate calculates the fraction of FPs divide by the total number of positives or target instances (Witten & Frank, 2005). The F-measure provides a relation between the precision and recall of the classification results (van Rijsbergen, 1979).

First, we analyzed the performance of the *kNN* classification method for the authentication of each of the selected local bee pollen types in isolation. These results are obtained by classifying the test dataset with the train one-class classifiers. The classifiers are trained without rejecting any training instance as outlier (rejection threshold equal to 0%). Table 2 shows the evaluation measures of the classifiers for each pollen type. By observing figures of Table 2 we can conclude that the FP rate is almost 0. This means that the classifiers are able to correctly identify all the outliers (non-local pollen types) without misclassifying them as local pollen types. The FN rate is higher than the FP rate although its value is low, 9% in the worst case (*Echium* pollen type).

Table 3 shows the performance measures of the final multi-classifier. In the first block of figures we show the accuracy and FP-FN rates when rejection threshold equals 0%. In the second block, a rejection threshold of 10% is used for obtaining the results. The *kNN* algorithm, with a rejection of 10%, is the model having the best results as it has low FP and FN rates, and has the highest accuracy.

Finally, Table 4 shows the confusion matrix of the best multi-classifier configuration with a *kNN* algorithm with a rejection of 10%. The reader can see that, in general, there is no miss-classification between the known pollen load types. There is just one miss-classified instance by the *kNN* multi-classifier. The highest error is in the right column. These are FNs because real pollen load type instances are classified as outliers (non-local pollen types). Nevertheless, as it can be seen in Table 2, the FN rate is low.

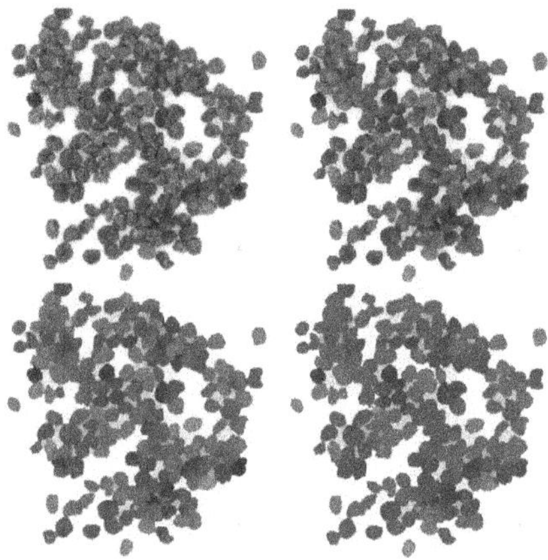

Figure 11. Original image and resulting images after applying the mean shift algorithm (varying spatial bandwidth $h_s$ to 7, 15, and 30, respectively).

From the results, we can also see that the tradeoff between FP and FN rates is better than in the isolated one-class classifiers. This fact comes from the confidence mechanism of the multi-classifier, able to discard instances classified by one of the one-class classifiers as local pollen types when they are not clearly confident about their decision. Although accuracy is obtained when classifying all the classes (four known pollen types and outliers), the FP and FN rates are calculated between the classification of the instance as known pollen type (one of the four known types) or as outlier. There is almost no error when classifying among local pollen types.

Finally and by comparing the validation measures of the multi-classifiers and the one-class classifiers, it can be concluded that the overall results of the multi-classifier are better than the independent one-class classifiers. This fact indicates a good behavior of the whole fusion scheme for detecting the known pollen type classifiers.

### 3.3.4. *Validation of the complete prototype*

The proposed software prototype includes the software methods to manage and create local pollen types, train the models, and validate the whole process. The software prototype was programmed in MATLAB using some functions of the DD tools library (Tax, 2011). Initially, the user needs to train the system with the known local samples. The user must manually separate

Figure 12. Pollen load samples separated by an expert to train the system.

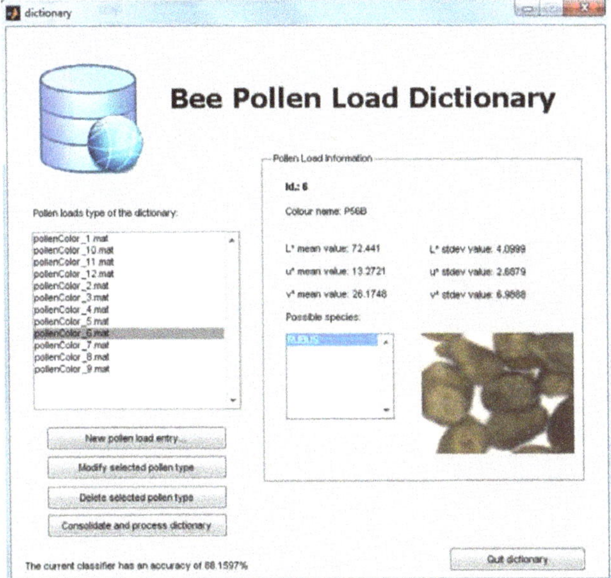


Figure 13. Screenshot of the prototype pollen dictionary. The ambiguity discovery algorithm checks if there is an existing pollen type having identical features compared to the new sample.

pollen types as shown in Figure 12. By using the dictionary tool of the software prototype, the user can also save her/his local (known) types. See for instance Figure 13 where an entry called *Rubus* is introduced into the system. After performing the training phase of the system, a new incoming pollen samples can be processed to start with the automatic pollen loads separation by color. Some screenshots of the software application are provided in Figure 14. In these images, one can see the graphical user interface with the list of separated pollen loads by type, the entries of the dictionary, and final output of the multi-classification algorithm.

### *3.4. Concluding remarks*

The propose standard methodology used a chain of methods based on computer vision and classification techniques for authentication the origin of bee pollen loads. This process can be used to check and detect fraudulent samples. A one-class multi-classification scheme based on *kNN* and its merging scheme improve the accuracy of isolated classification methods and is able to provide a final global output. The best multi-classifier configuration, formed by *kNN* one-class classifiers, is able to achieve a 94% accuracy when detecting local pollen types. The model has been validated in 1946 colors instances for the authentication of four Spanish pollen types against different outlier samples. The use of the presented standard methodology drastically reduces the time and effort spent by experts to several seconds and can be used as a standard method for macroscopically rejecting unknown pollen loads. Future work can be devoted to apply a more interpretable multi-classification system such as linguistic rule-based classifiers. In that way, the users of the standard method could understand the reason why a sample is rejected as known local pollen type.

## 4. Evaluation and classification of corbicular pollen morphology

The evaluation of quality of the corbicular pollen is one of the areas which lacks unified and standardized evaluation systems. However, descriptors used in the evaluation of genetic resources of cultivated species of plants could be a suitable system for the evaluation and characterization of various types of corbicular pollen.

Among other analyzes, the technical parameters of the corbiculae, may be an important marker for assessing the geographical and botanical origin and subsequently the nutritional quality of corbicular pollen. The industrial production, however, requires methods for the identification of corbicular pollen that can be systematically included into the production chain. What the plant species a corbicula is composed of, and what the plant composition and their proportionality is in the overall pollen lay, are the indicators of the geographic and botanical origin of the corbicular pollen (Carrión et al., 2003).

Briefly, the pollen color can distinguish the types of corbicular pollen and it mainly depends on factors such as pollen maturity, moisture, drying, pigments and the actual plant source (Nôžková, Brindza et al., 2010). The authors suggest a variety of colors ranging from white, yellow, orange, red, with a greenish tint, to black and its shades. The works also present a common fact: the color of the flower pollen varies with the color of the corbicular pollen of the same plant species (Chica & Campoy, 2012; Reiter, 1947).

The information regarding the size of corbiculae is very scarce in the literature. The authors mention the size only indirectly – either using the parameter of weight, or the number of pollen grains in one corbicula (Rodríguez-Damián et al., 2006). The size, shape and surface of the pollen grains, the corresponding botanical origin, size of the openings in the pollen trap, beekeeper's handling of the corbiculae after harvesting, etc., are all factors that have a decisive effect on the shape and

(a)

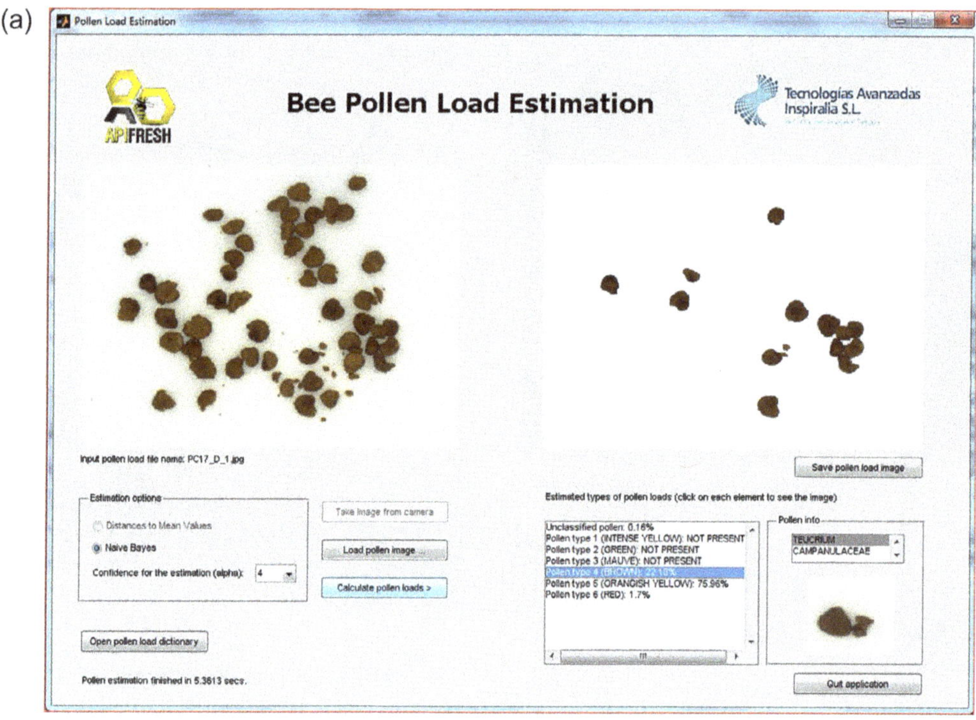


(b)

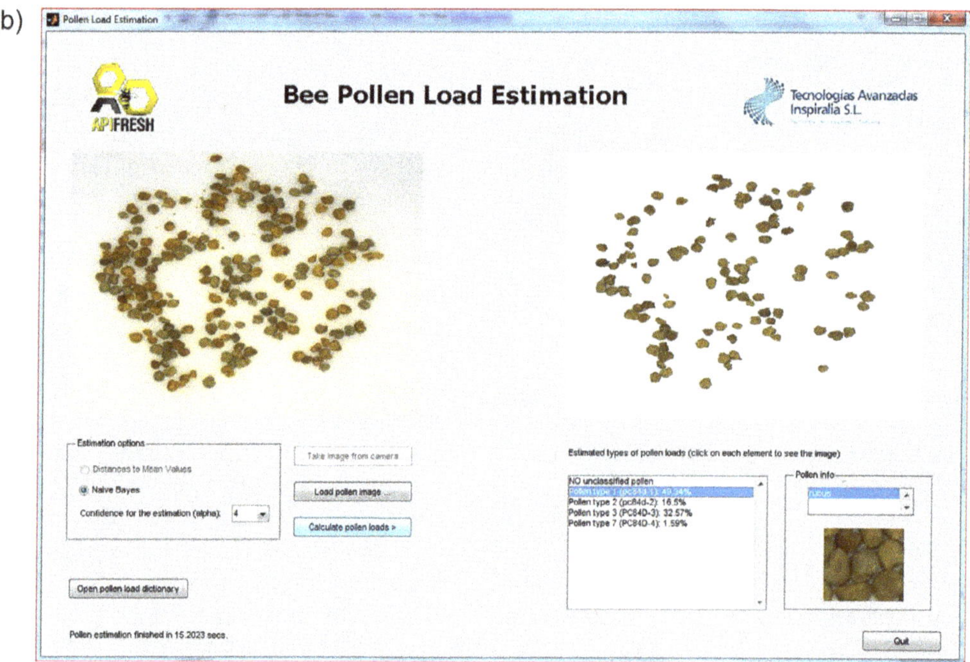


Figure 14. (a and b). Two screenshots of the prototype software detecting a local pollen type in the pollen load images.

size of the corbiculae, and so also on its the quality. The ornamentation of the exina of the pollen grains can cause differences in the structure of the corbicula surface.

### *4.1. Characteristics of corbicular pollen samples*

The selected samples of corbicular pollen, frozen and store at the temperature of −11 °C to −18 °C, immediately after collection.

### *4.2. Methodology for assessing the morphometric characters of corbicular pollen*

Quantitative and qualitative characteristics will be under evaluation of morphological parameters of corbicular pollen collected by the honey bees. For each characteristic is designed and defined the methodology of evaluation. Then a database of detailed image recordings is created. The corbiculae in each sample is placed on a glass slide using a fully automatic magnifying glass, for instance, Zeiss Discovery with a digital camera

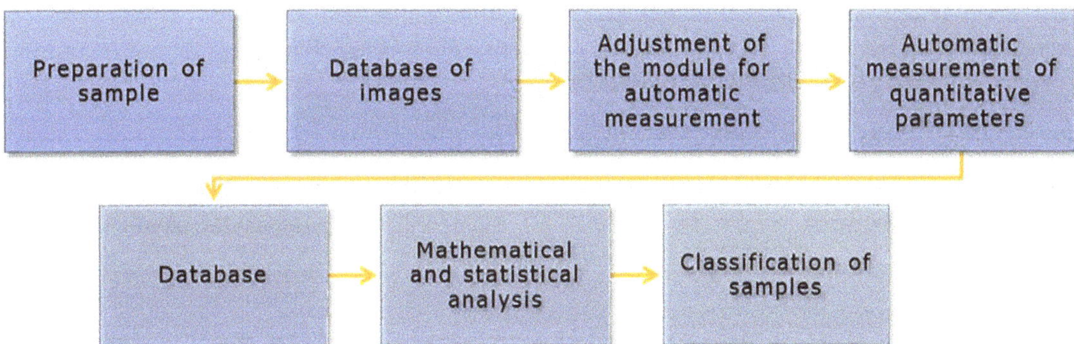


Figure 15. Flowchart of image analysis.

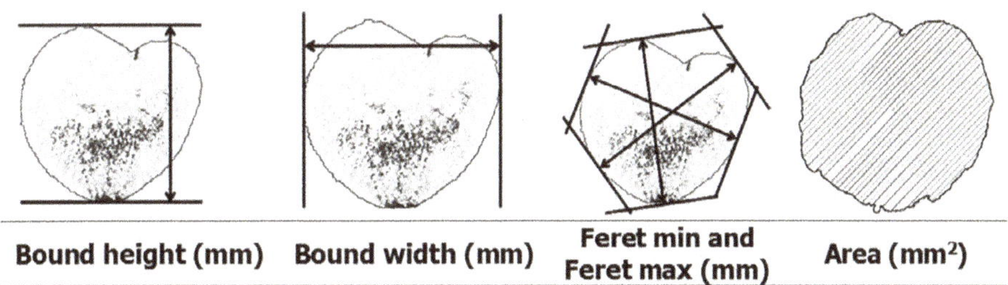


Figure 16. Evaluation of corbicula size characteristics.

AxioCam, or similar. The corbiculae are placed on the glass slide in the same direction. The upper part of the corbiculae in the image is always the depression created by placing the pollen around the bee leg inside the pollen basket.

*4.2.1. Methodological procedures for evaluating the quantitative characteristics*

In the selected quantitative traits evaluate the corbiculae by using the image analysis software Zeiss Axiovision or open source software like ImageJ with the corresponding modules (plugins) for automatic measurement (Figure 15). In the case of corbicular weight, conduct the measurement on precision analytical scales. The following section presents the assessment methodology of quantitative characteristics, which were published in Nôžková, Ostrovský et al. (2010).

1. Corbicula size. The size of corbiculae determines eight quantitative traits – bound height (mm), bound width (mm), Feret minimum (mm), Feret maximum (mm), shape index, symmetry, diameter (mm) and area ($mm^2$). The methodological procedure for the assessment of these characteristics is based on the manual for the Axiovision software.
    a. Corbicula bound height. Indicates the height (in the y axis) of a bound square for the selected area. The square is drawn parallel to the x and y axis (Figure 16). Unit of measurement: mm.
    b. Corbicula bound width. Indicates the width (in the x axis) of a bound square for the selected area. The square is drawn parallel to the x and y axis (Figure 16). Unit of measurement: mm.
    c. Feret maximum and Feret minimum. This trait is based on the measurement of a minimum and a maximum distance. Two parallel lines are place on the opposite sides of the object in 32 positions and evenly rotated by the respective angle (Figure 16). The corresponding distance is measured in each position. The maximum value is Feret maximum and the minimum value is Feret minimum. Unit of measurement: mm.
    d. Shape index (bound width/bound height). This parameter is determined by calculating the two values. It is not expressed by any unit. It can be used only if the width and height of the corbicula has been correctly specified. The shape index helps us express the shape of the object. If the value is $> 1$, the object is flatter. If the value is $< 1$, the object is taller. If the value approaches 1, the object has a round shape – it is equally as wide as it is tall.
    e. Symmetry or Feret index (Feret min/Feret max). This is a parameter determined by calculating the two measured values. It is not expressed by any unit. With this parameter, it is possible to express the shape of the object. Possible values range from 0 to 1. The values approaching 1 indicate the presence of compressed or round objects. Feret min and Feret max have approximately the same value. If the parameter has a low value, the objects are long, elongate or oblong.
    f. Diameter. It is assumed that the measure area is a circle. This surface is then use as a basis to calculate the corresponding diameter. Unit of measurement: mm.

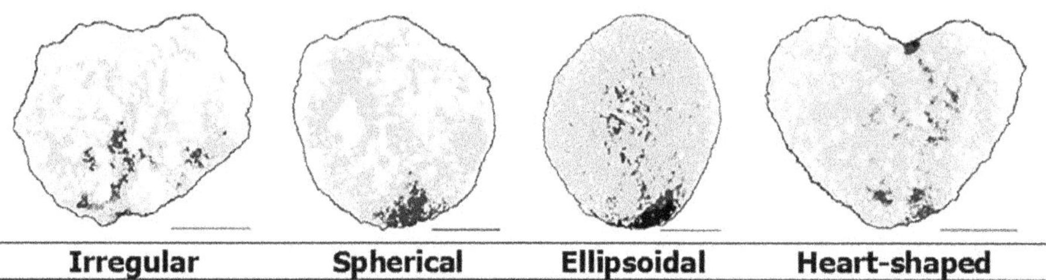


Figure 17. Evaluation of corbicula shape.

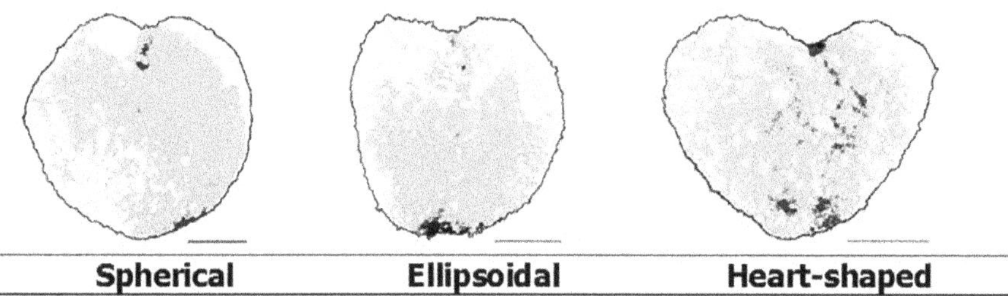


Figure 18. Evaluation of corbicula heart shape.

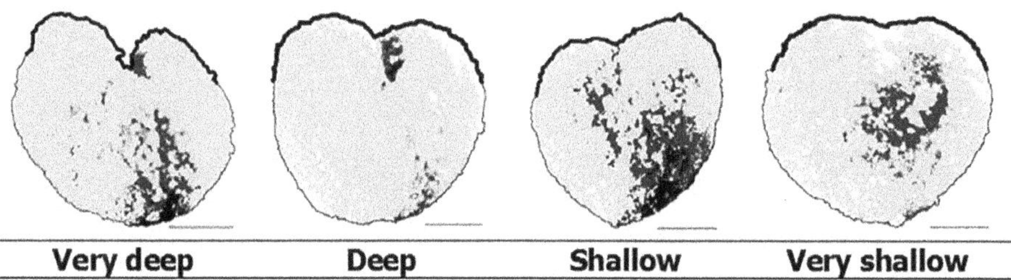


Figure 19. Evaluation of corbicula cutout depth.

g. Area. This is determined by the boundaries of the object by distinguishing the limit color of the object from the background (Figure 16). Unit of measurement: $mm^2$.

2. Corbicula weight (mg). Each corbicula is weighed on accurate laboratory scales.

From each sample select healthy and typical corbiculae. The minimum number is 5 and the maximum is 100 corbiculae per sample.

*4.2.2. Methodological procedures for evaluating the qualitative characteristics*

For the evaluation of qualitative traits, the set of corbiculae should be as big as possible, that it will cover the widest variability of the select characteristics. The qualitative characteristics of the corbiculae can be evaluated visually in digitalized images. From each sample select healthy and typical corbiculae. A database of detailed recordings, should be done with a fully automatic macro magnifying glass, for instance, Zeiss Discovery with a digital camera AxioCam (or similar). The following section presents the assessment methodology of the qualitative characteristics (Nôžková, Brindza et al., 2010).

1. Corbicula shape. For the evaluation of the corbicula shape, from three characteristics can be chosen: shape, heart shape and cutout depth. The variability in these characteristics is shown in Figures 17 and 18.
2. Corbicula surface. The variability in this characteristic is shown in the Figures 18 and 20.
3. Corbicula color. The color of the corbicula can only be assessed on samples of homogeneous color, i.e., samples of monofloral pollen. Color evaluate by using a visual color grading scale by Royal Horticultural Society (RHS) (2004) (or similar), which represents a universal color system with the respective color codes.

*4.2.3. List of descriptors and classification*

***4.2.3.1. General structure of descriptors list.*** The list of descriptors designed for the evaluation of genetic resources of cultivated plant species have always been specifically attuned to a particular plant species or genus.

Each list of descriptors consists of three main parts – the manual, passport descriptors and descriptors for evaluation and characterization. The manual contains the information on the structure and use of the descriptors, as well as the general principles for assessing the samples. The passport descriptors are designed according to the nature of the evaluated object and provide identification and passport data on the assessed sample.

Figure 20. Evaluation of corbicula surface.

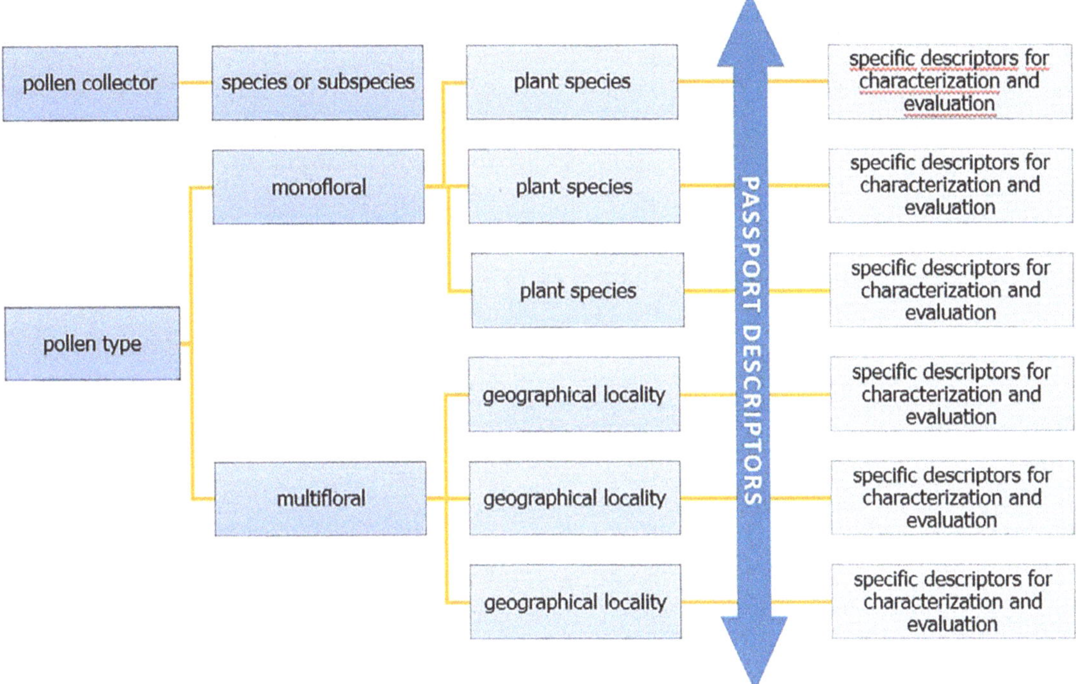


Figure 21. Flow chart of the descriptors list structure.

The descriptors for evaluation and characterization are designed based on the study of variation in quantitative and qualitative characteristics. Using these descriptors, the morphological, biological and economic characteristics are evaluated. When creating a list of descriptors for the corbicular pollen, it will always be necessary to specify the type of insect involved in the creation of the corbiculae and the descriptors for characterization and evaluation will be designed specifically for the different types of monofloral pollen, the predominant plant species and especially in the case of multifloral corbicular pollen, according to the geographical origin (Figure 21).

***4.2.3.2. Passports descriptors.*** The passport descriptors provide general information about the accession included in the collection and the management of growth (characterize by the collection method, storage and handling of the accession). The passport descriptors should define the mandatory information to identify the originality of the accession (Haussmann & Parzies, 2009). In the case of genetic plant resources, the passport data are exchange, and for the individual institutions to be able to communicate with each other, an international standardize list of Multi-crop passport descriptors (FAO/IPGRI, 2001) was created.

To evaluate the quality of corbicular pollen, it is also necessary to design the passport descriptors, which will guarantee the authenticity and identify the specific properties of the evaluated sample. What we consider to be the most important step is that the passport descriptors also determine the botanical and geographical origin. This means, it should be clear whether the pollen in question is monofloral or multifloral. In the case of multifloral pollen, the geographic origin should be identified, which ensures the representation of specific plant species.

Similar to plant species, even the corbicular pollen should have passport descriptors universal for all types of pollen. Despite the very high heterogeneity between the types of corbicular pollen, they should be designed in a way that they can be used to collect the necessary information be it in monofloral or multifloral pollen.

The following section contains a draft structure of passport descriptors (Figure 22). The information on

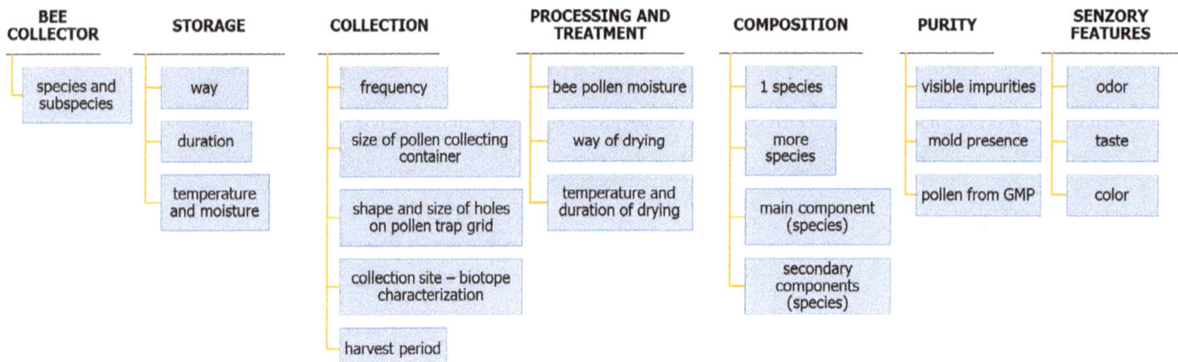


Figure 22. Descriptor structure.

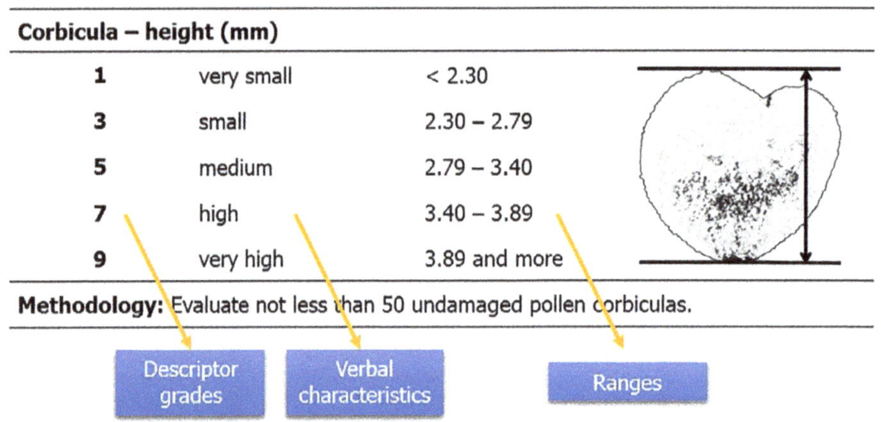


Figure 23. Visual representation of the line for the limit points $x_{min} = 1.81$; $x_{max} = 4.26$ and the coordinate "$x_4$" to the point "$y_4$."

the insect species or subspecies creating the corbiculae should come first. It should be followed by information on the collection by the beekeepers, i.e., the collection frequency, pollen trap characteristics, growing season and habitat characteristics. What is also important is the information on storage (method, length, temperature and humidity) and pollen processing (moisture, drying method). They should also include information on purity, i.e., presence of visible impurities, mold or whether the pollen comes from genetically modified plants (GMPs). The sensory characteristics such as flavor, color and taste, are also an important identifier of the specific characteristics and authenticity of pollen.

***4.2.3.3. General structure of characterization and evaluation descriptors.*** A descriptor is an independent rating system for one specific characteristic. We use them to describe quantitative or qualitative characteristics.

The structure of each descriptor is the same (Figure 24), which is a big advantage ensuring their universal use. Descriptors consist of a maximum of 9 classification grades (Figure 23). The minimum number of classification grades is two. Their number is proposed according to the observed variability in the characteristic.

For each classification grade, the respective verbal description is proposed. In the case of quantitative characteristics, the individual classification grades should come with intervals (ranges). An important part of the descriptors is the assessment methodology with concise diagrams and pictures that describe the evaluation in detail.

***4.2.3.4. Methods and techniques for creating and modifying the descriptors.*** A descriptor, which was included in the list of descriptors, covers the known variability of morphological, biological or economic characteristics and properties. If a new characteristic or trait is discovered in the future, which better characterizes and identifies the evaluated objects, the methods and techniques for creating new or modifying the existing descriptors should be observed. In the next section we present the procedures and rules for the creation and design of various parts of the descriptor for the purpose of characterization and evaluation.

a. Completeness. The descriptor grades should be designed to allow for any measured value or trait in the characterized sample to be included into the respective grade. This rule applies to both quantitative and qualitative characteristics (Table 5). In the case of quantitative characteristics, the intervals must be clearly defined. Also, a specification whether the range is open or closed should be included (Table 6).

b. Unambiguity and uniformity. A descriptor is a ranking system with a clear identification of what, how, when, and to what extent it evaluates. An adequately designed methodology, pictures, diagrams and classification levels helps maintain the overall unambiguity. The descriptor should be used uniformly to evaluate the given characteristic. This means that all evaluators should use the same procedure for characterization and classification.

c. Measurement ranges. In order to ensure the unambiguity in the evaluation of quantitative characteristics (in the case of a continuous scale), the classification grades are recommended to be supplemented by ranges. By specifying the respective ranges, misinterpretations in the characterization of samples are avoided. One of the ways to create the evaluation scales is using the principle of membership function in the fuzzy set (Nôžková et al., 2011).

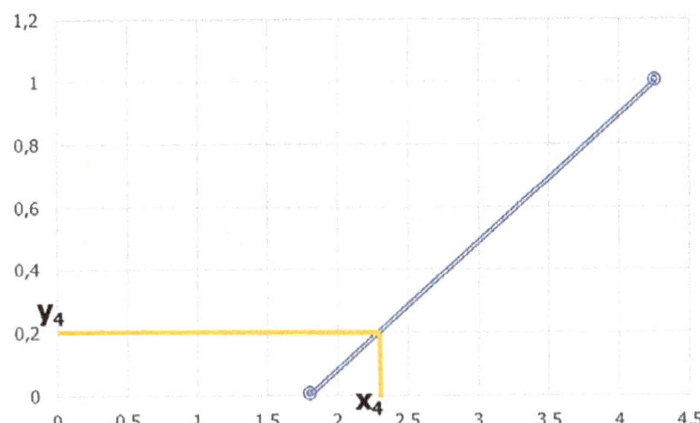


Figure 24. Flowchart of the passport data structure.

Table 5. Descriptor for quantitative character – principle of completeness.

| Corbicula – height (mm) | | |
|---|---|---|
| 1 | Very small | < 2.30 |
| 3 | Small | 2.30 – 2.79 |
| 5 | Medium | 2.79 – 3.40 |
| 7 | High | 3.40 – 3.89 |
| 9 | Very high | 3.89 and more |

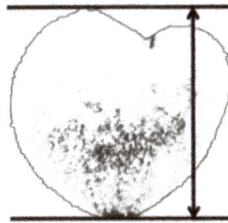

Table 6. Structure of open and closed intervals in descriptor grades.

| Corbicula – height (mm) | | | |
|---|---|---|---|
| 1 | Very small | < 2.30 | (0;2.30) |
| 3 | Small | 2.30–2.79 | (2.30;2.79) |
| 5 | Medium | 2.79–3.40 | (2.79;3.40) |
| 7 | High | 3.40–3.89 | (3.40;3.89) |
| 9 | Very high | 3.89 and more | (3.89;∞) |

a. The first important step is to determine the limit values in the given characteristic. This means that it is necessary to determine the minimum ($x_{min}$) and maximum limit ($x_{max}$), which can be achieved in the evaluated characteristic independent of the environmental conditions.

b. Based on the defined limits, the equation is expressed as follows:
The straight line is uniquely determined by two points ($x_{min}$, $y_{min}$); ($x_{max}$, $y_{max}$).

[1]

$$y = y_{min} + \frac{y_{max} - y_{min}}{x_{max} - x_{min}}(x - x_{min})$$

In our case, $x_{min}$ = minimum limit and $x_{max}$ = maximum limit, $y_{min} = 0$ and $y_{max} = 1$ (Figure 23). Thus, the equation for the line is

[2]

$$y = \frac{1}{x_{max} - x_{min}}(x - x_{min})$$

The above equation is used to express the variable "x" necessary to calculate the limit points of the individual ranges

[3]

$$x = x_{min} + (x_{max} - x_{min})y$$

c. The values of "y1" to "y19" (0.05, 0.1, 0.15 to 0.95) are then sequentially assigned to variable "y." The variable "x" represents the coordinate we are trying to find to the respective point "y" (Figure 23). The value of the variable "x" calculated according to Equation (2.3) is the limit point for the respective descriptor range (Figure 24).
An illustrative example of 5 levels in a descriptor for the height characteristic where $x_{min} = 1.81$; $x_{max} =$ 4.26 and the value of "$y_1$" to "$y_{19}$" (0.05, 0.1, 0.15 to 0.95) were attributed to variable "y" (Tables 7–9).

Table 7. Calculated value of variable "x."

| x1 | x2 | x3 | x4 | x5 | x6 | x7 | x8 | x9 | x10 | x11 | x12 | x13 | x14 | x15 | x16 | x17 | x18 | x19 |
|---|---|---|---|---|---|---|---|---|---|---|---|---|---|---|---|---|---|---|
| 1.93 | 2.06 | 2.18 | 2.30 | 2.42 | 2.55 | 2.67 | 2.79 | 2.91 | 3.04 | 3.16 | 3.28 | 3.40 | 3.53 | 3.65 | 3.77 | 3.89 | 4.02 | 4.14 |

Table 8. Determination of limit points in ranges for five descriptor levels.

| x1 | x2 | x3 | x4 | x5 | x6 | x7 | x8 | x9 | x10 | x11 | x12 | x13 | x14 | x15 | x16 | x17 | x18 | x19 |
|---|---|---|---|---|---|---|---|---|---|---|---|---|---|---|---|---|---|---|
| 1.93 | 2.06 | 2.18 | 2.30 | 2.42 | 2.55 | 2.67 | 2.79 | 2.91 | 3.04 | 3.16 | 3.28 | 3.40 | 3.53 | 3.65 | 3.77 | 3.89 | 4.02 | 4.14 |

4. Grading scales for qualitative characteristics. In the qualitative characteristics with a nominal range (color, shape, etc.), the classification grades are consecutive. Their number depends on the observed variability in the characteristic. The classification grade "9 – Other" is open to the newly detected variability in the characteristic (Table 10). In the case of ordinal scales (color intensity, etc.), the classification grades are not consecutive but follow the polarity of verbal characteristics (Table 11). In alternative characteristics, grade "1" refers to the absence of the given feature in the characteristic, and grade "9" refers to the full presence of the feature (Table 12).

Table 9. Ranges of 5 descriptor levels for height.

| Grade | Verbal characteristics | Intervals |
|---|---|---|
| 1 | Very small | < 2.30 |
| 3 | Small | 2.30–2.79 |
| 5 | Medium | 2.79–3.40 |
| 7 | High | 3.40–3.89 |
| 9 | Very high | 3.89 and more |

Table 10. Descriptor for qualitative character – structure of grades.

| Corbicula – heart shape | |
|---|---|
| 1 | Elliptical |
| 2 | Spherical |
| 3 | Heart-shaped |
| 9 | Other |

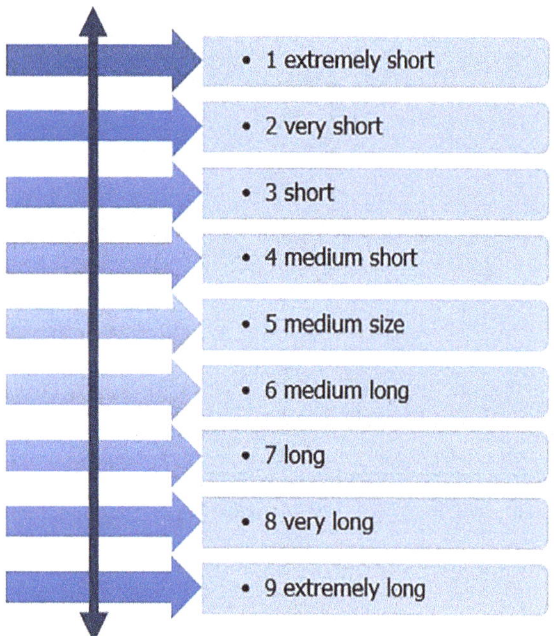


Figure 25. Polarity in verbal description.

Table 11. Descriptor for qualitative character – structure of grades in the case of ordinal scale.

| Corbicula – cutout depth | |
|---|---|
| 1 | Very deep |
| 3 | Deep |
| 7 | Shallow |
| 9 | Very shallow |

Table 12. Structure of alternative character.

| Corbicula – presence of visible impurities | |
|---|---|
| 1 | Absent |
| 9 | Present |

5. Polarity of verbal description. In the descriptors for quantitative characteristics, it is necessary to follow verbal polarization (Figure 25). This means that the classification grade 1 represents something extremely small and grade 9 something extremely large.
6. Descriptor name. The name of the descriptor should be concise, it should accurately determine the evaluation area and it should be integral. The name should employ correct terminology and its structure should be uniform throughout the list of descriptors (Biodiversity International, 2007).
7. Evaluation methodology. The methodology should contain the evaluation details, i.e., how much material from the sample is to be assessed in order to get a representative result, what the assessment period is, as well as other information specific to the given characteristic (Table 13).
8. Drawings and pictures. It is appropriate that the methodology be supplemented with concise diagrams, drawings and pictures. These help the evaluator to clearly and consistently determine the correct classification grade in the descriptor and to avoid confusion caused by undesirable side effects.

## 5. Identification of the floral origin by microscopy

Pollinic analysis by microscopy is one of the first steps and one of the very important because once the pollen morphology is identified, it is possible to infer the characteristics of the geographical regions where the BP was obtained, as well as the habitat of forest, fields and crops (Barth et al., 2010; Corbi et al., 1985; Dias et al., 2012; Rodríguez-Damián et al., 2006). This method is based on two stages: the first consists in the identification of the pollen grains that are under observation and the second one aims to count them (http://www.apiservices.com/abeille-de-france/articles/pollinic_analysis.htm). Pollinic analysis allows scientists to infer the vegetation present in the area and to date and ascertain any biodiversity changes, as for example, the presence and distribution of invasive or exotic plants. Also, allows the scientists to distinguish if the pollen is monofloral, bifloral or multifloral, or if the pollen grains present different color means that belongs to different botanical species (Nogueira et al., 2012). The application of Pollinic analysis in paleoclimatic reconstruction has been reported (Li & Flenley, 1999; Rodríguez-Damián et al., 2006). Some other studies revealed that BP analysis provides important information about the plant preferences of honeys bees (Chaturvedi, 1973). Therefore, BP

Table 13. Complete descriptor with methodology and pictures.

| Grade | Verbal characteristics | Figures |
|---|---|---|
| 3 | Smooth |  |
| 5 | Moderately rough | |
| 7 | Rough | |



Methodology: Evaluate not less than 50 undamaged pollen corbiculas.

Table 14. Descriptor for characteristic – corbicula – height (mm).

| Grade | Verbal characteristics | Range | Figure |
|---|---|---|---|
| 1 | Very low | | 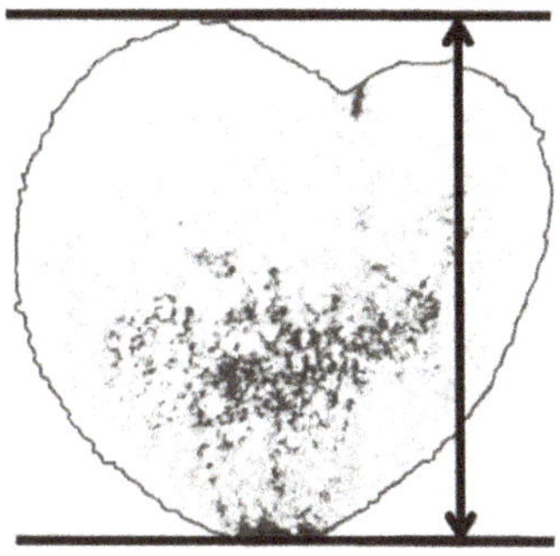 |
| 3 | Low | | |
| 5 | Moderately high | | |
| 7 | High | | |
| 9 | Very high | | |

**Methodology:** Evaluate at least 50 undamaged corbiculae, to two decimal places. The Figure shows the height (in the direction of axis *y*) of a bounded object.

Table 15. Descriptor for characteristic – corbicula – width (mm).

| Grade | Verbal characteristics | Range | Figure |
|---|---|---|---|
| 1 | Very narrow | | 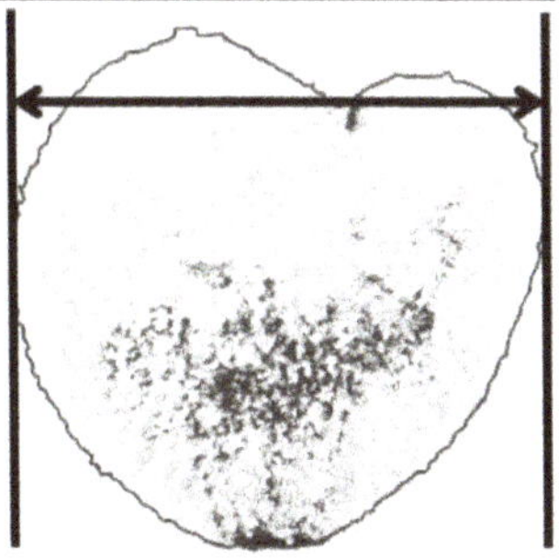 |
| 3 | Narrow | | |
| 5 | Moderately wide | | |
| 7 | Wide | | |
| 9 | Very wide | | |

**Methodology:** Evaluate at least 50 undamaged corbiculae, to two decimal places. The Figure shows the width (in the direction of axis *x*) of a bounded object.

Table 16. Descriptor for characteristic – corbicula – Feret maximum (mm).

| Grade | Verbal characteristics | Range | Figure |
|---|---|---|---|
| 1 | Very small | | 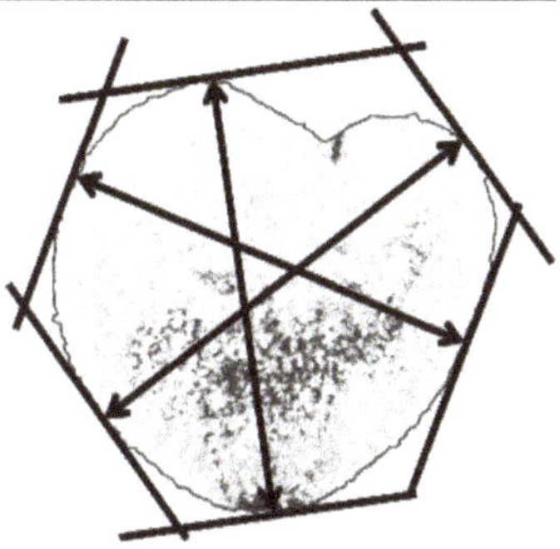 |
| 3 | Small | | |
| 5 | Medium | | |
| 7 | Large | | |
| 9 | Very large | | |

**Methodology:** Evaluate at least 50 undamaged corbiculae.

Table 17. Descriptor for characteristic – corbicula – Feret minimum (mm).

| Grade | Verbal characteristics | Range | Figure |
|---|---|---|---|
| 1 | Very small | | 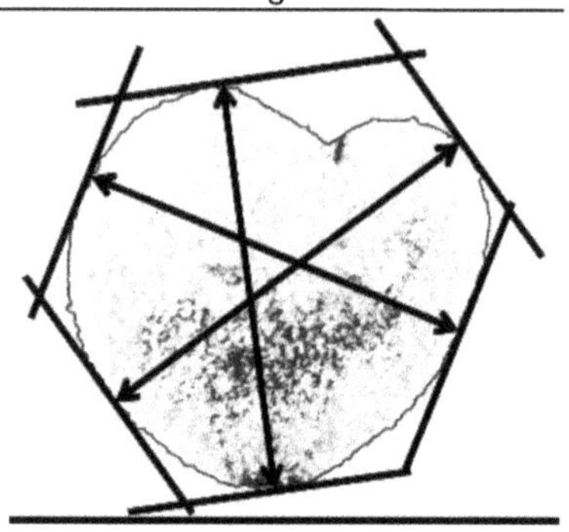 |
| 3 | Small | | |
| 5 | Medium | | |
| 7 | Large | | |
| 9 | Very large | | |

**Methodology:** Evaluate at least 50 undamaged corbiculae.

Table 18. Descriptor for characteristic – corbicula – shape index (width/height).

| Grade | Verbal characteristics | Range |
|---|---|---|
| 3 | ellipsoidal | <0.9 |
| 5 | spherical | 0.9–1.2 |
| 7 | heart shaped | 1.2 and more |

**Methodology:** Evaluate at least 50 undamaged corbiculae.

Table 19. Descriptor for characteristic – corbicula – symmetry (Feret min/Feret max).

| Grade | Verbal characteristics | Range |
|---|---|---|
| 1 | Unsymmetrical | 0 – 0.81 |
| 9 | Symmetrical | 0.81 – 1 |

**Methodology:** Evaluate at least 50 undamaged corbiculae.

load analysis indicates the behavioral pattern and sense of selection of plants for food (Sharma, 1970). Moreti et al. (2002) referred that allows the identification of the major pollen sources used by the bees, as well as the periods of pollen production in the field and possible times of shortage. By another side, Arroyo et al. (1986) referred that the determination of the flora where the bee products are originating is certainly of great interest to plan rationally the exploiting of natural resources. This knowledge allows us to know the geographic origin of these products, so avoiding counterfeits of origin (Ricciardelli, 1982). While these analyzes (Melissopalynology) are important to the aspects mentioned above, it's important to highlight that the process of counting and classification of pollen grains are much laborious and time consuming due to the fact that they are done manually by highly skilled experts. Therefore, a high level of training is required for obtain reliable identification results (Rodríguez-Damián et al., 2004, 2006). The classification of the different aspects of the structure used to the analysis of pollen by microscopy will be referred in Section 5.3 and complement the full characteristics. Accordingly to the *Dictionary.com* (http://dictionary.reference.com/browse/pollinic), pollen is defined as "male grain gametophyte generation of seed-bearing plants. In gymnosperms, each pollen grain also contains two sterile cells *(called prothallial cells),* thought to be remnants of the vegetative tissue of the male gametophyte."

### *5.1. Structure of the pollen grain*

Pollen is a granular mass of male reproductive cells produced in the anthers of a flower. A pollen grain is a living cell surrounded by two protective coats, the intine and exine. The cell contains cytoplasm and nucleus. On the surface are apertures or germinal pores or furrows. The detail structure of a pollen grain and pollen grain wall is shown in Figures 26 and 27. The inner layer or intine is thin, delicate and very elastic. It is semi-permeable membrane and does not stain. When the grain is seen in section, the intine can be recognized as a thin, clear line surrounding the cell contents. The outer layer exine is thicker, more brittle, and often variously sculptured or provided with various modifications such as spines, outgrowths or reticulations. The exine is made of an extremely durable material called sporopollenin and is composed by four layers, as listed below:

1. *Base:* This is a clear, uniform layer, the outer part of which can be stained to reveal a dark line in the optical section of the grain;
2. *Rods or columns:* These are arranged radially from the base;
3. *Tectum:* This layer forms a roof over the rods. It may be an incomplete layer, leaving some of the rods free-standing;
4. *Ornamentation.* This is provided by a layer of spines, outgrowths, reticulation, and other processes on the tectum.

Usually, all the layers are not present. Those which are present, show many modifications that are very useful in identifying a particular pollen grain.

Apertures are present on the surface of pollen grain. These are formed by the thinning or the absence of same of the layers of the exine. They are called furrows

Table 20. Descriptor for characteristic – corbicula – area ($mm^2$).

| Grade | Verbal characteristics | Range | Figure |
|---|---|---|---|
| 1 | Very small | | |
| 3 | Small | | |
| 5 | Medium | | |
| 7 | Large | | |
| 9 | Very large | | |

**Methodology:** Evaluate at least 50 undamaged corbiculae, to two decimal places. The area is determined by the boundaries of the object by distinguishing the limit color of the object from the background.

Table 21. Descriptor for characteristic – corbicula – weight (mg).

| Grade | Verbal characteristics | Range |
|---|---|---|
| 1 | Very small | |
| 3 | Small | |
| 5 | Medium | |
| 7 | Large | |
| 9 | Very large | |

**Methodology:** Evaluate at least 50 undamaged corbiculae.

Table 22. Descriptor for characteristic – corbicula – texture.

| Grade | Verbal characteristics | Figures |
|---|---|---|
| 3 | Smooth | 3 |
| 5 | Moderately rough | 5 |
| 7 | Rough | 7 |

**Methodology:** Evaluate not less than 50 undamaged pollen corbiculas.

Table 23. Descriptor for characteristic – corbicula – shape.

| Grade | Verbal characteristics | Figures |
|---|---|---|
| 1 | Irregular | 1 |
| 2 | Spherical | 2 |
| 3 | Ellipsoidal | 3 |
| 4 | Heart shaped | 4 |
| 9 | Other | |

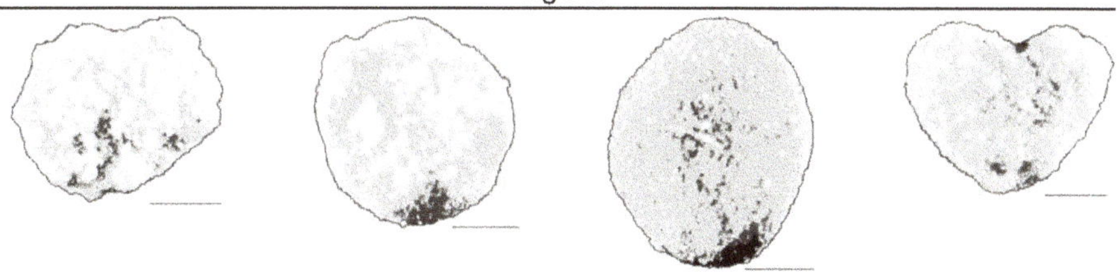

**Methodology:** Evaluate not less than 50 undamaged pollen corbiculas

Table 24. Descriptor for characteristic – corbicula – heart shape.

| Grade | Verbal characteristics | Figures |
|---|---|---|
| 1 | Ellipsoidal | 1 |
| 2 | Spherical | 2 |
| 3 | Heart shaped | 3 |
| 4 | Other | |

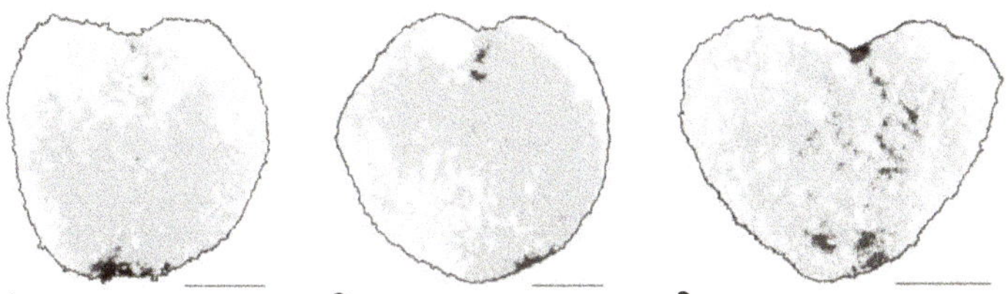

**Methodology:** Evaluate not less than 50 undamaged pollen corbiculas.

Table 25. Descriptor for characteristic – corbicula – cutout depth.

| Grade | Verbal characteristics | Figures |
|---|---|---|
| 1 | Very shallow | |
| 3 | Shallow | |
| 7 | Deep | |
| 9 | Very deep | |

1 3 7 9

**Methodology:** Evaluate not less than 50 undamaged pollen corbiculas.

Table 26. Evaluation of corbicular color (Nôžková, Brindza et al., 2010).

| Sample no. | Color (Visual assessment) | Color group name (RHS) | Color (RHS) |
|---|---|---|---|
| 2 | brownish yellow | gray-yellow | 162 A |
| 5 | grayish green | gray-green | 194 A |
| 66 | brown brick | gray-orange | 165 B |
| 80 | lemon yellow | yellow | 8 A |
| 81 | brownish yellow | gray-yellow | 162 A |
| 90 | brownish yellow | gray-yellow | 162 A |
| 98 | brownish yellow | gray-yellow | 162 A |
| 99 | brownish yellow | gray-yellow | 162 A |
| 100 | carrot orange | orange | 24 A |
| 113 | brownish yellow | gray-orange | 164 B |

or colpi, if they are elongated and tapering toward the end or pores when they are round or oval (Figure 28). The furrows run longitudinally with their ends toward the polar areas. The apertures are generally situated on the equator but pores are often found scattered over the whole grain surface. Pores and furrows often occur together, but at different levels of the exine. Apertures allow the grain to dry or to absorb water and thus to change into an expanded state. Also provide an easy outlet for the pollen tube when the grain germinates. Therefore, they are also called germ pores or germinal furrows.

### *5.2. Floral origin*

Considering the bee pollen loads, pollen may be classified as:

1. Unifloral, also named monofloral, when this product is derived from a single botanical source (Campos et al., 2008);
2. Bifloral when obtained from two species of bee plants;
3. Multifloral, also named heterofloral or plurifloral, when the pollen originates from several botanical species (Barth, 2004; Campos et al., 2008; Sharma, 1970; Stanley & Linskens, 1974).

### *5.3. Features of pollen grains*

Each pollen grain has its own peculiar structural features (size, shape, apertures type, exine sculpture (exine ornamentation) and surface and other features), which helps us recognize its origin. Scientists-palynologists Sawyer (1981) and Vorwohl (1990) have different opinions of these features. Some of them will be show along the text.

#### *5.3.1. Size*

The size (diameter) of the pollen grain is expressed in microns or micrometers (μm), where 1 μm = 1 / 1 000 mm. The maximum diameter of a pollen grain includes the spines or other ornamentation. Sawyer (1981) used five size classes:

1. Very small (<20 μm);
2. Small (20–30 (μm);
3. Medium (30–50 μm);
4. Large (50–100 (μm);
5. Very Large (>100 μm).

However, not all pollen grains are round. Many are either elliptical or triangular, even elongated. Vorwohl (1990) suggested that instead of defining size by the diameter, it should be defined by the length and breadth and defined nine classes of length and width (Table 28).

#### *5.3.2. Shape*

Shape is the outline; the optical section seen under the microscope and has been considered as an important feature in pollen identification (Sawyer, 1981). A pollen grain can be round, oval, elongated, triangular, semi-circular or boat-shaped, and irregular or multi-sided. The shape plays a very minor role in the identification of pollen grains, because it varies depending upon the

Table 27. Descriptor for characteristic – corbicula -color.

| Grade | Verbal characteristics |
|---|---|
| 1 | Black |
| 2 | Purple |
| 3 | Brown |
| 4 | Green |
| 5 | Orange |
| 6 | Yellow |
| 7 | |
| 8 | |
| 9 | Other |

**Figures**

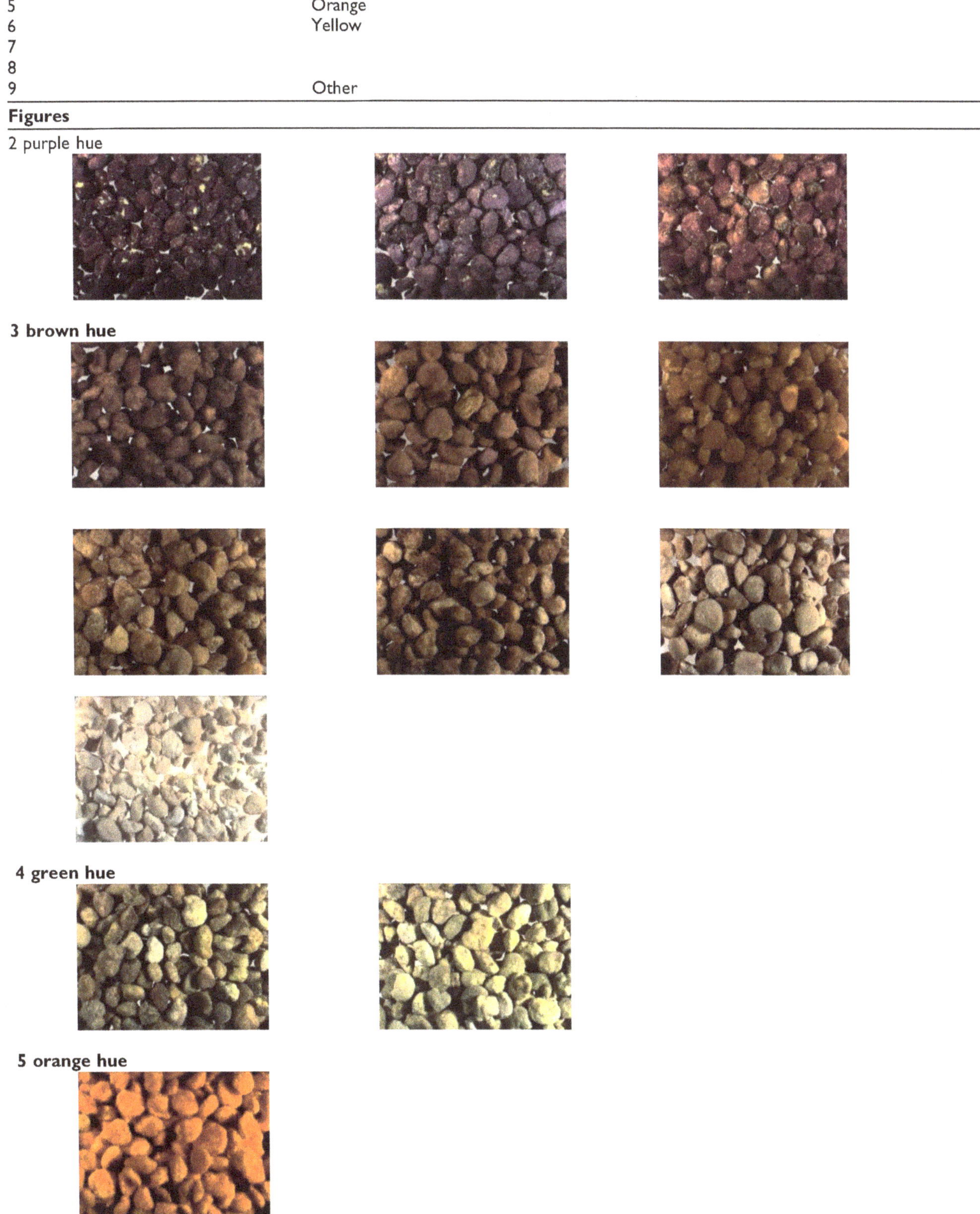

2 purple hue

**3 brown hue**

**4 green hue**

**5 orange hue**

*(Continued)*

Table 27. (*Continued*).

| Grade | Verbal characteristics |
|---|---|
| **6 yellow hue** | |
|  |  |

**Methodology:**

1. Visual assessment: the most frequent corbicula color is taken into account and the corbicula is assigned its respective classification grade.
2. Evaluation using RHS: the methodological rules set out in this classification system are observed. After determining the color of the sample, the respective code (e.g., 164B) is recorded.

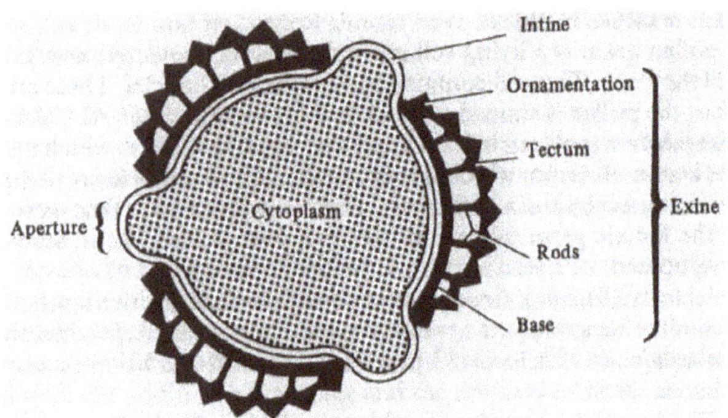


Figure 26. Diagrams showing the structure of a pollen grain in optical section (Sawyer, 1981).

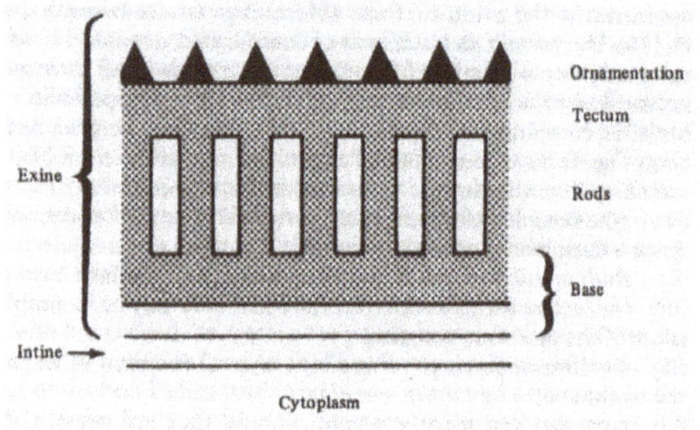


Figure 27. Diagrams showing details of pollen grain walls (Sawyer, 1981).

position in which the grain lies and upon the viewing aspect (polar view or equator view). Many grains appear round in one particular aspect. If they also show special characteristic shapes, these have been given preference in the key. These alternative shapes should be used for identification. In addition to the definitely triangular or tri-lobed appearance of many grains, other 3-furrowed grains may show a shape intermediate between rounds and triangular. Both these classifications should then be considered.

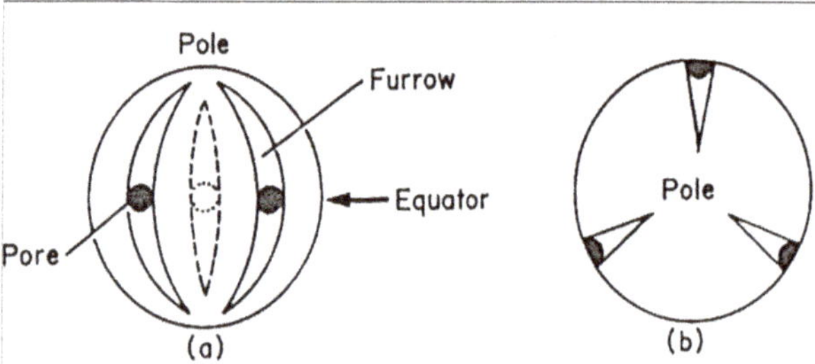


Figure 28. Diagrams illustrating the morphological regions of a pollen grain and pollen grain apertures: (a) equatorial view; (b) polar view (Sawyer, 1981).

### 5.3.3. Apertures

Pollen grains have apertures on the surfaces. The shape (type) and the numbers of apertures are important characteristics of a pollen grain and can be used in identification. Furrows containing pores are counted as one aperture. In compound pollen grains, the numbers refer to each single grain sub-unit. Nine classes of aperture number (as given in Table 28) can be used in pollen identification. Regarding aperture type, Vorwohl (1990) used the following five classes in pollen identification (Table 28). Whereas Sawyer (1981) recognized only four classes of aperture types to be important in pollen identification as:

1. Pores only;
2. Furrows only;
3. United pores and furrows;
4. Irregularly occurring furrows.

### 5.3.4. Exine sculpture (exine ornamentation) and surface

The exine, the outermost coat of the pollen grain, is either smooth or various ornamented with different outgrowths or projections, such as granules, various kinds of spines, and reticulum or window-like structures. The ornamentation of the exine has a major

Table 28. Diagnostic pollen grain features used for identification (Vorwohl, 1990).

| 1. | Length and breadth | 1 | <10 μm |
|---|---|---|---|
| | | 2 | 10–15 μm |
| | | 3 | 15–20 μm |
| | | 4 | 20–25 μm |
| | | 5 | 25–30 μm |
| | | 6 | 30–35 μm |
| | | 7 | 35–40 μm |
| | | 8 | 40–50 μm |
| | | 9 | >50 μm |
| 2. | Aperture number | 1 | |
| | | 2 | |
| | | 3 | |
| | | 4 | |
| | | 5 | |
| | | 6 | |
| | | 7 | |
| | | 8 | >7 |
| | | 9 | Absent or not clear |
| 3. | Aperture type | 1 | Pore |
| | | 2 | Colpa |
| | | 3 | Heterocolpa |
| | | 4 | Syncolpate |
| | | 5 | Heterocolporate |
| 4. | Exine Sculpture | 1 | Psilate, Faveolate, Fossulate |
| | | 2 | Scabrate, Verrucate, Gennuate |
| | | 3 | Echinate |
| | | 4 | Clavate, Bacculate |
| | | 5 | Rugulate, Striate |
| | | 6 | Reticulate |
| | | 7 | Fenestrate |
| 5 | Aggregation | 1 | 1 |
| | | 2 | 2 |
| | | 3 | 3 |
| | | 4 | 4 |
| | | 5 | 5 |
| | | 6 | 6 |
| | | 7 | 7 |
| | | 8 | >7 |

effect on the surface of pollen grain as viewed under the microscope. The surface of pollen grains, as seen under the microscope, can sometimes change with a change in focus. For example, the pollen grain of *Luffa cylindrica* may appear granular under one focus and reticulate under another.

1. Nine classes of exine ornamentation can be identify: thin; medium, no rods; medium with spaced rods or beaded; medium or thick with coarse external rods; layer of closed thin rods; long thin spines; large broad-based spines; small or very small spines; or warts and other projections.
2. Five classes of surface as: smooth or indefinite; granular or warts; striated; net or pitted and Isolated dots due to spines or other projections. (exine ornamentation can also be divided into seven classes of exine sculpture – Table 28):
   a. *Psilate, Faveolate and Fossulat:* When exines are smooth, they are known as psilate, e.g., *Betula* spp and *Pyrus* spp. Exine sculpture with little pits is reported *as* faveolate and is known as fossulate with relatively larger pits. In viewing the surface, the pollen grains with psilate, faveolate, and fossulate exine sculpture appear smooth or sometimes indefinite;
   b. *Scabrate, Verrucate and Gemuate:* Exines with little warts and diameters of not more than one pm are scabrate, e.g., some species of *Quercus*. In verrucates, the diameter of the warts is more than one pm, e.g., members of the *Ranunculaceae* family, some *Asteraceae,* and *Nigella* spp. Gennuate is an exine sculpture with round warts which are narrower at the base, for example, *Ilex* spp. In optical sections, this kind of exine appears to have small projections. Pollen grains with this kind of exine sculpture appear granular in viewing the surface;
   c. *Echinate:* Exines provided with different kinds of spines, e.g., *Malvaceae, Cucurbitaceae, Campanulaceae,* and many *Asteraceae* are echinate. Pollen grains with echinate exine sculpture appear to have dots on the surface;
   d. *Clavate and Bacculate:* Exines with rods which have thicker ends are clavate and those with stick-like rods are bacculate. Pollen grains with clavate or bacculate exine sculptures are very rare. On examining the surface, the pollen grains with this kind of exine sculpture appear granular (they have larger grains than scabrates, verrucates, and gennuates);
   e. *Rugulate and Striate:* Exine sculptures with striations, e.g., species of Acer, *Prunus, Datura,* and *Fragaria,* belong to these categories. In optical section, the exines appear to have close, thin rods. On examining the surface, the pollen grains also appear striated;
   f. *Reticulate:* This term is used when the exine sculpture is reticulate or net-like. This category is divided into fine reticulate, or beaded, or coarse reticulate. Examples are *Lilium* spp, *Hedera helix,* and *Ligustrum* spp. In optical sections, reticulate exines appear to have either thin or coarse rods, or they appear to be beaded. Pollen grains with reticulate exine sculpture appear reticulate on the surface also;
   g. *Fenestrate:* Exines in which there are window-like holes between the ribs of the exine belong to this category. Important examples include *Taraxacum officinale* (Ap.: and members of *Amaranthaceae).* On the surface, the pollen grains appear to have window-like holes.

### *5.3.5. Other features*

*According to* Vorwohl (1990), aggregation of pollen grains is one of the important features in identifying pollen grains. Pollen grains of some plant species have other diagnostic features. For example, the pollen grains of many plant species in the family's *Ericaceae* and

Figure 29. Photos to illustrate the variety of colors of pollen loads of different species of plants: (a) gray – *Vicia faba*; (b) lemon yellow – *Brassica napus*; (c) dark gray – *Papaver rhoeas*; (d) blue – *Phacelia tanacetifolia*; (e) mixed yellow pollen – mixed pollen from fruit trees (*Prunus, Pyrus, Malus*) and bright orange – *Taraxacum officinale*.

*Leguminosae* (sub-family, *Mimosoidae*) are polyads, i.e., compound. In *Ericaceae,* four grains are aggregated to form a tetrad, e.g., *Rhododendron* spp., whereas in the *Leguminosae,* eight, twelve, or sixteen pollen grains are aggregated to form a polyad, e.g., *Acacia* spp, *Albizia* spp, *Calliandra* spp, and *Mimosa* spp. Also another important *feature* for pollen determining as: Air sacs; Thickened or projecting edges to apertures; A cap or streak on the apertures; Granules or projections scattered on the apertures; Intine swollen below the apertures; Intines thick or very thick and cell content's granular has been reported by Sawyer (1981).

### 5.3.6. Irregular grains

In the study of pollen found abnormal, not typical, grains. Sawyer (1981) allocates such grains in the following groups:

1. Shriveled or abortive grains. Particularly in *Rubus* (blackberry), *Trifolium repens* (white clover) and many garden cultivars (e.g., *Saxifraga* and *Oxalis)* we find shriveled grains which do not expand when mounted. Such grains may even form the majority in a sample;
2. Giant and dwarf grains. Many pollen samples show the occasional giant grain; much larger than usual and often irregular. Others, for example *Plantago* (plantain) and *Fagopyrum* (buckwheat), produce a number of unusually small perfect grains among their pollen;
3. Aperture numbers. Variants with 4 furrows are occasionally found among 3-furrowed grains. This is particularly noticeable in some of the Rosaceae (e.g., blackberry and fruit blossom), *Rumex* (dock) and *Hippophae* (sea buckthorn). Regular variation of the number of pores is found in *Carpinus* (hornbeam) (3, 4 or 5); *Alniis* (alder) (4, 5 or 6) and *Epilobium angustifolium* (rosebay willowherb) (3 or 4);
4. Abnormal furrows. An interesting variation occurs in *Veronica* (speedwell) and occasionally in other Scrophulariaceae. This results in a square grain (Pl. 188). Irregular and united furrows are common in *Hypericum* while the 'furrows' of *Berberis* (barberry), *Crocus* and *Populus* (poplar) appear as irregular breaks in the exine rather than true furrows (*Crocus*). The study of atypical pollen grains used for palinoindication of environment. (See also the paragraph regarding to Palinoindication of environment).

### 5.3.7. Pollen "types"

Although the pollen from each species of plant differs from that of all others, the differences are often slight and with the multitude of plant species and hybrids in the world, it will be appreciated that we cannot expect to identify each one by its pollen alone. Here they are group according to their "type."

The pollen of some species is unique in its characters. Thus, *Trifolium repens* (white clover) and *Trifolium pratense* (red clover) may be distinguished from one another and from all others. But more often we find a "type" of pollen common to a genus or other group of plants. The pollen of all *Tilia* (lime) is unmistakable and although careful study may indicate the particular species, the classification "*Tilia* type" is usually sufficient in limiting.

### *5.4. Colors of pollen*

The colors of bulk pollen, dry from the anthers or the pollen loads of bees, have been arranged in spectral order. They show considerable variation but may at times provide a valuable clue to identity (Hodges, 1952; Kirk, 1994; Partap, 1997; Sawyer, 1981). Hodges (1952) and Kirk (1994) consider that the first steps for identification of pollen loads are

1. sorting by color;
2. selecting one or two pollen loads from each color group;
3. pollen loads are put on white paper and lightly patted to make a uniform flat surface. This is necessary in order to find an exact match to a known color chart.

The nice color charts in Hodges (1952) and Kirk (1994) illustrate the different color pollen loads of different plants. The Figure 29 presents an example of the different colors of pollen loads. Important is the lighting, in which the identification is carried out. Pollen loads that are to be identified must therefore be matched to the color chart under these same standard conditions. If the chart is used in direct sunlight or under electric light, the result may well be different. According to Kirk (1994), many plant species have pollen loads with very similar colors, so the particular species from which a pollen load has been collected can rarely be identified from just the pollen color. However, a combination of the color, flowering time and knowledge of local plants can usually narrow down the possibilities to one or a few species. However, this identification cannot be accurate. Therefore, for more exact identification microscopic examination pollen grains is necessary.

For instance, it has been proved that the color of pollen influences the color of honey; pollen and pollen loads from *Taraxacum* and *Helianthus annus are* bright orange – the color of pure honey with this plants – also is bright orange; pollen from *Phacelia* – blue, pure honey from *Phacellia* as a pale-blue color, pollen loads from *Aesculus hippocastanum give a* burgundy, honey with reddish color, etc. Most of the loads collected by honey bees consist only of pollen, but bees also collect propolis and occasionally fungal spores and various dusts (Hodges, 1952).

### *5.5. Methods of pollinic pollen analysis*

#### *5.5.1. Types of microscopy examination*

The exact definition of the botanical origin of bee products, especially honey and bee pollen, requires the use of microscopic examination. The literature describes three methods of microscopic procedures.

***5.5.1.1. Light microscope.***

- Use a light microscope at a magnification of which is 100, 500, 1000 makes it possible to determine the external characteristics of the pollen grain: the shape, size, surface sculpture, and the presence of pores, fissures, etc.). And identify it up to the family, genus and species (Tripathi et al., 2003). But this method has its drawbacks: the internal structure of the grain is unattainable because of the opacity of the outer shell (exine) and pollen grains of some plants are too large and therefore it is impossible to get a clear image.
- Preparation of pollen analysis for microscopic preparations. There are several techniques, which differ mainly in the medium fixing pollen grains turpentine, alcohol, honey solution or acetolized method (Dustman, 1992). Since the methods of preparation of pollen in microscopy greatly affect the appearance of pollen grains and noted the difference in the results of analysis, it is advisable to reference preparations of pollen grains and pollen description made in a relatively similar procedure.

Louveaux et al. (1978) proposed an affordable and simple method for the preparation of dust melisopalinology research involves the use of glycerol-gelatin. This method is used in many countries.

***5.5.1.2. Transmission electron microscope (TEM).***

- Microscopic examination of pollen using a transmission electron microscope, which allows the study the ultrathin sections of pollen grains;
- The image has pollen with unlimited depth transparency (X-ray allows us to study the internal structure of an object is significantly different than the other methods) (Kedves et al., 2002, 2003; Laere et al., 1969).

***5.5.1.3. Scanning electron microscope (SEM) (three-dimensional image).*** Accuracy to distinguish varieties even within the same species of plants. Especially for minor differences in the size and structure of the surface of pollen grains of different varieties of fruit crops (Brindza & Brovarskyi, 2014; Fogle, 1977; Klungness & Peng, 1983; Lindtner, 2014).

At the present stage of research melisopalinologiy widely practiced a combination of all methods of microscopy (Tripathi et al., 2003). For carrying pollen analysis must have reference preparations of pollen grains of the plants that could potentially occur in the samples of pollen, keys and atlases pollen. Help in defining various botanical origin keys pollen grains photographs, drawings. Modern pollen atlas published in 2004 by German melisopalinology. It contains pictures and descriptions of pollen about 170 species of plants and honey flora characterizes his country (Bucher et al., 2004). Several studies focused on computerized texture analysis of Scanning Electron Microscopy (SEM) images have been published (Langford et al., 1990; Li & Flenley, 1999).

### 5.6. *Microscopic techniques*

Identification of pollen is carried out in 3 stages.

#### *5.6.1. Preparation of pollen grain slides from pollen loads for microscopic research*

Method of preparation of pollen grain slides based on the method of mellisopalinology Louveaux et al. (1978).

1. Take 1 mg of bee pollen loads with tweezers and transfer to a glass slide, add a drop of distilled water, which cement pollen grains will be separate each other. The amount of water is such that pollen grains freely melt away in a single layer on a slide.
2. Evaporate the water.
3. Apply one drop of warm (+40 °C), heat by thermal table glycerin-gelatin and cover with cover slip.
4. Preparation of glycerin jelly. Weight 10 g of gelatin and place them place for two hours in distilled water (gelatin absorbs 34.2 ml of water) and add 66.3 g of glycerin. This ratio of ingredients allows to save pollen grain shape unchanged. To prevent microbial growth in glycerol-gelatin add 0.5 g of carbonic acid (phenol) and to get a clear picture of the structures of pollen grains – 0.1 g Basic Fuchsine. 12 hours later for the prevention of further drying up of the slides aids its conservation – edge of the coverslip treat with decorative lacquer (nail polish).

#### *5.6.2. Identification of pollen under microscope*

Determination of botanical species of pollen can be done using the computer program identification of pollen LUCIA (Rodlauer & Hüttinger, 1992) (or similar). As an example, Digital camera Sony with light microscope Nicon transfer image of pollen on the computer screen. The operator determines the size, shape, number of pores, fissures and other parameters of the pollen grain – creates a digital key for determine (details in Introduction). Using this key, another computer program (electronic database of pollen) "Pollen Data Bank," provides a search and similar pollen grains select and begins clarification of affiliation pollen to a particular plant species. For comparison also use own standard slides and/or slides from other collections as, for example, the collection of the Austrian Institute of Beekeeping.

Determination of the botanical origin of pollen will be performed by comparing investigational pollen grains with descriptions and photographs of existing atlases and reference books, as for instance: Ambruster and Oenike (1929), Burmistrov and Nikitina (1990), Faegri et al. (1993), Fossel and Pechhacker (2001), Hodges (1952), Maurizio and Louveaux (1965), Sawyer (1981) and Zander (1935).

Microscopic magnification of X 450 (*Zea mays*) and ×1000. For the magnification of ×1000 used oil immersion (cedar oil). A calibration line representing 10 μm should be add to each photograph.

#### *5.6.3. Description of pollen grains*

Description submitted by the following scheme:

1. Latin name of the plant;
2. The family of plant;
3. Plant life form (tree, shrub, herb; an annual or perennial plant);
4. Morphological characteristics of pollen grains:
   a. size of the diameter and the equator;
   b. number of furrows, pores;
   c. surface and exine *sculpture*;
5. Color of bee pollen loads.

### 5.7. *Palinoindication of environmental pollution*

The knowledge of the morphology of pollen grains from different species (plants) allows simultaneous pollen analysis of bee products, also to carry out environmental monitoring, determine ecologically disadvantaged regions. It has been proved that the formation and development of pollen is very sensitive to external factors (radiation, volcanic activity, solar activity, heavy metal contamination, etc.) (Bogdanov et al., 2005). The result is a significant amount of pollen grains with modified morphological structures. Research shows that in harsh conditions, plants may produce large amounts of atypical imperfect, irregular and abnormal pollen grains - the worse the state of the environment, the greater the percentage of abnormal pollen and vice versa. (Bessonov, 1992; Dzyuba, 2006, 2007; Dzyuba & Tarasovich, 2001; Gilber & Lisk, 1978; Glazunov, 2001; Muszynska & Warakomska, 1999).

### 5.8. *Concluding remarks*

Research using light and scanning electron microscopy show the changes in size and shape of the pollen grains, the number, the shape and type of apertures, their size and location relative to each other. Main thing changes the most stable structure of pollen grains – sporoderm surface structure (shell), the number and thickness of the layers. Russian scientists have developed a method for determining the quality of the environment by the number of abnormal, atypical pollen grains (Dzyuba, 2006). In selected samples investigated at least 200 pollen grains, each of which is examined by the following parameters:

1. The form of the pollen grain;
2. The size of the pollen grain;
3. The number of apertures;
4. The nature of the location and type of apertures;
5. The thickness and number of layers sporoderm;
6. The nature and type of sculpture sporoderm;

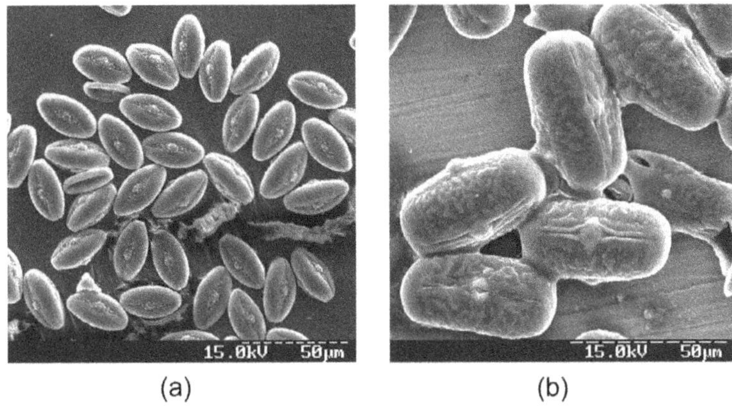


Figure 30. Typical (normal) pollen grains (scanning electron microscopy): (a) *Aesculus hippocastanum*; and (b) *Vicia faba*.

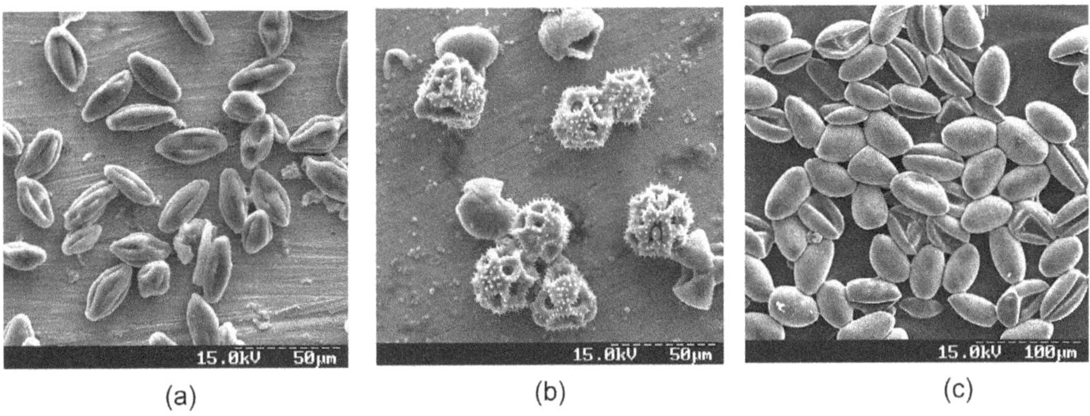


Figure 31. Pollen grains with changed morphological structures (scanning electron microscopy): (a) *Phacelia tanacetifolia*; (b) *Taraxacum officinale*; and (c) *Scilla bifolia*.

7. Symmetry breaking pollen grain.

Thus, each sample was fixed in the presence/absence teratomorfic (pathologically developed) pollen grains, their quantity and analyzed morphological features. The Figure 30 illustrates the typical, normal pollen grains. Pollen grains have no deviations in the morphological parameters (size and shape of grains, number, shape and type of apertures, their size and location relative to each other, the sculpture's surface the number and thickness of layers). In the Figure 30 presents the photographs of pollen grains morphological features which correspond to norm. Photos made with the help a scanning electron microscope (Figure 30) and the Figure 31 shows some pollen grains (scanning electron microscope) with changed morphological structures. Particularly large number of abnormal pollen grains in *Phacelia tanacetifolia* (atypical shape and size) (Figure 31a). Pollen grains of *Taraxacum officinale* observed symmetry breaking (Figure 31b) (Lokutova, 2014).

Therefore, numerous experiments have shown that during pollen analysis of bee products to determine their botanical and geographical origin there is a possibility at the same time estimate the state of the environment. According the quantitative content abnormal pollen grains can draw conclusions about environmental situation in the region, where these bee products were received.

## 6. Phenolic and polyphenolic profiles as fingerprints of floral origin of bee pollen

Rapid and easier method developed by Campos et al. (1997) use High Pressure Liquid Chromatography/ Diode Array Detector (HPLC/DAD) profiles, of phenolic and polyphenolic compounds, made with hydroalcoholic extracts of pollen and bee pollen loads. It is easy to do, reliable, highly effective, and requires minimum sample preparation, as indicated by the several reports, from worldwide, in which the method has been used to analyze the phenolic composition of pollen. This method allows the identification of the genus and the species (*sp.*) and provides valuable information related to chemical structure of the constituents under analysis. The phenolic acids and polyphenolics in pollen are expressed as specie-specific, which represents an important tool for taxonomy studies. It should be do a screening of different color pellets from a mixture of bee pollen and the identification of the potential floral source using a data base build with floral hand collect pollen or herbarium specimens. The first approach to the flavonoid structures can be based on Campos and Markham theory (2007).

### 6.1. Materials and methods

#### 6.1.1. Sample management

Sample management is an important issue for the analysis. It is essential to specify details about where and when specific pollen samples are collected, sample storage, handling, and preparation, as details of any of these aspects might affect the phenolic composition of pollen. More details found in Section 2 of this chapter.

#### 6.1.2. Sample collection

***6.1.2.1. Pollen hand collected – reference samples.*** Reference samples should be collected from the specimens at the Herbariums if they show a very good preservation or they can be hand collected from the stamens of flowers at the field and bring to the laboratory. The correct taxonomic identification should be assured. The HPLC/DAD fingerprints of the phenolic and polyphenolic compounds of these well-known samples will be used as a data base for match of the unknown samples of bee pollen. For that, it is important to:

1. record location of harvesting;
2. record date of sampling;
3. collect the respective voucher specimen, which must ~~to~~ be deposited at herbarium to be identified or authenticated.

***6.1.2.2. Bee pollen pellets/loads.*** Samples collected directly from bee pollen traps or purchased from beekeepers during the collection period. Information about the apiary locations and harvesting data is always welcome, as well the processing of drying, if it was applied.

#### 6.1.3. Methods

***6.1.3.1. HPLC/DAD analyses.***

*6.1.3.1.1. Sample preparation (hand collected pollen and/or pollen from herbarium).*

1. Weight ten milligrams of dried pollen of each sample. If the sample is from herbarium lower amounts can be used but the proportion 10 mg to 1000 µl should be maintained.
   Note: Using always the same concentration among samples makes easy to compare the approximate amount of phenolic and polyphenolic compounds in the fingerprints, without the need of quantification. If need, a rigorous determination of the total amounts of each compound can be done, using a calibration with standard solutions.
2. Add an ethanol (EtOH)-water solution (1 ml, 50% V/V).
3. Sonicate for 60 min.
4. Centrifuge the resultant extract at 5000 rpm for10 min.
5. Take out the supernatants and do a microfiltration with a 0.45 µm membrane for HPLC/DAD analysis as previously described (Campos et al., 1996).

*6.1.3.1.2. Sample preparation (bee pollen).* Usually bee pollen samples are made up of different color loads/pellets and set up a "multifloral bee pollen." However, "monofloral" samples are also found in the market and/or produced with this purpose according to Campos et al. (2008).

1. Separate pollen loads by colors (Figure 32).

Note: commonly each color corresponds to a botanical origin and, for each one, it will be carried out a HPLC/DAD profile (fingerprint of the *taxon*). This fingerprint will be compared to others in the data base to obtain the match with a similar one.

2. Weight one pellet, keeping the same ratio for extraction (10 mg/ml EtOH 50%), for instance, if the weight is 6.78 mg the solvent will be 0.678 ml of EtOH 50% (V/V).
3. Proceed the same way for mixtures of bee pollen. The extraction should keep the same ratio. If necessary, du an aliquot that represents the entire weight of the sample under analysis, for example, 10 g/l EtOH 50% (V/V). The portion of the mixture to be weighed is removed after sample homogenization.
4. Proceed as explained in points 3–5 in sSection 6.1.3.1.1.

Note: In addiction a standard of quercetin-3,5,7,3′,4′-*O*-pentamethyleted can be add to samples for HPLC/DAD injection, as an internal standard (Campos & Markham, 2007).

*6.1.3.1.3. HPLC/DAD analysis.* For the HPLC/DAD pollen analysis several steps must be performed rigorously to have a correct identification of the compounds in the different samples with different origins:

1. HPLC/DAD system with a RP18 column;
2. Eluent formed by acidified water with ortho-phosphoric acid to pH 2.4 (solvent A) and acetonitrile HPLC grade (solvent B), combined in a linear gradient according to the following: starting with 100% A, decreasing to 91% over the next 12 min, to 87% over the next 8 min, to 67% over the next 12 min, to 57% over the next 10 min, and holding that last proportion until the end of analysis (a total of 60 min).

Note: This gradient was optimized to avoid sample purification, which decrease time of preparation, sample manipulation and possible loose or degradation of compounds to be used on the fingerprints for each *taxon*.

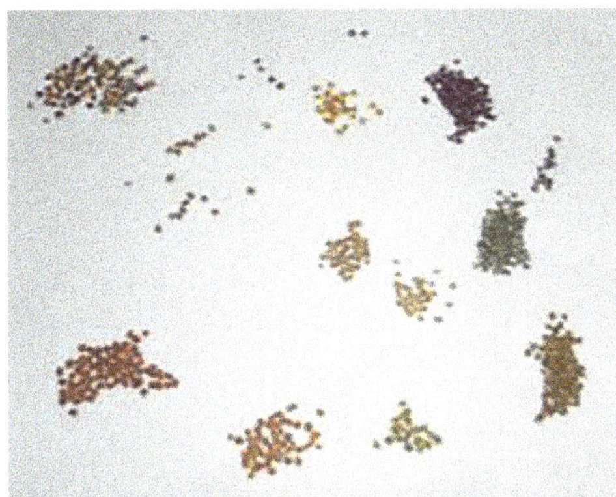

Figure 32. Separation of bee pollen by color pellets from a sample that include different floral origin pollens.

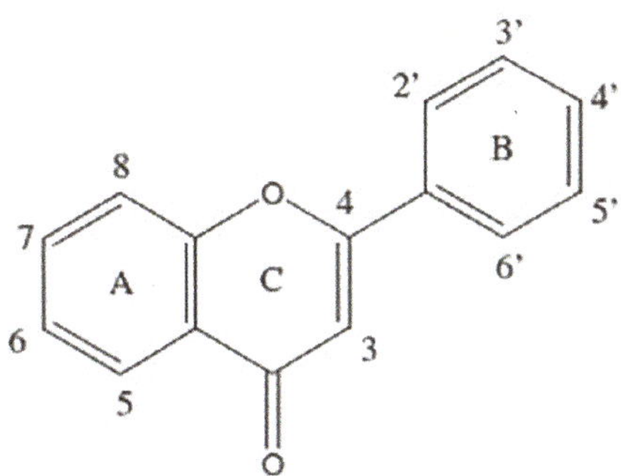


Figure 33. Basic flavonoid structure in order to show a better understanding of the explanation of the UV spectra related to the position of the substitution in the molecule (From Campos & Markham, 2007. Reproduced with permission).

Among phenolic and polyphenolic compounds many other polar constituents will be extracted with the hydroalcoholic solvent used; however, the contamination does not interfere with the HPLC analyze because at the beginning a water gradient is used and the compounds that will be separated stay in the top (head) of the reverse phase column choose, allowing the more polar, i.e., amino acids, proteins, sugars, hydrosoluble vitamins and minerals to elute in the first minutes without interfering in the fingerprint. Phenolic and polyphenolic compounds only start to be eluted 15 min later with the increase of acetonitrile in the gradient;

3. Flow rate 0.8 ml/min;
4. Temperature stated at 24 °C;
5. Injection of the samples extracts (20 to 100 µl, depending on phenolic concentration);
6. Standard chromatograms plotted at wavelengths 260 and 340 nm (the choice of these two wavelengths consist in the knowledge that flavonols and flavones glycosides, the main compounds in pollens, have two characteristic bands of ultraviolet – absorption around these values; phenolic acids and its derivatives can also be identified using the same range);
7. Spectral data for all peaks is accumulated in the range 220–400 nm (the phenolic/flavonoid spectra are all in this display) using DAD;
8. Apparatus: HPLC system, Diode Array Detector with respective software;
9. The phenolic/polyphenolic profile of each sample is made up with all compounds resolved in its respective chromatogram;
10. The structural information for each compound is made by direct comparison of retention time and ultraviolet absorption spectra with standards, and according to the theoretical rules developed and compiled by Campos and Markham (2007).

Note: the compounds can also be isolate and the structures determined by Nuclear Magnetic Resonance (NMR) and/or LC-MSMS (Campos et al., 1996, 1997). Briefly: The extracts are applying in sheets Whatman 3 MM chromatography paper for 1D separation. Each compound is prepared for further studies by chromatography on a C-18 reversed phase column using water: acetonitrile gradient and checked for purity by HPLC/DAD, according to Markham and Campos (1996). For the full structure elucidation by NMR spectra can be carry out in Bruker AC apparatus at 300 MHz (or similar) with the isolated compounds, solved in DMSO-$d_6$, with TMS as intern standard at 30 °C (Markham & Geiger, 1993).

### *6.2. Theoretical rules for structural determination of the main phenolic acids and flavonoids*

The identification of the compounds is carried out by direct comparison of retention time and ultraviolet absorption spectra (Campos & Markham, 2007). In Figure 33 flavonoid structures are shown to a better understanding of the explanation of the UV spectra related to the position of the substitution in the molecule. The majority of the flavonoids in pollen are derivatives of flavones and/flavonols. The majority of the flavones found in pollens are apigenin or luteolin derivatives, and the compounds with a flavonol nucleus, are mainly kaempferol, quercetin or isorhamnetin. In the most samples these compounds appear as glycoside derivatives. Examples of different flavonol and flavone structures and its respective UV-spectra are show in Figure 34(a–g). Isoflavones, as daidzein, glycitein and genistein derivatives can also be found in certain pollens but are not so common.

Examples; of rules developed in Campos and Markham (2007) for identification of details in the UV-spectra:

1. In all of them can be observed a clear existence of two intense bands of absorption which indicate the presence of the rings A, B and C from flavonoids. In Figure 34a are pointed the three important absorption points:

    a. Band I, usually with absorption around 340–390 nm. Flavones 340–350 nm; Flavonols-3-*O*-derivatized (Table 29) (as glycosides, for instance)

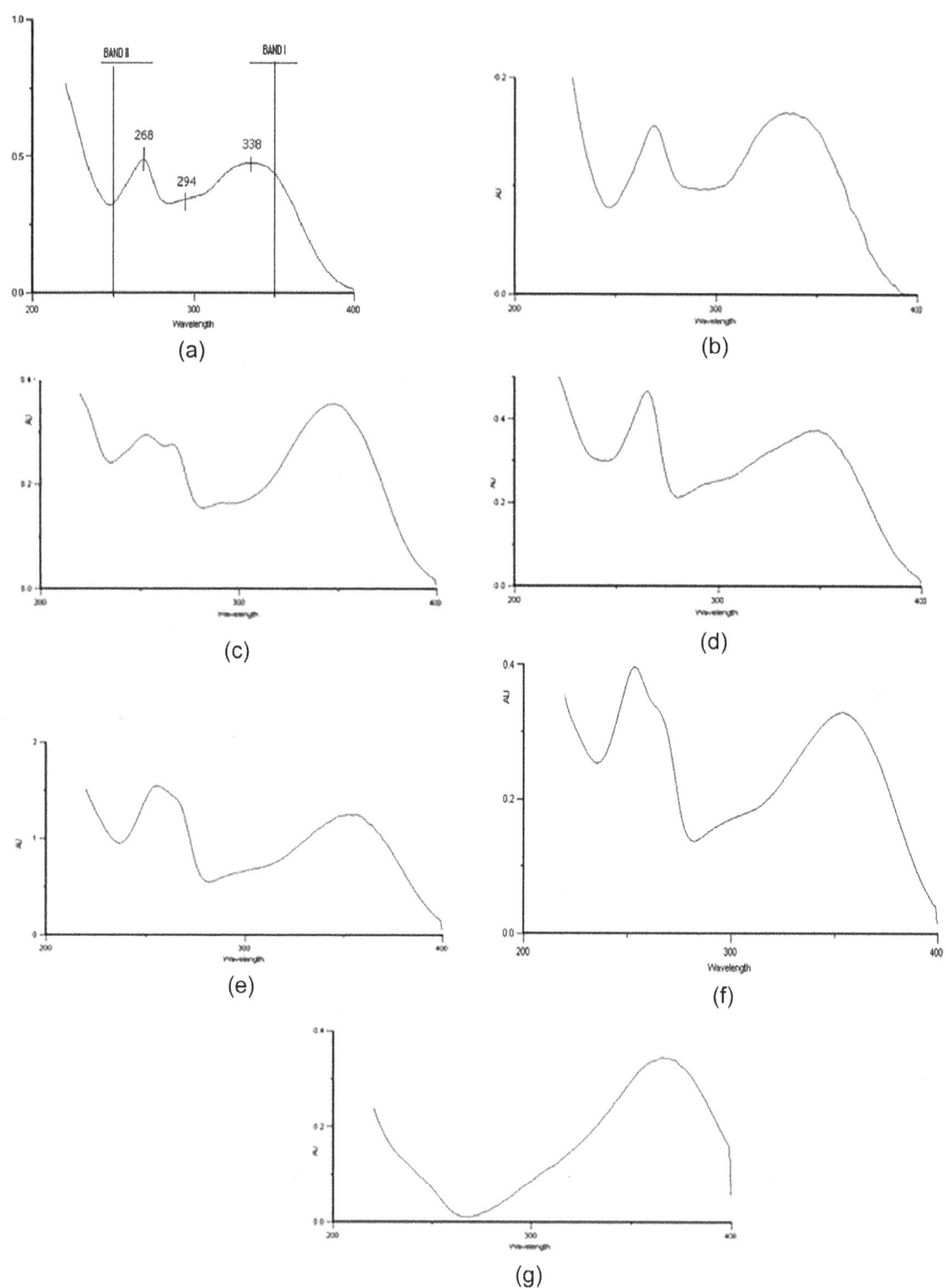


Figure 34. Examples of the more common UV- spectra found in bee pollens: (a) apigenin; (b) apigenin-7-*O*-derivative; (c) luteolin; (d) kaempferol-3-*O*-derivative; (e) quercitin-3-*O*-derivative; (f) isorhamnetin-3-*O*-derivative; (g) chalcone.

350–360 nm; Flavonols with the C-3-*O*-H free 370–390 nm;

b. Band II, usually with absorption around 240–280 nm;
c. shoulder between 280 and 310 nm – substitution in C7;

The absence of the shoulder between 280 and 310 nm is indicative of a substitution in C7-OH. Example, flavones as Apigenin (Figure 34a) and apigenin-7-*O*-derivative (Figure 34b). This –OH is free in Apigenin itself with no substitution, and the shoulder remain in the UV-spectra.

Example, flavonols as Kaempferol-3-*O*-derivatives (Figure 34d) and Kaempferol-3,7-*O*-di-derivatives (substitution in C7-OH) (Figure 34e). The same absence of the shoulder at 280–310 nm can also be seen for Quercetin-3,7-*O*-di-derivative (Figure 34g). When the C7-OH is free the shoulder is present at 280–310 nm (a flat region in this part of the spectrum), examples of luteolin, Kaempferol-3-*O*-derivative and Quercetin-3-*O*-derivative (Figures 34c, d and f, respectively – all have the C7-OH free);

2. Differences between isorhamnetin and quercetin: Comparing isorhamnetin-3-*O*-and quercetin-3-*O*-derivative, the difference in the structure is in B-ring with a methyl group in C3' hydroxyl in the first. This implies an intensification of Band IIb and IIa (wavelengths of UV absorption for both around 255 and 265shoulder nm). Quercetin show a smoother absorption among these two bands;
3. To distinguish both structures, flavones or flavonol 3-*O*-substituted the relative intensity of absorption among Band I and Band II should be analyzed. If Band I is higher than Band II, usually are flavones but in the inverse case, Band I lower then Band II is a high probability to be a flavonol-3-*O*-substituted (glycosylated in the majority of the times) (examples in Figure 34);

Note: More details for UV interpretation of the spectra in Campos and Markham (2007).

Table 29. Amplitude of the Wavelengths in Band I of flavones and flavonols UV spectra.

| Flavonoid type | $\lambda$ (nm) |
|---|---|
| Flavones | 304–350 |
| Flavonols (OH-3 derivative) | 328–357 |
| Flavonol (OH-3 free) | 352–385 |

Table 30. Wavelengths of Band I in UV spectra of flavonols with a different pattern of oxygenation in B-ring.

| | Oxigenation pattern | | |
|---|---|---|---|
| Flavonols | Ring-A and C | Ring-B | $\lambda_{max.}$ (nm) |
| Galangin | 3,5,7 | – | 359 |
| kampferol | 3,5,7 | 4′ | 367 |
| Quercetin | 3,5,7 | 3′, 4′ | 370 |
| Myricetin | 3,5,7 | 3′, 4′, 5′ | 374 |

Table 31. Wavelengths of Band II in UV spectra of flavones with oxygenation only in A-ring.

| Flavone | Oxygenation pattern of A-ring | $\lambda_{max.}$ (nm) |
|---|---|---|
| Flavone | | 250 |
| 5-hydroxiflavone | 5 | 268 |
| 7-hydroxiflavone | 7 | 252 |
| 5,7-di-hydroxiflavone | 5.7 | 268 |
| Baicalein | 5,6,7 | 274 |
| Norwogonin | 5,7,8 | 281 |

4. In Tables 30 and 31, the wavelengths related to the oxygenation pattern, for the more common flavones and flavonols can be consulted and there, a prevision of the bathochromic effect (relating to a shift to a longer wavelength in the absorption spectrum of a compound) in band I with the increase of the oxygenation pattern on B-ring in these structures. However, when these changes occur, usually the UV-absorption of Band II remain the same (Table 31).

### *6.3. Quantification of the major floral specie (floral source) in a mixture*

With this method it is also possible the identification of the major *taxon* source in mixtures of various floral species (different color loads/pellets in general corresponding to various floral sources), if one of them is in a high percentage.

1. Analyze the phenolic HPLC/DAD profiles (fingerprints) for the different *taxon* found in the mixture.
2. In each *taxon* select the compound that can be used as a marker for this specie. The compound used as marker should not be found in the other floral species in the full mixture. Note: This approach is supported by microscopic analysis. The flavonoid/phenolic HPLC/DAD fingerprint of the full mixture should be the same as the profile corresponding to the major floral specie, *i.e.*, if the major floral source in the bee pollen sample (batch) is *Salix atrocinerea*, the fingerprint match with the one obtained with same floral source in the data base made up with the hand/herbarium pollen sample.
3. To achieve the percentage of this major monofloral identification in a bee pollen mixture:

    a. Run HPL/DAD profile of the mixture and compare to the data base to verify if it is the same as one of them;
    b. Once the floral specie is identified choose one of the major compounds as a marker and its quantification will correspond to 100%. Note: analyze the fingerprints obtained with the various pellets (other floral origins) in the mixture and choose a compound that does not belong to them;
    c. Quantify the same compound in the mix. The amount determined will be a % of the marker and corresponds to the percentage of the major specie in the entire batch. In Figure 35 an example involving a mixture of bee pollen with about 60% of *Eucalyptus globulus* is shown. In this case, quercetin-3-*O*-sophoroside (di-glycoside) was chosen for quantification and the calibration curve was made with rutin, which is a similar di-glycoside of quercetin as the one used as marker for this *taxon*.

Note: The majority of the flavonoids in pollen are kaempferol, quercetin, isorhamnetin, apigenin or luteolin

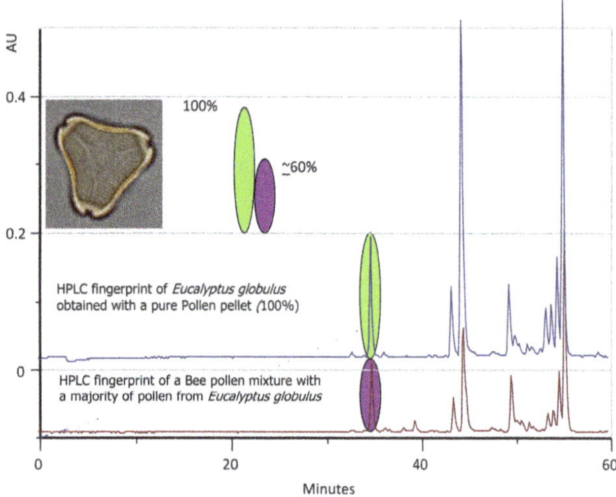


Figure 35. Fingerprint of *Eucalyptus globulus* obtained with a pollen pellet (100% pure) and a mixture of bee pollen with about 60% of *Eucalyptus globulus* pollen pellets.

derivatives, mainly glycosides. However, many different structures can be found and is not difficult found a good marker for each *taxon*. A combination of two or more markers can also be used if necessary.

## 7. Determination of phenolic profile of bee pollen extracts by LC and LC/MS

Polyphenols are known to interact with compounds present in foods, like carbohydrates, proteins or lipids (Jakobek, 2015; Ozdal et al. 2013). Comprehensive analytical methods such as LC and LC/MS can be used to obtain more detailed information on phenolic profile of bee pollen as well as on interactions in food systems. Standard procedures for the extraction of phenolic compounds from bee pollen, pre-concentration of phenolic extracts prior to analysis, LC and LC/MS techniques for their determination will be described.

### *7.1. Polyphenolics: importance, structure and classification*

The phenolic compounds in pollen grains are important as a part of reproductive process in plants. They protect pollen load from destructive exterior factors such as UV-radiation (LeBlanc et al., 2009). During the germination of pollen load, the formation of two protective membranes – exine and intine is occurred. Both membranes are rich in content of phenolics. Campos et al. (1997) have shown that it is possible to establish the correlation between the polyphenols and palynological origin of bee-collected pollen.

There are thousands phenolic structures of substances present in plants (Boudet, 2007). It can be the simple molecular structure with one substituted benzenoic ring or complex polyphenolic structure with two or more interconnected aromatic rings. Natural phenolic compounds

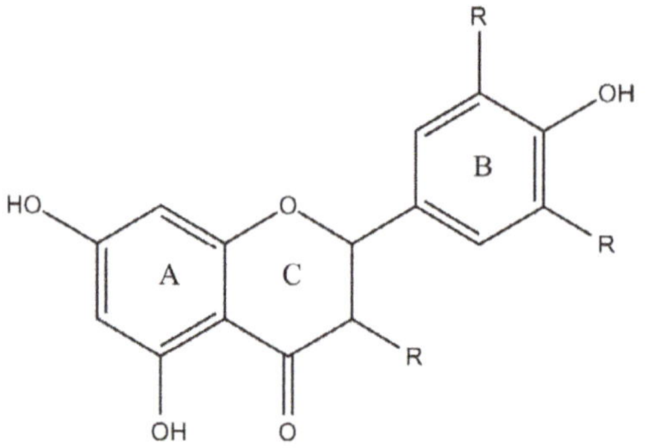

Figure 36. General formula of flavonoids.

can be classified according to their structure in two major classes:

1. non-flavonoids polyphenols (*Lignans and lignins*, phenolic acids, coumarins, stilbene oligomers, gallotannins, ellagitannins and procyanidins (Soto-Vaca et al., 2012)) (Table 32);
2. flavonoids (large group of molecules with complex structure derived from *p*-coummaric acid. Basically, it consists of two aromatic rings with different alkyl and (or) *O*-substituents and with multiple chiral centers (Yáñez et al., 2013). Benzene rings A and B are connected with $C_3$ chain which are often closed and form heterocyclic ring C (Figure 36).

There are a significant number of subclasses of flavonoids based on different reactions that occur on the rings over the lateral groups – hydroxylation, prenylation, alkalinisation and glycosylation (Stalikas, 2007): chalcones (without formed C ring), aurones, flavanols (includes catechins, epicatechins and procyanidins), flavonols, flavones flavanones, isoflavones and antocyanins. The general structures of these subclasses of flavonoids are presented in Table 33. Presence of sugar unit (in form of glycoside) and OH-groups increase water solubility of these molecules, meanwhile the presence of alkyl groups make them lipophilic. The most abundant form of flavonoids is the glycosidic form with different sugars linked (glucose, ramnose, etc.). If there is no bonded sugar unit, flavonoids are in aglycone form.

### *7.2. Preparation of pollen extracts for determination of phenolic compounds*

Bee pollen can be considered as a complex mixture of different compounds which differ, both, in chemical composition as well as in their solubility. The complexity of the structure of the pollen grains are also reflected in the fact that it is protected by two membranes with the main outer, exine, that consists of polymer sporopollenin. For this reason, isolation and extraction of material from pollen requires prior softening or maceration of pollen grain so that membranes

Table 32. Model of chemical structures of subclasses or representative molecule of subclasses of non-flavonoids phenolic compounds.

| Subclass | Model of chemical structure or model of representative molecule | Name |
|---|---|---|
| Lignans | 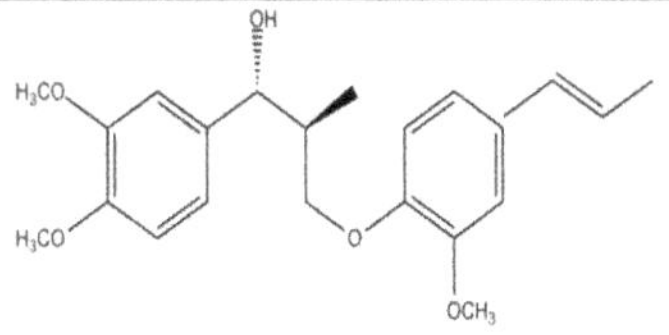 | **virolin** |
| Lignins | 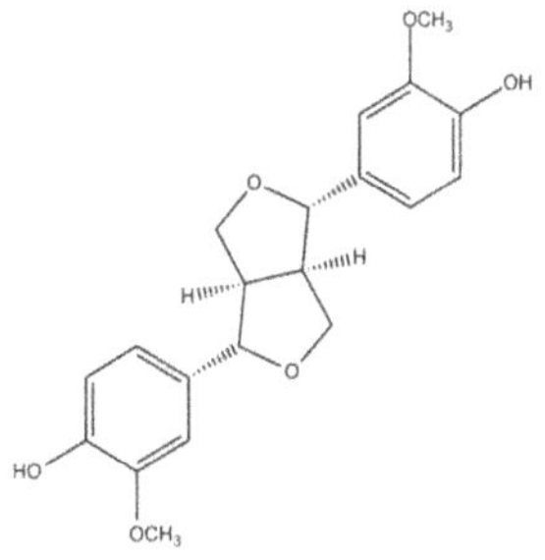 | **pinoresinol** |
| Phenolic acids | 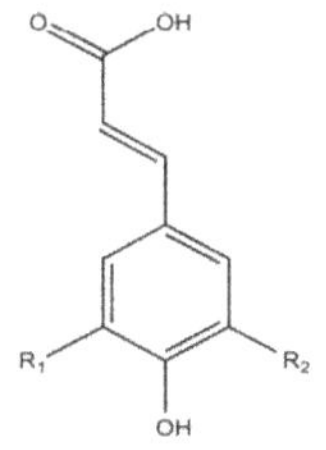 | **hydroxycinnamic acids** |
| | 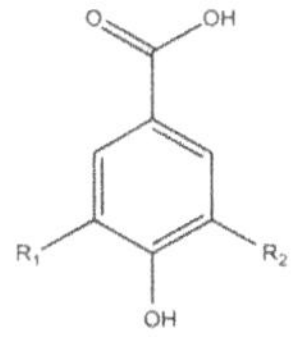 | **hydroxybenzoic acids** |
| Coumarins | 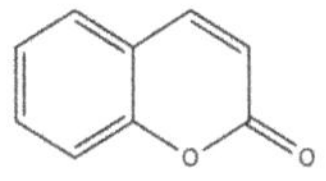 | **coumarin** |
| Stilbenes | 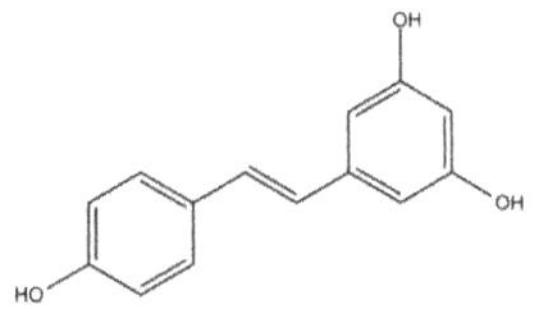 | **resveratrol** |
| Tannins | 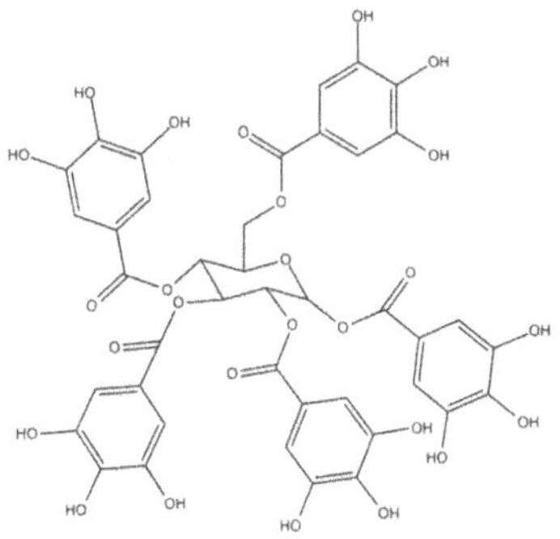 | **gallotannin** |

slightly relaxed or their physical disruption (by using ultrasound or sonification) in order to increase the extraction efficiency. Application of ultrasound (which increases penetration of solvent's molecules into the pollen matrix thereby enhance contact of solvent and the respective soluble components therein) is especially recommended due to the presence of thermolabile components in plant materials (Wang & Weller, 2006), including pollen. Generally, during the entire extraction process application of higher temperature should be avoided in order to preserve the original composition of the compounds present in the pollen

Table 33. Model of chemical structures of flavonoids.

| Subclass | Model of chemical structure |
|---|---|
| Chalcones | 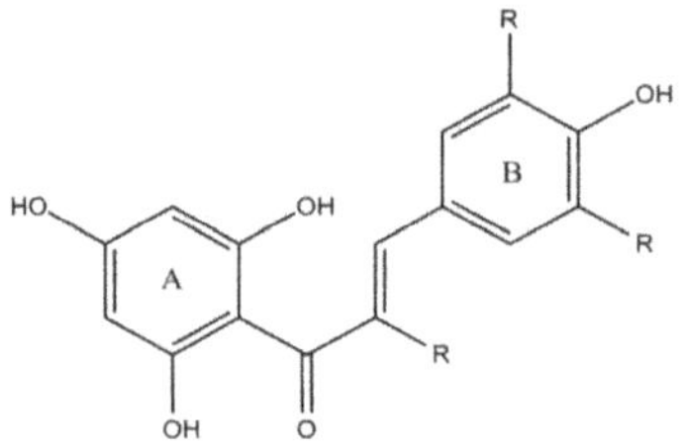 |
| Aurones | 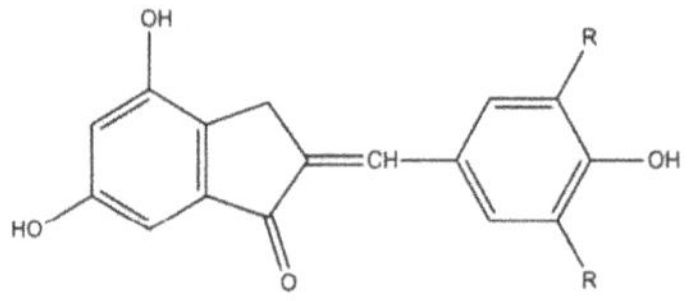 |
| Flavanols | 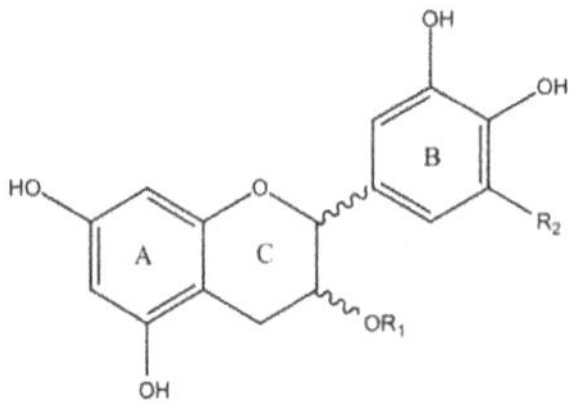 name R$_1$ R$_2$<br>Catechin H H<br>(OR$_1$/ B ring - cis)<br>Epicatechin H H<br>(OR$_1$/ B ring - trans)<br>Gallocatechin H OH<br>(OR$_1$/ B ring - cis)<br>Epigallocatechin H OH<br>(OR$_1$/ B ring - trans) |
| Flavonols | 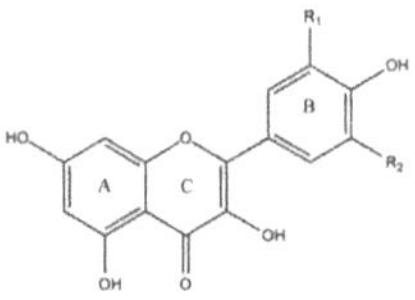 name R$_1$ R$_2$<br>Kaempeferol H H<br>Quercetin H OH<br>Myricetin OH OH<br>Isorhamnetin $OCH_3$ H |
| Flavones | 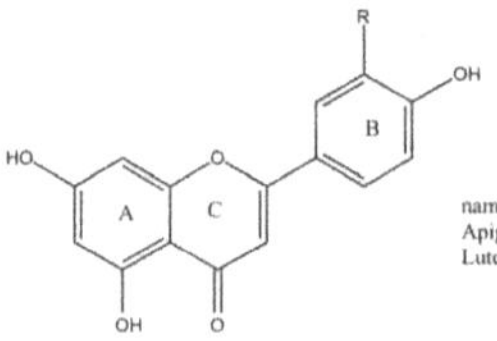 name R<br>Apigenin H<br>Luteolin OH |
| Flavanones | 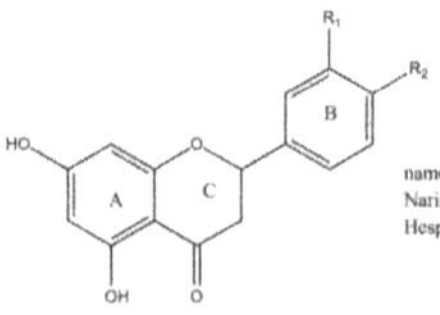 name R$_1$ R$_2$<br>Naringenin H OH<br>Hesperetin OH $OCH_3$ |

(*Continued*)

Table 33. (*Continued*).

| Subclass | Model of chemical structure |
|---|---|
| Isoflavones | 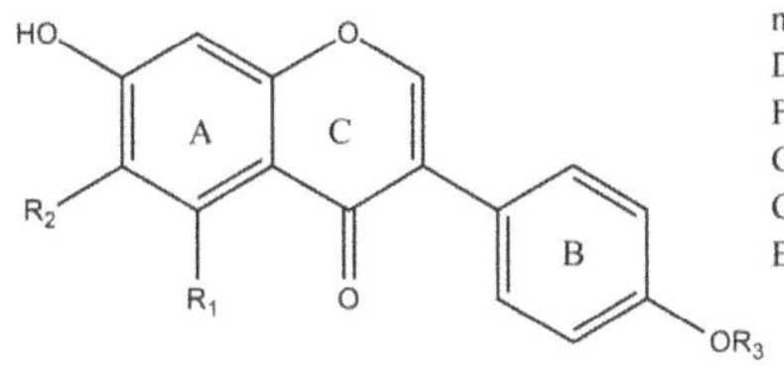<br>name R$_1$ R$_2$ R$_3$<br>Daidzein H H H<br>Formononentin H H OCH$_3$<br>Glycitein H OCH$_3$ H<br>Genistein OH H H<br>Biochanin OH H OCH$_3$ |
| Antocyanins | 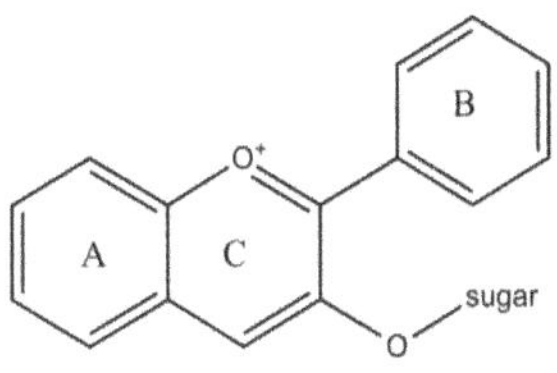 |

because some of them are volatile at elevated temperatures (lipids, phenolic compounds, etc.) or they can lose their native structure (free amino acids, proteins, polysaccharides). All phenolic compounds present in plants can be regarded as more or less water soluble (flavonoids, most of the phenolic acids, gallotannins and ellagitannins) and water insoluble (coumarine, stilbenes, proantocyanidins, some phenolic acids bonding in the cell wall, lignins). Solubility/Insolubility in water and similar solvents depends primary on the presence of side groups attached to the aromatic rings.

*7.2.1. Extraction of phenolic compounds from pollen – general procedure*

Based on the aforementioned the following procedure for the extraction of phenolic compounds from pollen can be proposed:

1. Measure the appropriate mass of pollen sample (e.g., 0.5 g);
2. Add 20 cm$^3$ (calculated for 0.5 g of pollen) of heptane in order to remove lipid components;
3. Shake thoroughly sample on Vortex and place it in an ultrasound bath for 15 min at 30 °C;
4. Centrifuge sample (5000 rpm) to separate non-polar, lipid fraction;
5. Add 20 cm$^3$ (calculated for 0.5 g of pollen) of 50% ethanol on the residual undissolved portion of pollen and place it in an ultrasound bath for 15 min at 30 °C;
6. Centrifuge sample (5000 rpm) in order to separate liquid and solid phases;
7. Evaporate ethanol from supernatant by vacuum-rotary evaporator at temperature of 35 °C;
8. Add ethyl acetate to the remaining aqueous solution (volumes of solution and extraction solvent should be equal) in order to perform the extraction of the phenolic compounds and their separation from the remaining organic substances (proteins and sugars);

A phenolic fraction, obtained in this way, is prepared for further concentration using the solid-phase extraction (SPE) method.

*7.2.2. Preparation of enzymatic hydrolysates from pollen*

Besides the common extracts for determination of phenolic profile of pollen, some authors (Rzepecka-Stojko, Pilawa et al., 2012; Rzepecka-Stojko, Stec et al., 2012) have prepared two types of enzymatic hydrolysates from pollen: pepsin hydrolysates from pollen (PEP) and ethanol extracts of pepsin-digested bee pollen (EEPP). On the other hand, Smirnova et al. (2012) prepared enzymatic hydrolysate of pollen's exine by using feruloyl esterase.

***7.2.2.1. Preparation of PEP.***

1. Measure 20 g of pollen samples.
2. Add distilled water (volume of distilled water should be five times the volume of sample) and acidify with conc. HCl until reach pH 2.
3. Add pepsin to obtain its concentration of 1%.
4. Incubate sample at 37 °C for 48 h to perform hydrolysis.
5. Boil sample for 10 min to stop enzymatic activity.
6. Filter the obtained enzymatic hydrolysate under reduced pressure.
7. Centrifuge the filtrate for 10 min at 10,000 rpm.
8. Evaporate the supernatant at rotary vacuum-evaporator to yield dry extract.

Table 34. Some publications on LC and LC/MS analysis of polyphenolic compounds from bee pollen samples.

| Compounds | Method | Stationary phase | Mobile phase A | Mobile phase B | t (°C) | Flow rate (mL/min) | λ (nm) | References |
|---|---|---|---|---|---|---|---|---|
| **Flavonoids** | HPLC-PAD | C18 11.9 cm × 0.4 cm, 5 μm | Orthophosphoric acid pH 2.5 | Acetonitrile | 24 | 0.8 | 220-400 | Campos et al. (1997) |
| **Flavonoids (aglycons)** | HPLC-DAD | C18 4.6 mm × 250 mm, 10 μm | Orthophosphoric acid pH 2.6 | Methanol | | 2 | 278-282 278-350 | Serra-Bonvehí et al. (2001) |
| **Phenolic acids, flavonoids** | Nano-LC-UV | C18 100 μm I.D. × 10 cm | 0.5% formic acid | Methanol | | 500 nl/min | 200, 280 | Fanali et al. (2013) |
| **Anthocyanins** | HPLC-DAD-ESI-MS/MS | C18 150 mm × 4.6 mm, 5 μm | 0.1% trifluoracetic acid | Acetonitrile | 35 | 0.5 | 520 | Di Paola-Naranjo et al. (2004) |
| **Flavonoids** | HPLC-PAD-ESI-MS/MS | C18 250 mm × 4 mm, 5 μm | 1% formic acid | Methanol | | 1 | 350, 520 | Ferreres et al. (2010) |
| **Flavonoids, hydroxycinnamic acid amide derivatives** | HPLC-PAD-ESI-MS/MS | C18 250 mm × 4.6 mm, 5 μm | 0.1% acetic acid | Methanol | 28 | 0.5 | 270, 350 | Negri et al. (2011) |
| **Flavonoid aglycones** | HPLC-DAD-APCI/MS | C8 150 mm × 4.6 mm, 5 μm | 0.1% formic acid in 5% acetonitrile | Acetonitrile | 30 | 1 | 258 | Lv et al. (2015) |

Table 35. Mass spectral data of compounds detected in extracts of bee pollen.

| No | $[M+H]^+$, m/z | $[M-H]^-$, m/z | MS/MS positive ion mode | MS/MS negative ion mode | Proposed identification | References |
|---|---|---|---|---|---|---|
| 1 | 685.0 | | | | Rosmarinic acid dihexoside derivative | Negri et al. (2011) |
| 2 | 627.2 | | 481.0, 319.0 | | Myricetin-3-*O*-rhamnosyl-glucoside | Negri et al. (2011) |
| 3 | 435.0 | | 303.0 | | Quercetin-3-*O*-arabinoside | Negri et al. (2011) |
| 4 | 641.2 | | 479.0, 317.0 | | Isorhamnetin-3-*O*-diglucoside | Negri et al. (2011) |
| 5 | 595.0 | | 449.0, 287.0 | | Kaempferol-3-*O*- rhamnosyl-glucoside | Negri et al. (2011) |
| 6 | 771.2 | | 625.1, 479.1, 317.1 | | Isorhamnetin-3-*O*-(2'',3''-dirhamnosyl)glucoside | Negri et al. (2011) |
| 7 | 611.1 | | 303.1 | | Rutin | Negri et al. (2011) |
| 8 | 625.2 | | 479.0, 317.0 | | Isorhamnetin-3-*O*- rhamnosyl-glucoside | Negri et al. (2011) |
| 9 | 449.0 | | 287.0 | | Kaempferol-3-*O*-glucoside | Negri et al. (2011) |
| 10 | 641.2 | | 495.0, 333.0 | | Patuletin-3-*O*- rhamnosylglucoside | Negri et al. (2011) |
| 11 | 449.2 | | 303.1 (100%) | | Quercitrin | Negri et al. (2011) |
| 12 | 381.4 | | 261.0, 235.2, 147.2 | | Chalcone | Negri et al. (2011) |
| 13 | 761.3 | | 615.3, 381.1, 287.0 | | Kaempferol-7-*O*-rhamnosyl-3-*O*-galloyl glucuronide | Negri et al. (2011) |
| 14 | 584.3 | | 438.2, 420.2 | | N' ,N'' ,N'''-*tris*-*p*-coumaroylspermidine | Negri et al. (2011) |
| 15 | 674.3 | | 498.2 , 480.3 | | N' ,N'' ,N'''-*tris*-*p*-feruloylspermidine | Negri et al. (2011) |
| 16 | 465 | | 303 | | Delphinidin-3-*O*-glucoside | Di Paola-Naranjo et al. (2004) |
| 17 | 611 | | 303, 465 | | Delphinidin-3-*O*-rutinoside | Di Paola-Naranjo et al. (2004) |
| 18 | 449 | | 287 | | Cyanidin-3-*O*-glucoside | Di Paola-Naranjo et al. (2004) |
| 19 | 595 | | 287, 449 | | Cyanidin-3-*O*-rutinoside | Di Paola-Naranjo et al. (2004) |
| 20 | 479 | | 317 | | Petunidin-3-*O*-glucoside | Di Paola-Naranjo et al. (2004) |
| 21 | 625 | | 317, 479 | | Petunidin-3-*O*-rutinoside | Di Paola-Naranjo et al. (2004) |
| 22 | 609 | | 301, 463 | | Peonidin-3-*O*-rutinoside | Di Paola-Naranjo et al. (2004) |
| 23 | 639 | | 331, 493 | | Malvidin-3-*O*-rutinoside | Di Paola-Naranjo et al. (2004) |
| 24 | 535 | | 287, 449 | | Cyanidin-3-(6''-malonylglucoside) | Di Paola-Naranjo et al. (2004) |
| 25 | | 625 | | 463, 445, 300 | Quercetin-3- sophoroside | Ferreres et al. (2010) |
| 26 | | 623 | | 315 | Isorhamnerin-3- rutinoside | Ferreres et al. (2010) |
| 27 | | 609 | | 463, 445, 300 | Quercetin-3- neohesperidoside | Ferreres et al. (2010) |
| 28 | | 609 | | 447, 429, 285 | Kaempferol-3- sophoroside | Ferreres et al. (2010) |
| 29 | | 739 | | 593, 575, 284 | Kaempferol-3-(4- rhamnosyl)-neohesperidoside | Ferreres et al. (2010) |
| 30 | | 593 | | 447, 429, 284 | Kaempferol- 3- neohesperidoside | Ferreres et al. (2010) |
| 31 | | 635 | | 593, 489, 471, 284 | Kaempferol-3-(acetyl)-neohesperidoside | Ferreres et al. (2010) |
| 32 | | 739 | | 593, 285 | Kaempferol-3- neohesperidosyl-7- rhamnoside | Ferreres et al. (2010) |
| 33 | | 447 | | 285 | Kaempferol-3- glucoside | Ferreres et al. (2010) |
| 34 | | 593 | | 285 | Kaempferol-3- rutinoside | Ferreres et al. (2010) |
| 35 | | 635 | | 593, 489, 471, 284 | Kaempferol- 3- (acetyl)-neohesperidoside isomer | Ferreres et al. (2010) |

9. Complete drying in a laboratory incubator (desiccator) at 38 °C.
10. Measure the dry extract and dissolve it in distilled water to prepare PEP-solution with concentration of 2 mg/cm$^3$.

#### 7.2.2.2. Preparation of EEPP.

1. Extract the supernatant obtained after pepsin-treatment of pollen with 200 cm3 of 50% ethanol for 60 min at room temperature with intensive shaking.
2. Filter the extract under reduced pressure.
3. Centrifuge the filtrate for 10 min at 10,000 rpm.
4. Evaporate the supernatant at rotary vacuum-evaporator to yield dry extract.
5. Complete drying in a laboratory incubator (desiccator) at 38 °C.
6. Measure the dry extract and dissolve in 50% ethanol solution to prepare EEPP-solution with concentration of 2 mg/cm$^3$.

**7.2.2.3. Preparation of feruloyl esterase hydrolysate from pollen's exine.** For the preparation of this type of hydrolysate, Smirnova et al. (2012) used enzyme feruloyl esterase derived from the mold of genus *Aspergillus* sp. according to the following procedure:

1. Incubate the exine of pollen in acetate buffer solution pH 4.5 containing enzyme feruloyl esterase in quantity of 5 g/dm3 for 24 h at 50 °C;
2. Thoroughly wash the digested material with distilled water;
3. Incubate in 1% sodium-dodecylsulfate (SDS) solution;

Table 36. Primer Sequences with indexes SA501 – SB712 (adapted from Kozich et al., 2013).

| Forward | |
|---|---|
| Name | Sequence |
| SA501 | AATGATACGGCGACCACCGAGATCTACAC **ATCGTACG** CCTGGTGCTG GT ATGCGATACTTGGTGTGAAT |
| SA502 | AATGATACGGCGACCACCGAGATCTACAC **ACTATCTG** CCTGGTGCTG GT ATGCGATACTTGGTGTGAAT |
| SA503 | AATGATACGGCGACCACCGAGATCTACAC **TAGCGAGT** CCTGGTGCTG GT ATGCGATACTTGGTGTGAAT |
| SA504 | AATGATACGGCGACCACCGAGATCTACAC **CTGCGTGT** CCTGGTGCTG GT ATGCGATACTTGGTGTGAAT |
| SA505 | AATGATACGGCGACCACCGAGATCTACAC **TCATCGAG** CCTGGTGCTG GT ATGCGATACTTGGTGTGAAT |
| SA506 | AATGATACGGCGACCACCGAGATCTACAC **CGTGAGTG** CCTGGTGCTG GT ATGCGATACTTGGTGTGAAT |
| SA507 | AATGATACGGCGACCACCGAGATCTACAC **GGATATCT** CCTGGTGCTG GT ATGCGATACTTGGTGTGAAT |
| SA508 | AATGATACGGCGACCACCGAGATCTACAC **GACACCGT** CCTGGTGCTG GT ATGCGATACTTGGTGTGAAT |
| SB501 | AATGATACGGCGACCACCGAGATCTACAC **CTACTATA** CCTGGTGCTG GT ATGCGATACTTGGTGTGAAT |
| SB502 | AATGATACGGCGACCACCGAGATCTACAC **CGTTACTA** CCTGGTGCTG GT ATGCGATACTTGGTGTGAAT |
| SB503 | AATGATACGGCGACCACCGAGATCTACAC **AGAGTCAC** CCTGGTGCTG GT ATGCGATACTTGGTGTGAAT |
| SB504 | AATGATACGGCGACCACCGAGATCTACAC **TACGAGAC** CCTGGTGCTG GT ATGCGATACTTGGTGTGAAT |
| SB505 | AATGATACGGCGACCACCGAGATCTACAC **ACGTCTCG** CCTGGTGCTG GT ATGCGATACTTGGTGTGAAT |
| SB506 | AATGATACGGCGACCACCGAGATCTACAC **TCGACGAG** CCTGGTGCTG GT ATGCGATACTTGGTGTGAAT |
| SB507 | AATGATACGGCGACCACCGAGATCTACAC **GATCGTGT** CCTGGTGCTG GT ATGCGATACTTGGTGTGAAT |
| SB508 | AATGATACGGCGACCACCGAGATCTACAC **GTCAGATA** CCTGGTGCTG GT ATGCGATACTTGGTGTGAAT |
| Reverse | |
| Name | Sequence |
| SA701 | CAAGCAGAAGACGGCATACGAGAT **AACTCTCG** AGTCAGTCAG CC TCCTCCGCTTATTGATATGC |
| SA702 | CAAGCAGAAGACGGCATACGAGAT **ACTATGTC** AGTCAGTCAG CC TCCTCCGCTTATTGATATGC |
| SA703 | CAAGCAGAAGACGGCATACGAGAT **AGTAGCGT** AGTCAGTCAG CC TCCTCCGCTTATTGATATGC |
| SA704 | CAAGCAGAAGACGGCATACGAGAT **CAGTGAGT** AGTCAGTCAG CC TCCTCCGCTTATTGATATGC |
| SA705 | CAAGCAGAAGACGGCATACGAGAT **CGTACTCA** AGTCAGTCAG CC TCCTCCGCTTATTGATATGC |
| SA706 | CAAGCAGAAGACGGCATACGAGAT **CTACGCAG** AGTCAGTCAG CC TCCTCCGCTTATTGATATGC |
| SA707 | CAAGCAGAAGACGGCATACGAGAT **GGAGACTA** AGTCAGTCAG CC TCCTCCGCTTATTGATATGC |
| SA708 | CAAGCAGAAGACGGCATACGAGAT **GTCGCTCG** AGTCAGTCAG CC TCCTCCGCTTATTGATATGC |
| SA709 | CAAGCAGAAGACGGCATACGAGAT **GTCGTAGT** AGTCAGTCAG CC TCCTCCGCTTATTGATATGC |
| SA710 | CAAGCAGAAGACGGCATACGAGAT **TAGCAGAC** AGTCAGTCAG CC TCCTCCGCTTATTGATATGC |
| SA711 | CAAGCAGAAGACGGCATACGAGAT **TCATAGAC** AGTCAGTCAG CC TCCTCCGCTTATTGATATGC |
| SA712 | CAAGCAGAAGACGGCATACGAGAT **TCGCTATA** AGTCAGTCAG CC TCCTCCGCTTATTGATATGC |
| SB701 | CAAGCAGAAGACGGCATACGAGAT **AAGTCGAG** AGTCAGTCAG CC TCCTCCGCTTATTGATATGC |
| SB702 | CAAGCAGAAGACGGCATACGAGAT **ATACTTCG** AGTCAGTCAG CC TCCTCCGCTTATTGATATGC |
| SB703 | CAAGCAGAAGACGGCATACGAGAT **AGCTGCTA** AGTCAGTCAG CC TCCTCCGCTTATTGATATGC |
| SB704 | CAAGCAGAAGACGGCATACGAGAT **CATAGAGA** AGTCAGTCAG CC TCCTCCGCTTATTGATATGC |
| SB705 | CAAGCAGAAGACGGCATACGAGAT **CGTAGATC** AGTCAGTCAG CC TCCTCCGCTTATTGATATGC |
| SB706 | CAAGCAGAAGACGGCATACGAGAT **CTCGTTAC** AGTCAGTCAG CC TCCTCCGCTTATTGATATGC |
| SB707 | CAAGCAGAAGACGGCATACGAGAT **GCGCACGT** AGTCAGTCAG CC TCCTCCGCTTATTGATATGC |
| SB708 | CAAGCAGAAGACGGCATACGAGAT **GGTACTAT** AGTCAGTCAG CC TCCTCCGCTTATTGATATGC |
| SB709 | CAAGCAGAAGACGGCATACGAGAT **GTATACGC** AGTCAGTCAG CC TCCTCCGCTTATTGATATGC |
| SB710 | CAAGCAGAAGACGGCATACGAGAT **TACGAGCA** AGTCAGTCAG CC TCCTCCGCTTATTGATATGC |
| SB711 | CAAGCAGAAGACGGCATACGAGAT **TCAGCGTT** AGTCAGTCAG CC TCCTCCGCTTATTGATATGC |
| SB712 | CAAGCAGAAGACGGCATACGAGAT **TCGCTACG** AGTCAGTCAG CC TCCTCCGCTTATTGATATGC |
| Index and Read | |
| Name | Sequence |
| Read1 | CCTGGTGCTG GT ATGCGATACTTGGTGTGAAT |
| Read2 | AGTCAGTCAG CC TCCTCCGCTTATTGATATGC |
| Index | GCATATCAATAAGCGGAGGA GG CTGACTGACT |

Index sequences indicated in bold.

4. Wash again with distilled water;
5. Dry at 55 °C until constant mass has been reached.

*7.2.3. Application of SPE for pre-concentration of phenolic compounds for further LC and LC/MS determination*

Solid-phase extraction (Figure 37) is one of the methods of sample preparation for the further analysis. It was developed during the 1980s. Currently, SPE is applied for pre-concentration of samples or for their isolation and purification (Bertoncelj et al., 2011; Gašić et al., 2014a; 2014b; Kečkeš et al., 2013; Santana et al., 2009). It is based on interaction of some components from liquid phase with solid (sorbent) phase. The desired compound retains on column, meanwhile, all the unwanted components have been washed and removed with rinsing of column in convenient medium. The analyte is than recovered with small quantity of appropriate solvent. In this way, apart from the isolation of the substance, its pre-concentration is achieved which facilitate

Table 37. International data on the protein content (% D.M) of bee pollen.

| Crude protein content (% D.M) | | | |
|---|---|---|---|
| Type of samples | Number of samples | Values | References |
| Multifloral | 21 | 21.0–29.3 | Herbert & Shimanuki, 1978 |
| Multifloral | 20 | 12.6–18.2 | Serra-Bonvehí & Escolá Jordá, 1997 |
| Monofloral | 194 | 9.2–37.4 | Somerville, 2001 |
| Multifloral | 15 | 9.8–16.5 | Villanueva et al., 2002 |
| Monofloral | 29 | 13.6–31.9 | Andrada & Telleria, 2005 |
| Multifloral | 10 | 21 ± 4* | Almeida-Muradian et al., 2005 |
| Multifloral | 6 batches | 23.59** | Melo et al., 2005 |
| Monofloral | 123 | 13.0–24.5 | Szczêsna, 2006a |
| Multifloral (Poland) | 13 | 15.8–24.1 | Szczêsna, 2006b |
| Multifloral (Korea) | 9 | 17.6–24.5 | Szczêsna, 2006b |
| Multifloral (China) | 5 | 17.8–26.1 | Szczêsna, 2006b |
| Monofloral | 42 | 6.3–26.3 | Forcone et al., 2011 |
| Multifloral | 154 | 12.2–27.0 | Martins et al., 2011 |
| Multifloral | 14 | 18.4–22.4 | Balkanska & Ignatova, 2012 |
| Multifloral | 7 | 15.2–28.5 | Stanciu et al., 2012 |
| Monofloral | 12 | 14.2–28.9 | Yang et al., 2014 |
| Multifloral | 7 batches | 23.38 ± 1.24* | (Arruda et al., 2013) |
| Multifloral | 106 | 13.65–26.5 | (Liolios et al., 2014) |
| Monofloral | 58 | 9.7–30.11 | (Liolios et al., 2014) |

*Average and standard deviation.
**Average.

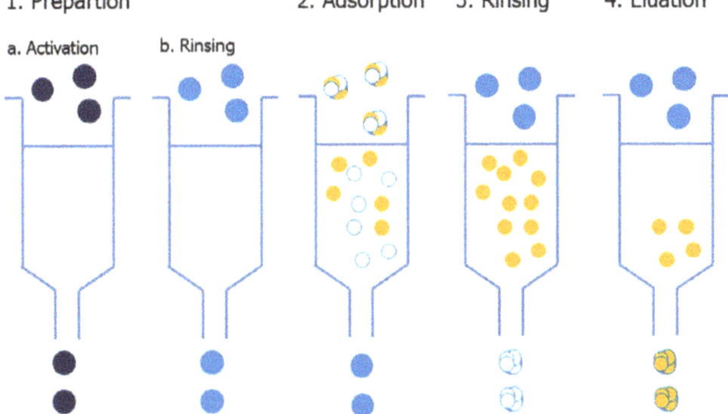


Figure 37. Solid phase extraction.

further analysis. Also, the use of large amounts of organic solvents for extraction can be avoided (Rouessac & Rouessac, 2007). This is important due to three reasons – medical, environmental and economic.

Typical solid phase extraction involves several steps (Rouessac & Rouessac, 2007) (Figure 37):

1. activation and rinsing of the sorbent before using it – performed by passing a certain volume of solvent similar polarity as well as samples that will be loaded to the cartridge;
2. adsorption of the analyte by loading of known volume of sample;
3. elimination of all interfering compounds by rinsing cartridge with adequate solvent;
4. desorption of an analyte by percolation of cartridge.

The solid phase in cartridge is made of similar materials like solid phase in LC-columns depending of type of analyte. A number of sorbents may be used such as silica gel, cellulose, aluminum oxide, polyamide and bonded silica gels (Milojković-Opsenica et al., 2015). Polar compounds are isolated from cartridge with non-polar matrix. If the analyzed sample contains non-polar components, they can be isolated from cartridge with polar matrix made of C-18 derivatized silica gel. This gel has attached a hydrocarbonic $C_{18}$-chain which increases its hydrophobicity. They were the first type of sorbents applied for SPE pre-concentration of phenolic compounds from water samples (Mußmann et al., 1994; Rodríguez et al., 2000; Tian et al., 2005). Interactions between phenolic compounds and this type of sorbent, based on Van der Waal's forces, can be weak if phenols are in ionized states, so it is necessary to adjust the pH at the proper value. Also, since silica materials are sensitive on pH value, it is possible to substitute this type of sorbent with some other, more polar matrices such as modified polystyrene-divinylbenzene polymeric (XAD) resins (Fontanals et al., 2007; Li et al., 2002; Wissiak & Rosenberg, 2002). The procedure is similar to those applied for C-18 derivatized silica gel but adsorption area is larger and presence of some aromatic parts in polymeric structure made it more suitable for phenols extraction (Santana et al., 2009).

A typical cartridge SPE manifold can accommodate up to 24 cartridges. Most SPE manifolds are equipped with exit for connection to vacuum or water pump. Application of reduced pressure speeds up the extraction process by pulling the liquid sample through the stationary phase. The analytes are collected in sample tubes inside or below the manifold after they pass through the stationary phase. Before the loading on the cartridge samples are usually passed through a disk for extraction −0.5 mm thickness and 25 to 90 mm in diameter (Rouessac & Rouessac, 2007). It serves to remove the low quantity of organic substances which are possibly present in the tested aqueous solution.

***7.2.3.1. Application of SPE for pre-concentration of phenolic compounds from pollen.*** The literature for pre-concentration of phenolic compounds from pollen by SPE is very limited (Di Paola-Naranjo et al., 2004). According to our laboratory practice and experience the following procedure for pre-concentration of phenolic compounds from pollen phenolic fraction prepared by general procedure can be proposed:

1. Condition and activate the SPE cartridge by passing 10 cm3 of methanol and 10 cm3 of 0.5% hydrochloric acid, successively. Do not allow air to enter the cartridge packing;
2. Load 2 cm3 of pollen phenolic fraction by passing slowly (no faster than one drop per second) into the cartridge. Do not allow air to enter in to the resin bed. Collect all cartridge effluent in clean tubes;
3. Wash the cartridge by passing 10 cm3 of 0.5% HCl (for purpose of elimination of potentially present sugars or some other polar substances). Collect all cartridge effluent in clean tubes;
4. Eluate the phenolic compounds by slowly passing (less than one drop per sec) of 5 cm3 of acidic solution of methanol (add 0.5% HCl). Do not allow air to enter into the resin bed. Collect phenolic solution in clean tubes;
5. Tap tubes and pack it into the refrigerator or freezer until further analysis.

### 7.3. LC and LC/MS techniques

#### 7.3.1. LC techniques

High-performance liquid chromatography or high-pressure liquid chromatography (HPLC) is highly improved form of column chromatography. It is based on the principle of forcing solvent through column, packed with stationary phase consisting of small particles, under high pressures (up to 400 atmospheres). Nano-liquid chromatography method with UV-Vis detection is also developed for analysis of phenolic compounds in bee pollen (Fanali et al., 2013). The main advantages of this method are short analysis time and reduced costs due to the small dimensions of the column.

***7.3.1.1. LC-DAD technique.*** The photodiode array detector has been extensively used for the detection of polyphenolics due to their absorption in the UV region (Robards & Antolovich, 1997). Detection is based on the absorptive measurement at characteristic wavelengths. Phenolic acids are usually detected at wavelengths between 210 and 320 nm, anthocyanidins and anthocyanins (glycosides or acylglycosides of anthocyanidins) between 500 and 530 nm, flavanols between 210 or 280 nm, flavanones at 280 nm, and possibly at 252, 285, 290 and 365 nm, flavones at 360 or 370 nm, flavonols and their glycosides at 270, 365, and 370 nm and isoflavones at 260 nm, and possibly between 230 and 280 nm. UV spectra of flavonoids, for example, have two absorption bands. Band I, with maximum absorption in the range of 300 to 550 nm, arises from the A aromatic ring and band II, with maximum absorption in the range of 240 to 285 nm, comes from the B ring. In LC-DAD system, polyphenolic standards are used to generate characteristic UV-Vis spectra and retention times. Identification of individual polyphenolic compounds is achieved by comparison with retention times and UV-Vis spectra of standards. Quantification of these compounds is performed from the peak areas by comparison to calibration curves obtained with standard solutions. As general procedure for determination of polyphenolic compounds from bee pollen extracts can be HPLC-DAD method developed by Campos et al. (1997).

***7.3.1.2. LC-MS, MS/MS techniques.*** Although HPLC-DAD method is the reference method for the analysis of phenolic compounds, certain problems, like co-elution and similar UV-Vis spectra of many compounds, can occur. As well, most glycosides and acyl residues are poor chromophores, so no further distinguishing can be achieved by means of DAD UV detection. In that case mass spectrometry (MS), more precisely, tandem mass spectrometry (MS/MS) detection is suggested for further characterization and identification of eluted compounds.

The MS principle consists in ionizing chemical compounds to generate charged molecules or molecule fragments and measuring their mass-to-charge ratios (Sparkman, 2000). LC-MS provides structural information about eluted compounds but is rarely used for full structure characterization. Using LC-MS/MS more fragmentation of the precursor and product ions is formed, providing additional structural information for the identification of phenolic compounds like details about the aglycone moiety, the types of carbohydrates (mono-, di-, tri- or tetrasaccharides and hexoses, deoxyhexoses or pentoses) or other substituents present, interglycosidic linkages, attachment points of the substituents to the aglycone, etc. Aforementioned useful structural information could be obtained directly and easily applying electrospray ionization (ESI) and atmospheric pressure chemical ionization (APCI) techniques, and so, many purification procedures can be circumvented. Both are the soft ionization techniques and the most widely used in analysis of polyphenolic compounds in bee pollen and other food. ESI is gentler method and is generally used for polar compounds that can be ionized in solution, while APCI is used for less polar molecules that can undergo acid-based reactions in the gas phase (Biesaga & Pyrzynska, 2009). Both positive and negative ionization are applied. Negative ion mode provides the highest sensitivity in flavonoid analysis and fragmentation is limited (Cuyckens & Claeys, 2002; de Rijke et al., 2003; Fabre et al., 2001; Wang et al., 2007), while mass spectrum in the positive ion mode shows different fragmentation pattern. Analyte responses can vary considerably so combined use of both EI and PI modes can prove complementary information's which are helpful in the structure determination, especially for the minor compounds (Cuyckens & Claeys, 2004; de Rijke et al., 2006). The eluent system also has a distinct influence on the ionization efficiency and therefore on structural information. The most common additives are formic acid, trifluoroacetic acid, ammonium acetate and ammonium formate. They are volatile and thus compatible with LC-MS. Combination of methanol-ammonium formate at pH 4 gives the highest response in negative mode APCI (de Rijke et al., 2003), while the best sensitivity in ESI is achieved with acidic ammonium acetate buffer and organic component consisted of methanol and acetonitrile as mobile phase (Rauha et al., 2001). Formic acid has slight negative effects on the ionization efficiency but improves chromatographic separation and peak shapes (Prokudina et al., 2012). For trifluoroacetic acid is reported that suppresses the ionization due to ion-pairing and surface-tension effects (Costa et al., 2000).

Chromatographic retention times provide useful structural information. In reversed-phase HPLC separation of polyphenolics on the basis of their polarity, the elution order of polyphenolics may be predicted Glycosylation increases polarity of flavonoids and therefore their mobility in the reversed-phase system. Acylation and methylation have the opposite effect but their position have significant influence on the retention time (Greenham et al., 1995; Harborne & Boardley, 1984). In case of flavonoid glycosides rutinosides precede neohesperidosides, galactosides precede glucosides (Robards & Antolovich, 1997), glucosides precede arabinosides and arabinosides precede rhamnosides (Harborne & Boardley, 1984; Schieber et al., 2002). Linkage position can have an influence on the retention, too. The elution order for different classes of flavonoids is flavanones, flavonols and flavones, for compounds with equivalent substitution pattern. The elution order for hydroxycinnamic acids is caffeic acid, *p*-coumaric acid, sinapic acid, ferulic acid and cinnamic acid while for benzoic acids order is gallic acid, protocatechuic acid, vanillic acid and syringic acid (Rodríguez-Delgado et al., 2001).

Quality and reliability of analysis depend on the combination of the mass analyzer with the detector. Analyzers can belong to magnetic sector, time-of-flight (TOF) or Fourier transform (FT) generic type, depending on the physics of mass analysis. There are analyzers that can perform MS and MS/MS ($MS^2$) analysis, sometimes to a high $MS^n$ stage or different analyzers could be combined together into one hybrid instrument. Mass analyzers used for analyzing polyphenolic compounds are usually quadrupole and ion-trap instruments. Ion-trap instruments can perform $MS^n$ experiments and that is its main advantage (de Rijke et al., 2003). Using ion-trap mass spectrometry Ferreres et al. (2010) reported the presence of non-colored flavonoids in bee pollen from *Echium plantagineum* and Lv et al. (2015) identified and quantified flavonoid aglycones in rape bee pollen collected from Qinghai-Tibetan Plateau. In order to optimize signals and to obtain maximal structural information from ion of interest, the different collision energy values should be applied. Methods based on low-energy collision induced dissociation permit the characterization of most structural features of flavonoid glycosides. High-energy CID methods provide additional structural information through more fragmentation processes. Collision energy directly affects abundance ratio of the various fragment ions. Optimum CE varies for different compounds. Gu et al. (2012) showed that a CE of 70 eV is sufficient to cause abundant product ions for structural elucidation of flavonoids and 30 eV for generation characteristic fragment ions for chlorogenic acid (Gu et al., 2012). Recent publications on LC and LC/MS analysis of polyphenolic compounds from bee pollen samples are listed in Table 34.

In the past, fragmentation behavior of phenolic compounds, especially flavonoids, have been extensively investigated. A great number of rules for structural characterization of unknown compounds, even without the reference standards, were summarized (de Rijke et al., 2006). As experience shows, fragmentation pathways are largely independent of the ionization mode (ESI or APCI) and the type of instrument (triple quadrupole or ion trap) used (de Rijke et al., 2003; Rauha et al., 2001) so that rules can be applied for resolving spectral data obtained with modern instruments. Characteristic *m/z* values of fragment ions from mass spectral data used for identification of certain polyphenolic compounds from bee pollen are listed in Table 35.

### *7.3.2. UHPLC-HESI-MS/MS technique for determination of phenolic compounds*

The LTQ-Orbitrap spectrometer supports a wide range of applications. Its MS/MS capabilities make this mass spectrometer extremely powerful, among other things, for unambiguous detection of polyphenolic compounds (Kečkeš et al., 2013; Natić et al., 2015; Pantelić et al., 2014; Ristivojević et al., 2014). Some chromatograms and mass spectra from our previous research are shown on Figures 38–40. Instrumentation and

Table 38. Determination of the protein content (%) using different volumes of the $H_3BO_3$ solution.

| Volume of $H_3BO_3$ (ml) | Protein content (%) n = 5 |
|---|---|
| 30 | 20.49 (± 0.07)[a] |
| 40 | 20.44 (± 0.06)[a] |

Analysis performed under the same conditions except the volume of the $H_3BO_3$ solution used.

conditions for determination of phenolic compounds using this method are as follows:

1. Pollen extracts (Section 7.2);
2. Thermo Scientific liquid chromatography system equipped with:
   a. Quaternary Accela 600 pump;
   b. Accela Autosampler;
   c. Linear ion trap-Orbitrap hybrid mass spectrometer (LTQ OrbiTrap XL, Thermo Fisher Scientific, Bremen, Germany) with heated electrospray ionization (HESI);
   d. Hypersil gold C18 column (50 × 2.1 mm, 1.9 µm) from Fisher Scientific.
3. Mobile phases:
   a. water + 0.1% formic acid;
   b. acetonitrile + 0.1% formic acid.
4. Linear gradient: 0–5 min from 5% to 95% (B), 5–6 min 95% (B) then 5% (B) for 3 min;
   a. HESI-source parameters:
   b. Source voltage 4 kV;
   c. Capillary voltage −47 V;
   d. Tube lens voltage −159.11 V;
   e. Capillary temperature 275 °C;
   f. Sheath and auxiliary gas flow (N2) 25 and 8 (arbitrary units);
   g. 100–1500 m/z;
   h. Collision energy 35 eV.
5. Injection volume 5 µl;
6. Flow rate of 0.400 ml/min.

## 8. Standard method for identification of bee pollen mixtures through meta-barcoding

Pollen analysis is a central part of bee ecology research (Carvell et al., 2006; Danner et al., 2014; Köppler et al., 2007). Identification of plant species origin of bee collected pollen traditionally relies on light microscopy and discrimination based on morphological differences of pollen grains (Mullins & Emberlin, 1997). This is labor- and time-intensive (Galimberti et al., 2014), requires expert knowledge (Keller et al., 2015) and lacks discriminative power at lower taxonomic levels (Galimberti et al., 2014; Williams & Kremen, 2007), which means that pollen from closely related plant species often has to be analyzed at the family or genus level. Recently, meta-barcoding has emerged as a suitable alternative for pollen analysis (Keller et al., 2015; Kraaijeveld et al., 2015; Richardson et al., 2015; Valentini et al., 2010). However, due to a missing consensus on the best marker for plant species identification and the variety of DNA sequencing platforms available, different methods and protocols exist (e.g., Bruni et al., 2015; Galimberti et al., 2014; Keller et al., 2015; Kraaijeveld et al., 2015; Richardson et al., 2015), which makes it difficult to compare independent studies. We here present a detailed protocol of the method described recently (Sickel et al., 2015) as a research standard that is highly cost-efficient and overcomes those limitations. It is based on ITS2-meta-barcoding, which has been validated for plant barcoding (Chen et al., 2010) and for which a comprehensive database has been established (Ankenbrand, 2015). A variant of this method has been recently developed also for the *rbcl* gene, and analyzing both markers in parallel with this workflow is recommended for most applications (Bell et al., 2019). Beside the laboratory process, we also provide information on data processing and analysis.

### 8.1. Meta-barcoding protocol

#### 8.1.1. Required materials

***8.1.1.1. Reagents.*** DNA isolation kit suitable for pollen grains (e.g., Macherey-Nagel NucleoSpin Food, Düren, Germany); PCR grade water; Ethanol (96–100%); Primers as given in Table 1; Polymerase with proofreading ability including dNTPs, GC buffer and co-factors (e.g., 2 × Phusion Master Mix); Agarose, suitable buffer (e.g., TAE), intercalating dye (e.g., Midori Green Advance, Biozym Scientific GmbH, Hessisch Oldendorg, Germany), 6 × loading dye, DNA ladder (e.g., FastRuler Low Range DNA Ladder, Life Technologies, Carlsbad, CA, USA); SequalPrep™ Normalisation Kit 96 wells (Invitrogen, Carlsbad, CA, USA); Bioanalyzer High Sensitivity DNA Chip (Agilent Technologies, Santa Clara, CA, USA); dsDNA High Sensitivity Assay (Life Technologies, Carlsbad, CA, USA); MiSeq Reagent Kit v2 2 × 250 bp (Illumina Inc., San Diego, CA, USA); 1 N NaOH (stock solution); PhiX Sequencing Control v3 (Illumina Inc., San Diego, CA, USA)

***8.1.1.2. Laboratory equipment.*** Microliter pipettes and tips; Microcentrifuge tubes; Electronic pestle; Bead mill; Incubator; Vortexer; Table centrifuge; 96 well PCR plates and PCR foils; 96 well plate cooling block; 96 well plate centrifuge; Thermal cycler; Agarose gel former; microwave, gel electrophoresis chamber; UV illuminator; Bioanalyzer, chip vortexer; Qubit Fluorometer; Access to an Illumina MiSeq desktop sequencer with MiSeq Control Software version 2.2 or later.

#### 8.1.2. Pollen acquisition

Pollen sampling should be performed as described in the pollination chapter of the COLOSS *BEEBOOK* Vol. I, chapters 3.1. and 4.1.1 (Delaplane et al., 2013). For long-term storage, we recommend lyophilization before freezing at −80 °C.

#### 8.1.3. Laboratory workflow

***8.1.3.1. DNA extraction.*** For the DNA extraction step, we recommend using the Macherey-Nagel (Düren, Germany) NucleoSpin Food Kit and following the supplementary guidelines for pollen samples, but equivalent extraction procedures may also be comparable. The DNA extraction steps are as follows:

1. Take 2 g of pollen and add 4 ml bidest $H_2O$;
2. Homogenize the sample with an electronic pestle;
3. Take 200 μl (~50 mg pollen) of the emulsion and grind it in a bead mill;
4. Add 400 μl Buffer CF (preheated to 65 °C) and 10 μl Proteinase K and mix carefully;
5. Incubate at 65 °C for 30 min;
6. Centrifuge the mixture for 10 min (>10,000 × g);
7. Transfer the supernatant into a new microcentrifuge tube and add 1 vol Buffer C4 and 1 vol ethanol;
8. Vortex for 30 s;
9. Pipette 700 μl mixture onto a NucleoSpin Food Column placed in a Collection Tube;
10. Centrifuge for 1 min at 11,000 × g;
11. Discard the flow-through;
12. Repeat steps 9-1;
13. Add 400 μl Buffer CQW onto the spin column;
14. Centrifuge for 1 min at 11,000 × g;
15. Discard the flow-through;
16. Add 700 μl Buffer C5 onto the spin column;
17. Centrifuge for 1 min at 11,000 × g.
18. Discard the flow-through;
19. Add 200 μl Buffer C5 onto the spin column;
20. Centrifuge for 2 min at 11,000 × g;
21. Place the spin column into a new 1.5 ml microcentrifuge tube;
22. Add 100 μl Elution Buffer CE (pre-heated to 70 °C) onto the membrane;
23. Incubate for 5 min at room temperature (18–25 °C);
24. Centrifuge for 1 min a 11,000 × g;
25. Proceed with amplification or keep frozen until further processing.

***8.1.3.2. Amplification.*** This protocol utilizes a dual-indexing strategy (Kozich et al., 2013) amplifying the ITS2 region, using the primers ITS-S2F (Chen et al., 2010) and ITS4R (White et al., 1990), but can be adapted to amplify other markers (Bell et al., 2019). The primer sequences are as follows: forward: 5′-AATGATACGGCGACCACCGAGATCTACAC XXXXXXXX CCTGGTGCTG GT ATGCGATACTTGGTGTGAAT-3′; reverse: 5′-CAAGCAGAAGACGGCATACGAGAT XXXXXXXX AGTCAGTCAG CC TCCTCCGCTTATTGATATGC-3′, where XXXXXX indicates the variable index sequences (Table 36). The detailed protocol is described below:

1. Sample index combinations should be planned beforehand according to the scheme in Figure 41;

Table 39. Protein content (%) using different distillation times.

| Distillation time (min) | Protein content (%) n = 5 |
|---|---|
| 4 | 20.48 (± 0.07)[a] |
| 3 | 20.29 (± 0.25)[a] |
| 2 | 20.41 (± 0.04)[a] |

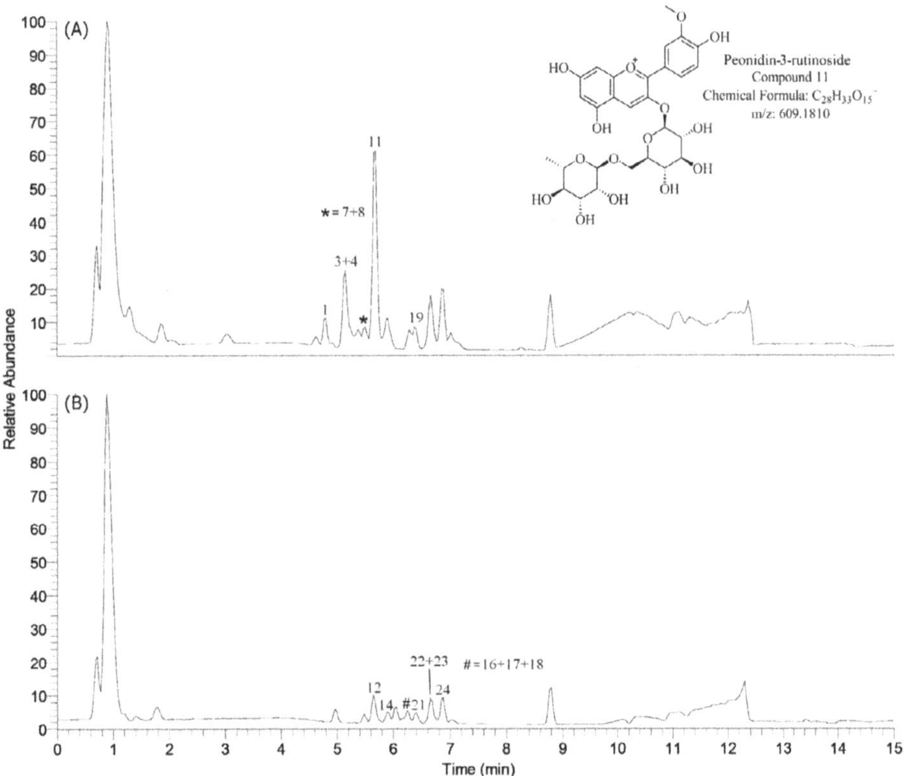


Figure 38. Base peak chromatograms of (a) cherry wine and (b) Cabernet Sauvignon-Central Serbia in positive ion mode. (Pantelić et al., 2014).

2. Prepare 3 × 10 µl reaction mixes for each sample containing (also see PCR sample design below for details):
   a. 5 µl 2 × Phusion Master Mix (New England Biolabs, Ipswich, MA, USA) or equivalent;
   b. 0.33 µM each of the forward and reverse primers (sample-specific combinations of forward and reverse index sequences);

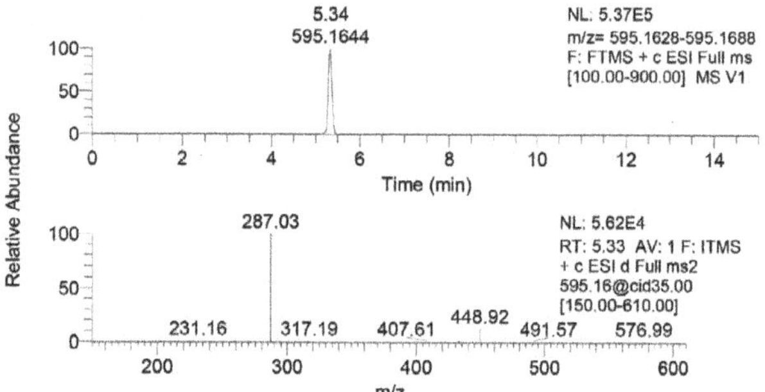


Figure 39. Extracted ion chromatograms and MS/MS spectra of cyanidin-3-rutinoside (Pantelić et al., 2014).

   c. 3.34 µl PCR grade water;
   d. 1 µl DNA template.
3. Carry out the PCR with a program of:
   a. 95 °C for 4 min, then;
   b. 37 cycles of 95 °C for 40 sec;
   c. 49 °C for 40 sec;
   d. 72 °C for 40 sec. and
   e. a final extension at 72 °C for 5 min.
4. Combine the triplicate PCR reactions of each sample and mix well.

For quality control purposes, successful amplification can be checked on a 1% agarose gel using 5 µl of the combined PCR product.

*The design used in 96-well PCR sample is:*

**Design 1:** Well-equipped laboratories with pipetting robots or 96-channel pipettes can directly fill each well with a different sample and generate three replicates of these. This will result in 4 × 3 replicate

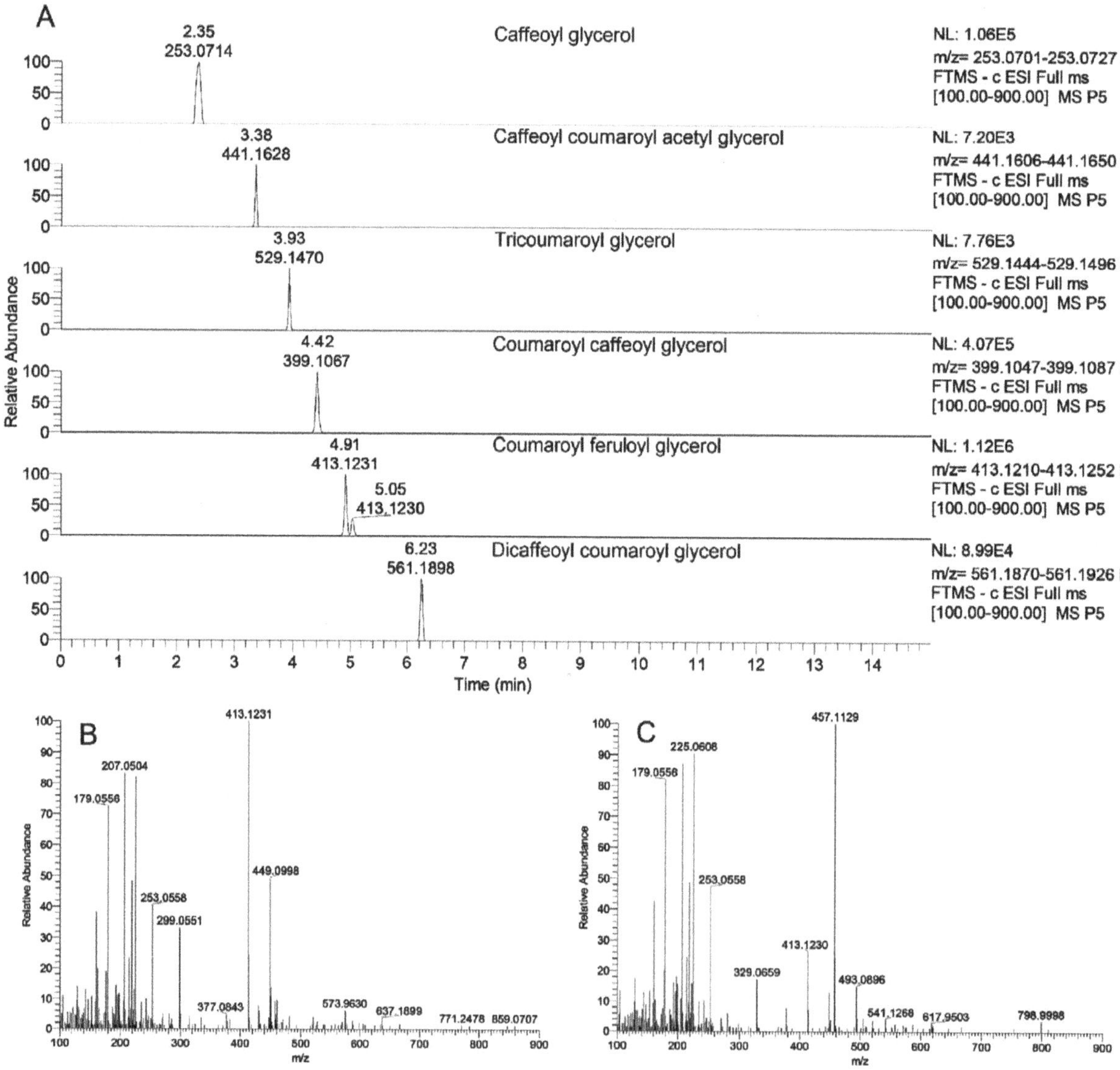

Figure 40. (a) Extracted ion chromatograms of phenolic glyceride derivatives with their retention times and accurate masses; (b and c) mass spectra of two phenolic glyceride derivatives with the same accurate masses, m/z 413 (Ristivojević et al. 2014).

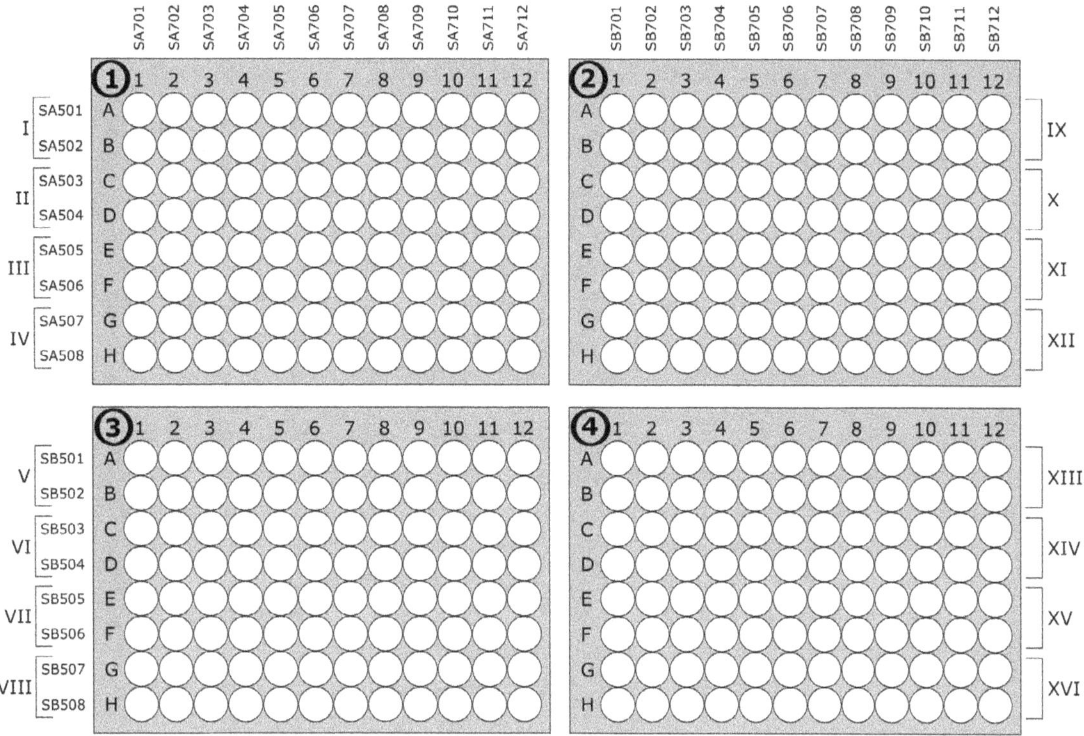


Figure 41. Planning scheme for samples and the corresponding index-combinations. Roman numbers indicate PCR plate numbers, bold Arabian numbers on 96 well plates indicate normalization plate number.

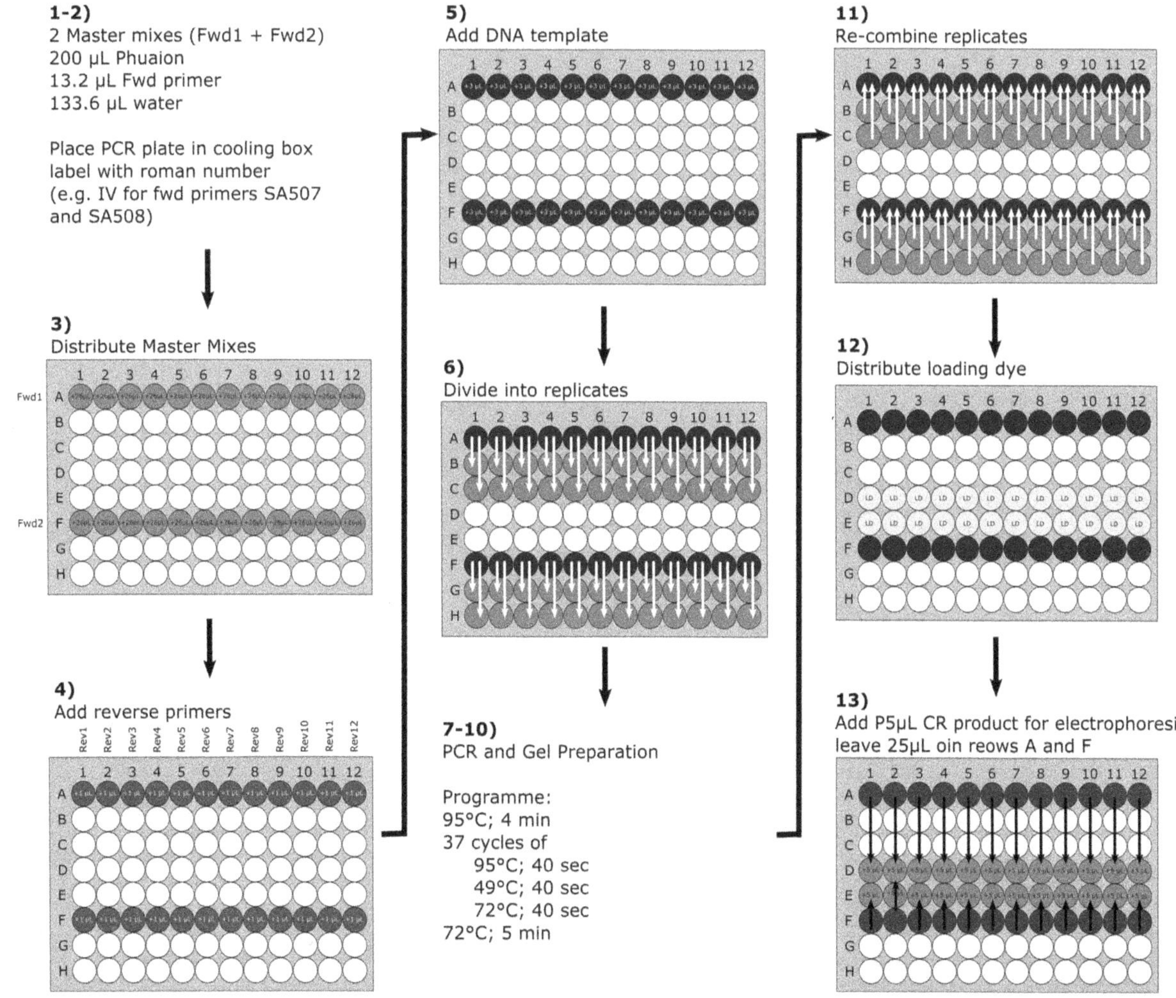

**1-2)**
2 Master mixes (Fwd1 + Fwd2)
200 µL Phuaion
13.2 µL Fwd primer
133.6 µL water

Place PCR plate in cooling box
label with roman number
(e.g. IV for fwd primers SA507 and SA508)

**3)**
Distribute Master Mixes

Fwd1
Fwd2

**4)**
Add reverse primers

Rev1 Rev2 Rev3 Rev4 Rev5 Rev6 Rev7 Rev8 Rev9 Rev10 Rev11 Rev12

**5)**
Add DNA template

**6)**
Divide into replicates

**7-10)**
PCR and Gel Preparation

Programme:
95°C; 4 min
37 cycles of
95°C; 40 sec
49°C; 40 sec
72°C; 40 sec
72°C; 5 min

**11)**
Re-combine replicates

**12)**
Distribute loading dye

**13)**
Add P5µL CR product for electrophoresis;
leave 25µL oin reows A and F

Figure 42. Detailed workflow (schematic), suitable for laboratories with limited access to equipment for automated pipetting. Bold numbers indicated step number of Design 2 in Section 8.1.3.2.

96-well plates according to Figure 41 used for amplification. After amplification one can proceed with normalization.

**Design 2:** For laboratories with little equipment for automated pipetting, the workflow described above is impractical, since manual pipetting in that format is time-intensive and pipetting errors can be easily introduced. To facilitate the process, we recommend to work with all triplicates but only 24 samples on one 96 well plate (Figure 42). This way, 16 PCR plates will be produced, but pipetting effort is minimized. PCR plate labeling is therefore of utter importance, for example with roman numbers, I – XVI to be able to map the samples back to the scheme in Figure 41. The complete workflow is shown schematically in Figure 42 and described in the following:

1. Prepare two PCR master mixes, each containing one forward primer, corresponding to the samples you want to amplify; each master mix contains:
   a. 200 µl 2 × Phusion Master Mix (New England Biolabs, Ipswich, MA, USA) or equivalent.
   b. 13.2 µl forward primer (10 µM).
   c. 133.6 µl PCR grade water.
2. Place a new PCR plate into a cooling block;
3. Distribute 26 µl of the master mixes into row A (Master Mix 1) and F (Master Mix 2);
4. Add 1 µl of the correct reverse primer;
5. Add 3 µl of the correct DNA template;
6. Using a pipette set to 10 µl, pipette up and down to mix and distribute 10 µL each into the two rows below: from row A into rows B + C; from row F into rows G + H;
7. Seal with a foil, spin down briefly;
8. Perform PCR;
9. Prepare a 1% agarose gel;
10. After PCR, briefly spin down again;
11. Lift the foil carefully and combine the triplicate reactions, pipette up and down to mix;
12. For gel electrophoresis, add 1 µl of 6× loading buffer into the so far unused rows D + E;
13. Add 5 µl PCR product to the loading buffer;
14. Briefly spin down;
15. Load the gel, add a DNA ladder;
16. Run the gel (e.g., 25 min, 120 V);
17. Check under UV illuminator for successful PCR amplification;
18. Freeze PCR product until further processing;
19. Continue with normalization.

***8.1.3.3. Normalisation.*** To ensure more equalized library sizes, DNA amounts in each PCR product are normalized using the SequalPrep™ Normalisation Kit (Invitrogen, Carlsbad, CA, USA). For 384 samples, four normalization plates are needed. After normalization, samples from each plate will be combined in "plate pools" for the following quality control.

**Design 1:** Pool the samples of all three replicates together by keeping the sample scheme. Transfer 25 µl of PCR products onto the normalization plates. Proceed with the normalization as described below.

**Design 2:** For normalization, PCR plates I – IV; V – VIII; IX – XII and XIII – XVI will be combined to normalization plates 1, 2, 3 and 4. The pipetting scheme is as follows:

1. Thaw the PCR plates.
2. Briefly spin down.
3. Use four normalization plates and add 25 µL of PCR product into the wells following this scheme.

   a. Normalization Plate 1: PCR plates I –IV
   b. *PCR plate I:* row A →row A; row F → row B
   c. *PCR plate II:* row A →row C; row F → row D
   d. *PCR plate III:* row A →row E; row F → row F
   e. *PCR plate IV:* row A →row G; row F →row H
   f. Repeat analogous for the other three normalization plates.
   g. Proceed with the normalization as described below.

**Design 1 & 2:** Continue for both designs with the normalization:

1. Add 25 µl of Binding buffer;
2. Mix by pipetting up and down or seal the plate with foil tape, vortex to mix and briefly centrifuge the plate;
3. Incubate for 1 hour at room temperature; alternatively leave to incubate overnight;
4. Aspirate liquid from wells, do not scrape the well sides;
5. Add 50 µl Wash buffer, mix by pipetting up and down;
6. Completely aspirate the buffer from wells, you may need to invert and tap the plate on paper towels;
7. Add 20 µl of Elution buffer;
8. Mix by pipetting up and down or seal the plate with foil tape, vortex and briefly spin down;
9. Incubate for 5min at room temperature;
10. Combine 5 µl of each sample (plate-wise) in a new microcentrifuge tube, mix well;
11. Prepare 1:10 dilutions of each plate pool;

***8.1.3.4. Quality control and quantification.*** Quality control is performed on a Bioanalyzer High Sensitivity DNA Chip (Agilent Technologies, Santa Clara, CA, USA) to ensure that the correct fragment size (peak at approximately 450 bp; target plus adapters) has been amplified. Additionally, libraries are quantified using the dsDNA High Sensitivity Assay on the Qubit fluorometer (both Life Technologies GmbH, Darmstadt, Germany) in order to combine the four plate pools equimolarly to the final sequencing library. We recommend preparing three independent concentration measurements per plate pool.

***Bioanalyzer:***

1. Prepare a Bioanalyzer Chip according to the protocol;
2. Allow all reagents to equilibrate to room temperature;
3. If not ready, prepare a gel-dye mix;
4. Add 15 μl of the dye concentrate (blue lid) to a gel matrix vial (red lid);
5. Vortex well and spin down, transfer to spin filter;
6. Centrifuge at 2240 × g for 10 min;
7. Protect solution from light, store at 4 °C, use within 6 weeks;
8. Put a new chip on the chip priming station;
9. Pipette 9 μl gel-dye mix into the well marked with a white "G";
10. Close the chip priming station, with the plunger at position 1ml;
11. Press plunger until held by the clip;
12. Wait for 60 s then release clip;
13. After 5 s slowly pull back the plunger to the 1 ml position;
14. Open the priming station, pipette 9 ml gel-dye mix in the wells marked with black 'G's;
15. Pipette 5 μl marker (green lid) into all sample wells and the ladder wells;
16. Pipette 1 μl of ladder (yellow lid) in the well-marked with a ladder symbol;
17. In each sample well, pipette 1 μl of sample (concentrated and diluted Plate pools) or 1 μl marker (unused wells);
18. Put the chip horizontally in the adapter and vortex for 1 min at 2400 rpm;
19. Run the chip within 5 min;
20. The samples are of sufficient quality, if the electropherograms show a single peak at approximately 450bp; this peak can be rather wide due to different lengths of the ITS2 region, a minor peak shortly after the lower marker is acceptable and corresponds to left-over primer dimers, which will not interfere with sequencing.

#### *8.1.3.5. Quantification.*

1. Measure concentrations of plate pools with the dsDNA High Sensitivity Assay on the Qubit Fluorometer.
2. Mix 1 × n μl Qubit reagent with 199 × n μl Qubit buffer (working solution).
3. For each measurement, mix 180–199 μl working solution with 1–20 μl sample.
4. Vortex and incubate at room temperature for 2 min.
5. Combine plate pools to final library equimolarly, starting with the least concentrated library of which take 20 μl.
6. Quantify the final pool and dilute to 2 nM, if final pool contains less than 2 nM proceed without dilution.

### *8.1.4. Sequencing*

Before sequencing, a sample sheet has to be prepared. This can be done at the MiSeq instrument using the Illumina Experiment Manager (IEM). However, due to the high number of samples, we recommend to prepare the sample sheet in advance, which can be done on a different computer with the IEM installed, and load it on to the MiSeq instrument when starting the sequencing procedure. Due to using custom index sequences, a new assay has to be added to the IEM, this is described in the Supplemental Material of Kozich et al. (2013).

For library dilution, we follow the Illumina Sample Preparation Guide for a 2 nM library, with some modifications. In order to increase read quality, 5% PhiX control is added to the sample library. Additionally, the reagent cassette of the sequencing kit (e.g., Illumina MiSeq Reagent Kit v2 2 × 250 bp) is spiked with the custom Read1, Read2 and index primers (for primer sequences, see Table 1).

#### *8.1.4.1. Sample library.*

1. Remove Buffer HT1 from freezer.
2. Prepare a fresh dilution of 0.15 N NaOH (less than a week old).
3. Mix 5μl of the sample library with 5 μl of 0.15 N NaOH.
4. Vortex briefly and centrifuge at 280 × g for 1 min.
5. Incubate at room temperature for 5 min.
6. Add 990 μl Buffer HT1 (10 pM library).
7. Mix 480 μl of 10 pM library and 120 μl Buffer HT1 (8 pM library).

#### *8.1.4.2. PhiX control.*

1. Thaw PhiX control at room temperature.
2. Mix 2 μl 10 nM PhiX control with 3 μl $H_2O$ (4 nM PhiX).
3. Add 5 μl 0.15 N NaOH.
4. Vortex briefly and centrifuge at 280 × g for 1 min.
5. Incubate at room temperature for 5 min.
6. Add 990 μl Buffer HT1 (20 pM PhiX).
7. Mix 375 μl of 20 pM PhiX and 225 μl Buffer HT1 (12.5 pM PhiX).
8. Mix 570 μl 8 pM library with 30 μl 12.5 pM PhiX.

#### *8.1.4.3. Preparing reagent cassette and loading the sample.*

1. Remove the reagent cassette from the freezer.
2. Rename the sample sheet to match the barcode of the reagent cassette.
3. Place in water bath, do not fill higher than maximum water line.
4. Prepare 3 μl each of Read1, Read2 and index primers in new microcentrifuge tubes.
5. Remove cassette from water bath, dry with paper towel.
6. Invert the cassette several times to mix.

7. Inspect wells, make sure all reagents are thawed and there are no precipitates.
8. Gently tap the cassette on the bench to remove air bubbles.
9. With a 1000 µl pipette tip, break the foils over wells 12–14 and well 17.
10. With a 100 µl pipette set to 75 µl, transfer the read and index primers to the following wells of the reagent cartridge: Read1 → Well 12; Index → Well 13; Read2 → Well14, mix well by pipetting up and down.
11. Load 600 µl of the spiked library to well 17.
12. Load the cassette, PR2 bottle and flow cell as prompted by the instrument.
13. Sequence.

### 8.2. Bioinformatics

#### 8.2.1. Required software

a. up to date Linux or Unix-based OS.
b. usearch, version v11 (Edgar, 2010), if necessary add location to your system PATH.
c. SeqFilter, https://github.com/BioInf-Wuerzburg/SeqFilter
d. usearch additional python scripts, https://drive5.com/python

Table 40. Protein content (%) analyzing different quantities of pollen.

| Pollen quantity (g) | Protein content (%) n = 5 |
|---|---|
| 0.5 | 19.12 (± 0.16)[b] |
| 1 | 20.59 (± 0.03)[a] |
| 2 | 20.64 (± 0.06)[a] |

Different letter shows statistically significant changes (Duncan's multiple range test, $\alpha = 0.005$).

#### 8.2.2. Classification

##### 8.2.2.1. Reference databases.

a. Reference database for direct classification

We recommend to first use a reference database that is most specific to your project design and limited in represented taxa. This helps strongly with the quality of the assignments. Examples for restrictions are species lists for the biogeographic region or specific taxonomy groups of interest. Such databases can be created with the BCdatabaser software (Keller et al., 2020). If you have unpublished sequences of your laboratory that should be considered as well, add them into this file.

a. Reference database for hierarchical classification

All remaining sequences not classified with the direct database can be additionally classified to a hierarchical taxonomic level as deep as possible. For this it is important to have a database that covers all taxonomic groups. One such database could be for ITS2 accessible under the DOI https://doi.org/10.5281/zenodo.3339028 which was also created with the BCdatabaser.

Table 42. Total lipid content (%) by using different amount of pollen for the analysis.

| Pollen amount (g) | Total lipid content % (n = 5) |
|---|---|
| 1 | 2.24 (±0.113)[a] |
| 2 | 2.87 (±0.056)[b] |
| 3 | 2.83 (±0.098)[b] |

The elution time was equal (5 h) and the petroleum ether served as solvent. Different Latin letter indicates statistically significant differences, according to Duncan's multiple range test (a = 0.05).

Table 43. Total lipid content (%) in pollen samples, by using different elution times (3 h, 5 h, 7 h).

| Elution time (h) | Total lipid content % (n = 5) |
|---|---|
| 3 | 2.56 (±0.080)[a] |
| 5 | 2.95(±0.040)[b] |
| 7 | 2.96 (±0.025)[b] |

The pollen amount was 2 g and the petroleum ether served as solvent. Different Latin letter indicates statistically significant differences, according to Duncan's multiple range test (a = 0.05).

Table 41. International data on the lipid content (% DM) of bee pollen.

| Crude lipid content (%D.M) | | | |
|---|---|---|---|
| Type of samples | Number of samples | Values | References |
| Multifloral | 21 | 4.0–5.5 | (Herbert & Shimanuki, 1978) |
| Multifloral | 3 | 2.0–18.0 | (Echigo et al., 1986) |
| Multifloral | 20 | 4.8–7.1 | (Serra-Bonvehí & Escolá Jordá, 1997) |
| Monofloral | 172 | 0–11.2 | (Somerville, 2001) |
| Multifloral | 15 | 3.6–8.9 | (Villanueva et al., 2002) |
| Multifloral | 10 | 7 ± 2* | (Almeida-Muradian et al., 2005) |
| Multifloral | 6 batches | 4.97** | (Melo et al., 2005) |
| Multifloral (Korea) | 9 | 3.3–9.8 | (Szczêsna, 2006a) |
| Multifloral (China) | 5 | 3.5–8.8 | (Szczêsna, 2006b) |
| Multifloral (Poland) | 13 | 6.74–10.99 | (Szczêsna, 2006c) |
| Multifloral | 154 | 4.0–13.3 | (Martins et al., 2011) |
| Multifloral | 7 | 4.3–5.0 | (Stanciu et al., 2012) |
| Multifloral | 14 | 6.3–8.7 | (Balkanska & Ignatova, 2012) |
| Monofloral | 12 | 0.6–6.5 | (Yang et al., 2014) |
| Multifloral | 7 batches | 5.39 ± 0.60* | (Arruda et al., 2013) |
| Multifloral | 99 | 1.15–6.97 | (Liolios et al., 2014) |
| Monofloral | 43 | 0.4–13.6 | (Liolios et al., 2014) |

*Average and standard deviation; ** Average.

***8.2.2.2. Preparation and classification of sequencing data.*** The sequence reads created in step 2.3.5 have to be joined, quality filtered and classified. For this purpose:

1. Copy all R1 and R2 fastq files and the reference database folders into a single folder.
2. Navigate on the command line shell to this folder.
3. Merge forward and reverse reads:
```
usearch -fastq_mergepairs *R1*.fastq \
  -relabel @ \
  -fastq_minmergelen 150 -fastq_maxmergelen 350 \
  -fastqout analyse_merged.fq -fastq_maxdiffs 15
```
4. Quality filtering and format parsing:
```
usearch -fastq_filter analyse_merged.fq
  -fastq_maxee 1 -fastaout reads.fa
sed "s/^>\([a-zA-Z0-9-]*\)\.\([0-9]*\)/>1_\2;barcodelabel=\1;/g"\
reads.fa>reads_bc.fa
```
5. Cluster amplicon sequence variants:
```
usearch -fastx_uniques reads_bc.fa \
  -sizeout -minuniquesize 3
  -fastaout reads_bc.derep.fa
usearch -sortbysize reads_bc.derep.fa \
  -fastaout reads_bc.derep.sort.fa
usearch -unoise3 reads_bc.derep.sort.fa \
  -zotus reads_bc.zotus.fa \
  -tabbedout unoise3.txt
```
6. Classify with direct reference database:
```
usearch -usearch_global reads_bc.zotus.fa \
  -db PATH-TO-DIRECT-DB/sequences.tax.fa \
  -id 0.97 -uc reads_bc.zotus_direct.uc -strand both
```
7. Classify remaining unclassified sequences with hierarchical reference database:
```
grep "^N" reads_bc.zotus_manRefs.uc | cut -f 9 > nohits
SeqFilter reads_bc.zotus.fa --ids nohits \
  -o reads_bc.zotus.nohits.fa
usearch -sintax reads_bc.zotus.nohits.fa \
  -db PATH-TO-HIERARCHICAL-DB/sequences.tax.fa \
  -tabbedout reads_bc.zotus.sintax -strand plus \
  -sintax_cutoff 0.8
```
8. Combine classifications:
```
cut -f1,4 reads_bc.zotus.sintax>  reads_bc.zotus.sintax.cut
cat reads_bc.zotus.sintax.cut | \
    sed -E -e "s/\_[0-9]+//g" -e "s/,s:.*$//" \
    > reads_bc.zotus.sintax.cutx
cut -f 9,10 reads_bc.zotus_ direct.uc | \
    grep -v "*" | sed "s/[A-Za-z0-9]*;tax=//" \
    > reads_bc.zotus.cutx
echo ",kingdom,phylum,order,family,genus,species" > header.cutx
cat header.cutx reads_bc.zotus.cutx \
    reads_bc.zotus.sintax.cutx>taxonomy.data
```
9. Count read abundance in the sample data:
```
usearch -usearch_global reads_bc.fa -strand plus \
    -db reads_bc.zotus.fa -id 0.97 -uc zotus.uc
python2.7 PATH-TO-PYTHON-SCRIPTS/uc2otutab.py sample_zotus.uc \
    > combined.txt
sed -e "s/OTUId//" combined.txt>combined.out
```
   This procedure will end with the following final files used for further analyses:
   (a) `combined.out` (the abundance of each taxonomic unit in the samples)
   (b) `taxonomy.data` (taxonomy assignments)

Table 44. Total lipid content (%) in pollen samples, by using different solvents (acetone, petroleum ether, hexane).

| Solvent | Total lipid content % n = 5 |
|---|---|
| Acetone | 2.99 (±0.155)[a] |
| Petroleum ether | 2.91 (±0.084)[a] |
| Hexane | 2.9 (±0.268)[a] |

The pollen amount was 2 g and the elution time was 5 h.

Table 45. Literature data on the carbohydrate content of bee pollen calculated as: 100 – (g fat + g protein + g ash + g water) (g/100 g DM).

| Type of samples | Number of samples | Values | References |
|---|---|---|---|
| Monofloral (*Aloe greatheadii* var. *davyana*) | 6 | 59.5 (±1.3)* | Human and Nicolson (2006) |
| Multiflora | 22 | 67.7 (±2.6)* 61.2–70.6** | Feas et al. (2012) |
| Multifloral | 8 | 69.68–84.25** | Nogueira et al. (2012) |
| Monoflora (*Brasica napus* L., *Citrullus lanatus* L., *Camellia japonica* L., *Dendranthema indicum* L., *Fagopyrum esculentum* L., *Helianthus annus* L., *Nelumbo nucifera* Gaertn., *Papaver rhoeas* L., *Rosa rugosa*, *Schisandra chinensis*, *Vicia faba* L., *Zea mays* L.) | 12 | 59.43–77.82** | Yang et al. (2014) |
| Multifloral | – | 13–55** | Bogdanov (2015) |
| Multifloral | 26 | 64.42–81.84** | Kostić et al. (2015) |

*Average and standard deviation; **min – max.

Table 46. Literature data on the content of sugars in bee pollen (g/100 g DM).

| Type of samples/ Country of origin | Number of samples | Sugars | Values | Method used References |
|---|---|---|---|---|
| Multiflora (with a predominance of *Cistus landaniferus*)/ Spain | 20 | Fructose | 15.20–23.90* | GC-FID |
| | | Glucose | 10.86–18.19* | Serra-Bonvehí & Escolá |
| | | Saccharose | 4.20–9.40* | Jordá (1997) |
| | | Maltose | 0.79–3.20* | |
| | | F/G**** | 1.13–1.53* | |
| Multifloral/Poland | 95 | Fructose | 13.38–21.80* | GC-FID |
| | | Glucose | 5.99–18.17* | Szczęsna et al. (2002) |
| | | Saccharose | 3.74–7.59* | |
| | | Maltose | 0.96–3.32* | |
| | | F/G**** | 1.03–2.51* | |
| Multifloral/ Poland, South Korea, China | 27 | Fructose | 9.74–22.06* | HPLC-RID |
| | | Glucose | 8.45–20.06* | Szczęsna (2007) |
| | | Saccharose | 1.13–4.84* | |
| | | Maltose | 1.07–3.51* | |
| | | F/G**** | 1.02–1.63* | |
| Multifloral/Israel, China, Romania, Spain | 5 | Fructose | 15.9–19.9* | HPLEC-PAD |
| | | Glucose | 8.2–13.1* | Qian et al. (2008) |
| | | Saccharose | 14.8–18.4* | |
| | | F/G**** | 1.33–1.66* | |
| Multifloral/Romania | -*** | Fructose | 19.31** | HPLC-RID |
| | | Glucose | 17.86** | Bobis et al. (2010) |
| | | Saccharose | 0.85** | |
| | | Maltose | 0.64** | |
| Multifloral/Brazil | 154 | Fructose | 12.59–23.62* | HPLC-RID |
| | | Glucose | 6.99–21.85* | Martins et al. (2011) |
| | | F/G**** | 1.01–2.24 | |
| Multifloral/Romania | 16 | Fructose | 8.44–15.39* | HPLC-RID |
| | | Glucose | 4.37–16.14** | Mărgăoan |
| | | F/G**** | 0.78–1.44* | et al. (2012)***** |
| Monofloral (*Cucurbita pepo* Thunb, *Phoenix dactylifera* L., *Helianthus annus* L., *Medicago sativa* L., *Brasica napus* L.)/ Saudi Arabia | -*** | Fructose | 17.13–21.30* | HPLC-RID |
| | | Glucose | 15.44–17.06* | Taha (2015) |
| | | F/G**** | 1.07–1.25* | |

*min – max; **average; ***no information; ****fructose to glucose ratio; *****results for fresh pollen.

Table 47. Methods used to evaluate the vitamin content of bee pollen.

| Method | Vitamins analyzed | Country | Reference | Vitamin content (mean) |
|---|---|---|---|---|
| Open column chromatography and microfluorimetric method | C, beta-carotene as provitamin A | Brazil | Almeida-Muradian et al., 2005 | Absence of vitamin C and beta-carotene (dried bee pollen) |
| Open column chromatography and titrimetric method | E, C and beta-carotene as provitamin A | Brazil | Oliveira et al., 2009 | 13,5 μg/g (E); 56,3–198,9 μg/g (beta-carotene); 273,9–560,3 μg/g (C) (fresh bee pollen) |
| HPLC, open column chromatography and titrimetric method | E, C and beta-carotene as provitamin A | Brazil | Melo et al., 2009 | 114–340 μg/g (C); 16,27–38,64 μg/g (E); 3,14–77,88 μg/g (beta carotene) (dried bee pollen) |
| HPLC, open column chromatography and titrimetric method | E, C and beta-carotene as provitamin A | Brazil | Melo & Almeida-Muradian, 2010 | 14–119 ug/g (C); 19.43–37.19 μg/g (E); 5.95–99.27 μg/g (beta-carotene) (fresh bee pollen) |
| HPLC method with fluorescence detection. | B1, B2, B6 and PP | Brazil | Arruda et al., 2013a; Arruda et al., 2013b | 0.64 mg/100 g ($B_1$); 0.60 mg/100 g ($B_2$); 12.18 mg/100 g (PP); 0.55 mg/100 g ($B_6$) |
| HPLC and titrimetric method | E (alpha, beta, gamma and delta tocoperol), C (ascorbic acid) and provitamin A (alpha and beta-carotene) | Brazil | Sattler, 2013 | 60–797 μg/g (C); 0.57–11.7 mg/100g (E); 0–179.21 μg/g of beta-carotene; 0–828.43 μg/g of alpha-carotene |
| HPLC method with fluorescence detection. | | Brazil | Souza, 2014 | 0.46–1.57 mg/100g ($B_1$); 0.40–1.86 mg/100g ($B_2$); 2.67–7.74 mg/100g (PP or $B_3$);0.78–7.13 mg/100g ($B_6$) |

Table 48. Calibration curves obtained for echimidine, lycopsamine and intermedine.

| Echimidine or Lycopsamine or Intermedine (μg/ml) | Echimidine (intensity) | Intermedine (intensity) | Lycopsamine (intensity) |
|---|---|---|---|
| 0.02 | 2950 | 3130 | 3670 |
| 0.1 | 13660 | 15540 | 16200 |
| 0.5 | 63940 | 62050 | 71200 |
| 2 | 243800 | 200400 | 277000 |
| Calibration curve* | y = 122755x + 686 | y = 104410x + 1900 | y = 138740x + 1143 |

*Weighted by 1/x.

These files can be loaded directly into R with the package phyloseq (McMurdie & Holmes, 2013). In addition, also the results of the intermediate steps are retained and can be used for troubleshooting, archiving or further analyses.

### 8.3. Data analysis

#### 8.3.1. Required software

a. Up to date R distribution (R Core Team, 2014).
b. R package: phyloseq (McMurdie & Holmes, 2013); https://joey711.github.io/phyloseq

#### 8.3.2. Prepare sample meta-data

Additionally, to the sequencing data, phyloseq is intended to read in a meta-data file. This can be a simple spreadsheet in csv format with any number of variables to be investigated. For example, continuous vectors like "altitude" or "temperature" or categorical factors as "bee species" or "site" can be used. For this, open the file with your preferred text-editor or spreadsheet application and add columns according to the sampling design. Save the file again in tab-separated format.

#### 8.3.3. Importing data

The data generated above can be directly imported into R as a phyloseq class object. This allows a variety of analytical procedures and is recommended. However, other software tools handling community datasets may be equally well used for the task of analyses. The following are R scripts, which can be directly used on the console:

1. Load packages and data:
```
library(phyloseq) # load the package
setwd("<PATH-TO-DATA>") # data folder
tax <- tax_table(as.matrix(read.table("taxonomy.data",
        header=T,row.names = 1,fill=T,sep=",")))
otu <- otu_table(read.table("zotu_table.txt", header=T, row.names = 1),
        taxa_are_rows=T)
map <- sample_data(read.table("microbiome_DATA_SamBoff.csv", sep=",",
        header=T, row.names = 1,fill=T))
data <- merge_phyloseq(otu, tax, map) # create phyloseq object
```
2. Adjust for missing taxa names in hierarchical classification:
```
tax_table(data)[tax_table(data)[,"phylum"]=="","phylum"]<- paste(
        tax_table(data)[tax_table(data)[,"phylum"]=="","kingdom"],
        "_spc",sep="")
tax_table(data)[tax_table(data)[,"order"]=="","order"]<- paste(tax_table(
        data)[tax_table(data)[,"order"]=="","phylum"],
        "_spc",sep="")
tax_table(data)[tax_table(data)[,"family"]=="","family"]<- paste(
        tax_table(data)[tax_table(data)[,"family"]=="","order"],
        "_spc",sep="")
tax_table(data)[tax_table(data)[,"genus"]=="","genus"]<- paste(
        tax_table(data)[tax_table(data)[,"genus"]=="","family"],
        "_spc",sep="")
```
3. Combine Amplicon Sequence Variants to species or genus level:
```
(data.genus <- tax_glom(dataset.comp, taxrank="genus"))
(data.species <- tax_glom(dataset.comp, taxrank="species"))
```
4. Transform to relative abundances and filter low abundance taxa:
```
data.genus.rel=transform_sample_counts(data.genus,
        function(x) x/sum(x))
otu_table(data.genus.rel)[otu_table(data.genus.rel)<0.01 ]<-0
```

Table 49. Retention and mass characteristics of known *Echium*-type and *Eupatorium*-type PAs (free bases/-*N*-oxides and isomers).

| PA peak n° | Compound Name | RT (min) | Chemical formula | $[M+H]^+$ experimental | $[M+H]^+$ calculated | Error (mDa) |
|---|---|---|---|---|---|---|
| | • *Echium vulgare* | | | | | |
| 1 | Echimidine | 2.28 | C20H31NO7 | 398.2184 | 398.2179 | 0.5 |
| | Echimidine-*N*-oxide | 2.27 | C20H31NO8 | 414.2126 | 414.2128 | −0.2 |
| | Acetylechimidine | 2.74 | C22H33NO8 | 440.2282 | 440.2284 | −0.2 |
| 3 | Acetylechimidine-*N*-oxide | 2.71 | C22H33NO9 | 456.2234 | 456.2234 | 0.0 |
| | Echivulgarine | 3.81 | C25H37NO8 | 480.2575 | 480.2575 | 0.0 |
| 5 | Echivulgarine-*N*-oxide | 3.83 | C25H37NO9 | 496.3402 | 496.3405 | −0.3 |
| | Vulgarine | 2.33 | C20H31NO7 | 398.2174 | 398.2179 | −0.5 |
| 2 | Vulgarine-*N*-oxide | 2.43 | C20H31NO8 | 414.2133 | 414.2128 | 0.5 |
| | Acetylvulgarine | 2.88 | C22H33NO9 | 440.2285 | 440.2284 | 0.1 |
| 4 | Acetylvulgarine-*N*-oxide | 2.87 | C22H33NO9 | 456.2243 | 456.2234 | 0.9 |
| | • *Eupatorium cannabinum* | | | | | |
| 6 | Lycopsamine | 1.12 | C15H25NO5 | 300.1807 | 300.1811 | −0.4 |
| 7 | Intermedine | 1.16 | C15H25NO5 | 300.1810 | 300.1811 | −0.1 |
| 8 | Lycopsamine-*N*-oxide or Intermedine-*N*-oxide* | 1.27 | C15H25NO6 | 316.1756 | 316.1760 | 0.4 |
| 9 | Lycopsamine-*N*-oxide or Intermedine-*N*-oxide* | 1.35 | C15H25NO6 | 316.1758 | 316.1760 | 0.2 |

*Distinction between the *N*-oxides of intermedine and lycopsamine not possible due to the lack of standards.

Table 50 MS/MS fragment ions for PA-N-oxides from *E. vulgare* or *E. cannabinum*.

| Pyrrolizidine alkaloid | $[M+H]^+$ | MS/MS fragment ions |
|---|---|---|
| • *Echium vulgare* | | |
| Echimidine-*N*-oxide | 414.2126 | 396.2029, 352.1764, 338.1608, 254.1403, 220.1350, 120.0817 |
| Acetylechimidine-N-oxide | 456.2234 | 438.2131, 396.2025, 338.1610, 254.1393, 220.1339 |
| Echivulgarine-*N*-oxide | 496.3402 | 478.2448, 396.2030, 338.1613, 254.1397, 220.1341, 120.0816 |
| Vulgarine-*N*-oxide | 414.2133 | 396.2021, 314.1612, 256.1189, 172.0976, 138.0923, 136.0764 |
| Acetylvulgarine-*N*-oxide | 456.2243 | 438.2132, 356.1714, 298.1295, 214.1081, 180.1030, 120.0818 |
| • *Eupatorium cannabinum* | | |
| Lycopsamine/Intermedine-*N*-oxide | 316.1756 | 172.0973, 155.0947, 138.0917, 111.0684 |

After completion of the tasks above, the dataset is in a condition where individual analyses can be started. The tutorials at the repository of phyloseq ((McMurdie & Holmes, 2013); https://joey711.github.io/phyloseq) provide a good starting point for this.

#### *8.3.4. Recommended packages for further analysis*

Whilst phyloseq provides basic tools suited for most purposes, the modularity of R packages allows a variety of more and deeper analyses. It is not possible to discuss all the features here, yet we provide a list some of the major packages relevant for community ecology and pollination studies:

a. vegan: comprehensive community ecology package;
b. picante: phylogenetic diversity indices;
c. bipartite: interaction network ecology;
d. edgeR: sequencing library size correction.

## 9. Standard method for the determination of protein content of pollen using the Kjeldahl method

### *9.1. Introduction*

The knowledge of bee pollen chemical composition increases the commercial value of the product. Especially, among the chemical constituents of bee pollen, the proteins' concentration has been the most studied (Table 37). There seems to be a considerable variation among the results regarding the protein concentration, probably because of the different number of the analyzed samples and whether the pollen samples are unifloral or polyfloral.

For the determination of the total protein content, the majority of authors use the Kjeldahl method modified for bee pollen analysis by Rabie et al. (1983). However, the method of analysis has not been investigated. Regarding the calculation of the nitrogen content in this type of samples, several authors used different factors (6.25 or 5.60). Indeed, Rabie et al. (1983), recommended the factor 5.60, while Serra-Bonvehí and Escolá Jordá (1997) stated that the use of factor 6.25 would overestimate the protein content of analyzed samples. The follow methods elucidate the optimum conditions, in order to minimize the reagents' consumption and the sample quantity.

### *9.2. Determination of protein content by Kjeldahl method*

1. Weigh 1 g of bee pollen.
2. Heat the sample with 20 ml of sulfuric acid (95–97%) at the presence of a catalyst (potassium sulfate, copper sulfate) for about 4 h until the solution becomes clear and blue-green in color.
3. Neutralize the digested sample with 90 ml NaOH (30%) and distillate the solution for 2 min using for the trapping of $NH_3$, 30 ml of $H_3BO_3$ solution (4%).
4. Titrate with HCl solution (0.1 M).

The nitrogen content is determined by the consumption of the HCl solution (0.1 M) during the titration. The crude protein content is estimated using the most suitable factor 5.60, which was calculated and suggested by Rabie et al. (1983) after amino-acids analysis in different pollen species.

### *9.3. Development of the method*

For the development and validation of the suggested method, bee pollen of *Papaver rhoeas* was used. In order to find the optimum conditions for the determination of the protein content of bee pollen tests were applied regarding the volumes of NaOH, and $H_3BO_3$ solutions, the quantity of sample and the distillation time.

#### *9.3.1. Volume of NaOH solution*

During the distillation, the NaOH solution converts any nitrogen in the bee pollen into $NH_3$, while the appearance of the blue color demonstrates a sufficient volume of the solution for this formation. Volumes of 50 ml, 60 ml, 70 ml, 80 ml, 100 ml and 110 ml were tested. No change of color was observed in volumes of 50 ml to 80 ml, while in volume of 90 ml, the color turned from light green into dark blue indicating the end of the reaction. The volumes of 100 ml and 110 ml were rejected, to avoid the unnecessary consumption and to reduce the cost of the analysis. Thus, the suitable volume of the NaOH solution (30%) was determined at 90 ml.

*9.3.2. Volume of $H_3BO_3$ solution*

Boric acid is used for the capture of $NH_3$, during distillation. Volumes of 30 ml and 40 ml were tested without statistically significant changes to be observed between the results (Duncan's multiple range test, $\alpha = 0.05$) (Table 38). Eventually, the volume of 30 ml of the $H_3BO_3$ solution (4%) was chosen for the analysis.

*9.3.3. Distillation time*

Different distillation periods of time (2, 3, 4 min) were used in order to choose the most appropriate. In all cases, 1 g of sample, 90 ml NaOH and 30 ml $H_3BO_3$ were used. According to the statistical process of the results, no significant changes were observed in the protein content regarding to the distillation time (Table 39). For this reason, the shortest time of 2 min was chosen for the analysis.

### 9.4. Sample preparation

For the chemical analysis of bee pollen of a single plant, the pollen pellets should be separated from the mixture according to its color, shape and texture and this is a tedious and time consuming process. Consequently, the minimum amount of pollen for chemical analysis will reduce the needed time and facilitate the analysis.

1. Different quantities of pollen 0.5 g, 1 g and 2 g were tested with five replications in each case, maintaining the other parameters constant.
2. The volumes of 90 ml NaOH and 30 ml $H_3BO_3$ were used, while the distillation time was 2 min.
3. Statistically significant changes were observed regarding to the protein content (Table 43). The minimum quantity of 0.5 g gave significantly less amount of proteins, while no differences were found when one or two grams were used.
4. After that, we propose the use of one gram sample for the analysis of proteins in bee pollen samples.

### 9.5. Validation of the method

For the determination of the repeatability of the method:

1. eight pollen samples of *Papaver rhoeas* of the same quantity (1 g) were analyzed;
2. samples were treated equally during the three steps of the Kjeldahl method, as described previously;
3. average protein content of *Papaver rhoeas* pollen was found 20.6%, while the low value of RSD (0.81%) indicated the very good repeatability of the method.

### 9.6. Application of the method

Pollen from different botanical species has different nutritional value for honey bees, affecting the development and the productivity of the bee colony. To investigate qualitative and quantitative differences on pollen composition we collected bee pollen samples from the apiary of Aristotle University in consecutive years during the full season and samples from beekeepers all over Greece. The samples were analyzed as a mixture or as separate pollen pellets that are classified according to their color, shape and texture. For the identification of the pollen microscopic method is applied. Monofloral and multifloral bee pollen samples were analyzed using the method described above. Some of the results regarding the differences based on plant origin are presented below, in Figure 43. The lowest value was observed for *Actinidia chinensis* (9.7%) and the highest for *Crocus sativa* (26.1%). The range of 16.4%, confirms the large variation among pollen from different plants. The protein content of mixed bee pollen samples was also examined during the whole beekeeping season, in order to find out differences based on the harvesting period (Figure 44).

The mixed pollen samples collected from March till the middle of June indicated that the nutritional requirements of bees are very well met since the protein content ranged from 19.7% to 26.6%. Herbert (1992) mentions that a protein content of 20–23% in pollen substitutes is ideal for the dietary requirements of honey bees. As opposed to that, pollen collected from the middle of June till the middle of September had average protein content below 20%, including the lowest observed value (14.8%) in the end of August. During autumn, the protein content of pollen increased again and exceeded 20%. According to these results, it is obvious that the protein content of mixed pollen ranges according to the season and this could affect the colony's development and the quality of the bee product for sale.

## 10. Standard method for determination of lipid content of pollen using Soxhlet extraction

### 10.1. Introduction

The lipid fraction is one of the basic constituents of bee pollen and varies significantly among the different botanical species (Somerville, 2001; Yang et al., 2014). Several authors have studied the lipid concentration in pollen, whose results are referred in Table 41. The variations among the research observed, could be attributed to the different number of the analyzed samples and to variable botanical origin of the samples. The most common method for the determination of the total lipid content of foods is the extraction method described by Soxhlet. According to this procedure, oil and fat are extracted by repeated elution with an organic solvent (extraction cycle). Nevertheless, in case of the extraction of lipids in pollen, there is vagueness about the optimum methodology concerning the reagent's consumption, the sample quantity and the elution time. The following methods clarify these parameters and provide experimental results that should be implemented to produce reliable analytical data on lipids analysis of pollen.

### 10.2. Methodology for total lipid content determination

The total lipid content is determined as crude fat from a solvent extract.

1. weigh 2 g of bee pollen and place in a porous cellulose thimble (25 × 100 mm).
2. place the thimble in an extraction chamber, which is suspend above a round bottom flask containing petroleum ether as solvent and below a condenser.
3. heat the flask – evaporate the solvent and move up into the condenser where it is converted into a liquid that trickle into the extraction chamber containing the sample. This extraction cycle is repeated many times for 5 hours. In order to remove the excess solvent.
4. fat residue is dried in an oven at 80 °C for 45 min until to constant weight.
5. place the flask for 15 minutes in a desiccator which contains $CaCl_2$, in order to cool down without absorbing moisture.
6. estimate the fat content as the difference in weight of the flask before and after the extraction. Express as percent of dry matter.

### *10.3. Development of the method*

For the development and validation of the previous method, pollen of *Papaver rhoeas* was chosen.

1. Pollen samples: collected from traps placed in bee hives (in this example they were located in the surrounding area of Thessaloniki). The specific plant *taxon* was chosen, because of its abundance in the area where the bee hives were located and its regular foraging by the bees. Moreover, its pollen loads possess a distinct black color, so they can be separated easily from mixed pollen samples.
2. Determination of the optimum quantities of the materials and the appropriate analysis' conditions of the method – tests implemented regarding the pollen amount (1 g, 2 g, 3 g), extraction time (3 h, 5 h, 7 h) and the type of the solvent (acetone, petroleum ether, hexane). Statistically significant differences were found in total lipid content (Duncan's multiple range test, $\alpha = 0.05$), using different amounts of pollen (Table 42) and different elution times (Table 43), while no significant differences were observed by using different solvents (Table 44). Note: Considering that the separation of pollen samples based on color is a tedious and time consuming process, the quantity of pollen should minimize as possible. As a consequence, according to the Table 2, the amount of 2 g was chosen for the analysis.
3. Elution time the period of 5 h was chosen. The time of 7 h was rejected in order to reduce the time consumption, while when 3 h elution was applied statistically significant differences were observed in the total lipid content.
4. Solvent extraction – no statistically significant differences were found among the solvents and petroleum ether is widely referred in bibliography for the extraction of the lipids in pollen this liquid was chosen as the appropriate solvent.

### *10.4. Validation of the method*

For the determination of the repeatability of the method, *Papaver rhoeas* pollen was separated, homogenized and eight subsamples of 2 g analyzed under the same conditions (extraction time: 5 h, solvent: petroleum ether). The average liquid content of *Papaver rhoeas* was found 2.47%, while the relative standard deviation (RSD) was 4.66% indicating a satisfactory precision.

### *10.5. Application of the method*

The method described above was applied in monofloral and polyfloral pollen samples collected from University apiary. Some of the results regarding the differences based on plant origin are presented in Figure 45. The identification of pollen taxa was based on the microscopic characteristics (Lau et al., 2018; Louveaux et al., 1978). The lowest value was observed for *Erica manipuliflora* (0.40%) and the highest for *Echinops ritro* (13.60%). The average of the samples examined was 4.61% and the range was 13.2%.

The lipid content of mixed pollen samples was also examined during the whole beekeeping season, in order to find differences based on the harvesting time (Figure 46). The higher lipid content of mixed pollen samples was observed in samples collected between 19/9 until 15/10. Pollen collected early in autumn (13/9) has very low lipid content (2.15%) but after that, the lipid concentration increases considerably to reach the high level of 6.22% on the day 15/10. One factor that may contribute to this increase is the second period of blooming of *Sisymbrium irio* which is very reach in lipids (7.22%). The fluctuation of lipid curve of pollen mixtures collected in different period of year (Figure 46) indicates the effect of various blossoms to the final chemical composition of pollen.

## 11. Determination of sugar content in pollen using HPLC-RID

### *11.1. Introduction*

Carbohydrates are the sugars, dietary fiber, and starch which constitute the main fraction of pollen dry matter and represent important components from the nutritional and energy point of view. Collected pollen provides bees a valuable source of simple sugars and other carbohydrates which are essential for proper nutrition and advisable for certain physiological conditions. The generally accepted calculation for total carbohydrate content in pollen is: 100 – (g fat + g protein + g ash + g water) (Bogdanov, 2015; Feas et al., 2012; Human & Nicolson, 2006; Kostić et al., 2015; Nogueira et al., 2012; Yang et al., 2014) (Table 45). This calculation is

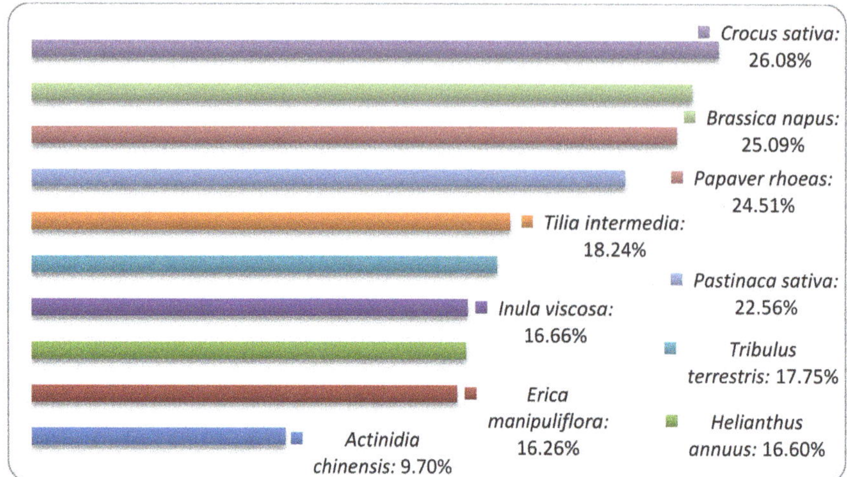


Figure 43. Crude protein content (% D.M) of bee pollen from selected botanical origins.

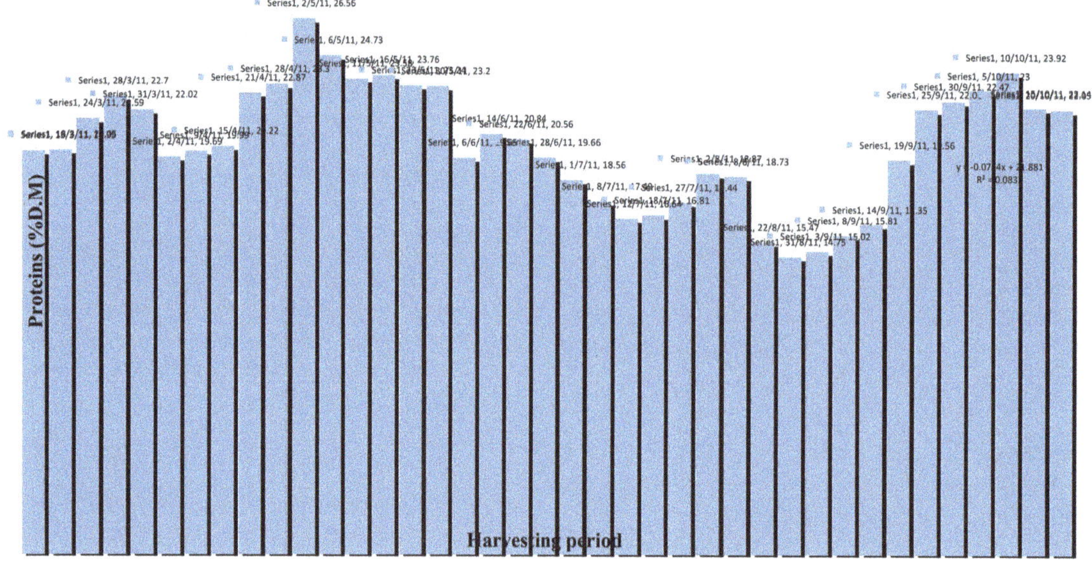


Figure 44. Protein content (% D.M) of mixed pollen samples harvested during the beekeeping season of 2011.

much higher than results received through HPLC or GC methods because they do not determine such components as dietary fiber, starch, and cell wall material. Studies dealing with the composition of sugars in pollen were concerned with the total content of reducing and non-reducing sugars (Cirnu et al., 1969; Rosenthal et al., 1969; Solberg & Remedios, 1980; Szczęsna et al., 1995a, 1995b; Youssef et al., 1978).

Advances in chromatographic techniques with different detections (TLC, HPLC-RID, HPTLC, HPLEC-PAD, HPLEC coupled to LC-ESI-MS, GC-FID; GC-MS) made it possible to analyze, in detail, the composition of sugars in bee pollen (Aratchi et al., 2018; Bobis et al., 2010; Echigo et al., 1986; Mărgăoan et al., 2012; Martins et al., 2011; Qian et al., 2008; Serra-Bonvehí et al., 1986; Serra-Bonvehí & Escolá Jordá, 1997; Szczęsna, 2007; Szczęsna et al., 2002; Taha, 2015). The quality and quantity composition of pollen sugars was found to be closely related to the plant species of origin as well as to the methods of analysis. Researchers identified fourteen different sugars in pollen of which fructose, glucose, and sucrose occurred in the greatest amounts, followed by maltose (Table 46). The following composition of basic sugars in bee pollen was determined to be: fructose 44–46%, glucose 35–37%, sucrose 8–30%,and maltose 5–10% of the total sugar content. Mono-, di- and trisaccharides furanose, arabinose, ribose, trehalose, isomaltose, turanose, kojibiose, gentiobiose, melibiose and melezitose accounted for about 1% (Aratchi et al., 2018; Serra-Bonvehí et al., 1986; Serra-Bonvehí & Escolá Jordá, 1997; Szczęsna, 2007; Szczęsna et al., 2002). The metabolome analysis of bee-collected pollen, especially myo-inositol (predominant physiologica form of six-carbon sugar alcohol) and pentose sugars (furanose) is useful in the study of plant-pollinators mutualism (Aratchi et al., 2018). Tetrasaccharide stachyose is also determined in pollen samples (Quian et al., 2008). The fructose-to-glucose ratio (F/G) of bee-collected pollen samples varies considerably from 1.01 in Brazil (Martins et al., 2011) to

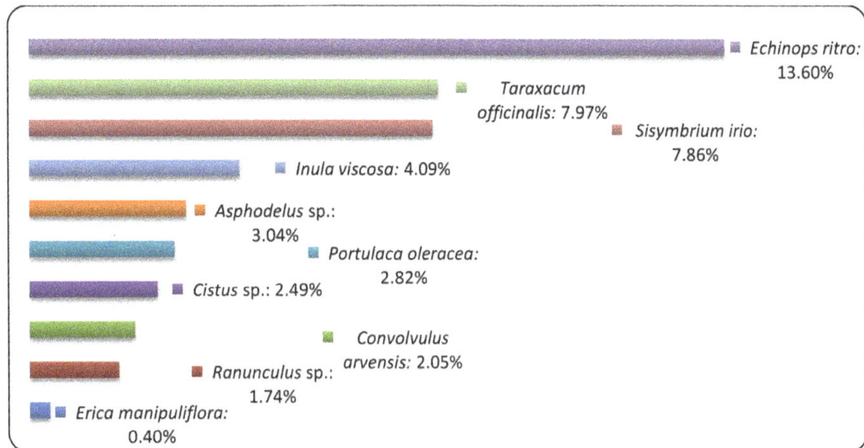


Figure 45. Crude lipid content (% D.M) of bee pollen from selected botanical origins.

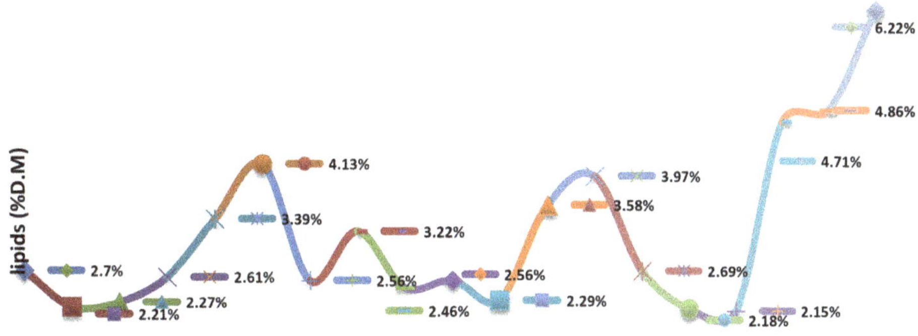


Figure 46. Lipid content (% D.M) of mixed pollen samples harvested during the beekeeping season of 2013.

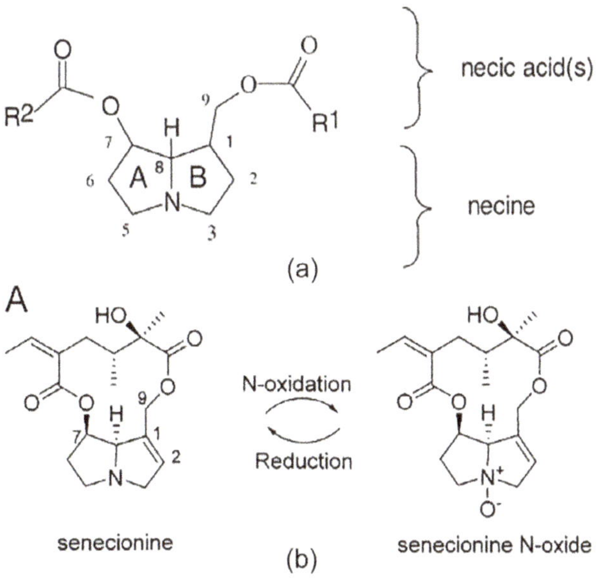

Figure 47. (a) Structure of PAs: Necinic acids are connected to the necine base by ester bond in position 1 and/or 7. Necinic acids can be connected to form a macrocyclic ester; (b) Example of PA free base and N-oxide macrocyclic esters (Hartmann & Witte, 1995).

2.51 in Poland (Szczęsna et al., 2002) (Table 46). The sugar content of pollen has not been normalized, and in fact only Poland, Switzerland and Argentina currently have a standard for it. The International Honey Commission proposes 4 g/100 g minimum sugar content for dried bee pollen (Campos et al., 2008). The HPLC technique with refractive index detection (RID) has generally been used for the determination of sugars in bee-collected pollen (Bobis et al., 2010; Mărgăoan et al., 2012; Martins et al., 2011; Szczęsna, 2007; Taha, 2015).

### 11.2. Method for identification and quantification of sugars by HPLC-RID

#### 11.2.1. Principle

This method is based on the procedure elaborated by Bogdanov et al. (1997) for the determination of honey sugars and adopted by Szczęsna (2007). Briefly, the sugars are extracted from pollen samples and then, determined by HPLC with a refractive index detector (RID) using a normal-phase column containing amine-modified silica gel with a 5–7 μm particle size. The analyzed sugars are: fructose, glucose, saccharose, maltose, turanose, trehalose, izomaltose, erlose, melezitose, and rafinose. The sugars are identified through the comparison of individual sugars' retention times of the reference and the analyzed solution. The quantitative assays are done by the external standard method comparing those sugars' peak surfaces.

#### 11.2.2. Reagents

a. deionized water (18.3 MΩ).
b. methanol, acetonitrile for HPLC (J.T. Baker or comparable purity from another company).
c. analytical standards of: fructose, glucose, saccharose, maltose, turanose, trehalose, izomaltose, erlose, melezitose, rafinose (SIGMA-ALDRICH or comparable purity from another company).
d. Carrez solution I: 15 g of potassium hexacyanoferrate (II) trihydrate, $K_4Fe(CN)_6 \cdot 3H_2O$ dissolved in 100 ml of deionized water.
e. Carrez solution II: 30 g of zinc acetate dihydrate, $Zn(CH_3COO)_2{\cdot}2H_2O$ dissolved in 100 ml of deionized water.

#### 11.2.3. Equipment

a. High Performance Liquid Chromatograph operated by computer software and equipped with: pump, degasser, autosampler, column-oven, and refractive index detector (RID).
b. amine-modified silica gel HPLC column with 5–7 μm particle size (250 × 4.6 mm × 5 μm) (Phenomenex or another company).
c. analytical balance.
d. mechanical shaker.
e. automatic dispenser.
f. analytical grade filter paper.
g. PVDF membrane filter, pore size 0.45 μm (Karl Roth or another company).
h. volumetric flask, 100 ml.
i. Eylenmeyer flask, 100 ml.
j. beaker, 50 ml.
k. autosampler vial.

#### 11.2.4. Preparation of stock standard solution and working standard solutions of sugars

1. weigh each sugar to prepare a standard solution, with an accuracy of 0.001 g. Weights of individual sugars: (a) fructose: 1.000 g, (b) glucose: 0.750 g, (c) saccharose, maltose: 0.500 g, (d) turanose, trehalose, izomaltose, erlose, melezitose, rafinose: 0.075g.
2. dissolve the standard sugars in a small amount of deionized water and transferred quantitatively to a volumetric 100 ml flask. Add 25 ml of methanol with the use of an automatic dispenser, and poured into a volumetric flask up to the mark with deionized water.
3. filter the standard solution with 0.45 μm membrane filter and stored at −18 °C. A stock standard sugar solution kept at this temperature is stable for six months.

Note: a calibration curve is used for the external standard method of the quantitative analysis of sugar in pollen. The calibration curve is prepared for three different concentrations of each analyzed sugar. For this purpose, the working standard solutions of those sugars are prepared on the day of analysis from the stock standard solution.

#### 11.2.5. Sample preparation

1. weigh a pollen sample of 5 g ± 0.001 g and put into a 100 ml Eylenmeyer flask.
2. add about 40 ml of deionized water, 5 ml of Carrez solution I, and 5 ml of Carrez solution II and mix for 2 h at 30 °C using a mechanical shaker.
3. transfer the solution quantitatively to a 100 ml volumetric flask, add 25 ml of methanol and filled up to the notch, with deionized water.
4. filter the sample solution through qualitative filter paper into a beaker. The first 10 ml of the filtrate are discarded. The solution is passed through a 0.45 μm membrane filter into an autosampler vial and the filtrate injected in the HPLC-RID system.

#### 11.2.6. HPLC-RID analysis

The HPLC system is run using computer software in the following conditions:

1. flow rate: 1.5 ml/min.
2. mobile phase: acetonitrile: water (80:20 V/V).
3. column and detector temperature: 30 °C.
4. injection volume: 20 μl.

After the HPLC System is stabilized, the working standard solutions of the sugars and then pollen sample are injected and analyzed under the same conditions. The external standard method is used to calculate the results using a calibration curve determine for three different concentrations of each sugar standard. The results are express as g/100 g to one decimal place.

### 11.3. Validation of the method

The method described above can be applied for routine analysis of fructose, glucose, saccharose, maltose, turanose, trehalose, izomaltose, erlose, melezitose, and rafinose in pollen samples. The detection limit, working range, linearity, repeatability and within-laboratory reproducibility were determined for this method. The detection limit was set at 0.2 g/100 g DM for fructose and glucose and at 0.5 g/100 g DM for saccharose, maltose, turanose, trehalose, izomaltose, erlose, melezitose, and rafinose. This procedure is used for the routine determination of sugars in pollen, in the range of: 5–40 g/100 g for fructose and glucose; 0.5–30 g/100 g for saccharose, maltose, trehalose, turanose, and isomaltose; 0.5–10 g/100 g for erlose, rafinose and melecytoze. The correlation coefficient of the linear dependence of sugar concentration and peak area in the working range, reached the value of over 0.995 for all the analyzed sugars. A series analysis for three pollen samples showed that the variation coefficient of the repeatability and the within-laboratory reproducibility of the elaborated method did not exceed 10% for fructose and glucose, and 20% for both other analyzed disaccharides (saccharose, maltose, turanose, trehalose, izomaltose) and trisaccharides (erloze, melezitose). Thus, sufficient accuracy and precision of the methods were demonstrated.

### 11.4. Application of the method

The results on the content of the sugars analyzed by the HPLC-RID method, in multifloral pollen samples originating from Poland, South Korea and China are presented Szczęsna, (2007) in Table 46 and show that the sugar content of pollen dry matter averages 40%. The fructose ranged from 9.74 to 22.06 DM, and the average was 15.94 g/100 g DM. The glucose content was lower by several g/100 g and ranged from 8.45 to 20.06 g/100 g; the average being 14.02 g/100 g DM. Of the tested disaccharides, saccharose occurred in the highest concentrations, ranging from 1.13 to 4.84 g/100 g DM; the average being 3.20 g/100 g DMmaltose followed ranging from 1.07 to 3.51 g/100 g DM; the average being 2.47 g/100 g DM. Those four sugars varied substantially within each of the samples. The variation was related to their botanical origin. In the majority of samples, the content of other such analyzed disaccharides as trehalose and turanose was at the detection limit of 0.2 g/100 g DM.

Bobis et al. (2010), Martins et al. (2011), Mărgăoan et al. (2012) and Taha (2015) used the same method as Szczęsna (2007) to quantify the sugar content of sugars and confirmed the predominance of fructose followed by glucose in pollen samples (Table 48). Bobis et al. (2010) found fructose and glucose to be on average at concentrations of 19.31 and 17.86 g/100 g, respectively. They investigated disaccharides were in similar quantities as those investigated by Szczęsna (2007). Martins et al. (2011) determined the fructose concentration to be ranged between12.59 and 23.62 g/100 g and the average 18.99 g/100 g, while the glucose concentration between 6.99 and 21.85 g/100 g and the average 14.89 g/100 g. According to Mărgăoan et al. (2012), fructose content ranged between 8.44 and 15.39 g/100 g, and glucose content – between 4.37 and 16.14 g/100 g. The results received by Taha (2015) for a few monofloral pollens (*Cucurbita pepo* Thunb, *Phoenix dactylifera* L., *Helianthus annus* L., *Medicago sativa* L., *Brasica napus* L.), showed that fructose content ranged from 17.13 g/100 g for rape (*Brassica napus* L.) pollen to 21.30 g/100 g for date palm pollen (*Phoenix dacttylifera* L.), and glucose content from 15.44 g/100 g for sunflower (*Helianthus annus* L.) pollen to 17.06 g/100 g for date palm (*Phoenix dacttylifera* L.) pollen.

With the above-described method, the Bee Products Quality Testing Laboratory, the Apiculture Division of the National Institute of Horticultural Research, Poland produced results for multifloral pollen samples were in the same range as the results received by the above-mentioned authors. Fructose content ranged between 15.40 and 22.54 g/100 g DM, 17.01 g/100 g DM, on average. Glucose content ranged between 10.42 and18.54 g/100g DM, 14.01 g/100 g DM, on average. Saccharose content ranged between 2.80 and 7.40 g/100 g DM, with an average value of 5.70 g/100 g DM.

## 12. Standard method for vitamin analysis in bee pollen

### 12.1. Introduction

Bee pollen contains vitamins of B complex, vitamin C, E and carotenes with provitamin A function (alpha and beta-carotene) (Arruda et al., 2013a, 2013b; Melo et al., 2009; Melo & Almeida-Muradian, 2010; Oliveira et al., 2009; Sattler et al., 2015; Souza et al., 2018). Vitamins are classified in terms of solubility. The lipid-soluble vitamins (vitamins A, D, E and K) are a group of chemical substances with different structure which are soluble in organic solvents and are stored in the body fat, but can be toxic when consumed in excess. Water-soluble vitamins (vitamins $B_1$, $B_2$, $B_6$, $B_{12}$, folic acid, pantothenic acid, niacin, biotin and vitamin C) are not normally stored in significant amounts in the body, which leads to the need for a daily supply of these vitamins (Ball, 1998). Methods used to evaluate the vitamin content can be seen in Table 47.

### 12.2. B complex vitamins

#### 12.2.1. Methods for B complex vitamins

Before vitamin analysis, it is necessary to determine the moisture content, so the results can be expressed in dry basis. From the current methods for moisture determination in dried bee pollen samples, the drying out process with infrared light at 85 °C and lyophilization is recommended (Melo & Almeida-Muradian, 2010, 2011). The following procedures are based mainly in the on the work of Arruda et al. (2013a) and Arruda et al. (2013b) and Souza et al. (2018).

***12.2.1.1. Simultaneous extraction of vitamins $B_1$, $B_2$, vitamers from vitamin $B_6$ and niacin.*** For the simultaneous extraction of B vitamins, it is used an acid hydrolysis followed by enzymatic treatment with fungal diastase like this:

In an Erlenmeyer flask, approximately 5 g of dried bee pollen is mixed with 50 ml of HCl 0.1 mol/l and kept in boiling water bath for 30 min. After cooling to room temperature, the solutions have their pH adjusted to 4.6 with sodium acetate 2.5 M. After adding 0.5 g of the diastase (from fungi), the solutions are incubated for 2 h at 42 °C in a water bath. After the enzymatic treatment and cooling at room temperature, the solutions are transferred to 100 ml volumetric flasks; the volumes should be completed with deionized water, homogenized and filtered through filter paper and in 0.45-mm membrane for injection into the chromatographic system.

***12.2.1.2. Chromatographic conditions for the analysis of the B complex vitamins in dried bee pollen.*** The chromatographic conditions vary for each vitamin. Vitamin $B_1$ and $B_2$ are analyzed under the same conditions, while vitamin $B_6$ and PP are quantified in distinct conditions. Vitamins $B_2$ and $B_6$ are detected directly by fluorescence, while vitamins $B_1$ and PP underwent pre- and post-column reaction, respectively, to be also detected by fluorescence.

***12.2.1.3. Determination of thiamin (vitamin B1), with pre-column reaction.*** For the reaction of thiamin into thiochrome, 1 ml of standard solution is diluted from the sample extract is pipetted into 10 ml amber volumetric flasks, 2 ml of deionized water and 3 ml of freshly prepared alkaline potassium ferricyanide solution, [0.25% (w/v) in 15% NaOH]. The mixture is homogenized and put to rest for 2 min to allow the reaction. After this period, 450 ml of 85% orthophosphoric acid is added. The solution is then cooled, and the final volume is completed with deionized water. These solutions are filtered in a 0.45-mm membrane and injected into the chromatographic system immediately after preparation. For the separation, 20 ml is injected into a $C_{18}$ reversed-phase column (RP-18 spherical 5 μmm/250 × 4.6 mm) with pre-column (5 μm/10 × 4.6 mm). The mobile phase is composed of phosphate buffer pH 7.2 (10 mM $KH_2PO_4.3H_2O$) and dimethylformamide (85:15); flow rate: 1 ml/min and detection by fluorescence (Ex 368 nm; Em 440 nm).

***12.2.1.4. Determination of riboflavin (vitamin $B_2$).*** 20 ml is injected into $C_{18}$ reversed-phase column (RP-18 spherical 5 μm/250 × 4.6 mm ) with pre-column μm/10 × 4.6 mm). The mobile phase is composed of phosphate buffer pH 7.2 (10 mM $KH_2PO_4.3H_2O$) and dimethylformamide (85:15); flow rate of 1 ml/min and detection by fluorescence (Ex 450 nm; Em 530 nm).

***12.2.1.5. Determination of vitamin PP through niacin and niacinamide vitamers.*** Vitamin PP is detected by fluorescence after reaction under ultraviolet (UV). For the separation, 20 ml is injected into $C_{18}$ reversed-phase column (Luna $C_{18}$, spherical 5 μm/250 × 4.6 mm) with pre-column (5 μm/2 × 8 mm). The mobile phase is composed of 0.07 M potassium dihydrogen phosphate, $KH_2PO_4$ solution (pH 4.5) of phosphate buffer of 75 mM hydrogen peroxide and 0.1% of 5 mM copper sulfate solution; flow: 1.5 ml/min; reactor composed of 12 m of tetrafluoroethylene (TFE) tuve (0.5 mm internal diameter, wrapped around a black lamp) that emits radiation from 300 to 400 nm; fluorescence detection (Ex 322 nm; Em 380 nm).

***12.2.1.6. Determination of vitamin $B_6$ through the vitamers pyridoxol, pyridoxal and pyridoxamine.*** For the separation, 20 ml is injected into a $C_{18}$ reversed-phase column (LiChrospherI 100 RP-18 end-capped 5 μm/250 × 4.0 mm), with pre-column (5 μm/4 × 4 mm Lichrospher 100 RP-18). The mobile phase is composed of phosphate buffer pH 2.5 (39 mM $KH_2PO_4$) with ion pair (0.63 mM PIC 7) and acetonitrile (96:4); flow: 0.6 ml/min; detection by fluorescence (Ex 296 nm; Em 390 nm).

### 12.3. Methods recommended to analyze antioxidant vitamins of bee pollen

The following procedures are based mainly in the *on the work of* Melo and Almeida-Muradian (2010, 2011) and Sattler et al. (2013).

#### *12.3.1. Determination of vitamin C*

The method is a titrimetric method, based on the reduction of 2,6-dichlorophenol-indophenol (DCPIP) by ascorbic acid (also recommended by AOAC, 1995).

#### *12.3.2. Determination of vitamin E*

This vitamin is analyzed using high-performance liquid chromatography (HPLC) using silica column and fluorescence detector (Ex. 295; Em. 330); mobile phase: hexane/isopropyl alcohol (99:1); flow rate: 1.5 ml/min.

#### *12.3.3. Determination of β-carotene (provitamin A)*

This carotenoid, as well as other carotenoids can be analyzed using both open column chromatography (OCC) (Almeida-Muradian et al., 2005; Oliveira et al., 2009) or HPLC, using $C_{18}$ column (25 cm, 5 um), diode array detection (450 nm); mobile phase: methanol, ethyl acetate, acetonitrile (70:20:10); flow rate: 1.2 ml/min.

### *12.4. Concluding remarks*

Note that the dehydration process, which is favorable for the preservation of bee pollen and the increase of shelf life of the product, do not interfere in the content of complex B vitamins ($p < 0.05$). In general, it can be stated that the concentration of vitamins is dependent on storage time and not on the conditions in which the dry pollen is stored ($p < 0.05$). When stored at room temperature, vitamin C is rapidly lost, probably due to oscillations in temperature, and, after this initial sharp loss, it becomes quite stable during the storage process. In relation to vitamin E (Melo & Almeida-Muradian, 2010), the storage conditions do not account for drastic losses in its concentrations.

## 13. Standard methods for mineral and trace elements

### *13.1. Introduction*

Insufficient data are available on the different composition and variability of minerals and trace elements in bee pollen between plant species. In each region, the collected bee pollen generally came from just a few spontaneous plant species and agricultural crops in each region play also an important role as pollen sources. Among the micronutrients vitamins, polyphenols and minerals are the more important for human well-being. Nevertheless, some minerals can be toxic if the intake is higher or over a long period, (e.g., Zn, Se, Mn and Mo). Others at excessive levels may exhibit high toxicity (Villanueva & Marquina, 2001; Serra-Bonvehí & Escolá Jordá, 1997; Yang et al, (2014). According FAO (2001) the mineral elements can be classified as "nutritionally essential major elements" (e.g., Ca, Cl, K, Mg, N, Na, P and S), "nutritionally essential minor elements" (e.g., Fe, B, Br, I and Si) and those regarded as "toxic" at excessive levels and may exhibit toxicity (e.g., Zn, Se, Mn, Mo). Given the importance and variability of different minerals founded in bee pollen, Campos et al. (2008) proposed the inclusion in label of bee pollen, for sale, additional information related to different constituents and also minerals, to complement the label and improve the value of the product.

### *13.2. Materials and methods*

Usually the determination of mineral and trace elements on bee pollen is carried out on pollen ash, most frequently by atomic absorption. The main mineral is K (about 60% of total mineral content), Mg constitutes about 20%, and Na and Ca 10% (Szczęsna & Rybak-Chmielewska, 1998).

#### *13.2.1. Sample preparation*

According to the analytical method used, different sample preparations methodologies can be performed.

*Method 1: for Flame and Electrothermal Atomic Absorption Spectrometry* (Stanciu, Marghitas, Dezmirean and Campos, 2012).

1. weigh 2 g of pollen;
2. ash in a muffle furnace at 450 °C overnight;
3. digest by treating with hydrochloric acid 6 M (5 ml) and $H_2O_2$ (3 ml) and then evaporate;
4. dissolve with 25 ml nitric acid 0.1 M. A blank digest was also carried out;
5. Lanthanum nitrate [$La(NO_3)_3x6H_2O$] were used as matrix modifier for the determination of potassium, calcium and magnesium.

*Method 2: for flame atomic absorption spectrometry and analyzing honey and pollen (Grembecka & Szefer, 2013):*

1. weigh 1 g (±0.0001 g);
2. treat with 9 ml 65% $HNO_3$ and digest in an automatic microwave digestion system: 250 W, 48 s; 0 W, 48 s; 250W, 6 min 24 s; 400 W, 4 min; 650 W, 4 min;
3. cycle consists of five food samples and one blank sample (9 ml 65% $HNO_3$);
4. every digested sample is dissolved in up to 10 ml deionized water.

*Method 3 (Loper* et al., *1980):*

1. weigh 1g of pollen;
2. digest with alternate aliquots of concentrated nitric acid and hydrogen peroxide;
3. boil and dry at 130 °C until the residual ash will be white or light gray. Three cycles were sufficient;
4. taken up residues in dilute HCl and determine directly for Zn and Pb. For Mg, Ca and K analyses, samples are diluted 50–100 times to provide concentrations in the proper absorption range.

*Method 4: for Inductively coupled argon plasma-atomic emission spectrometry (ICP-AES): Paulo* et al. *(2012, 2014)*

1. 200 mg of dry bee pollen digest with a mixture of nitric acid (10% v/v) and hydrogen peroxide (30% v/v) at 100 °C in a block digestion system (Digiprep MS) until complete digestion (12 h).
2. final residue is dilute to 50 ml with nitric acid (10% v/v) and filter.

*Method 5: Yang* et al. *(2014)*

1. 1.0 g of digested bee pollen with 10 ml of concentrate nitric acid (65%) and 1 ml of hydrogen peroxide (30%) in a close polytetrafluoroethylene-stopper vessel for 8 h at 170 °C.

*13.2.2. Methods of analysis*

The methods that are usually used for mineral composition determination in food products were the Flame and Electrothermal Atomic Absorption Spectrometry and the inductively coupled plasma atomic emission spectroscopy with or without mass detector. However, some easier methods could be use if only the determination of calcium will be required. Kelina et al. (2013) determine the concentration of calcium in bee pollen samples from Lambayeque – Peru using complexometric method. In this method 0.1 g of ash were used to adding a few drops of hydrochloric acid, the results being expressed as percentage on dry basis.

***13.2.2.1. Flame and electrothermal atomic absorption spectrometry.*** Flame and Electrothermal Atomic Absorption Spectrometry is a very common method of food mineral analyses applied mostly for the determination of alkali and alkaline metals.

*Equipment*

(1) Device equipped with deuterium lamp for background correction and hallow-cathode lamps for each of the elements studied and air/acetylene flame (Stanciu et al., 2011, 2012). (2) Flame atomic spectrometric measurements (F-AES/AAS) using the optimal instrumental parameters for each element with the following wavelengths: 766.5 nm (K); 422.7 nm (Ca); 285.2 nm (Mg); 248.3 nm (Fe) and 213.9 nm (Zn). Good linearity for the method is observed and the calibration curves present a very good accuracy with an $r^2$ ranged between 0.9990 and 0.9999.

Grembecka and Szefer (2013) analyzed the concentration of elements (Mg, Ca, K, Na, Zn, Cu, Fe, Cr, Co, Ni, Mn, Pb and Cd) in pollen and honey in an air–acetylene flame by AAS method using deuterium-background correction. These authors added Cs to samples, for Na and K determinations and standards as an ionization buffer at a concentration of 0.2%w/v. In the case of Ca and Mg measurements, La was used as a releasing agent at a concentration of 0.4%w/v.

***13.2.2.2. Inductively coupled plasma atomic emission spectroscopy (ICP-AES) methods.*** Method I (Paulo et al., 2012, 2014):

1. All experiments were carried out using an inductively coupled plasma-optical emission spectrometry (ICP-OES- Activa M, Horiba Jobin Yvon), operating at 1000 W plasma power;
2. 15 l $min^{-1}$ plasma gas flow, 0.02 l $min^{-1}$ nebulizer Air flow and 1.0 bar air pressure;
3. The analytical wavelengths (nm) were set of the following: Cd (228.802), Cr (205.571), Cu (327.395), Fe (259.940), Mn (257.611), Pb (283.305) and Zn (213.857);
4. Results expressed in $mg.kg^{-1}$ on a dry weight basis.

Method II (Yang et al., 2014):

1. Plasma gas flow rate (Ar), 16 $l.min^{-1}$;
2. auxiliary flux, 1.0 $l.min^{-1}$;
3. nebulized pressure, 25 psi; sample flush time, 30 s; delay time, 30 s; solution uptake rate, 1.60 $ml.min^{-1}$;
4. radio frequency, 27.12 MHz; power, 1.05 kW;
5. absorbance data for each element recorded at the following wavelengths: Al ($\lambda$ = 308.2 nm), As ($\lambda$ = 189.0 nm), Ca ($\lambda$ = 317.9 nm), Cd ($\lambda$ = 228.8 nm), Co ($\lambda$ = 228.6 nm), Cr ($\lambda$ = 283.5 nm), Cu ($\lambda$ = 324.7 nm), Fe ($\lambda$ = 259.9 nm), Ge ($\lambda$ = 265.1 nm), Hg ($\lambda$ = 184.9 nm), K ($\lambda$ = 766.4 nm), Mg ($\lambda$ = 279.5 nm), Mn ($\lambda$ = 257.6 nm), Mo ($\lambda$ = 202.0 nm), Na ($\lambda$ = 589.5 nm), Ni ($\lambda$ = 231.6 nm), P ($\lambda$ =213.6 nm), Pb ($\lambda$ = 220.3 nm), Se ($\lambda$ = 196.0 nm) and Zn ($\lambda$ = 213.8 nm);
6. Results expressed in micrograms per gram on a dry weight basis.

To be underlining that ICP-MS is a powerful, but expensive technique for determination of trace mineral. It is also used for other bee products too as honey, for example Caroli et al. (1999), Madejczyk and Baralkiewicz (2008) and Chudzinska and Baralkiewicz (2011).

### *13.3. Concluding remarks*

Some preliminary results from the composition of minerals in bee pollen can give a good idea of the amounts and the identification of the main of these compounds on this crude material. Still the information remains scarce relatively to different floral sources and in special correlating minerals and regions of the globe. The countries should do similar assays to collect the most data possible to validate the amounts and the contaminants as this content could be a bio-indicator of the environmental condition or pollutants.

The amount of minerals in bee pollen from the various data collected in the published papers can be found in Table 49. Stanciu et al. (2012) analyzed pollen loads samples obtained by pollen traps installed in *Apis mellifera* (L.) hives located in the northwest and central counties of Transylvania area (Romania). These authors confirmed that the examined bee pollen is a natural source of nutritionally essential minerals and can contribute for a better balanced diet or for special therapeutic applications. In their work, the authors reported that the main macro elements were potassium (2483–7620 mg/kg), calcium (553–2798 mg/kg) and magnesium (3555–205 mg/kg), while the main microelements were iron (18.8–135 mg/kg) and zinc 18.8–60.5 $mg.kg^{-1}$. Variations in the mineral levels of the analyzed monofloral bee pollens were due to differences in the floral origin.

Loper et al. (1980) use similar methodology, with flame absorption spectrometry for mineral determination (e.g., Mg, Ca, K, Zn and Pb) in fresh pollen collected by hand from flowers, bee pollen removed from the bees' legs and collected from the pollen traps from California. Potassium is the main compound in bee pollen. Highest values of potassium were reported by Stanciu et al. (2011) in *Salix* sp. pollen (8686 mg/kg), in *Helianthus annuus* L. pollen (5421 mg/kg) and by Stanciu et al. (2012) in *Knautia arvensis* (L.) (7619.57 mg/kg), in *Onobrychis viciifolia* Scop. (6748.72 mg/kg), *Crataegus monogyna* Jacq. (6224.17 mg/kg) and *Inula helenium* L. (6078.28 mg/kg), *Taraxacum officinale* Web. (2483.51 mg/kg), *Carduus* sp. (3865.27 mg/kg) and among others pollen samples analyzed.

Different authors reported different values for pollen potassium concentration, namely: 4000 mg/kg (Villanueva & Marquina, 2001), 5530 mg/kg (Somerville & Nicol, 2002), 2843–5976 mg/kg (Szczęsna, 2007) and 4950–5131 mg/kg (Salamanca et al., 2008) (Table 49). These results confirm the higher variability of pollen mineral composition related to the different origin of pollen. Calcium followed by magnesium were the next highest mineral content in pollen. All authors founded significant differences related to the higher concentration of potassium compared with the other minerals. Yang et al. (2014) analyzed twelve common varieties of monofloral bee pollen collected from China's main producing regions. On average the values obtained were the following: P 5946 mg/kg, K 5324 mg/kg, Ca 2068 mg/kg, Mg 1449 mg/kg, sodium 483 mg/kg, aluminum 129 mg/kg, iron 119 mg/kg, magnesium 70 mg/kg, zinc 45 mg/kg and cupper 17 mg/kg.

Paulo et al. (2012) analyzing *Cistus ladanifer* L., *Rubus ulmifolius* Schott and *Calluna vulgaris* (L.) found different composition of micronutrients (Cu, Fe, Mn and Zn) in pollen related to different botanical origins but not from the same species in different region. These authors concluded that the determination of micronutrients could be suitable for the identification of botanical species. The contents of Cd, Pb, As and Hg were usually lower or not detected in pollen samples. In Campos et al. (2008), the limits for this trace elements were Cd $\leq$ 0.1 mg/kg, Pb $\leq$ 0.5 mg/kg, As $\leq$ 0.5 mg/kg and Hg $\leq$ 0.03 mg/kg.

## 14. Standard methods for pyrrolizidine alkaloid profiling by UHPLC-HRMS

### 14.1. Introduction

Pyrrolizidine alkaloids (PAs) are secondary metabolites produced by some plants as defense mechanism against herbivores and phytophagous insects. They can occur as *N*-oxides or free-bases/tertiary PAs. In plants, the *N*-oxides are often found in higher concentrations as compared to the corresponding free-bases/tertiary PAs (Hartmann & Toppel, 1987; Hartmann & Witte, 1995). PA containing plant species mainly belong to the families of Boraginaceae (all genera), Asteraceae (mainly genera of Senecioneae and Eupatorieae), Fabaceae (mainly genus of *Crotalaria*) and Apocynaceae (Hartmann & Witte, 1995). PAs can get into bee products, such as honey and pollen, when bees collect nectar or pollen of PA containing plants. PAs have been detected in plant pollen (Boppré et al., 2005; Kempf et al., 2010). High concentrations of PAs have also previously been reported in bee collected pollen (Dübecke *et al.*, 2011; Kempf et al., 2010) presenting a food safety concern for the consumers of pollen as a nutritional supplement. PAs are classified as esters of hydroxylated methyl pyrrolizidines, consisting of a necine base (1,2-saturated or unsaturated) and one or more necinic acids (Hartmann & Witte, 1995) (Figure 47a,b). Esters of 1,2 unsaturated retronecine- and otonecine type PAs are toxic for humans and animals (Prakash et al., 1999; Rösemann et al., 2014). The structural diversity of more than 400 known PAs (Boppré, 2011; Kempf et al., 2011) represents an analytical challenge and no standardized method has been established so far to determine the PA content in bee products. Many PAs are difficult to quantify, due to the lack of standard materials.

Two main analytical approaches are commonly used: a sum parameter GC-MS method (Kempf et al., 2008, 2010, 2011) and a targeted LC-MS method allowing the individual identification of PAs and PA-*N*-oxides (Betteridge et al., 2005; Boppre et al., 2005; Kempf et al., 2011). The GC-MS method covers most PAs, except the otonecine-type PAs, but derivatization is required and structural information of the original PAs is lost. On the other hand, the targeted LC-MS method identifies individual PAs, but unknown PAs are not detected and the chromatographic separation of many PA isomers is also not easily obtained. To at least partially overcome these limitations, ultra-high pressure liquid chromatography coupled to high resolution mass spectrometry (UHPLC-HRMS) as a way to profile both known and unknown PAs can be used (Avula et al., 2015). Compared to conventional HPLC, UHPLC enables a more rapid separation with higher resolution and is more efficient at separating closely related molecules, e.g., positional isomers (Plumb et al., 2004). UHPLC relies on the use of smaller particles (sub-2µm) and instrumentations able to resist higher pressure, typically up to 1000 or 1500 bars. HRMS systems, such as time-of-flights (TOF) or electrostatic trap (Orbitrap™) represent an attractive way to record data from UHPLC separation at high frequency and in a non-targeted manner. HRMS measures ions with high mass and spectral accuracies and allows for the determination of elemental compositions, which in turn may assist the identification of unknowns. Furthermore, recent instruments are fully compatible with quantitative or semi-quantitative analysis (Rochat, 2012). Here we describe a UHPLC-HRMS-based method for the profiling of PAs in plant pollen from *Echium* and *Eupatorium* species, which may be applied to other plant genera with minor adaptation.

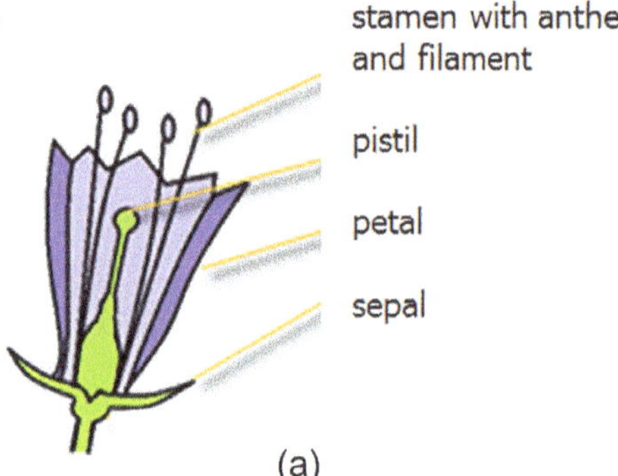


(a)

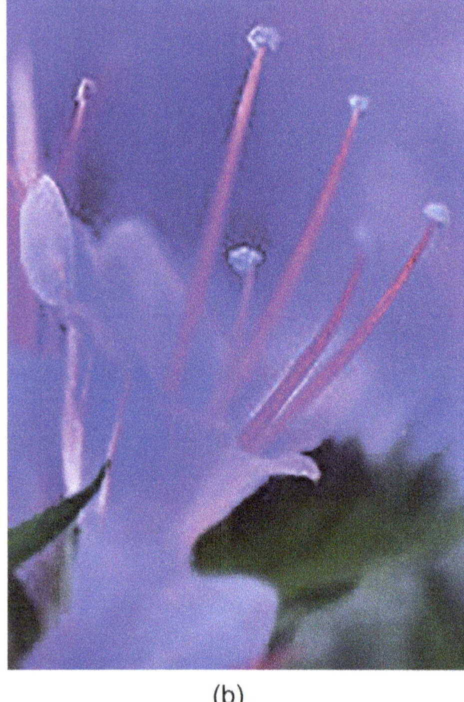

(b)

Figure 48. (a and b) Structure of a flower of *E. vulgare*.

## *14.2. Materials and methods*

### *14.2.1. Equipment and chemicals*

a. Steel-made forceps from A. Dumont (#0208-55-PO)
b. 1.5 ml and 2 ml tubes from Eppendorf (#0030120086, #0030120094)
c. 2 ml glass LC-MS tubes (screw top) from Kinesis (#KVP6112)
d. polystyrene Petri dishes (Ø 10 cm) from Greiner (#663102)
e. glass beads (Ø 2 mm) from Sigma-Aldrich
f. nitrile gloves (powder free) from Kimberly-Clark (#52003M)
g. Solvents: cyclohexane SupraSolv (#102817), methanol SupraSolv (#106011) and formic acid for Analysis Emsure (#100264) from Merck, Acetonitrile ULC/MS (#012041) and Water ULC/MS (#232141) from Biosolve
h. Standards: echimidine (CAS #520-68-3) and lycopsamine (CAS #10285-07-1) from Phytolab (#89553, #89726).

### *14.2.2. Collection of plant pollen from PA-containing plants*

Two examples: collection of pollen from *E. vulgare* (Boraginaceae) and from *E. cannabinum* (Asteraceae).

***14.2.2.1. Pollen collection from* E. vulgare.** Plant pollen from *E. vulgare*, located as blue/violet dust on the anthers of the flower can be collected using steel-made surgery forceps or using pure cyclohexane (Figure 48a,b and Figure 49). Rainy days are not recommended for collection, since pollen can be washed out from the anthers.

1. Verify that most of the flowers are well developed and fully open.
2. Tightly bag the plants on the field with fine mesh nets to avoid the pollen harvest from insects (at least one day before the collection).
3. Spread a layer of paraffin around the lower part of the stem to avoid that non-flying insects are climbing the plant to collect pollen.
4. Plant pollen can be collected two days after bagging the plant and is stored at −20 °C until extraction. It is suggested to bring a cooler box container filled with dry ice on the field in order to place the freshly collected samples immediately on dry ice.

*14.2.2.1.1. Collection with forceps.*

1. Wear nitrile gloves during the collection.
2. Using forceps, scratch the pollen from the surface of the anthers of the flowers, carefully avoiding PA contamination by the stamen tissues.
3. Collect the pollen using pre-weighed 1.5 ml tubes. Collect 1 mg of pollen in each tube.
4. Insert the forceps carefully inside the tube and release the pollen on the walls of the vial. Try to avoid any electrostatic charge of the tube that could compromise the collection.

*14.2.2.1.2. Collection with cyclohexane.*

1. Remove the stamens from the flower.
2. Immerge the stamens in a pre-weighed LC-MS glass tube containing cyclohexane.
3. Shake the stamens delicately into the cyclohexane to release the pollen.
4. If possible, bring the tubes back to the laboratory to evaporate the cyclohexane under laminar-flow hood. Determine the weight of the dry pollen.
5. Store the dry pollen at −80 °C until extraction.

**Pros:** This technique allows a faster collection of pollen when compared to the method by forceps. Both collection methods give similar concentrations of PAs as we have determined in our laboratory.

### *14.2.3. Pollen collection from* E. cannabinum

Plant pollen from *E. cannabinum* (Figures 50 and 51), located along the stamens, can't be collected using forceps due to the small dimensions of the tubular floral units (approximately 7 mm).

Figure 49. An *E. vulgare* plant bagged with a net and supported by a metal structure.

*Collection on Petri dishes*

1. Bag the plant and apply the paraffin layer on the lower part of *E. cannabinum*.
2. After two days, shake the entire floral head of *E. cannabinum* over a Petri dish in order to collect the pollen that is released by this procedure.
3. Remove the impurities from the dish using forceps.
4. Collect the pollen into a pre-weighted 1.5 ml tube.
5. Store the tube at −80 °C until extraction.

**Pros:** This method allows to collect rapidly a high amount of pollen from *E. cannabinum*.

### *14.3. Extraction of PAs from plant pollen*

PAs are polar organic compounds with basic characteristics. They are soluble in polar organic solvents or in mixtures of solvents and acidified water.

1. Transfer 1 mg of plant pollen in a pre-weighed 2 ml microcentrifuge tube using a metal spatula.
2. Dissolve the pollen in 100 μl of extraction solvent consisting of 70% methanol, 29.5% ultra-pure water and 0.5% formic acid.
3. Add 4–8 glass beads to the tube.
4. Shake the tube at 30.0 Hz for 4 minutes.
5. Centrifuge the tube at 18,407 *g* for 4 minutes.
6. Collect the supernatant (extract) in a new tube.
7. Transfer with a pipette 5 μl of the extract and dilute 5–20 times with the extraction solvent into a glass LC-MS vial containing a conical glass insert.

**Pros:** This technique can be used for extracting plant pollen from many plant genera.

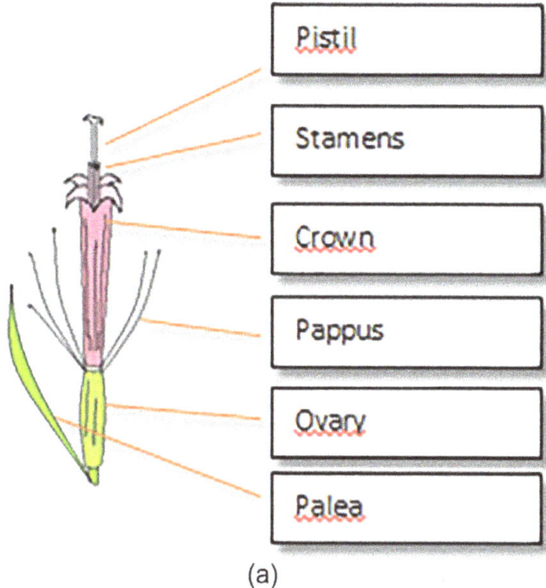



(a)

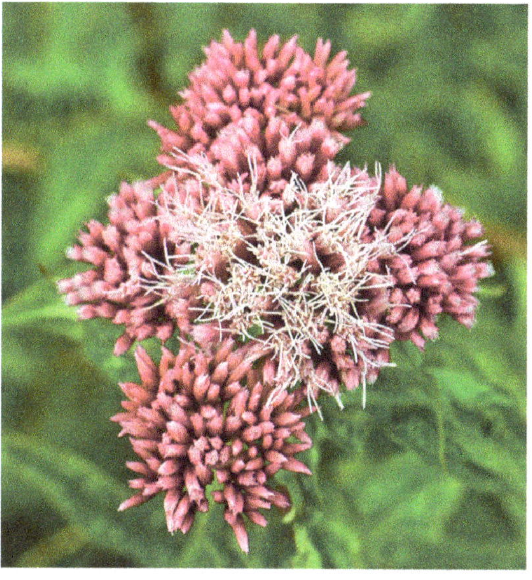

(b)

Figure 50. Structure of a floret and a floral head of *E. cannabinum*.

### *14.4. Profiling of PAs in plant pollen with UHPLC-HRMS*

Non-targeted analysis using LC-HR-MS system allows the detection of alkaloids found in *Echium vulgare* (echimidine/-*N*-oxide, vulgarine/-*N*-oxide, acetyl-echimidine/-*N*-oxide, acetylvulgarine/-*N*-oxide and echivulgarine/-*N*-oxide) and alkaloids found in *Eupatorium cannabinum* (intermedine/-*N*-oxide, lycopsamine/-*N*-oxide). Separation of the alkaloids is performed using an Acquity UPLC (Waters). The UPLC system is coupled to a Synapt G2 QTOF mass spectrometer (Waters). An Acquity BEH C18 column (50 × 2.1 mm i.d., 1.7 μm particle size, Waters) fitted with guard column (5 × 2.1 mm) of identical phase is utilized. The column is maintained at 30 °C and a binary gradient of separation is performed at flow rate of 0.4 ml min$^{-1}$. The mobile phase consists of 0.05% formic acid in water (solvent A) and 0.05% formic acid in acetonitrile (solvent B). The gradient program used is: 0–4 min 5–40% B, 4–6 min 40–100% B, 6–9 min 100% B, 9.1–10.5 min 5% B. Injection volume is 1 µl. The autosampler needle is washed with methanol/acetonitrile/isopropanol (1:1:1) followed by 0.05% formic acid in water between injections to eliminate carryover. The QTOF operates in electrospray positive mode over a mass range of 50–600 Da. MS conditions are: Capillary voltage +2800 V, cone voltage +30 V, source temperature 120 °C, desolvation gas temperature and flow 350 °C and 800 l/h, respectively, scan time 0.4 s. A leucine-enkephaline solution at 400 ng/ml is infused throughout the analysis to ensure high mass accuracy (<2 ppm). Fragmentation spectra are recorded in separate analyses in MSe mode using a collision energy ramp of 10–30 eV. Data are recorded using Masslynx 4.1 and PAs are identified on the basis of their retention times, exact mass fragmentation characteristics and comparison with the existing literature and databases containing information on PAs known in *Echium* and *Eupatorium* genus. External calibration for the quantification of the PAs is made using echimidine, intermedine and lycopsamine as standards. Five calibration points are made: 0.02, 0.1, 0.5, 2, and 10 µg/ml. Linear responses are obtained from 0.02 to 2 µg/ml (Table 48). The limit of detection (LOD) and limit of quantitation (LOQ) for echimidine, lycopsamine and intermedine are 1.5 ng/ml and 3 ng/ml, respectively. As only a limited number of reference standards are available, a number of other PAs and PA-*N*-oxides commonly found in *Echium* (echimidine-*N*-oxide, vulgarine/-*N*-oxide, acetyl-echimidine/-*N*-oxide, acetylvulgarine/-*N*-oxide, echivulgarine/-*N*-oxide) are indirectly quantified as echimidine equivalents. PAs and their PA-*N*-oxides and isomers in pollen of *E. vulgare* and *E. cannabinum* determined with LC-MS analysis are reported in Table 49. Two examples of extracted ion chromatograms of PAs are shown in Figure 52.

Figure 51. An *E. cannabinum* plant bagged with a net.

Pollen samples of *E. vulgare* and *E. cannabinum* analyzed contain mainly PA-*N*-oxides while free bases were only present as traces. Table 50 shows the MS/MS fragments of PA-*N*-oxides detected in pollen samples. In addition, PAs not previously reported to be present in pollen of *E. vulgare* and *E. cannabinum* can also be detected by UHPLC-HRMS. They are tentatively identified through the determination of their molecular formula based on high mass and spectral accuracy measurements by the high resolution QTOF mass spectrometer (Table 51).

### *14.5. Data processing*

The following procedure is used for the detection and quantification of PAs. Provided that the identity of the PA is known, this procedure may also be applied to other types of alkaloids from other plant species.

1. To identify the different alkaloids, generate an extracted ion chromatogram (EIC) using a mass window of ± 0.01 Da around the values provided in Table 49.
2. Verify the mass accuracy of the ions compared to the theoretical values.
3. In the EICs integrate the peaks and calculate concentrations in ng/g or µg/g of pollen using your own

Table 51. PAs which have not been previously reported from *E. vulgare* and *E. cannabinum*.

| Compound name | RT (min) | Chemical formula | [M + H]$^+$ experimental | [M + H]$^+$ calculated | Error (mDa) |
|---|---|---|---|---|---|
| • *Echium vulgare* | | | | | |
| Curassavine-*N*-oxide* | 1.54 | C16H29NO5 | 316.2125 | 316.2124 | 0.1 |
| • *Eupatorium cannabinum* | | | | | |
| Leptantine* | 1.41 | C15H27NO6 | 318.1922 | 318.1917 | 0.5 |
| Amabiline* | 1.48 | C15H25NO4 | 284.1865 | 284.1862 | 0.3 |
| Uplandicine* | 1.71 | C17H27NO7 | 358.1868 | 358.1866 | 0.2 |

*Tentative identification.

Table 52. Frequency of each plant genus as predominant pollen, secondary pollen, important minor pollen and minor pollen in the 134 pollen samples mixture analyzed.

| Genus | Predominant pollen (>45%) | Secondary pollen (15–45%) | Important minor pollen (3–15%) | Minor pollen (<3%) |
|---|---|---|---|---|
| *Eucalyptus* spp. | 13 | 21 | 21 | 21 |
| *Prunus* spp. | 8 | 4 | 4 | 4 |
| *Rubus* spp. | 8 | 11 | 8 | 8 |
| *Castanea* spp. | 15 | 34 | 23 | 8 |
| *Trifolium* spp. | 0 | 6 | 11 | 19 |
| *Cistus* spp. | 53 | 17 | 6 | 4 |
| *Echium* spp. | 21 | 17 | 11 | 4 |
| *Kiwi* spp. | 2 | 0 | 4 | 11 |
| *Leontondon* spp. | 0 | 2 | 17 | 11 |
| *Quercus* spp. | 0 | 6 | 11 | 15 |
| *Erica* spp. | 2 | 4 | 6 | 2 |
| *Thymus* spp. | 2 | 0 | 11 | 2 |
| *Genista* spp. | 0 | 6 | 6 | 2 |
| *Lavandula* spp. | 0 | 4 | 13 | 0 |
| *Cytisus* spp. | 11 | 23 | 11 | 2 |

Table 53. Statistics of the 134 sample of bee pollen analyzed.

| Parameter | Mean ± $\sigma$ | Min – max | Coefficient of variation |
|---|---|---|---|
| Total phenolic compounds | 23.9 ± 4.16 | 16.1–35.5 | 17.4 |
| Total flavonoids | 4.1 ± 0.96 | 2.2–5.9 | 23.2 |
| $EC_{50}$ | 3.5 ± 1.05 | 2.0–6.4 | 39.1 |
| ABTS | 244.9 ± 24.40 | 203.8–298.6 | 10.0 |

Table 54. Results for calibration and cross-validation.

| Calibration | | | | | | |
|---|---|---|---|---|---|---|
| Parameter | Spectral range ($cm^{-1}$) | Pre-processing | Rk | $r^2$ | RMSEE | RPD |
| Total phenolic compounds | 2787 – 2459 + 2135 – 1809 | 2ndDer | 8 | 92.6 | 1.16 | 3.7 |
| Total flavonoids | 2858 – 2509 + 2164 – 1470 + 777 – 430 | VecNor | 9 | 85.9 | 0.373 | 2.7 |
| $EC_{50}$ | 3055 – 2731 + 2087 – 1119 + 799 – 474 | VecNor | 10 | 89.7 | 0.351 | 3.1 |
| ABTS | 2557 – 1838 + 1479 – 399 | 1stDer | 10 | 91.3 | 7.57 | 3.4 |
| Cross-validation | | | | | | |
| Parameter | Spectral range($cm^{-1}$) | Pre-processing | Bias | $r^2$ | RMSECV | RPD |
| Total phenolic compounds | 2787 – 2459 + 2135 – 1809 | 2ndDer | 0.0044 | 79.3 | 1.88 | 2.2 |
| Total flavonoids | 2858 – 2509 + 2164 – 1470 + 777 – 430 | VecNor | −0.0020 | 76.0 | 0.469 | 2.0 |
| $EC_{50}$ | 3055 – 2731 + 2087 – 1119 + 799 – 474 | VecNor | 0.0185 | 77.9 | 0.492 | 2.1 |
| ABTS | 2557 – 1838 + 1479 – 399 | 1stDer | 0.2470 | 72.5 | 12.7 | 1.9 |

2ndDer – seconded derivative; VecNor – vector normalization; 1stDer – first derivative; $r^2$ – coefficient of determination; Rk – number of PLS components; RMSEE – root mean square error of calibration; RPD – residual prediction deviation; Bias – mean value of deviation, also called systematic error; RMSECV – root mean square error of cross-validation.

calibration curves. Examples of equations are provided in Table 48.

In this study we obtained a total PA concentration of 7.43 (± 2.45) mg/g in pollen of *E. vulgare* and 5.32 (± 1.98) mg/g in pollen of *E. cannabinum*. Pollen samples were collected in Switzerland in 2013.

### 14.6. Concluding remarks

Liquid chromatography approaches are frequently chosen for the identification and quantification of PAs in plants and for food analysis, allowing the determination of individual PAs and PA-*N*-oxides. This is particularly important for determining the exact type of PAs which are present in a given plant species and also in food analysis, where the spectrum of PAs may help deduce the plant species that contribute to PA contamination. Additionally, the sample preparation is faster and does not need derivatization as required for gas chromatography analysis. However, many types of PAs are still not available on the market as reference substances. Therefore, a number of PAs are quantified using a reference substance that is closely related but not identical to the analyzed PAs. Hence, quantification may not be entirely accurate, since the response factor of the

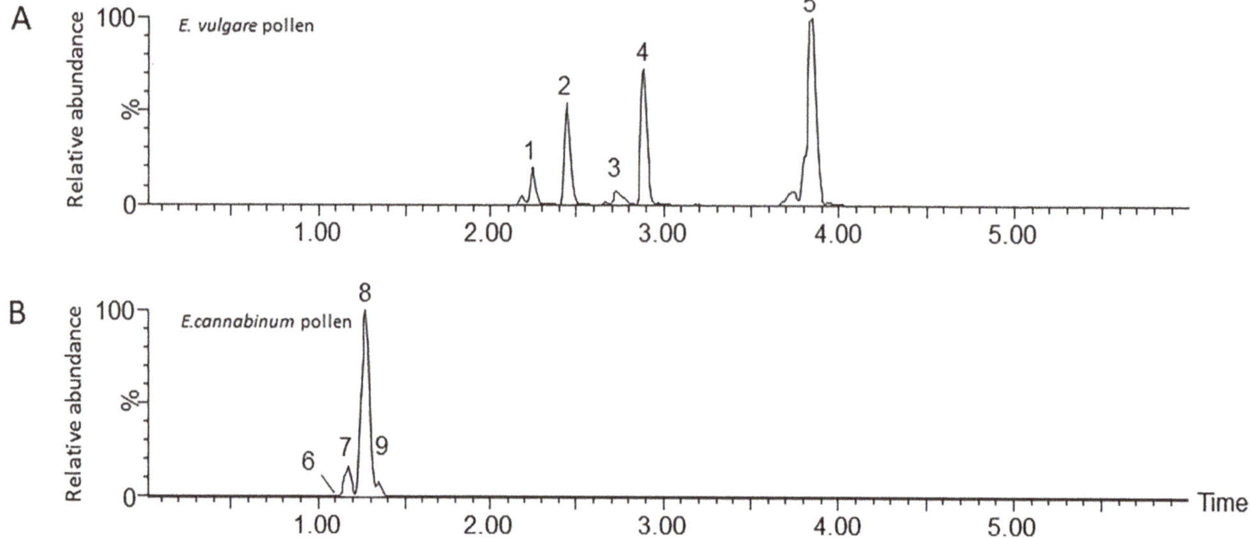


Figure 52. Extracted ion chromatograms of PAs identified in plant pollen of *E. vulgare* (A) and *E. cannabinum* (B). Peak numbers correspond to the PAs described in Table 51 (1: echimidine-*N*-oxide; 2: vulgarine-*N*-oxide; 3: acetylechimidine-*N*-oxide; 4: acetylvulgarine-*N*-oxide; 5: echivulgarine-*N*-oxide; 6: lycopsamine; 7: intermedine; 8-9: lycopsamine-*N*-oxide or intermedine-*N*-oxide).

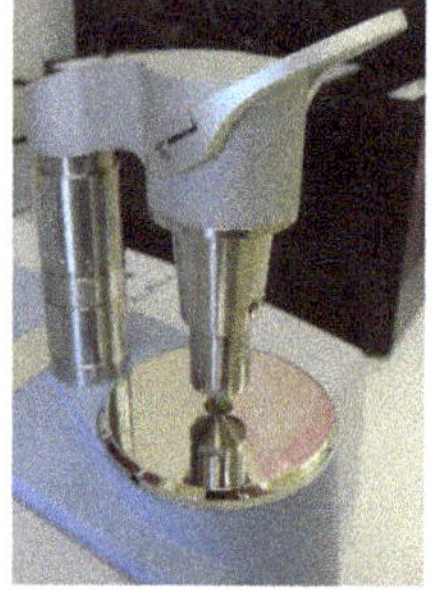

Figure 53. (a–d) Scheme of sample preparation for analysis in FTIR-ATR equipment.

detector may differ between PAs, even when they are closely related to each other. Furthermore, positional isomers, such as lycopsamine and intermedine, are also not easy to separate. Finally, while some plant species have been relatively well characterized regarding their PA content (e.g., *E. vulgare*), little is known for many other PA-producing plant species. Recently, ultra-high performance liquid chromatography coupled with high resolution mass spectrometry has been described as a selective and sensitive method for the analysis of PAs in plants (Avula et al., 2015; Crews et al., 2009, 2010; Xiong et al., 2009). This method provides an accurate mass information based on which unknown PAs may also be detected in an untargeted manner without the need for reference standards for each PA, as has been shown in the present study. In the future we anticipate that UHPLC-HRMS will become the method of choice for the profiling of both known and unknown PAs in plants.

## 15. Standard methods for total phenolic compounds, total flavonoids and antioxidative capacity bee pollen evaluation by FTIR-ATR spectroscopy

### 15.1. Introduction

FTIR-ATR techniques have been applied in several different crud materials in order to simplify the analysis and have results quickly. Gok et al. (2015) used FTIR-ATR spectroscopy in automated and highly sensitive botanical origin estimation of honey. According Anjos et al. (2017) FTIR-ATR spectroscopy technique in combination with partial least squares regression (PLS-R) can be used to evaluate the ability to quantify total phenolic compounds, total flavonoids, free radical scavenging ($EC_{50}$) and antioxidant capacity (ABTS) in bee pollen. The method described in this chapter is based in this previously paper was PLS-Regression analysis was used to predicted for total phenolic compounds, total flavonoids, EC50 and ABTS. The calibrations model founded present a good accuracy for total phenolic compounds ($r^2 = 92.6\%$; RPD = 3.7), total flavonoids ($r^2 = 85.9\%$; RPD = 2.7); $EC_{50}$ ($r^2 = 89.7\%$; RPD = 3.1) and ABTS ($r^2 = 91.3\%$; RPD = 3.4) of bee pollen).

Table 55. Condition variations of the DPPH* assay on bee pollen according to different authors.

| Number of report | Reference | Initial DPPH* concentration | Reaction time | Expression of activity | DPPH* volume/extract volume |
|---|---|---|---|---|---|
| 1 | Campos et al. (1997) | 5.91 mg/250 ml in ethanol | 10 min | $EC_{50}$ | 2.5 ml-DPPH* solution + 1 to 10 µl extract |
| 2 | Kroyer and Hegedus (2001) | $6 \times 10^{-5}$ M in methanol | 60 min | % inhibition | 2.7 ml-DPPH* solution + 0.3 ml extract |
| 3 | Nagai et al. (2002) | 1 mM | NI | % activity | 0.3 ml-DPPH* solution + 2.4 ml ethanol + 0.3 ml extract |
| 4 | Campos et al. (2003) | $6 \times 10^{-5}$ M in methanol | 10 min | $EC_{50}$ | 2 ml-DPPH* solution + appropriate amount of extract |
| 5 | Almaraz-Abarca et al. (2004) | 3.4 µg/ml in 50% etanol (v/v) | 10 min | $EC_{50}$ | 1600 to 1990 µl-DPPH* solution + 10 to 400 µl extract. Final volume: 2 ml |
| 6 | Nagai et al. (2005) | 1 mM | NI | % activity | 0.3 ml-DPPH* solution + 2.4 ml etanol + 0.3 ml extract |
| 7 | Silva et al. (2006) | $6 \times 10^{-5}$ M in ethanol | 30 min | $EC_{50}$ | 2 ml-DPPH* solution + appropriate amount of extract |
| 8 | Leja et al. (2007) | 0.1 mM in ethanol | 30 min | % inhibition | 0.083% pollen in the reaction mixture |
| 9 | Baltrušaitytė et al. (2007) | $6.5 \times 10^{-5}$ M in methanol | 16 min | % inhibition | 2 ml-DPPH* solution + 50 µl extract |
| 10 | Carpes et al. (2008) | 0.5 mM in ethanol | 100 min | % inhibition, $EC_{50}$ | 0.3 ml-DPPH* solution + 3 ml ethanol + 0.5 ml extract |
| 11 | Silva et al. (2009) | 100 µM in ethanol | 30 min | $EC_{50}$ | 2 ml-DPPH* solution + appropriate amount of extract |
| 12 | LeBlanc et al. (2009) | 51 mg/100 ml in methanol | 60 min | % activity, $EC_{50}$ | 242.5 µl DPPH* solution + 280 µl ethanol + 72 µl extract |
| 13 | Lee et al. (2009) | 10 mM | 10 min | % scavenging effect | 1 ml-DPPH* solution + 4 ml extract |
| 14 | Mărghitas et al. (2009) | 0.02 mg/ml | 15 min | % inhibition | 200 µl-DPPH* solution + 40 µl extract |
| 15 | Wang et al. (2010) | 0.5 mM in ethanol | 30 min | % activity | 0.5 ml-DPPH* solution + 0.5 ml acetate buffer (100 mM, pH 5.5) + 0.5 ml extract |
| 16 | Menezes et al. (2010) | 0.5 mM in ethanol | 30 min | % activity | 4 ml-DPPH* solution + 100 µl extract |
| 17 | Morais et al. (2011) | $6 \times 10^{-5}$ M in methanol | 60 min | %-DPPH* discoloration, $EC_{50}$ | 2.7 ml-DPPH* solution + 300 µl extract |
| 18 | Negri et al. (2011) | 20 µg/ml in methanol | 20 min | % activity | 3.9 ml-DPPH* solution + 0.1 ml extract |
| 19 | Graikou et al. (2011) | 70 µM in methanol | Until a plateau was reached | % activity, $EC_{50}$ | 1 ml-DPPH* solution + 50 µl extract |
| 20 | Feás et al. (2012) | $6 \times 10^{-5}$ M in methanol | 60 min | % absorbance, $EC_{50}$ | 2.7 ml-DPPH* solution + 0.3 ml extract |
| 21 | Chantarudee et al. (2012) | 0.15 mM in methanol | 60 min | % activity, $EC_{50}$ | 50 µl-DPPH* solution + 50 µl extract |
| 22 | Mejías and Montenegro (2012) | 0.02 mg/mL in methanol | NI | µg ascorbic acid equivalents/g sample | 1.5 ml-DPPH* solution + 750 µl extract |
| 23 | Melo et al. (2012a) | NI | NI | % activity | NI |
| 24 | Melo et al. (2012b) | NI | NI | % activity | NI |
| 25 | Melo et al. (2012c) | NI | 30 min | % inhibition | NI |
| 26 | Melo et al. (2012d) | NI | 30 min | % inhibition | NI |
| 29 | Arruda (2013) | $6 \times 10^{-5}$ M in ethanol | 30 min | $EC_{50}$ | 1.5 ml-DPPH* solution + 2.0 ml etanol + 0.5 ml extract |
| | Arruda et al. (2013) | NI | NI | $EC_{50}$ | NI |
| 27 | Montenegro et al. (2013) | 0.02 mg/ml | 15 min | µg ascorbic acid equivalents/g pollen | 1.5 ml-DPPH* solution + 0.75 ml extract |
| 28 | Cheng et al. (2013) | 0.1 mM in methanol | 30 min | % inhibition | 4 ml-DPPH* solution + variable volumes of extract. Final volume: 10 ml |
| 30 | Ulusoy and Kolayli (2014) | 0.1 mM in methanol | 50 min | $SC_{50}$ | 1.5 ml-DPPH* solution + 1.5 ml extract |
| 31 | Melo et al. (2013) | NI | 30 min | $IC_{50}$ | NI |
| 32 | Barriada-Bernal et al. (2014) | 40 µg/ml in ethanol | 30 min | $EC_{50}$ | 600 to 990 µl-DPPH* solution + 10 to 400 µl extract. Final volume: 1 m |

NI = Not indicated.

Table 56. Advantages and disadvantages of some techniques and methods used for anti-microbial analysis.

| Methods | Advantages | Disadvantages | References |
|---|---|---|---|
| Agar Diffusion | • The technique works well with defined inhibitors;<br>Flexibility in selection of disks for testing;<br>Its simplicity of implementation and it is the least costly of all susceptibility methods;<br>For testing of antifungal susceptibility its relative ease;<br>Does not require for specialized equipment;<br>Has intrinsic flexibility by being able to test the drugs the laboratory chooses;<br>The provision of categorical results easily interpreted by all clinicians;<br>The disk test has been standardized for testing not all fastidious or slow growing bacteria but, for some as streptococci, *Haemophilus influenzae*, and *N. meningitides*. | • Less sensitive than micro-dilution techniques;<br>Requires much quantity of sample for analysis;<br>The antimicrobial effect may be inhibited or increased by extrinsic factors or contaminants as the type of agar and his concentration, salt concentration, incubation temperature, and molecular size of the antimicrobial component;<br>There are substances in the culture medium that influence the test susceptibility as thymidine that antagonizes the activity of sulfonamides and trimetroprinas, leading to a false result of resistance to such agents;<br>This technique also does not distinguish between bactericidal and bacteriostatic effects;<br>The pH of the culture medium should be compatible with the microbial growth because the increase in acidity of the medium decreases the antibacterial activity. By another side, increases the activity of acidic substances such as penicillin;<br>The minimum inhibitory concentration (MIC) cannot be determined;<br>Is difficult to examine the susceptibility of fastidious and slow-growing bacteria;<br>It is labor-intensive and time-consuming;<br>Can represent an expensive approach if more than a few drugs are tested;<br>The lack of mechanization or automation of the test. | Hewitt and Vincent (1989); Marsh and Goode (1994); Eloff (1998); Scorzoni et al. (2007); Rex et al. (2001); Ostrosky et al. (2008); Barry and Thornsberry (1991); Klancnik et al. (2010); Wilkins and Thiel (1973); Jiang (2011); Jorgensen and Ferraro (2009). |
| Macrodilution | -Generation of a quantitative result (i.e., the MIC);<br>-Ease of performance;<br>-Economy;<br>-More rapid results, therefore their development could be highly desirable. | • Requires large volumes of antibiotic;<br>It is not recommend using for most routine procedures;<br>Time consuming;<br>It is tedious;<br>Manual task of preparing the antibiotic solutions for each test;<br>The possibility of errors in preparation of the antibiotic solutions;<br>Relatively large amount of reagents and space;<br>Generate large amounts of waste. | Jorgensen and Ferraro (2009); CLSI (2008); Wiegand et al. (2008); Ostrosky et al. (2008). |
| Micro-dilution *in microtiter plate (96 wells)* | • The method is robust, is not expensive;<br>Gives reproducible results and convenience of having pre-prepared panels; | • The lack of or poor growth of many anaerobic microorganisms;<br>Testing some fastidious anaerobes gives | Eloff (1998); Ostrosky et al. (2008); Clinical and Laboratory Standards Institute (CLSI) (2009); Jiang (2011); Jorgensen and Ferraro (2009). |

(*Continued*)

Table 56. (*Continued*).

| Methods | Advantages | Disadvantages | References |
|---|---|---|---|
| | Is *ca.* 32 times more sensitive than other methods used in the literature; Requires a small/minimal quantity of reagents, samples and space that occurs due to the miniaturization of the test; Can be used for large numbers of samples; Can test the susceptibility of multiple antimicrobials at the same time; Enables a larger number of replicates, increasing the ability of confident results; Does not require high levels of skill; Leaves a permanent record; It decreases the intensive labor and time cost compared with agar-based method; The considerable savings in media usage; Also allow assistance in generating computerized reports if an automated panel reader is used; Works better with antibiotics because no precipitate forms; Compared with agar-based method, broth microdilution can decrease much labor and time; Is conveniently used for routine antimicrobial susceptibility testing in clinical laboratory; Is a suitable and fast screening method for MIC determination compared to other methods. | inconsistent and unreliable results because of poor growth of strains due to excessive exposure to oxygen during the set-up procedure; Some inflexibility of drug selections available in standard commercial panels. | |

Table 57. Summary of some antimicrobial effects of bee pollen.

| Pharmacological activity | Susceptible microorganisms | Methods of determinations | References |
|---|---|---|---|
| | *E. coli, S. aureus, B. cereus* and *P. aeruginosa* | Agar well diffusion | Abouda et al. (2011) |
| | **Fungi:** *Aspergillus fumigatus, Aspergillus niger* and *Aspergillus flavus.* **Yeasts:** *Candida albicans (C. albicans), Candida glabrata, Candida krusei* and *Rhodotorula mucilaginosa.* **Bacteria:** *E. coli, S. aureus, Salmonella entérica (S. entérica), P. aeruginosa* and *Listeria monocytogenes (L. monocytogenes)* | " | Kacániová et al. (2012) |
| | *L. monocytogenes, S. entérica, E. coli, S. aureus* and *P. aeruginosa* | " | Fatrcová-Šramková et al. (2013) |
| | *Clostridium butyricum, Clostridium perfringens, Clostridium hystoliticum, Clostridium intestinale* and *Clostridium ramosum* | " | Kacániová et al. (2014) |
| | *Paenibacillus larvae, Bacillus subtilis (B. subtilis), B. cereus , E. coli* and *P. aeruginosa* | " | Barbosa et al. (2006) |
| | *S. aureus* and *Streptococcus pyogenes* | " | Cabrera and Montenegro (2013) |
| | *Enterobacteriaceae* genera | " | Hleba et al. (2013) |
| | *S. aureus* and *Staphylococcus epidermidis* | " | Baltrušaityte et al. (2007) |
| | *Viridans streptococci* | " | Tichy and Novak (2000) |
| | *Agrobacterium tumefaciens, P. syringae pv. tomato, X. axonopodis pv. Vesicatória, E. amylovora, P. corrugata, R. solanacearum, X. campestris pv. Campestris, A. vitis, C. michiganensis subsp. Michiganensis, E. carotovora pv. carotovora, P. savastanoi pv. savastanoi, P. syringae pv. Phaseolicola, P. syringae pv. syringae* | Paper Disk diffusion | Basim et al. (2006) |
| | *S. aureus, B. subtilis, P. aeruginosa, Klebsiella sp.* and *B. cereus* | " | Carpes et al. (2007) |
| | *S. aureus, B. Cereus, B. Subtillis, P. aeruginosa* and *K. Pneumoniae* | " | Basuny et al. (2013) |
| | *E. coli, S. aureus, B. subtilis, P. aeruginosa, K. pneumonia, Shigella flexneri, Salmonella typhi* and *Proteus mirabilis* | " | Eswaran and Bhargava (2014) |
| | *S. thyphimurium* and *E. coli.* | " | Yerlikaya (2014) |
| | *S. aureus* [ATCC 29213], *E. coli* [ATCC 27853], *Enterococcus faecalis* [ATCC 29212], or *P. aeruginosa* [ATCC 25922] | Microdilution in Microtiter plate | Eloff (1998) |
| | *S. aureus, S. epidermidis, E. coli, Enterobacter cloacae, Klebsiella pneumonia, P. aeruginosa, C. albicans, C. tropicalis* and *C. glabrata* | " | Graikou et al. (2011) |
| | *Salmonella typhi, S. aureus, B. cereus, E. coli, Zygosacharomyces mellis, Zygosacharomyces bailii, Zygosacharomyces rouxii* and *Candida magnolia* | " | Morais et al. (2011) |
| | *S. aureus, P. aeruginosa, E. coli* and *C. glabrata* | " | Pascoal et al. (2014) |

### 15.2. Materials and methods

#### 15.2.1. Sample characterization and preparation

The pollen samples are checked for the botanical origin based on the microscopic analysis, according the methodology described in Section 5.

For FTIR-ATR analysis a representative part of each samples are carefully homogenizing and reduce to a fine powder (Figure 53). A small amount of sample is put in the apparatus with diamond solid ATR crystal properly mounter, and press against crystal (Anjos et al., 2017).

#### 15.2.2. Reference analytical methods

**15.2.2.1. Total phenolic and flavonoids content analysis.** The total phenolic content of the extracts was determined using the Folin–Ciocalteu method as described by Moreira et al. (2008). Total phenols content is expressed as mg of Galic Acid equivalents per g of bee pollen (GAEs).

For flavonoids contents the aluminum chloride method was used. Total flavonoids content is express as mg of catechin equivalents per g of bee pollen (CAEs).

**15.2.2.2. Antioxidant activity analysis.** The antioxidant activity was assessed by calculating the 2,2-diphenyl-1-picrylhydrazyl (DPPH*) free radical scavenging activity the trolox equivalent antioxidant capacity (ABTS) according the methodology proposed in Section 16.

The concentration of the extract that induced a 50% inhibition ($EC_{50}$) was calculated from the graphic of the percentage effect of eliminating radicals as a function of the concentration of the sample extract solution.

For trolox results were expressed in millimol of equivalent trolox per gram of sample.

#### 15.2.3. FTIR-ATR collection and data processing

1. The data acquisition and processing was described by Anjos et al. (2017). Pollen spectra are acquiring in a Bruker FTIR spectrometer with a resolution of 4 $cm^{-1}$ in the wavelength region 4000–400 $cm^{-1}$ (or similar) using a diamond single reflection attenuated total reflectance (ATR).
2. obtain a duplicate spectrum per sample, for example, with 32 scans per spectrum and calculate the average for each sample.
3. Note: A background measurement is made before each ten-sample analyzing using air. ATR crystal is carefully clean before and between each analysis with MilliQ water and isopropanol and the ATR crystal should be dried with soft tissue paper. All experiments are carried out at room temperature.
4. In a first screening all abnormal spectra are remove and if it was necessary a new sample spectra acquisition is performed.
5. PLS regression is done by using OPUS 7.5.18 BRUKER software. Firstly, the spectral data from the calibration samples are regress against the measured parameters and by means of full cross-validation with one sample omitted in order to obtain a significant number of PLS components (rank).

Different preprocess spectra could be tested for PLS analysis with the following methods:

a. multiplicative scatter correction (MSC);
b. minimum maximum normalization (MinMax);
c. vector normalization (VecNor);
d. straight line subtraction (SLS);
e. constant offset elimination (ConOff);
f. first derivative (1stDer);
g. second derivative (2ndDer);
h. combination of the previously methods namely 1stDer + MSC, 1stDer + VecNor, 1stDer + SLS.

For PLS analysis models with no spectral data processing (None) are tested. All models can be calculated to a maximum rank of 10 and the results of the cross-validation test for the better values of coefficient of determination ($r^2$), root mean square error of cross-validation (RMSECV) and ratios of performance to deviation (RPD) (Anjos et al., 2018).

### 15.3. Concluding remarks

These developed models have acceptable accuracy for the screening analysis for total phenolic compounds, total flavonoids and antioxidative capacity of bee pollen. They are fast and easy to handle and could give a good evaluation of the properties of bee pollen moisture available in the market. In the future is important produce more robust models with a higher number of samples, produced in different countries to have a better chemical characterization of the bee pollen to use in the labeling.

A total of 134 pollen samples mixture was analyzed and the frequency of each plant genus as predominant pollen, secondary pollen, important minor pollen and minor pollen is represented on Table 52. The variability of genus observed in all samples is given by the richness of floral resources of the region where the bee pollen samples were collected. In a higher amount of samples, the *Cistus* spp and *Echium* spp pollen were the predominant. As secondary pollen the *Castanea* spp pollen was the highest representative genus found in 34 samples, followed by the *Cytisus* spp, *Eucalyptus* spp, *Cistus* spp and *Echium* spp.

*Castanea* spp pollen was found in 60% of the samples from as Predominant pollen to minor pollen. The same distribution was observed for *Cistus* spp pollen that was presented in 61% of the analyzed samples.

The spectra for pollen are very similar, like it can be observed in Figure 54, which shows the average spectrum of different types of analyzed samples. All the spectra analyses are characterized by bands between 1600 and 900 $cm^{-1}$, which correspond to the signal

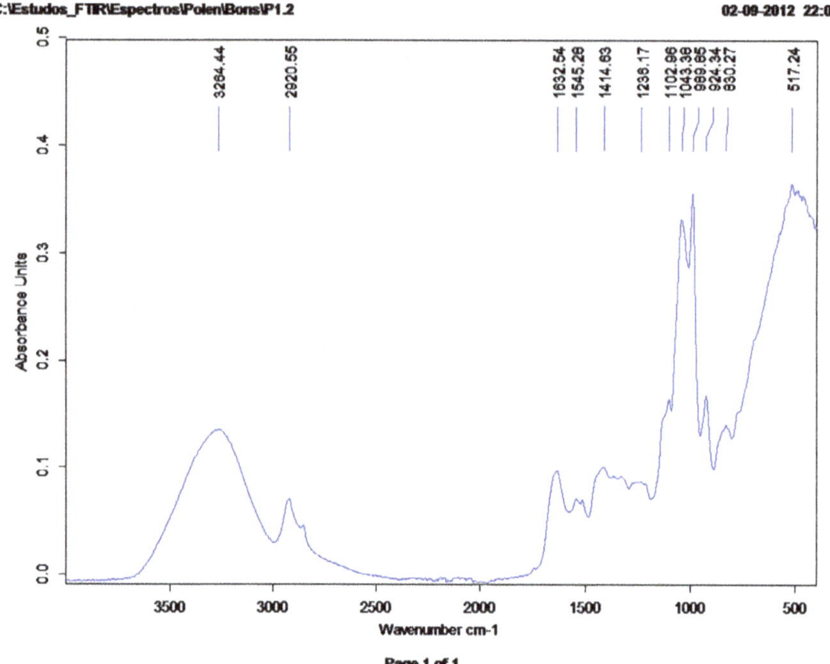


Figure 54. Average ATR spectrum of pollen samples.

characteristic of the phenols compounds and is due to the C-C and C-O vibration of phenolic and flavonoids compounds. An absorbance band appeared at 1632 $cm^{-1}$, which could be assigned to the carboxyl group in asymmetric stretching, presented also in the honey samples (Anjos et al., 2015). The band from 3400 to 3000 $cm^{-1}$ is usually associated to the water content in the samples because is due to OH groups but this groups are also important in the identification of phenolic compounds, alcohols and carboxylic groups and also the hydrogen vibration of amide N-H groups from olive wastewater samples (Droussi et al., 2009). The water content in this kind of samples usually varies between 6 and 8%. The other important bands that were observed in bee pollen samples are described in Anjos et al. (2017).

Table 53 summarizes the statistical data for each parameter evaluated. The higher values of coefficient of variation denote the variability of data important for good model calibration. The samples data must be divided into two parts, one for calibration (30% to 50% of data) and the other for validation set (70% to 50% of data). PLS multivariate analysis was done with spectral data in order to obtain calibration models for each analyzed parameter in pollen samples. The calibration models determination of total phenolic compounds, total flavonoids and antioxidant capacity in pollen mixtures were developed based on the highest $r^2$, lowest standard error of calibration and cross-validation, lower number of factors used in the calculation and higher residual prediction deviation. Table 54 represents the results for optimal calibration established. Values of $r^2$ for selected models were greater than 0.859 for calibration and greater than 0.72 for the Cross validation. The worst model founded was for the ABTS with a lower RPD for cross-validation. In calibration all models present values of RPD higher than 2.7. However, more accurate model could be obtained with more samples, from different regions and countries.

## 16. Standard methods for determination of antioxidant bioactivity

### 16.1. Introduction

Bee pollen is considered as an important source of antioxidant flavonoids and phenolic acids for human consumption. Given the complex composition (view the first sections of this chapter), the natural variation of that composition, the storage conditions (Campos et al., 2003; Melo & Almeida-Muradian, 2010), the different ways to prepare the pollen extracts, and the different assay conditions to evaluate the antioxidant capacity, the determination and comparison of antioxidant capacity is difficult to assess. A great deal of methods for assessing the antioxidant capability of foods and beverages have been developed, and reviews on this topic have been published (Prior et al., 2005). In this section, a proposal for standardizing some methods to assess the antioxidant levels of bee pollen is done. As the antioxidant capacity varies with the type of oxidants, more than one assay is needed (Niki & Noguchi, 2000). - DPPH* free radical scavenging activity and the oxygen radical absorbance capacity (ORAC) methods are proposed for standardization of the routine assessment of bee pollen antioxidant capacity; the former because of its wide use in bee pollen, easiness of doing it, and good correlation with other assays like $\beta$-carotene bleaching (Morais et al., 2011) among others; and the second one because of it has been proposed to be considered for standardization for foods, botanicals, nutraceuticals, and other dietary supplements (Prior et al., 2005), and this would allow to compare this property between bee pollen and any other food.

### 16.2. Methods

*16.2.1. 2,2-diphenyl-1-picrylhydrazyl (-DPPH*) free radical scavenging activity principle*

-DPPH* is a stable organic nitrogen free radical. It has a spare electron, which being delocalized in the molecule, confers a deep violet color and makes the compound no capable to form dimers (Molyneux, 2004) neither reacts with oxygen (Ionita, 2005). In ethanolic or methanolic solutions, -DPPH* absorbs between 515 and 528 nm (Almaraz-Abarca et al., 2004; Molyneux, 2004; Sánchez-Moreno, 2002), but when it is mixture with a reducing substance, the reduced form of -DPPH* (-DPPH*-H, 2,2-diphenyl-1-picrylhydrazine) is formed (Ionita, 2005), then the deep purple color becomes to a pale purple and eventually, this becomes to a pale yellow. The wide range of maximum absorbance reported for the alcoholic -DPPH* solution indicates that the $\lambda_{max}$ must to be determined under a given set of particular conditions. Stoichiometric determinations have indicated that two moles of -DPPH* are reduced by one mole of water soluble vitamin E analogue (Trolox) (Leong & Shui, 2002), but the interaction of a potential antioxidant with -DPPH* depends on its structural conformation (Kaur & Kapoor, 2001). In a -DPPH* assay, this molecule represents the radical formed in a system suffering oxidation and the antioxidant to evaluate represents the reductant compound. Although the -DPPH* scavenging activity can be evaluated by electron spin resonance (Wasek et al., 2001; Yu, 2001), the evaluation of the absorbance reduction at 515 to 528 nm when in contact with a hydrogen donator compound by using visible spectroscopy, is the most common manner of evaluation. A quantitative estimation of the activity using visible spectroscopy is based on the registration of the differences between the initial concentration of -DPPH* in the reaction medium and its diminution with increased amounts of antioxidant.

The -DPPH* scavenging activity of one sample can be expressed as the percentage of -DPPH* free radical scavenging remaining in the reaction (Equation (1)).

$$\%\text{DPPH}^{*}\text{remaining} = [[\text{DPPH}^{*}]\text{remaining}/[\text{-DPPH}^{*}]\text{initial}] \times 100 \qquad (1)$$

That manner has been used to express the -DPPH* scavenging activity in around 65.6% of reports concerning bee pollen. However, this activity can also be expressed in terms of the efficient concentration at 50% ($EC_{50}$), defined as the amount of antioxidant needed for reducing by 50% the initial concentration of -DPPH*, the larger the $EC_{50}$ the less efficient the antioxidant.

Sánchez-Moreno et al. (1998) have proposed that besides the concentration of antioxidant, the reaction time to reach the plateau of the scavenging reaction should be taken in account for estimating this activity, according to Equation (2). Results from quercetin evaluations suggest that the increasing in the reductor compound do not affect the time to reach the plateau (Campos, 1997).

$$AE = 1/(EC_{50})(T_{EC50}) \qquad (2)$$

where AE means the antiradical efficiency, $EC_{50}$ the efficient concentration at 50% and $T_{EC50}$ the time needed to reach the steady state with $EC_{50}$.

As both -DPPH* and phenols are highly soluble in methanol and ethanol, the antiradical activity is evaluated in either of those solvents without apparent interference with the reaction; other solvents, like water and acetone seem to under estimate the extent of reduction (Guo et al., 2020). The use of buffered methanolic solutions has been reported for the evaluation of antiradical activity of bee pollen (Wang et al., 2010); that solvent can affect the estimation of $IC_{50}$ in methanol and buffered methanol for antioxidants like BHT, but the effect is lower for phenolic antioxidants like propyl gallato, which bears several hydroxyl groups, being the $IC_{50}$ values comparable in both kinds of solvents (Sharma & Bhat, 2009).

A great variation in the initial concentration of -DPPH* solution can be found in many reports of the antiradical activity. According to Molyneux (2004), that concentration should be in the range of 50 to 100 μM in order to get solutions giving absorbance values less than 1, thus, observing the normal practice in spectrophotometry to reach accurate measurements. The range given by Sharma and Bhat (2009) was 25 to 70 μM.

Concerning the stability of the -DPPH* solution, its spectrum (200 to 700 nm) did not reveal neither molecular nor quantitative alterations after eight hours in light (Campos, 1997). For longer periods, the free radical activity of the -DPPH* solution is reduced about 2 to 4% per week (Molyneux, 2004).

Among the disadvantage of -DPPH* assay are the reversibility of the initial step of the reaction (Bondet et al., 1997) and the fact of some substance, like proteins, most likely present in total extracts of plants, can precipitate in the alcoholic reaction medium (Sánchez-Moreno, 2002). Prior et al. (2005) informed about the following three disadvantages of this assay, the not easy interpretation when the test compounds have spectra overlapping -DPPH* $\lambda_{max}$; the overestimation of the activity for small molecules, having better access to the radical site (steric accessibility); and the inaccurate interpretation of the antioxidant activity for a given sample because -DPPH* solutions are decolorized by both reducing and hydrogen atom transfers. Contrary, among the advantage of the -DPPH* free radical scavenging activity are that it is a rapid method, does not have to be generated previously to the proper radical scavenging as in other assays, the compound is commercially available, the method is accurate and reproducible, and needs only a UV-Vis spectrophotometer or a colorimeter to perform. Also, -DPPH* assay is independent of sample polarity (Koleva et al., 2002). Besides, good correlations between this assay and TEAC (Mărghitaş et al., 2009) and with $\beta$-carotene bleaching assay (Morais et al., 2011) have been reported, meaning that -DPPH* test may represent a good approaching to evaluate the antioxidant activity for complex samples like bee pollen extracts.

There is a high correlation between the decreasing of DPPH* concentration and the increasing of flavonoid or phenolics concentration in the reaction medium (Almaraz-Abarca et al., 2004, 2006, 2008; Barriada-Bernal et al., 2014). However, a clear correlation between the DPPH* scavenging activity and the flavonoid or phenol concentration in different samples has not been found in all systems (Morais et al., 2011; Negri et al., 2011). This suggests that the types of phenolics present in a particular combination in a given sample are involved in the manifestation of the antioxidant properties.

Sánchez-Moreno (2002) informed about a long list of compounds from different sources analyzed with the -DPPH* assay, which could be update with the reports concerning the bee pollen activity.

***16.2.1.1. Variations in the -DPPH* assay for bee pollen.*** The main variations of the -DPPH* assay found in the several reports concerning bee pollen refer to the -DPPH* concentration of the solution used to carry out the analysis, reaction time, and the proportion of volume of -DPPH* solution related to volume of extract or sample to analyze (Table 55). Differences are also found in the form of expression of the activity. Under this situation important difficulties are to be faced when comparisons of scavenging activity of different bee pollen samples are done. Thus, the expression of the antiradical activity in terms of $IC_{50}$ would represent a more accurate activity than the expression in terms of % activity, which is estimated at one extract concentration selected in an arbitrary manner and mentioned in the reports only in few cases.

***16.2.1.2. Material and methods for -DPPH* assay.***

1. Prepare a solution of -DPPH* in ethanol or methanol (depending on the solubility of the compounds to be tested), and calculate the UV-Vis absorption that should be close to 1 (Absorption Unities – AU). The concentration of the free radical depends according to the method choose (see Table 55).
2. Prepare a test solution with the sample. For bee pollen a ratio of 10 mg/ml of solvent can be used as a start point. After a preliminary assay, if needed, different concentration of the simple solution can be done.
3. Measure 1 or up to 4 ml of -DPPH* according to the method choose (see Table 55) for at least 5 test tubes.
4. Add to each different concentration of the sample solutions and adjust with solvent for an equal final volume in each test tube. Example: 1 ml of -DPPH* solution + (50 µl extract + 950 µl of solvent), to complete a final volume of 1 ml.
5. Wait the time necessary for the complete reaction, according to the method choose (see Table 55)
6. Measure the absorption at 517 nm in the visible range with a spectrophotometric system.
7. Do a chart with the concentration of the sample and the respective consume of -DPPH*. Example: O µg of sample + 1 ml of -DPPH* correspond to 0.96 AU. These values correspond to zero activity. When the concentration of the sample increase (if active) the -DPPH* solution absorption decreases and then is possible to determine the concentration of the sample necessary to decrease this AU to 50% corresponding to the $EC_{50}$.
8. Proceed as above utilizing a standard (ascorbic acid, trolox, various flavonoids or other polyphenols, depending of the simple under evaluation) for comparison of the results and to validate the accuracy of the assay.

*16.2.2. Antioxidant capacity using ORAC (oxygen radical absorbance capacity) assay principle*

The ORAC method measures the sequestrating capability of an antioxidant against the induced radical 2,2′-azo-bis(2-amidinopropane)dihydrochloride (AAPH) in reactions which are temperature sensitive; 37 °C is the most frequent temperature of assay. The radical reacts with a fluorescent compound forming a nonfluorescent product. The protector effect of antioxidant is estimated by calculating the area under a fluorescence decreasing curve over time, comparing with a reference substance, which is not an antioxidant. ORAC method was developed by Glazer (1990), using phycoerythrin, but currently fluorescein and dichlorofluorescein are preferred because of their stability and less reactiveness (Prior et al., 2005). At present, ORAC assay is highly automated. Special plates of fluorometers are used. Arruda (2013) used plate wells for 300 $\mu$l, and it was used for the external ones 300 $\mu$l of distilled water. Other holes were filled with 25 $\mu$l of phosphate buffer (blank) and the others with 25 $\mu$l of the extract dilutions. Then it was injected 150 $\mu$l of fluoresceine, waits for 30 minutes, and then injected 25 $\mu$l of the AAPH solution to initiate the reaction of peroxyl radical with the fluorescein (oxygen radical absorbance capacity). ORAC values are usually expressed as Trolox equivalents/g sample, then a standard curve using three to five concentrations of Trolox® (25 $\mu$l) vs. the fluorescence intensity is constructed. ORAC unit is equivalent of a protection against oxidation offered by a 1 µmol of Trolox. The fluorescence must be monitored every minute, using as work temperature 37 °C. It is used extinction wavelength at 493 nm (485/20 filter) and at 515 nm (filer 528/20) of emission. For calculation, it is used the sample area under the curve of the testing compounds minus the area of the blank. The same calculation is made for the Trolox curve. The area under the curve (AUC) is calculated after 30 minutes, when it is injected 25 $\mu l$ of the AAPH solution to start the reaction, this delay in the starting of reaction avoids underestimation of antioxidant activity due to effects of secondary potential antioxidant products (Prior et al., 2005).

The area under the curve (AUC) is calculated using Equation (3).

$$AUC = 1 + \sum_{i=1}^{N} \frac{f_i}{f_0}$$

where $f_0$ = initial fluoresence at time zero; $f_i$ = fluorescence registration after i

At the beginning of the reaction kinetics there is a time in which the fluorescence is maintained, this is the lag phase, in it there is no consume of peroxide by fluorescent probe due to the presence of the antioxidants and consequently enhance the area under the curve. Prior et al. (2005) consider that methods like ORAC, which measures the antioxidant capacity by AUC, take in account both inhibition time and degree; this gives a more accurate estimation of the action of an antioxidant than the methods which are based on the evaluation at a fixed time or inhibition degree.

***16.2.2.1. Material and methods for antioxidant activity (ORAC) assay.*** The ORAC procedure recommended use an automated plate reader (Synergi HT, Bio TeK, USA) with 96-well plates (Arruda et al., 2013; Ou et al., 2001), as follows:

1. Conduct all the analyses in a pH 7.4 (75 mM) phosphate buffer at 37 °C under condition with a blank sample in parallel.
2. Dilute the extracts with sodium phosphate buffer (75 mM, pH 7.4). Peroxyl radical will generated using 2, 2′-azobis (2-amino-propane) dihydrochoride aqueous solution – that will be prepared for each run (153 mM). Fluorescein will be used as substrate (40 nM).
3. Program the ORAC analyzer to record the fluorescence of FL every minute after addition of AAPH. All fluorescent measurements express relatively to the initial reading. Fluorescence analysis conditions are performed as follows: excitation at 493 nm (filter 485/20) and 515 nm (filter 528/20). The standard curve must be linear between 6.25 and 100 µM Trolox®.
4. Calculate the final results using the differences of areas under the FL decay curves between the blank and a sample and express as µM TE/g bee pollen.

### 16.3. Concluding remarks

The great diversity of bee pollen that can be found worldwide, varying in the antioxidant capabilities, encourages to propose standard evaluations for this property, allowing a comparison among bee pollens from different geographical and botanical origins. Standards evaluations also would be useful for understanding the effects of several environmental factors on the antioxidant properties, and for monitoring bee pollen quality.

## 17. Standard methods for anti-inflammatory activities in bee pollen

### 17.1. Introduction

The inflammation process has been defined by several pathologists with different approach. Most pathologists would probably agree that inflammation represents a response of living tissue to local injury; that it leads to the local accumulation of blood cells and fluid (Ryan & Majno, 1977). Briefly, inflammation is a physiological response to the damage of tissues or cells that is caused by physical or biological agents and also free radicals involving different reactions intended to remove the cause and repair the damage (Bogdanov, 2015). Paulino et al. (2008) defined the inflammation as the process that involves production and/or release of mediators from neurons or damaged tissues, which are responsible for different responses including pain.

For determination of standard methods on anti-inflammatory activities much methodology *in vitro* and *in vivo* has been developed by several researchers worldwide. Recently Pascoal et al. (2014) following the methodologies *in vitro* described by da Silva et al. (2012), present a method based on the measurement of the inhibitory effect of bee pollen (BP) on the reactions catalyzed by hyaluronidase was evaluated.

### 17.2. Determination of standard methods

*17.2.1. Anti-inflammatory activities of BP in vitro*

***17.2.1.1. BP extract preparation.*** For BP extract preparation the procedures described by Pascoal et al. (2014) are described below regarding to the procedures.

*17.2.1.1.1. Materials and reagents.*

1. BP sample(s).
2. Methanol ($CH_3OH$).
3. Whatman filter paper No. 4.
4. Flasks.
5. Rotary vacuum evaporator.

*17.2.1.1.2. Procedure.* Note: The sample concentration and diluents used varies from the author to other:

1. BP and Methanol (MeOH) were mixed (1:2) (w/v).
2. Macerate for 72 h at room temperature.

3. Then, the solution was filtered by Whatman filter paper No. 4.
4. The solid residue was re-extracted.
5. Mix the two filtered into a flask, thus obtaining the methanol extracts of bee pollen.
6. Evaporate the extracts on a rotary evaporator at 45 °C for approximately 1 h.
7. Finally, the dried extract of pollen was stored in the dark at room temperature, for further analysis.

***17.2.1.2. BP sample preparation.*** *17.2.1.2.1. Materials and reagents.*

1. BP extract.
2. Hyaluronidase enzyme (Type IV-S: bovine testes).
3. Calcium chloride (CaCl2).
4. Hyaluronic acid sodium salt.
5. Potassium tetraborate ($K_2B_4O_7{\cdot}4H_2O$).
6. Water-bath.
7. *p*-dimethylaminebenzaldehyde (DMSO).
8. Spectrophotometer.

*17.2.1.2.2. Procedure.*

1. Prepare BP extract concentrations [5mg/ml – 25 mg/ml] using dimethyl sulfoxide (DMSO) as solvent.
2. Prepare a mixture of 50 $\mu$l of BP' extract and 50 $\mu$l (350 units) of hyaluronidase enzyme (Type IV-S: bovine testes).
3. Incubate at 37 °C for 20 min.
4. Add 1.2 $\mu$l of calcium chloride ($2.5 \times 10^{-3}$ M/l) to activate the enzyme.
5. Incubate at 37 °C for 20 min.
6. Add 0.5 ml of hyaluronic acid sodium salt (0.1 M/l) to start the reaction.
7. Incubate the mixture at 37 °C for 40 minutes.
8. Then add 0.1 ml of potassium tetraborate 0.8 M.
9. Incubate in boiling water bath for 3 minutes.
10. Place the mixture at 10 °C.
11. Add 3 ml of (DMSO).
12. Afterwards incubate at 37 °C for 20 minutes.
13. Measured the absorbance at 585 nm using water as control.

*17.2.2. Anti-inflammatory activities of BP in vivo*

***17.2.2.1. BP extract preparation.*** For BP extract preparation the procedures described by Maruyama et al. (2010) are suggest.

*17.2.2.1.1. Materials and reagents.*

1. BP sample(s)
2. Ethanol (EtOH) and others.

*17.2.2.1.2. Procedure.*

1. BP (200 g) extracted with water (Water BP) or 95% ethanol (EtOH BP).
2. Stir for 4 h at 60 °C (Water BP) or stir for 16 h (EtOH BP) at room temperature.
3. Take the filtrate to dryness and powdered.
4. The yields are stored for further use.

*17.2.3. Animals*

For the animal experimental protocol procedures, the Guide for the Care and Use of Laboratory Animals Eighth Edition (2011) is suggested.

### *17.3. Concluding remarks*

The high anti-inflammatory effects of BP compared to such nonsteroidal anti-inflammatory drugs as naproxen, analgin, phenylbutazone, or indomethacin also has been reported (Pascoal et al., 2014). The mechanism of the inflammatory effect is about inhibiting the activity of enzymes (cyclooxygenase and lipoxygenase) which are responsible for turning arachidonic acid into such toxic compounds as prostaglandin and leukotrienes, inducing acute and chronic inflammatory conditions in tissues. Flavonoids and phenolic acids are mainly responsible for such actions, but fatty acids and phytosterols also take part in this process (Akkol et al., 2010; Choi, 2007; Komosinska-Vassev et al., 2015; Pascoal et al., 2014). Kaempferol it is a compound present in BP and has been attributed ability to inhibit the activity of two enzymes: hyaluronidase, which is the enzyme catalyzing depolymerization of hyaluronic acid, and elastase, which hydrolyzes elastin, strengthens the connective tissue and seals blood vessels. This results in decreased transudates, inflammatory reactions and swellings. Blood circulations in the vessels improve and, therefore, skin becomes moistened and tight. Quercetin another flavonoid compound with several bioactivities as anti-inflammatory, antiedematous and analgesic action, may result by inhibiting the activity of histidine decarboxylase, lowers the histamine level in the organism (Kanashiro et al., 2007; Lee et al., 2009; Sahin et al., 2011). Studies done by Lee et al., 2009 in order to elucidated the anti-inflammatory activities of several subclass flavonoids such as the antioxidant property, inhibition of NO production, and inhibition of inflammatory enzymes (hyaluronidase, collagenase, LOX, and COXs), found that Kaempferol, quercetin, myricetin, and rutin inhibited hyaluronidase reaction, while apigenin, baicalin, and baicalein showed specific inhibition to collagenase reaction and catechin, which is flavon-3-ols-subclass, had suppressed both hyaluronidase and collagenase reaction. The degradation of the hyaluronic acid by the hyaluronidase enzyme may cause bone loss, inflammation and pain. As consequence, the determination of the hyaluronidase enzyme is an indirect way to assess the anti-inflammatory activity (da Silva et al., 2012). Choi (2007) evaluated the antinociceptive and anti-inflammatory activity of pine (*Pinus densiflora*) pollen in mice. The antinociceptive activity was determined using acetic acid-induced abdominal constriction and formalin-induced licking, and the hot plate test and anti-inflammatory effects were evaluated using carrageenan and formalin-induced paw edema, and arachidonic acid-induced ear edema. The obtained results showed that ethanol extract of pine pollen (100 and 200 mg/kg) produced a significant inhibition of both phases of the formalin pain test in mice. Also, the pine pollen extract produced a significant inhibition of carrageenan and formalin-induced paw edema as well as arachidonic acid-induced ear edema in mice. These activities could be due to different polyphenols found in pine pollen related with antinociceptive and anti-inflammatory actions. Among other flavonoids, Chrysin has been reported as to possess anti-inflammatory activity, antioxidant properties and promotes cell death by perturbing cell cycle progression (Pichichero et al., 2011). Another studies developed by Maruyama et al. (2010) in order to investigate the anti-inflammatory effect of bee pollen from *Cistus* sp. of Spanish origin by a method of carrageenan-induced paw edema in rats, and to investigate the mechanism of anti-inflammatory, found that the ethanol extract of bee pollen show a potent anti-inflammatory activity and its effect acts *via* the inhibition of nitric oxide (NO) production, besides the inhibitory activity of COX-2. Maruyama et al. (2010) and, reported that some flavonoids found in plant foods included also in BP may partly participate in some of the anti-inflammatory action. Therefore, the anti-inflammatory effects of BP may be due to their flavonoids compounds as Kaempferol, quercetin, myricetin, and rutin that inhibit the inflammatory enzymes such as hyaluronidase, collagenase, LOX, and COXs, helping the retardation of inflammatory responses.

## 18. Standard methods for antimicrobial activities in bee pollen

### 18.1. Introduction

In order to investigate new antimicrobial agents from extracts of plants and other natural products to be used in pharmaceutical and cosmetic products, several methods have been developed by several researchers for evaluate the antibacterial and antifungal activity of the plant extracts. Nowadays, the most known assays have been based on diffusion in agar, macro and microdilution methods (Ostrosky et al., 2008). Table 56 shows the advantages and disadvantages of these methods with references. Microdilution methods have been reported as the most widely used testing methods that use commercially marketed materials and devices, and the manual methods that provide flexibility and possible cost savings include the disk diffusion and gradient diffusion methods (Jorgensen & Ferraro, 2009). Some methods provide quantitative results and all provide qualitative assessments using the categories susceptible, intermediate or resistant. Also, it's important to highlight that all of these methods have strengths and weaknesses, including organisms that may be accurately tested by the method; but in general way, these current testing methods provide accurate detection of common antimicrobial resistance mechanisms (Jorgensen & Ferraro, 2009).

According to the information available the diffusion in agar technique is widely used but problems associated with this have been reported. However, the *Microdilution technique* in microtiter plate (96 wells) has been proved to be quick provides reproducible results and was useful in screening plants for antimicrobial activity. All of these methods have been reported that each of them has strengths and weaknesses. Also, it is important to highlight that those methods are widely used in the worldwide by several researchers for antimicrobial activities determinations but sometimes suffer minor modifications (Estevinho et al., 2008).

### 18.2. Materials and methods

#### 18.2.1. BP sample collection

Accordingly, with Morais et al. (2011), for getting BP samples from beekeepers, the selected beehives were equipped with bottom-fitted pollen traps. After dried the BP, the samples were delivered to the Microbiology Lab, where they were stored in a dark place at room temperature (−20 °C) for further analysis.

#### 18.2.2. Extract preparation of BP for antimicrobial analysis

Nowadays, many experiments *in vitro* and *in vivo* are performed with Pollen Methanolic Extracts (PME) and Pollen Ethanolic Extracts (PEE) to confirm its antimicrobial activity. According to the revised bibliography, once the pollen is received in the laboratories from beekeepers the methods for samples preparation commonly used are based on Methanol (MeOH) (Graikou et al., 2011; Kacániová et al., 2012; Pascoal et al., 2014) and Ethanol (EtOH) (Eswaran & Bhargava, 2014; Hleba et al., 2013; Kacániová et al., 2014) pollen extracts. In dependence of the final concentration desired, amounts of BP are weighted and dissolved in methanol or ethanol aqueous solutions. These methodologies have been used by several researchers with some modifications. The BP extracts could be prepared accordingly with Pascoal et al. (2014) as described below regarding to the procedures:

1. The sample concentration and diluents used varies from the author to other.
2. Place the mixture under magnetic stirring at 150 rpm for 1 h.
3. Filtration using Whatman No. 4 paper.
4. Repeat this procedure once more.
5. Mix the two filtered into a flask, thus obtaining the methanol extracts of BP.
6. Evaporate the extracts on a rotary evaporator at 45 °C for approximately 1 h.
7. Keep extracts in dark and in refrigeration until their further use.

8. Preparing different sample concentrations using dimethyl sulfoxide (DMSO) as solvent.

*18.2.3. Antimicrobial analysis by agar disk diffusion*

***18.2.3.1. Preparation of inoculum suspension.*** The preparation of inoculum could be accordingly with Matuschek et al. (2014) with bit modifications as described below regarding to the procedures:

1. Inoculate the bacteria in broth of Muller Hinton Agar or Brain Heart Infusion Broth (BHIA); and Sabouraud dextrose Agar or potato dextrose agar for yeast.
2. Incubate for 18 to 24 hours for bacteria and 48 h for yeast.
3. Inoculate 3 to 4 colonies of bacteria in 5 ml Tryptic Soy Broth (TSB) and use Yeast Extract Peptone Dextrose (YPD) for yeasts.
4. Incubate at 37 °C for bacteria and 25 °C for yeast until to reach an optical density 0.5 of Marc Farland standard, which is equivalent approximately from 1–2 × 108 CFU/ml for bacteria and 1–5 × $10^6$ UFC/ml for yeasts.

Note: Alternatively, may prepare directly from the culture medium with agar 18 to 24 hours for bacteria and 48 hours for yeast, suspending the colony in saline or sterile nutrient broth until an optical density of 0.5 Marc Farland standard, to perform this step properly, either a photometric device should be used or, if done visually, adequate light is needed to visually compare the inoculum tube and the 0.5 Marc Farland standard against a card with a white background and contrasting black lines. For 0.5 Marc Farland standard preparations, see Cavalieri et al. (2005) on page 220.

5. With swab, remove a small portion of inoculum is evenly spread on plates containing Muller-Hinton broth by passing the swab over the entire surface in three directions.
6. Let dry for 5 minutes.
7. Apply 10 µl of the antimicrobial solution on filter paper disks of 6 mm in diameter, at different concentrations to be tested.
8. Place the antibiotic disks on the agar surface using sterile forceps, by pressing the agar.
9. Let the plates 15 minutes at room temperature, in order to initiate diffusion of antibiotics.
10. Incubate the plate upside down at 37 °C for 18 to 24 hours.
11. Measure the diameter of the inhibition zones with a ruler. To facilitate the measurement, the plates should be placed in a dark background surface. Figure 55 shows the inhibition zone around the disk.

*18.2.4. Antimicrobial analysis by agar well diffusion*

***18.2.4.1. Preparation of inoculum suspension.*** The preparation of inoculum could be accordingly with (Clinical and Laboratory Standards Institute (CLSI), 2012) with bit modifications as described below regarding to the procedures:

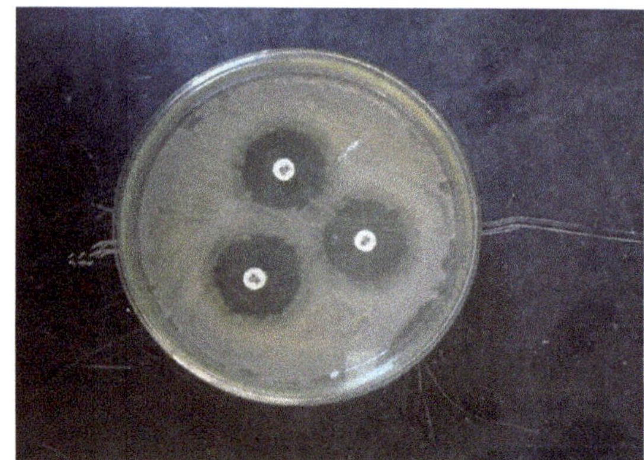

Figure 55. Inhibition zones around the disks.

1. Please, see proceedings described in points 1 to 6 regarding to agar Disk diffusion, then, follow the next step.
2. Forming wells in the culture medium previously inoculated from 6 to 8 mm, with the aid of stainless-steel cylinders.
3. Fill the wells with 100 µl of the substances to be analyzed.
4. Incubate the plate upside down at 37 °C for 18 to 24 hours for bacteria and yeast to 25 °C for 48 h.
5. Measure the diameter of the inhibition zones with a ruler. To facilitate the measurement, the plates should be placed in a dark background surface.

Note: Use a positive control (may be an antibiotic) and a negative control (solvent used for dissolving the dry extract in different concentrations); The disks are positioned at 15 mm plate edge and widely separated from each other, so there is no overlap of zones of inhibition (approximately 20 mm). The pH of the culture medium should be between 7.2 and 7.4, and the recommended depth is approximately 4 mm (Barry & Thornsberry, 1991). Figure 56 shows the inhibition zone around the well diffusion.

*18.2.5. Antimicrobial analysis by micro-dilution in microtiter plate (96 wells)*

***18.2.5.1. Preparation of inoculum suspension.*** The preparation of inoculum could be accordingly with Cavalieri et al. (2005) in Manual of Antimicrobial Susceptibility Testing, with bit modifications as described below.

*Procedure*

1. In sterile 96 well microplate, placing 200 µl of the sample (pollen extract) in the wells of the first column.

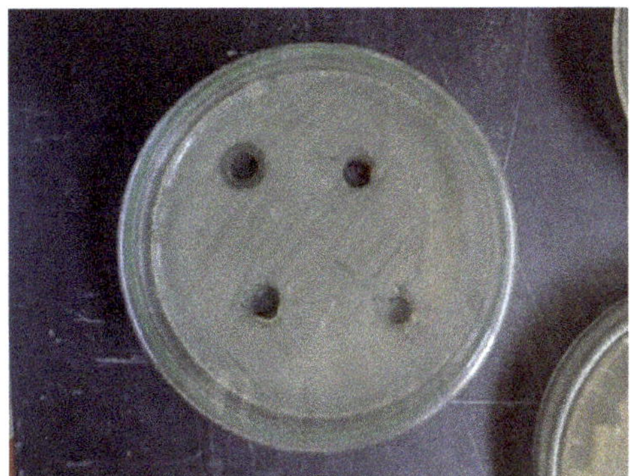

Figure 56. Inhibition zone around the well diffusion.

2. In wells 2 to 8 and 11 to 12 put 100 μl medium (nutrient broth), the wells 11 and 12 being the positive control.
3. In wells 9, 10 (negative control), putting 200 μl of medium.
4. Perform serial dilutions until the well 8.
5. Please, see proceedings described in points 1 to 4 regarding to agar Disk diffusion, then, follow the next step.
6. Then, within 15 minutes of adjusting the inoculum to the 0.5 Marc Farland turbidity standard, mix the suspension and dilute it so that the final concentration in each well is 5 x $10^5$ CFU/ml. Deliver 2.0 ml of the original suspension into 38 ml of water (1:20 dilution). From 1:20 dilution, prepare another 1:10 dilution that will be the inoculum.
7. Add 100 μl of inoculum in each well except for the columns 1, 9 and 10. Inoculate MIC panel carefully to avoid splashing from one well to another. Failure to adjust the inoculum and to dilute it within 15 minutes may adversely affect the concentration of organisms and test results.
8. Cover the microplate with a breathable film.
9. Incubate the yeast to 48 hours at 25 °C and the bacteria at 37 °C for 24 hours.
10. After this period, all wells except the first column, add 20 μL of TTC (2, 3, 5-triphenyl tetrazolium chloride) at 1%.
11. Incubate plates again for another 3 hours at 37 °C and 25 °C for bacteria and yeast, respectively.
12. Reading: See the color change (red/pink), indicating the reduction of TTC by microorganisms. The minimum inhibitory concentration (MIC) is defined as the lowest concentration of pollen extract that is capable of inhibiting visible growth, as indicated by TTC staining. Figure 57 shows the color change indicating the reduction of TTC by microorganisms in the wells of microplates.

#### *18.2.5.2. Calculating the lethal concentrations.*

1. After reading the microplates at 48 hours.
2. Seeding in plates (*Mueller-Hinton* agar for bacteria and Sabouraud for yeasts) in duplicate 20 μl of each of the wells in which there has been no microbial growth and the last in which growth was obtained.
3. Then mix thoroughly the contents of each well.
4. Once homogenized, inoculate 20 μl of each well in duplicate plates in their respective medium.
5. Incubate during 24 and 48 hours for bacteria and yeast respectively.
6. Then make counts.
7. If they are <10 or >10 CFU means that the substance has bactericidal and bacteriostatic activity, respectively.

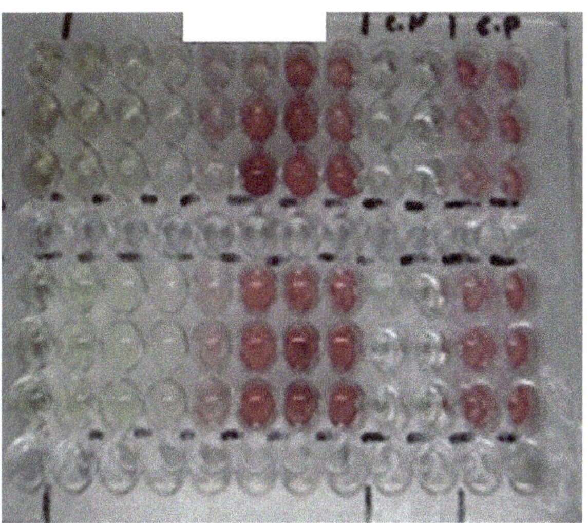

Figure 57. Microtiter plate (96 wells), showing the color change indicating the reduction of TTC by microorganisms.

### *18.3. Concluding remarks*

The Agar well diffusion techniques (Table 57) are used widely, problems associated with this technique are reported in Table 56). Therefore, Morais et al. (2011), Graikou et al. (2011) and Pascoal et al. (2014) used a Micro-dilution technique because is quick, worked well with *Staphylococcus aureus, Enterococcus faecalis, Pseudomonas aeruginosa (P. aeruginosa),* and *Escherichia coli* (E. coli) also this method gave reproducible results; provided a permanent record of the results and was useful in screening plants for antimicrobial activity (Eloff, 1998). Although BP has antimicrobial properties, it is important to highlight that these properties are depending on the type of pollen, concentrations applied and the inhibitions can occur against different species of bacteria and inhibition ratios can also vary (Carpes et al., 2007; Erkmen & Ozcan, 2008). For elucidate these affirmations, Campos et al. (1998) and Erkmen and Ozcan (2008) in their studies when investigated the antibacterial activities of BP on some bacteria, yeasts and molds, the obtained results showed that, the used concentrations had no antimicrobial effect on the bacteria and fungi. Carpes et al. (2007) in their studies found similar results as previously mentioned researchers. In general way the results obtained by all of cited researchers even if the methodologies applied were different showed that the Gram-negative bacteria were more resistant to the BP action than the Gram-positive bacteria, what may be related with the additional outer layer membrane, impermeable to most molecules, that consists of phospholipids, proteins and lipopolysaccharides (Silici & Kutluca, 2005).

## 19. Determination of standard methods for anti-mutagenic activities in bee pollen

### 19.1. Introduction

Some published studies, reported the beneficial effects of BP to human health such as the prevention of prostate problems (Shoskes, 2002) and tumors (Zhang et al., 1995). The increase of the of the large number of mutagens that has widespread in the world, led to researchers to investigate and discovering anti-mutagenic materials that limit and stop the harmful effects of mutations on the physiology of the organism. The most important of these materials has been reported as a plant extracts (Jaiswal et al., 2014). Over the past decade, some studies have lauded the anti-mutagenic properties of diverse beekeeping products, such as BP, propolis, honey mixtures and royal jelly, against the influence of some chemical and physical mutagens (Bariliak et al., 1996). Recently, studies done by Abdella et al. (2009) and Tohamy et al. (2014) when studied the protective efficiency of Egyptian BP extracts and propolis, treated mice with cisplatin [cis-diammine-dichloro-platinum(II)] (CDDP), observed that BP extracts presented potential chemoprotective against CDDP induced genotoxicity in mice by another side found that the chemoprotective frequency of BP extracts was much greater than propolis. Cisplatin (CDDP) is known as an inorganic platinum compound with a broad spectrum anti-neoplastic activity against different types of human tumors as testicular, ovarian, breast, lung, bladder, head and neck cancer (Khynriam & Prasad, 2003; Pabla et al., 2008; Siddik, 2003). The genotoxicity action of CDDP is due to its ability to bind with DNA, block and prolong the cell division in the G2 phase that is related to the inhibition of chromatin condensation (Aly et al., 2003). Bariliak et al. (1996) when studied apiculture products (propolis, beebread, honey and queen bee milk) observed a chemoprotective effect of these beehive products against the influence of some chemical and physical mutagens. BP was able to reduce the chromosome damage induced by the three cancer drugs used, thus supporting their use as a chemoprotective agent (Shi et al., 2010).

### 19.2. Determination of standard methods for anti-mutagenic activities in BP

#### 19.2.1. Preparation of BP extracts

Different methods for preparations of BP extracts have been reported (Carpes et al., 2007; Kaur et al., 2015; Pascoal et al., 2014; Pinto et al., 2010).

Procedure (according with Kaur et al., 2015).

1. The sample concentration and diluents used varies from the author to other.
2. Milling 1 g of pollen grains.
3. Homogenize with the help of needle in Petri dish.
4. Add 7.5 ml of 70% ethanol to the pollen mixture.
5. Keep at 70 °C for 1 h with agitation of 1 min after every 10 min interval.
6. Separate the supernatant.
7. Re-extract solid residue with 70% ethanol.
8. Repeat the process 3–4 times.
9. Group supernatants in a 25 ml conical flask.
10. The collected supernatant is considered as 100% extract.
11. Store these extracts at 5 °C till further analysis.
12. Prepare different concentrations *viz.*, 25%, 50%, 75% and 100% of pollen extracts using double distilled water.

#### 19.2.2. Anti-mutagenic activities in vitro

##### 19.2.2.1. Materials and reagents.

a. BP.
b. Ethyl methanesulfonate (EMS) and other chemicals.
c. Methanol ($CH_3OH$) and ethanol ($CH_3CH_2OH$).
d. Folin–Ciocalteu reagent chloroform ($CHCl_3$).
e. Sodium carbonate ($Na_2CO_3$).
f. Gentaminice and fluconazole.
g. Gallic acid and (+)-catechin.
h. The bovine testicular hyaluronidase (350 units).
i. Potassium salt of human umbilical cord hyaluronic acid.
j. The culture mediums.
k. The TTC solution (235-Triphenyl-2H-tetrazolium chloride).
l. High purity water (18 MX cm) and other chemicals.

##### 19.2.2.2. Procedure.

1. Yeast cells (D7 diploid strain of *Saccharomyces cerevisiae* ATCC 201137), according Pascoal et al. (2014).
2. Prior to each experiment the *S. cerevisiae* D7 strain must be tested for the frequency of spontaneous convertants at the tryptophan (trp) locus and revertants at the isoleucine (ilv) locus.
3. Cells from a culture with low spontaneous gene conversion and reverse point mutation frequencies were grown in a liquid medium at 28 °C until they reached the stationary growth phase.
4. Yeast cells pelleted and re-suspended in a volume of 0.1 M sterile potassium phosphate buffer, pH 7.4, to obtain a final mixture of $2 \times 10^8$ cell/ml.
5. The mixture (4 ml) contain 1 ml of cell suspension, potassium phosphate buffer and methanolic extracts of BP, in order to reach final concentrations of 0.00, 0.25, 0.50 and 0.75 mg/L.
6. Incubate the mixture under shaking for 2 h at 37 °C.

7. Cells are plated in complete and selective media to ascertain survival, *trp* convertants and *ilv* revertants.
8. Ethyl methanesulfonate (EMS), a mutagenic compound, is used as control.

*19.2.3. Anti-mutagenic activities of BP in vivo*

***19.2.3.1. Materials and reagents.***

a. BP.
b. Cis-platin (CDDP) (cis-diamminedicholorplatinum).
c. Ethyl methanesulfonate (EMS) and other chemicals.

***19.2.3.2. Experimental animals.*** For animal experimentation, we recommend to proceed according to the Guide for the Care and Use of Laboratory Animals Eighth Edition (2011) in order to perform as per international ethical standards. The used animals are usually mice (Abdella et al., 2009; Tohamy et al., 2014).

***19.2.3.3. Preparation of BP extracts.*** The preparation of BP extract can be according with Orsolić et al. (2005), Yamaguchi et al. (2007), Abdella et al. (2009) and Tohamy et al. (2014).

*Procedure*

1. BP (280 mg) was suspended in 10 ml of distilled water.
2. Mix vigorously.
3. Keep the suspension overnight in dark.
4. Centrifuge at 10,000 rpm in a cooling centrifuge for 45 minutes at 10 °C.
5. Collect and filtrate the supernatant fraction.
6. Keep the filtrated in freeze condition at −20 °C until use.

### 19.3. Concluding remarks on anti-mutagenic effects of BP

Studies published by Pinto et al. (2010) suggest that some BP preparations could be employed as future chemoprotective and/or chemopreventive agents, especially for mammary or other estrogen-dependent cancers because in their studies found that at least one preparation of both species showed a marked anti-genotoxic effect against the action of tested anticancer drug. These anti-mutagenic properties of BP have been attributed to their composition rich in phenolic compounds that justify the higher protection (Pinto et al., 2010). This author also reported that the most effective species of BP from *Salix alba* in inducing inhibitory activity of E2 (17β-estradiol) and protected against the three mutagenic drugs and these results could be due its higher flavonoid content. Some of two studied samples were rich in luteolin, quercetin, chrysin, and kaempferol compounds that have been previously demonstrated to be antiestrogenic (Oh & Chung, 2006; Pinto et al., 2008; van Meeuwen et al., 2007), also exhibit anti-mutagenic and genotoxic activity (Lautraite et al., 2002; Nagy et al., 2009) whereas Galangin, pinocembrin and chrysin, as well as isorhamnetin 3-*O*-neohesperidoside, a derivative of isorhamnetin were proven to effectively contrast the DNA-damaging activity of several mutagens/carcinogens without exhibiting any genotoxic activity (Heo et al., 2001; Snyder & Gillies, 2002). Similar studies revealed that phenolic and flavonoids compounds as luteolin, chrysin, quercetin and taxifolin showed anti-mutagenic and genotoxic activity (Benkovic et al., 2008; Dutta et al., 2007; Makena et al., 2009; Uhl et al., 2003). Therefore, the anti-mutagenic effects, can be affected by total phenolic compound content as well as variation in their composition, and the higher phenolic compound content found in *S. alba* would justify the higher protection elicited toward the mutagenic drugs (Pinto et al., 2010).

The specific mechanism of protection of the BP extract is reported to involve the scavenging of potentially toxic and mutagenic electrophiles and free radicals and the modification of antioxidant pathways (Ohta et al., 2007). By another side, Nagy et al. (2009) referred that the anti-mutagenic effect of most phenolic compounds could be clearly ascribed to their - DPPH* (1,1-diphenyl-2-picrylhydrazyl) scavenging activity, substitution patterns and lipophilicity. Barzin et al. (2011) in their studies evaluated the anti-mutagenic effects of pollen grams of *Phoenix dactylifera* using Ames assay and found that had 46% anti-mutagenic response. According to Ames et al. (1973, 1976) when the percent (46% of anti-mutagenic response), is more than 40, mutagenesis effect of the test sample is strong. Also, Bariliak et al. (1996) observed the ability of apiculture products to decrease the mutagenic effects of some chemical and physical mutagens. In general way, the bioactivity of the pollens has been reported to be due to presence of compounds such as proteins, carbohydrates, amino acids, lipids, vitamins carotenoids, steroids, terpenes, flavonoids and polyphenols (Leja et al., 2007; Kao et al., 2020; Kroyer & Hegedus, 2001).

## Acknowledgements

The COLOSS (Prevention of honey bee COlony LOSSes) Association aims to explain and prevent massive honey bee colony losses. It was funded through the COST Action FA0803. COST (European Cooperation in Science and Technology) is a unique means for European researchers to jointly develop their own ideas and new initiatives across all scientific disciplines through trans-European networking of nationally funded research activities. Based on a pan-European intergovernmental framework for cooperation in science and technology, COST has contributed since its creation more than 40 years ago to closing the gap between science, policy makers and society throughout Europe and beyond. COST is supported by the EU Seventh Framework Program for research, technological development and demonstration activities *(Official Journal L 412, 30 December 2006)*. The European Science Foundation as implementing agent of COST provides the COST Office through an EC Grant Agreement. The Council of the European Union provides the COST Secretariat. The COLOSS network is now supported by the Ricola Foundation – Nature & Culture. Figures 26–28 are reproduced from Sawyer (1981) with the permission of the publishers University College Cardiff Press and Northern Bee

Books. Lidia Barreto and J Nordi wish to thank the Apiculture Research Center of Taubate University (UNITAU-SP/Brazil) and Agriculture Secretary of Bahia State (SEAGRI-BA/BRAZIL). Maria Campos wishes to thank (UI0204): UIDB/00313/2020, Center of Chemistry from Faculty of Sciences and Technology of University of Coimbra, Portugal. Ofélia Anjos wishes to thank to Forest Research Centre, a research unit funded by Fundação para a Ciência e a Tecnologia I.P. (FCT), Portugal (UIDB/00239/2020), and to Centro de Biotecnologia de Plantas da Beira Interior for the OPUS software availability. Norma Almaraz Abarca thanks to the Instituto Politecnico Nacional for financial and logistic support. Manuel Chica and Pascual Campoy wish to thank the APIFRESH project. APIFRESH has been co-funded by the European Commission under the R4SMEs 7th Framework Program. Olena Lokutova thanks the Austrian Institute of beekeeping, Dr H Pehhacker and the same members of the Institute H Hagel and E Hüttinger for conducting photomicroscopic studies and pollen analysis Ukrainian samples of pollen loads, which were the basis of the atlas of pollen "honey plants" of Ukraine. Olena is grateful also to Polish colleagues Z Warakomska (Department of Botany University of Lublin) and D Teper (Polish Institute of beekeeping) for their professionalism and consultations to determine the botanical origin of some Ukrainian honey. Also thanks to their scientific advisers' academician G Bogdanov (National Academy of Agrarian Sciences of Ukraine), Prof. V Polishuk (Department of beekeeping National University of Life and Environmental Sciences of Ukraine) and O Martynyuk (M.G. Kholodny Institute of Botany, Kiev, Ukraine) for his helpful co-operation in the field of beekeeping and palinology. Janka Nozkova wishes to thank the Operational Program Research and Development of the European Regional Development Fund in the frame of the project "Support of technologies innovation for special bio-food products for human healthy nutrition" (ITMS 26220220115) and also by the Excellence Center for Agrobiodiversity Conservation and Benefit – project implemented under the Operational Program Research and Development financed by European Fund for Regional Development ITMS 26220120015 (Slovak Republic) and all colleagues from Institute of Biodiversity and Biosafety, Slovak University of Agriculture in Nitra for their help with image analysis. Ananias Pascoal, Georgina Tolentino and Letícia Estevinho would like to thank Fundação para a Ciência e Tecnologia (FCT), Programa Operacional Pontencial Humano (POPH) and European Union (EU) for his Postdoctoral grant SFRH/BPD/91380/2012. Wiebke Sickel, Markus Ankenbrand, Gudrun Grimmer, Frank Förster, Ingolf Steffan-Dewenter and Alexander Keller thank the financial support by the DFG Collaborative Research Center 1047, Insect Timing. MJA was further supported by a grant of the German Excellence Initiative to the Graduate School of Life Sciences of the University of Würzburg. They are grateful to the members of the Departments of Animal Ecology and Tropical Biology; Bioinformatics; and Human Genetics, University of Würzburg, for constructive input on the design of the workflow. Additionally thank to the Department of Human Genetics, especially S. Rost, for granting access to the Illumina MiSeq device. Živoslav Tešić, Mirjana Mosić, Aleksandar Kostić, Mirjana Pešić, Dušanka Milojković-Opsenica thank the Ministry of Education, Science and Technological Development of Serbia, Grants 172017 and TR 31069. Gina Tolentino would like to thank the Mountain Research Center (CIMO), Agricultural College of Bragança, Polytechnic Institute of Bragança for his research grant in the project titled " Development of new bee products in biological production way."

## Disclosure statement

No potential conflict of interest was reported by the authors.

## ORCID

*Maria G. Campos* http://orcid.org/0000-0003-1012-6240
*Ofélia Anjos* http://orcid.org/0000-0003-0267-3252
*Manuel Chica* http://orcid.org/0000-0002-4717-1056
*Pascual Campoy* http://orcid.org/0000-0002-9894-2009
*Norma Almaraz-Abarca* http://orcid.org/0000-0003-1603-4865
*Leticia M. Estevinho* http://orcid.org/0000-0002-9249-1948
*Ananias Pascoal* http://orcid.org/0000-0003-3823-0402
*Luis G. Dias* http://orcid.org/0000-0002-1210-4259
*Živoslav L. j. Tešić* http://orcid.org/0000-0002-5162-3123
*Aleksandar Ž. Kostić* http://orcid.org/0000-0002-1012-4029
*Mirjana B. Pešić* http://orcid.org/0000-0001-9779-1686
*Dušanka M. Milojković-Opsenica* http://orcid.org/0000-0001-6274-4222
*Wiebke Sickel* http://orcid.org/0000-0002-0038-1478
*Markus J. Ankenbrand* http://orcid.org/0000-0002-6620-807X
*Ingolf Steffan-Dewenter* http://orcid.org/0000-0003-1359-3944
*Alexander Keller* http://orcid.org/0000-0001-5716-3634
*Frank Förster* http://orcid.org/0000-0003-4166-5423
*Chrysoula H. Tananaki* http://orcid.org/0000-0002-0056-483X
*Vasilios Liolios* http://orcid.org/0000-0001-7353-2364
*Dimitrios Kanelis* http://orcid.org/0000-0001-7736-4783
*Maria-Anna Rodopoulou* http://orcid.org/0000-0003-1643-4615
*Andreas Thrasyvoulou* http://orcid.org/0000-0001-6731-4184
*Christina Kast* http://orcid.org/0000-0002-5673-6848
*Ligia Bicudo de Almeida-Muradian* http://orcid.org/0000-0001-5322-1358
*Teresa Szczęsna* http://orcid.org/0000-0002-0161-0287
*Norman L. Carreck* http://orcid.org/0000-0001-7779-9736

## References

Abdella, E. M., Tohamy, A., & Ahmad, R.R. (2009). Antimutagenic activity of Egyptian propolis and bee pollen water extracts against cisplatin-induced chromosomal abnormalities in bone marrow cells of mice. *Iranian Journal of Cancer Prevention*, *2*(4), 175–181.

Abouda, Z., Zerdani, I., Kalalou, I., Faid, M., & Ahami, M. T. (2011). The antibacterial activity of Moroccan bee bread and bee-pollen (fresh and dried) against pathogenic bacteria. *Research Journal of Microbiology*, *6*(4), 376. http://.dx.doi.org/10.3923/jm.2011.376.384

Akkol, E. K., Orhan, D. D., Gurbuz, I., & Yesilada, E. (2010). In vivo activity assessment of a "honey-bee pollen mix" formulation. *Pharmaceutical Biology*, *48*(3), 253–259. https://doi.org/10.3109/13880200903085482

Allen, G. (2006). *An automated pollen recognition system* [Master's thesis]. Institute of information Sciences and Technology, Massey University.

Almaraz-Abarca, N., Campos, M. d. G., Ávila-Reyes, J. A., Naranjo-Jimenez, N., Herrera-Corral, J., & Gonzalez, V. L. S. (2004). Variability of antioxidant activity among honey bee-collected pollen of different botanical origin. *Interciencia*, *29*(10), 574–578. doi: 0378-1844/04/10/574-05.

Almaraz-Abarca, N., Campos, M. D. G., Delgado-Alvarado, A., Ávila-Reyes, J. A., Naranjo-Jiménez, J. A., & Herrera Corral, H. (2006). Variación intrapoblacional en el Perfil Fenolico del polen de *Stenocactus multicostatus* subsp. *zacatecasensis* (Cactaceae). *Boletín Nakari*, *17*(3), 59–64.

Almaraz-Abarca, N., Campos, M. G., Ávila-Reyes, J. A., Naranjo-Jiménez, N., Herrera Corral, J., & González-Valdez, L. S. (2007). Antioxidant activity of polyphenolic extract of monofloral honey bee – collected pollen from mesquite (*Prosopis juliflora, Leguminosae*). *Journal of Food Composition and Analysis*, *20*(2), 119–124. https://doi.org/10.1016/j.jfca.2006.08.001

Almaraz-Abarca, N., Campos, M. G., Delgado-Alvarado, A., Ávila-Reyes, J. A., Naranjo-Jiménez, N., Herrera Corral, J., Tomatas, A. F., Almeida, A. J., & Vieira, A. (2007). Fenoles del polen de *Stenocactus, Echinocereus y Mammillaria* (Cactaceae). *Polibotanica*, *23*, 37–55. http://www.herbario.encb.ipn.mx/pb/pdf/pb23/3Fenoles.pdf

Almaraz-Abarca, N., Campos, M. G., Delgado-Alvarado, E. A., Ávila-Reyes, J. A., Herrera-Corral, J., González-Valdez, L. S., Naranjo-Jiménez, N., Frigerio, C., Tomatas, A. F., Almeida, A. J., Vieira, A., & Uribe-Soto, J. N. (2008). Pollen flavonoid/phenolic acid composition of four species of Cactaceae and its taxonomic significance. *American Journal of Agricultural and Biological Science*, *3*(3), 534–543. https://doi.org/10.3844/ajabssp.2008.534.543

Almaraz-Abarca, N., González-Elizondo, M. S., Campos, M. G., Ávila-Sevilla, Z. E., Delgado-Alvarado, E. A., & Ávila, R. J. A. (2013). Variability of the foliar phenol profiles of *Agave victoriae-reginae* complex (Agavaceae). *Botanical Sciences*, *91*(3), 295–306. https://doi.org/10.17129/botsci.9

Almaraz-Abarca, N., Rivera-Rodríguez, D. M., Arráez-Román, D., Segura-Carretero, A., Sánchez-González, J. J., Delgado-Alvarado, A., Ávila-Reyes, J. A. (2013). Los fenoles del polen del género *Zea*. *Acta Botanica Mexicana*, *105*, 59–85. http://www.ibiologia.unam.mx/sociedad/www/pdf/BSBM91_3/Almaraz-Abarca_91_3_295-306.pdf

Almeida-Muradian, L. B., Arruda, V. A. S., & Barreto, L. M. R. C. (2012). *Manual de Controle de Qualidade do Pólen Apícola* (p. 28). APACAME.

Almeida-Muradian, L. B., Pamplona, L. C., Coimbra, S., & Barth, O. M. (2005). Chemical composition and botanical evaluation of dried bee pollen pellets. *Journal of Food Composition and Analysis*, *18*(1), 105–111. https://doi.org/10.1016/j.jfca.2003.10.008

Aly, M. S., Ashour, M. B., EL-Nahas, S. M., & Abo Zeid, M. A. F. (2003). Genotoxicity and cytotoxicity of the anticancer drugs gemcitabine and cisplatin, separately and in comination: in vivo studies. *Journal of Biological Sciences*, *3*(11), 961–972. https://doi.org/10.3923/jbs.2003.961.972[Mismatch] .

Ambruster, L., & Oenike, G. (1929). *Die pollenformen als mittel zur honig herkunftsbestimmung* (p. 116). Wachholtz.

Ames, B. N., Durston, W. E., Yamasaki, E., & Lee, F. D. (1973). Carcinogens are mutagens: A simple test system combining liver homogenates for activation and bacteria for detection. *Proceedings of the National Academy of Sciences of the United States of America*, *70*(8), 2281–2285. https://doi.org/10.1128/AAC.04918-14.

Andrada, A. C., & Telleria, M. C. (2005). Pollen collected by honey bees (*Apis mellifera* L.) from south of Caldén district (Argentina): Botanical origin and protein content. *Grana*, *44*(2), 115–122. https://doi.org/10.1080/00173130510010459

Anjos, O., Campos, M. G., Ruiz, P. C., & Antunes, P. (2015). Application of FTIR-ATR spectroscopy to the quantification of sugar in honey. *Food Chemistry*, *169*, 218–223. https://doi.org/10.1016/j.foodchem.2014.07.138

Anjos, O., Santos, A. J. A., Dias, T., Estevinho, L. M. (2017). Application of FTIR-ATR Spectroscopy on the bee pollen characterization. *Journal of Apicultural Research*, *56*(3), 210–218. https://doi.org/10.1080/00218839.2017.1289657

Anjos, O., Santos, A. F. A., Paixão, V., Estevinho, M. L. (2018). Physicochemical Parameters of *Lavandula spp.* honey accessed by FT-RAMAN. *Talanta*, *178*, 43–48. https://doi.org/10.1016/j.talanta.2017.08.099

Ankenbrand, M. J. (2015). The ITS2 Database V – Twice as much. *Molecular Biology and Evolution*, 32, 3030–3032.

AOAC. (1995). *Official methods of analysis* (*16th Ed.*). Association of Official Analytical Chemists.

Arroyo, J., Devesa, J. A., Herrera, P., Ortiz, P. L., & Lozano, S. T. (1986). Resumen del proyecto de investigación: La flora melitófila en Andalucía occidental. *Vida Apícola*, *18*, 33–39.

Arruda, V. A. S. (2013). *Dried bee pollen: Physicochemical, microbiological quality, phenolic and flavonoid compounds, antioxidant activity and botanical origin* [PhD thesis]. Pharmaceutical Science School, University of São Paulo.

Arruda, V. A. S., Freitas, A. S., B., O. M., Estevinho, L. M., & Almeida-Muradian, L. B. (2013). Propriedades biológicas do pólen apícola de coqueiro, coletado na Região Nordeste do Brasil. *Magistra*, *25*, 1–6.

Arruda, V. A. S., Santos Pereira, A. A., de Freitas, A. S., Barth, O. M., & de Almeida-Muradian, L. B. (2013b). Dried bee pollen: B complex vitamins, physicochemical and botanical composition. *Journal of Food Composition and Analysis*, *29*(2), 100–105. https://doi.org/10.1016/j.jfca.2012.11.004

Arruda, V. A. S., Santos Pereira, A. A., Estevinho, L. M., & de Almeida-Muradian, L. B. (2013a). Presence and stability of B complex vitamins in bee pollen using different storage conditions. *Food and Chemical Toxicology*, *51*, 143–148. https://doi.org/10.1016/j.fct.2012.09.019

Avula, B., Satyanarayanaraju, S., Wang, Y., Zweigenbaum, J., Wang, M., & Khan, I. A. (2015). Characterization and screening of pyrrolizidine alkaloids and *N*-oxides from botanicals and dietary supplements using UHPLC-high resolution mass spectrometry. *Food Chemistry*, *178*, 136–148. https://doi.org/10.1016/j.foodchem.2015.01.053

Balkanska, R. G., Ignatova, M. M. (2012). Chemical composition of multifloral bee pollen from Bulgaria. In *Proceedings of 6th Central European Congress on Food*, CE Food 2012 (pp. 375–378).

Ball, G. (1998). *Bioavailability and analysis of vitamins in foods.* Chapman & Hall.

Baltrušaitytė, V., Venskutonis, P. R., & Čeksterytė, V. (2007b). Radical scavenging activity of different floral origin honey and beebread phenolic extracts. *Food Chemistry*, *101*(2), 502–514. https://doi.org/10.1016/j.foodchem.2006.02.007

Barbosa, S. I. T. R., Silvestre, A. J. D., Simões, M. M. Q., & Estevinho, M. L. (2006). Composition and antibacterial activity of the lipophilic fraction of honey bee pollen from native species of Montesinho Natural Park. *International Journal of Agricultural Research*, *1*(5), 471–749.

Bariliak, I. R., Berdyshev, G. D., & Dugan, A. M. (1996). The antimutagenic action of apiculture products. *The Journal of Cytology and Genetics*, *30*, 48–55.

Barreto, L. M. R. C. (2004). Pólen Apícola: beneficiamento, armazenamento e legislação. Botucatu, 2004, 150 (Tese Doutorado) – Faculdade de Medicina Veterinária e Zootecnia – Universidade Estadual Paulista.

Barreto, L. M. R. C., Dib, A. P. S., & Peão, G. F. R. (2006). *Higienização e Sanitização da Produção Apícola* (p. 150). Cabral Editora e Livraria Universitária.

Barreto, L. M. R. C., Funari, S. R. C., Orsi, R. O., & Dib, A. P. S. (2006). *Produção de Pólen no Brasil* (p. 100). Cabral Editora e Livraria Universitária.

Barriada-Bernal, L. G., Almaraz-Abarca, N., Delgado-Alvarado, E. A., Gallardo-Velázquez, T., Ávila-Reyes, J. A., Torres-Morán, M. I., González-Elizondo, M. d. S., & Herrera-Arrieta, Y. (2014). Flavonoid composition and antioxidant capacity of the edible flowers of *Agave durangensis* (Agavaceae). *Cyta – Journal of Food*, *12*(2), 105–114. https://doi.org/10.1080/19476337.2013.801037

Barry, A. L., & Thornsberry, C. (1991). Susceptibility tests: Diffusion test procedures. In A. Balows, W. J. Hauser, K. L. Hermann, H. D. Isenberg, & H. J. Shamody (Eds.), *Manual of clinical microbiology* (Vol. 5, pp. 1117–1125). American Society for Microbiology.

Barth, O. M. (2004). Melissopalynology in Brazil: A review of pollen analysis of honeys, propolis and pollen loads of bees. *Scientia Agricola*, *61*(3), 342–350. https://doi.org/10.1590/S0103-90162004000300018

Barth, O. M., Freitas, A. S., Oliveira, E. S., Silva, R. A., Maester, F. M., Andrella, R. R., & Cardozo, G. M. (2010). Evaluation of the botanical origin of commercial dry bee pollen load batches using pollen analysis: A proposal for technical standardization. *Anais da Academia Brasileira de Ciencias*, *82*(4), 893–902. https://doi.org/10.1590/s0001-37652010000400011

Barzin, G., Entezari, M., Hashemi, M., Hajiali, S., Ghafoori., & Gholami, M. (2011). Survey of Antimutagenicity and Anticancer effect of *Phoenix dactylifera* pollen grains. *Advances in Environmental Biology*, *5*(12), 3716–3718.

Basim, E., Basim, H., & Ozcan, M. (2006). Antibacterial activities of Turkish pollen and propolis extracts against plant bacterial pathogens. *Journal of Food Engineering*, *77*(4), 992–996. https://doi.org/10.1016/j.jfoodeng.2005.08.027

Bassi, E. A. (2000). Programa de qualidade total na produção de mel. In: *Congresso brasileiro de apicultura*, 13, 2000. Sonopress. 1CD.

Basuny, A. M., Arafat, S. M., & Soliman, H. M. (2013). Chemical analysis of olive and palm pollen: Antioxidant and antimicrobial activation properties. *Herald Journal of Agriculture and Food Science Research*, *2*(3), 091–097.

Bell, K. L., Burgess, K. S., Botsch, J. C., Dobbs, E. K., Read, T. D., & Brosi, B. J. (2019). Quantitative and qualitative assessment of pollen DNA metabarcoding using constructed species mixtures. *Molecular Ecology*, *28*, 431–455. https://doi.org/10.1111/mec.14840.

Benkovic, V., Knezevic, A. H., Dikic, D., Lisicic, D., Orsolic, N., Basic, I., Kosalec, I., & Kopjar, N. (2008). Radioprotective effects of propolis and quercetin in gamma-irradiated mice evaluated by the alkaline comet assay. *Phytomedicine*, *15*(10), 851–858. https://doi.org/10.1016/j.phymed.2008.02.010

Bertoncelj, J., Polak, T., Kropf, U., Korošec, M., & Golob, T. (2011). LC-DAD-ESI/MS analysis of flavonoids and abscisic acid with chemometric approach for the classification of Slovenian honey. *Food Chemistry*, *127*(1), 296–302. https://doi.org/10.1016/j.foodchem.2011.01.003

Bessonov, V. N. (1992). Status of pollen as an indicator of pollution by heavy metals. *Ecology. – Ekaterinburg*, *1992*(3), 45–50. (Article in Russian).

Betteridge, K., Cao, Y., & Colegate, S. M. (2005). Improved method for extraction and LC-MS analysis of pyrrolizidine alkaloids and their N-oxides in honey: Application to *Echium vulgare* honeys. *Journal of Agricultural and Food Chemistry*, *53*(6), 1894–1902. https://doi.org/10.1021/jf0480952

Biesaga, M., & Pyrzynska, K. (2009). Liquid chromatography/tandem mass spectrometry studies of the phenolic compounds in honey. *Journal of Chromatography. A*, *1216*(38), 6620–6626. https://doi.org/10.1016/j.chroma.2009.07.066

Biodiversity International. (2007). *Guidelines for the development of crop descriptor lists. Biodiversity Technical Bulletin Series* (xii + 72 pp.). Biodiversity International.

Bobis, O., Marghitas, L., Al, Dezmirean, D., Morar, O., Bonta, V., & Chirila, F. (2010). Quality parameters and nutritional value of different commercial bee products. *Bulletin UASVM Animal Science and Biotechnologies*, *67*(1-2), 91–96.

Bogdanov, S. (2015). *Pollen: Production, nutrition and health: A review.* Bee Product Science. www.bee-hexagon.net

Bogdanov, S., Martin, P., & Lüllmann, C. (1997). Harmonized methods of the European Honey Commission. Determination of sugars by HPLC. *Apidologie*, *28*, 42–44.

Bondet, V., Brand-Williams, W., & Berset, C. (1997). Kinetics and mechanisms of antioxidant activity using the DPPH* free radical method. *LWT – Food Science and Technology*, *30*(6), 609–615. https://doi.org/10.1006/fstl.1997.0240

Boppré, M. (2011). The ecological context of pyrrolizidine alkaloids in food, feed and forage: An overview. *Food Additives & Contaminants. Part A, Chemistry, Analysis, Control, Exposure & Risk Assessment*, *28*(3), 260–281. https://doi.org/10.1080/19440049.2011.555085

Boppré, M., Colegate, S. M., & Edgar, J. A. (2005). Pyrrolizidine alkaloids of *Echium vulgare* honey found in pure pollen. *Journal of Agricultural and Food Chemistry*, *53*(3), 594–600. https://doi.org/10.1021/jf0484531

Boucher, A., Hidalgo, P. J., Thonnat, M., Belmonte, J., Galan, C., Bonton, P., & Tomczak, R. (2002). Development of a semi-automatic system for pollen recognition. *Aerobiologia*, *18*(3/4), 195–201. https://doi.org/10.1023/A:1021322813565

Boudet, A. M. (2007). Evolution and current status of research in phenolic compounds. *Phytochemistry*, *68*(22-24), 2722–2735. https://doi.org/10.1016/j.phytochem.2007.06.012

Brasil. (1997). Ministério da Agricultura e do Abastecimento. Portaria no 368, de 4 de setembro de 1997. Regulamento Técnico sobre as Condições Higiênico-Sanitárias e de Boas Práticas de Elaboração para Estabelecimentos Elaboradores/Industrializadores de Alimentos. Diário Oficial da União da Republica Federativa do Brasil, Brasília.

Brasil. (2001). Instrução Normativa do Ministério da Agricultura n° 3, de 19 de janeiro de 2001. (DOU – 23/01/01, Secção I, pg. 18-23) Regulamentos técnicos de identidade e qualidade de apitoxina, cera de abelhas, geléia real, geléia real Liofilizada, pólen apícola, própolis e extrato de própolis. Diário Oficial (DO), Brasília, DF.

Brindza, J., & Brovarskyi, V. (2014). *Pollen and bee pollen of some plant species* (p. 138). Kyiv: Korsunskiy vidavnichiy dim Vsesvit.

Bruni, I., Galimberti, A., Caridi, L., Scaccabarozzi, D., De Mattia, F., Casiraghi, M., & Labra, M. (2015). A DNA barcoding approach to identify plant species in multiflower honey. *Food Chemistry*, *170*, 308–315. https://doi.org/10.1016/j.foodchem.2014.08.060

Bucher, E., Rjfler, V., Vorwoht, G., & Zieger, E. (2004). *Das pollebild der sudtiroler honige* (p. 676). Biologischer labor.

Burmistrov, N., & Nikitina, V. A. (1990). *Honey plants and their pollen* (p. 192). Rosagropromizdat. (Book on Russian).

Cabrera, C., & Montenegro, G. (2013). Pathogen control using a natural Chilean bee pollen extract of known botanical origin. *Ciencia e Investigación Agraria*, *40*(1), 223–230. https://doi.org/10.4067/S0718-16202013000100020

Campos, M. G. (1997). *Caracterização do pólen apícola pelo seu perfil em compostos fenólicos e pesquisa de algumas actividades biologicas* [PhD thesis]. Faculty of Pharmacy, University of Coimbra.

Campos, M. G., Bogdanov, S., Almeida-Muradian, L. B., Szczesna, T., Mancebo, Y., Frigerio, C., & Ferreira, F. (2008). Pollen composition and standardization of analytical methods. *Journal of Apicultural Research*, *47*(2), 154–161. https://doi.org/10.3896/IBRA.1.47.2.12[10.1080/00218839.2008.11101443]

Campos, M. G., Frigerio, C., Lopes, J., & Bogdanov, S. (2010). What is the future of bee pollen? *Journal of ApiProduct and ApiMedical Science*, 2(4), 131–144. https://doi.org/10.3896/IBRA.4.02.4.01

Campos, M. G., & Markham, K. R. (2007). *Structure information from HPLC and on-line measured absorption spectra: Flavones, flavonols and phenolic acids.* Imprensa da Universidade de Coimbra. https://doi.org/10.14195/978-989-26-0480-0

Campos, M. G., Markham, K. R., Mitchell, K. A, & da Cunha, A. P. d. (1997). An approach to the characterization of bee pollens via their flavonoid/phenolic profiles. *Phytochemical Analysis*, *8*(4), 181–185. https://doi.org/10.1002/(SICI)1099-1565(199707)8:4 < 181::AID-PCA359 > 3.0.CO;2-A

Campos, M. G., Markham, K., & Proença da Cunha, A. (1996). Quality assessment of bee-pollens using flavonoid/phenolic profiles. *Bull. Groupe Polyphenols*, *18*, 54–55.

Campos, M. G., Mitchel, K., Cunha, A., & Markham, K. R. (1997). A systematic approach to the characterization of bee pollens via their flavonoid/phenolic profiles. *Phytochemical Analysis*, *8*(4), 181–185. 2-A https://doi.org/10.1002/(SICI)1099-1565(199707)8:4 < 181::AID-PCA359 > 3.0.CO,[10.1002/(SICI)1099-1565(199707)8:4 < 181::AID-PCA359 > 3.0.CO;2-A]

Campos, M. G., Webby, R. F., Markham, K. R., Mitchell, K. A., & da Cunha, A. P. (2003). Age-induced diminution of free radical scavenging capacity in bee pollens and the contribution of constituent flavonoids. *Journal of Agricultural and Food Chemistry*, *51*(3), 742–745. https://doi.org/10.1021/jf0206466

Caroli, S., Forte, G., Iamiceli, A. L., & Galoppi, B. (1999). Determination of essential and potentially toxic trace elements in honey by inductively coupled plasma-based techniques. *Talanta*, *50*(2), 327–336. https://doi.org/10.1016/S0039-9140(99)00025-9

Carpes, S. T., Begnini, R., DE Alencar, S. M., & Masson, M. L. (2007). Study of preparations of bee pollen extracts, antioxidant and antibacterial activity. *Ciência e Agrotecnologia*, *31*(6), 1818–1825. https://doi.org/10.1590/S1413-70542007000600032

Carpes, S. T., Prado, A. P., Moreno, I. A. M., Mourão, G. B., DE Alencar, S. M., & Masson, M. L. (2008). Avaliação do potencial antioxidante do pólen apícola producido na região sul do Brasil. *Química Nova*, *31*(7), 1660–1664. https://doi.org/10.1590/S0100-40422008000700011

Carrión, P., Cernadas, E., Gálvez, J. F., & Díaz-Losada, E. (2003). Determine the composition of honey bee pollen by texture classification. In F. J. Perales (Ed.), *IbPRIA 2003, LNCS 2652* (pp. 158–167). https://doi.org/10.1007/978-3-540-44871-6_19

Carvell, C., Westrich, P., Meek, W. R., Pywell, R. F., & Nowakowski, M. (2006). Assessing the value of annual and perennial forage mixtures for bumble bees by direct observation and pollen analysis. *Apidologie*, *37*(3), 326–340. https://doi.org/10.1051/apido:2006002

Cavalieri, S. J., Harbeck, R. J., Mccarter, Y. S., Ortez, J. H., Rankin, I. D., Sautter, R. L., Sharp, S. E., & Spiegel, C. A. (2005). Manual of antimicrobial susceptibility testing. In M. B. Coyle (Ed.), *American Society for Microbiology © 2005* (pp. 1–236). American Society for Microbiology.

Chandola, V., Banerjee, A., & Kumar, V. (2009). Anomaly detection: A survey. *ACM Computing Surveys*, *41*(3), 1–15. 58. https://doi.org/10.1145/1541880.1541882

Chantarudee, A., Phuwapraisirian, P., Kimura, K., Okuyama, M., Mori, H., Kimura, A., & Chanchao, C. (2012). Chemical constituents and free radical scavenging activity of corn pollen collected from Apis mellifera hives compared to floral corn pollen at Nam, Thailand. *BMC Complementary & Alternative Medicine*, *12*, 45.

Chaturvedi, M. (1973). An analysis of honey bee pollen loads from Banthra, Lucknow, India. *Grana*, *13*(3), 139–144. https://doi.org/10.1080/00173137309429890

Chen, S., Yao, H., Han, J., Liu, C., Song, J., Shi, L., Zhu, Y., Ma, X., Gao, T., Pang, X., Luo, K., Li, Y., Li, X., Jia, X., Lin, Y., & Leon, C. (2010). Validation of the ITS2 region as a novel DNA barcode for identifying medicinal plant species. *PLoS One*, *5*(1), e8613. https://doi.org/10.1371/journal.pone.0008613

Chica, M., & Campoy, P. (2012). Discernment of bee pollen loads using computer vision and one-class classification techniques. *Journal of Food Engineering*, *112*(1-2), 50–59. https://doi.org/10.1016/j.jfoodeng.2012.03.028

Choi, E. (2007). Antinociceptive and antiinflammatory activities of pine (*Pinus densiflora*) pollen extract. *Phytotherapy Research*, *21*(5), 471–475. https://doi.org/10.1002/ptr.2103

Chudzinska, M., & Baralkiewicz, D. (2011). Application of ICP-MS method of determination of 15 elements in honey with chemometric approach for the verification of their authenticity. *Food and Chemical Toxicology*, *49*(11), 2741–2749. https://doi.org/10.1016/j.fct.2011.08.014

Cirnu, I., Slusanschi, N., Maronescu, R., Filipescu, H., & Grosu, E. (1969). Compozitia chimică a polenului de porumb (*Zea mays* L.) si floarea-soarelui (*Helianthus annuus* L.) recoltat la diferite epoci, in cardul amestecului furajer-melifer. *Apicultura*, *22*(8), 22–25.

Clinical and Laboratory Standards Institute (CLSI). (2009). Performance standards for antimicrobial disk susceptibility tests; approved standard-tenth edition M02-A10. *National Committee for Clinical Laboratory Standards*, *29*. 76 pp

Clinical and Laboratory Standards Institute (CLSI). (2012). Methods for dilution antimicrobial susceptibility tests for bacteria that grow aerobically; approved standard – Ninth Edition M07 – A9. *National Committee for Clinical Laboratory Standards*, *32*(2), 1–64.

Comaniciu, D., & Meer, P. (2002). Mean shift: A robust approach toward feature space analysis. *IEEE Transactions on Pattern Analysis and Machine Intelligence*, *24*(5), 603–619. https://doi.org/10.1109/34.1000236

Costa, C. T., Dalluge, J. J., Welch, M. J., Coxon, B., Margolis, S. A., & Horton, D. (2000). Characterization of prenylated xanthones and flavanones by liquid chromatography/atmospheric pressure chemical ionization mass spectrometry. *Journal of Mass Spectrometry*, *35*(4), 540–549. https://doi.org/10.1002/(SICI)1096-9888(200004)35:4 < 540::AID-JMS966 > 3.0.CO;2-Y

Crane, E. E. (1990). *Bees and beekeeping: Science, practice and world resources* (p. 593). Cornstock Publ.

Crews, C., Berthiller, F., & Krska, R. (2010). Update on analytical methods for toxic pyrrolizidine alkaloids. *Analytical and Bioanalytical Chemistry*, *396*(1), 327–338. https://doi.org/10.1007/s00216-009-3092-2

Crews, C., Driffield, M., Berthiller, F., & Krska, R. (2009). Loss of pyrrolizidine alkaloids on decomposition of ragwort (*Senecio jacobaea*) as measured by LC-TOF-MS. *Journal of Agricultural and Food Chemistry*, *57*(9), 3669–3673. https://doi.org/10.1021/jf900226c

Cuyckens, F., & Claeys, M. (2002). Optimization of a liquid chromatography method based on simultaneous electrospray ionization mass spectrometric and ultraviolet photodiode array detection for analysis of flavonoid glycosides. *Rapid Communications in Mass Spectrometry*, *16*(24), 2341–2348. https://doi.org/10.1002/rcm.861

Cuyckens, F., & Claeys, M. (2004). Mass spectrometry in the structural analysis of flavonoids. *Journal of Mass Spectrometry*, *39*(1), 1–15. https://doi.org/10.1002/jms.585

Danner, N., Härtel, S., & Steffan-Dewenter, I. (2014). Maize pollen foraging by honey bees in relation to crop area and landscape context. *Basic and Applied Ecology*, *15*(8), 677–684. https://doi.org/10.1016/j.baae.2014.08.010

Dasarathy, B. (1991). *Nearest Neighbor (NN) norms: Nearest neighbor pattern classification techniques*. IEEE Computational Society.

de Rijke, E., Out, P., Niessen, W. M. A., Ariese, F., Gooijer, C., & Brinkman, U. A. (2006). Analytical separation and detection methods for flavonoids. *Journal of Chromatography. A*, *1112*(1-2), 31–63. https://doi.org/10.1016/j.chroma.2006.01.019

de Rijke, E., Zappey, H., Ariese, F., Gooijer, C., & Brinkman, U. A. (2003). Liquid chromatography with atmospheric pressure chemical ionization and electrospray ionization mass spectrometry of flavonoids with triple quadrupole and ion-trap instruments. *Journal of Chromatography. A, 984*(1), 45–58. https://doi.org/10.1016/S0021-9673(02)01868-X

Delaplane, K. S., Dag, A., Danka, R. G., Freitas, B. M., Garibaldi, L. A., Goodwin, R. M., & Hormaza, J. I. (2013). Standard methods for pollination research with *Apis mellifera*. In *V. Dietemann, J. D. Ellis, & P. Neumann (Eds.), The COLOSS* BEEBOOK, *Volume I: standard methods for* Apis mellifera *research. Journal of Apicultural Research*, 52(4). https://doi.org/10.3896/IBRA.1.52.4.12.

Di Paola-Naranjo, R. D., Sánchez, S. J., Paramás, A. M. G., & Rivas Gonzalo, J. C. (2004). Liquid chromatographic-mass spectrometric analysis of anthocyanin composition of dark blue bee pollen from Echium plantagineum. *Journal of Chromatography. A, 1054*(1-2), 205–210. https://doi.org/10.1016/j.chroma.2004.05.023

Droussi, Z., D'orazio, V., Provenzano, M. R., Hafidi, M., & Ouatmane, A. (2009). Study of the biodegradation and transformation of olive-mill residues during composting using FTIR spectroscopy and differential scanning calorimetry. *Journal of Hazardous Materials, 164*(2-3), 1281–1285. https://doi.org/10.1016/j.jhazmat.2008.09.081

Dübecke, A., Beckh, G., & Lüllmann, C. (2011). Pyrrolizidine alkaloids in honey and bee pollen. *Food Additives & Contaminants. Part A, Chemistry, Analysis, Control, Exposure & Risk Assessment, 28*(3), 348–358. https://doi.org/10.1080/19440049.2010.541594

Dustman, J. H. (1992). Honey, quality and its control. Abstracts of the 8th International Palynological Congress. France, p. 40.

Dutkiewicz, S. (1996). Usefulness of Cernilton in the treatment of benign prostatic hyperplasia. *International Urology and Nephrology, 28*(1), 49–53. https://doi.org/10.1007/BF02550137

Dutta, D., Devi, S. S., Krishnamurthi, K., Kumar, K., Vyas, P., Muthal, P. L., Naoghare, P., & Chakrabarti, T. (2007). Modulatory effect of distillate of *Ocimum sanctum* leaf extract (Tulsi) on human lymphocytes against genotoxicants. *Biomedical and Environmental Sciences, 20*(3), 226–234.

Dzyuba, O. F. (2006). Palinoindikatsiya environmental quality. SPb.: Publishing House "Nedra", 198 p. (Article in Russian).

Dzyuba, O. F. (2007). Teratomorfnye pollen grains in a modern and paleopalinologicheskih pollen spectra and some problems. *PalynostratigraphylOF DziubalIOil and Gas Technologies: Theory and Practice, 2007*(2), 1–22. (Article in Russian).

Dzyuba, O. F., & Tarasovich, V. F. (2001). The morphological features of pollen grains of Tilia Cordata Mill. In the modern metropolis. International Workshop: Pollen as an indicator of environmental and paleoecological reconstruction – St. Petersburg, pp. 79–90. (Article in Russian).

Echigo, T., Takenaka, T., & Yatsunami, K. (1986). Comparative studies on chemical composition of honey, royal jelly and pollen loads. *Bulletin of the Faculty of Agriculture Tamagawa University, 26*, 1–8.

Edgar, R. C. (2010). Search and clustering orders of magnitude faster than BLAST. *Bioinformatics, 26*(19), 2460–2461. https://doi.org/10.1093/bioinformatics/btq461

Eloff, J. N. (1998). A sensitive and quick microplate method to determine the minimal inhibitory concentration of plant extracts for bacteria. *Planta Medica, 64*(8), 711–713. https://doi.org/10.1055/s-2006-957563

Eraslan, G., Kanbur, M., & Silici, S. (2009). Effect of carbaryl on some biochemical changes in rats: The ameliorative effect of bee pollen. *Food and Chemical Toxicology, 47*(1), 86–91. https://doi.org/10.1016/j.fct.2008.10.013

Eraslan, G., Kanbur, M., Silici, S., Liman, B., Altinordulu, S., & Sarica, Z. S. (2009). Evaluation of protective effect of bee pollen against propoxur toxicity in rat. *Ecotoxicology and Environmental Safety, 72*(3), 931–937. https://doi.org/10.1016/j.ecoenv.2008.06.008

Erkmen, O., & Ozcan, M. M. (2008). Antimicrobial effects of Turkish propolis, pollen, and laurel on spoilage and pathogenic food-related microorganisms. *Journal of Medicinal Food, 11*(3), 587–592. https://doi.org/10.1089/jmf.2007.0038

Estevinho, L. M., Pereira, A. P., Moreira, L., Dias, L. G., & Pereira, E. (2008). Antioxidant and antimicrobial effects of phenolic compounds extracts of Northeast Portugal honey. *Food and Chemical Toxicology, 46*(12), 3774–3779. https://doi.org/10.1016/j.fct.2008.09.062

Eswaran, V. U., & Bhargava, H. R. (2014). Chemical analysis and anti-microbial activity of Karnataka bee bread of *Apis species. World Applied Sciences Journal, 32*(3), 379–385. https://doi.org/10.5829/idosi.wasj.2014.32.03.1006

Evans, J. D., Schwarz, R. S., Chen, Y.-P., Budge, G., Cornman, R. S., De La Rua, P., De Miranda, J. R., Foret, S., Foster, L., Gauthier, L., Genersch, E., Gisder, S., Jarosch, A., Kucharski, R., Lopez, D., Lun, C. M., Moritz, R. F. A., Maleszka, R., Muñoz, I., & Pinto, M. A. (2013). Standard methodologies for molecular research in *Apis mellifera*. In *V. Dietemann, J. D. Ellis & P. Neumann (Eds), The COLOSS* BEEBOOK, *Volume I: standard methods for* Apis mellifera *research. Journal of Apicultural Research, 52*(4), https://doi.org/10.3896/IBRA.1.52.4.11

Fabre, N., Rustan, I., Hoffmann, E., & Quetin-Leclercq, J. (2001). Determination of flavone, flavonol and flavanone aglycones by negative ion LC–ES ion trap mass spectrometry. *Journal of the American Society for Mass Spectrometry, 12*(6), 707–715. https://doi.org/10.1016/S1044-0305(01)00226-4

Faegri, K., Kaland P, E., & Krzywinski, K. (1993). *Bestimmungsschlussel fur die nordwesteuropaische pollenflora* (p. 85). G. Fischer.

Fanali, C., Dugo, L., & Rocco, A. (2013). Nano-liquid chromatography in nutraceutical analysis: determination of polyphenols in bee pollen. *Journal of Chromatography. A, 1313*, 270–274. https://doi.org/10.1016/j.chroma.2013.06.055

FAO/IPGRI. (2001). *Multi-crop passport descriptors*. http://www.bioversityinternational.org/fileadmin/bioversity/publications/pdfs/124.pdf

Fatrcová-Šramková, K., Nôžková, J., Kačániová, M., Máriassyová, M., & Kropková, Z. (2010). Microbial properties, nutritional composition and antioxidant activity of *Brassica napus* subsp. *napus* L. bee pollen used in human nutrition. *Ecological Chemistry and Engineering A, 17*(1), 45–54.

Fatrcová-Šramková, K., Nôžková, J., Kačániová, M., & Mariássyová, M. (2011). Antioxidant and antimicrobial properties of monofloral bee pollen usable in human nutrition. In *Chemické Listy, 105*(Suppl), 1005–1006. https://doi.org/10.1080/03601234.2013.727664

Fatrcová-Šramková, K., Nôžková, J., Kačániová, M., Máriássyová, M., Rovná, K., & Stričík, M. (2013). Antioxidant and antimicrobial properties of monofloral bee pollen. *Journal of Environmental Science and Health. Part. B, Pesticides, Food Contaminants, and Agricultural Wastes, 48*(2), 133–138. https://doi.org/10.1080/03601234.2013.727664

Feás, X., Vázquez-Tato, M. P., Estevinho, L., Seijas, J. A., & Iglesias, A. (2012). Organic bee pollen: Botanical origin, nutritional value, bioactive compounds, antioxidant activity and microbiological quality. *Molecules, 17*(7), 8359–8377. http://dx.doi.org/10.3390/molecules17078359

Forcone, A., Aloisi, P. V., Ruppel, S., & Munoz, M. (2011). Botanical composition and protein content of pollen collected by *Apis mellifera* L. in the north-west of Santa Cruz (Argentinean Patagonia). *Grana, 50*(1), 30–39. https://doi.org/10.1080/00173134.2011.552191

Ferreres, F., Pereira, D. M., Valentão, P., & Andrade, P. B. (2010). First report of non-coloured flavonoids in Echium plantagineum bee pollen: Differentiation of isomers by liquid chromatography/ion trap mass spectrometry. *Rapid Communications in Mass Spectrometry, 24*(6), 801–806. https://doi.org/10.1002/rcm.4454

Fogle, H. W. (1977). Identification of tree fruit species by pollen ultrastructure. *Journal – American Society for Horticultural Science*, *102*(5), 548–551.

Fontanals, N., Marcé, R. M., & Borrull, F. (2007). New materials in sorptive extraction techniques for polar compounds. *Journal of Chromatography. A*, *1152*(1-2), 14–31. https://doi.org/10.1016/j.chroma.2006.11.077

Forcone, A., Aloisi, P. V., Ruppel, S., & Munoz, M. (2011). Botanical composition and protein content of pollen collected by *Apis mellifera* L. in the north-west of Santa Cruz (Argentinean Patagonia). *Grana*, *50*(1), 30–39. https://doi.org/10.1080/00173134.2011.552191

Fossel, A.-M., & Pechhacker, H. (2001). *Bienen and blumen* (p. 676). Institut fur Bienenkunde.

Franco, B. D. G. M. (2002). Fatores intrínsecos e extrínsecos que controlam o desenvolvimento microbiano nos alimentos. In B. D. G. M. Franco & M. Landgraf (Eds.), *Microbiologia dos Alimentos* (pp. 13–26). Atheneu.

Fukunaga, K., & Hostetler, L. (1975). The estimation of the gradient of a density function, with applications in pattern recognition. *IEEE Transactions on Information Theory*, *21*(1), 32–40. https://doi.org/10.1109/TIT.1975.1055330

Galimberti, A., De Mattia, F., Bruni, I., Scaccabarozzi, D., Sandionigi, A., Barbuto, M., Casiraghi, M., & Labra, M. (2014). A DNA barcoding approach to characterize pollen collected by honeybees. *PLoS One*, *9*(10), e109363. https://doi.org/10.1371/journal.pone.0109363

Gašić, U., Kečkeš, S., Dabić, D., Trifković, J., Milojković-Opsenica, D., Natić, M., & Tešić, Z. (2014b). Phenolic profile and antioxidant activity of Serbian polyfloral honeys. *Food Chemistry*, *145*, 599–607. https://doi.org/10.1016/j.foodchem.2013.08.088

Gašić, U., Šikoparija, B., Tosti, T., Trifković, J., Milojković-Opsenica, D., Natić, M., & Tešić, Ž. Ž (2014a). Phytochemical fingerprints of lime honey collected in Serbia. *Journal of AOAC International*, *97*(5), 1259–1267. https://doi.org/10.5740/jaoacint.SGEGasic

Germano, M. I. S., Germano, P. M. L., Kamei, C. A. K., Abreu, E. S. d., Ribeiro, E. R., Silva, K. C. d., Lamardo, L. C. A., Rocha, M. F. G., Vieira, V. K. I., & Kawasaki, V. M. (2000). Manipuladores de alimentos: capacita? É preciso! Regulamentar? Será preciso? *HigieneAlimentar*, *14*(78/79), 18–22.

Gilber, M. D., & Lisk, D. J. (1978). Honey as an environmental indicator of radionuclide contamination. *Bulletin of Environmental Contamination and Toxicology*, *19*(1), 32–34. https://doi.org/10.1007/BF01685763

Gilliam, M., Prest, D. B., & Lorenz, B. J. (1989). Microbiology of pollen and bee bread: taxonomy and enzimology of molds. *Apidologie*, *20*(1), 53–68. https://doi.org/10.1051/apido:19890106

Glazer, A. N. (1990). Phycoerythrin fluorescence-based assay for reactive oxygen species . *Methods in Enzymology*, *186*, 161–168. https://doi.org/10.1016/0076-6879(90)86106-6

Glazunov, K. P. (2001 *Pollen as an indicator of negative environmental factors: embryological aspect* [Paper presentation]. International Seminar: Pollen as an Indicator of Environmental and Paleoecological Reconstruction, pp. 61–63. (Article in Russian).

Goh, K., Chang, E., & Cheng, K. (2001). *SVM binary classifier ensembles for image classification* [Paper presentation]. Proceedings of the ACM Conference Information and Knowledge Management, New York, pp. 395–402. https://doi.org/10.1145/502585.502652

Goh, K., Chang, E., & Li, B. (2005). Using one-class and two-class SVMs for multiclass image annotation. *IEEE Transactions on Knowledge and Data Engineering*, *17*(10), 1333–1346. https://doi.org/10.1109/TKDE.2005.170

Gok, S., Severcan, M., Goormaghtigh, E., Kandemir, I., & Severcan, F. (2015). Differentiation of Anatolian honey samples from different botanical origins by ATR-FTIR spectroscopy using multivariate analysis. *Food Chemistry*, *170*, 234–240. https://doi.org/10.1016/j.foodchem.2014.08.040

Gonzalez, R. C., & Woods, R. E. (2008). *Digital image processing* (3rd ed.). Prentice Hall.

Graikou, K., Kapeta, S., Aligiannis, N., Sotiroudis, G., Chondrogianni, N., Gonos, E., & Chinou, I. (2011). Chemical analysis of Greek pollen – Antioxidant, antimicrobial and proteasome activation properties. *Chemistry Central Journal*, *5*(1), 33–39. https://doi.org/10.1186/1752-153X-5-33

Greenham, J., Williams, C., & Harborne, J. B. (1995). Identification of lipophilic flavonols by a combination of chromatographic and spectral techniques. *Phytochemical Analysis*, *6*(4), 211–217. https://doi.org/10.1002/pca.2800060407

Grembecka, M., & Szefer, P. (2013). Evaluation of honeys and bee products quality based on their mineral composition using multivariate techniques. *Environmental Monitoring and Assessment*, *185*(5), 4033–4047. https://doi.org/10.1007/s10661-012-2847-y

Gu, D., Yang, Y., Abdulla, R., & Aisa, H. A. (2012). Characterization and identification of chemical compositions in the extract of *Artemisia rupestris* L. by liquid chromatography coupled to quadrupole time-of-flight tandem mass spectrometry. *Rapid Communications in Mass Spectrometry*, *26*(1), 83–100. https://doi.org/10.1002/rcm.5289

Guide for the Care and Use of Laboratory Animals Eighth Edition. (2011). The National Academies Press 500 Fifth Street, NW Washington, DC 20001. www.national-academies.org

Guo, J. T., Lee, H. L., Chiang, S. H., Lin, F. I., & Chang, C. Y. (2020). Antioxidant properties of the extracts from different parts of broccoli in Taiwan. *Journal of Food and Drug Analysis*, *9*(2), 96–101. https://doi.org/10.38212/2224-6614.2795

Harborne, J. B., & Boardley, M. (1984). Use of high performance liquid chromatography in the separation of flavonol glycosides and flavonol sulphates. *Journal of Chromatography A*, *299*(2), 377–385. https://doi.org/10.1016/S0021-9673(01)97853-7

Hartmann, T., & Toppel, G. (1987). Senecionine n-oxide. The primary product of pyrrolizidine alkaloid biosynthesis in root cultures of *Senecio vulgaris*. *Phytochemistry*, *26*(6), 1639–1643. https://doi.org/10.1016/S0031-9422(00)82261-X

Hartmann, T., & Witte, L. (1995). Chemistry, biology and chemoecology of the pyrrolizidine alkaloids. In S. W. Pelletier (Ed.), *Alkaloids: Chemical and biological perspectives* (Vol. 9, pp. 155–233). Pergamon Press. https://doi.org/10.1016/b978-0-08-042089-9.50011-5

Haussmann, B. I. G., & Parzies, H. K. (2009). Methodologies for generating variability: Part 1: Use of genetic resources in plant breeding. In S. Ceccarelli (Ed.), *Plant breeding and farmer participation* (pp. 107–128). FAO of the UN.

Heo, M. Y., Sohn, S. J., & Au, W. W. (2001). Anti-genotoxicity of galangin as a cancer chemopreventive agent candidate. *Mutation Research*, *488*(2), 135–150. https://doi.org/10.1016/S1383-5742(01)00054-0 http://www.mercola.com/article/diet/bee_pollen.htm.

Herbert, E. W. J. (1992). Honey bee nutrition. In J. E. Graham (Ed.), *The hive and the honey bee* (pp. 197–233). Dadant & Sons Inc.

Herbert, E. W. J., & Shimanuki, H. (1978). Chemical composition and nutritive value of bee-collected and bee-stored pollen. *Apidologie*, *9*(1), 33–40. https://doi.org/10.1051/apido:19780103

Hewitt, W., & Vincent, S. (1989). *Theory and application of microbiological assay*. Academic Press.

Hleba, L., Pochop, J., Felšöciová, S., Petrová, J., Čuboň, J., Pavelková, A., & Kačániová, M. (2013). Antimicrobial effect of bee collected pollen extract to *Enterobacteriaceae* Genera after application of bee collected pollen in their feeding. *Animal Science and Biotechnologies*, *46*(2), 108–113.

Hleba, L., Pochop, J., Felšöciová, S., Petrová, J., Čuboň, J. & Pavelková, A. (2013). Characterization of chemical composition of bee pollen in China. *Food Chemistry, 61*(3), 708–718. https://doi.org/10.1021/jf304056b

Hodges, D. (1952). *The pollen loads of the honey bee* (p. 154). Bee Research Association.

Human, H., Brodschneider, R., Dietemann, V., Dively, G., Ellis, J. D., Forsgren, E., Fries, I., Hatjina, F., Hu, F.-L., Jaffé, R., Jensen, A. B., Köhler, A., Magyar, J., Özikrim, A., Pirk, C. W. W., Rose, R., Strauss, U., Tanner, G., Tarpy, D. R., van der Steen, J. J. M., Vaudo, A., Vejsnaes, F., Wilde, J., Williams, G. R., & Zheng, H.-Q. (2013). Miscellaneous standard methods for *Apis mellifera* research. In *V. Dietemann, J. D. Ellis & P. Neumann (Eds), The COLOSS* BEEBOOK, *Volume I: standard methods for* Apis mellifera *research. Journal of Apicultural Research, 52*(4), https://doi.org/10.3896/IBRA.1.52.4.10

Human, H., & Nicolson, S., W. (2006). Nutritional content of fresh, bee-collected and stored pollen of *Aloe greatheadii* var. *davyana* (Assphodelaceae). *Phytochemistry, 67*(14), 1486–1492. https://doi.org/10.1016/j.phytochem.2006.05.023

Ibrahim, A., & Spivak, M. (2006). The relationship between hygienic behavior and suppression of mite reproduction as honey bee (*Apis mellifera*) mechanisms of resistance to *Varroa destructor. Apidologie, 37*(1), 31–40. https://doi.org/10.1051/apido:2005052

Ionita, P. (2005). Is DPPH stable free radical a good scavenger for oxygen active species? *Chemical Papers, 50*(1), 11–16.

Jaiswal, S., Mansa, N., Prasad, M. S., Jena, B. S., & Negi, P. S. (2014). Antibacterial and antimutagenic activities of (*Dillenia indica*) extracts. *Food Bioscience, 5*, 47–53. https://doi.org/10.1016/j.fbio.2013.11.005

Jakobek, L. (2015). Interactions of polyphenols with carbohydrates, lipids and proteins. *Food Chemistry, 175*, 556–567. https://doi.org/10.1016/j.foodchem.2014.12.013

Jiang, L. (2011). *Comparison of disk diffusion, agar dilution, and broth microdilution for antimicrobial susceptibility testing of five chitosans* [Thesis in Master of Science]. Faculty of the Louisiana State University and Agricultural and Mechanical College, Department of Food Science.

Jorgensen, J. H., & Ferraro, M. J. (2009). Antimicrobial susceptibility testing: A review of general principles and contemporary practices. *Clinical Infectious Diseases, 49*(11), 1749–1755. https://doi.org/10.1086/647952

Kacániová, M., Vuković, N., Chlebo, R., Haščík, P., Rovná, K., Cubon, J., Džugan, M., & Pasternakiewicz, A. (2012). The antimicrobial activity of honey, bee pollen loads and beeswax from Slovakia. *Archives of Biological Sciences, 64*(3), 927–934. https://doi.org/10.2298/ABS1203927K

Kacániová, M., Terentjeva, M., Petrová, J., Hleba, L., Shevtsova, T., Kluz, M., Rožek, P., & Vukovič, N. (2014). Antimicrobial activity of non-traditional plant pollen against different species of microorganisms. *Journal of Microbiology Biotechnology and Food Science, 4*(1), 80–82. http://dx.doi.org/10.15414/jmbfs.2014.4.1.80-82

Kao, Y. T., Lu, M. J., & Chen, C. (2020). Preliminary analysis of phenolic compounds and antioxidant activities in tea pollen extracts. *Journal of Food and Drug Analysis, 19*(4), 470–477. https://doi.org/10.38212/2224-6614.2177

Kaur, C., & Kapoor, H. C. (2001). Antioxidants in fruits and vegetables-the millennium's health. *International Journal of Food Science and Technology, 36*(7), 703–725. https://doi.org/10.1046/j.1365-2621.2001.00513.x

Kaur, R., Nagpal, A., & Katnoria, J. K. (2015). Exploration of antitumor properties of pollen grains of plant species belonging to Fabaceae family. *Journal of Pharmaceutical Sciences and Research, 7*(3), 127–129.

Kečkeš, S., Gašić, U., Veličković, T. Ć., Milojković-Opsenica, D., Natić, M., & Tešić, Ž. (2013). The determination of phenolic profiles of Serbian unifloral honeys using ultra-high-performance liquid chromatography/high resolution accurate mass spectrometry. *Food Chemistry, 138*(1), 32–40. https://doi.org/10.1016/j.foodchem.2012.10.025

Kedves, M., Kocsicska, I., & Priskin, K. (2003). Transmission electron microscopy of the connectives of the pollen grains of *Ginkgo biloba* L. *Plant Cell Biology and Development, Szeged, 15*, 28–42.

Kedves, M., Pardutz, A., & Horvath, A. (2002). Transmission electron microscopy of partially degraded pollen grains of *Ambrosia artemisiifolia* (ragweed). *Plant Cell Biology and Development, Szeged, 14*, 49–58.

Kelina, I., Saavedra, C., Consuelo Rojas, I., Guillermo, E., & Delgado, P. (2013). Pollinic characteristics and chemical composition of bee pollen collected in Cayalti (Lambayeque – Peru). *Revista Chilena de Nutricion, 40*(1), 71–78.

Keller, A., Danner, N., Grimmer, G., Ankenbrand, M., von der Ohe, K., von der Ohe, W., Rost, S., Härtel, S., & Steffan-Dewenter, I. (2015). Evaluating multiplexed next-generation sequencing as a method in palynology for mixed pollen samples. *Plant Biology, 17*(2), 558–566. https://doi.org/10.1111/plb.12251

Keller, A., Hohlfeld, S., Kolter, A., Schultz, J., Gemeinholzer, B., & Ankenbrand, M. J. (2020). BCdatabaser: On-the-fly reference database creation for (meta-)barcoding. *Bioinformatics, 36*(8), 2630–2631. https://doi.org/10.1093/bioinformatics/btz960

Keller, I., Fluri, P., & Imdorf, A. (2005). Pollen nutrition and colony development in honey bees: Part I. *Bee World, 86*(1), 3–10. https://doi.org/10.1080/0005772X.2005.11099641

Kempf, M., Beuerle, T., Bühringer, M., Denner, M., Trost, D., von der Ohe, K., Bhavanam, V. B. R., & Schreier, P. (2008). Pyrrolizidine alkaloids in honey: Risk analysis by gas chromatography-mass spectrometry. *Molecular Nutrition & Food Research, 52*(10), 1193–1200. https://doi.org/10.1002/mnfr.200800051

Kempf, M., Heil, S., Hasslauer, I., Schmidt, L., von der Ohe, K., Theuring, C., Reinhard, A., Schreier, P., & Beuerle, T. (2010). Pyrrolizidine alkaloids in pollen and pollen products. *Molecular Nutrition & Food Research, 54*(2), 292–300. https://doi.org/10.1002/mnfr.200900289

Kempf, M., Wittig, M., Reinhard, A., von der Ohe, K., Blacquière, T., Raezke, K.-P., Michel, R., Schreier, P., & Beuerle, T. (2011). Pyrrolizidine alkaloids in honey: Comparison of analytical methods. *Food Additives & Contaminants. Part A, Chemistry, Analysis, Control, Exposure & Risk Assessment, 28*(3), 332–347. https://doi.org/10.1080/19440049.2010.521772

Khynriam, D., & Prasad, S. B. (2003). Changes in glutathione-related enzymes in tumor-bearing mice after cisplatin treatment. *Cell Biology and Toxicology, 18*(6), 1573–6822.

Kirk, W. D. J. (1994). *A color guide to pollen loads of the honey bee* (p. 54). International Bee Research Association.

Klancnik, A., Piskernik, S., Jersek, B., & Mozina, S. S. (2010). Evaluation of diffusion and dilution methods to determine the antibacterial activity of plant extracts. *Journal of Microbiological Methods, 81*(2), 121–126. https://doi.org/10.1016/j.mimet.2010.02.004

Klungness, L. M., & Peng, Y.-S. (1983). A scanning electron microscopic study of pollen loads collected and stored by honey bees. *Journal of Apicultural Research, 22*(4), 264–271. https://doi.org/10.1080/00218839.1983.11100598

Koleva, I. I., VAN Beek, T. A., Linssen, J. P. H., De Groot, A., & Evstatieva, L. N. (2002). Screening of plant extracts for antioxidant activity: A comparative study on three testing methods. *Phytochemical Analysis, 13*(1), 8–17. https://doi.org/10.1002/pca.611

Komosinska-Vassev, K., Olczyk, P., Kafmierczak, J., Mencner, L., & Olczyk, K. (2015) Bee pollen: Chemical composition and therapeutic application. *Evidence-Based Complementary and Alternative Medicine, 2015*, 1–6. https://doi.org/10.1155/2015/297425.

Köppler, K., Vorwohl, G., & Koeniger, N. (2007). Comparison of pollen spectra collected by four different subspecies of the honey bee *Apis mellifera*. *Apidologie*, *38*(4), 341–353. https://doi.org/10.1051/apido:2007020

Kostić, A. Ž., Barać, M. B., Stanojević, S. P., Milojković-Opsenica, D. M., Tešić, Ž. L., Šikoparija, B., Radišić, P., Prentović, M., & Pešić, M. B. (2015). Physicochemical composition and techno-functional properties of bee pollen collected in Serbia. *LWT – Food Science and Technology*, *62*(1), 301–309. https://doi.org/10.1016/j.lwt.2015.01.0310023-6438

Kozich, J. J., Westcott, S. L., Baxter, N. T., Highlander, S. K., & Schloss, P. D. (2013). Development of a dual-index sequencing strategy and curation pipeline for analyzing amplicon sequence data on the MiSeq Illumina sequencing platform. *Applied and Environmental Microbiology*, *79*(17), 5112–5120. https://doi.org/10.1128/AEM.01043-13

Kraaijeveld, K., de Weger, L. A., Ventayol García, M., Buermans, H., Frank, J., Hiemstra, P. S., & den Dunnen, J. T. (2015). Efficient and sensitive identification and quantification of airborne pollen using next-generation DNA sequencing. *Molecular Ecology Resources*, *15*(1), 8–16. https://doi.org/10.1111/1755-0998.12288

Kroyer, G., & Hegedus, N. (2001). Evaluation of bioactive properties of pollen extracts as functional dietary food supplement. *Innovative Food Science & Emerging Technologies*, 2(3), 171–174. https://doi.org/10.1016/S1466-8564(01)00039-X

Langford, M., Taylor, G., & Flenley, J. (1990). Computerized identification of pollen grains by texture analysis. *Review of Palaeobotany and Palynology*, *64*(1-4), 197–203. https://doi.org/10.1016/0034-6667(90)90133-4

Lautraite, S., Musonda, A. C., Doehmer, J., Edwards, G. O., & Chipman, J. K. (2002). Flavonoids inhibit genetic toxicity produced by carcinogens in cells expressing CYP1A2 and CYP1A1. *Mutagenesis*, *17*(1), 45–53. https://doi.org/10.1093/mutage/17.1.45

LeBlanc, B. W., Davis, O. K., Boue, S., DeLucca, A., & Deeby, T. (2009). Antioxidant activity of Sonoran Desert bee pollen. *Food Chemistry*, *115*(4), 1299–1305. https://doi.org/10.1016/j.foodchem.2009.01.055

Lee, K.-H., Kim, A.-J., & Choi, E.-M. (2009). Antioxidant and anti-inflammatory activity of pine pollen extract in vitro. *Phytotherapy Research*, *23*(1), 41–48. https://doi.org/10.1002/ptr.2525

Leja, M., Mareczek, A., Wyżgolik, G., Klepacz-Baniak, J., & Czekońska, K. (2007). Antioxidative properties of bee pollen in selected plant species. *Food Chemistry*, *100*(1), 237–240. https://doi.org/10.1016/j.foodchem.2005.09.047

Leong, L. P., & Shui, G. (2002). An investigation of antioxidant capacity of fruits in Singapore market. *Food Chemistry*, *76*(1), 69–75. https://doi.org/10.1016/S0308-8146(01)00251-5

Li, A., Zhang, Q., Zhang, G., Chen, J., Fei, Z., & Liu, F. (2002). Adsorption of phenolic compounds from aqueous solutions by a water-compatible hypercrosslinked polymeric adsorbent. *Chemosphere*, *47*(9), 981–989. https://doi.org/10.1016/S0045-6535(01)00222-3

Li, P., & Flenley, J. R. (1999). Pollen texture identification using neural networks. *Grana*, *38*(1), 59–64. In https://doi.org/10.1080/001731300750044717

Lindtner, P. (2014). *Garden plants for honey bees* (p. 389). Wicwas Press LLC.

Liolios, V., Tananaki, C., Dimou, M., Kanelis, D., Goras, G., Karazafiris, E., Thrasyvoulou, A. (2014). Predict the chemical composition of pollen. In *Proceedings of International Symposium on Bee Products*, 3rd edition, September 28th – October 1st, p. 28.

Lokutova, O. (2014). Quality of bee products in context of environmental safety. Zborník vedeckých prác a výsledkov z riešenia výskumného projektu Podpora inovácie technológií špeciálnych výrobkov biopotravín pre zdravú výživu ľudí – Slovenská polnohospodárska univerzita v Nitre, pp. 182–188.

Loper, G. M., Standifer, L. N., Thompson, M. J., & Gilliam, M. (1980). Biochemistry and microbiology of bee collected almond (*Prumus dulcis*) pollen and bee bread. I Fatty acids, vitamins and minerals. *Apidologie*, *11*(1), 63–73. https://doi.org/10.1051/apido:19800108

Louveaux, J., Maurizio, A., & Vorwohl, G. (1978). Methods of melissopalynology. *Bee World*, *59*(4), 139–157. https://doi.org/10.1080/0005772X.1978.11097714

Lucchese, L., Mitra, S. (2001). Color image segmentation: A state-of-the-art survey. In *Proceedings of the Indian National Science Academy (INSA-A)*, pp. 207–221.

Lv, H., Wang, X., He, Y., Wang, H., & Suo, Y. (2015). Identification and quantification of flavonoid aglycones in rape bee pollen from Qinghai-Tibetan Plateau by HPLC-DAD-APCI/MS. *Journal of Food Composition and Analysis*, *38*, 49–54. https://doi.org/10.1016/j.jfca.2014.10.011

Madejczyk, M., & Baralkiewicz, D. (2008). Characterization of Polish rape and honeydew honey according to their mineral contents using ICP-MS and F-AAS/AES. *Analytica Chimica Acta*, *617*(1-2), 11–17. https://doi.org/10.1016/j.aca.2008.01.038

Makena, P. S., Pierce, S. C., Chung, K. T., & Sinclair, S. E. (2009). Comparative mutagenic effects of structurally similar flavonoids quercetin and taxifolin on tester strains Salmonella typhimurium TA102 and Escherichia coli WP-2 uvrA. *Environmental and Molecular Mutagenesis*, *50*(6), 451–459. https://doi.org/10.1002/em.20487

Mărgăoan, R., Mărghitas, L., Al, Dezmirean, D. S., Bobis, O., & Mihai, C. M. (2012). Physical-Chemical composition of fresh bee pollen from Transylvania. *Bulletin UASVM Animal Science and Biotechnologies*, *69*(1-2), 351–355.

Mărghitaş, L. A., Stanciu, O. G., Dezmirean, D. S., Bobiş, O., Popescu, O., Bogdanov, S., & Campos, M. G. (2009). *In vitro* antioxidant capacity of honey bee-collected pollen of selected floral origin harvested from Romania. *Food Chemistry*, *115*(3), 878–883. https://doi.org/10.1016/j.foodchem.2009.01.014

Markham, K., & Campos, M. G. (1996). 7- and 8-*O*-methylherbacetin-3-*O*-sophoroside from bee pollens and some structure/activity observations. *Phytochemistry*, *43*(4), 763–767. https://doi.org/10.1016/0031-9422(96)00286-5

Markham, K. R., & Geiger, H. (1993). 1H nuclear magnetic resonance spectroscopy of flavonoids and their glycosides in hexadeuterodimethylsulphoxide. In J. B. Harborne (Ed.), *The flavonoids: Advances in research since 1986*. Chapman and Hall, pp. 441–496.

Markowicz Bastos, D. H., Barth, M. O., Rocha, C. I., Cunha, I. B. S., Carvalho, P. O., Torres, E. A. S., & Michelan, M. (2004). Fatty acid composition and palynological analysis of bee (*Apis*) pollen loads in the states of São Paulo and Minas Gerais, Brazil. *Journal of Apicultural Research*, *43*(2), 35–39. https://doi.org/10.1080/00218839.2004.11101107

Marsh, J., & Goode, J. A. (Eds.) (1994). *Antimicrobial Peptides [Ciba Foundation Symposium 186]*. John Wiley & Sons.

Martins, M. C. T., Morgano, M. A., Vincente, E., Baggio, S. R., & Rodriguez-Amaya, D. B. (2011). Physicochemical composition of bee pollen from eleven Brazilian states. *Journal of Apicultural Science*, *55*(2), 107–116.

Maruyama, H., Sakamoto, T., Araki, Y., & Hara, H. (2010). Anti-inflammatory effect of bee pollen ethanol extract from *Cistus sp* of Spanish on carrageenan-induced rat hind paw edema. *BMC Complementary and Alternative Medicine*, *10*(1), 11. https://doi.org/10.1186/1472-6882-10-30

Matuschek, E., Brown, D. F. J., & Kahlmeter, G. (2014). Development of the EUCAST disk diffusion antimicrobial susceptibility testing method and its implementation in routine microbiology laboratories. *Clinical Microbiology and Infection*, *20*(4), O255–O266. https://doi.org/10.1111/1469-0691.12373

Maurizio, A., & Louveaux, J. (1965). *Pollens de plantes melliferes d'Europe* (p. 148). Union des Groupements apicoles Français.

McMurdie, P. J., & Holmes, S. (2013). phyloseq: An R package for reproducible interactive analysis and graphics of microbiome census data. *PLoS One*, *8*(4), e61217. https://doi.org/10.1371/journal.pone.0061217

Mejías, E., & Montenegro, G. (2012). The antioxidant activity of Chilean honey and bee pollen produced in the Llaima volcano's zonez. *Journal of Food Quality, 35*(5), 315–322. https://doi.org/10.1111/j.1745-4557.2012.00460.x

Melo, E. A., Filho, J. M., & Guerra, N. B. (2005) Characterization of antioxidant compounds in aqueous coriander extract (*Coriandrum sativum* L.). *LWT—Food Science and Technology, 38*, 15–19. https://doi.org/10.1016/j.lwt.2004.03.011.

Melo, I. L. P., & Almeida-Muradian, L. B. (2010). Stability of antioxidants vitamins in bee pollen samples. *Química Nova, 33*(3), 514–518. https://doi.org/10.1590/S0100-40422010000300004

Melo, I. L. P., & Almeida-Muradian, L. B. (2011). Comparison of methodologies for dried bee pollen's moisture determination. *Ciência e Tecnologia de Alimentos, 31*(1), 194–197. v. https://doi.org/10.1590/S0101-20612011000100029

Melo, I. L. P., Freitas, A. S., Barth, O. M., & Almeida-Muradian, L. B. (2009). Relação entre a composição nutricional e a origem floral de pólen apícola desidratado. [Correlation between nutritional composition and floral origin of dried bee pollen]. *Revista Do Instituto Adolfo Lutz, 68*, 346–353.

Melo, A. A. M., Meira, D. F. S., Chan, J., Sattler, J. A. G., & Almeida-Muradian, L. B. (2012a). Comparação entre métodos de extração de compostos antioxidantes do pólen apícola desidratado. In: *11° Encontro de Química de Alimentos, Bragança. 11° Encontro de Química de Alimentos – Qualidade dos alimentos: novos desafios*/Resumos, p. 65–65.

Melo, A. A. M., Meira, D. F. S., Sattler, J. A. G., & Almeida-Muradian, L. B. (2012b). Atividade antioxidante do pólen apícola desidratado avaliada por diferentes métodos de extração. Mensagem Doce (Associação Paulista de Apicultores, Criadores de Abelhas Melificas Européias), São Paulo, p. 59, 22 maio.

Melo, A. A. M., Meira, D. F. S., Sattler, J. A. G., & Almeida-Muradian, L. B. (2012c). *Antioxidant activity of bee pollen produced in two Brazilian states* [Paper presentation]. 5th European Conference of Apidology, Halle an Der Saale. EurBee 5 – 5th European Conference of Apidology (p. 223).

Melo, A. A. M., Meira, D. F. S., Sattler, J. A. G., Almeida-Muradian, L. B. (2012d). Antioxidant activity of dehydrated bee pollen produced in Rio Grande do Sul state, Brazil. In *II International Symposium on Bee Products/Annual Meeting of IHC, Bragança*/Book of Abstracts (p. 75).

Melo, A. A. M., Nagai, M. K., Chan, J., & Almeida-Muradian, L. B. (2013). Chemical composition, total phenolic compounds and antioxidant activity of dehydrated bee pollen produced in four Brazilian states. In *XXXXIII International Apicultural Congress*. Apimondia – Scientific Program/Posters.

Menezes, J. D. S., Maciel, L. F., Miranda, M. S., & Druzian, J. I. (2010). Compostos bioactivos e potencial antioxidante de pólen apícola producido por abelhas africanizadas (*Apis mellifera* L.). *Revista do Instituto Adolfo Lutz, 69*(2), 233–242.

Molyneux, P. (2004). The use of the stable free radical diphenylpicryl-hydracyl (DPPH) for estimating antioxidant activity. *Songklanakarin Journal of Science and Technology, 26*, 211–219.

Montenegro, G., Pizarro, R., Mejías, E., & Rodríguez, S. (2013). Evaluación biológica de polen apícola de plantas de Chile. *Revista Internacional de Botánica Experimental, 82*, 7–14.

Morais, M., Moreira, L., Feás, X., & Estevinho, L. M. (2011). Honeybee-collected pollen from five Portuguese Natural Parks: Palynological origin, phenolic content, antioxidant properties and antimicrobial activity. *Food and Chemical Toxicology, 49*(5), 1096–1101. https://doi.org/10.1016/j.fct.2011.01.020

Moreira, L., Dias, L. G., Pereira, J. A., & Estevinho, L. (2008). Antioxidant properties, total phenols and pollen analysis of propolis samples from Portugal. *Food and Chemical Toxicology, 46*(11), 3482–3485. https://doi.org/10.1016/j.fct.2008.08.025

Moreti, A. C. C. C., Marchini, L. C., Souza, V. C., & Rodrigues, R. R. (2002). Atlas do pólen de plantas apícolas. Papel Virtual.

Moya, M., Koch, M., Hostetler, L. (1993). One-class classifier networks for target recognition applications. In *Proceedings on World Congress on Neural Networks, International Neural Network Society* (pp. 797–801).

Mullins, J., & Emberlin, J. (1997). Sampling pollens. *Journal of Aerosol Science, 28*(3), 365–370. https://doi.org/10.1016/S0021-8502(96)00439-9

Mußmann, P., Levsen, K., & Radeck, W. (1994). Gas-chromatographic determination of phenols in aqueous samples after solid phase extraction. *Fresenius' Journal of Analytical Chemistry, 348*(10), 654–659. https://doi.org/10.1007/BF00325568

Muszynska, J., & Warakomska, Z. (1999). Honey bees in the monitoring of the environment's contamination: I. The numerousness of the monitoring apiary giving samples of pollen loads, which are representative for a neighborhood rich in pollen flow. Pszczelnicze scientific notebooks – 1999. – Rok XLIII, pp. 197–208.

Nagai, T., Inoue, R., Inoue, H., & Suzuki, N. (2002). Scavenging capacities of pollen extracts from *Cistus ladanifeus* on autoxidation, superoxide radicals, hydroxyl radicals, and DPPH radicals. *Nutrition Research, 22*(4), 519–526. https://doi.org/10.1016/S0271-5317(01)00400-6

Nagai, T., Nagashima, T., Suzuki, N., & Inoue, R. (2005). Antioxidant activity and angiotensin I-converting enzyme inhibition by enzymatic hydrolysates from bee bread. *Zeitschrift fur Naturforschung. C, Journal of Biosciences, 60*(1-2), 133–138. https://doi.org/10.1515/znc-2005-1-224

Nagy, M., Križková, L., Mučaji, P., Kontšeková, Z., Šeršeň, F., & Krajčovič, J. (2009). Antimutagenic activity and radical scavenging activity of water infusions and phenolics from ligustrum plants leaves. *Molecules, 14*(1), 509–518. https://doi.org/10.3390/molecules14010509

Natić, M. M., Dabić, D., Č; Papetti, A., Fotirić Akšić, M. M., Ognjanov, V., Ljubojević, M., & Tešić, Ž L. J. (2015). Analysis and characterisation of phytochemicals in mulberry (Morus alba L.) fruits grown in Vojvodina, North Serbia. *Food Chemistry, 171*, 128–136. https://doi.org/10.1016/j.foodchem.2014.08.101

Negri, G., Teixeira, E., Alves, M. L. T. F., Moreti, A. C. C. C., Otsuk, I. P., Borguini, R. G., & Salatino, A. (2011). Hydroxycinnamic acid amide derivatives, phenolic compounds and antioxidant activities of extracts of pollen samples from Southeast Brazil. *Journal of Agricultural and Food Chemistry, 59*(10), 5516–5522. https://doi.org/10.1021/jf200602k

Niki, E., & Noguchi, N. (2000). Evaluation of antioxidant capacity. What capacity is being measured by which method? *IUBMB Life, 50*(4-5), 323–329. https://doi.org/10.1080/15216540051081119

Nogueira, C., Iglesias, A., Feas, X., & Estevinho, L. M. (2012). Commercial bee pollen with different geographical origins: A comprehensive approach. *International Journal of Molecular Sciences, 13*(9), 11173–11187. https://doi.org/10.3390/ijms130911173

Nôžková, J., Brindza, J., Ostrovský, R., Stehlíková, B., & Fatrcová-Šramková, K. (2010). Hodnotenie kvantitatívnych a kvalitatívnych znakov včelieho pelu a ich klasifikácia podľa navrhnutých deskriptorov: Evaluation of quantitative and qualitative traits of bee pollen and their classification according to proposed list of descriptors. In Potravinárstvo. Združenie HACCP Consulting 2007: Nitrianske Hrnčiarovce, Slovensko. 4, 204–216.

Nôžková, J., Ostrovský, R., Brindza, J., & Molnárová, E. (2010). Morfologická charakteristika včelích pelových obnôžok. In B. Brovarskyi (Ed.), *Včelí obnôžkový peľ. SPU v Nitre* (pp. 101–113).

Nôžková, J., Pavelek, M., Bjelková, M., Brutch, N., Tejklová, E., Porokhovinova, E., & Brindza, J. (2011). *Descriptor List for Flax – Linum usitatissimum L, Nitra* (p. 104).

Oh, S. M., & Chung, K. H. (2006). Antiestrogenic activities of Ginkgo biloba extracts. *The Journal of Steroid Biochemistry and Molecular Biology, 100*(4-5), 167–176. https://doi.org/10.1016/j.jsbmb.2006.04.007

Ohta, S., Fujimaki, T., Uy, M. M., Yanai, M., Yukiyoshi, A., & Hirata, T. (2007). Antioxidant hydroxycinnamic acid derivatives isolated from Brazilian bee pollen. *Natural Product Research, 21*(8), 726–732. https://doi.org/10.1080/14786410601000047

Oliveira, K. C. L. S., Moriya, M., Azedo, R. A. B., Almeida-Muradian, L. B., Teixeira, E. W., Alves, M. L. T. M. F., & Moreti, A. C. C. C. (2009). Relationship between botanical origin and antioxidants vitamins of bee-collected pollen. *Química Nova, 32*(5), 1099–1102. https://doi.org/10.1590/S0100-40422009000500003

Orsolić, N., Kosalec, I., & Basić, I. (2005). Synergistic antitumor effect of polyphenolic components of water soluble derivative of propolis against Ehrlich ascites tumour. *Biological & Pharmaceutical Bulletin, 28*(4), 694–700. https://doi.org/10.1248/bpb.28.694

Ostrosky, E. A., Mizumoto, M. K., Lima, M. E. L., Kaneko, T. M., Nishikawa, S. O., & Freitas, B. R. (2008). Métodos para avaliação da actividade antimicrobiana e determinação da concentração mínima inibitória (CMI) de plantas medicinais. *Revista Brasileira de Farmacognosia, 18*(2), 301–307. https://doi.org/10.1590/S0102-695X2008000200026

Otero, M. P., & Losada, S. B. (2002). Método de determinación del origen geográfico del polen apícola comercial. *Lazaroa, 23*, 25–34.

Otsu, N. (1979). A threshold selection method from grey-level histograms. *IEEE Transactions on Systems, Man, and Cybernetics, 9*(1), 62–66. https://doi.org/10.1109/TSMC.1979.4310076

Ou, B., Hampsch-Woodill, M., & Prior, R. R. (2001). Development and validation of an improved oxygen radical absorbance capacity assay using fluorescein as the fluorescent probe. *Journal of Agricultural and Food Chemistry, 49*(10), 4619–4626. https://doi.org/10.1021/jf010586o

Ozcan, M., Ceylan, D. A., Unver, A., & Yetisir, R. (2003). Antifungal effect of pollen and propolis extracts collected from different regions of Turkey, Uludag. *Bee Journal, 3*, 27–34.

Ozdal, T., Capanoglu, E., & Altay, F. (2013) A review on protein-phenolic interactions and associated changes. *Food Research International, 51*, 954–970. https://doi.org/10.1016/j.foodres.2013.02.009.

Pabla, N., Huang, S., Mi, Q.-S., Daniel, R., & Dong, Z. (2008). ATR-Chk2 Signaling in p53 Activation and DNA Damage Response during Cisplatin-induced Apoptosis. *The Journal of Biological Chemistry, 283*(10), 6572–6583. https://doi.org/10.1074/jbc.M707568200

Panetta, J. C. (1998). O caráter educativo da Vigilância Sanitária. *Higiene Alimentar, 12*(57), 8–10.

Pantelić, M., Dabić, D., Matijašević, S., Davidović, S., Dojčinović, B., Milojković-Opsenica, D., Tešić, Ž., & Natić, M. (2014). Chemical characterization of fruit wine made from Oblačinska sour cherry. *The Scientific World Journal, 2014*, 454797–454799. https://doi.org/10.1155/2014/454797

Partap, U. (1997). *Bee flora of the Hindu Kush-Himalayas: Inventory and management* (p. 297). International Centre for Integrated Mountain Development.

Pascoal, A., Rodrigues, S., Teixeira, A., Feás, X., & Estevinho, L. M. (2014). Biological activities of commercial bee pollens: Antimicrobial, antimutagenic, antioxidant and anti-inflammatory. *Food and Chemical Toxicology, 63*, 233–239. https://doi.org/10.1016/j.fct.2013.11.010

Paulino, N., Abreu, S. R., Uto, Y., Koyama, D., Nagasawa, H., Hori, H., Dirsch, V. M., Vollmar, A. M., Scremin, A., Bretz, W. A. (2008). Anti-inflammatory effects of a bioavailable compound, Artepillin C, in Brazilian propolis. *European Journal of Pharmacology, 10;587*(1–3), 296–301. doi: 10.1016/j.ejphar.2008.02.067. Epub 2008 Feb 29. PMID: 18474366.

Paulo, L., Antunes, P., Campos, M. G., & Anjos, O. (2012). Utilização do Teor em Metais Pesados no Pólen Como Marcador Ambiental – Estudo Preliminar. *II Congresso Ibérico de Apicultura, 18-20 Setembro, Guadalajara*, pp. 75–76.

Paulo, L., Antunes, P., & Anjos, O. (2014). Mineral composition of pollen using inductively coupled plasma atomic emission spectroscopy. *Planta Medica, 80*(16), 2–16. https://doi.org/10.1055/s-0034-1394851

Petersen, J., Souza, E. M. P., Moreira, R. M., Pasin, L. E. V., Nordi, J. C., & Barreto, L. M. R. C. (2011). Comercialização do pólen apícola em 11 países da America Latina. *Magistra, Cruz das Almas-BA, 23*(número especial), 14–16.

Pichichero, E., Cicconi, R., Mattei, M., & Canini, A. (2011). Chrysin-induced apoptosis is mediated through p38 and Bax activation in B16-F1 and A375 melanoma cells. *International Journal of Oncology, 38*(2), 473–483. https://doi.org/10.3892/ijo.2010.876

Pinto, B., Bertoli, A., Noccioli, C., Garritano, S., Reali, D., & Pistelli, L. (2008). Estradiol-antagonistic activity of phenolic compounds from leguminous plants. *Phytotherapy Research, 22*(3), 362–366. https://doi.org/10.1016/j.fct.2013.11.010.

Pinto, B., Caciagli, F., Riccio, E., Reali, D., Sarić, A., Balog, T., Likić, S., & Scarpato, R. (2010). Antiestrogenic and antigenotoxic activity of bee pollen from Cystus incanus and Salix alba as evaluated by the yeast estrogen screen and the micronucleus assay in human lymphocytes. *European Journal of Medicinal Chemistry, 45*(9), 4122–4128. https://doi.org/10.1016/j.ejmech.2010.06.001

Plumb, R., Castro-Perez, J., Granger, J., Beattie, I., Joncour, K., & Wright, A. (2004). Ultra-performance liquid chromatography coupled to quadrupole-orthogonal time-of-flight mass spectrometry. *Rapid Communications in Mass Spectrometry, 18*(19), 2331–2337. https://doi.org/10.1002/rcm.1627

Prakash, A. S., Pereira, T. N., Reilly, P. E. B., & Seawright, A. A. (1999). Pyrrolizidine alkaloids in human diet. *Mutation Research/Genetic Toxicology and Environmental Mutagenesis, 443*(1-2), 53–76. https://doi.org/10.1016/S1383-5742(99)00010-1

Prior, R. L., Wu, X., & Schaich, K. (2005). Standardized methods for the determination of antioxidant capacity and phenolics in foods and dietary supplements. *Journal of Agricultural and Food Chemistry, 53*(10), 4290–4302. https://doi.org/10.1021/jf0502698

Prokudina, E. A., Havlíček, L., Al-Maharik, N., Lapčík, O., Strnad, M., & Gruz, J. (2012). Rapid UPLC–ESI–MS/MS method for the analysis of isoflavonoids and other phenylpropanoids. *Journal of Food Composition and Analysis, 26*(1-2), 36–42. https://doi.org/10.1016/j.jfca.2011.12.001

Qian, W. L., Khan, Z., Watson, D. G., & Fearnley, J. (2008). Analysis of sugars in bee pollen and propolis by ligand exchange chromatography in combination with pulse amperometric detection and mass spectrometry. *Journal of Food Composition and Analysis, 21*(1), 78–83. https://doi.org/10.1016/j.jfca.2007.07.001

Rabie, A. L., Wells, J. D., & Dent, L. K. (1983). The nitrogen content of pollen protein. *Journal of Apicultural Research, 22*(2), 119–123. https://doi.org/10.1080/00218839.1983.11100572

Rauha, J. P., Vuorela, H., & Kostiainen, R. (2001). Effect of eluent on the ionization efficiency of flavonoids by ion spray, atmospheric pressure chemical ionization, and atmospheric pressure photoionization mass spectrometry. *Journal of Mass Spectrometry, 36*(12), 1269–1280. https://doi.org/10.1002/jms.231

R Core Team. (2014). *R: A language and environment for statistical computing*. R Foundation for Statistical Computing. http://www.R-project.org/

Reiter, R. (1947). The coloration of anther and corbicular pollen. *The Ohio Journal of Science, 17*(4), 137–152. http://hdl.handle.net/1811/3589

Rex, J. H., Pfaller, M. A., Walsh, T. J., Chaturvedi, V., Espinel-Ingroff, A., Ghannoum, M. A., Gosey, L. L., Odds, F. C., Rinaldi, M. G., Sheehan, D. J., & Warnock, D. W. (2001).

Antifungal susceptibility testing: Practical aspects and current challenges. *Clinical Microbiology Reviews*, *14*(4), 643–658. https://doi.org/10.1128/CMR.14.4.643-658.2001

Royal Horticultural Society (RHS). 1966, 1986, 1995, 2004. *RHS Color Chart*. Royal Horticultural Society.

Ricciardelli, D. G. (1982). Analisi organolettico-microscopica dei prodotti apistici: un problema attuale e indilazionabile. *L'Ape Nosta Amica*, *4*, 19–21.

Richardson, R. T., Lin, C.-H., Sponsler, D. B., Quijia, J. O., Goodell, K., & Johnson, R. M. (2015). Application of ITS2 metabarcoding to determine the provenance of pollen collected by honey bees in an agroecosystem. *Applications in Plant Sciences*, *3*(1), 1400066. https://doi.org/10.3732/apps.1400066

Robards, K., & Antolovich, M. (1997). Analytical chemistry of fruit bioflavonoids. *The Analyst*, *122*(2), 11R–34R. https://doi.org/10.1039/a606499j

Rochat, B. (2012). Quantitative/qualitative analysis using LC-HRMS: the fundamental step forward for clinical laboratories and clinical practice. *Bioanalysis*, *4*(14), 1709–1711. https://doi.org/10.4155/bio.12.159

Rodlauer, R., & Hüttinger, E. (1992). A Computer-supported method of honey pollen analysis. Abstracts of the 8th International palynological Congress, France. p. 40.

Rodríguez, I., Llompart, M. P., & Cela, R. (2000). Solid-phase extraction of phenols. *Journal of Chromatography. A*, *885*(1-2), 291–304. https://doi.org/10.1016/S0021-9673(00)00116-3

Rodríguez-Damián, M., Cernadas, E., Formella, A., Fernández-Delgado, M., & Sá-Otero, P. D. (2006). Automatic detection and classification of grains of pollen based on shape and texture. *IEEE Transactions on Systems, Man, and Cybernetics, Part C: Applications and Reviews*, *36*, 531–542. https://doi.org/10.1109/TSMCC.2005.855426

Rodríguez-Damián, M., Cernadas, E., Formella, A., Sá-Otero, P. (2004). Pollen classification using brightness-based and shape-based descriptors. *IEEE Transactions on System, Man, and Cybernetics*.

Rodríguez-Damián, M., Formella, A., & Sá-Otero DE, P. (2004 *Pollen classification using brightness-based and shape-based descriptors* [Paper presentation]. Proceedings of the 17th International Conference on Pattern Recognition, ICPR 2004, Vol. 2, pp. 212–215. https://doi.org/10.1109/ICPR.2004.1334098

Rodríguez-Delgado, M. A., Malovaná, S., Pérez, J. P., Borges, T., & García-Montelongo, F. J. (2001). Separation of phenolic compounds by high-performance liquid chromatography with absorbance and fluorimetric detection. *Journal of Chromatography. A*, *912*(2), 249–257. https://doi.org/10.1016/S0021-9673(01)00598-2

Rösemann, G. M., Botha, C. J., & Eloff, J. N. (2014). Distinguishing between toxic and non-toxic pyrrolizidine alkaloids and quantification by liquid chromatography–mass spectrometry. *Phytochemistry Letters*, *8*, 126–131. https://doi.org/10.1016/j.phytol.2014.03.002

Rosenthal, C., Petrescu, A., & Caragiani, S. (1969). Cercetări ascupra compozitiei biochimice si valorii biologice a unui produs inlocuitor de pollen. *Apicultura*, *22*(8), 18–21.

Rouessac, F., & Rouessac, A. (2007). Sample preparation: Solid phase extraction (SPE). In Francis, A. Rouessac, & S. Brooks (Trans.), *Chemical analysis – Modern instrumentation methods and techniques* (2nd ed., chapter 21 pp. 487–490). John Wiley & Sons, Ltd.

Rzepecka-Stojko, A., Pilawa, B., Ramos, P., & Stojko, J. (2012). Antioxidative properties of bee pollen extracts examined by EPR spectroscopy. *Journal of Apicultural Science*, *56*(1), 23–30. https://doi.org/10.2478/v10289-012-0003-0

Rzepecka-Stojko, A., Stec, M., Kurzeja, E., Gawrońska, E., Pawlowska-Goral, K. (2012). The effect of storage of bee pollen extracts on polyphenol content. *Polish Journal of Environmental Studies*, *21*(4), 1007–1011. http://www.pjoes.com/pdf/21.4/Pol.J.Environ.Stud.Vol.21.No.4.1007-1011.pdf

Sahin, H., Aliyazicioglu, R., Yildiz, O., Kolayli, S., Innocenti, A., & Supuran, C. T. (2011). Honey, pollen, and propolis extracts show potent inhibitory activity against the zinc metalloenzyme carbonic anhydrase. *Journal of Enzyme Inhibition and Medicinal Chemistry*, *26*(3), 440–444. https://doi.org/10.3109/14756366.2010.503610

Salamanca, G., Moreno, C., Quiralte, J., Barber, D., Villalba, M., & Rodriguez, R. (2008). Immunological characterisation of Ole e 11, a major olive pollen allergen. *Allergy*, 63, 500.

Sánchez, E. T. (2004). *Polen: Producción Envasado y Comercialización*.

Sánchez-Moreno, C. (2002). Methods used to evaluate the free radical scavenging activity in foods and biological systems. *Food Science and Technology International*, *8*(3), 121–137.

Sánchez-Moreno, C., Larrauri, J. A., & Saura-Calixto, F. (1998). A procedure to measure the antiradical efficiency of polyphenols. *Journal of the Science of Food and Agriculture*, *76*(2), 270–276. https://doi.org/10.1002/(SICI)1097-0010(199802)76:2 < 270::AID-JSFA945 > 3.0.CO;2-9

Santana, C. M., Ferrera, Z. S., Torres Padrón, M. E., & Santana Rodríguez, J. J. (2009). Methodologies for the extraction of phenolic compounds from environmental samples: New approaches. *Molecules*, *14*(1), 298–320. https://doi.org/10.3390/molecules14010298

Sattler, J. A. G. (2013). *Quantificatio of antioxidant vitamins E (alpha-, beta-, gama- and delta-tocopherol), C (ascorbic acid), provitamin A (alpha- and beta-carotene) and chemical composition of dehydrated bee pollen produced in georeferenced apiaries of southern Brazil* [Master dissertation]. University of São Paulo. Retrieved July 7, 2014, from http://www.teses.usp.br/teses/disponiveis/9/9131/tde-18032014-151137/

Sawyer, R. (1981). *Pollen identification for beekeepers* (p. 112). University College Cardiff Press.

Sawyer, R. (1988). *Honey identification* (p. 115). Cardiff Academic Press.

Schieber, A., Keller, P., Streker, P., Klaiber, I., & Carle, R. (2002). Detection of isorhamnetin glycosides in extracts of apples (*Malus domestica* cv. "Brettacher") by HPLC-PDA and HPLC-APCI-MS/MS. *Phytochemical Analysis*, *13*(2), 87–94. https://doi.org/10.1002/pca.630

Scorzoni, L., Benaducci, T., Almeida, A. M. F., Silva, D. H. S., Bolzani, V. S., & Gianinni, M. J. (2007). The use of standard methodology for determination of antifungal activity of natural products against medical yeasts *Candida* sp. and *Cryptococcus* sp. *Brazilian Journal of Microbiology*, *38*, 391–397. https://doi.org/10.1590/S1517-83822007000300001.

Serra-Bonvehí, J., & Escolá Jordá, R. (1997). Nutrient composition and microbiological quality of honey bee-collected pollen in Spain. *Journal of Agricultural and Food Chemistry*, *45*(3), 725–732. https://doi.org/10.1021/jf960265q

Serra-Bonvehí, J., Gonello Galindo, J., & Gomez Pajuelo, A. (1986). Estudio de la composition y caracteristicas fisicoquimicas del pollen de abejas. *Alimentaria*, *23*(176), 63–67.

Serra-Bonvehí, J., Torrentó, S. M., & Lorente, C. E. (2001). Evaluation of polyphenolic and flavonoid compounds in honey bee – Collected pollen produced in Spain. *Journal of Agricultural and Food Chemistry*, *49*(4), 1848–1853. https://doi.org/10.1021/jf0012300

Sharma, M. (1970). An analysis of pollen loads of honey bees from Kangra, India. *Grana*, *10*(1), 35–42. https://doi.org/10.1080/00173137009429855

Sharma, O. P., & Bhat, T. K. (2009). DPPH antioxidant assay revisited. *Food Chemistry*, *113*(4), 1202–1205. https://doi.org/10.1016/j.foodchem.2008.08.008

Shaw, D. E. (1990). The incidental collection of fungal spores by bees and the collection of spores in lieu of pollen. *Bee World*, *71*(4), 158–176. https://doi.org/10.1080/0005772X.1990.11099059

Shi, H. B., Kong, M., Chen, G., Zhao, J., Shi, H. L., Chen, Y., & Rowan, F. G. (2010). Compound pollen protein nutrient increases serum albumin in cirrhotic rats. *Gastroenterology Research*, *3*(6), 253–261. https://doi.org/10.4021/gr240e

Shoskes, D. A. (2002). Phytotherapy in chronic prostatitis. *Urology, 60*(6 Suppl), 35–37. https://doi.org/10.1016/S0090-4295(02)02383-X

Sickel, W., Ankenbrand, M. J., Grimmer, G., Holzschuh, A., Härtel, S., Lanzen, J., Steffan-Dewenter, I., & Keller, A. (2015). Increased efficiency in identifying mixed pollen samples by meta-barcoding with a dual-indexing approach. *BMC Ecology, 15*, 20. https://doi.org/10.1186/s12898-015-0051-y

Siddik, Z. (2003). Cisplatin: mode of cytotoxic action and molecular basis of resistance. *Oncogene, 22*(47), 7265–7279. https://doi.org/10.1038/sj.onc.1206933

Silici, S., & Kutluca, S. (2005). Chemical composition and antibacterial activity of propolis collected by three different races of honeybees in the same region. *Journal of Ethnopharmacology, 99*(1), 69–73. https://doi.org/10.1016/j.jep.2005.01.046

Silva, E. N. (1999). Tecnologia de alimentos. Higiene na indústria de alimentos. Unicamp- Campinas. 5.

Silva, J. A. (2000). Tópicos da tecnologia de alimentos. *São Paulo: Livraria Varela, 229*. 82 pp.

Silva, T. M. S., Camara, C. A., da Silva Lins, A. C., Maria Barbosa-Filho, J., da Silva, E. M. S., Freitas, B. M., & de Assis Ribeiro dos Santos, F. (2006). Chemical composition and free radical scavenging activity of pollen loads from stingless bee *Melipona subnituda* Ducke. *Journal of Food Composition and Analysis, 19*(6-7), 507–511. https://doi.org/10.1016/j.jfca.2005.12.011

Silva, T. M. S., Camara, C. A., Lins, A. C. S., Agra, M. F., Silva, E. M. S., Reis, I. T., & Freitas, B. M. (2009). Chemical composition, botanical evaluation and screening of radical scavenging activity of collected pollen by the stingless bees Melipona rufiventris (Uruçu-amarela)). *Anais da Academia Brasileira de Ciencias, 81*(2), 173–178. https://doi.org/10.1590/s0001-37652009000200003

Silva, C. I., Bordon, N. G., da, R., Filho, L. C., & Garófalo, C. A. (2012). The importance of plant diversity in maintaining the pollinator bee, Eulaema nigrita (Hymenoptera: Apidae) in sweet passion fruit fields. *Revista de Biologia Tropical, 60*(4), 1553–1565.

Smirnova, V. A., Timofeyev, N. K., Breygina, M. A., Matveyeva, N. P., & Yermakov, P. I. (2012). Antioxidant properties of the pollen exine polymer matrix. *Biophysics, 57*(2), 174–178. https://doi.org/10.1134/S0006350912020224

Snyder, R. D., & Gillies, P. J. (2002). Evaluation of the clastogenic, DNA intercalative, and topoisomerase II-interactive properties of bioflavonoids in Chinese hamster V79 cells. *Environmental and Molecular Mutagenesis, 40*(4), 266–276. https://doi.org/10.1002/em.10121

Solberg, Y., & Remedios, G. (1980). Chemical composition of pure and bee-collected pollen. *Scientific Reports Agricultural University, Norway, 59*(18), 1–13.

Somerville, D. C. (2001). *Nutritional value of bee collected pollens, a report for the Rural Industries Research and Development Corporation # 01/047* (pp. 57–58). NSW Agriculture Goulburn.

Somerville, D. C., & Nicol, H. I. (2002). Mineral content of honey bee collected pollen from southern New South Wales. *Australian Journal of Experimental Agriculture, 42*(8), 1131–1136. https://doi.org/10.1071/EA01086

Soto-Vaca, A., Gutierrez, A., Losso, J. N., Xu, Z., & Finley, J. W. (2012). Evolution of phenolic compounds from color and flavor problems to health benefits. *Journal of Agricultural and Food Chemistry, 60*(27), 6658–6677. https://doi.org/10.1021/jf300861c

Souza, B. R. (2014). *Quantification of B complex vitamins (B1, B2) and vitamers of vitamin B3 and B6 in dehydrated bee pollen samples from Southern Brazil* [Master dissertation]. University of São Paulo.

Souza, E. H., Souza, F. V. D., Rossi, M. L., Packer, R. M., Cruz-Barros, M. A. V., & Martinelli, A. P. (2018) Pollen morphology and viability in Bromeliaceae. *Annals of the Brazilian Academy of Sciences*, 89(4), 3067–3082. http://dx.doi.org/10.1590/0001-3765201720170450.

Sparkman, O. D. (2000). *Mass spectrometry desk reference*. Global View Pub. http://www.amazon.com/exec/obidos/ASIN/0966081323/102-8754827-3860108

Stalikas, C. D. (2007). Extraction, separation, and detection methods for phenolic acids and flavonoids. *Journal of Separation Science, 30*(18), 3268–3295. https://doi.org/10.1002/jssc.200700261

Stanciu, O. G., Marghitas, L. A., Dezmirean, D., & Campos, M. G. (2011). A comparison between the mineral content of flower and honey bee collected pollen of selected plant origin (*Helianthus annuus* L. and *Salix* sp.). *Romanian Biotechnological Letters, 16*(4), 6291–6296.

Stanciu, O. G., Marghitas, L. A., & Dezmirean, D. (2012). Nutritional and biological values of bee pollen and beebread harvested from Romania. In *Proceedings of International Conference of Agricultural Engineering CIGR-AGENG*. http://journals.usamvcluj.ro/index.php/zootehnie/article/viewFile/9391/7820

Stanciu, O. G., Marghitas, L. A., Dezmirean, D., & Campos, M. G. (2012). Specific distribution of minerals in selected unifloral bee pollen. *Food Science and Technology Letters, 3*(1), 27–31.

Stanley, R. G., & Linskens, H. F. (1974). *Pollen – Biology, biochemistry, management* (p. 307). Springer-Verlag. https://doi.org/10.1007/978-3-642-65905-8

Szczêsna, T. (2006a). Long – Chain fatty acids composition of honey bee-collected pollen. *Journal of Apicultural Science, 50*, 65–78.

Szczêsna, T. (2006b). Protein content and amino acid composition of bee-collected pollen from selected botanical origins. *Journal of Apicultural Science, 50*, 81–90.

Szczêsna, T. (2006c). Protein content and amino acids composition of bee-collected pollen originating from Poland, South Korea and China. *Journal of Apicultural Science, 50*, 91–99.

Szczęsna, T. (2007). Study on the sugar composition of honey-bee-collected pollen. *Journal of Apicultural Science, 51*(1), 15–21.

Szczęsna, T., & Rybak-Chmielewska, H. (1998). Some properties of honey bee collected pollen. In Polnisch-Deutsches Symposium Salus *Apis Mellifera*, new demands for honey bee breeding in the 21st century. *Pszczelnicze Zeszyty Naukowe, 42*(2), 79–80.

Szczęsna, T., Rybak-Chmielewska, H., & Chmielewski, W. (2002). Sugar composition of pollen loads harvested at different periods of the beekeeping season. *Journal of Apicultural Science, 46*(2), 107–115.

Szczęsna, T., Rybak-Chmielewska, H., & Skowronek, W. (1995a). Wpływ utrwalania na wartość biologiczną obnóży pyłkowych. *Pszczelnicze Zeszyty Naukowe, 39*(1), 177–187.

Szczęsna, T., Rybak-Chmielewska, H., & Skowronek, W. (1995b). Zmiany w składzie chemicznym obnóży pyłkowych zachodzące podczas ich przechowywania w różnych warunkach. I. Cukry, tłuszcze, popiół. *Pszczelnicze Zeszyty Naukowe, 39*(2), 145–156.

Taha, E. A. (2015). Chemical composition and amounts of mineral elements in honey bee-collected pollen in relation to botanical origin. *Journal of Apicultural Science, 59*(1), 75–81. https://doi.org/10.1515/JAS-2015-0008

Tax, D. (2011). DDtools, the data description toolbox for Matlab. Version 1.9.0.

Thakur, M., & Nanda, V. (2020). Composition and functionality of bee pollen: A review. *Trends in Food Science & Technology, 98*, 82–106. https://doi.org/10.1016/j.tifs.2020.02.001

Tian, S., Nakamura, K., Cui, T., & Kayahara, H. (2005). High-performance liquid chromatographic determination of phenolic compounds in rice. *Journal of Chromatography. A, 1063*(1-2), 121–128. https://doi.org/10.1016/j.chroma.2004.11.075

Tichy, J., & Novak, J. (2000). Detection of antimicrobials in bee products with activity against *viridans streptococci*. *Journal of Alternative and Complementary Medicine, 6*(5), 383–389. https://doi.org/10.1089/acm.2000.6.383

Tohamy, A. A., Abdella, E. M., Ahmed, R. R., & Ahmed, Y. K. (2014). Assessment of anti-mutagenic, anti-histopathologic and antioxidant capacities of Egyptian bee pollen and propolis extracts. *Cytotechnology, 66*(2), 283–297. https://doi.org/10.1007/s10616-013-9568-0

Tripathi, S. K. M., Kumar, M., Kedves, M., & Varga, B. (2003). LM, SEM and TEM investigations on partially degraded pollen grains of Cycas rumphii Miq. from India. *Plant Cell Biology and Development, Szeged, 15*, 28–42.

Uhl, M., Ecker, S., Kassie, F., Lhoste, E., Chakraborty, A., Mohn, G., & Knasmüller, S. (2003). Effect of chrysin, a flavonoid compound, on the mutagenic activity of 2-amino-1-methyl-6-phenylimidazo[4,5- b]pyridine (PhIP) and benzo(a)-pyrene (B(a)P) in bacterial and human hepatoma (HepG2) cells. *Archives of Toxicology, 77*(8), 477–484. https://doi.org/10.1007/s00204-003-0469-4

Ulusoy, E., & Kolayli, S. (2014). Phenolic composition and antioxidant properties of anzer bee pollen. *Journal of Food Biochemistry, 38*(1), 73–82. http://.dx.doi.org/10.1111/jfbc.12027 https://doi.org/10.1111/jfbc.12027

Valentini, A., Miquel, C., & Taberlet, P. (2010). DNA barcoding for honey biodiversity. *Diversity, 2*(4), 610–617. https://doi.org/10.3390/d2040610

van Laere, O., Lagasse, A., & de Mets, M. (1969). Use of the scanning electron microscope for investigating pollen grains isolated from honey samples. *Journal of Apicultural Research, 8*(3), 139–145. https://doi.org/10.1080/00218839.1969.11100230

van Meeuwen, J. A., Korthagen, N., De Jong, P. C., Piersma, A. H., & van den Berg, M. (2007). (Anti)estrogenic effects of phytochemicals on human primary mammary fibroblasts, MCF-7 cells and their co-culture. *Toxicology and Applied Pharmacology, 221*(3), 372–383. https://doi.org/10.1016/j.taap.2007.03.016

van Rijsbergen, C. (1979). *Information retrieval*. Butterworths.

Villanueva, G. (1999). Pollen resources used by European and Africanized honey bees in the Yucatan peninsula. *Mexico. Journal of Apicultural Research, 38*(1-2), 105–111. https://doi.org/10.1080/00218839.1999.11101001

Villanueva, M. T. O., Díaz Marquina, A., Bravo Serrano, R., & Blazquez Abellán, G. (2002). The importance of bee-collected pollen in the diet: A study of its composition. *International Journal of Food Sciences and Nutrition, 53*(3), 217–224. https://doi.org/10.1080/09637480220132832

Vorwohl, G. (1990). Progress, problems and future tasks of Melissopalynologie. *Apidologie, 21*(5), 383–389. (Article in German). https://doi.org/10.1051/apido:19900501

Wang, L., & Weller, C. L. (2006). Recent advances in extraction of nutraceuticals from plants. *Trends in Food Science & Technology, 17*(6), 300–312. https://doi.org/10.1016/j.tifs.2005.12.004

Wang, Q., Garrity, G. M., Tiedje, J. M., & Cole, J. R. (2007). Naive Bayesian classifier for rapid assignment of rRNA sequences into the new bacterial taxonomy. *Applied and Environmental Microbiology, 73*(16), 5261–5267. https://doi.org/10.1128/AEM.00062-07

Wang, S., Xie, B., Yin, L., Duan, L., Li, Z., Eneji, A. E., Tsuji, W., & Tsunekawa, A. (2010). Increased UV-B radiation affects the viability, reactive oxygen species accumulation and antioxidant enzyme activities in maize (*Zea mays* L.) pollen. *Photochemistry and Photobiology, 86*(1), 110–116. https://doi.org/10.1111/j.1751-1097.2009.00635.x

Wang, W., Hu, J., Cheng, J. (1987). Biological effect of pollen from beehives radioprotective effect on hematopoietic tissues of irradiated mice. In *Proceedings of the 31st International Apicultural Congress Apimondia*, p. 176.

Wasek, M., Nartowska, J., Wawer, I., & Tudruj, T. (2001). Electron spin resonance assessment of the antioxidative potential of medicinal plants. Part 1. Contribution of anthocyanosides and flavonoids to the radical scavenging ability of fruits and herbal teas. *Acta Poloniae Pharmaceutica, 58*(4), 283–288.

Wiegand, I., Hilpert, K., & Hancock, R. E. W. (2008). Agar and broth dilution methods to determine the minimal inhibitory concentration (MIC) of antimicrobial substances. *Nature Protocols, 3*(2), 163–175. https://doi.org/10.1038/nprot.2007.521

Wilkins, T. D., & Thiel, T. (1973). Modified broth-disk method for testing the antibiotic susceptibility of anaerobic bacteria. *Antimicrobial Agents and Chemotherapy, 3*(3), 350–356. https://doi.org/10.1128/AAC.3.3.350

Williams, N. M., & Kremen, C. (2007). Resource distributions among habitats determine solitary bee offspring production in a mosaic landscape. *Ecological Applications, 17*(3), 910–921. https://doi.org/10.1890/06-0269

Witten, I. H., & Frank, E. (2005). *Data mining: Practical machine learning tools and techniques*. (2nd ed.). Morgan Kaufmann. https://doi.org/10.1016/b978-0-12-374856-0.00023-7

Xiong, A., Yang, L., He, Y., Zhang, F., Wang, J., Han, H., Wang, C., Bligh, S. W. A., & Wang, Z. (2009). Identification of metabolites of adonifoline, a hepatotoxic pyrrolizidine alkaloid, by liquid chromatography/tandem and high-resolution mass spectrometry. *Rapid Communications in Mass Spectrometry, 23*(24), 3907–3916. https://doi.org/10.1002/rcm.4329

Yamaguchi, M., Uchiyama, S., & Nakagawa, T. (2007). Preventive effects of bee pollen Cistus iadaniferus extract on bone loss in ovariectomized rats in vivo. *Journal of Health Science, 53*(5), 571–575. https://doi.org/10.1248/jhs.52.268[10.1248/jhs.53.571]

Yáñez, Y. A., Remsberg, C. M., Takemoto, J. K., Vega-Villa, K. R., Andrews, P. K., Sayre, C. L., Yang, K., Wu, D., Liu, D., Chen, J., & Sun, P. (2013). Characterization of chemical composition of bee pollen in China. *Journal of Agricultural and Food Chemistry, 61*(3), 708–718. https://doi.org/10.1021/jf304056b

Yang, W. Z., Qiao, X., Bo, T., Wang, Q., Guo, D. A., & Ye, M. (2014). Low energy induced homolytic fragmentation of flavonol 3-O-glycosides by negative electrospray ionization tandem mass spectrometry. *Rapid Communications in Mass Spectrometry, 28*(4), 385–395. https://doi.org/10.1002/rcm.6794

Yerlikaya, O. (2014). Effect of bee pollen supplement on antimicrobial, chemical, rheological, sensorial properties and probiotic viability of fermented milk beverages. *Mljekarstvo, 64*(4), 268–279. https://doi.org/10.15567/mljekarstvo.2014.0406

Youssef, A. M., Farag, R. S., Ewies, M. A., & El-Shakaa, S. M. A. (1978). Chemical studies on pollen collected by honey bees in Giza region, Egypt. *Journal of Apicultural Research, 17*(3), 110–113. https://doi.org/10.1080/00218839.1978.11099914

Yu, L. (2001). Free radical scavenging properties of conjugated linoleic acids. *Journal of Agricultural and Food Chemistry, 49*(7), 3452–3456. https://doi.org/10.1021/jf010172v

Zhang, X., Habib, F. K., Ross, M., Burger, U., Lewenstein, A., Rose, K., & Jaton, J. (1995). Isolation and characterization of a cyclic hydroxamic acid from a pollen extract, which inhibits cancerous cell growth *in vitro*. *Journal of Medicinal Chemistry, 38*(4), 735–738. https://doi.org/10.1021/jm00004a019

## Appendix: authors

**1. Introduction**
Maria Graça R Campos, Ofélia Anjos and Norman L Carreck

**2. Producing bee pollen from harvest to storage**
Lidia M R C Barreto, Maria G R Campos and João C Nordi

**3. Inexpensive pollen load authentication by means of computer vision and machine learning**
Manuel Chica and Pascual Campoy

**4. Evaluation and classification of corbicular pollen morphology**
Janka Nozkova

**5. Identification of the floral origin of bee pollen by microscopy**
Olena Lokutova, Maria G R Campos, Ananias Pascoal, Letícia M Estevinho

**6. Phenolic and polyphenolic profiles as fingerprints of floral origin of bee pollen**
Maria G R Campos, Norma Almaraz-Abarca, Lígia Muradian-Bicudo and Ofélia Anjos

**7. Determination of phenolic profile of bee pollen extracts by LC and LC/MS**
Živoslav Lj Tešić, Mirjana D Mosić, Aleksandar Ž Kostić, Mirjana B Pešić, Dušanka M Milojković-Opsenica

**8. Standard method for identification of bee pollen mixtures through meta-barcoding**
Wiebke Sickel, Markus J Ankenbrand, Gudrun Grimmer, Frank Förster, Ingolf Steffan-Dewenter and Alexander Keller

**9. Standard method for the determination of protein content of pollen using the Kjeldahl method**
Chrysoula H Tananaki, Vasilios Liolios, Dimitrios Kanelis, Maria-Anna Rodopoulou and Andreas Thrasyvoulou

**10. Standard method for determination of lipid content of pollen using Soxhlet extraction**
Chrysoula H Tananaki, Vasilios Liolios, Dimitrios Kanelis, Maria-Anna Rodopoulou and Andreas Thrasyvoulou

**11. Determination of the content of sugars in pollen, using HPLC-RID**
Teresa Szczęsna

**12. Standard method for vitamin analysis in bee pollen**
Ligia Bicudo de Almeida-Muradian

**13. Standard methods for mineral and trace elements in bee pollen**
Ofélia Anjos, Luísa Paulo and Maria Graça Campos

**14. Standard methods for pyrrolizidine alkaloid profiling in pollen by UHPLC-HRMS**
Matteo A Lucchetti, Christina Kast and Gaetan Glauser

**15. Standard methods for total phenolic compounds, total flavonoids and antioxidative capacity bee pollen evaluation by FTIR-ATR Spectroscopy**
Ofélia Anjos and Letícia M Estevinho

**16. Standard methods for determination of antioxidant bioactivity of bee pollen**
Norma Almaraz-Abarca, Ligia Bicudo de Almeida-Muradian and Maria G Campos

**17. Standard methods for anti-inflammatory activities in bee pollen**
Ananias Pascoal, Vanessa Branco Paula and Letícia M Estevinho

**18. Standard methods for antimicrobial activities in bee pollen**
Ananias Pascoal, Vanessa Branco Paula and Letícia M Estevinho

**19. Determination of standard methods for anti-mutagenic activities in bee pollen**
Ananias Pascoal, Vanessa Branco Paula and Letícia M Estevinho

www.ingramcontent.com/pod-product-compliance
Lightning Source LLC
Chambersburg PA
CBHW041153220326
41598CB00045B/7420

* 9 7 8 1 9 1 3 8 1 1 0 5 1 *